Digital Optical Communications

Digital Optical Communications

Le Nguyen Binh

CRC Press
Taylor & Francis Group
Boca Raton   London   New York

CRC Press is an imprint of the
Taylor & Francis Group, an **informa** business

MATLAB® is a trademark of The MathWorks, Inc. and is used with permission. The MathWorks does not warrant the accuracy of the text or exercises in this book. This book's use or discussion of MATLAB® software or related products does not constitute endorsement or sponsorship by The MathWorks of a particular pedagogical approach or particular use of the MATLAB® software.

FIRST INDIAN REPRINT, 2012

CRC Press
Taylor & Francis Group
6000 Broken Sound Parkway NW, Suite 300
Boca Raton, FL 33487-2742

© 2009 by Taylor & Francis Group, LLC
CRC Press is an imprint of Taylor & Francis Group, an Informa business

No claim to original U.S. Government works

Printed at Chennai Micro Print (P) Ltd, Chennai - 29.

International Standard Book Number-13: 978-1-4200-8205-0

Library of Congress Cataloging-in-Publication Data

Binh, Le Nguyen.
   Digital optical communications / Le Nguyen Binh.
      p. cm.
   Includes bibliographical references and index.
   ISBN 978-1-4200-8205-0
   1. Optical communications. 2. Digital communications. 3. Computer networks. I. Title.

TK5103.59.B52 2008
621.382'7--dc22   2008043512

Visit the Taylor & Francis Web site at
http://www.taylorandfrancis.com

and the CRC Press Web site at
http://www.crcpress.com

Dedication

To the memory of my father

To my mother who will not understand a word

To Phuong and Lam

Contents

Chapter 3

Optical Receivers and Transmission Performances 65

Chapter 4

Coherent Photonic Transmission 131

Chapter 5

Differential Phase Shift Keying Photonic Systems 171

Chapter 6

Chapter 8

Frequency Discrimination Receiver for Optical MSK..307

Chapter 9

Partial Responses and Single-sideband Modulation Formats 331

Chapter 10

Chapter 11

Chapter 12

Optical Soliton Transmission 481

Chapter 13

OFDM Optical Transmission Systems 533

Chapter 14

Comparison of Modulations Formats for Long-Haul Optically Amplified Transmission Systems and Networks 547

Chapter 15

Preface

The trends in long-haul terrestrial and intercontinental telecommunications systems and networks are reaching towards longer transparent distance and increased information capacity with economical implementation of the communications systems.

One of the possibilities is the use of modulation formats rather than the on–off keying intensity modulation and direct detection as traditionally employed in present optical communications networks. That is, the use of the amplitude, phase, and frequency of the lightwave carrier and formats of return-to-zero, non-return-to-zero with and without suppression of the carrier, to represent the information for effective transmission. These modulation formats must provide a strong robustness towards noises and distortion, nonlinear impairments, and high spectral efficiency.

Thus, we have witnessed the merging of the fields of digital communications and optical communications in advanced transmission techniques for long-haul optically amplified communications in global systems networks. Within this book, digital modulation techniques are applied to optical communications systems, including binary and quadrature amplitude shift keying, phase shift keying, differential shift keying, continuous phase shift keying, frequency shift keying, multilevel modulation as well as the use of multi-subcarriers, such as orthogonal frequency division multiplexing. Further, the formats of the modulation with pulse shape of return to zero or non-return to zero and the suppression of the carrier can be employed to combat the nonlinear impairment due to the enhancement of the signal energy and reduction of unused carrier energy. Furthermore, with the advancement of digital signal processing with ultra-high sampling rate, lightwave modulated signals can be detected coherently, thus the phase of the signals can be recovered, which can then be processed via digital signal processors. This digital signal processing will overcome the difficulty of the recovery of the signal phases if analog circuitry is employed. Thus, we have now witnessed the revival of coherent communications that were developed during the 1980s to significantly improve the optical signal-to-noise ratio for extending the repeaterless distance.

The modulation and transmission are treated theoretically, experimentally, and by simulation. MATLAB® (is a registered trademark of the MathWorks, Inc. For product information, please contact: The MathWorks, Inc., 3 Apple Hill Drive, Natick, MA 01760-2098 USA. Tel: 508 647 7000; Fax: 508-647-7001; E-mail: info@mathworks.com; Web: www.mathworks.com) Simulink® is extensively described in the chapters, as this modeling platform is now universally available in research laboratories in industries and universities around the world. So, it is expected that the models provided in this book will assist researchers and engineers to further develop specific optical communications systems employing new modulation formats in MATLAB. Simulink is preferred as the software package within MATLAB, and is based on block sets for ease of use and shortening of learning and development time. Furthermore, the use of MATLAB Simulink would require users' understanding of the principles of digital communications with various communications and mathematical blocks. There are no such optical communications block sets in MATLAB Simulink, so one of the main objectives of this book is to provide the operational principles of optical communications blocks as examples for users who wish to model their systems.

Our objectives in the chapters of the book are: (i) to provide the principles of digital communications within the context of optical and photonic transmission technology; (ii) to present the optical transmission of various modulation formats over optically amplified fiber links within the contexts of experimental works, theoretical and by modeling; and (iii) to identify the roles of frequency spectra of digital modulation formats for effective transmission of signals, to combat impairments, and to increase the communications capacity.

Acknowledgments

I am grateful to the Faculty of Engineering of Monash University for the availability of laboratories and facilities to carry out the work. The chapters were finalized while I was on sabbatical at the Christian Albretchs University in Kiel, Germany, as a visiting professor associated with the Chair of Communications. Therefore, I would like to thank Professor Dr. Werner Rosenkranz for his kind invitation and several fruitful discussions.

I am grateful to Dr. Thomas Lee of SHF AG of Berlin, Germany, and formerly of Nortel Networks, for his advice and the loan of various optical transmitters and receivers and the purchase of 40 Gb/s BERT for the experiments on ASK and DPSK using various formats conducted since 2004. I am also indebted to my undergraduate and postgraduate students over the years for the development and research of optical communications, including Dr. S.V. Chung of Corning Optical Systems (Australia), Professor John Ngo of Nayang Technological University of Singapore, Dr. Kai-fu Chang, Dr. T.L. Huynh, Dr. W.J. Lai, Nhan Nguyen, Calvin Li, H.S. Chong, H.C. Chong, and H.Q. Lam of the Department of Electrical and Computer Systems Engineering of Monash University. While associated with the Chair of Communications, I participated in several interesting discussions with the following people: Dr. Jochen Lebrich, Abdulamir Ali, Chumin Xia, Stefan Schoemann, Liu Yun, and also discussed the doctoral theses of Dr. C. Wree and Dr. Lebrich, as well as the lecture notes on high-speed systems and networks of Professor Rosenkranz.

Last, but not least, I thank my wife Phuong and my son Lam who have supported me over the years and put up with my hours of writing these chapters. Over the years, my parents have given me the education of *"learning for life"*, and I thank them for their lifetime of encouragement and their sacrifices to educate their children. Thus, I quote their education philosophy in Vietnamese:

"hoc.sau, thau.gon, chon.dung, dung.hay"

Le Nguyen Binh, PhD, C. Eng.
Kiel, Schleswig Holstein, Germany

Author

Dr. Le N. Binh received a Baccalaureate (Parts I and II) in Vietnam and was also awarded the Colombo Plan Scholarship of the Australian Government to continue his university education in Australia. In 1975 and 1980, he received a BE (Hons) and PhD in Electronic Engineering and Integrated Photonics, respectively, from the University of Western Australia, Nedlands, Western Australia. In 1980, he joined the Department of Electrical Engineering at Monash University after a three-year period with CSIRO Australia as a research scientist developing parallel microcomputing systems. In 1995, he was appointed reader of Monash University.

Dr. Binh has also worked for the Department of Optical Communications of Siemens AG Central Research Laboratories in Munich, Germany, and Advanced Technology Centre of Nortel Networks in Harlow, UK. From 2007 to 2008, he was a visiting professor of the Faculty of Engineering of Christian Albretchs University of Kiel, Germany.

Dr. Binh has published more than 250 papers in leading journals and refereed conferences in the fields of linear and nonlinear integrated photonics, optical communications, and photonic signal processing. He has one book published by CRC Press in the field of *Photonic Signal Processing* (2007). He is a member of the Institute of Electrical and Electronic Engineers (USA), the Optical Society of America (SPIE), and the Institution of Engineers and Technologist (UK). He has served on the technical programs committee of numerous international conferences and as a reviewer for several leading international journals, numerous Australian and international research grants.

Dr. Binh is well experienced in delivering lectures on subjects of signal processing, physical electronics, electromagnetic propagation, and optical communications to undergraduate and postgraduate students of electrical, electronic, and computer systems engineering.

Abbreviations

AM	Amplitude Modulation/Modulator
ASE	Amplified Spontaneous Emission
ASK	Amplitude Shift Keying
BDPSK	Binary Differential Phase Shift Keying
CD	Chromatic Dispersion
CPFSK	Continuous Phase Frequency Shift keying
CS-RZ	Carrier Suppressed Return to Zero format
DBM	Duo-Binary Modulation
DCF	Dispersion Compensating Fiber
DCM	Dispersion Compensating Module
DD	Direct Detection
Demux	Demultiplexer
DFB	Distributed Feedback (Laser)
DI	Delay Interferometer
DPSK	Differential Phase Shift Keying
DQPSK	Differential Quadrature Phase Shift Keying
DSF	Dispersion Shifted Fiber
DSP	Digital signal processing (processor)
DuoB	Duo-Binary
EDFA	Erbium-doped fiber amplifier
FWM	Four-wave Mixing
GVD	Group Velocity Dispersion
IF	Intermediate frequency
IM	Intensity Modulation/Modulator
IM/DD	Intensity Modulation/Direct Detection
I-Q	In-phase and Quadrature
ITU	International Telecommunications Union
LO	Local oscillator
MADPSK	Multi-level (M-ary) Amplitude-Differential Phase Shift Keying
MMF	Multi-mode optical fibers
MSK	Minimum Shift Keying
Mux	Multiplexers
MZDI	Mach–Zehnder Delay Interferometer
MZI	Mach–Zehnder Interferometer

MZIM	Mach–Zehnder Interferometer Modulator or Mach–Zehnder Interferometric Intensity Modulator
NF	Noise figure
NLPN	Nonlinear Phase Noise
NLSE	Non-Linear Schroedinger Equation
NRZ	Non-Return to Zero
NZDSF	Non-zero Dispersion Shifted Fiber (ITU-655)
OA	Optical Amplifier
O-DPSK	Offset Differential Phase Shift Keying
OFDM	Orthogonal Frequency Division Multiplexing
OOK	On–Off Keying or Amplitude Shift keying (ASK)
OPLL	Optical phase locked loop
OSNR	Optical Signal-to-Noise Ratio
Pdf	Probability density function
PLL	Phase locked loop
PM	Phase modulator
PMD	Polarization Mode Dispersion
PMF	Polarization maintaining fiber
QAM	Quadrature Amplitude Modulation
ROA	Raman Optical Amplifier
RZ	Return-to-Zero format
RZ33	RZ pulse of width of 33% of bit period format
RZ50	RZ pulse of width of 50% of bit period format
RZ67	RZ pulse of width of 67% of bit period format (normally CSRZ)
SMF	Single Mode Fiber
SPM	Self Phase Modulation
SSMF	Standard Single Mode Fiber (ITU-652)
STAR-QAM	Star quadrature amplitude modulation
TOD	Third Order Dispersion
XPM	Cross Phase Modulation

NOTATIONS

α_L	Attenuation (linear scale)
αdB	Attenuation factor in dB
β	Propagation constant of fiber
β_1	First order differentiation of the propagation with respect to the angular frequency – propagation delay
β_2	Second order differentiation of the propagation constant – group velocity dispersion (GVD)
β_3	Third order differentiation of the propagation constant – dispersion slope.

b_0, b_1	energy of a "1" transmitted and received at the front end of a photodetector
δ	Delta-factor – equivalent to the Q-factor
f_c	optical carrier frequency (center)
h	Plank's constant
L	Fiber length
η	Quantum efficiency
N	Number of fiber spans
P	Responsivity
λ	Wavelength
q	Electronic charge
B_e	Electrical bandwidth
B_o	Optical bandwidth
i_{Neq}	total equivalent noise current at the input of an electronic amplifier
i_{Ns}	shot noise current
i_d	Dark current noise of a photodetector
$<i_s(t)>$	Average current generated by signal power $P_s(t)$
S_N	Noise spectral density (A^2/Hz)
y	admittance parameters of the small signal model of electronic device

b_{th} energy of a "1" transmitted and received at the front end of a photodetector
δ   Delta-factor—equivalent to the Q-factor
f   optical ring frequency/report
h   Planck's constant
L   Fiber length
η   Quantum efficiency
N   Number of fiber spans
ρ   Reflectivity
λ   Wavelength
q   Electronic charge
B_e Electrical bandwidth
B   Optical bandwidth
I_{tot} total equivalent noise current at the input of an electronic amplifier
I_{sc} shot pulse current
i   Dark current noise of a photodetector
σ^2_{SPN} Average current generated by signal photons P(t)
S^2 Noise spectral density i^2/Hz
y   admittance parameters of the small signal model of electronic device

1 Introduction

1.1 DIGITAL OPTICAL COMMUNICATIONS AND TRANSMISSION SYSTEMS: CHALLENGING ISSUES

Starting from the proposed dielectric waveguides by Kao and Hockham* in 1966, the first research phase attracted intensive interest in the early 1970s in the demonstration of fiber optics. Optical communications have progressed tremendously over the last three decades. The first generation lightwave systems were commercially deployed in 1983, and operated in the first wavelength window of 800 nm over multi-mode optical fibers (MMF) at transmission bit rates of up to 45 Mb/s [1,2]. After the introduction of ITU-G652 standard single-mode fibers (SSMF) in the late 1970s [2,3], the second generation of lightwave transmission systems became available in the early 1980s [4,5]. Operating wavelengths were shifted to the second window of 1300 nm, which offered much lower attenuation for silica-based optical fibers compared with the previous 800-nm region, and particularly because the chromatic dispersion (CD) factor was almost zero. These second-generation systems could operate at bit rates of $\leq$1.7 Gb/s, and had a repeater-less transmission distance of about 50 km [6]. Further research and engineering efforts were put into the improvement of receiver sensitivity by coherent detection techniques, and the repeater-less distance reached 60 km in installed systems with a bit rate of 2.5 Gb/s. Optical fiber communications evolved to third-generation transmission systems that utilized the lowest attenuation 1550-nm wavelength window, and operated at a bit rate of up to 2.5 Gb/s [6,7]. These systems were commercially available in 1990 with repeater spacing of about 60–70 km [6,8]. At this stage, the generation of optical signals was mainly based on direct modulation of the semiconductor laser source and direct detection. Since the invention of erbium-doped fiber amplifiers (EDFA) in the early 1990s [9–11], lightwave systems rapidly evolved to wavelength division multiplexing (WDM) and, shortly after that, dense WDM (DWDM), optically amplified transmission systems capable of transmitting multiple 10 Gb/s channels. This is because loss is no longer a major issue for external optical modulators that normally suffer an insertion loss of $\leq$3 dB. These modulators allow preservation of the narrow linewidth of distributed feedback lasers (DFB). These high-speed and high-capacity systems extensively exploited external modulation in their optical transmitters. Present optical transmission systems are considered fifth generation, having a transmission capacity of a few Tb/s [6].

Coherent detection (homodyne or heterodyne) was the focus of extensive research and development during the 1980s and early 1990s [12–17]. Coherent detection was the main detection technique in the first three generations of lightwave transmission systems. At that time, the main motivation for development of coherent optical systems was to improve receiver sensitivity, commonly by 3–6 dB [13, 16]. Repeater-less transmission distance could be extended to >60 km of SSMF (with an attenuation factor of 0.2 dB/km). However, coherent optical systems suffer severe performance degradation due to impairment of fiber dispersion. In addition, phase coherence for lightwave carriers of the laser source and local laser oscillator was very difficult to maintain. The incoherent detection technique minimizes linewidth obstacles of the laser source and the local laser

---

* Kao, C. and G. Hockham "Dielectric surface waveguides for optical frequencies", IEE Proceedings, July 1966. C. Kao "40th Anniversary of Kao and Hockham paper", IET Communications Engineer, Vol. 4, No.1, Feb/March 2006, p.5.

oscillator, thereby relaxing the requirement of phase coherence. Moreover, incoherent detection mitigates the problem of polarization control in the mixing of transmitted lightwaves and the local laser oscillator at multi-THz optical frequency range. The invention of EDFAs, which can produce optical gains of ≥ 20 dB, has greatly contributed to the progress of incoherent digital photonic transmission systems up till now.

Recent years have seen a huge increase in demand for broadband communications driven mainly by the rapid growth of multi-media services, peer-to-peer networks, and IP streaming services, in particular Internet Protocol Television (IP TV). It is most likely that such tremendous growth will continue in the coming years. This is the main driving force for local and global telecommunications service carriers to develop high-performance and high-capacity next-generation optical networks. The overall capacity of WDM or DWDM optical systems can be boosted by increasing the base transmission bit rate of each optical channel, multiplexing more channels in a DWDM system or, preferably, by combining both of these schemes. However, while implementing these schemes, optical transmission systems encounter a number of challenging issues, as discussed in the following paragraphs (Figure 1.1).

Current 10 Gb/s transmission systems employ intensity modulation (IM), also known as on-off keying (OOK) and utilize non-return-to-zero (NRZ) pulse shapes. The term "OOK" can also be used interchangeably with amplitude shift keying (ASK)[†]. Moving towards high bit-rate transmission such as 40 Gb/s, the performance of OOK photonic transmission systems is severely degraded due to fiber impairments, including fiber dispersion and fiber nonlinearities. Fiber dispersion is classified into CD and polarization mode dispersion (PMD), causing inter-symbol interference (ISI) problems. Conversely, severe deteriorations on system performance due to fiber nonlinearities result from high-power spectral components at the carrier and signal frequencies of OOK-modulated optical signals. It is also of concern that existing transmission networks comprise millions of kilometers of SSMF that were installed approximately two decades' ago. These fibers do not have the advanced properties of the state-of–the-art fibers used in recent laboratory "hero" experiments, and have degraded after many years of use.

Total transmission capacity can be enhanced by increasing the number of multiplexed DWDM optical channels. This can be carried out by reducing the frequency spacing between these optical channels (e.g., from 100 GHz down to 50 GHz, or even 25 GHz to 12.5 GHz) [18,19]. Reduction of channel spacing also results in narrower bandwidths for optical multiplexers (mux) and demultiplexers

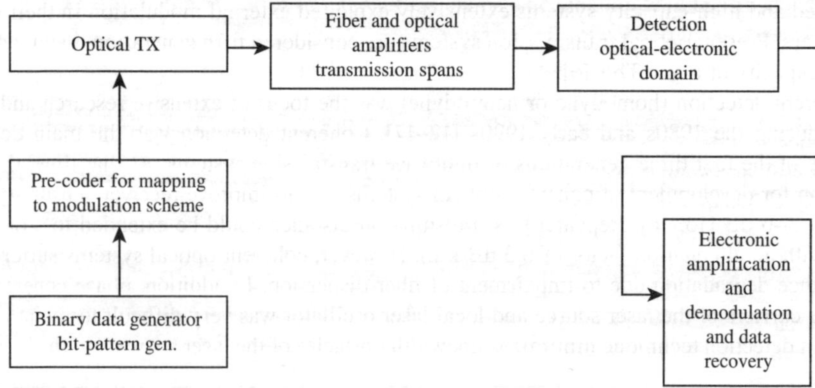

FIGURE 1.1 Modulation and electronic detection and demodulation of an advanced modulation format optical communications system (schematic).

---

† The OOK format simply implies the on-off states of the lightwaves where only the optical intensity is considered. On the other hand, the ASK format is a digital modulation technique representing the signals in the constellation diagram by both the amplitude and phase components.

(demux). Passing through these narrowband optical filters, signal waveforms are distorted and optical channels suffer inter-channel cross-talks. Narrowband filtering problems are getting more severe at high data bit rate (e.g., 40 Gb/s), significantly degrading system performance.

Together with the demand for boosting total system capacity, another challenge for service carriers is to find cost-effective solutions for upgrading. These cost-effective solutions should require minimum renovation to the existing photonic and electronic sub-systems (i.e., upgrading should take place only at the transmitter and receiver ends of an optical transmission link). Another possible cost-effective solution is to significantly extend the uncompensated reach of optical transmission links, i.e., without using dispersion compensation fibers (DCF), thereby reducing many required inline EDFAs. This network configuration has recently received much interest from the photonic research community and service carriers.

The principal motivation of this book is to provide a description of the employment of digital communications in modern optical communications. The fundamental principles of digital communications (coherent and incoherent transmission and detection techniques) are described, with the focus on technological development and limitations in the optical domain. The enabling technologies, research results, and demonstration in laboratory experimental platforms of the development of high-performance and high-capacity next-generation optical transmission systems impose significant challenges to the engineering of optical transmission systems in the near future and techniques for network monitoring.

1.2 ENABLING TECHNOLOGIES

1.2.1 MODULATION FORMATS AND GENERATION OF OPTICAL SIGNALS

Modulation is a process of facilitating information transfer over a medium. In optical communications, the process of converting information so that it can be successfully sent through an optical fiber is termed "optical modulation".

There are three basic types of digital modulation techniques: Amplitude Shift Keying (ASK), Frequency Shift Keying (FSK), Phase Shift Keying (PSK) in which the amplitude, frequency or phase of the carrier are manipulated to represent the information. Digital modulation is a process of mapping such that the digital data of "1" and "0" or symbols of "1" and "0" that converts it into some aspects of the carrier of either amplitude or/and phase. They are then transmitted. At the receiver, the phase or/and amplitude of the carrier is remapped back to a near copy of the information data.

1.2.1.1 Binary Level

Modulation is a process that facilitates information transport over a medium. In this book, the medium is the optical-guided fiber and associate photonic components. In digital communications, there are three basic types of digital modulation techniques: ASK, PSK and FSK. All of these techniques vary a parameter of a sinusoidal carrier to represent the information "1" and "0".

In ASK, the amplitude of the lightwave carrier, usually generated by a narrow-linewidth laser source, is changed in response to the digital data, and everything else is kept fixed. That is, bit "1" is transmitted by the lightwave carrier of a particular amplitude. To transmit "0", the amplitude is changed, keeping the frequency unchanged (Figure 1.2). NRZ or RZ can be assigned depending on the occupation of the state "1" during the time length of a bit period. For RZ, usually only half of the bit period is occupied by digital data.

In addition to NRZ and RZ formats, in optical communications, the carrier can be suppressed under these formats so as to achieve non-return-to-zero carrier suppression (NRZ-CS) and return-to-zero carrier suppression (RZ-CS). This is usually generated by biasing the optical modulator so that the carriers passing through the two parallel paths of an interferometric modulator are π phase shift difference with each other. Thus, the carrier at the center frequency is suppressed, but at the sidebands of the modulated signals.

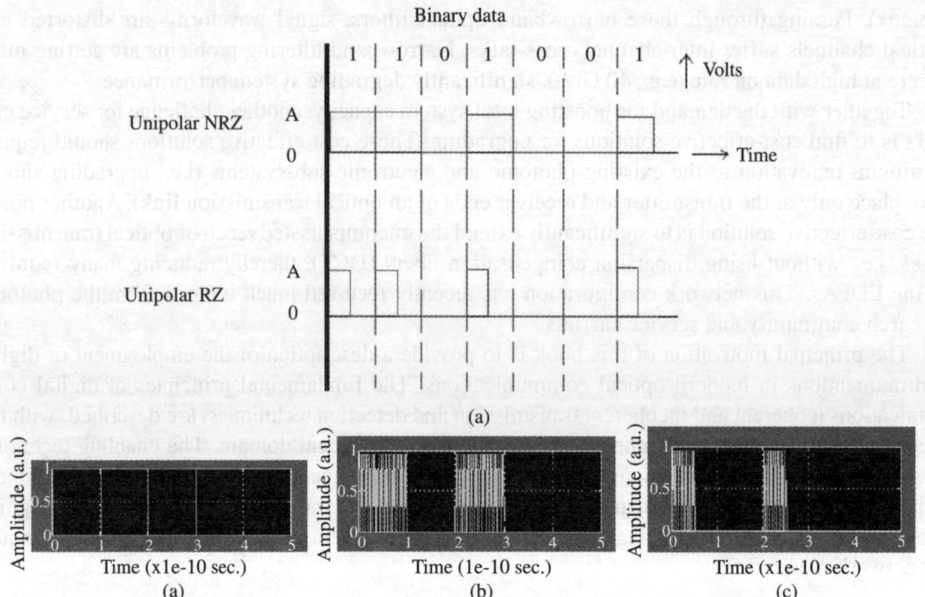

FIGURE 1.2 (a) NRZ and RZ Pulse amplitude-modulated formats for a sequence of {1 0 1 0 1 0 1 0 1 0 1 0} (b) Generated ASK signals with carrier (not to scale)-data and carrier-modulated NRZ and NZ formats.

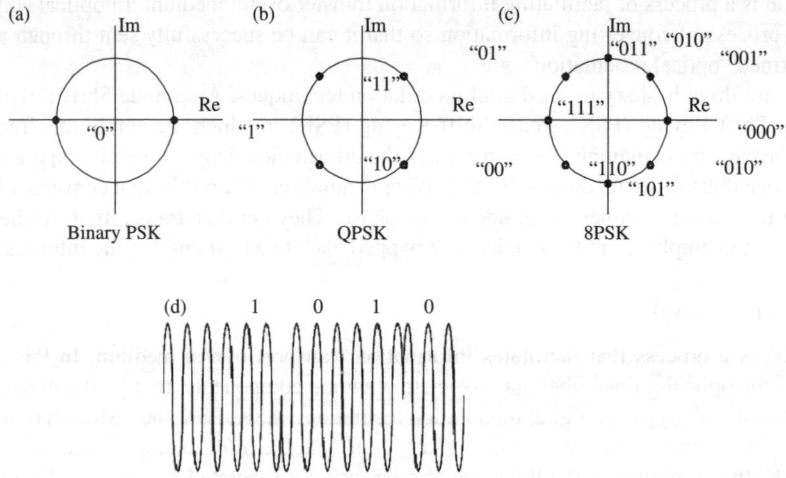

FIGURE 1.3 Signal-space constellation of discrete PM (a) binary PSK (b) quadrature binary PSK (c) 8-PSK. (d) Phase of the carrier under modulation with π phase shift of the BPSK at the edge of the pulse period.

In PSK, the phase of the lightwave carrier is changed to represent the information. The phase in this context is the shift of angle at the start at which the sinusoidal carrier starts. To transmit a "0", the phase would be shifted by π and a "1" without a change of phase. The phase angle can be changed and take a value of a set of phases corresponding to the mapping of the symbols (Figure 1.3).

In FSK, the frequency of the carrier represents the digital information. One particular frequency is assigned to a "1" and another frequency is assigned to the "0" (Figure 1.4). An FSK can be considered as continuous-phase modulation, e.g., the continuous phase modulation minimum phase

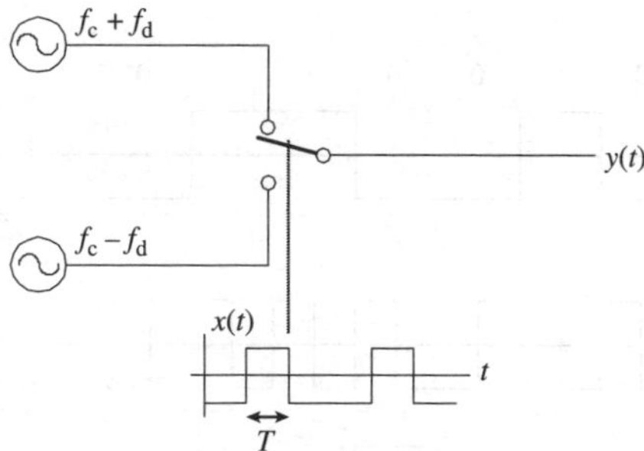

FIGURE 1.4 FSK transmitter (schematic). f_d is the deviation frequency.

shift keying (MSK) whose frequency separation f_d is selected such that the signals carried by these frequencies are orthogonal.

ASK can be combined with PSK to create a hybrid modulation scheme such as quadrature amplitude modulation (QAM) in which the phase and amplitude of the carrier are changed simultaneously. The carriers are expected to follow a similar pattern to that of the DPSK in Figure 1.3d, but with different frequencies of the carrier under the envelope of the bit "0" and "1". For MSK signals, the carrier frequency is chirped up or down depending on the "0" or "1". That is, the phase of the carrier is continuously varied during the bit period, and the carrier frequencies of the bits are such that there is an orthogonality of carriers and signal envelope.

1.2.1.2 Binary and Multi-level

Additional degree of freedom for detection can be used to effectively enhance the capacity due to effective equivalence of the multi-level and symbol rate, thence the detection of received optical signals. A widely used and mature detection scheme for optical signals is direct detection in which the optical power $P = [E]^2$, the square of a complex optical field amplitude. The photodetector would not be able to distinguish between a "0" or "π" phase shift of the carrier lightwave embedded within the pulse. The carrier phase could be possibly extracted only if photonic processing is used to extract the phase at the front-end of the receiver. Thus, a $+$ or $-$ field complexity would be seen as identical in the photodetector.

This ambiguity of the phase detection process would allow one to shape the optical spectra of optical signals to induce a modulation format more resilient to the distortion effects accumulated during the transmission process.

Formats making use of the tri-level are illustrated in Figure 1.5. They could be termed "pseudo-multilevel", "tri-level" or "poly-binary" signals. These tri-level signals can be represented in terms of the phase or frequency of the lightwave carrier.

The use of more than two symbols to encode a single bit of information leads to the symbol rate B_s. Under optical transmission, the -1 and $+1$ can be coded in terms of the phase of the carrier because there is no negative intensity representation unless the field of the lightwaves is used. The tri-level uses $\{+IEI, -IEI, \text{and } 0\}$ and its equivalent phase representation can be mapped to $\{0, IEI^2\}$ at the optical receiver (e.g., duo-binary form; see later chapter). A phase difference of π and 0 between the three levels would minimize pulse dispersion because they are propagating along the fiber due to the relative phase difference of π, hence the destructive interference of any pulse spreading due to dispersion.

This tri-level must not be confused with the truly multi-level signaling in which $\log_2 M$ bits are encoded on N-symbols, and then transmitted at a reduced rate $B_s/\log_2 N$. Multi-level amplitude or ASK and DQPSK are multi-level optical modulation techniques. Multi-level modulation formats are discussed in Chapter 7.

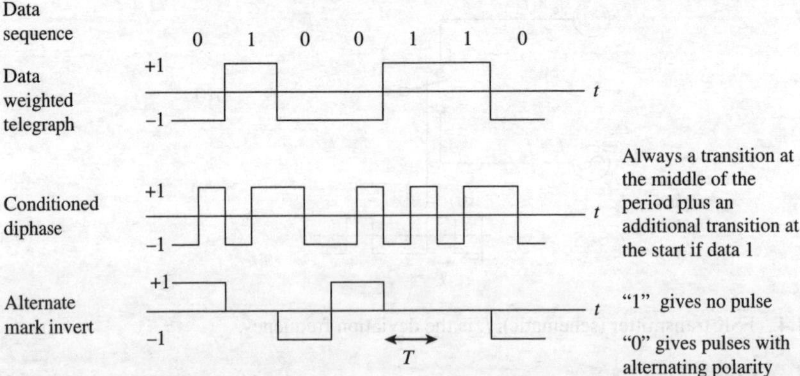

FIGURE 1.5 Pseudo-multi-level or polybinary baseband signals. Binary data sequence, weighted signals, diphase RZ and alternate mark inversion formats.

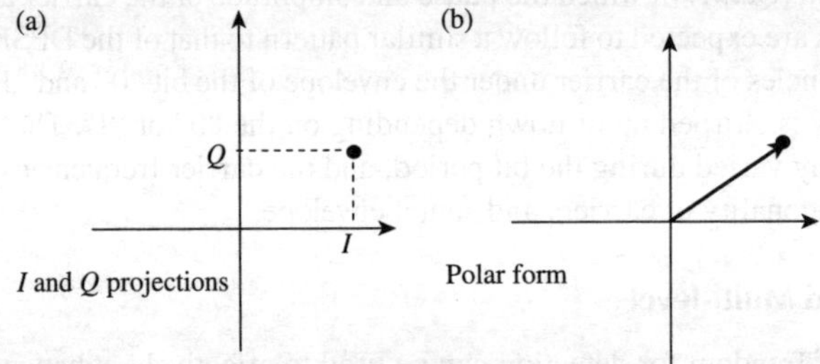

FIGURE 1.6 Signal vectors plotted in signal space (a) Cartesian coordinate (b) Polar form.

1.2.1.3 In-phase and Quadrature Phase Channels

Another form of modulation that would enhance the capacity of transmission is use of the orthogonal channels, in which the information can be coded into the in-phase and quadrature as shown in its polar or Cartesian coordinates (Figure 1.6).

QPSK is the most commonly used in the differential phase modulation; and I and Q components are used extensively. It is an extension of the binary PSK signals, but with the phase change of only $\pi/2$ instead of π. The signal $s(t)$ can be expressed mathematically as

$$s(t) = A_c p_s(t) \cos\left(2\pi f_c t + \frac{2\pi i}{M}\right),$$

where $p_s(t)$ is the pulse shaping of the data, M is the quantized level or the total number of phase states of the modulation, and I is the phase modulation index. QPSK can be combined with ASK to generate QAM, in which the phase and amplitude can be used to map a symbol of data information into one of the points on the signal space.

1.2.1.4 External Optical Modulation

External modulation is the essential technique for modulating lightwaves so that linewidth narrowness is preserved, and only the sidebands of the modulation scheme dominate the spectral property of the generated passband characteristics. Figure 1.7 and Figure 1.8 show the typical structure of optical transmitters for generation of NRZ and RZ optical signals. The laser is always switched on and its lightwaves are modulated via the electro-optic modulator using the principles of interferometric

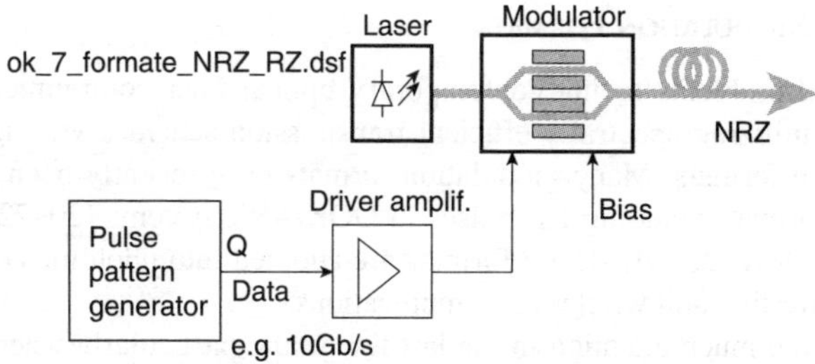

FIGURE 1.7 Generation of optical signals of format NRZ using an external modulator.

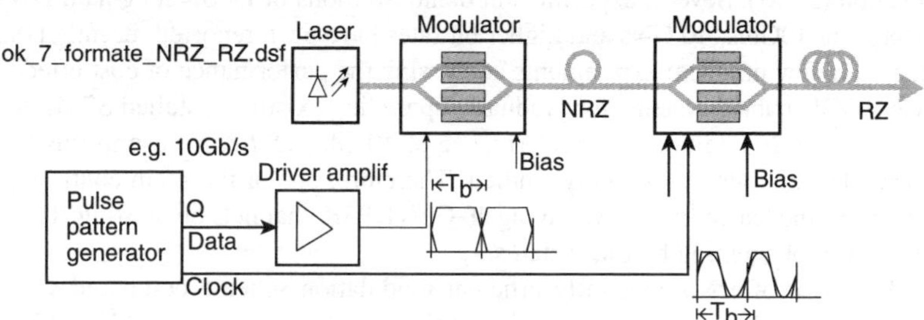

FIGURE 1.8 Generation of optical signals of format RZ.

constructive and destructive interference to represent the ON and OFF states of the lightwaves. The RZ can be similarly generated, but with an additional optical modulator that would generate periodic optical pulses whose width is half that of the bit period. Phase and frequency modulation can also be generated using these electro-optic modulators by biasing conditions and control of the amplitudes of electrical pulses; these optical transmitters are described in Chapter 2. The fiber that connects between the two modulators of Figure 1.8 must be the polarization-maintaining (PM) type, otherwise there would be polarization fluctuation and hence reduction of the coupling of lightwave power from one to the other.

The laser source would usually be a narrow-linewidth laser that is turned on continuously to preserve narrowness characteristics. Lightwaves are generated and coupled to the optical modulator via the pigtails of both devices. The modulator would be driven by data pulse sequence output of a bit pattern generator condition to the appropriate driving level required by the V_π and phase variation of the carrier if phase modulation is needed. If a modulation format is necessary, an electronic pre-coder is required to code the serial sequence to appropriate coding. The pre-coder can be a differential coding, multi-level coding or IFFT to generate orthogonal data sub-channels in case of the OFDM.

The pulses of all modulation formats can take the form of NRZ or RZ. For RZ format, an additional optical modulator is required to generate or condition the "1" NRZ to RZ (Figure 1.8). The second modulator can exchange its position with that of the other modulator without affecting generation of the modulation formats. This second modulator is usually called the "pulse carver".

If the RZ modulator is biased such that the phase difference at the biasing condition is π phase difference, we would have carrier suppression of the carriers at the central location of the generate spectra, but at the sidebands of the optical signals. The bandwidth of the modulator determines the rise time and fall time of edges of the pulse sequence shown in Figure 1.7 and Figure 1.8. Details of optical transmitters for different modulation formats are given in Chapter 2 and sections of other chapters in this book.

1.2.2 ADVANCED MODULATION FORMATS

The above-described problems facing contemporary optical fiber communications can be effectively overcome by utilizing spectrally efficient transmission schemes via the implementation of advanced modulation formats. Many modulation formats have recently been reported as alternatives for the OOK format, including RZ pulses in OOK/ASK systems [20–22], DPSK [18,20–25] and, more recently, MSK [26–33]. These formats are adopted into photonic communications from the knowledge of wire-line and wireless communications.

DPSK has received much attention in the last few years, particularly when combined with RZ pulses. The main advantages of RZ DPSK are: (i) a 3-dB improvement in receiver sensitivity over the OOK format by using an optical balanced receiver; [21,34,35] and (ii) high resilience to fiber nonlinearities [6,20,34,36] such as intra-channel self-phase modulation (SPM) and inter-channel cross phase modulation (XPM). Several experimental demonstrations of DPSK long-haul DWDM transmission systems for 10Gb/s, 40 Gb/s and higher bit rates have been reported recently [18,23,37–39]. However, there are few practical experiments addressing the performance of cost-effective 40 Gb/s DPSK–10 Gb/s OOK hybrid systems for gradually upgrading existing installed SSMF transmission infrastructure [40,41]. In addition, the performance of 40 Gb/s DPSK for use in this hybrid transmission scheme has not been thoroughly studied. Therefore, one of the main contributions of this research is to prove the feasibility of overlaying 40 Gb/s DPSK channels on existing 10 Gb/s network infrastructure for implementing hybrid systems.

The MSK format offers a spectrally efficient modulation scheme compared with its DPSK counterpart at the same bit rate. As a subset of continuous phase frequency shift keying (CPFSK), MSK possesses spectrally efficient attributes of the CPFSK family. The frequency deviation of MSK is equal to one-quarter of the bit rate, and this frequency deviation is also the minimum spacing to maintain orthogonality between two FSK-modulated frequencies. Conversely, MSK can be considered a particular case of offset differential quadrature phase shift keying (ODQPSK) [42–46], which enables MSK to be represented by I and Q components on the signal constellation. The advantageous characteristics of the optical MSK format can be summarized as follows: (i) compact spectrum, which is of particular interest for spectrally efficient and high-speed transmission systems; this also provides robustness to tight optical filtering; (ii) high suppression of spectral side lobes in the optical power spectrum compared with DPSK. The roll-off factor follows f^{-4} rather than f^{-2}, as in the case of DPSK; this also reduces the effects of inter-channel cross-talks; (iii) no high-power spectral spikes in the power spectrum, thus reducing fiber nonlinear effects compared with OOK; (iv) as a subset of CPFSK or ODQPSK, MSK can be detected incoherently based on the phase or frequency of the lightwave carrier, or coherently based on the popular I-Q detection structure; and (v) constant envelop property, which eases the measure of the average optical power.

Several studies have been conducted recently investigating the generation and direct detection of externally modulated optical MSK signals [26–29]. However, there are few reports investigating the performance of an externally modulated MSK format for digital photonic transmission systems, particularly at high bit rates such as 40 Gb/s [26,47,48]. Furthermore, if MSK can be combined with a multi-level modulation scheme, transmission baud rate would be reduced in addition to the spectral efficiency of MSK formats. This is of great interest for long-haul and metropolitan optical networks, and provides the main motivation for proposing dual-level MSK modulation format in this research. In addition, the potential of the optical dual-level MSK format transmission has yet to be explored. Therefore, another main contribution of this research is to provide comprehensive studies on the performance of MSK and dual-level MSK modulation formats for long-haul and metropolitan optical transmission systems.

1.2.3 INCOHERENT OPTICAL RECEIVERS

The modulation formats studied in this research, optical DPSK and MSK-based formats, can be demodulated incoherently by using an optical balanced receiver that employs a Mach–Zehnder delay interferometer (MZDI). In the optical DPSK format, MZDI is used to detect differentially coded phase information between every two consecutive symbols [6,34,35,49]. This detection is carried out in the photonic domain because it is beyond the speed of the electrical domain, particularly at very high bit rates of ≥40 Gb/s. The MZDI balanced receiver is also used for incoherent detection of optical MSK signals [26,28,29] by also detecting the differential phase of MSK-modulated optical pulses. However, using the MZDI-based detection scheme, it is found that optical MSK provides slight improvement for CD tolerance compared with DPSK and OOK [28,30].

As a subset of the CPFSK family, MSK-modulated lightwaves can also be incoherently detected based on the principles of optical frequency discrimination. Thus, an optical frequency discrimination receiver (OFDR) employing dual narrowband optical filters and an optical delay line (ODL) is proposed in this research. This receiver scheme effectively mitigates CD-induced ISI effects and enables breakthrough CD tolerances for optical MSK transmission, as reported in my first-authored papers [30,31]. In addition, the feasibility of this novel receiver is based on recent advances in the design of optical filters, in particular the micro-ring resonator filters. Such optical filters have very narrow bandwidths, e.g., <2 GHz (3-dB bandwidth) and they have been realized commercially by Little Optics [50–53]. This research provides comprehensive study of this OFDR scheme, from operational principles to analysis of receiver design, and on the performance of OFDR-based MSK optical transmission systems.

1.2.4 COHERENT OPTICAL RECEIVERS

Coherent detection and transmission techniques were extensively exploited in the mid-1980s to extend the repeater-less distance a further 20–40 km of SSMF with an expected improvement of receiver sensitivity of 10–20 dB depending on the modulation format and receiver structure using phase or polarization diversity.

In general, a coherent receiver would operate on the beating of received optical signals and that of the field of a local laser oscillator. The beating optical signals are then detected by a photodetector with the phase of the carrier preserved, permitting detection of the phase of the carriers. Hence, phase modulation and continuous phase or frequency modulation signals can be processed in the electronic domain. With advancement of digital electronic processor, processing of the received signals in the intermediate frequency (IF) or base band of heterodyne and homodyne detection, respectively, can be processed to determine the phase of the modulated and transmitted signals. Coherent receivers for different modulation formats are described in appropriate sections of the chapters of this book, but particularly in Chapter 4.

The three type of coherent receivers (homodyne, heterodyne, intradyne) detection techniques are dependent on the frequency difference of zero, IF greater or smaller than the passband of the signals between the local oscillator laser and that of the signal carrier. With modern advanced optically amplified fiber communications, broadband ASE noises are always present and, under coherent detection, the beating between the local laser source and ASE dominate the electronic noise of the receiver at the front-end. These noise considerations are described in Chapter 4.

1.2.5 TRANSMISSION OF ULTRA-SHORT PULSE SEQUENCE

The bit rate of all optical transmission systems and networks has been proposed to reach 100 Gb/s and 160 Gb/s. At this rate, the processing speed of electronics will encounter severe difficulty. The generation of data pulse sequence at this speed can be derived from ultra-short pulses of mode-locked fiber lasers with a width of sub-picoseconds. To deal with the dispersiveness of these ultra-short pulses, equalization can be done using the principles of temporal imaging or effectively focusing and defocusing of time-domain pulses using optical quadratic phase modulation.

1.2.6 ELECTRONIC EQUALIZATION

Electronic equalizers have recently become one of the most promising solutions for future high-performance optical transmission systems. The Si—Ge technological development has enhanced electron mobility and hence shorter rise and fall time of pulse propagation, which has increased processing speed. Sampling rate can now reach several Giga-samples/s, enabling processing of 10 Gb/s bit rate data channel without difficulty. It is very probable that electronic processing and equalization can be implemented in real systems in the very near future (Figure 1.9).

The channel is single-mode optical fibers of dispersive in the negative or positive factors with some residual dispersion. Thus, pulse distortion is purely phase distortion before detection by the photodiode, which follows the square law rule for direct detection. Conversely, for coherent detection, the beating between the local oscillator and the signal in the photodetection device would lead to phase preservation, and could be considered a pure phase distortion. In direct detection, after square law detection, phase distortion is transferred to the amplitude distortion. To conduct the equalization process, it is important to know the impulse and step responses of the fiber channel $h_F(t)$ and $s_F(t)$. We then give the fundamental aspects of equalization using feed forward equalization, decision feedback equalization with maximum mean square error (MMSE) or maximum likelihood sequence estimation of Viterbi algorithm (MLSE).

1.2.6.1 Feed Forward Equalizer

FFE is a linear equalization process which has been widely studied. A transversal filter would offer such linear processing of the distorted signals for correction of the distortion process prior to decision making. The decision-feedback equalizer (DFE) is illustrated in Figure 1.11. It consists of m coefficients and m delay taps. Each tap spacing equals to the bit duration T. From Figure 3.2b we can see that the received signal sequence goes through the forward filter firstly. After making decisions on previously detected symbols, the feedback filter provides information from the previously detected symbols for present estimating. The structure of a FFE transversal filter consists of a cascade delay of the input sample and, at teach delay, the signal is tapped and multiplied with coefficients. These tapped signals, whose delay tap time is the bit period, are then summed to give the output sample. The coefficients of the transversal filter must take values that match with the channel, so the convolution of the channel impulse response and that of the filter result to unity so as to achieve complete equalized pulse sequence at the output. Figure 1.10a and Figure 1.10b show the linear equalization scheme that uses feed forward

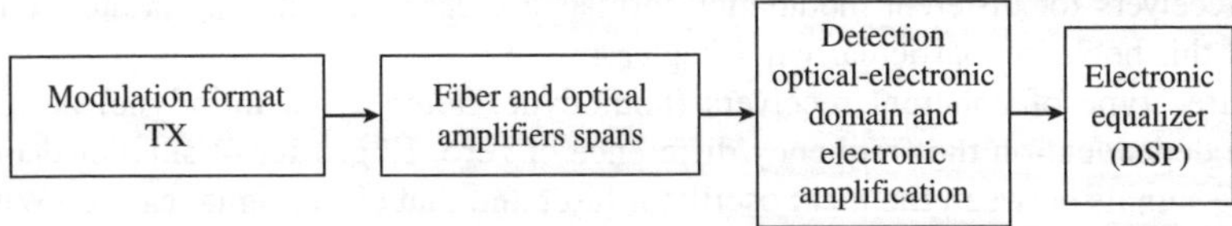

FIGURE 1.9 Location of the electronic equalizer at the receiver of an advanced optical communications system (schematic). DSP = digital signal processor.

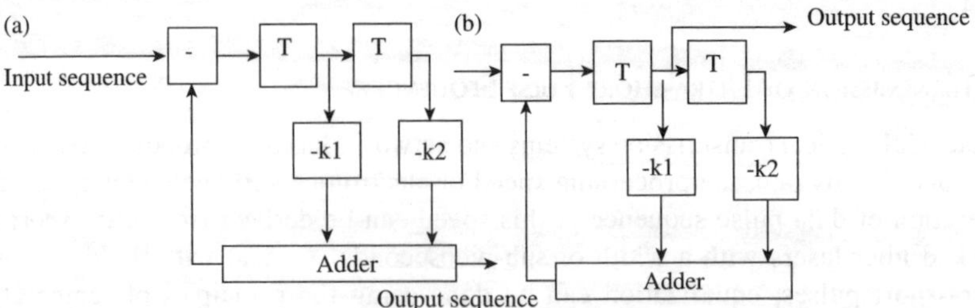

FIGURE 1.10 Linear (a) feed forward equalization (transversal equalization) and (b) feedback equalization scheme.

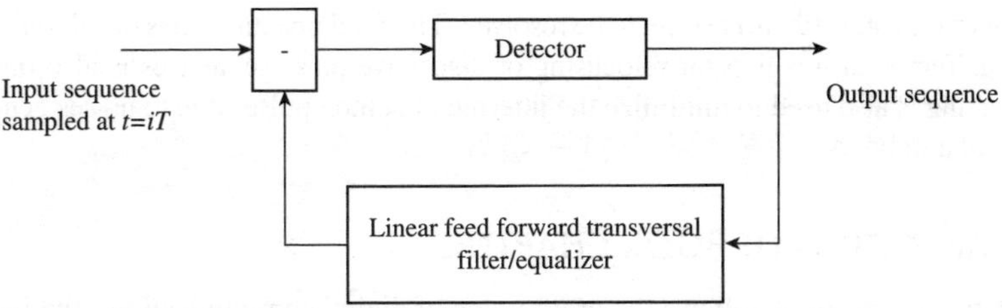

FIGURE 1.11   Receiver using nonlinear equalization by decision-directed cancellation of ISI (schematic).

or feedback equalization. The difference between these two schemes is the tapped signals at the output of the transversal filter or at the output of the feedback that minimizes the input sequence.

Conversely, a decision feedback equalizer (DFE) differs from that of the linear equalizer with a decision detector that would determine the signal amplitude required for feed back to the difference error at the input (Figure 1.11).

1.2.6.2   Decision Feedback Equalization

The feedforward equalization method is based on the use of a linear filter with adjustable coefficients. The equalization method that exploits the use of previous detected symbols to suppress the ISI in the present symbol being detected, is termed as decision-feedback equalization.

1.2.6.3   Minimum Mean Square Error Equalization

Consider the general case where the linear equalizer is adjusted to minimize the mean square error due to ISI and noise; this is called the MMSE.

1.2.6.4   Placement of Equalizers

Linear and nonlinear equalization are possible because they are well known in signal processing. The principles of equalization with equalizers placed at the transmitter, the receiver, or sharing between the transmitter and receiver, are critical for practical networks.

1.2.6.5   MLSE Electronic Equalizers

Among electronic equalization techniques, MLSE (which can be implemented effectively with the Viterbi algorithm) has attracted considerable research interest [54–59]. However, most of the studies on MLSE equalizers have focused on ASK or DPSK formats [54,55,58,60]. Apart from a co-authored paper reported recently [32], there has not been a study on the performance of MLSE equalizers for optical MSK transmission, particularly when OFDR is used as the detection scheme. Performance of OFDR-based MSK optical transmission systems is significantly enhanced with the incorporation of post-detection MLSE electronic equalizers because the ISI problem caused by fiber dispersion impairments or tight optical filtering effects is effectively mitigated. Therefore, this decision process is included as a case study in a chapter of this book to comprehensively investigate the performance of MLSE equalizers for OFDR-based MSK optical transmission systems.

1.2.7   Ultra-short Pulse Transmission

The need for ultra-high speed transmission beyond the processing speed of electronic devices drives the interest in transmission techniques in the photonic domain using an ultra-short pulse with sufficient energy to interplay with the nonlinear SPM effects for balancing the linear dispersion, the solitons. Because the first installed soliton long-haul transmission system is in operation between South and Western Australia, practical demonstration has proved the superiority of solitons. However, challenges regarding environmental effects have yet to be resolved. The fundamental issues of soliton transmission are described. Temporal imaging for refocusing of dispersive pulse sequences is also outlined. This temporal imaging is also used to minimize the jittering of soliton pulses due to noises generated from cascade optical amplifiers.

1.3 ORGANIZATION OF BOOK CHAPTERS

The presentation of this book follows the progression of digital communications and integration of modulation techniques in optical communications, as described below.

In Chapter 2, optical transmitter configurations based on the principles of operation of interferometric effects for generation of phase and frequency modulation, either CPFSK format or I-Q structure of ODQPSK format, are proposed for generating optical MSK signals. Conversely, the generation scheme of optical dual-level MSK format comprises two MSK optical transmitters connected in parallel. The spectral characteristics of these formats are also discussed in detail. In an optical transmitter, data modulation is implemented by using external electro-optic phase modulators (EOPM) or MZIMs. Phasor principles are extensively applied in this chapter to derive the modulation of the carrier phase and amplitude.

Chapter 3 provides an insight into various advanced optical modulation formats, including RZ and NRZ pulse shaping, MSK and its multi-level version, and dual-level MSK. The focus of this chapter is the generation and detection schemes of these formats. Chapter 3 also describes the modeling of critical optical sub-systems of a digital photonic transmission system. System modeling is described sequentially from the optical transmitter, onto the fiber channel, and to the optical receiver. Lightwaves generated from the optical transmitter and propagating along the optical fiber are degraded by fiber dispersion impairments and fiber nonlinearities. Detailed descriptions of these impairments (particularly fiber CD) are presented. These fiber dynamics are embedded in the nonlinear Schrodinger equation (NLSE), which is solved numerically by symmetric split step Fourier method (SSFM). Chapter 3 also describes the modeling of amplified spontaneous emission (ASE) noise and receiver noise sources, as well as modeling of optical and electrical filters. Key signal quality metrics and methods used to evaluate the performance of optical transmission systems are described in detail. The final part of Chapter 3 discusses the advantages of the MATLAB® Simulink® modeling platform developed for advanced digital photonic transmission systems.

A fast method for evaluation of the statistical properties of the distribution of the received eye diagrams is described. This enables measurement of the bit error rate (BER) from the received eye diagram rather than the Monte Carlo Method, which would consume considerable time for computing the errors.

Chapter 2 and Chapter 3 provide an insight on various advanced optical modulation formats studied in this research, including return-to-zero (RZ), carrier suppressed RZ (CSRZ) pulse shaping, differential phase shift keying (DPSK) and DQPSK (quadrature) with discrete phase shift keying. Modulation with continuity of the phase at the transition of the logical bits is also demonstrated, such as continuous phase frequency shift keying (CPFSK) and minimum shift keying (MSK) and its multi-level version, dual-level MSK. The focus of these chapters is the generation and detection schemes of these formats. Two optical transmitter configurations which are based on principles of the CPFSK format or I-Q structure of the ODQPSK format are proposed for generating optical MSK signals. Conversely, the generation scheme of optical dual-level MSK format comprises two MSK optical transmitters connected in parallel. The spectral characteristics of these formats are also discussed in detail. Further details of transmitters and receivers can be found in sections of later chapters addressing advanced modulation format optical transmission systems.

Why should Simulink® be developed while there are various commercial simulation packages (e.g., VPI systems) available? Simulink is a modular tool box attached to MATLAB that is now used as a universal computing tool in universites worldwide. Its communications blockset contains many tool boxes for digital communications in the base band and passband. Thus, these toolboxes can be used as a modular set for generation and detection of lightwaves signals. Because the frequency of the lightwaves is much higher than base band signals, it is unlikely that the carrier is contained under the envelope of the signals because an extremely high sampling rate would be required, which is currently not possible in computing systems. Thus, complex envelope signals are used to represent the carrier phase, allowing simulation models for digital optical communications to be modeled very easily and accurately due to the modularity of Simulink. From chapter to chapter, Simulink models are given and an appendix is dedicated to the modeling techniques of digital optical communications. Simulink techniques and model samples may be made available as a toolbox in MATLAB in the future so that laboratories worldwide can equip and facilitate teaching and research activities. Thus, the Simulink models described in this book could be easily integrated into a standard MATLAB software package in university, research and engineering laboratories. This is particularly relevant because the toolset "Digital Optical Communications" is the most universally affordable package for students of advanced level telecommunications engineering.

Chapter 4 provides the fundamental techniques of coherent optical communication; the optical receivers and associated noises in these receiving systems are described. The principal motivation of the introduction in this chapter is the emerging technological developments in photonic and optoelectronic components, and digital signal processors. The limitation and obstacles due to the linewidth of the laser source is no longer a major hurdle; it is now used in the transmitters and as a local oscillator at the receiver. The high sampling speed of electronic digital processors enable estimation of the phase of the signal carrier after signals beating with the local oscillator. Coherent techniques must therefore be described for applications in modern digital optical communication systems. These high-speed digital signal processors are also employed as electrical equalization systems to compensate for disturbance or residual dispersion in optical transmission systems. Equalization techniques in the electrical domain of digital optical communications are described in Chapter 10. In coherent receivers, the noise sources due to beating between the oscillator and ASE noise sources would dominate electronic noises at the front-end of the receiver. Analysis and simulation are described in this chapter.

Chapter 5 presents the theoretical and experimental results for studying performance characteristics of 40 Gb/s optical DPSK, with the focus on its robustness to tight optical filtering. This chapter introduces phase discrete modulation, especially differential phase coding. It is to show the feasibility of implementing cost-effective hybrid optical transmission systems in which 40 Gb/s DPSK channels can be co-transmitted simultaneously with 10 Gb/s OOK channels over the existing 10 Gb/s network infrastructure. These experiments were conducted in collaboration with commercial and industrial engineering partners SHF Communications Technology AG (Germany) and Research Laboratories-Telstra Corporation (Australia).

Chapter 6 continues with the phase modulation of the carrier, but continuously rather than discrete. This is different to the description in Chapter 5, in which minimum shift keying (MSK) is described extensively because it is the most efficient continuous phase modulation due to the orthogonal properties of the spectra of the modulated signals at the two distinct carriers.

Chapter 7 introduces multi-level in amplitude and phase modulation formats. The driving conditions for a dual-drive MZIM for generation of multi-level ADPSK are derived, as well as the models in MATLAB Simulink platform. Dual-level MSK signals are also given. A 16-ADPSK and 16-star quadrature amplitude modulation (QAM) are studied as two typical case studies that would allow reduction of the symbol rate from 100 Gb/s to 25 Gb/s, at which electronic processing could assist in detection using direct or coherent detection.

A novel incoherent detection scheme, the optical frequency discrimination receiver (OFDR), is proposed in Chapter 8 for optical MSK systems aiming to extend significantly uncompensated optical links. Receiver design is optimized by analyzing the significance of key sub-system optical components on receiver performance. These receiver design guidelines are verified with analytical and simulation results. Performance of OFDR-based optical MSK systems are evaluated numerically for receiver sensitivities, CD and PMD tolerances, resilience to SPM non-linear effects, and transmission limits due to high PMD coefficient of old fiber vintages.

Chapter 9 introduces equalization processes in the electronic domain so as to equalize the transmitted pulse sequence suffered from impairments of the transmission media, the quadratic phase optical fibers. Both linear and non-linear equalization processes are described. Noise contributions in the equalization processes are stated and derived for the BER. Transmission examples are given for duo-binary modulation and MSE with Viterbi trellis tracing. This chapter briefly revises the principles of MLSE equalization technique, Viterbi algorithm, and state trellis structure. Detailed explanations of state-based Viterbi-MLSE equalizers for optical communications are given. This chapter then investigates the performance of MLSE equalizers for 40 Gb/s OFDR-based optical MSK systems. In this receiver scheme, OFDR serves as the optical front-end and is integrated with a post-detection MLSE equalizer. The CD tolerance performance of Viterbi-MLSE and template-matching MLSE equalizers is studied. The performance limit of Viterbi-MLSE equalizers with 2^4 to 2^{10} states is carried out based on maximum uncompensated transmission distances. The performance of 16-state Viterbi-MLSE equalizers for PMD equalization is also investigated. This number of state reflects the feasibility of high-speed electronic signal processing in the near future. The significance of multi-sample sampling schemes (two and four samples per one bit period) over the conventional single-sample sampling technique is also highlighted.

Chapter 10 introduces multi-sub-carrier modulation with its carriers orthogonal to eachother, the orthogonal frequency division multiplexing (OFDM), in which the FFT and IFFT digital signal processors are extensively exploited to combat dispersion compensation in long-haul optically amplified transmission systems. Thus, there is a possibility that dispersion-compensating fiber sections can be discarded, allowing reduction of the number of optical amplifiers in spans, therefore reduction of ASE noises, permitting longer transmission reach.

Chapter 11 introduces impulse and step responses of quadratic phase-guided optic media, the single-mode optical fiber. This chapter gives an equivalent temporal imaging in the time domain for pulse multiplication and compensation of ultra-short pulse transmission. Simulation results are given to demonstrate the photonic equalization of ultra-short pulse propagation.

Chapter 12 completes the picture of ultra-short pulse transmission by presenting the transmission of solitons over optically amplified transmission spans. Although soliton transmission was developed in the late twentieth century, the first soliton 10 Gb/s was installed between Adelaide and Perth of Australia in the late 1990s, and proved to be robust in practice. This chapter reactivates soliton transmission.

Chapter 13 deals with multi-sub-carriers carrying information data with the orthogonal property of adjacent sub-channels, the orthogonal frequency division multiplexing (OFDM). The lowering of data rates of sub-channels assists the mitigation of linear and nonlinear impairments of the fiber channel. The principles and performances of the scheme are described.

Chapter 14 summarizes the key achievements of digital optical communications in the transmission of several Tera-bits/s capacity over single-mode optical fibers. A brief comparison of the modulation formats is given incorporating experimental data. The roles of digital communications in emerging advanced photonic transmission systems and networks are identified. Furthermore, emerging ultra-high-speed electronic processors have reached the possibility for real-time equalization of distortion and other disturbance effects in long-haul ultra-high-speed optical communications.

Practice problems are given (if appropriate) at the end of each chapter.

REFERENCES

1. Kaminow, I. P., and T. Li. 2002. *Optical fiber communications, volume IVB*. USA: Elsevier.
2. Sanferrare, R. S. 1987. Terrestrial lightwave systems. *AT&T Technology Journal* 66: 95–107.
3. Lin, C., H. Kogelnik, and L. G. Cohen. 1980. Optical pulse equalization and low dispersion transmission in single-mode fibers in the 1.3–1.7 μm spectral region. *Optics Letters* 5: 476–78.
4. Korotky, S., A. Gnauck, B. Kasper, J. Campbell, J. Veselka, J. Talman, and A. McCormick. 1987. 8-Gbit/s transmission experiment over 68 km of optical fiber using a Ti:LiNbO$_3$ external modulator. *IEEE J. Lightwave Technology* 5 (10): 1505–09.
5. Kogelnik, H. 1985. High-speed lightwave transmission in optical fibers. *Science* 228: 1043–48.

6. Agrawal, G. P. 2002. *Fiber-optic communication systems.* 3rd ed. New York: Wiley.

7. Chraplyvy, A. R., A. H. Gnauck, R. W. Tkach, and R. M. Derosier. 1993. 8 × 10 Gb/s transmission through 280 km of dispersion-managed fiber. *IEEE Photonics Technology Letters* 5: 1233–35.

8. Kogelnik, H. 2000. High-capacity optical communications: Personal recollections. *IEEE Journal on Selected Topics in Quantum Electronics* 6 (6): 1279–86.

9. Giles, C. R., and E. Desurvire. 1991. Propagation of signal and noise in concatenated erbium-doped fiber amplifiers. *IEEE Journal of Lightwave Technology* 9 (2): 271–83.

10. Becker, P. C., N. A. Olsson, and J. R. Simpson. 1999. *Erbium-doped fiber amplifiers, fundamentals and technology.* San Diego, CA: Academic Press.

11. Farries, M. C., P. R. Morkel, R. I. Laming, T. A. Birks, D. N. Payne, and E. J. Tarbox. 1989. Operation of erbium-doped fiber amplifiers and lasers pumped with frequency-doubled Nd:YAG lasers. *IEEE Journal of Lightwave Technology* 7 (10): 1473–77.

12. Okoshi, T. 1982. Heterodyne and coherent optical fiber communications: Recent progress. *IEEE Transactions on Microwave Theory and Techniques* 82 (8): 1138–49.

13. Okoshi, T. 1987. Recent advances in coherent optical fiber communication systems. *IEEE Journal of Lightwave Technology* 5 (1): 44–52.

14. Salz, J. 1986. Modulation and detection for coherent lightwave communications. *IEEE Communications Magazine* 24 (6): 38–49.

15. Okoshi, T. 1986. Ultimate performance of heterodyne/coherent optical fiber communications. *IEEE Journal of Lightwave Technology* 4 (10): 1556–62.

16. Henry, P. S. 1990. *Coherent lightwave communications.* New York: IEEE Press.

17. Elrefaie, A. F., R. E. Wagner, D. A. Atlas, and A. D. Daut. 1988. Chromatic dispersion limitation in coherent lightwave systems. *IEEE Journal of Lightwave Technology* 6 (5): 704–10.

18. Charlet, G., E. Corbel, J. Lazaro, A. Klekamp, W. Idler, R. Dischler, S. Bigo, et al. 2005. Comparison of system performance at 50, 62.5 and 100 GHz channel spacing over transoceanic distances at 40 Gbit/s channel rate using RZ-DPSK. *Electronics Letters* 41 (3): 145–46.

19. Cho, P. S., V. S. Grigoryan, Y. A. Godin, A. Salamon, and Y. Achiam. 2003. Transmission of 25-Gb/s RZ-DQPSK signals with 25-GHz channel spacing over 1000 km of SMF-28 fiber. *IEEE Photonics Technology Letters* 15 (3): 473–75.

20. Ishida, K., T. Kobayashi, J. Abe, K. Kinjo, S. Kuroda, and T. Mizuochi. 2003. A comparative study of 10 Gb/s RZ-DPSK and RZ-ASK WDM transmission over transoceanic distances. In *Proceedings of OFC'03*, vol. 2, 451–53.

21. Atia, W. A., and R. S. Bondurant. 1999. Demonstration of return-to-zero signaling in both OOK and DPSK formats to improve receiver sensitivity in an optically preamplified receiver. In *Proceedings of IEEE LEOS '99*, vol. 1, 226–27.

22. Bosco, G., A. Carena, V. Curri, R. Gaudino, and P. Poggiolini. 2002. On the use of NRZ, RZ, and CSRZ modulation at 40 Gb/s with narrow DWDM channel spacing. *IEEE Journal of Lightwave Technology* 20 (9): 1694–1704.

23. Gnauck, A. H., G. Raybon, P. G. Bernasconi, J. Leuthold, C. R. Doerr, and L. W. Stulz. 2003. 1-Tb/s (6/spl times/170.6 Gb/s) transmission over 2000-km NZDF using OTDM and RZ-DPSK format. *IEEE Photonics Technology Letters* 15 (11): 1618–20.

24. Zhu, B., L. E. Nelson, S. Stulz, A. H. Gnauck, C. Doerr, J. Leuthold, L. Gruner-Nielsen, M. O. Pedersen, J. Kim, and R. L. Lingle, Jr. 2004. High spectral density long-haul 40-Gb/s transmission using CSRZ-DPSK format. *IEEE Journal of Lightwave Technology* 22 (1): 208–14.

25. Hirano, A., Y. Miyamoto, and S. Kuwahara. 2003. Performances of CSRZ-DPSK and RZ-DPSK in 43-Gbit/s/ch DWDM G.652 single-mode-fiber transmission. In *Proceedings of OFC'03*, vol. 2, 454–56, Anaheim, CA.

26. Binh, L. N., and T. L. Huynh. 2007. Linear and nonlinear distortion effects in direct detection 40 Gb/s MSK modulation formats multi-span optically amplified transmission. *Optics Communications* 237 (2): 352–61.

27. Mo, J., D. Yi, Y. Wen, S. Takahashi, Y. Wang, and C. Lu. 2005. Optical minimum-shift keying modulator for high spectral efficiency WDM systems. In *Proceedings of ECOC'05*, vol. 4, 781–82.

28. Mo, J., Y. J. Wen, Y. Dong, Y. Wang, and C. Lu. 2005. Optical minimum-shift keying format and its dispersion tolerance. In *Proceedings of OFC'05, Paper JThB12*, Anaheim, CA, 1–3.

29. Ohm, M., and J. Speidel 2004. Optical minimum-shift keying with direct detection (MSK/DD). In *Proceedings of SPIE on Optical Transmission, Switching and Systems*, vol. 5281, 150–61.

30. Huynh, T. L., T. Sivahumaran, L. N. Binh, and K. K. Pang. 2007. Narrowband frequency discrimination receiver for high dispersion tolerance optical MSK systems. In *Proceedings of Coin-Acoft'07, Paper TuA1-3*, June, Melbourne, Australia, 1–3.

31. Huynh, T. L., T. Sivahumaran, L. N. Binh, and K. K. Pang. 2007. Sensitivity improvement with offset filtering in optical MSK narrowband frequency discrimination receiver. In *Proceedings of Coin-Acoft '07, Paper TuA1-5*, June, Melbourne, Australia, 1–5.

32. Sivahumaran, T., T. L. Huynh, K. K. Pang, and L. N. Binh. 2007. Non-linear equalizers in narrowband filter receiver achieving 950 ps/nm residual dispersion tolerance for 40 Gb/s optical MSK transmission systems. In *Proceedings of OFC'07, Paper OThK3*, Anaheim, CA, 1–3.

33. Sakamoto, T., T. Kawanishi, and M. Izutsu. 2005. Optical minimum-shift keying with external modulation scheme. *Optics Express* 13: 7741–47.

34. Gnauck, A. H., and P. J. Winzer. 2005. Optical phase-shift-keyed transmission. *IEEE Journal of Lightwave Technology* 23 (1): 115–30.

35. Lazaro, J. A., W. Idler, R. Dischler, and A. Klekamp. 2004. BER depending tolerances of DPSK balanced receiver at 43Gb/s. In *Proceedings of IEEE/LEOS Workshop on Advanced Modulation Formats 2004*, 15–16.

36. Kim, H., and A. H. Gnauck. 2003. Experimental investigation of the performance limitation of DPSK systems due to nonlinear phase noise. *IEEE Photonics Technology Letters* 15 (2): 320–22.

37. Bhandare, S., D. Sandel, A. F. Abas, B. Milivojevic, A. Hidayat, R. Noe, M. Guy, and M. Lapointe. 2004. 2/spl times/40 Gbit/s RZ-DQPSK transmission with tunable chromatic dispersion compensation in 263 km fibre link. *Electronics Letters* 40 (13): 821–22.

38. Mizuochi, T., K. Ishida, T. Kobayashi, J. Abe, K. Kinjo, K. Motoshima, and K. Kasahara. 2003. A comparative study of DPSK and OOK WDM transmission over transoceanic distances and their performance degradations due to nonlinear phase noise. *IEEE Journal of Lightwave Technology* 21 (9): 1933–43.

39. Xu, C., X. Liu, L. F. Mollenauer, and X. Wei. 2003. Comparison of return-to-zero differential phase-shift keying and ON–OFF keying in long-haul dispersion managed transmission. *IEEE Photonics Technology Letters* 15 (4): 617–19.

40. Binh, L. N., and T. L. Huynh. 2007. Phase-modulated hybrid 40 Gb/s and 10 Gb/s DPSK DWDM long-haul optical transmission. In *Proceedings of OFC '07, Paper JWA94*, Anaheim, CA, 1–3.

41. Ito, T., K. Sekiya, and T. Ono. 2002. Study of 10 G/40 G hybrid ultra long haul transmission systems with reconfigurable OADM's for efficient wavelength usage. In *Proceedings of ECOC'02, Paper 1.1.4*, Copenhagen, Denmark, 1–4.

42. Iwashita, K., and N. Takachio. 1989. Experimental evaluation of chromatic dispersion distortion in optical CPFSK transmission systems. *IEEE Journal of Lightwave Technology* 7 (10): 1484–87.

43. Proakis, J. G. 2001. *Digital communications*. 4th ed. New York: McGraw-Hill.

44. Proakis, J. G., and M. Salehi. 2002. *Communication systems engineering*. 2nd ed., 522–24. Englewood Cliffs, NJ: Prentice Hall.

45. Pang, K. K. 2005. *Digital transmission*. Melbourne, Australia: Mi-Tec Media.

46. Iwashita, K., and T. Matsumoto. 1987. Modulation and detection characteristics of optical continuous phase FSK transmission system. *IEEE Journal of Lightwave Technology* 5 (4): 452–60.

47. Mo, J., Y. J. Wen, and Y. Wang. 2007. Performance evaluation of externally modulated optical minimum shift keyed data. *Optical Engineering* 46 (3): 035001.

48. Huynh, T. L., L. N. Binh, and K. K. Pang. 2006. Optical MSK long-haul transmission systems. In *Proceedings of SPIE APOC'06, Paper 6353–86, Thu9a*. Asia Pacific Optical Communications APOC'06, Kwangju, Korea, September 2006.

49. Elrefaie, A. F., and R. E. Wagner. 1991. Chromatic dispersion limitations for FSK and DPSK systems with direct detection receivers. *IEEE Photonics Technology Letters* 3 (1): 71–73.

50. Little, B. E. 2003. Advances in microring resonator. *Integrated Photonics Research Conference 2003*, Invited Paper.

51. Van, V. B. E. Little, S. T. Chu, and J. V. Hryniewicz. 2004. Micro-ring resonator filters. In *Proceedings of LEOS'04*, vol. 2, 571–72.

52. Absil, P. P., S. T. Chu, D. Gill, J. V. Hryniewicz, F. Johnson, O. King, B. E. Little, F. Seiferth, and V. Van. 2004. Very high order integrated optical filters. In *Proceedings of OFC'04*, vol. 1, Anaheim, CA, 1–4.

53. Brent, L., C. Sai, C. Wei, C. Wenlu, H. John, G. Dave, K. Oliver, et al. 2006. Advanced ring resonator based PLCs. In *IEEE Lasers & Electro-Optics Society*, 751–52.

54. Alic, N., G. C. Papen, R. E. Saperstein, L. B. Milstein, and Y. Fainman. 2005. Signal statistics and maximum likelihood sequence estimation in intensity modulated fiber optic links containing a single optical preamplifier. *Optics Express* 13 (12): 4514–25.

55. Alic, N., G. C. Papen, and Y. Fainman. 2004. Performance of maximum likelihood sequence estimation with different modulation formats. In *Proceedings of IEEE/LEOS Workshop on Advanced Modulation Formats*, 49–50.

56. Curri, V., R. Gaudino, A. Napoli, and P. Poggiolini. 2004. Electronic equalization for advanced modulation formats in dispersion-limited systems. *IEEE Photonics Technology Letters* 16 (11): 2556–58.
57. Downie, J. D., M. Sauer, and J. Hurley. 2006. Flexible 10.7 Gb/s DWDM transmission over up to 1200 km without optical in-line or post-compensation of dispersion using MLSE-EDC. In *Proceedings of OFC'06, Paper THB5*, Anaheim, CA, 1–4.
58. Agazzi, O. E., M. R. Hueda, H. S. Carrer, and D. E. Crivelli. 2005. Maximum-likelihood sequence estimation in dispersive optical channels. *IEEE Journal of Lightwave Technology* 23 (2): 749–62.
59. Napoli, A. 2006. Limits of maximum-likelihood sequence estimation in chromatic dispersion limited systems. In *Proceedings of OFC'06, Paper JThB36*, Anaheim, CA.
60. Haunstein, H. 2004. PMD and chromatic dispersion control for 10 and 40 Gb/s systems. In *Proceedings of OFC'04, Invited Paper, ThU3*, Anaheim, CA, 1–4.

53. Carr, V.M., Gaudino, A., Napoli, J.M.P., Popescu, ... "Thermose equalization for advanced graphics information ... in dispersion flattened systems, 2006. Photonics Technology Letters, 18 (1): 32, 38–38.

54. Donetz, ..., Shaw, D.J., Earley, 2006, ... PMC PMC 16-, DWDM transmission degree up to 1200 km, compensation in-line or post-compensation of dispersion using ..., ... OFC06, Paper OWX, Anaheim, CA, 1–4.

55. Aktas, ..., Ibraim, H.S. Garren and D.L. Galwell, 2006, ... non-linear in-line compensation dispersion optical channel ... IEEE Journal of Lightwave Technology, v. ... 797–799-A ...

57. Gnauk, ..., 2006, Limits of maximum likelihood sequence estimation in chromatic dispersion limited systems, in Proceedings of OFC06, Paper OWB30, Anaheim, CA.

60. Haunstein, H. 2004, Principle and performance of dispersion ... control for 10 and 40 Gb/s ..., in Proceedings of Invited Paper ..., Anaheim, CA, 1–4.

2 Photonic Transmitters

2.1 OPTICAL MODULATORS

Modulation of lightwaves via an external optical modulator can be classified into three types depending on the special effects that alter the property of lightwaves, especially the intensity or phase of the lightwave carrier. In an external modulator, the intensity is usually manipulated by changing the phase of the carrier lightwaves guided in one path of an interferometer. The Mach–Zehnder interferometric structure is the commonest type [1,2].

The electro-absorption effect (EA) modulator employs the Franz and Keldysh effect, which is observed as lengthening the wavelength of the absorption edge of a semiconductor medium under the influence of an electric field [3,5]. In quantum structure (e.g., multi-quantum well structure), this effect is called the Stark effect or the EA effect. The EA modulator can be integrated with a laser structure on the same integrated circuit chip. For a $LiNbO_3$ modulator, the device is externally connected to a laser source via an optical fiber.

The total insertion loss of semiconductor intensity modulator is about 8–10 dB, including fiber-waveguide coupling loss, which is rather high. However, this loss can be compensated by a semiconductor optical amplifier (SOA) that can be integrated on the same circuit. Compared with $LiNbO_3$, its total insertion loss is about 3–4 dB, which can be affordable because erbium-doped fiber amplifier (EDFA) is now readily available.

The driving voltage for an EA modulator is usually lower than that required for $LiNbO_3$. However, the extension ratio is not as high as that of the $LiNbO_3$ type, which is about 25 dB, compared with 10 dB for an EA modulator. This feature contrasts with the operating characteristics of $LiNbO_3$ and EA modulators. Although the driving voltage for an EA modulator is about 2–3 V and is 5–7 V for $LiNbO_3$, the former would be preferred for intensity or phase modulation formats due to the extinction ratio, which offers much lower "zero" noise level and hence a high quality factor.

2.1.1 PHASE MODULATORS

A phase modulator is a device that manipulates the "phase" of optical carrier signals under the influence of an electric field created by an applied voltage. When voltage is not applied to the RF-electrode, the number of periods of the lightwaves, n, exists in a certain path length. When voltage is applied to the RF-electrode, one or a fraction of one period of the wave is added, i.e., $(n+1)$ waves exist in the same length. In this case, the phase has been changed by 2π, and the half voltage of this is called the "driving voltage". In long-distance optical transmission, the waveform is susceptible to degradation due to non-linear effects, such as self-phase modulation. A phase modulator can be used to alter carrier phase to compensate for this degradation. The magnitude of the change of the phase depends on the change of the refractive index created via the electro-optic effect, which in turn depends on the orientation of the crystal axis with respect to the direction of the established electric field by the applied signal voltage.

An integrated optic phase modulator operates in a similar manner, except that the lightwave carrier is guided via an optical waveguide with diffused or ion-exchanged confined regions for $LiNbO_3$, and rib-waveguide structures for semiconductor types. Two electrodes are deposited so that an electric field can be established across the wave-guiding cross section, which enables a change of refractive index via the electro-optic or EA effect (Figure 2.1). For ultra-fast operation, one of the electrodes is a travelling wave type or hot electrode, and the other is a ground electrode. The travelling wave electrode must be terminated with matching impedance at the end to avoid wave reflection. Usually, impedance of one-quarter wavelength is used to match the impedance of the travelling wave electrode to that of the 50-Ω transmission line.

Phasor representation of a phase-modulated lightwave can be by circular rotation at a radial speed of ω_c. Thus, the vector with an angle ϕ represents the magnitude and phase of the lightwave.

2.1.2 Intensity Modulators

The basic structure of a LN modulator comprises: (i) two waveguides; (ii) two Y-junctions; and (iii) RF/DC traveling wave electrodes. Optical signals coming from the lightwave source are launched into the LN modulator through the polarization-maintaining fiber. They are then split equally into two branches at the first Y-junction on the substrate. If no voltage is applied to the RF electrodes, the two signals are recombined constructively at the second Y-junction and coupled into a single output. In this case, output signals from the LN modulator are recognized as "ONE". When voltage is applied to the RF electrode, due to the electro-optic effects of LN crystal substrate, the waveguide refractive index is changed, and the carrier phase in one arm is advanced (though retarded) in the other arm. Thence the two signals are recombined destructively at the second Y-junction, transformed into higher-order mode, and radiated at the junction. If the phase retarding is in multiple odd factor of π, the two signals are completely out of phase, the combined signals are radiated into the substrate and the output signal from the LN modulator is recognized as a "ZERO". The voltage difference that induces this "ZERO" and "ONE" is called the "driving voltage" of the modulator, and is one of the important parameters in deciding the performance of a modulator.

2.1.2.1 Phasor Representation and Transfer Characteristics

Consider an interferometic intensity modulator consisting of an input waveguide which splits into two branches and then recombines to a single output waveguide. If the two electrodes are initially biased with voltages V_{b1} and V_{b2}, then the initial phases exerted on the lightwaves would be $\phi_1 = \pi V_{b1}/V_\pi = - \phi_2$, which are indicated by the bias vectors shown in Figure 2.3b. From these

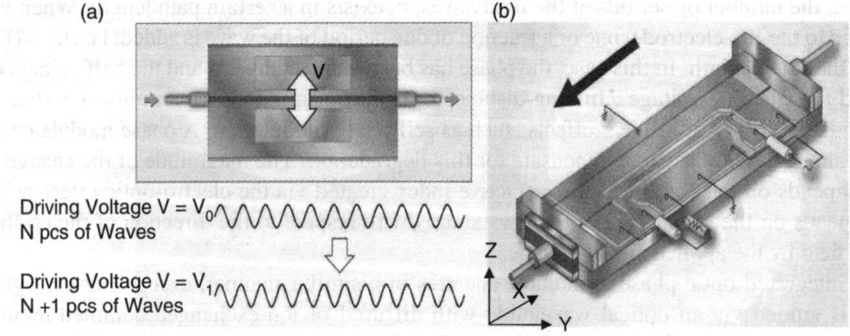

FIGURE 2.1 Electro-optic phase modulation in an integrated modulator using LiNbO3. Electrode impedance matching is not shown. (a) schematic diagram (b) integrated optic structure.

positions, the phasors are swinging according to the magnitude and sign of the pulse voltages applied to the electrodes. They can be switched to two positions: constructive or destructive. The output field of the lightwave carrier can be represented by

$$E_0 = \frac{1}{2} E_{iRMS} e^{j\omega_c t} \left(e^{j\phi_1(t)} + e^{j\phi_2(t)} \right),$$ (2.1)

where ω_c is the carrier radial frequency, E_{iRMS} is the root mean square value of the magnitude of the carrier, and $\phi_1(t)$ and $\phi_2(t)$ are the temporal phase generated by the two time-dependent pulse sequences applied to the two electrodes. Figure 2.2 illustrates the evolution of the fileds in Equation 2.1 under the cases when the applied traveling electric field causes 0 or pi phase shift so as to switch ON or OFF the lightwaves at the output port. With the voltage levels varying according to the magnitude of the pulse sequence, one can obtain the transfer curve shown in Figure 2.3a. This phasor representation can be used to determine exactly the biasing conditions and magnitude of the RF or digital signals required for driving the optical modulators to achieve 50%, 33% or 67% bit period pulse shapes.

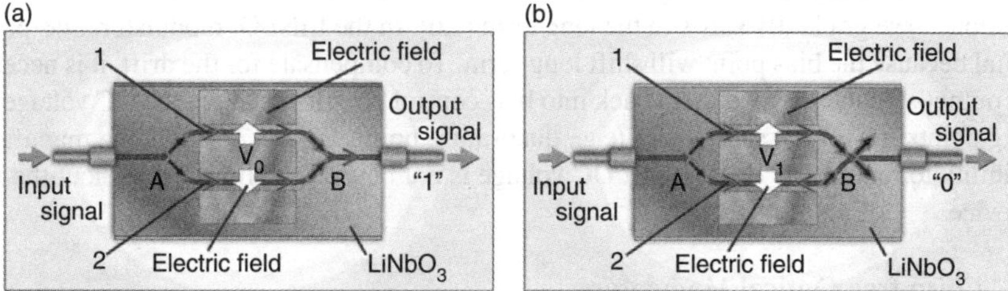

FIGURE 2.2 Intensity modulation using interferometric principles in guide wave structures in LiNbO₃ (a) ON – constructive interference mode (b) OFF – destructive interference mode. Optical-guided wave paths 1 and 2. Electric field is established across the optical waveguide.

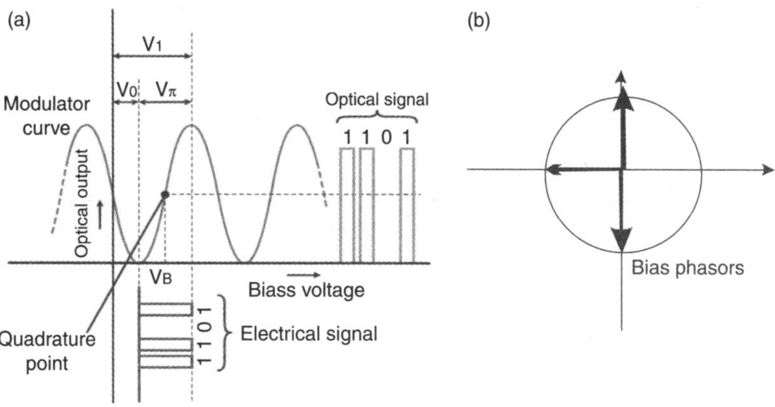

FIGURE 2.3 Electrical-to-optical transfer curve of an interferometric intensity modulator.

The power transfer function of Mach–Zehnder modulator is expressed as[*]

$$P_0(t) = \alpha P_i \cos^2 \frac{\pi V(t)}{V_\pi},\qquad(2.2)$$

where $P_0(t)$ is the output transmitted power, α is the modulator total insertion loss, P_i is the input power (usually from the laser diode), $V(t)$ is the time-dependent signal-applied voltage, and V_π is the driving voltage so that a π phase shift is exerted on the lightwave carrier. It is necessary to set the static bias on the transmission curve through the bias electrode. It is common practice to set the bias point at 50% transmission point or a $\pi/2$ phase difference between the two optical waveguide branches, the quadrature bias point. As shown in Figure 2.3, electrical digital signals are transformed into optical digital signal by switching voltage to both ends of quadrature points on the positive and negative (Figure 2.3).

2.1.2.2 Bias Control

The drift of the bias voltage affects modulator performance. For the Mach–Zehnder interferometric modulator (MZIM) type, it is very critical when it is required to bias at the quadrature point or at minimum or maximum locations on the transfer curve. DC drift is the phenomenon that occurs in $LiNbO_3$ due to the build-up of charges on the surface of the crystal substrate. Under this drift, the transmission curve gradually shifts in the long term [1,6]. In the $LiNbO_3$ modulator, bias point control is vital because the bias point will shift long-term. To compensate for the drift, it is necessary to monitor output signals and feed them back into bias control circuits to adjust the DC voltage, so that operating points stay at the same point (e.g., quadrature point; Figure 2.4). It is the manufacturer's responsibility to reduce DC drift so that DC voltage is not beyond the limit throughout the lifetime of the device.

2.1.2.3 Chirp-free Optical Modulators

Due to the symmetry of the crystal refractive index of the uniaxial anisotropy of the class m of $LiNbO_3$, the crystal cut and propagation direction of the electric field affect modulator efficiency (denoted as driving voltage) and modulator chirp. The uniaxial property of $LiNbO_3$ is shown in Figure 2.5.

In a Z-cut structure, because a hot electrode is placed on top of the waveguide, RF field flux is more concentrated, resulting in improvement of overlap between RF and the optical field (Figure 2.6).

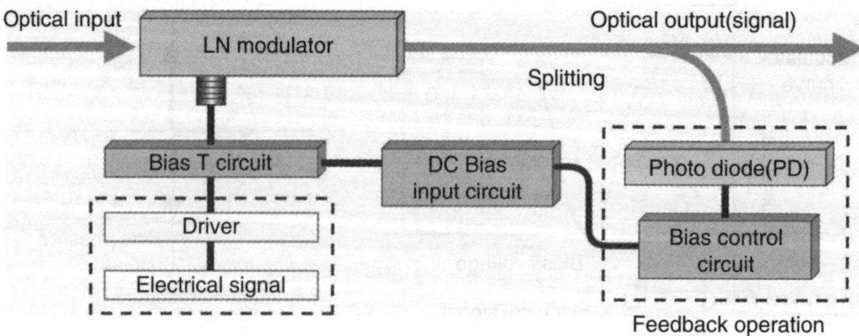

FIGURE 2.4   Arrangement of bias control of integrated optical modulators.

---

[*] Note that this equation is representative for single-drive MZIM; it is identical for dual-drive MZIM provided that the bias voltages applied to the two electrodes are equal and opposite in sign. The transfer curve of the field representation would have half the periodic frequency of the transmission curve shown in Figure 2.3.

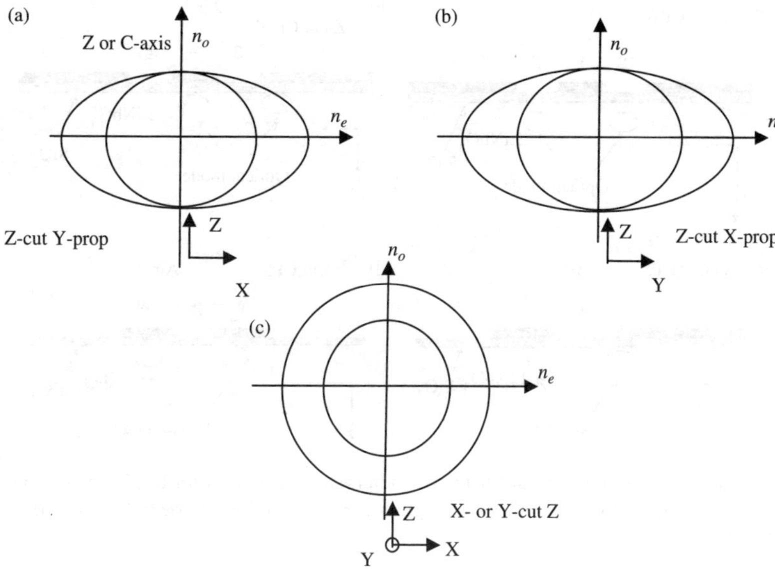

FIGURE 2.5 Refractive index contours of LiNbO3 uniaxial crystal with Z- or C- denoting the principal axis (a) Lightwaves propagation in the Z-axis polarize along Y-cut $LiNbO_3$ (b) prop- direction Z-axis and X-cut crystal (c) Y-prop and Y-cut crystal.

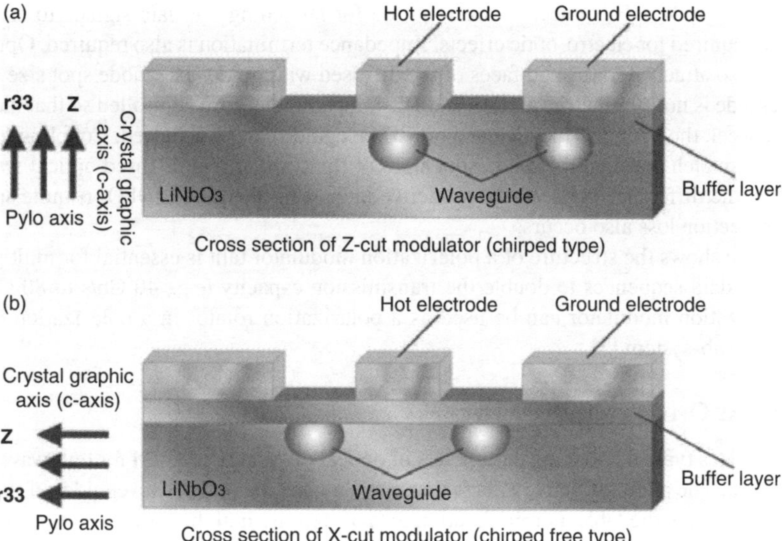

FIGURE 2.6 Different crystal cuts of $LiNbO_3$-integrated structures (a) Integration of electrodes and optical waveguides in Z-cut (b) X-cut.

However, overlap between RF in ground electrode and waveguide is reduced in the Z-cut structure so that overall improvement of driving voltage for the Z-cut structure compared with X-cut is approximately 20%. The different overlapping area for the Z-cut structure results in a chirp parameter of 0.7, whereas X-cut and Z-propagation has almost zero-chirp due to its symmetric structure. Various commonly arranged electrode and waveguide structures to maximize interaction between the travelling electric field and optical guided waves are shown in Figure 2.7. Furthermore, a buffer layer, (usually SiO_2) is used to match the velocities between these waves to optimize optical modulation bandwidth.

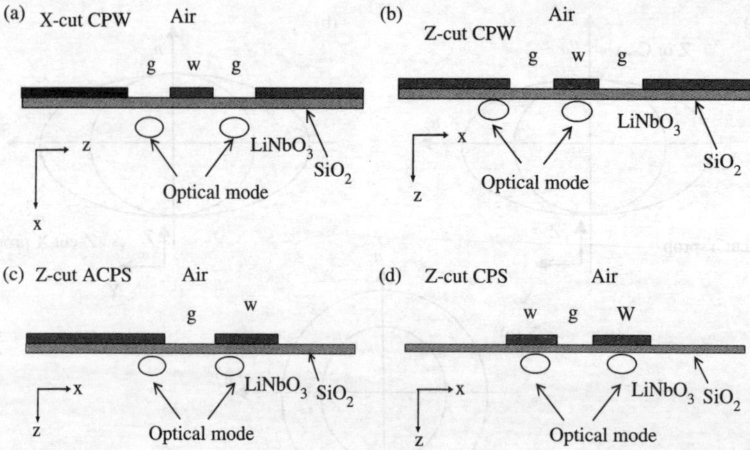

FIGURE 2.7 Commonly used electrode structure and crystal orientation for inteferometric modulation to maximize the overlap integral between the optical guided mode and the electric field distribution for largest electro-optic coefficients.

2.1.3 STRUCTURES OF PHOTONIC MODULATORS

Figure 2.8a and Figure 2.8b show the structure of a MZ intensity modulator using single- and dual-electrode configurations, respectively. The thin line electrode is called the "hot" electrode or "traveling wave" electrode. RF connectors are required for launching RF data signals to establish the electric field required for electro-optic effects. Impedance termination is also required. Optical fibers pig tails are also attached to the end faces of the diffused waveguide. The mode spot size of the diffused waveguide is not symmetric, and some diffusion parameters are controlled so that maximizing coupling between the fiber and the diffused or rib waveguide can be achieved. Coupling loss occurs due to this mismatch between the mode spot sizes of the circular and diffused optical waveguides. Furthermore, the difference between the refractive indices of fiber and $LiNbO_3$ is quite substantial, so Fresnel reflection loss also occurs.

Figure 2.8c shows the structure of a polarization modulator taht is essential for multiplexing of two polarized data sequences to double the transmission capacity (e.g., 40 Gb/s to 80 Gb/s). This type of polarization modulator can be used as a polarization rotator in a polarization dispersion compensating sub-system [7].

2.1.4 TYPICAL OPERATIONAL PARAMETERS

Table 2.1 tabulates typical operating parameters of optical modulators MZIM for lightwaves at wavelength 1550 nm. The photorefractive effects that may damage the optical waveguides of the modulator are not included in the table. For modulation formats require high linearity. The electrical signals must swing within the linear range of the power-applied voltage characteristics of the modulator.

2.2 RETURN-TO-ZERO OPTICAL PULSES

2.2.1 GENERATION

Figure 2.9 shows the conventional structure of a RZ-ASK transmitter in which two external $LiNbO_3$ MZIMs can be used. The MZIM shown in this transmitter can be a single- or dual-drive (push–pull) type. Operational principles of the MZIM were presented in Section 2.2 of the previous chapter. The optical OOK transmitter would usually consist of a narrow-linewidth laser source to generate lightwaves whose wavelength satisfies the ITU grid standard.

TABLE 2.1

Typical Operational Parameters of Optical Intensity Modulators

Parameters	Typical Values	Definition/Comments
Modulation speed	10 Gb/s	Capability to transmit digital signals
Insertion loss	Max 5 dB	Optical power loss within the modulator
Driving voltage	Max 4 V	RF voltage required to have a full modulation
Optical bandwidth	Min 8 GHz	3-dB roll-off in efficiency at the highest frequency in the modulated signal spectrum
ON/OFF extinction ratio	Min 20 dB	Ratio of maximum optical power (ON) and minimum optical power (OFF)
Polarization extinction ratio	Min 20 dB	Ratio of two polarization states (TM- and TE- guided modes) at the output

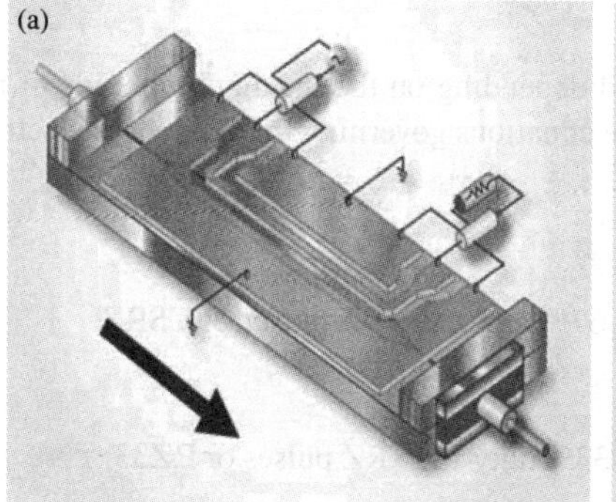
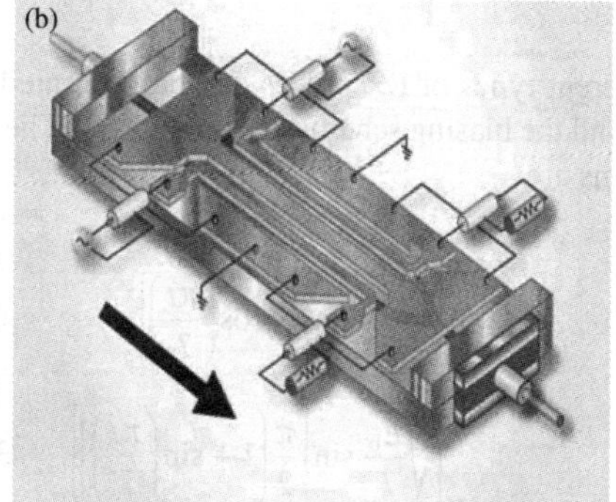
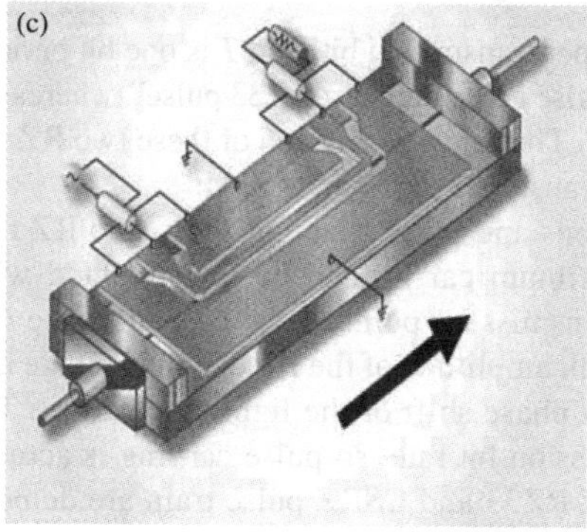

FIGURE 2.8 Intensity modulators using LiNbO$_3$. (a) single drive electrode (b) dual-electrode structure (c) electro-optic polarization scrambler using LiNbO$_3$.

The first MZIM (commonly known as the "pulse carver") was used to generate the periodic pulse trains with a required return-to-zero (RZ) format. The suppression of the lightwave carrier can also be carried out at this stage if necessary (commonly known as the "carrier-suppressed RZ" (CSRZ)). Compared with other RZ types, CSRZ pulse shape is found to have attractive attributes for long-haul WDM transmissions, including π phase difference of adjacent modulated bits, suppression of the optical carrier component in optical spectra, and narrower spectral width.

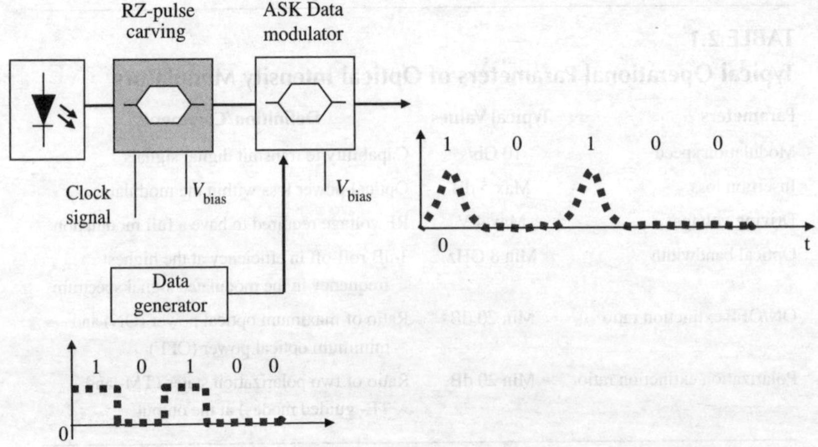

FIGURE 2.9 Conventional structure of an OOK optical transmitter utilizing two MZIMs.

Different types of RZ pulses can be generated depending on the driving amplitude of the RF voltage and the biasing schemes of the MZIM. The equations governing the RZ pulses electric field waveforms are

$$
E(t) = \begin{cases} \sqrt{\dfrac{E_b}{T}} \sin\left[\dfrac{\pi}{2}\cos\left(\dfrac{\pi t}{T}\right)\right] & \text{67\% duty-ratio RZ pulses or CSRZ,} \\[3em] \sqrt{\dfrac{E_b}{T}} \sin\left[\dfrac{\pi}{2}\left(1+\sin\left(\dfrac{\pi t}{T}\right)\right)\right] & \text{33\% duty-ratio RZ pulses or RZ33,} \end{cases}
\tag{2.3}
$$

where E_b is the pulse energy per transmitted bit, and T is one bit period.

The 33% duty-ratio RZ pulse is denoted as "RZ33 pulse", whereas the 67% duty cycle RZ pulse is known as the "CSRZ type". The art in generation of these two RZ pulse types stays at the difference of biasing point on the transfer curve of an MZIM.

The bias voltage conditions and pulse shape of these two RZ types, the carrier suppression and non-suppression of maximum carrier can be implemented with the biasing points at the minimum and maximum transmission point of the transmittance characteristics of the MZIM, respectively. The peak-to-peak amplitude of the RF driving voltage is $2V_\pi$ where V_π is the required driving voltage to obtain a π phase shift of the lightwave carrier. The RF signal is operating at only one-half of the transmission bit rate, so pulse carving is actually implementing frequency doubling. The generations of RZ33 and CSRZ pulse train are demonstrated in Figure 2.10a and Figure 2.10b.

The pulse carver can also utilize a dual-drive MZIM that is driven by two complementary sinusoidal RF signals. This pulse carver is biased at $-V_{\pi/2}$ and $+V_{\pi/2}$ with the peak-to-peak amplitude of $V_{\pi/2}$. Thus, a π phase shift is created between the state "1"and "0" of the pulse sequence, and hence the RZ with alternating phase 0 and π. If carrier suppression is required, the two electrodes are applied with voltages V_π and swing voltage amplitude of V_π.

Although RZ modulation offers improved performance, RZ optical systems usually require more complex transmitters than those in the NRZ optical systems. Compared with only one stage for modulating data on NRZ optical signals, two modulation stages are required for generation of RZ optical pulses.

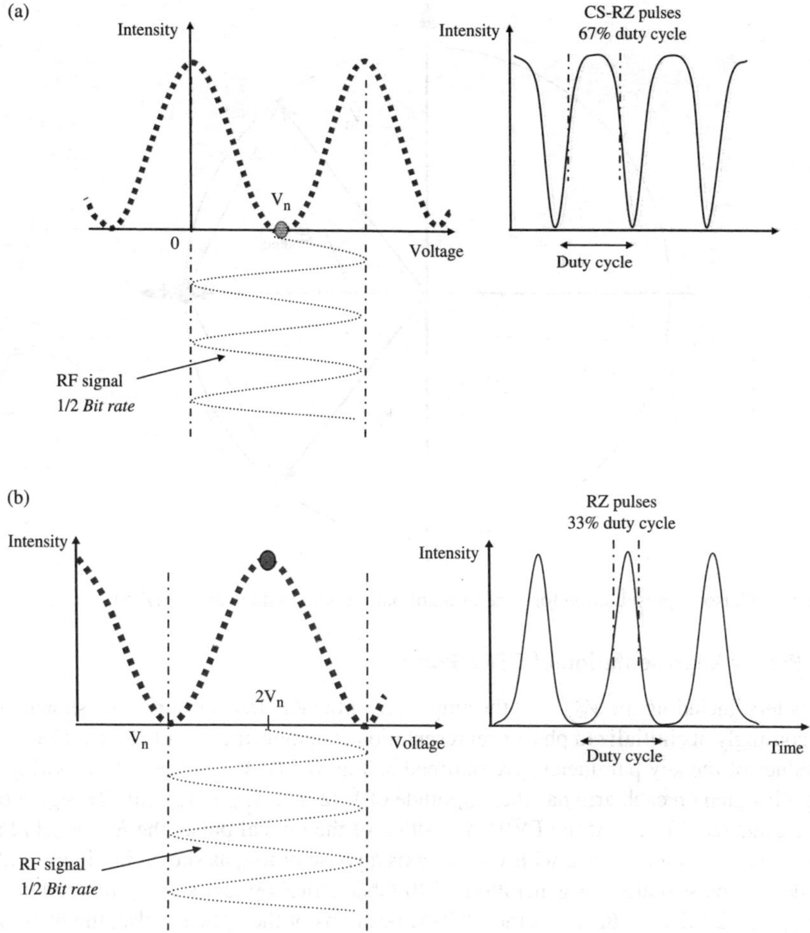

FIGURE 2.10 Bias point and RF driving signals for generation of (a) CSRZ and (b) RZ33 pulses.

2.2.2 PHASOR REPRESENTATION

Recalling Equation 2.1 we have

$$E_o = \frac{E_i}{2}\left[e^{j\varphi_1(t)} + e^{j\varphi_2(t)}\right] = \frac{E_i}{2}\left[e^{j\pi v_1(t)/V_\pi} + e^{j\pi v_2(t)/V_\pi}\right]. \tag{2.4}$$

The modulating process for generation of RZ pulses can be represented by a phasor diagram (Figure 2.10). This technique gives a clear understanding of the superposition of the fields at the coupling output of two arms of the MZIM. Here, a dual-drive MZIM is used, i.e., the data-driving signals $[V_1(t)]$ and inverse data ($\overline{data}$: $V_2(t) = -V_1(t)$) are applied into each arm of the MZIM, respectively, and the RF voltages swing in inverse directions. Applying the phasor representation as shown in Figure 2.11, vector addition and simple trigonometric calculus, the process of generation RZ33 and CSRZ is explained in detail and verified.

The width of these pulses are commonly measured at the position of full-width half maximum (FWHM). Measured pulses are intensity pulses, whereas we are considering the addition of the fields in the MZIM. Thus, the normalized E_o field vector has the value of $\pm 1\sqrt{2}$ at the FWHM intensity pulse positions, and the time interval between these points gives the FWHM values.

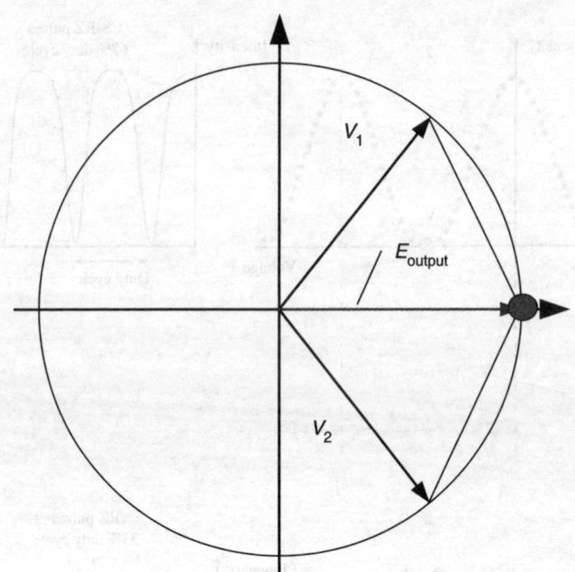

FIGURE 2.11   Phasor representation for generation of output field in dual-drive MZIM.

2.2.2.1   Phasor Representation of CSRZ Pulses

Key parameters, including the V_{bias}, and the amplitude of the RF driving signal, are shown in Figure 2.12a. Accordingly, its initialized phasor representation is demonstrated in Figure 2.12b.

The values of the key parameters are outlined as follows: (i) V_{bias} is $\pm V_\pi /2$; (ii) swing voltage of driving RF signal on each arm has the amplitude of $V_\pi/2$ (i.e., $V_{\mathrm{p\text{-}p}} = V_\pi$); (iii) RF signal operates at half of bit rate ($B_\mathrm{R}/2$); (iv) At the FWHM position of the optical pulse, the $E_{\mathrm{out}} = \pm 1\sqrt{2}$ and the component vectors V_1 and V_2 form with vertical axis a phase of $\pi/4$, as shown in Figure 2.13.

Considering the scenario for generation of 40 Gb/s CSRZ optical signal, the modulating frequency is f_m ($f_\mathrm{m} = 20$ GHz $= B_\mathrm{R}/2$). At the FWHM positions of the optical pulse, the phase is given by the following expressions:

$$\frac{\pi}{2}\sin(2\pi f_\mathrm{m}) = \frac{\pi}{4} \Rightarrow \sin 2\pi f_\mathrm{m} = \frac{1}{2} \Rightarrow 2\pi f_\mathrm{m} = \left(\frac{\pi}{6}, \frac{5\pi}{6}\right) + 2n\pi. \tag{2.5}$$

Thus, calculation of TFWHM can be carried out and the duty cycle of the RZ optical pulse can be obtained, as given in the following expressions:

$$T_{\mathrm{FWHM}} = \left(\frac{5\pi}{6} - \frac{\pi}{6}\right)\frac{1}{R2\pi} = \frac{1}{3}\pi \times \frac{1}{R} \Rightarrow \frac{T_{\mathrm{FWHM}}}{T_{\mathrm{BIT}}} = \frac{1.66 \times 10^{-4}}{2.5 \times 10^{-11}} = 66.67\%. \tag{2.6}$$

The result obtained in Equation 2.6 clearly verifies the generation of CSRZ optical pulses from phasor representation.

2.2.2.2   Phasor Representation of RZ33 Pulses

Key parameters, including the V_{bias}, the amplitude of driving voltage and its correspondent initialized phasor representation, are shown in Figure 2.14a and Figure 2.14b respectively.

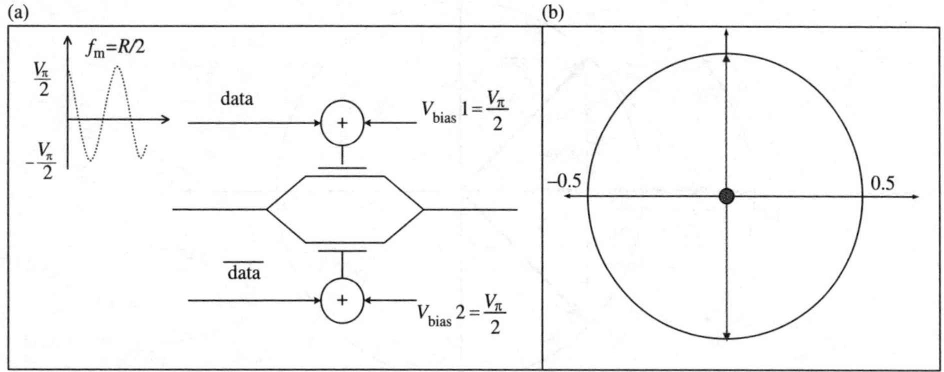

FIGURE 2.12 Initialized stage for generation of CSRZ pulse. (a) RF driving signal and bias voltages (b) Initial phasor representation.

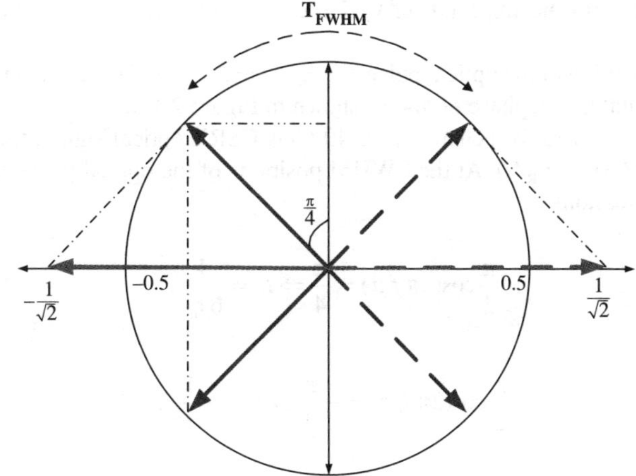

FIGURE 2.13 Phasor representation of CSRZ pulse generation using dual-drive MZIM.

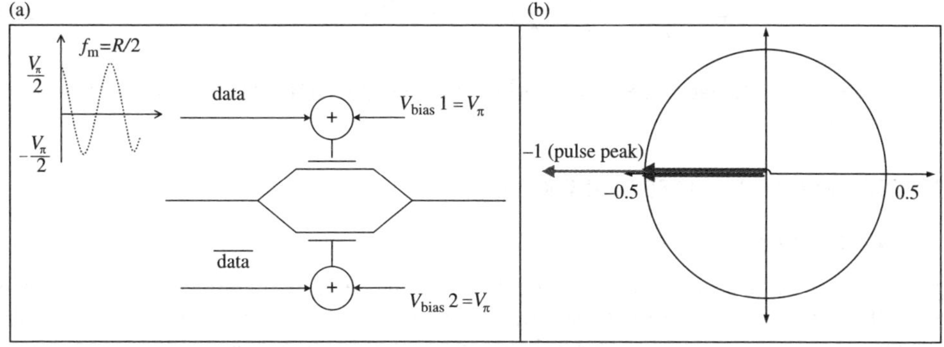

FIGURE 2.14 Initialized stage for generation of RZ33 pulse: (a) RF driving signal and the bias voltage; (b) Initial phasor representation.

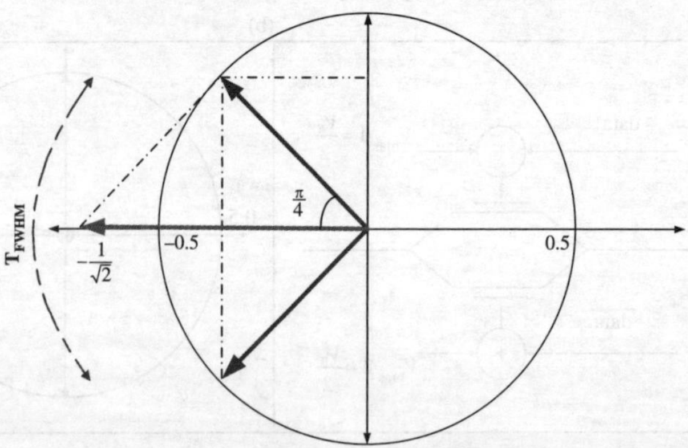

FIGURE 2.15 Phasor representation of RZ33 pulse generation using dual-drive MZIM.

The values of the key parameters are (i) V_{bias} is V_π for both arms; (ii) swing voltage of driving RF signal on each arm has the amplitude of $V_\pi/2$ (i.e., $V_{p-p} = V_\pi$); (iii) RF signal operates at half of bit rate ($B_R/2$).

At the FWHM position of the optical pulse, the $E_{output} = \pm 1\sqrt{2}$ and the component vectors V_1 and V_2 form with horizontal axis a phase of $\pi/4$, as shown in Figure 2.15.

Considering the scenario for generation of 40 Gb/s CSRZ optical signal, the modulating frequency is f_m ($f_m = 20$ GHz $= B_R/2$). At the FWHM positions of the optical pulse, the phase is given by the following expressions:

$$\frac{\pi}{2}\cos(2\pi f_m t) = \frac{\pi}{4} \Rightarrow t_1 = \frac{1}{6 f_m}, \tag{2.7}$$

$$\frac{\pi}{2}\cos(2\pi f_m t) = -\frac{\pi}{4} \Rightarrow t_2 = \frac{1}{3 f_m}. \tag{2.8}$$

Thus, the calculation of TFWHM can be carried out and the duty cycle of the RZ optical pulse can be obtained, as given in the following expressions:

$$T_{FWHM} = \frac{1}{3 f_m} - \frac{1}{6 f_m} = \frac{1}{6 f_m} \quad \therefore \quad \frac{T_{FWHM}}{T_b} = \frac{1/6 f_m}{1/2 f_m} = 33\%. \tag{2.9}$$

The result obtained in Equation 2.8 clearly verifies the generation of RZ33 optical pulses from phasor representation.

2.3 DIFFERENTIAL PHASE SHIFT KEYING

2.3.1 BACKGROUND

Digital encoding of data information by modulating the phase of the lightwave carrier is termed "optical phase shift keying" (PSK). In the early days, optical PSK was studied extensively for coherent photonic transmission systems. This technique requires the manipulation of the absolute phase of the lightwave carrier. Thus, precise alignment of the transmitter and demodulator center frequencies for the coherent detection is required. These coherent optical PSK systems face severe obstacles (e.g., broad linewidth and chirping problems of the laser source). The differential phase shift keying (DPSK) scheme overcomes these problems because DPSK optically-modulated signals can be detected incoherently. This technique requires only the coherence of the lightwave carriers over a one-bit period for comparison of the differentially coded phases of consecutive optical pulses.

A binary "1" is encoded if the present input bit and the previous encoded bit are of opposite logic, whereas a binary "0" is encoded if the logics are similar. This operation is equivalent to an XOR logic operation, so an XOR gate is employed as a differential encoder. NOR can also be used to replace XOR operation in differential encoding (Figure 2.16a). In DPSK, the electrical data "1" indicates a π phase change between the consecutive data bits in the optical carrier, whereas the binary "0" is encoded if there is no phase change between the consecutive data bits. Hence, this encoding scheme gives rise to two points located exactly at π phase difference with respect to each other in the signal constellation diagram. For continuous phase shift keying, such as minimum shift keying, the phase evolves continuously over a quarter of the section, thus a phase change of $\pi/2$ exists between one phase state and the other. This is indicated by the inner bold circle in Figure 2.16b.

2.3.2 OPTICAL DPSK TRANSMITTER

Figure 2.17 shows the structure of a 40 Gb/s DPSK transmitter in which two external LiNbO$_3$ MZIMs are used. Operational principles of a MZIM were presented above. The MZIMs shown in Figure 2.17 can be of single- or dual-drive type. The optical DPSK transmitter also consists of a narrow-linewidth laser to generate a lightwave whose wavelength conforms to the ITU grid.

The RZ optical pulses are then fed into the second MZIM through which RZ pulses are modulated by the pre-coded binary data to generate RZ-DPSK optical signals. Electrical data pulses are differentially pre-coded in a pre-coder using the XOR coding scheme. Without pulse carver, the structure

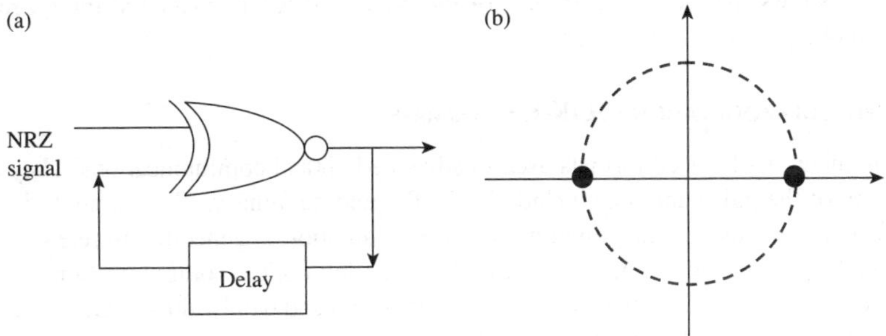

FIGURE 2.16   (a) DPSK pre-coder (b) Signal constellation diagram of DPSK [20].

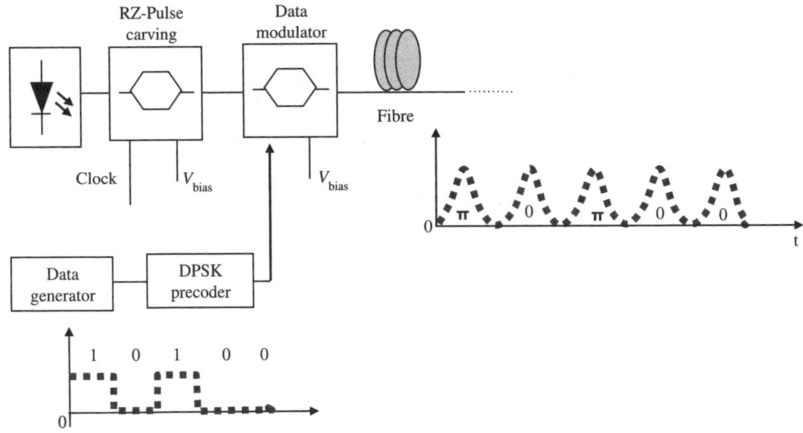

FIGURE 2.17   DPSK optical transmitter with RZ pulse carver.

shown in Figure 2.17 is an optical NRZ-DPSK transmitter. Figure 2.18 shows the swing level of the electrical signals in the modulator transfer characteristic in order to obtain the phase difference of pi. In data modulation for DPSK format, the second MZIM is biased at the minimum transmission point. The pre-coded electrical data have peak-to-peak amplitude equal to $2V_\pi$, and operates at the transmission bit rate. The modulation principles for generation of optical DPSK signals are demonstrated in Figure 2.9.

The electro-optic phase modulator may also be used for generation of DPSK signals instead of MZIM. Using an optical phase modulator, the transmitted optical signal is chirped, whereas using MZIM, especially the X-cut type with Z-propagation, chirp-free signals can be produced. However, in practice, a small amount of chirp may be useful for transmission [4]. Refs [21–28] give detailed description of the electro-optic modulators and applications in system transmission.

2.4 GENERATION OF MODULATION FORMATS

Modulation is the process facilitating the transfer of information over a medium, e.g., wireless or optical environment. Three basic types of modulation techniques are based on the manipulation of a parameter of the optical carrier to represent the information digital data. These types are amplitude shift keying (ASK), phase shift keying (PSK) and frequency shift keying (FSK). In addition to manipulation of the carrier, the occupation of the data pulse over a single period would also determine the amount of energy concentrates and the speed of the system required for transmission. The pulse can remain constant over a bit period or return to zero level within a portion of the period. These formats would be named non-return-to-zero (NRZ) or return-to-zero (RZ). They are combined with modulation of the carrier to form various modulations formats presented in this section.

Figure 2.19 shows the base band signals of the NRZ and RZ formats and the corresponding block diagram of a photonic transmitter.

2.4.1 AMPLITUDE-MODULATION OOK-RZ FORMATS

There are a number of advanced formats used in advanced optical communications. They are based on the intensity of the pulse and may include NRZ, RZ and duobinary. These amplitude shift keying (ASK) formats can also be integrated with phase modulation to generate discrete or continuous phase NRZ or RZ formats. Currently, most 10-Gb/s installed optical communication systems have been developed with NRZ due to its simple transmitter design and bandwidth-efficient characteristic. However, RZ format has higher robustness to fiber nonlinearity and polarization mode dispersion

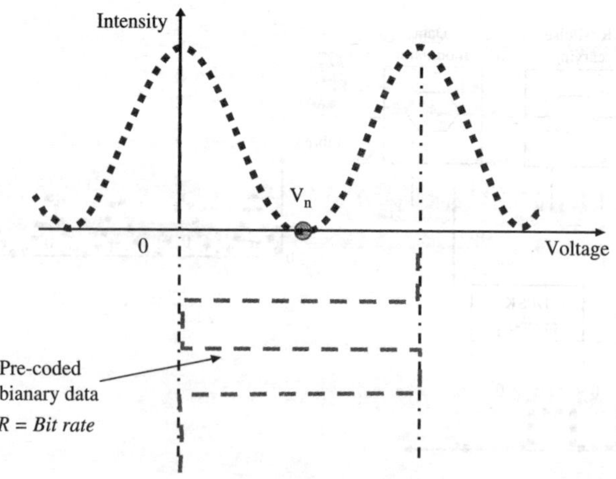

FIGURE 2.18 Bias point and RF driving signals for generation of optical DPSK format.

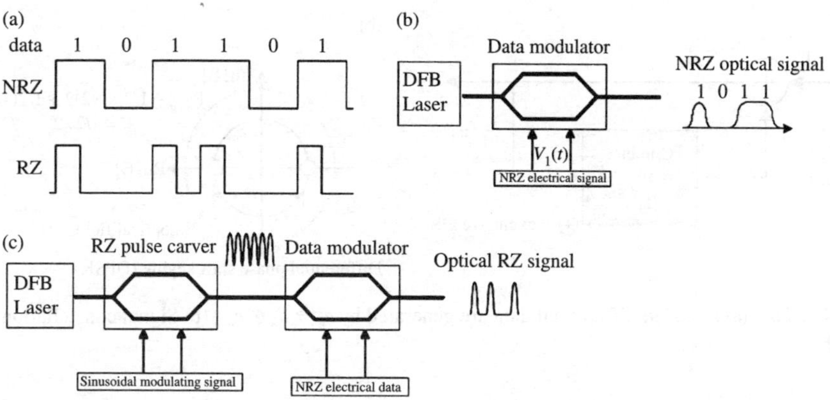

FIGURE 2.19   (a) Baseband NRZ and RZ line coding for 101101 data sequence. (b) Block diagram of NRZ photonics transmitter. (c) RZ photonics transmitter incorporating a pulse carver.

TABLE 2.2

Summary of RZ Format Generation and Characteristics of Single-drive MZIM Based on Biasing Point, Drive Signal Amplitude and Frequency

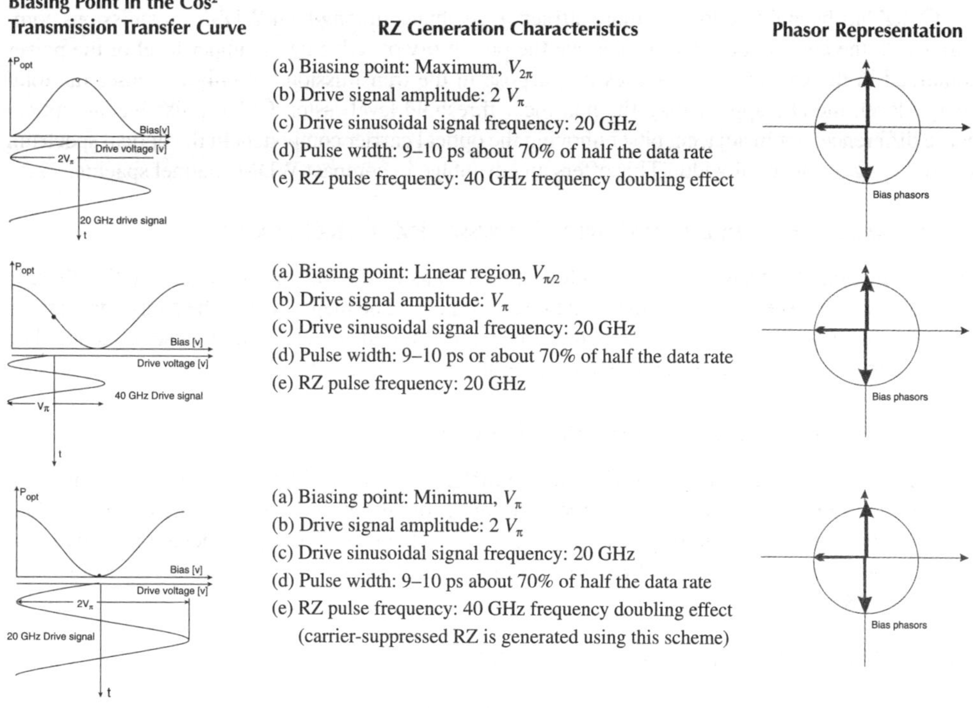

Biasing Point in the Cos² Transmission Transfer Curve	RZ Generation Characteristics	Phasor Representation
P_{opt}; Bias[v]; Drive voltage [v]; $2V_\pi$; 20 GHz drive signal	(a) Biasing point: Maximum, $V_{2\pi}$ (b) Drive signal amplitude: $2\,V_\pi$ (c) Drive sinusoidal signal frequency: 20 GHz (d) Pulse width: 9–10 ps about 70% of half the data rate (e) RZ pulse frequency: 40 GHz frequency doubling effect	Bias phasors
P_{opt}; Bias [v]; Drive voltage [v]; V_π; 40 GHz Drive signal	(a) Biasing point: Linear region, $V_{\pi/2}$ (b) Drive signal amplitude: V_π (c) Drive sinusoidal signal frequency: 20 GHz (d) Pulse width: 9–10 ps or about 70% of half the data rate (e) RZ pulse frequency: 20 GHz	Bias phasors
P_{opt}; Bias [v]; Drive voltage [v]; $2V_\pi$; 20 GHz Drive signal	(a) Biasing point: Minimum, V_π (b) Drive signal amplitude: $2\,V_\pi$ (c) Drive sinusoidal signal frequency: 20 GHz (d) Pulse width: 9–10 ps about 70% of half the data rate (e) RZ pulse frequency: 40 GHz frequency doubling effect (carrier-suppressed RZ is generated using this scheme)	Bias phasors

(PMD). In this section, the RZ pulse is generated by MZIM, commonly known as "pulse carver", as arranged in Figure 2.20 and Figure 2.21. There are a number of variations in RZ format based on the biasing point in the transmission curve (Table 2.2). The phasor representation of the biasing and driving signals are shown in Table 2.2.

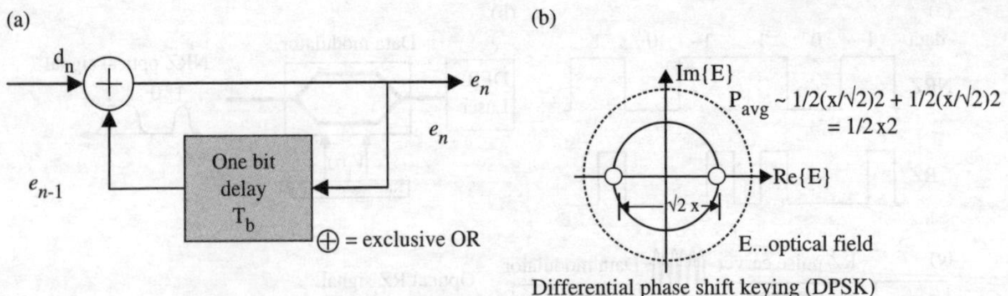

FIGURE 2.20   (a) Encoded differential data are generated by $e_n = d_n \oplus e_{n-1}$ (b) Signal constellation diagram of DPSK.

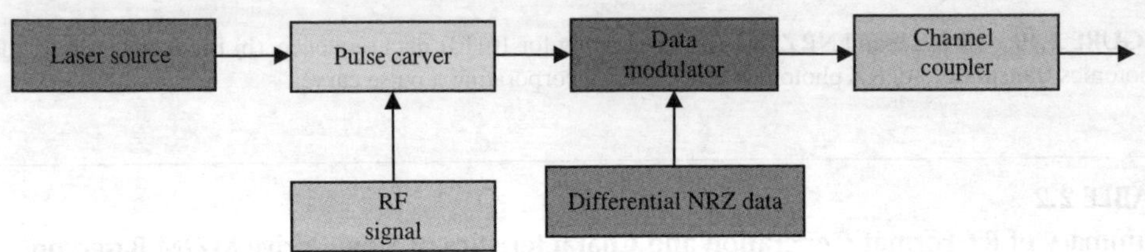

FIGURE 2.21   Block diagrams of RZ-DPSK transmitter.

CSRZ has been found to have more attractive attributes in long-haul WDM transmissions compared with the conventional RZ format due the possibility of reducing the upper level of the power contained in the carrier, which serves no purpose in the transmission but only increases the total energy level, thereby approaching the nonlinear threshold level faster. CSRZ pulse has an optical phase difference of π in adjacent bits, removing the optical carrier component in the optical spectrum and reducing the spectral width. This offers an advantage in compact WDM channel spacing.

2.4.2   Amplitude-modulation Carrier-Suppressed RZ (CSRZ) Formats

The suppression of the carrier can be implemented by biasing the MZ interferometer so that there is a π phase shift between the two arms of the interferometer. The magnitude of the sinusoidal signals applied to an arm or both arms would determine the width of the optical output pulse sequence. The driving conditions and phasor representation are shown in Table 2.2.

2.4.3   Discrete Phase-modulation NRZ Formats

The term "discrete phase modulation" refers to differential phase shift keying, whether DPSK or quadrature DPSK (DQPSK), to indicate the states of the phases of the lightwave carrier are switched from one distinct location on the phasor diagram to the other state. Examples include from 0 to π or $-\pi/2$ to $-\pi/2$ for binary PSK (BPSK), or even more evenly spaced phase shift keying levels such as M-ary PSK.

2.4.3.1   Differential Phase Shift Keying (DPSK)

Information encoded in the phase of an optical carrier is commonly referred to as optical phase shift keying. In the early days, PSK required precise alignment of the transmitter and demodulator center frequencies. Hence, PSK system is not widely deployed. When the DPSK scheme was introduced, coherent detection was not critical because DPSK detection required only source coherence over one bit period by comparison of two consecutive pulses.

A binary "1" is encoded if the present input bit and the previous encoded bit are of opposite logic, and a binary 0 is encoded if the logic is similar. This operation is equivalent to an XOR logic operation. Hence, an XOR gate is usually employed in a differential encoder. NOR can also be used to replace the XOR operation in differential encoding (Figure 2.20).

In optical applications, electrical data "1" is represented by a π phase change between the consecutive data bits in the optical carrier, while state "0" is encoded with no phase change between consecutive data bits. Hence, this encoding scheme gives rise to two points located exactly at π phase difference with respect to each other in the signal constellation diagram (Figure 2.20b).

A RZ-DPSK transmitter consists of an optical source, pulse carver, data modulator, differential data encoder and a channel coupler. Channel coupler model is not developed in simulation by assuming no coupling losses when optical RZ-DPSK modulated signal is launched into the optical fiber. This modulation scheme has combined the functionality of dual-drive MZIM modulator of pulse carving and phase modulation.

The pulse carver, usually a MZ interferometric intensity modulator, is driven by a sinusoidal RF signal for single-drive MZIM and two complementary electrical RF signals for dual-drive MZIM, to carve pulses out from optical carrier signal forming RZ pulses. These optical RZ pulses are fed into a second MZ intensity modulator in which RZ pulses are modulated by differential NRZ electrical data to generate RZ-DPSK. This data phase modulation can be done using a straight-line phase modulator, but the MZ waveguide structure has several advantages over the phase modulator due to its chirpless property. Electrical data pulses are differentially pre-coded in a differential pre-coder (Figure 2.20a). Without pulse carver and sinusoidal RF signal, the output pulse sequence follows NRZ-DPSK format, i.e., the pulse would occupy 100% of the pulse period and there is no transition between the consecutive "1s".

2.4.3.2 Differential Quadrature Phase Shift Keying (DQPSK)

This differential coding is similar to DPSK except that each symbol consists of two bits which are represented by the two orthogonal axial discrete phases at $(0, \pi)$ and $(-\pi/2, +\pi/2)$ as shown in Figure 2.22, or two additional orthogonal phase positions, as located on the imaginary axis of Figure 2.20b.

2.4.3.3 NRZ-DPSK

Figure 2.23 shows the block diagram of a typical NRZ-DPSK transmitter. Differential pre-coder of electrical data is implemented using the logic explained in the previous section. In phase modulating of an optical carrier, MZ modulator known as "data phase modulator" is biased at minimum point and driven by data swing of $2V_\pi$. The modulator showed excellent behaviour; the phase of the optical carrier will be altered by exactly π when the signal transmits the minimum point of the transfer characteristic.

2.4.3.4 RZ-DPSK

The arrangement of RZ-DPSK transmitter is essentially similar to that of RZ-ASK (Figure 2.24) with the data intensity modulator replaced with a data phase modulator. The difference between them is the biasing point and the electrical voltage swing. Different RZ formats can also be generated.

2.4.3.5 Generation of M-ary Amplitude Differential Phase Shift Keying (M-ary ADPSK) Using One MZIM

As an example, a 16-ary MADPSK signal can be represented by the constellation shown in Figure 2.25. It is a combination of a 4-ary ASK and a DQPSK scheme. At the transmitting end, each group of four bits $[D_3D_2D_1D_0]$ of user data are encoded into a symbol, among them two least significant bits $[D_1D_0]$ are encoded into four phase states $[0, \pi/2, \pi, 3\pi/2]$ and the other two most significant bits, $[D_3D_2]$, are encoded into four amplitude levels. At the receiving end, because MZ Delay Interferometers (DI) are used for phase comparison and detection, a diagonal arrangement of the signal constellation shown

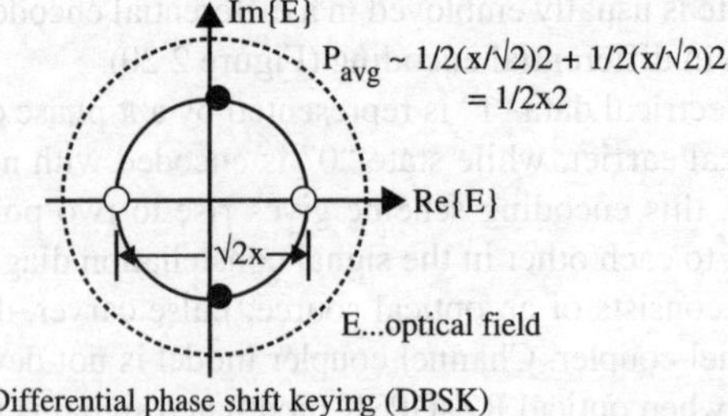

FIGURE 2.22 Signal constellation diagram of DQPSK. Two bold dots are orthogonal to the DPSK constellation.

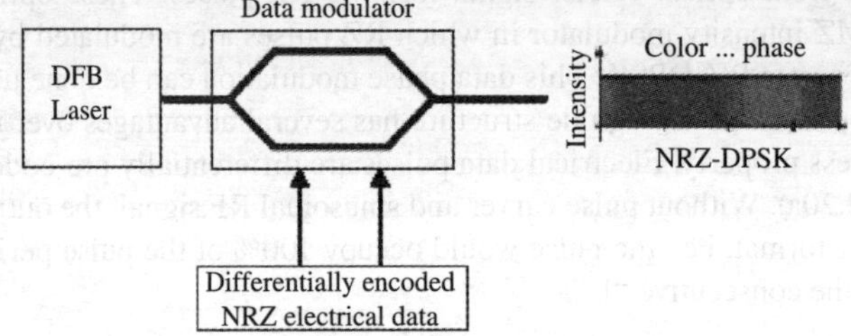

FIGURE 2.23 Block diagram of a NRZ-DPSK photonics transmitter.

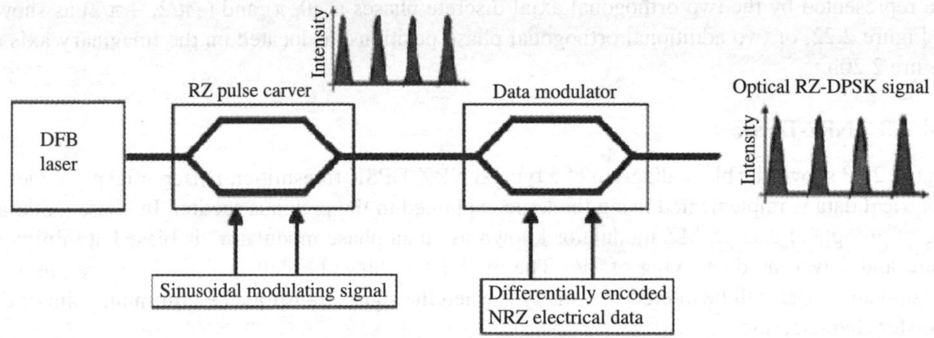

FIGURE 2.24 Block diagram of a RZ-DPSK photonics transmitter.

in Figure 2.25a is preferred. This simplifies the design of transmitter and receiver and minimizes the number of phase detectors, leading to high receiver sensitivity.

To balance the bit error rate (BER) between ASK and DQPSK components, the signal levels corresponding to four circles of the signal space should be adjusted to a reasonable ratio, which depends on the noise power at the receiver. For example, if this ratio is set to $[I_0/I_1/I_2/I_3] = [1/1.4/2/2.5]$, where I_0, I_1, I_2 and I_3 are the intensity of the optical signals corresponding to circle 0, circle 1, circle 2, and circle 3, respectively, then by selecting E_i equal to the amplitude of the circle 3 and V_π equal to 1, the driving voltages should have the values given in Table 2.3. Conversely, one can set the outermost level such that its peak value is below the non-linear SPM threshold, so that the voltage level of the outer most circle would be determined. The innermost circle is limited to the condition that the largest signal-to-noise ratio (SNR) should be achieved. This is related to the optical SNR

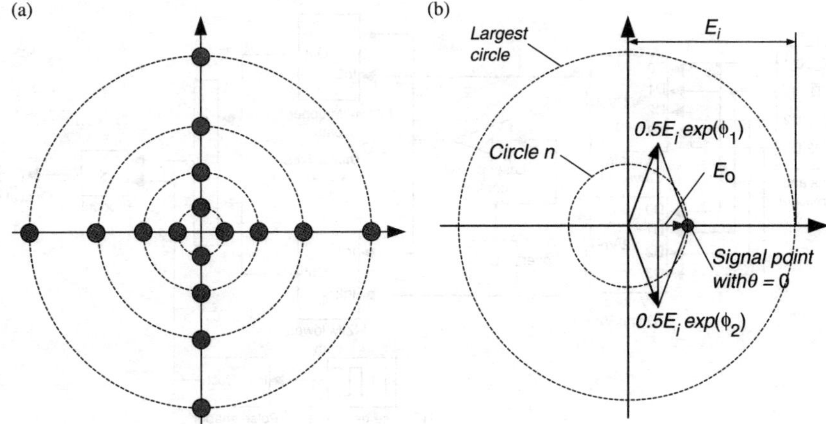

FIGURE 2.25 Signal constellation of 4-ary ADPSK format and phasor representation of a point on the constellation point for driving voltages applied to dual-drive MZIM.

TABLE 2.3
Driving Voltages for 16-ary MADPSK Signal Constellation

Circle 0			Circle 1			Circle 2			Circle 3		
Phase	$V_1(t)$ volt	$V_2(t)$ volt	Phase	$V_1(t)$ volt	$V_2(t)$ volt	Phase	$V_1(t)$ volt	$V_2(t)$ volt	Phase	$V_1(t)$ volt	$V_2(t)$ volt
0	0.38	−0.38	0	0.30	−0.30	0	0.21	−0.21	0	0.0	0.0
$\pi/2$	0.88	0.12	$\pi/2$	0.80	0.20	$\pi/2$	0.71	0.29	$\pi/2$	0.5	0.5
π	−0.62	0.62	π	−0.7	0.70	π	−0.79	0.79	π	−1.0	1.0
$3\pi/2$	−0.12	−0.88	$3\pi/2$	−0.20	−0.8	$3\pi/2$	−0.29	−0.71	$3\pi/2$	−0.5	−0.5

required for a certain BER. Thus, from the largest amplitude level and smallest amplitude level, we can then design the other points of the constellation.

Furthermore, to minimize the effect of inter-symbol interference, 66%-RZ and 50%-RZ pulse formats are also used as alternatives to the traditional NRZ counterpart. These RZ pulse formats can be created by a pulse carver that precedes or follows the dual-drive MZIM modulator. Mathematically, waveforms of NRZ and RZ pulses can be represented by the following equations, where E_{on}, $n = 0, 1, 2, 3$ are peak amplitude of the signals in the circle 0, circle 1, circle 2, and circle 3 of the constellation, respectively:

$$p(t) = \begin{cases} E_{on} & \text{for NRZ,} \\[2ex] E_{on}\ \cos\left(\dfrac{\pi}{2}\cos^2\left(\dfrac{1.5\pi t}{T_s}\right)\right) & \text{for } 66\% \text{ - RZ,} \\[2ex] E_{on}\ \cos\left(\dfrac{\pi}{2}\cos^2\left(\dfrac{2\pi t}{T_s}\right)\right) & \text{for } 50\% \text{ - RZ.} \end{cases} \tag{2.10}$$

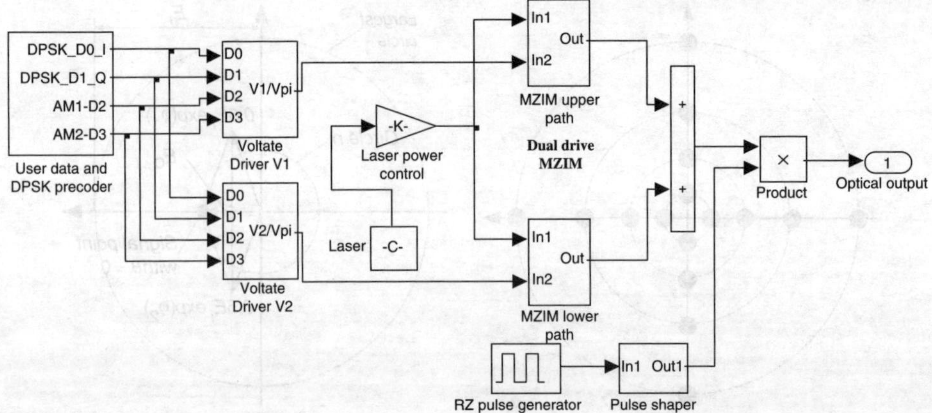

FIGURE 2.26 MATLAB® SIMULINK® model of a MADPSK photonic transmitter. The MZIM is represented by two phase-shifter blocks.

Figure 2.26 shows a MATLAB® Simulink® model of the pre-coder for mapping the binary data pulse sequence into multi-level data symbols for modulating the optical modulator MZIM.

2.4.4 Continuous Phase-modulation PM-NRZ Formats

In the previous section, the optical transmitters for discrete PSK modulation formats were been described. Obviously, the phase of the carrier has been used to indicate the digital states of the bits or symbols. These phases are allocated in a non-continuous manner around a circle corresponding to the magnitude of the wave. Alternatively, the phase of the carrier can be continuously modulated and the total phase changes at the transition instants, usually at the end of the bit period, would be the same as those for discrete cases. Because the phase of the carrier continuously varies during the bit period, this can be considered as a FSK modulation technique, except that the transition of the phase at the end of one bit and the beginning of next bit would be continuous. The continuity of the carrier phase at these transitions would reduce the signal bandwidth and hence be more tolerable to dispersion effects and higher energy concentration for effective transmission over the optical guided medium. An example of the reduction of the phase at the transition is the offset DQPSK, which is a minor (but important) variation on the QPSK or DQPSK. In OQPSK, the Q-channel is shifted by half a symbol period so that the transition instants of I and Q channel signals do not occur at the same time. The result of this simple change is that the phase shifts at any one time are limited, and hence the offset QPSK is more constant envelope that the normal QPSK.

The enhancement of the efficiency of the bandwidth of the signals can be further improved if the phase changes at these transitions are continuous. In this case, the change of the phase during the symbol period is continuously changed by using half-cycle sinusoidal driving signals with the total phase transition over a symbol period that is a fraction of π, depending on the levels of this phase shift keying modulation. If the change is $\pi/2$, then we have a minimum shift keying (MSK) scheme. The orthogonality of the I- and Q-channels will also reduce further the bandwidth of the carrier-modulated signals. In this section, we describe the basic principles of optical MSK and photonic transmitters for these modulation formats. Ideally, the driving signal to the phase modulator should be a triangular wave so that a linear phase variation of the carrier in the symbol period is linear. However, when a sinusoidal function is used, there are some nonlinear variations, we thus term this type of MSK a "nonlinear MSK format". This nonlinearity contributes to some penalty in the optical signal-to-noise ratio (OSNR) which will be explained in a later chapter. Furthermore, the MSK as a special form of ODQPSK is also described for optical systems.

2.4.4.1 Linear and Non-linear MSK

MSK is a special form of continuous phase FSK (CPFSK) signal in which the two frequencies are spaced so that they are orthogonal and hence have minimum spacing between them, defined by

$$s(t) = \sqrt{\frac{2E_b}{T_b}}\, \cos\left[2\pi f_1 t + \theta(0)\right] \text{ for symbol 1,} \tag{2.11}$$

$$s(t) = \sqrt{\frac{2E_b}{T_b}}\, \cos\left[2\pi f_2 t + \theta(0)\right] \text{ for symbol 0.} \tag{2.12}$$

As shown by the equations above, the signal frequency change corresponds to higher frequency for data-1 and lower frequency for data-0. Both frequencies, f_1 and f_2, are defined by

$$f_1 = f_c + \frac{1}{4T_b}, \tag{2.13}$$

$$f_2 = f_c - \frac{1}{4T_b}. \tag{2.14}$$

Depending on the binary data, the phase of signal changes: data-1 increases the phase by $\pi/2$, while data-0 decreases the phase by $\pi/2$. The variation of phase follows paths of sequence of straight lines in phase trellis (Figure 2.27), in which the slopes represent frequency changes. The change in carrier frequency from data-0 to data-1, or vice versa, is equal to half the bit rate of incoming data [8]. This is the minimum-frequency spacing that allows the two FSK signals representing symbols 1 and 0 to be coherently orthogonal in that they do not interfere with one another in the process of detection.

A MSK signal consists of both I and Q components, which can be written as

$$s(t) = \sqrt{\frac{2E_b}{T_b}}\, \cos[\theta(t)]\cos[2\pi f_c t] - \sqrt{\frac{2E_b}{T_b}}\, \sin[\theta(t)]\sin[2\pi f_c t]. \tag{2.15}$$

The in-phase component consists of half-cycle cosine pulse defined by

$$s_I(t) = \pm\sqrt{\frac{2E_b}{T_b}}\, \cos\left(\frac{\pi t}{2T_b}\right), \quad -T_b \le t \le T_b. \tag{2.16}$$

While the quadrature component would take the form

$$s_Q(t) = \pm\sqrt{\frac{2E_b}{T_b}}\, \sin\left(\frac{\pi t}{2T_b}\right), \quad 0 \le t \le 2T_b. \tag{2.17}$$

During even bit interval, the I-component consists of positive cosine waveform for phase of 0, while negative cosine waveform for phase of π; during odd bit interval, the Q- component consists of positive sine waveform for phase of $\pi/2$, while negative sine waveform for phase of $-\pi/2$ (Figure 2.27). Any of four states can arise: 0, $\pi/2$, $-\pi/2$, π. However, only state 0 or π can occur during any even

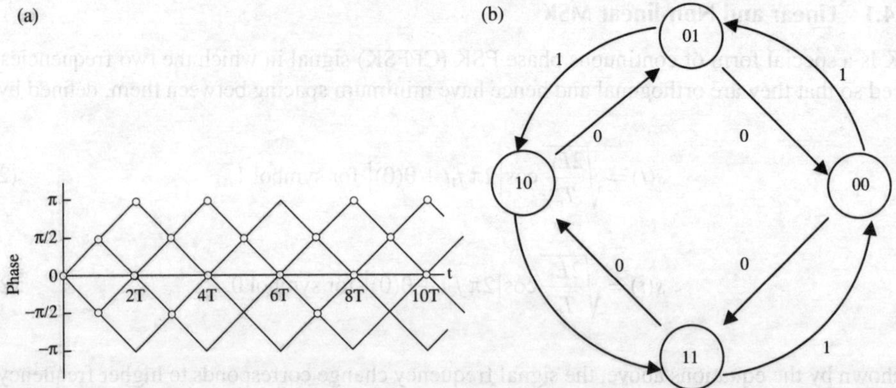

FIGURE 2.27   (a) Phase trellis for MSK (b) State diagram for MSK (K.K. Pang, "Digital Transmission," Melbourne: Mi-Tec Publishing, 2002, p. 58).

bit interval, and only $\pi/2$ or $-\pi/2$ can occur during any odd bit interval. The transmitted signal is the sum of I and Q components, and its phase is continuous with time.

Two important characteristics of MSK are that each data bit is held for two bit period, meaning the symbol period is equal two bit period ($h = 1/2$) and the I and Q-components are interleaved. I- and Q-components are delayed by one-bit period with respect to each other. Therefore, only I- or Q-component can change at a time (when one is at zero-crossing, the other is at maximum peak). The pre-coder can be combinational logic (Figure 2.28).

A truth table can be constructed based on the logic state diagram and combinational logic diagram above. For positive half cycle cosine wave and positive half cycle sine wave, the output is 1; for negative half cycle cosine wave and negative half cycle sine wave, the output is 0. Then, a K-map can be constructed to derive the logic gates of the pre-coder, based on the truth table. The following three pre-coding logic equations are derived:

$$S_0 = \overline{b_n}\,\overline{S_0'}\ \overline{S_1'} + b_n\overline{S_0'}\ S_1' + \overline{b_n}S_0'\ S_1' + b_nS_0'\ \overline{S_1'}\,, \tag{2.18}$$

$$S_1 = \overline{S_1'} = \overline{b_n}S_1' + b_n\overline{S_1'}\,, \tag{2.19}$$

$$\text{Output} = \overline{S_0}. \tag{2.20}$$

The resultant logic gates construction for the precoder is as shown in Figure 2.28.

2.4.4.2 MSK as a Special Case of Continuous Phase FSK (CPFSK)

CPFSK signals are modulated in upper and lower side band frequency carriers f_1 and f_2 as expressed in Equation 2.16 and Equation 2.17.

$$s(t) = \sqrt{\frac{2E_b}{T_b}}\,\cos[2\pi f_1 t + \theta(0)] \quad \text{symbol 1}, \tag{2.21}$$

$$s(t) = \sqrt{\frac{2E_b}{T_b}}\,\cos[2\pi f_2 t + \theta(0)] \quad \text{symbol 0}, \tag{2.22}$$

where $f_1 = f_c + 1/4T_b$ and $f_2 = f_c - 1/4T_b$ with T_b is the bit period.

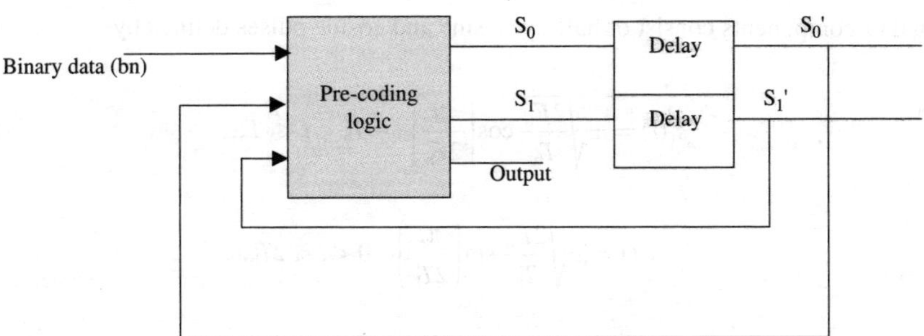

FIGURE 2.28 Combinational logic, the basis of the logic for constructing the pre-coder.

The phase slope of lightwave carrier changes linearly or nonlinearly with the modulating binary data. In linear MSK, the carrier phase linearly changes by $\pi/2$ at the end of the bit slot with data "1", while it linearly decreases by $\pi/2$.with data "0". The variation of phase follows paths of well-defined phase trellis in which the slopes represent frequency changes. The change in carrier frequency from data-0 to data-1, or *vice versa*, equals half the bit rate of incoming data [9]. This is the minimum frequency spacing that allows the two FSK signals representing symbols 1 and 0 to be coherently orthogonal in the sense that they do not interfere with one another in the process of detection. MSK carrier phase is always continuous at bit transitions. The MSK signal in Equation 2.14 and Equation 2.15 can be simplified as:

$$s(t) = \sqrt{\frac{2E_b}{T_b}} \cos[2\pi f_c t + d_k \frac{\pi t}{2T_b} + \Phi_k], \quad kT_b \leq t \leq (k+1)T_b \tag{2.23}$$

and the base-band equivalent optical MSK signal is represented as:

$$\tilde{s}(t) = \sqrt{\frac{2E_b}{T_b}} \exp\left\{ j\left[d_k \frac{\pi t}{2T} + \Phi(t,k) \right] \right\}, \quad kT \leq t \leq (k+1)T$$

$$= \sqrt{\frac{2E_b}{T_b}} \exp\left\{ j\left[d_k 2\pi f_d t + \Phi(t,k) \right] \right\}, \tag{2.24}$$

where $d_k = \pm 1$ are the logic levels; f_d is the frequency deviation from the optical carrier frequency; and $h = 2f_d T$ is defined as the frequency modulation index. In optical MSK, $h = 1/2$ or $f_d = 1/(4T_b)$.

2.4.4.3 MSK as Offset Differential Quadrature Phase Shift Keying (ODQPSK)

Equation 2.16 can be rewritten to express MSK signals in form of I-Q components as:

$$s(t) = \pm \sqrt{\frac{2E_b}{T_b}} \cos\left(\frac{\pi t}{2T_b} \right) \cos[2\pi f_c t] \pm \sqrt{\frac{2E_b}{T_b}} \sin\left(\frac{\pi t}{2T_b} \right) \sin[2\pi f_c t]. \tag{2.25}$$

The I- and Q-components consist of half-cycle sine and cosine pulses defined by

$$s_I(t) = \pm\sqrt{\frac{2E_b}{T_b}}\cos\left(\frac{\pi t}{2T_b}\right) \quad -T_b < t < T_b, \tag{2.26}$$

$$s_Q(t) = \pm\sqrt{\frac{2E_b}{T_b}}\sin\left(\frac{\pi t}{2T_b}\right) \quad 0 < t < 2T_b. \tag{2.27}$$

During even bit intervals, the in-phase component consists of positive cosine waveform for phase of 0, while negative cosine waveform for phase of π; during odd bit interval, the Q-component consists of positive sine waveform for phase of $\pi/2$, while negative sine waveform for phase of $-\pi/2$. Any of four states can arise: 0, $\pi/2$, $-\pi/2$, π. However, only state 0 or π can occur during any even bit interval, and only $\pi/2$ or $-\pi/2$ can occur during any odd bit interval. The transmitted signal is the sum of I- and Q-components, and its phase is continuous with time.

Two important characteristic of MSK are that each data bit is held for two bit period, meaning the symbol period is equal two bit period ($h = 1/2$) and the I-component and Q-component are interleaved. I- and Q-components are delayed by one bit period with respect to each other. Therefore, only I- or Q-component can change at any time (when one is at zero-crossing, the other is at maximum peak).

2.4.4.4 Configuration of Photonic MSK Transmitter Using Two Cascaded Electro-optic Phase Modulators

Electro-optic phase modulators and interferometers operating at high frequency using resonant-type electrodes have been studied [10–12]. In addition, high-speed electronic driving circuits evolved with the ASIC technology using 0.1-μm GaAs P-HEMT or InP HEMTs [1] enables realization of the optical MSK transmitter structure. The base-band equivalent optical MSK signal is represented in Equation 2.4.

The first electro-optic phase modulator (E-OPM) enabled the frequency modulation of data logics into upper side bands (USB) and lower side bands (LSB) of the optical carrier with frequency deviation of f_d. Differential phase pre-coding is not necessary in this configuration due to the nature of the continuity of the differential phase trellis. By alternating the driving sources $V_d(t)$ to sinusoidal waveforms for simple implementation, or combination of sinusoidal and periodic ramp signals (first proposed by Amoroso in 1976) [13], different schemes of linear and non-linear phase shaping MSK transmitted sequences can be generated (see spectra in Figure 2.40).

The second E-OPM enforces the phase continuity of the light wave carrier at every bit transition. The delay control between the E-OPMs is usually implemented by the phase shifter shown in Figure 2.29. The driving voltage of the second E-OPM is pre-coded to fully compensate the transitional phase jump at the output $E_{01}(t)$ of the first E-OPM. Phase continuity characteristic of the optical MSK signals is determined by the algorithm in Equation 2.19 to Equation 2.21.

$$\Phi(t,k) = \frac{\pi}{2}\left(\sum_{j=0}^{k-1} a_j - a_k I_k \sum_{j=0}^{k-1} I_j\right), \tag{2.28}$$

where $a_k = \pm1$ are the logic levels; $I_k = \pm1$ is a clock pulse whose duty cycle is equal to the period of the driving signal $V_d(t)$; f_d is the frequency deviation from the optical carrier frequency; and $h = 2f_d T$ is previously defined as the frequency modulation index. In optical MSK, $h = 1/2$ or $f_d = 1/(4T)$. The phase evolution of the continuous phase optical MSK signals are explained in Figure 2.29.

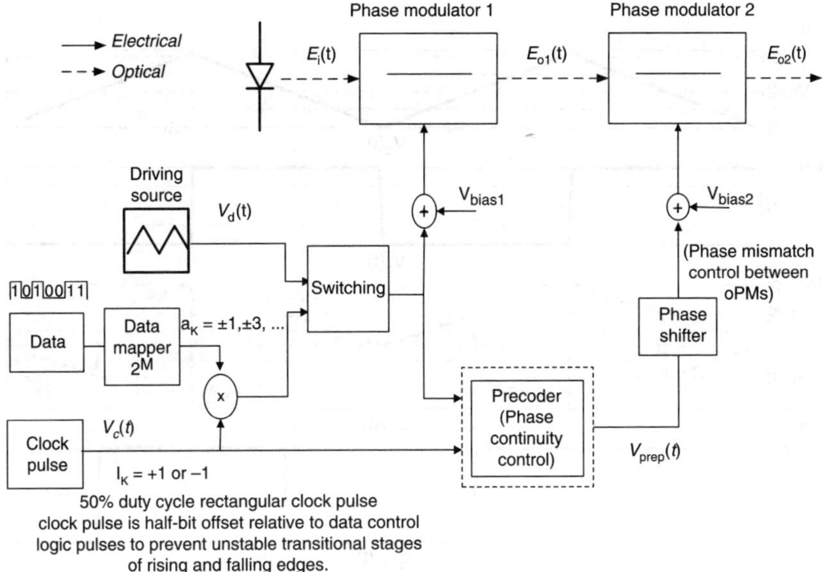

FIGURE 2.29 Block diagram of optical MSK transmitter employing two cascaded optical phase modulators.

To mitigate the effects of unstable stages of rising and falling edges of the electronic circuits, the clock pulse $V_c(t)$ is offset with the driving voltages $V_d(t)$.

2.4.4.5 Configuration of Optical MSK Transmitter Using Mach–Zehnder Intensity Modulators: I-Q Approach

The evolution of the phase and amplitude in the time domain of MSK signals is shown in Figure 2.30. Table 2.4 shows the logical states of the symbols of the MSK signals. The conceptual block diagram of an optical MSK transmitter is shown in Figure 2.31. The transmitter consists of two dual-drive electro-optic MZM modulators generating chirpless I and Q components of MSK-modulated signals; this is considered as a special case of staggered or offset QPSK. The binary logic data is pre-coded and de-interleaved into even and odd bit streams which are interleaved with each other by one-bit duration offset.

Two arms of the dual-drive MZM modulator are biased at $V_\pi/2$ and $-V_\pi/2$, and driven with data and $\overline{\text{data}}$. Phase-shaping driving sources can be a periodic triangular voltage source in linear MSK generation or simply a sinusoidal source for generating a nonlinear MSK-like signal which also produces a linear phase trellis property but with small ripples introduced in the magnitude. The magnitude fluctuation level depends on the magnitude of the phase-shaping driving source. High spectral efficiency can be achieved with tight filtering of the driving signals before modulating the electro-optic MZMs. Three types of pulse shaping filters are investigated, including Gaussian, raised cosine, and squared-root raised cosine filters. The optical carrier phase trellis of linear and non-linear optical MSK signals are shown in Figure 2.32.

2.4.5 Single Side Band (SSB) Optical Modulators

2.4.5.1 Operating Principles

A SSB modulator can be formed using a primary interferometer with two secondary MZM structures. The optical Ti-diffused waveguide paths that form a nested primary MZ structure are shown in Figure 2.33. Each of the two primary arms contains a MZ structure. Two RF ports are for RF modulation, and three DC ports are for biasing the two secondary MZMs and one primary MZM.

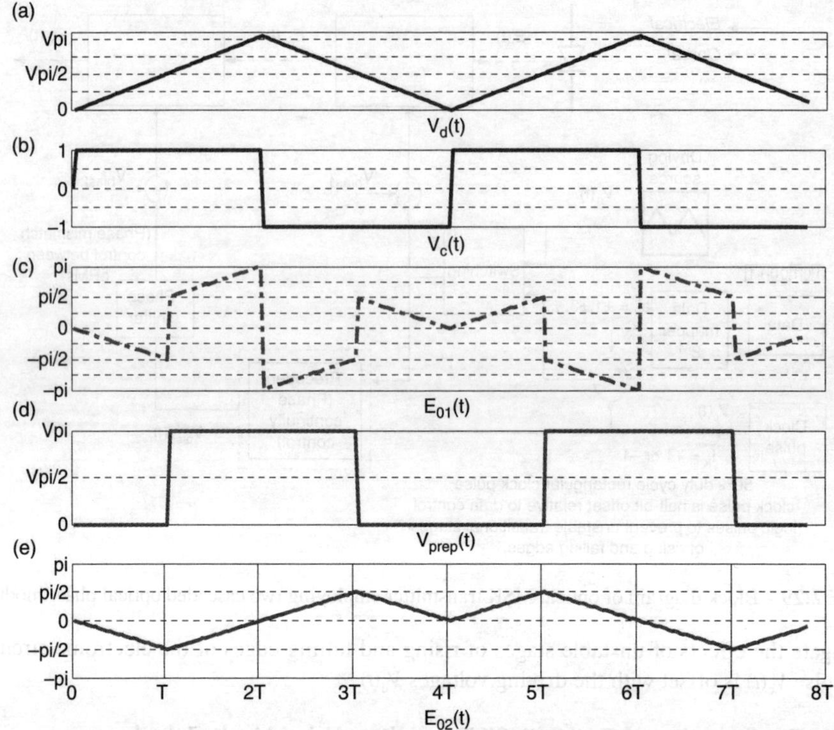

FIGURE 2.30 Evolution of time-domain phase trellis of optical MSK signal sequence [–1 1 1 –1 1 –1 1 1] as inputs and outputs at different stages of the optical MSK transmitter. The notation is denoted in Figure 2.29 accordingly: (a) $V_d(t)$: periodic triangular driving signal for optical MSK signals with duty cycle of 4 bit period, (b) $V_c(t)$: the clock pulse with duty cycle of $4T$, (c) $E_{01}(t)$: phase output of oPM1 (d) $V_{prep}(t)$: pre-computed phase compensation driving voltage of oPM2 and (e) $E_{02}(t)$: phase trellis of a transmitted optical MSK sequences at output of oPM2.

TABLE 2.4
Truth Table Based on MSK State Diagram

$B_n S_0' S_1'$	$S_0' S_1'$	Output
100	01	1
001	00	1
100	01	1
101	10	0
010	01	1
101	10	0
110	11	0
111	00	1
000	11	0
011	10	0

The modulator consists of X-cut Y-propagation LiNbO$_3$ crystal, in which you can produce a SSB modulation just by driving each MZ. DC voltage is supplied to produce the π phase shift between upper and lower arms. DC bias voltages are also supplied from DC$_B$ to produce the phase shift

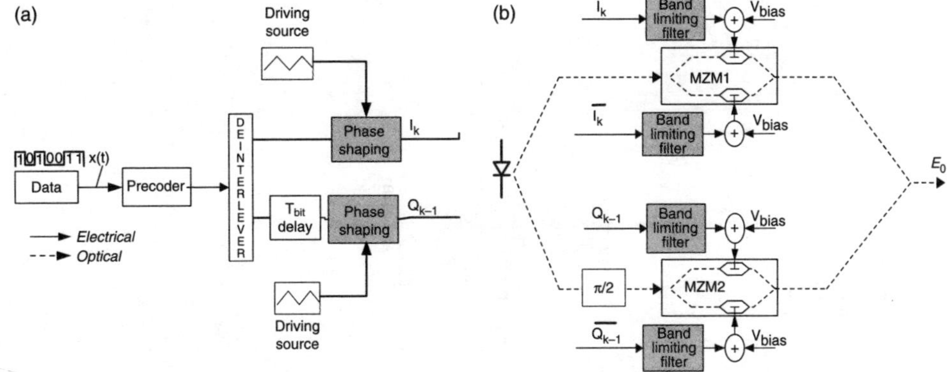

FIGURE 2.31 Block diagram configuration of band-limited phase-shaped optical MSK.

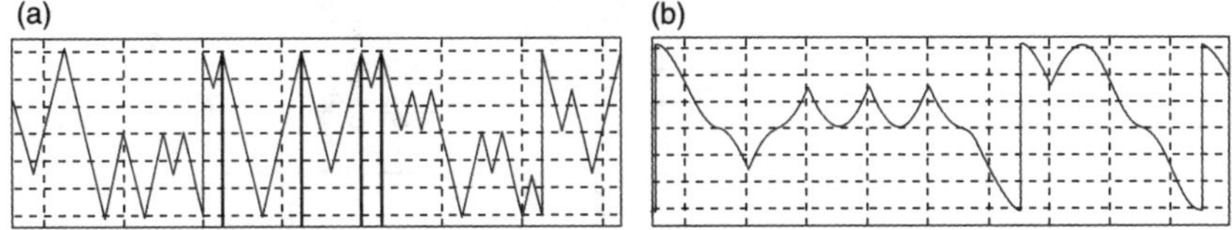

FIGURE 2.32 Phase trellis of linear and non-linear MSK transmitted signals.

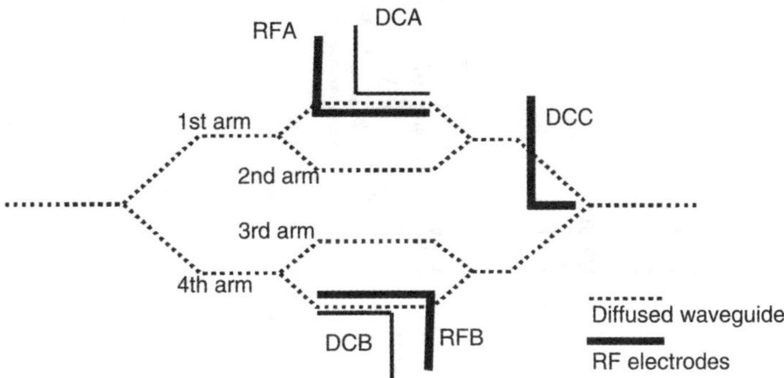

FIGURE 2.33 Schematic diagram (not to scale) of a single sideband (SSB) FSK optical modulator formed by nested MZ modulators.

between third and fourth arms. A DC bias voltage is supplied from DC_C to achieve a $\pi/2$ phase shift between $MZIM_A$ and $MZIM_B$. The RF voltage applied $\Phi_1(t) = \Phi \cos\omega_m t$ and $\Phi_2(t) = \Phi \sin\omega_m t$ are inserted from RF_A and RF_B, respectively, by using a wideband $\pi/2$ phase shifter. Φ is the modulation level, ω_m is the RF angular frequency.

2.4.5.2 Optical RZ-MSK

The RZ format of the optical MSK modulation scheme can also be generated. A bit is used to generate the ASK-like feature of the bit. A Simulink® structure of such transmitters is shown in Figure 2.24. The encoder, as shown in the far left of the model, provides two outputs, one for MSK signal generation and the other for amplitude modulation for generation of the RZ or NRZ format. Figure 2.34 shows the schematic of the transmission system using multiple spans setup for evaluation of performance of MSK modulation formats. The amplitude and phase of the RZ signals at the receivers are shown in Figure 2.35 after one 100-km span transmission and full compensation. The phase of the RZ MSK must be non-zero to satisfy the continuity of the phase between one state and the other.

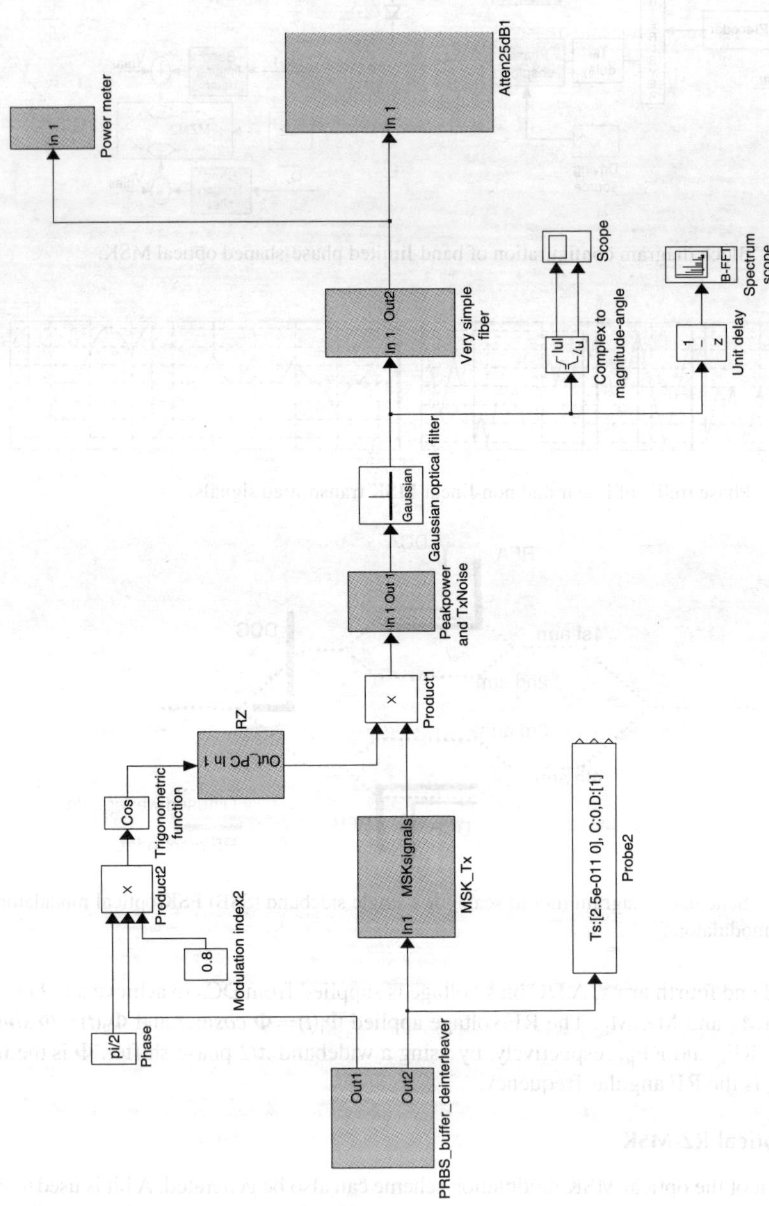

FIGURE 2.34 MATLAB SIMULINK model of a RZ-MSK optical transmission system.

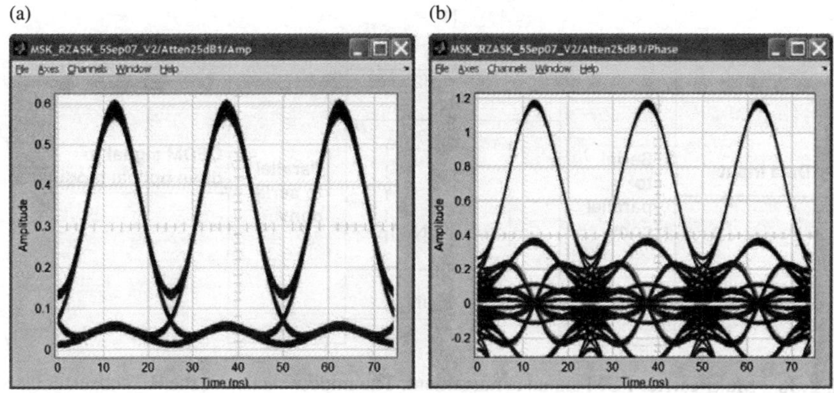

FIGURE 2.35   RZ eye diagram at output of the amplitude receiver (a) and phase detection (b).

2.4.6 MULTI-CARRIER MULTIPLEXING (MCM) OPTICAL MODULATORS

Another modulation format that can offer much higher single channel capacity and flexibility in dispersion and non-linear impairment mitigation is the employment of multi-carrier multiplexing. When these sub-carrier channels are orthogonal, the term "orthogonal frequency division multiplexing" (OFDM) is used.

Our motivation in the introduction of OFDM is due to its potential as a ultra-high capacity channel for the next generation Ethernet, the optical internet. The network interface cards for 1- and 10-Gb/s Ethernet are readily available commercially. Traditionally, the Ethernet data rates have grown in 10-Gb/s increments, so the data rate of 100 Gb/s can be expected as the speed of the next generation of Ethernet. The 100-Gb/s all-electrically time-division-multiplexed (ETDM) transponders are becoming increasingly important because they may be able to meet the speed requirements of the new generation of Ethernet. Despite the recent progress in high-speed electronics, ETDM [14] modulators and photodetectors are not widely available, so that alternative approaches to achieving 100-Gb/s transmission using commercially available components and QPSK is very attractive.

OFDM is a combination of multiplexing and modulation. The data signal is first split into independent sub-sets and then modulated with independent sub-carries. These sub-channels are then multiplexed to OFDM signals. OFDM is thus a special case of FDM but is a combination of several small streams into one bundle.

A schematic signal flow diagram of a MCM is shown in Figure 2.36. The basic OFDM transmitter and receiver configurations are given in Figure 2.37a and Figure 2.37b, respectively. Data streams (e.g., 1 Gb/s) are mapped into a two-dimensional signal point from a point signal constellation such as QAM. The complex-valued signal points from all sub-channels are considered as the values of the discrete Fourier transform (DFT) of a multi-carrier OFDM signal. The symbol interval length in an OFDM system is the symbol-interval length in a single-carrier system. By selecting, a number of sufficiently large sub-channels, the OFDM symbol interval can be made much larger than the dispersed pulse width in a single-carrier system, resulting in an arbitrary small intersymbol interference (ISI). The OFDM symbol shown in Figure 2.38 is generated under software processing as follows: input QAM symbols are zero-padded to obtain input samples for inverse fast Fourier transform (IFFT), the samples are inserted to create the guard band, and the OFDM symbol is multiplied by the window function (raised cosine function can be used). The purpose of cyclic extension is to preserve the orthogonality among sub-carriers even when the neighboring OFDM symbols partially overlap due to dispersion.

A system arrangement of the OFDM for optical transmission in laboratory demonstration is shown in Figure 2.38 [15]. Each individual channel at the input would carry the same data rate sequence. These sequences can be generated from an arbitrary waveform generator. The multiplexed channels are then combined and converted to time domain using the IFFT module, and then converted to the analog version via two digital-to-analog converters. These orthogonal data sequences are used to modulate

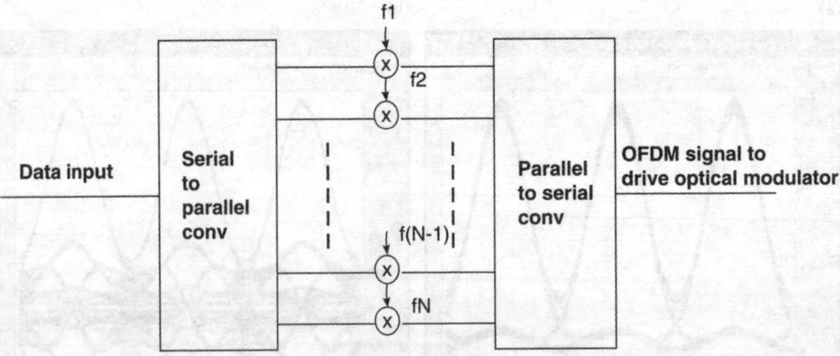

FIGURE 2.36 Multi-carrier FDM signal arrangement. The middle section is the RF converter (as shown in Figure 2.26).

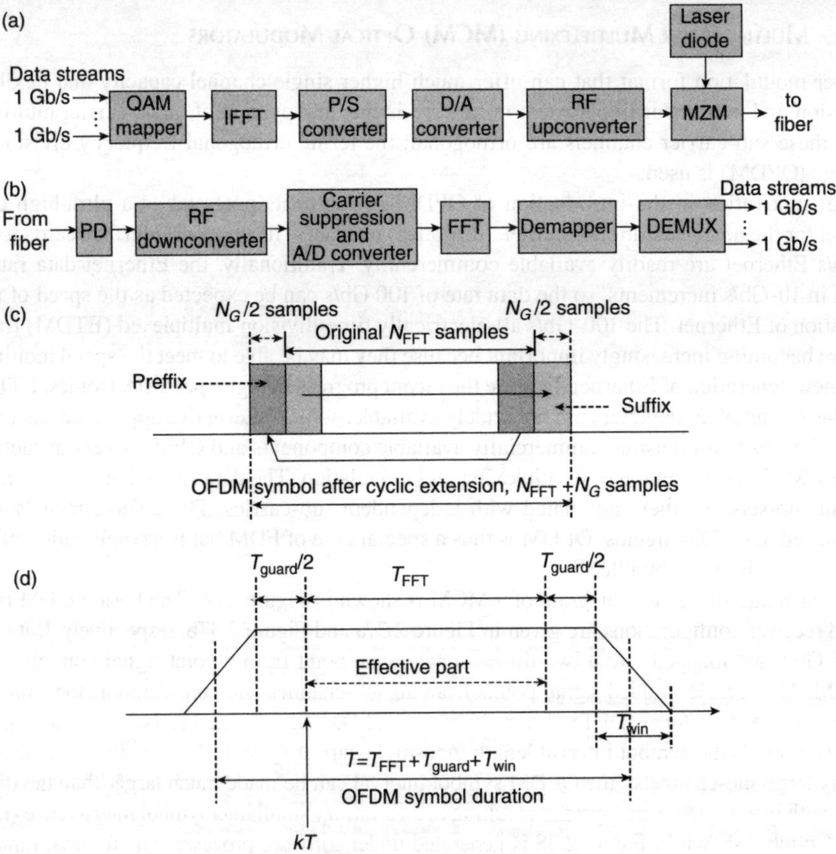

FIGURE 2.37 Schematic diagram of the principles of generation and recovery of OFDM signals (a) Electronic processing and optical transmitter and (b) opto-electronic and receiver configurations; (c) OFDM symbol cyclic extension; (d) OFDM symbol after windowing (extracted from [7]).

I- and Q- optical waveguide sections of the electro-optical modulator to generate the orthogonal channels in the optical domain. Similar decoding of I- and Q- channels is carried out in the electronic domain after optical transmission and optical–electronic conversion via the photodetector and electronic amplifier.

In OFDM, the serial data sequence, with a symbol period of T_s and a symbol rate of $1/T_s$, is split up into N-parallel sub-streams (sub-channels). Figure 2.39 shows the schematic of the OFDM

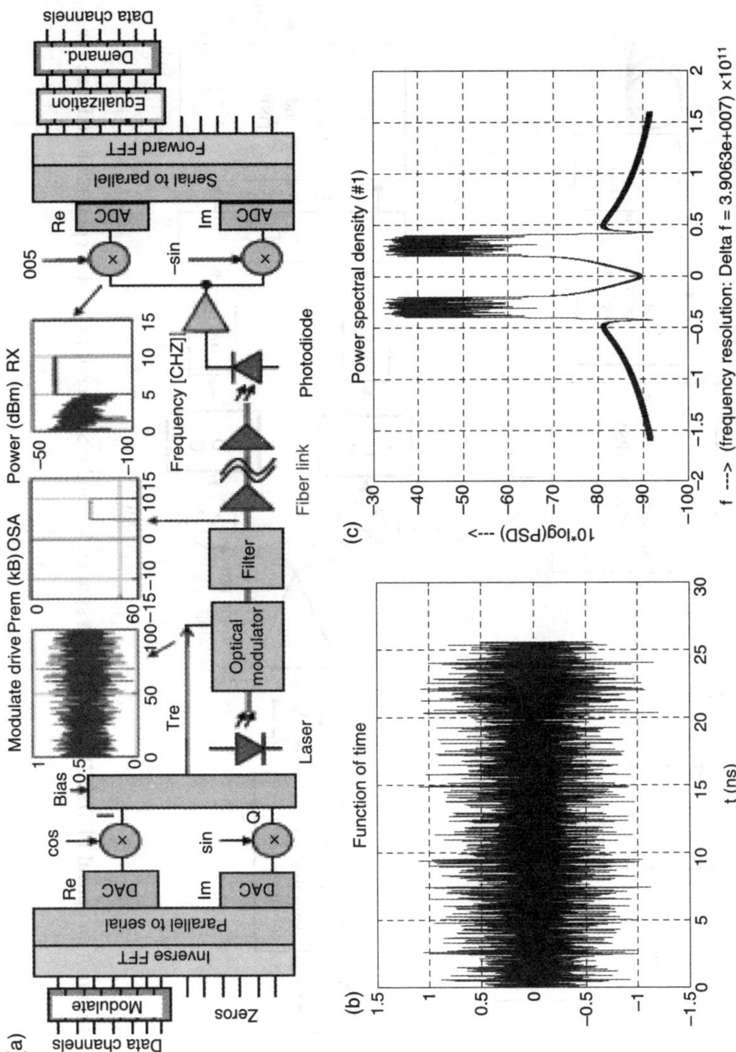

FIGURE 2.38 An optical FFT/IFFT based (a) OFDM system including representative waveforms and spectra (extracted from [14]). (b) Typical time domain OFDM signals (c) Power spectral density of OFDM signal with 512 sub-carriers at a shift of 30 GHz for the line rate of 40 Gb/s and QPSK modulation. (From IEEE. 2006. IEEE Optical Fiber Conference, Chapter 2, Ref. 7. With permission.)

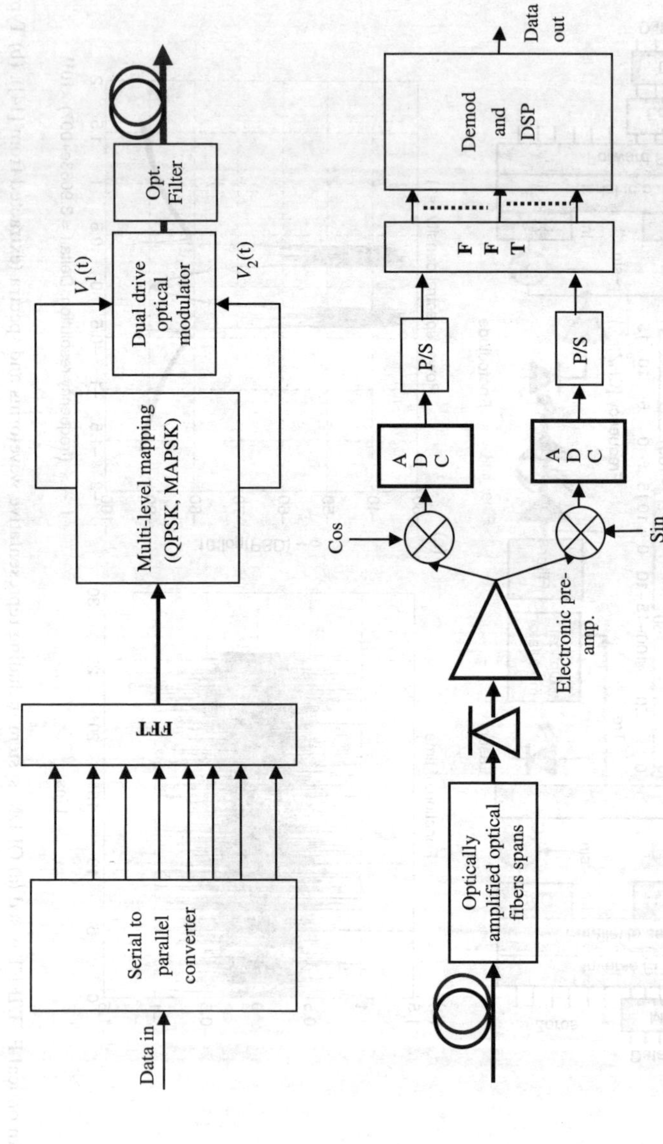

FIGURE 2.39 Schematic diagram of an optical FFT/IFFT based OFDM system. S/P and P/S = serial to parallel conversion and vice versa.

signals at the optical transmitting and receiving sides with the arrangement of the FFT and IFFT sub-systems.

2.4.7 SPECTRA OF MODULATION FORMATS

Utilizing this double-phase modulation configuration, different types of linear and non-linear CPM phase-shaping signals including MSK, weakly-non-linear MSK and linear-sinusoidal MSK can be generated. A third scheme was introduced by Amoroso [15] and its side lobes decay with a factor of f-8 compared to f-4 of MSK. The simulated optical spectra of DBPSK and MSK schemes at 40 Gb/s are contrasted in Figure 2.40. Table 2.5 outlines the characteristics and spectra of different modulation schemes.

2.5 SPECTRAL CHARACTERISTICS OF DIGITAL MODULATION FORMATS

Figure 2.41 shows the power spectra of the DPSK-modulated optical signals with various pulse shapes, including NRZ, RZ33 and CSRZ.

For convenience of comparison, the optical power spectra of the return-to-zero OOK counterparts are also shown in Figure 2.40.

Several key notes observed from Figure 2.42 and Figure 2.41 are outlined as follows: (i) The optical power spectrum of the OOK format has high-power spikes at the carrier frequency or at signal modulation frequencies; this contributes significantly to the severe penalties caused by the nonlinear effects. DPSK optical power spectra do not contain these high-power frequency components; (ii) RZ pulses are more sensitive to the fiber dispersion due to their broader spectra. In particular, RZ33 pulse type has the broadest spectrum at the point of –20 dB down from the peak. This property of the RZ pulses thus leads to faster spreading of the pulse when propagating along the fiber. Thus, the peak values of the optical power of these CSRZ or RZ33 pulses decrease much faster than the NRZ counterparts. As the result, the peak optical power quickly becomes lower than the nonlinear threshold of the fiber, which means that the effects of fiber nonlinearity are significantly reduced; and (iii) However, NRZ optical pulses have the narrowest spectrum, so are expected to be most robust to the fiber dispersion. As a result, there is a trade-off between RZ and NRZ pulse types. RZ pulses are much more robust to nonlinearity, but less tolerant to the fiber dispersion. The RZ33/CSRZ-DPSK optical pulses are proven to be more robust against impairments, especially self-phase modulation and polarization mode dispersion compared with the NRZ-DPSK and the CSRZ/RZ33-OOK counterparts.

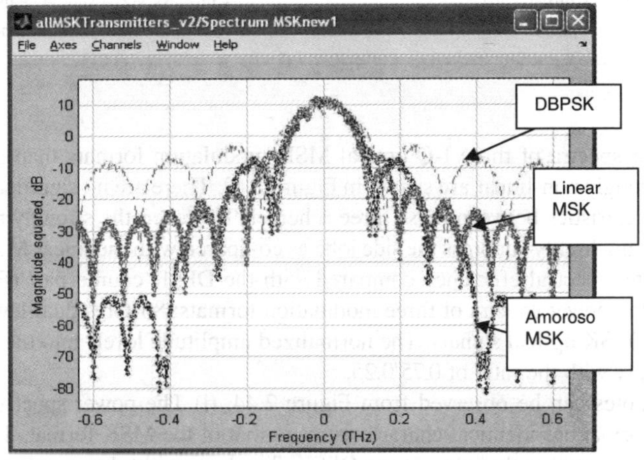

FIGURE 2.40   Spectra of 40-Gbps DBPSK, and linear and non-linear MSK.

TABLE 2.5

Typical Parameters of Optical Intensity Modulators for Generation of Modulation Formats

Modulation Techniques	Spectra	Formats	Definition/Comments
Amplitude modulation ASK–NRZ	DSB + carrier	ASK–NRZ	Biased at quadrature point or offset for pre-chirp
AM–ASK–RZ	DSB + carrier	ASK–RZ	Two MZIMs required: one for RZ pulse sequence and other for data switching
ASK–RZ-Carrier suppressed	DSB– CSRZ	ASK–RZCS	Carrier suppressed, biased at π phase difference for the two electrodes
Single sideband	SSB + carrier	SSB NRZ	Signals applied to MZIM are in phase quadrature to suppress one side band; alternatively an optical filter can be used
CSRZ DSB	DSB– carrier	CSRZ–ASK	RZ pulse carver is biased such that there a π phase shift between the two arms of the MZM to suppress the carrier, and then switch on and off or phase modulation via a data modulator
DPSK–NRZ DPSK–RZ, CSRZ–DPSK		Differential BPSK RZ or NRZ/RZ-carrier suppressed	
DQPSK		DQPSK–RZ or NRZ	Two bits per symbol
MSK	SSB equivalent	Continuous phase modulation with orthogonality	Two bits per symbol and efficient bandwidth with high side-lobe suppression
Offset-DQPSK			
MCM (multi-carrier multiplexing e.g., OFDM)	Multiplexed bandwidth – base rate per sub-carrier		
Duo-binary	Effective SSB		Electrical low pass filter required at the driving signal to the MZM
FSK			
Continuous-phase FSK			
Phase modulation (PM)	Chirped carrier phase		Chirpless MZM should be used to avoid inherent crystal effects, hence carrier chirp

Optical power spectra of three I-Q optical MSK modulation formats that are linear, weakly nonlinear, and strongly non-linear are shown in Figure 2.43. There are no significant distinctions of the spectral characteristics between these three schemes. However, the strongly non-linear optical MSK format does not highly suppress the side lobe as compared with the linear MSK type. All three formats offer better spectral efficiency compared with the DPSK counterpart (Figure 2.44). This figure compares the power spectra of three modulation formats: 80-Gb/s dual-level MSK, 40-Gb/s MSK, and NRZ-DPSK optical signals. The normalized amplitude levels into the two optical MSK transmitters comply with the ratio of 0.75/0.25.

Several key notes can be observed from Figure 2.44: (i) The power spectrum of the optical dual-level MSK format has identical characteristics to that of the MSK format. The spectral width of the main lobe is narrower than that of the DPSK. The base width has a value of approximately

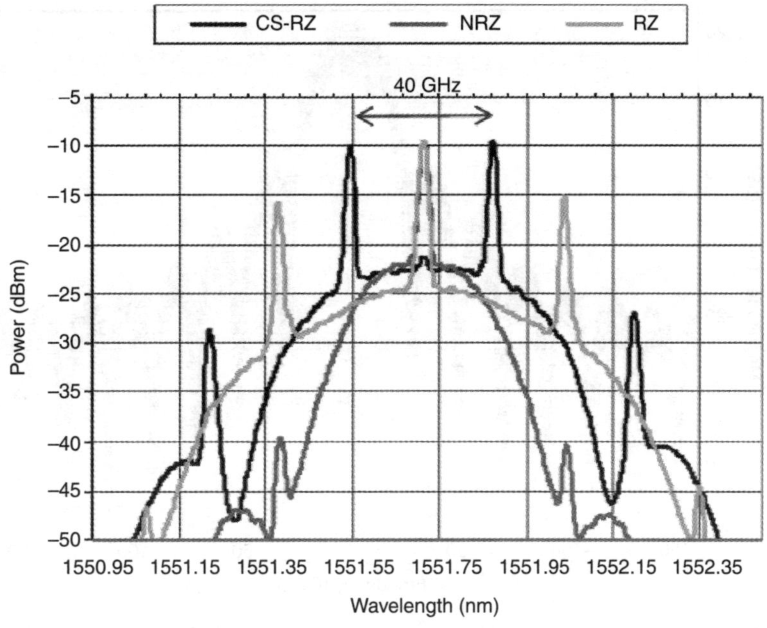

FIGURE 2.41 Spectra of CSRZ/RZ33/NRZ–DPSK modulated optical signals.

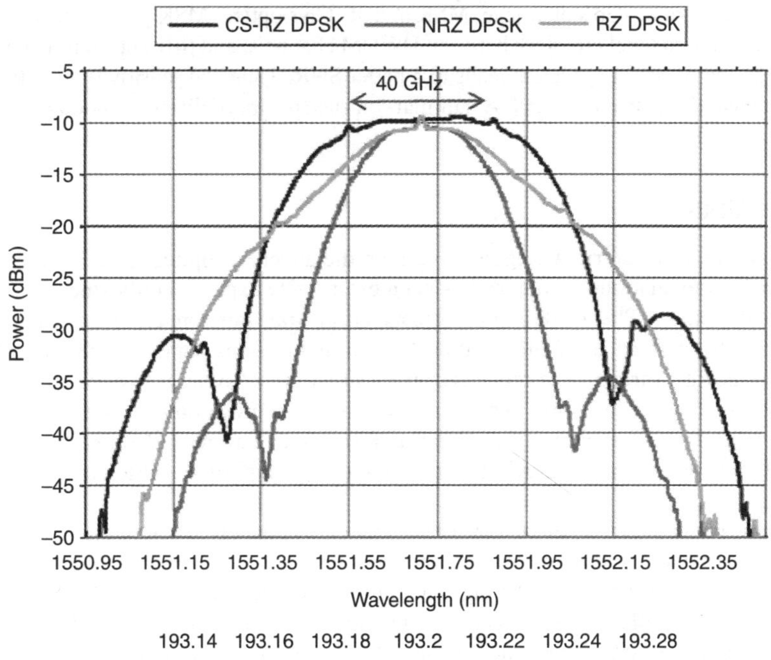

FIGURE 2.42 Spectra of CSRZ/RZ/NRZ–OOK modulated optical signals.

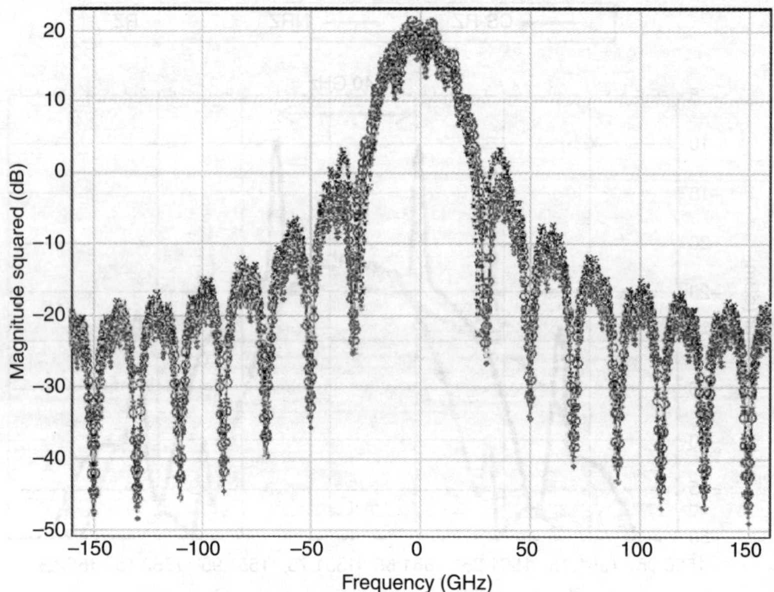

FIGURE 2.43 Optical power spectra of three types of I-Q optical MSK formats: linear (red), weakly non-linear (blue) and strongly non-linear (black).

± 32 GHz on either side compared to ± 40 GHz in the DPSK format. Hence, the tolerance to fiber dispersion effects is improved; (ii) High suppression of the side lobes with a value of approximately >20 dB in the case of 80-Gb/s dual-level MSK and 40-Gb/s optical MSK power spectra; thus, more robustness to inter-channel crosstalk between DWDM channels; and (iii) The confinement of signal energy in the main lobe of spectrum leads to a better SNR. Thus, the sensitivity to optical filtering can be significantly reduced [16–20]. A summary of the spectra of different modulation formats is given in Table 2.5.

2.6 REMARKS

Since the proposed dielectric waveguide and then the advent of optical circular waveguide, the employment of modulation techniques has been extensively exploited only since the availability of optical amplifiers. The modulation formats allow transmission efficiency, and hence the economy of ultra-high capacity information telecommunications. Optical communications have evolved significantly through several phases, from single-mode systems to coherent detection and modulation, which was developed with the main aim to improve optical power. Optical amplifiers defeated this main objective of modulation formats, and allowed the possibility of incoherent and all possible formats employing modulation of the amplitude, phase and frequency of the lightwave carrier.

Currently, photonic transmitters play a principal part in the extension of the modulation speed into several GHz range, and make possible the modulation of the amplitude, phase and frequency of optical carriers and their multiplexing. Photonic transmitters using $LiNbO_3$ have been proven in laboratory and installed systems. The principal optical modulator is the MZIM, which can be a single or a combined set of these modulators, allowing formation of binary or multilevel amplitude or phase modulation, and is even more effective for discrete or continuous phase shift keying techniques. The effects of modulation on transmission performance will be given in the coming chapters.

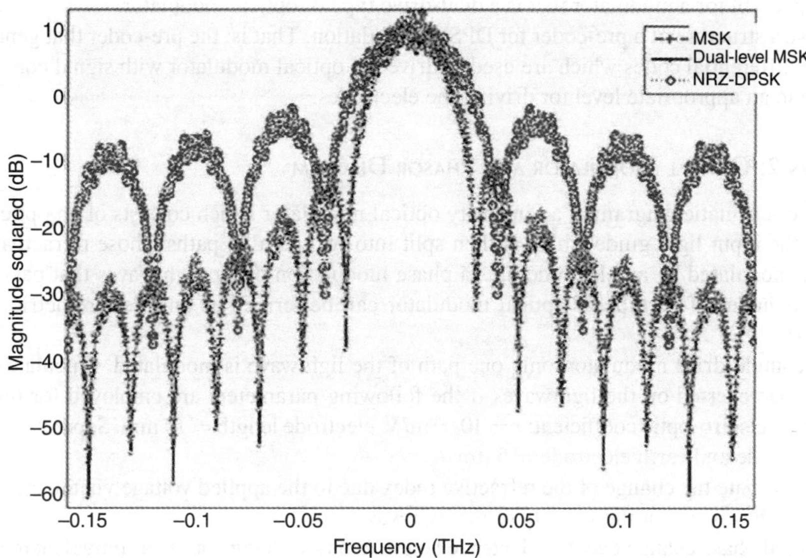

FIGURE 2.44 Spectral properties of three modulation formats: MSK (red, dash), dual-level MSK (black, solid) and DPSK (blue, dot).

Spectral properties of the optical 80-Gb/s dual-level MSK, 40-Gb/s MSK and 40-Gb/s DPSK with various RZ pulse shapes are compared. The spectral properties of the first two formats are similar. Compared with optical DPSK, the power spectra of optical MSK and dual-level MSK modulation formats have more attractive characteristics. These include high spectral efficiency for transmission, higher energy concentration in the main spectral lobe, and more robustness to inter-channel crosstalk in DWDM due to greater suppression of the side lobes. In addition, the optical MSK offers the orthogonal property, which may offer great potential in coherent detection in which the phase information is reserved via I and Q components of the transmitted optical signals. In addition, the multi-level formats would permit lowering of the bit rate and hence substantial reduction of the signal-effective bandwidth, and the possibility of reaching the highest speed limit of the electronic signal processing (digital signal processing) for equalization and compensation of distortion effects. The demonstration of ETDM receiver at 80 G and higher speed [29] would allow applications of these modulation-formatted schemes in ultra-high speed transmission.

PROBLEMS ADVANCED MODULATION FORMATS FOR LONG-HAUL TRANSMISSION SYSTEMS

QUESTION 1

A bit pattern "1 1 0 1 0 0 1 1 0 1" with a bit rate of 10 Gb/s is input into two separate modulators to generate ASK and BDPSK modulation format.

(a) Sketch the modulated bit pattern, including the carrier; this can be drawn with two or three periods within the duration of a bit period. Hence the phase distribution or the scatter diagram of the modulated sequence.

(b) A Mach–Zehnder intensity modulator (MZIM) is used as an optical modulator, and its Vπ is 5 V. Show how the data sequence can be conditioned to feed into the electrode port of the optical modulator so that ASK or BDPSK signals can be generated. Make sure that you show the biasing voltage.

(c) Repeat (b) for a modulator that is a dual-drive type of optical modulator.
(d) Give a structure of a pre-coder for DPSK modulation. That is, the pre-coder that generates
     the differential codes which are used to drive the optical modulator with signal condition-
     ing to an appropriate level for driving the electrodes.

QUESTION 2: OPTICAL MODULATOR AND PHASOR DIAGRAM

Sketch the schematic diagram of an intensity optical modulator which consists of an optical waveguide as the input light guide which is then split into two parallel paths whose refractive indices
would be modulated by an electrode, i.e., a phase modulation of the lightwaves that pass through
these waveguides. This type of optical modulator can be termed as an interferometric intensity
modulator.

For a single-drive modulator, only one path of the lightwave is modulated. Calculate the total
phase change exerted on the lightwaves if the following parameters are employed for the optical
modulation: electro-optic coefficient: $r = 10^{-11}$ m/V, electrode length $= 10$ mm. Separation between
active electrode and earth electrode $= 5$ μm.

First estimate the change of the refractive index due to the applied voltage via the electro-optic
effect, then the change of the velocity of the lightwave.

The total phase change over the electrode length (a phase change of 2π is equivalent to the slow-
ing down of one wavelength).

You may confirm that the phase change can be given by

$$\Delta\phi = \frac{c\,\Delta n}{n^2}\,L\,\frac{2\pi}{\lambda}.$$

Estimate the voltage required for applying to the electrode so that a π phase change occurs on
the lightwave carrier passing through it.

Write an expression that represents the lightwave at the input, thence those for the split waves
propagating in the two parallel lightpaths of the interferometer. Represent their phasors in a plane.

Then find the sum of the two phasors and hence project this total phasor vector on the real axis
and obtain the equation of the lightwave at the output of the modulator as a function of the applied
voltage V.

QUESTION 3A

The figure shown below shows the pulse sequence and its optical modulated spectrum for NRZ and
RZ formats.

What is the bit rate of the sequence?

Examine the spectra and identify their principal features, such as the lightwave carrier and its
power, and the 3-dB bandwidth of the passband spectrum. Thence compare the 3 dB of RZ and
NRZ modulation formats. Which modulation schemes can be derived from the spectra?

QUESTION 3B

Figure Question 3b. Note vertical scale should be from 10 dBm, and the 10 dB per division.
A carrier-suppressed RZ optical transmitter is shown below.

(i)   Sketch the time-domain pulse sequence over 10-bit period for a bit rate of 40 Gb/s at the
      output of the pulse pattern generator.
(ii) Give a brief description of the principles of the suppression of the carrier. Which compo-
      nent of the transmitter would implement the suppression?

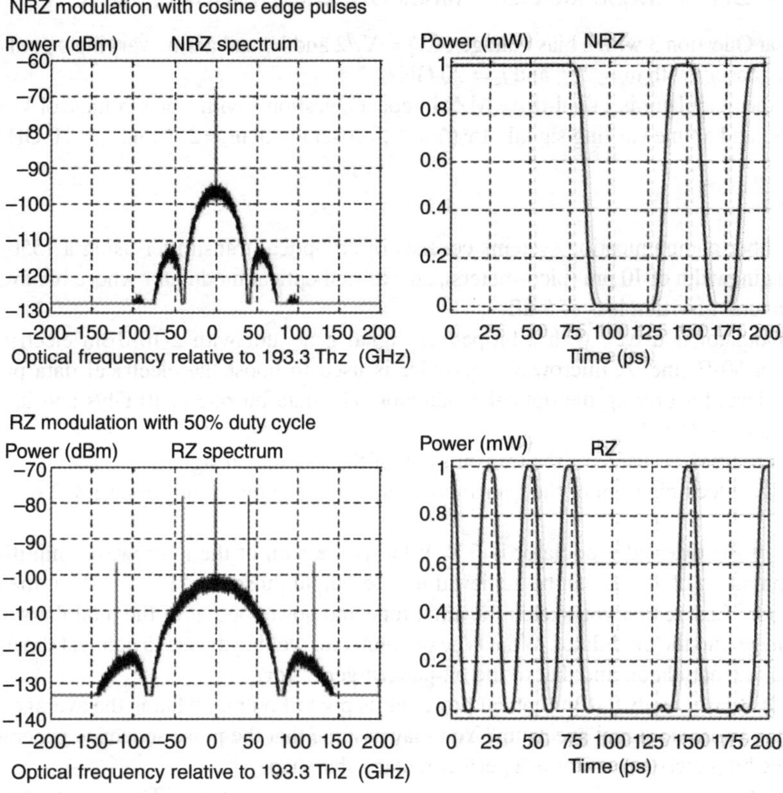

FIGURE QUESTION 3B

(iii) What are the functions of the laser, the modulator and the push–pull modulator? If the V_π for the two modulators is 5 volts, sketch the transfer characteristics of the modulators, i.e., the output power versus the input driving voltage. Ensure that you set appropriate biasing voltages for the modulators. The output power of the laser is 10 dBm and the total insertion loss for each modulator is 4 dB. For the pulse pattern generator, the output power at the output port data is 10 dBm and that at the clock output port is 2 Vp-p. All line impedances are 50 Ω.

(iv) Is it necessary to use a booster optical amplifier to increase to total average power launching into an optical fiber for transmission. If it does, what is the gain and noise of the optical amplifier? The nonlinear limit of a SSMF is about 5 dBm.

(v) Sketch the spectra at the outputs of the laser, the modulator and the push–pull modulator.

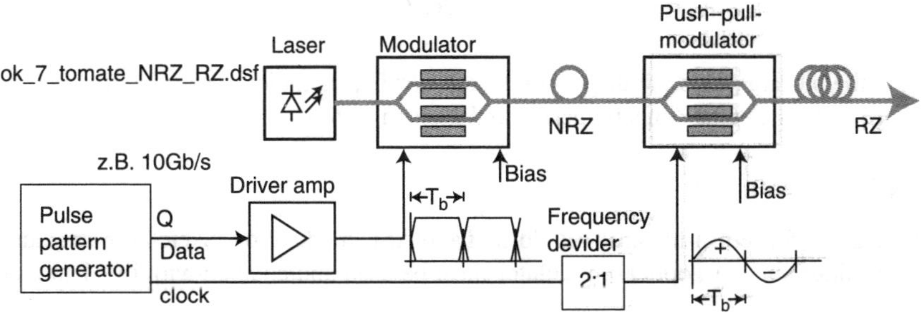

FIGURE QUESTION 3C

Question 4 Optical Modulator and Phasor Diagram: Dual-Drive MZM

(a) Repeat Question 3 with a bias voltage of $V_b = V_\pi/2$ and V_π and a time varying signal of $v_s(t)$ $= V_{\pi/2}\cos\omega_s t$ with $\omega_s = 2\pi f_s$ and $f_s = 20$ GHz.

(b) Now the modulator is a dual-drive MZM, repeat Question 1 with a bias voltage of $V_b = V_\pi/2$ and V_π and a time varying signal of $v_s(t) = V_{\pi/2}\cos\omega_s t$ with $\omega_s = 2\pi f_s$ and $f_s = 20$ GHz.

Question 5

An optical fiber communication systems consists of an optical transmitter using a 1550-nm DFB laser with a linewidth of 10 pm (pico-meters), an external optical modulator whose bandwidth is 20 GHz, and a total insertion loss of 5 dB.

The modulator is driven with a bit-pattern-signal generator with a 10-dBm electrical power output into a 50-Ω line. A microwave amplifier is used to boost the electrical data pulse to an appropriate level for driving the optical modulator. The data bit rate is 10 Gb/s and its format is non-return-to-zero (NRZ).

An 80-km SSMF is used for the transmission of the modulated signals.

Sketch the block diagram of the transmission system. You may refer to page 8–2 of the lecture notes.

The V_π of the external modulator is 5 V. What is the gain of the microwave amplifier so that an extension ration of 20 dB can be achieved for the output pulses of "1" and "0" at the output of the modulator? Ensure that you sketch the amplitude and power output of the modulator versus the driving voltage into the modulator. What type of connector would you use to connect the microwave amplifier to the modulator, and that to the bit-pattern-generator?

If the DFB laser emits 0 dBm optical power at its pig tail output, what is the average of optical power contained in the signal spectrum? You may assume that the pulse sequence generated at the output of the bit pattern generator is a perfect rectangular shape.

What is the effective 3-dB bandwidth of the signal power spectrum? Thence estimate the total pulse broadening of the pulse sequence at the end of the 80-km fiber length. Similarly, estimate the pulse sequence if the bit rate is 40 Gb/s.

If a dispersion compensating fiber of 20 km is used to compensate for the signal distortion in the 80-km fiber, what is the required dispersion factor of this fiber so that there would be no distortion? The loss of the dispersion compensating fiber is 1.0 dB/km at 1550 nm. Estimate the average optical power of the signal at the output of the dispersion-compensating fiber.

Based on the dispersion limit given below, plot the dispersion length as a function of the bit rate for NRZ format.

NOTE: The dispersion limit, under linear regime operation, can be estimated in the following equation (ref, Forgheti et al.[*] 1997)

$$L_D = \frac{c}{\lambda}\frac{\rho}{B_R{}^2 D}$$

where B_R is the bit rate , D is the dispersion factor (s/m²), ρ is the duty cycle ration, i.e., the ratio between the "ON" and "OFF" in a bit period and L_D is in meters.

Question 6

Repeat Question 1 for return-to-zero (RZ) format and ASK modulation. Sketch the structure of the optical transmitter (an extra optical modulator must be used and coupled with the data modulator

---

[*] Forghieri, F., P. R. Prucnal, R. W. Tkach, and A. R. Chraplyvy. 1997. RZ versus NRZ in nonlinear WDM systems. *IEEE Photonics Technology Letters* 9 (7): 1035–37.

of Question 1, the optical pulse carver). Give details of the pulse carver, including driving voltage, driving signal and synchronization with the data generator.

QUESTION 7: SPECTRAL EFFICIENCY

(a) A DWDM optical transmission system can transmit optical channels whose channel spacing is 100 GHz. What is the spectral efficiency if the bit rate of each channel is 40Gb/s and the modulation is NRZ-ASK?
(b) Repeat (a) for RZ-ASK modulation format.
(c) Repeat (a) and (b) for channel spacing of 50 GHz.

QUESTION 8

(a) Give the structure of an optical transmitter for generation of RZ-ASK modulation format. Ensure that you assign the optical power of lightwaves generated from the light source, and that at the output of the optical modulators a maximum of 10 dBm of optical power is launched into the SSMF so that it is below the nonlinear SPM effect limit.
(b) Describe the operation of the optical modulator and the pulse carver, so that it can generate periodic pulse sequence before feeding into the data generator. Ensure that you provide the amplitude- and intensity- versus the driving signal voltage levels which are used to drive the optical modulators.

QUESTION 9: NONLINEAR SPM EFFECT

The nonlinear refractive index coefficient of silica-based SSMF is $n_2 = 2.5 \times 10^{-20}$ m^2/W.

What is the effective area of the SSMF? You can refer to the technical specification of the Corning SMF-28 and its mode field diameter to estimate this area.

Estimate the change of the refractive index as a function of the average optical power. Hence estimate the total phase change due to this nonlinear effect after propagating through a length L (in km) of this fiber.

Hence estimate the maximum length L of the SSMF that the lightwaves can travel so that not more than 0.1 rad of the phase change on this lightwave carrier would occur.

Show how you can generate a format that would have a RZ format and a suppression of the lightwave carrier. Show that the width of the RZ pulse on this case is 67% of the bit period. Hint: you may represent the lightwaves in the path of the optical modulator, an optical interferometer by using phasors. First, sketch the phasor of the input lightwave. Then, those of the two paths and then the phase applied onto these paths. Then, sum up at the output to give the resultant output. For the pulse width, you can estimate the width over which the amplitude falls to 1/sqrt(2) of its maximum.

Now show you can generate RZ pulse sequence with 50% and 33% pulse width of the bit period.

QUESTION 10: BALANCED RECEIVER

Sketch the schematic diagram of an optical balanced receiver. A balanced receiver would consist of a delay interferometer and a back-to-back connected pair of photodetector with its output connected to the input of an optical pre-amplifier.

What is the functionality of the delay interferometer? What is the temporal length of the delay unit?

What are the roles of the two optical couplers and their ideal coupling coefficients?

What is the relationship between the two output ports of the delay interferometer?

Suppose that a sequence of 4 bits of a DPSK 10 Gb/s data channel is presented at the input of a balanced receiver. The phases of the lightwave carrier contained within these four bits are π π o π at the transition of the bit period.

Sketch the carrier wave and the pulse envelope. The lightwave has a wavelength of 1550 nm, but to illustrate the wave you are expected to sketch only a few periods of the waves contained within the bit period at the input of the receiver.

Sketch the electrical signal at the output of the electronic preamplifier (excluding noises).

Assume that an optical amplifier is used as an optical preamplifier and is placed at the input of the balanced receiver that would give an optical signal power of −10 dBm for the "0" and "1" of the DPSK sequence. The responsibility of the photodetector is 0.9; and the electronic pre-amplifier has a trans-impedance of 150 Ω, a total equivalent noise current spectral density of 2 pA/(Hz)$^{1/2}$ and a bandwidth of 15 GHz. Sketch the signal waveform at the output of the electronic preamplifier.

QUESTION 11: DUO-BINARY MODULATION FORMAT

Design a block diagram of a pre-coder that would generate tri-level modified duo-binary format signals. Make sure that the coefficients of the filters are specified. Hint: you may refer to pages 8–11 of the lectures notes. Hence derive the spectrum of the signals after the pre-coder of the modified duo-binary.

If possible, obtain the pre-coders for AMI and duo-binary and their frequency responses. Compare the frequency responses of the three modulation schemes.

Sketch the structure of the tri-level duo-binary pre-coder with its output of −1, 0, +1.

Show how to use the coded signals to drive a dual-drive MZIM to generate optical duo-binary signals.

QUESTION 12: DUO-BINARY MODULATION FORMAT

A modulation format that would allow detection of the modulated signals is duo-binary, which is a special case of partial response coding (see lecture notes on "digital communications").

Give a brief account of the principles of operation of this line code.

A duo-binary coder using a delay and add coding structures is shown below. If three-level duo-binary coded signals are required, design the pre-coder for this type of modulation.

Figure Question 12.

If setting the delay time Tb is that of a bit period, transform the structure into the z-transform diagram and hence obtain the transfer function of the coder in the z-domain, thence the frequency response of this coder. Plot the frequency response of the transfer function of the filter in continuous domain.

Find the impulse response of the coder, and hence the term partial response.

Sketch a block diagram that shows the functionality of pre-coding, coding, tri-level conversion (offset) and decoding.

A binary sequence $d(k) = \{0\ 0\ 1\ 1\ 0\ 1\ 0\ 0\ 1\}$ is applied to the input of the duo-binary coder. Determine the data sequences b(K), c(k) and c´(k) in the electrical domain which can be used to modulate an optical modulator.

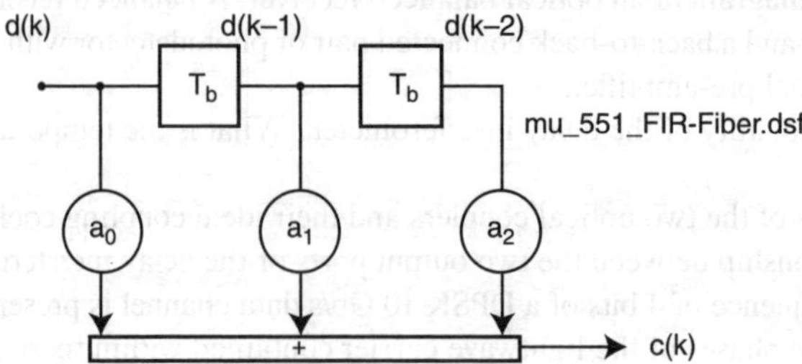

FIGURE QUESTION 12

Assuming that there is no dispersion in the transmission of the duo-binary data sequence, find the output pulse sequence at the output of the decoder. What is the physical realization of the decoder? Thence sketch the sequence at the output of a decision circuit.

The electrical signals are applied to a microwave amplifier that would condition the signals to appropriate signal levels so as to modulate the optical modulator. The measured spectra are recorded as shown in the diagrams below.

Determine where in the block diagram (as per attached diagram) that each of the spectra belongs to the points of the diagram of the transmission system.

QUESTION 13: DQPSK

DQPSK is a 2 bit per symbol modulation, i.e., 2 bits/symbol, thus the scheme is spectral-efficient.

Give a brief account of the modulation scheme DPSK and DQPSK.

Give a structure of a pre-coder for DPSSK, i.e., give a differential modulation with phase as the codes for "1" and "0".

Extend this pre-coder and the phase quadrature modulation technique for the structure of a DQPSK optical transmitter.

QUESTION 14: SSB AND DSB MODULATION

Refer to the diagrams shown below for generation of optical signals with SSB.

State the functionality of the Hilbert transformer. Hence, could you deduce a general principle for suppression of a sideband to generate single side band signals?

What is the role of the phase shifter $\pi/2$?

Explain the operation of the optical SSB transmitter, in both the time and frequency domain. Confirm that the spectrum is correct.

Figure for Question 14 (see below): SSB modulation and generation using (a) transform in optical domain (b) transform in electrical domain (c) realization of a SSB optical transmitter.

QUESTION 15: COHERENT OPTICAL COMMUNICATION SYSTEMS

Sketch the structure of an optical coherent receiver. Give a brief description of the roles of each component in the system.

What is the typical modern linewidth of the laser that acts as the local oscillator?

Distinguish between homodyne and heterodyne coherent systems.

A homodyne optical receiver has the parameters decribed below.

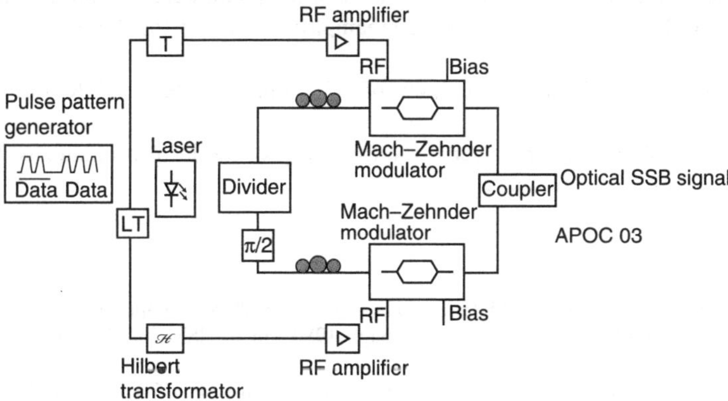

FIGURE QUESTION 14

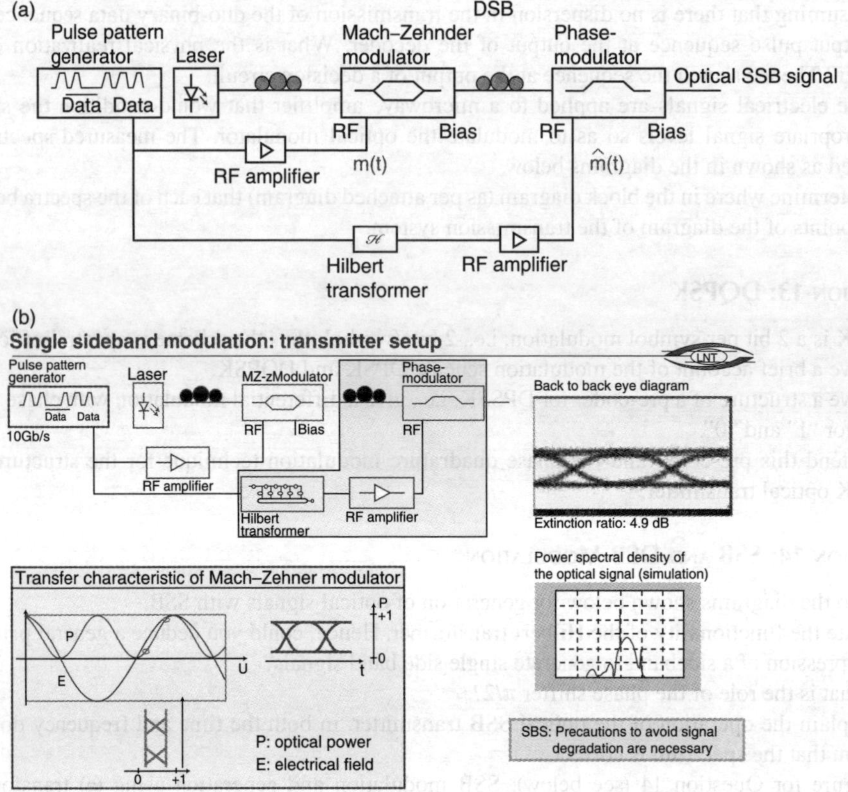

FIGURES FOR QUESTIONS 12, 13 AND 14

A photodetector with a responsibility of 0.9 which is followed by an electronic preamplifier whose total equivalent noise spectral density is 5 pA/(Hz)$^{1/2}$ and an electrical bandwidth of 15 GHz. The transmission bit rate is 10 Gb/s.

The local oscillator is a tunable laser source with a line width of 100 MHz. The wavelength in vacuum of the signals and the local oscillator is 1550.92 nm. The average optical power of the local oscillator coupled to the photodetector is 0 dBm.

Sketch the structure of the receiver and then its equivalent small signal circuit, which includes the generated electronic signal current at the output of the photodetector, and the total noise currents looking from the input of the electronic preamplifier. What is the dominant noise source in this receiver?

For an optical signal with an average power of –20 dBm, estimate the SNR at the output of the photodetector.

Re-calculate the SNR of the receiver if the frequency of the local oscillator is 20 GHz away from that of the signal carrier frequency.

REFERENCES

1. Gnauck, A. H., X. Liu, X. Wei, D. M. Gill, and E. C. Burrows. 2004. Comparison of modulation formats for 42.7-gb/s single-channel transmission through 1980 km of SSMF. *IEEE Photonics Technology Letters* 16 (3): 909–11.

2. Yamada, Y., H. Taga, and K. Goto. 2002. Comparison between VSB, CS-RZ and NRZ format in a conventional DSF based long haul DWDM system. In *Proceedings of ECOC'02*, vol. 4, 1–2.

3. Agrawal, G. P. 2002. *Fiber-optic communication systems*. 3rd ed. New York: Wiley.

4. Hirano, A., Y. Miyamoto, and S. Kuwahara. 2003. Performances of CSRZ-DPSK and RZ-DPSK in 43-Gbit/s/ch DWDM G.652 single-mode-fiber transmission. In *Proceedings of OFC'03*, vol. 2, 454–56.
5. Gnauck, A. H., G. Raybon, P. G. Bernasconi, J. Leuthold, C. R. Doerr, and L. W. Stulz. 2003. 1-Tb/s (6/spl times/170.6 Gb/s) transmission over 2000-km NZDF using OTDM and RZ-DPSK format. *IEEE Photonics Technology Letters* 15 (11): 1618–20.
6. Djordjevic, I. B., and B. Vasic. 2006. 100-Gb/s transmission using orthogonal frequency-division multiplexing. *IEEE Photonics Technology Letters* 18 (15): 1576–78.
7. Lowery, A. J., L. Du, and J. Armstrong. 2006. Orthogonal frequency division multiplexing for adaptive dispersion compensation in long haul WDM systems. In *OFC 2006 Postdeadline Sessions*, 9th March, Anaheim, CA, USA, Paper PDP39.
8. (a) Alferness, R. C. 1986. Optical guided-wave devices. *Science,* 234 (4778): 825–29; (b) Rizzi, M., and B. Castagnolo. 2002. Electro-optic intensity modulator for broadband optical communications. *Fiber and Integrated Optics* 21: 243–51; (c) Takara, H. 2001. High-speed optical time-division-multiplexed signal generation. *Optical and Quantum Electronics* 33 (7–10): 795–810; (d) Wooten, E. L., K. M. Kissa, A. Yi-Yan, E. J. Murphy, D. A. Lafaw, P. F. Hallemeier, D. Maack, et al. 2000. A review of lithium niobate modulators for fiber-optic communications systems. *IEEE Journal of Selected Topics in Quantum Electronics* 6 (1): 69–80; (e) Noguchi, K., O. Mitomi, H. Miyazawa, and S. Seki. 1995. A broadband Ti: LiNbO$_3$ optical modulator with a ridge structure. *Journal of Lightwave Technology* 13 (6): 1164–68.
9. Yariv, A., C. A. Mead, and J. V. Parker. 1966. *IEEE Journal of Quantum Electronics* QE-2: 243.
10. Kawanishi, T., S. Shinada, T. Sakamoto, S. Oikawa, K. Yoshiara, and M. Izutsu. 2005. Reciprocating optical modulator with resonant modulating electrode. *Electronics Letters* 41 (5): 271–72.
11. Krahenbuhl, R., J. H. Cole, R. P. Moeller, and M. M. Howerton. 2006. High-speed optical modulator in LiNbo$_3$ with cascaded resonant-type electrodes. *Journal of Lightwave Technology* 24 (5): 2184–89.
12. Painter, O., P. C. Sercel, K. J. Vahala, D. W. Vernooy, and G. H. Hunziker. 2002. Resonant optical modulators. US Patent No. WO/2002/050575, June 27, 2002.
13. Schubert, C., R. H. Derksen, M. Möller, R. Ludwig, C.-J. Weiske, J. Lutz, S. Ferber, A. Kirstädter, G. Lehmann, and C. Schmidt-Langhorst. 2007. Integrated 100-Gb/s ETDM receiver. *IEEE Journal of Lightwave Technology* 25 (1): 122–30.
14. Amoroso, F. 1976. Pulse and spectrum manipulation in the minimum frequency shift keying (MSK) format. *IEEE Transactions on Communications* 24: 381–84.
15. Lee, W. S. 80+ Gb/s ETDM systems implementation: An overview of current technology. In *Proceedings OFC 2006, Paper No. OTuB*, 1–3.
16. Amoroso, F. 1976. Pulse and spectrum manipulation in the minimum frequency shift keying (MSK) format. *IEEE Transactions on Communications* 24: 381–84.
17. Noda, Y. 1990. Electro-optic modulation method and device using the low-energy oblique transition of a highly coupled super-grid, United States Patent 5073809, 12/03/1990.
18. Suzuki, M., Y. Noda, H. Tanaka, S. Akiba, Y. Kuahiro, and H. Isshiki. 1987. Monolithic integration of InGaAsP/InP distributed feedback laser and electroabsorption modulator by vapor phase epitaxy. *IEEE Journal of Lightwave Technology* LT-5 (9): 127.
19. Ivan P. Kaminow, Tingye Li. 2002. Optical fiber telecommunications IV: Components. New York: Academic Press.
20. Stillman 1976, Dutta 1984, Noda 1985.
21. Nagata, H., Y. Li, W. R. Bosenberg, and G. L. Reiff. 2004. DC Drift of X-Cut LiNbO$_3$ Modulators. *IEE Photonics Technology Letters* 16(10), pp. 2233–335.
22. Nagata, H. 2000. DC drift failure rate estimation on 10 Gb/s X-cut lithium niobate modulators. *IEEE Photonics Technology Letters* 12 (11): 1477–79.
23. Epworth, R. E., K. S. Farley, and D. Watley. 2003. Polarization mode dispersion compensation. US Patent 398/152, 398/202, 398/65.
24. Langton, C. Basic concept of modulation. www.complextoreal.com.
25. Pang, K. K. 2002. *Digital transmission*. Melbourne, Australia: Mi-Tec Publishing, 58.
26. Redner, R., and H. Walker. 1984. *SIAM Review* 26 (2): 195–234.
27. Kaminow, I. P., and T. Li. 2002. *Optical fiber communications, volume IVA*. Chapter 16. USA: Elsevier.
28. Lach, E., and K. Schuh. 2006. Recent advances in ultrahigh bit rate ETDM transmission systems. *IEEE Journal of Lightwave Technology* 24 (12): 4455–67.
29. Lee, W. S. 2006. 80+ GBit/s ETDM systems implementation: An overview of current technologies. *Proceedings of the Optical Fiber Communication Conference, 2006 and the 2006 National Fiber Optic Engineers Conference. OFC 2006*, Annaheim, CA, USA, 5–10 March 2006, 1–3.

3 Optical Receivers and Transmission Performances

3.1 INTRODUCTION

Optical detection and the noise interference in such processes are critical to the performance of optical communications systems. Fundamental understanding of the noises and the sensitivity of the receiver, and the minimum optical power available at the photodetector, requires evaluation of the transmission performance of different modulation formats. Optical receivers have evolved from the binary digital direct detection to coherent mixing of signals and local lightwave oscillator and detection with optical pre-amplification. This chapter presents an overview of the noise process and mechanism in term of equivalent noise power and contribution.

A schematic diagram of a single-channel DPSK system is illustrated in Figure 3.1.

As the bit or symbol rate of the optical transmission systems is increased, the demand for modeling is intense, especially for a modeling platform that can truly structure the photonic sub-systems. In order to enhance the effective transmission capacity with minimum renovation of the photonic and electronic sub-systems, there are two possible solutions: (i) increasing the base bit rate; and (ii) employing multilevel modulation techniques such as M-ary amplitude and/or phase shift keying [1]. The later would be preferred because the symbol rate is lower than the bit rate, thus multi-level modulation formats offer much higher transmission capacity than the case in which the base rate would be increased. Increasing the base rate would face the complexity of the transmitters and receivers as well as the dispersion tolerance.

One possible technique that could offer insight into possible solutions that facilitate the simulation of ultra-high capacity and ultra-high bit rate transmission systems is to develop a comprehensive modeling platform. Furthermore, the modeling platform should take advantage of any user-friendly software platform that is popular and easy to use for further development. This platform would offer the research community of optical communication engineering a basis for extension, and enhance the linkages between research groups.

Thus, one of the principal objectives of this chapter is to present the development of a MATLAB® Simulink® 7.0 [2] for computer experiments of optical transmission systems under advanced modulation formats, especially the amplitude and/or phase shift keying modulation of binary or multi-level. Duo-binary can be employed in association with appropriate low pass filter and can offer multiple rates much higher than the basic rate. The advantage of the duo-binary is that the detection is much simpler than that of DPSK, DQPSK or M-ary PSK, in which the receiver is a direct detection type. In the photonic domain, the combination of phase and amplitude is taken into account for coding modulation schemes that involves tri- or higher-order levels. These features can be implemented with ease on the Simulink platform.

We base our system bit rate for 40 Gb/s per channel. The modulation formats of binary RZ, NRZ, CSRZ ASK DPSK are demonstrated with the modeling of transmission over standard single mode optical fibers (SSMF), as well as the photonic decoding of optical signaling. The transmission performance of some optical systems is also given in terms of the bit error rate (BER), the δ-factor (usually known as the Q-factor) and receiver sensitivity.

In this chapter, we present the first MATLAB Simulink-platform simulation test-bed for modeling of the transmission of amplitude and differential phase modulation incorporating RZ,

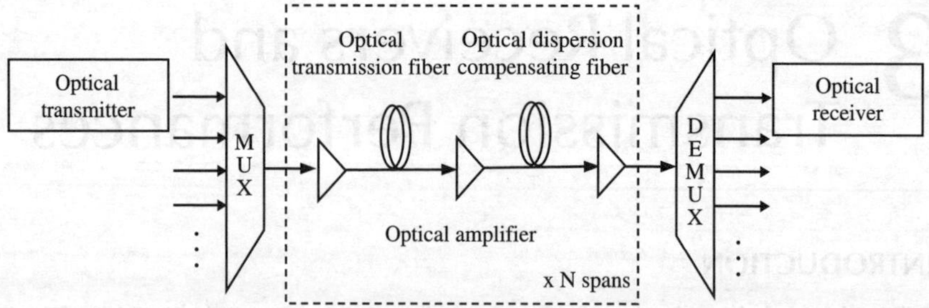

FIGURE 3.1 Schematic diagram of a single-wavelength channel optical transmission system.

NRZ, carrier-suppressed RZ formats for advanced high capacity and long-haul optical fiber transmission systems. A novel modified fiber propagation algorithm has been developed and optimized to minimize simulation processing time and enhance its accuracy. The performance of the optical transmission systems can be automatically and accurately evaluated by various methods. Analysis and methodologies for future development of the simulation test-bed are also presented.

In recent years, the capacity and reach distance of optical transmission systems have been dramatically increased due to the accelerating growth of data usage demand (e.g., Internet, peer-to-peer network). Therefore, the necessity of upgrading of current DWDM 10 Gb/s systems to DWDM 40 Gbps or even higher bit rates becomes crucial to telecommunications service providers. Accordingly, several modulation formats have been proposed and investigated as the alternatives for the current on–off keying (OOK) intensity modulation that is severely degraded at high bit rate due to dispersion and non-linear effects of the transmission fibers. Apart from the requirements of robustness to transmission impairments, the cost effectiveness for system upgrades is also significant. Among the candidates, differential binary and/or quadrature phase shift keying (DPSK/DQPSK) have recently attracted much attention due to the following advantages: (i) a 3-dB improvement on receiver sensitivity (if balanced receiving technique is used) [1,2]; (ii) high tolerance to fiber nonlinearities, especially to intra-channel non-linear effects, the cross phase modulation (XPM) and the four wave mixing (FWM) [2,3]; (iii) superior spectral efficiency (DQPSK), hence high tolerance to optical filtering [4]; and (iv) advantages in all-optical networks incorporating optical add-drop multiplexers or optical cross-connects [5]. Recently, several experimental demonstrations of DPSK/DQPSK long-haul transmission DWDM systems for 10 Gbps and 40 Gbps have been reported [6–8]. Therefore, a simulation test-bed is necessary for detailed design, investigation and verification on the benefits and shortcomings of these advanced modulation formats on fiber-optic transmission systems. In this chapter, we present the first MATLAB Simulink-based simulation package for photonic DWDM systems. The simulator is in the first phase of the development, whereby the focus is on a single-channel system, and it is continuously improved and updated.

This chapter is organized as follows: (i) in Section 3.2, receiver noise sources are outlined and noise analyses and performance of basic binary digital optical receiver are then given in Section 3.3 (ii) the fundamentals for the modeling of DPSK transmission system are given; (iii) Section 3.3 presents the architecture and operational principles of the simulator; (iv) Section 3.4 presents simulation results of the up-to-date simulator and corroboration of the developed models in comparison with experimental results presented in Section 3.5 for ASK and DPSK modulation formats. Finally, concluding remarks are given.

3.2　DIGITAL OPTICAL RECEIVERS

This section gives the fundamental concepts of receiver structure, sensitivity and noises. It is mainly focused on the direct detection that would be integrated into more complex receiver structure for different modulation formats.

3.2.1　PHOTONIC AND ELECTRONIC NOISE

Before proceeding to the systems calculations to determine the performance of optical receivers, the noises generated in the photodetector and the preamplifier front-end must be considered. One can consider investigating the system calculation and returning back to the noise calculation provided that they are taking either the total equivalent noise spectral density at the input of the detector or its noise figure. This section describes all noise mechanisms related to the photodetection process, including the electronic noise associated with the receiver.

3.2.1.1　Electronic Noise of Receiver

At the receiver, noise sources that are superimposed onto the electrical signals detected after photo-diodes include the electrical shot noise, dark current noise, and the equivalent noise current as seen from the input of the electronic pre-amplifier. The electronic noise sources are represented as the square of current which can be found by using the noise spectral density as a function of frequency and then integrating over the bandwidth of the system. They can be given as

$$i_{Nshot}^2 = 2q\Re P_{in} B_e \tag{3.1}$$

$$i_{ND}^2 = 2q I_D B_e \tag{3.2}$$

$$i_{EA}^2 = (I_{eq})^2 B_e. \tag{3.3}$$

3.2.1.2　Shot Noises

Shot noises and thermal noises are the two most significant noises in optical detection systems. Shot noises are generated by quantum process or electronic biasing. The noises are specified in noise spectral density, i.e., the square of noise current per unit frequency. Thus, noise spectral density must be integrated over the total amplifier bandwidth to obtain the equivalent noise currents.

Electrical shot noises are generated by the random generation of streams of electrons (current). In optical detection, shot noises are generated by: (i) biasing currents in electronic devices; and (ii) photo currents generated by the photodiode.

3.2.1.3　Biasing Current Shot Noises

Any biasing current I has a spectral current density S_I given by:

$$S_I = \frac{d\langle i_I^2 \rangle}{df} = 2qI \quad \text{in A}^2/\text{Hz} \tag{3.4}$$

where q is the electronic charge. The current i_I represents the noise current generated due to the biasing current I.

3.2.1.4 Quantum Shot Noise

The average current $<i_s^2>$ generated by the photodetector by an optical signal with an average optical power P_{in} is given by

$$S_Q = \frac{d\langle i_s^2 \rangle}{df} = 2q\langle i_s^2 \rangle. \tag{3.5}$$

This is the signal-dependent noise and it is a unique feature of optical communications. If the APD is used, then the noise spectral density is given by

$$S_Q = \frac{d\langle i_s^2 \rangle}{df} = 2q\langle i_s^2 \rangle\langle G_n^2 \rangle. \tag{3.6}$$

The dark currents generated by the photodetector as mentioned in Chapter 4 must be included to the total equivalent noise current at the input after it is evaluated. These currents are generated even in the absence of the optical signal. These dark currents can be eliminated by cooling the photodetector to at least below the temperature of liquid nitrogen (77 K).

3.2.1.5 Thermal Noise

At a certain temperature, the conductivity of a conductor varies randomly. The random movement of electrons generates a fluctuating current even in the absence of an applied voltage. The thermal noise of a resistor R is given by

$$S_R = \frac{d\left(i_R^2\right)}{df} = \frac{2k_B T}{R} \tag{3.7}$$

where k_B is the Boltzmann's constant, T is the absolute temperature (in K), R is the resistance in Ohms and i_R denotes the noise current due to resistor R.

3.2.1.6 ASE Noise of Optical Amplifier

The following formulation accounts for all noise terms that can be treated as Gaussian noise due to the optical amplifier

$$N_{ASE} = mn_{sp}h\nu(G_{op} - 1)B_o \tag{3.8}$$

where G_{op} = amplifier gain; n_{sp} = spontaneous emission factor; m = number of polarization modes (1 or 2); PN = mean noise in bandwidth; OSNR at the output of EDFA

3.2.1.7 Optical Amplifier Noise Figure

Amplifier Noise Figure (NF) is defined at the output of the optical amplifier as the ratio between the output OSNR at the output on that OSNR at the input of the EDFA.

$$F_N = \frac{\mathrm{OSNR}_{in}}{\mathrm{OSNR}_{out}} \approx 2n_{sp} \text{ for } G_{op} \gg 1. \tag{3.9}$$

Modern optical amplifiers are optimized and n_{sp} reaches unity, so the NF can reaches 3 dB.

3.2.1.8 Electronic Beating Noise

If an optical preamplifier is used in association with an optoelectronic receiver front-end, then there would be noises generated due to the signal-dependent shot noise, beat noises from the beating of the electronic currents generated from the signals and the random ASE noises, and the beating noise between the ASE electronic currents. The signal-ASE noise usually dominates this noise process.

The beating noises generated from the signal current and the ASE noise current dominates the detection process in an optical pre-amplifying optoelectronic receiver, and is given by

$$i^2_{\text{sig-ASE}} = 2(q\eta G_{\text{op}})^2 (2n_{\text{sp}})P_{\text{in}}\frac{B_{\text{e}}}{h\upsilon} \tag{3.10}$$

where B_{e} is the bandwidth of the electronic amplifier system.

Under direct detection, the ASE noises appear at the input of an electronic preamplifier would follow the square detection, i.e., the noise vector would be taken with its absolute value, then squared and multiplied with the responsibility of the detector to obtain the spectral density noise current.

Conversely, under coherent detection, the ASE noise superimposes on the electric field of the local oscillator and the signals. The local oscillator amplitude is much higher than that of the signal, and the beating noise between the local oscillator and the ASE noise dominates the noise source presented at the input of the electronic pre-amplifier. This is the significant difference between the optically amplified and non-optically amplified receiver for direct and coherent optical receiver, respectively.

3.2.1.9 Accumulated ASE Noise in Cascaded Optical Amplifiers

Long-haul optical communications would be structured with dispersion-managed and loss-equalized through tens or hundred of spans. It has been demonstrated in error-free transmission over several thousands of kilometers without repeaters. It is of extreme importance to account for the accumulated noise sources that ultimately limit the transmission distance. The ASE noise accumulated is the principal noise source that has been built-up over several optical amplifiers in cascade that would be originally from the photons generated randomly under spontaneous emission. The spontaneous noise factor is given as n_{sp}. Practical EDFAs have now reached mature stage with n_{sp} reaching unity. In this section, an effective noise factor at the end of the transmission spans or at the input of the optical receiver is derived.

Let the linear loss factor of the ith transmission fiber be $\alpha_{\text{L,I}}$ and G_i be the linear gain factor of the optical amplifier in the span. A recursive relation between the noise of the first stage and i^{th} stage can be written as

$$n''_{\text{sp},1} = n'_{\text{sp},1}$$

$$n''_{\text{sp},i} = \frac{n''_{\text{sp},i-1}G_{i-1}}{\alpha_{\text{L},i-1}} + n'_{\text{sp},i} \quad \text{for } 2 \leq i \leq N \tag{3.11}$$

where $n''_{\text{sp},1}$ is the noise as seen from the input of the amplifier and $n'_{\text{sp},1}$ is the equivalent noise factor at the output of the amplifier. The accumulated ASE noise over the transmission spans is given by the summation of all the noise sources of the amplifiers. At the output of the final N^{th} stage, without taking into account the other noise sources, the equivalent ASE accumulated noise is given as

$$n''_{sp,N} = \sum_{i=1}^{N} n''_{sp,i}$$ (3.12)

$$n^e_{sp,i} = \frac{n''_{sp,N} G_N}{\alpha_L} + n_{sp,i}$$

with the assumption that the gain of optical amplifiers has equalized all the losses of transmission and dispersion-compensating fibers.

If a optical pre-amplified receiver and two optical filters are placed before and after the optical preamplifier, then the effective ASE noise is given by

$$n^e_{sp} = \frac{n''_{sp,N} G_N \tau_2}{\alpha_L \tau_1} + n_{sp}$$ (3.13)

where τ_1 and τ_2 are the time constant of the two optical filters.

The noise figure of the equivalent accumulated noise is then given by

$$NF^e = 10 \mathrm{Log}_{10} n^e_{sp}.$$ (3.14)

This is also the amount of degradation of the OSNR due to accumulated ASE noise at the front of the photodetector.

3.3 PERFORMANCE EVALUATION OF BINARY AMPLITUDE MODULATION FORMAT

As described above, the noises of the optical receivers consist of the thermal noises and quantum shot noises due to the bias currents and the photocurrent generated by the photodetector with and without the optical signals. Thus, this quantum shot noise is strongly signal dependent. These noises degrade the sensitivity of the receiver [1] and thus a penalty can be estimated. Another source of interference that would also result in signal penalty is the intersymbol interference (ISI). The goal of this section is to obtain an analytical expression for the receiver sensitivity of the direct detection optical receiver on/off keying system.

3.3.1 RECEIVED SIGNALS

Assuming that the received signal power is

$$p(t) = \sum_j a_j h_p(t - jT_b).$$ (3.15)

The average output voltage at the output of the electronic preamplifier is thus given by

$$\langle v_o \rangle = \Re \langle G_n \rangle \left[\sum_j a_j \frac{1}{T_b} \int_{-T_b/2}^{T_b/2} h_p(t - jT_b) dt \right] R_1 A$$ (3.16)

where T_b is the bit period, and $h_p(t-jT_b)$ is the impulse response of the system evaluated at each time interval. R_I is the input resistance of the overall amplifier of the system, including the front-end and linear channel amplifier. It is assumed that the overall amplifier has a flat gain response A over the band width of the system.

$$a_j \bigg|_{t=0} = \begin{cases} b_0 \approx 0 \\ b_1 \end{cases} \tag{3.17}$$

where b_0 is the energy when a transmitted "0" is received, and b_1 is the energy when a transmitted "1" is received. The summation over a number of periods must be taken into account the contribution of adjacent optical pulses. We now have to distinguish between two cases when a "0" or a "1" is transmitted and received.

3.3.1.1 Case 1: OFF or a Transmitted "0" is Received

Using Equation 3.12 and Equation 3.13, we have:

$$\langle v_0 \rangle_0 = v_{00} = \Re \frac{b_0}{T_b} G R_I A \cong 0 \tag{3.18}$$

with the total equivalent noise voltage at the output, $v_{NT_0}^2$ is

$$v_{NT_0}^2 = v_{NA}^2 = i_{Neq}^2 R_I^2 A^2 \tag{3.19}$$

where i_{Neq} is the total equivalent current at the input of the electronic pre-amplifier. The appendix gives a method for estimating this noise current for any pre-amplifier whose equivalent Y-parameters are known.

3.3.1.2 Case 2: ON Transmitted "1" Received

In this case, the average signal voltage at the output is received as

$$\langle v_0 \rangle_1 = v_{01} = \Re \frac{b_1}{T_b} \langle G_n \rangle R_I A \tag{3.20}$$

where a total noise equivalent mean voltage at the output is given by

$$v_{NT1}^2 = v_{0SN}^2 + v_{NA}^2 \tag{3.21}$$

where v_{0SN}^2 is the signal-dependent shot noise. v_{NA}^2 is the amplifier noise at the output, and given by

$$v_{NA}^2 = i_{Neq}^2 R_I^2 A^2 B. \tag{3.22}$$

The signal-dependent noise is the quantum shot noise, and is given by

$$v_{0SN}^2 = \int_0^B 2q \langle i_s \rangle_\lambda \langle G_n^2 \rangle R_I^2 A^2 df \tag{3.23}$$

where B is the 3-dB bandwidth of the overall amplifier, $<i_s>_1$ is the average phototcurrent current received when a "1" was transmitted. This current can be estimated as follows

$$\langle i_s \rangle_1 = \sum_{-\infty}^{\infty} \Re \frac{b_1}{T_b} \int_{-T_b/2}^{T_b/2} h_p(t - jT_b)dt \tag{3.24}$$

or

$$\langle i_s \rangle_1 = \Re \frac{b_1}{T_b} \int_{-\infty}^{\infty} h_p(t)dt \tag{3.25}$$

with a normalization that is $\displaystyle\int_{-\infty}^{\aleph} h_p(t)dt = 1$.

Equation 3.21 becomes

$$\langle i_s \rangle_1 = \Re \frac{b_1}{T_b}. \tag{3.26}$$

3.3.2 Probability Distribution Functions

The optical systems under consideration are typical IM/DD systems, but are also suitable for further calculations for modulation formats in digital optical communications. Further development of statistical distribution is discussed in Section 3.4. In these systems, the optical energy of each pulse period is equivalent to that of at least a few hundred photons. This number is sufficiently large so that a Gaussian distribution of the probability density is warranted. The probability density function (pdf) of a "0" transmitted and received by the optical receiver is thus given by

$$p\left[v_0\big|\text{"0"}\right] = \frac{1}{\left(2\pi v_{NT0}^2\right)^{1/2}} \exp\left[\frac{-(v_0 - v_{00})^2}{2v_{NT0}^2}\right] \tag{3.27}$$

and similarly for a "1" transmitted

$$p[v_0|\text{"1"}] = \frac{1}{\left(2\pi v_{NT1}\right)^{1/2}} \exp\left[\frac{-(v_0 - v_{01})^2}{2v_{NT1}^2}\right]. \tag{3.28}$$

The total probability of error or a bit-error-rate (BER) is defined as

$$\text{BER} = p(1)p(0/1) + p(0)p(1/0) \tag{3.29}$$

where $p(1)$ and $p(0)$ are the probabilities of receiving "1" and "0", respectively, and $p(1/0)$ and $p(0/1)$ are the probabilities of deciding "1" when "0" is transmitted and *vice versa*. In an OOK bit stream, "1" and "0" are likely to occur equally, that is $p(1) = p(0) = 0.5$, then Equation 3.25 becomes

$$\text{BER} = \frac{1}{2}[p(0/1) + p(1/0)] \tag{3.30}$$

Thus, as equal probability of transmitting "0" and "1" is assumed. For a decision voltage level of d, the total probability of error P_{E} is the summation of the errors of deciding "0" or "1" by integrating the probability density function over the overlapping regions of the two pdf , gives

$$P_{\text{E}} = \frac{1}{2}\int\limits_{d}^{\infty} p[v_0|"0"]\mathrm{d}v_0 + \frac{1}{2}\int\limits_{-\infty}^{d} p[v_0|"1"]\mathrm{d}v_0. \tag{3.31}$$

Substituting for the probability distribution using Equation 3.27 and Equation 3.26 leading to

$$\text{BER} = \frac{1}{2\sqrt{\pi}}\int\limits_{\frac{d-v_{00}}{v_{\text{NT0}}}}^{\infty} e^{-\frac{x^2}{2}}\mathrm{d}x + \frac{1}{2\sqrt{\pi}}\int\limits_{\frac{v_{01}-d}{v_{\text{NT1}}}}^{d} e^{-\frac{x^2}{2}}\mathrm{d}x. \tag{3.32}$$

The functions in Equation 3.28 have the standard form of the complementary error function $Q(\alpha)$ defined as

$$Q(\alpha) = \frac{1}{\sqrt{2\pi}}\int\limits_{\alpha}^{\infty} e^{-x^2/2}\mathrm{d}x \tag{3.33}$$

then Equation 3.28 becomes

$$\text{BER} = \frac{1}{2}\left[Q\left(\frac{d}{v_{\text{NTO}}}\right) + Q\left(\frac{v_{01}-d}{v_{\text{NT1}}}\right)\right]. \tag{3.34}$$

3.3.3 Receiver Sensitivity

Again using the condition $p[v_0/"0"] = p[v_0/"1"]$, we have

$$\frac{d}{v_{\text{NTO}}} = \frac{v_{01}-d}{v_{\text{NT1}}} \equiv \partial o. \tag{3.35}$$

Assuming $v_{00} = 0$, hence

$$\text{BER} = Q(\delta). \tag{3.36}$$

The Marcum $Q(\delta)$ function is a standard function, and this curve is shown in Figure 3.2. For a $\text{BER} = 10^{-9}$, the value of δ is about 6, which is the normal standard for communications at bit rate of 140–40 Gb/s.

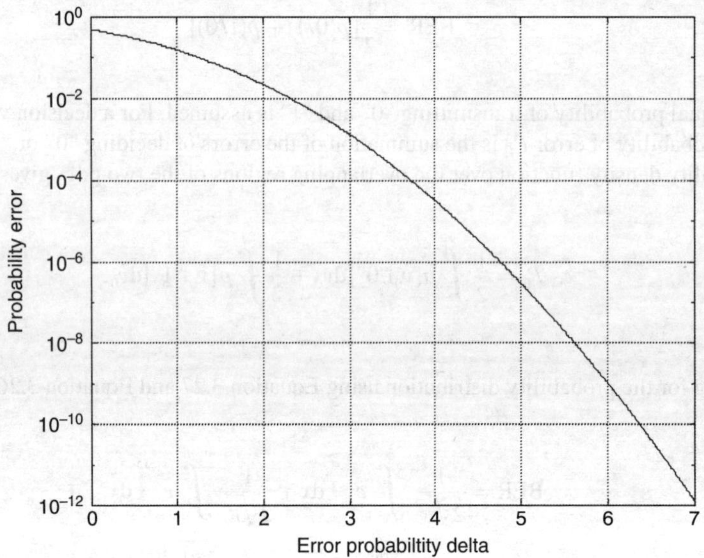

FIGURE 3.2 BER as a function of the δ parameter.

Thus, by eliminating the decision level variable d from Equation 3.31, we obtain

$$\partial = \frac{v_{01} - \partial v_{\mathrm{NT0}}}{v_{\mathrm{NT1}}} \tag{3.37}$$

or alternatively

$$v_{01} = \partial(v_{\mathrm{NT0}} + v_{\mathrm{NT1}}) \tag{3.38}$$

substituting v_{01}, v_{NT1}, v_{NT0}, we have

$$\Re \frac{b_1}{T_{\mathrm{b}}} \langle G_{\mathrm{n}} \rangle R_{\mathrm{I}} A = \partial \left\{ \left[2\Re q \frac{b_1}{T_{\mathrm{b}}} \langle G_n^2 \rangle B R_{\mathrm{I}}^2 A^2 + v_{\mathrm{NA}}^2 \right]^{1/2} + v_{\mathrm{NA}} \right\}. \tag{3.39}$$

However, the amplifier noise voltage v_{NA} can is given by $v_{\mathrm{NA}} = i_{\mathrm{Neq}} R_{\mathrm{I}} A\, B^{1/2}$. Thus, by substituting this noise voltage and eliminating $R_{\mathrm{I}} A$ and solving for the energy required for the "1" transmitted and received at the photodetector b_1 as

$$b_1 = q\Re \left(\delta^2 G + \frac{2 i_{\mathrm{Neq}} \delta T_{\mathrm{b}}}{qG} \right) \tag{3.40}$$

we have used the approximation $<G_n> = G$ and $<G_n^2> = G^{2+x}$ where x is the factor dependent on the ionization ration in an APD. For pin photodetector $G = 1$. The optical receiver sensitivity in dBm, denoted as RS in dBm, can thus be obtained as

$$\mathrm{RS} = 10\mathrm{Log}_{10} \frac{P_{\mathrm{av}}}{P_0} \text{ in dBm} \tag{3.41}$$

where $P_{av} = b_1/T_b$ and $P_0 = 1.0$ mW. This is the optical receiver sensitivity defined as the minimum optical power required for the receiver to operate reliably with a BER below a specific value. In Equation 3.36, there are two terms clearly specifying the dependence of the signal dependence δ and the amplifier noise contribution at a certain data rate and for a certain modulation format. The term b_1 represents the optical energy of the "1" required for the optical receiver to detect with a certain bit error rate. The term q/R in the RHS of Equation 3.36 is equivalent to the optical power required to generate one electron, or the number of photons required to generate one electron. Typical measurement for optical receiver is the number of photons required for it to operate with a specified BER.

The common term of the energy b_1 is q/R, which is equivalent to about one photon energy. We can rewrite Equation 3.36 by using $q/R = h\upsilon/\eta$

$$b_1 = \frac{h\upsilon}{\eta}\left(\delta^2 G + \frac{2i_{Neq}\delta T_b}{qG}\right). \tag{3.42}$$

Thus, when we have an ideal electronic amplifier, i.e., noiseless amplifier, then it requires a number of photon energy of $\delta^2 G$ for error-free detection. The second term in the bracket is the number of photons required to overcome the amplifier noise. Determination of the BER using the eye diagram when the pdf is not Gaussian leads to an inaccurate value. However, a much more accurate statistical analysis is developed and presented in Section 3.9 of this chapter.

3.3.4 Optical Signal-to-Noise Ratio and Noise Impact

3.3.4.1 OSNR

For an optically pre-amplified receiver, the ASE dominates the noise processes. An optical signal-to-noise ratio (OSNR) is commonly used, which is determined by measuring the noise and optical signals levels in both polarization directions under an optical filtering bandwidth B_0 of 0.1 nm [2].

$$\text{OSNR} = \frac{\text{optical} - \text{signal} - \text{power}}{\text{optical} - \text{noise} - \text{power}}. \tag{3.43}$$

For an average output power P_0 of the EDFA, the OSNR can be determined by

$$\text{OSNR} = \frac{P_0}{2n_{sp}hf_T(G-1)B_0} \tag{3.44}$$

where n_{sp} is the number of spontaneous emission of the EDFA, h is Planck's constant, and G is the linear power gain coefficient of the EDFA. If the signal-ASE beat noise dominates the noises of the detection process, a noise figure (NF) of the EDFA can be used to determine the ratio of the OSNR at the input and output because this is much more practical. Thus, the NF can be approximated by $F_n \sim 2n_{sp}$. Thus, the OSNR can be rewritten as

$$\text{OSNR} = \frac{P_0}{F_n hf_T(G-1)B_0}. \tag{3.45}$$

3.3.4.2 Determination of the Impact of Noise

One way to measure the impact of noises on the received signals of optically amplified systems is by attenuating the signal power at the input of the optical pre-amplifier and then obtain the BER. A BER versus received power will be obtained. Assuming that only one span is used and the optical gain is much greater than unity, then the OSNR can be written for an operating wavelength of 1550 nm

$$\text{OSNR} = 58(\text{dBm}) - F_n(\text{dB}) + P_{in}(\text{dBm}).$$ (3.46)

If an eye diagram is used, the eye-opening penalty (EOP) can be determined for the same BER. EOP can be written as

$$\text{EOP} = -10\text{Log}_{10}\frac{\text{EO}}{\text{EO}_n}\ \text{dB}.$$ (3.47)

3.4 QUANTUM LIMIT OF OPTICAL RECEIVERS UNDER DIFFERENT MODULATION FORMATS

Instead of using equivalent noise current density or NEP, optical receiver front-ends are sometimes also characterized in terms of their receiver sensitivity. While the receiver sensitivity is undoubtedly of high interest in optical receiver design, it comprises not only the degrading effects of noise, but also encompasses essential properties of the received signal, such as extinction ratio, signal distortions and intersymbol interference (ISI), generated within the transmitter or within the receiver. Thus, knowledge of receiver sensitivity alone does not allow trustworthy predictions on how the receiver will perform for other formats.

Electronics noise usually dominates shot noise, but can be squeezed to zero. The signal-dependent shot noise, however, is fundamentally present. The limit when only fundamental noise sources determine receiver sensitivity is called "quantum limit" in optical communications. Quantum limits make optical receiver design an exciting task because there is always a fundamental measure against which practically implemented receivers can be compared. Note, however, that each class of receivers in combination with each class of modulation formats has its own quantum limit.

From Equation 3.36, we can observe that when the amplifier is noiseless, the receiver would require $\delta^2 G$ number of photon energy for detection. This is the quantum limit of the receiver. For example, for a BER of $1e^{-9}$ for a pin detector under ASK modulation we would need 36 photon energy per bit for detection when both the "1" and "0" are 50 to 50 randomly received.

Noises play the major roles in the receiving end of communications systems. In optical communications, using coherent or non-coherent detection techniques noises are contributed by: (i) electronic noises of the electronic amplifications following the opto-electronic processes in the photodetector; (ii) quantum shot noises due to the electronic current generated by the optical signals; (iii) quantum shot noises due to the high power of the local oscillator (additional and dominant source of coherent detection); and (iv) beating of the local oscillator and the optical signals.

These noise sources vary from optical receivers to the others depending on their structures, i.e., whether they consist of a photodetector and electronic amplifiers with and without optical pre-amplifications under coherent or non-coherent detection. This section gives the fundamental issues of the noise processes and their impact on the sensitivity of the receiving systems. In particular, we examine the quantum limits of the optical receivers, i.e., when the electronic noises are considered nullified.

Schematic diagrams of coherent and incoherent direct detection receivers are shown in Figure 3.3a and Figure 3.3b without using an optical amplifier, while Figure 3.3c and Figure 3.3d show their counterparts with optical amplifiers. Figure 3.3f and Figure 3.3g shows the balanced and fiber version of the detection and receiving systems. The difference between these configurations is the noises generated after the photodetection and at the input of the electronic amplifier. Note that coherent systems are identified with the mixing of the optical signals and the local oscillator whose polarization directions are aligned with each other.

3.4.1 DIRECT DETECTION

Optical detection for optical fiber communications is in the form of direct modulation and direct detection. The latter is the most simple form of detection and requires only a photodetector followed by an electronic amplifier and the decision circuitry, clock recovery and data recovery. For the ASK

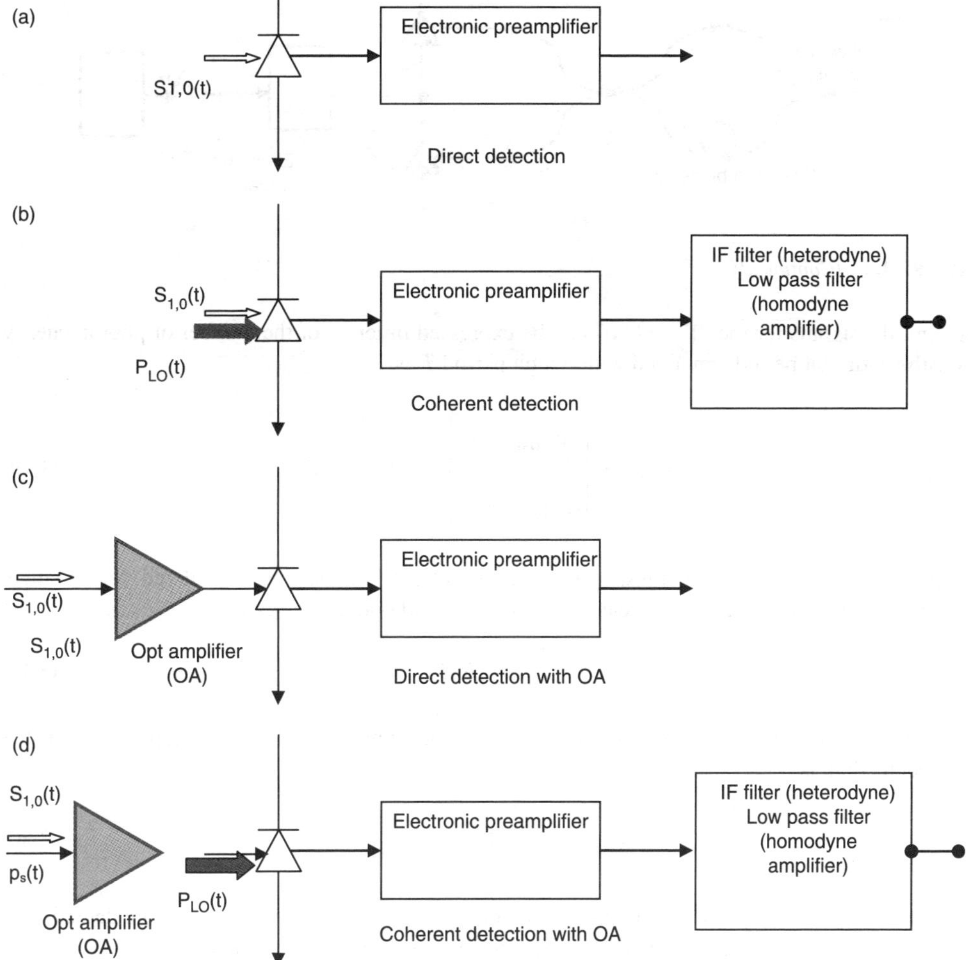

FIGURE 3.3 Schematic diagram of (a) direct detection and (b) coherent detection (c) direct detection with OA (d) coherent detection with OA (e) fiber version of coherent receiver (f) Single-ended and (g) balanced (h) self heterodyne balanced receiver structures. Following photodetection, various kinds of electronic signal-processing can be performed, including equalization. Detection of (a) (f) balanced receiver (g) balanced receiver with photonic phase comparator.

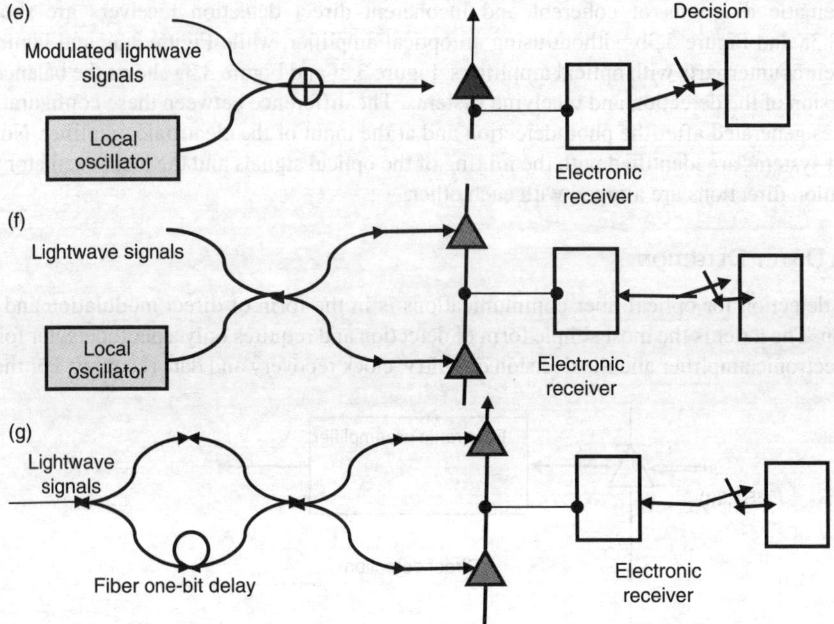

FIGURE 3.3 (*Continued*)

system, the signals for the "1" and "0" can be expressed in terms of the number of photon energy over the entire bit period contained within a bit period T as:

$$s_{1,0}(t) = \begin{cases} \dfrac{n_p}{T} & \text{for} \quad \text{"1"} \\[2mm] 0 & \text{for} \quad \text{"0"} \end{cases} \qquad 0 \leq t \leq T \tag{3.48}$$

where n_p is the number of photons, and the energy of the lightwave is normalized with a single photon energy at the operating wavelength. Thus, the total optical energy is

$$E_s = n_p h\upsilon T. \tag{3.49}$$

Thus, the electronic current generated after the photodetector, with R as the responsivity of the photodetection, is

$$\langle i_s \rangle = n_p h\upsilon T \Re = n_p h\upsilon T \frac{\eta q}{h\upsilon}. \tag{3.50}$$

Therefore, we could say that the number of photon energy per bit is n_p required for the detection of the "1" if there is no noise contributed by the electronic amplifier or detection.

If the probability of the "1" and "0" are equal, 50% then the probability of error of the detection is

$$P_e = \frac{1}{2} e^{-n_p} \tag{3.51}$$

Thus for a BER of 1e^9 then the argument $n_p = (3\sqrt{2}\,)^2$ or $n_p = 18$. with the allowance of the 1/2 factor of the single sided estimation so that $n_p = 20$ for the full detection error. This is the super quantum limit.

3.4.2 Coherent Detection

In coherent homodyne detection with a local oscillator whose optical power P_{LO} is very much larger than the signal average power, the quantum shot noise current dominates the noise process. The detected electronic current is the beating current between the local oscillator lightwave and that of the signal. Thus we have

$$\left\langle i_{N(LO)}^2 \right\rangle = 2qP_{LO}\frac{\eta q}{h\upsilon}\frac{1}{T} = 2q^2 P_{LO}\frac{\eta}{h\upsilon}\frac{1}{T} \tag{3.52}$$

and the signal to noise ratio is given by

$$\mathrm{SNR} = \frac{\left\langle i_s^2 \right\rangle}{\left\langle i_{N(LO)}^2 \right\rangle} = \frac{4\Re^2 p_s(t)P_{LO}/T}{2q\Re P_{LO}/T} = 2n_p \tag{3.53}$$

$$P_e = \mathrm{erfc}(2n_p). \tag{3.54}$$

Thus, for 10^{-9} BER, we have $2n_p = 3(2)^{1/2}$.

3.4.3 Coherent Detection with Matched Filter

It is assumed that a matched filter is inserted after the coherent receiver of Figure 3.3b and different modulation formats are used for transmission over long-haul optically amplified fiber systems.

In general and under the assumption that the noise process in the optical detection is Gaussian, the BER is given by

$$\mathrm{BER} = \mathrm{erfc}\left(\frac{d}{N_0}\right) \tag{3.55}$$

where d is the signal power separation between the average level of the "1" and "0" for the binary system and the equidistance between the constellation points of the modulation scheme (Figure 3.4). Let E_1 and E_0 be the field amplitude of the signals "1" and "0". The Euclidean distance d is given by

$$d^2 = E_1^2 + E_0^2 - 2\rho E_1 E_0 \quad \text{with} \quad \rho = \frac{2}{T}\int_0^T s_1(t)s_0(t)\mathrm{d}t \tag{3.56}$$

ρ is the correlation coefficient between the two logic levels or alternatively the Euclidean angle between the two vectors signals as represented on the scattering plane of constellation.

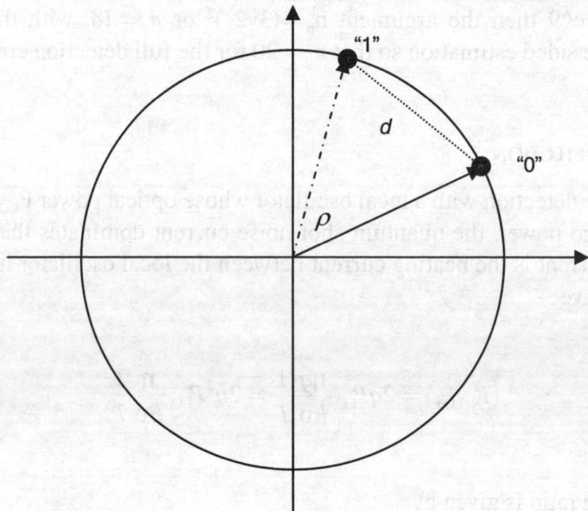

FIGURE 3.4   Signal constellation and energy level and the geometrical distance between "1" and "0".

3.4.3.1 Coherent ASK Systems

In heterodyne detection with an IF frequency region, the two bit signals as shown in Figure 3.5 are given by

$$s_{1,0}(t) = \left\{ \begin{array}{ll} \sqrt{\dfrac{2n_p}{T}}\,\cos\omega_{IF}t & \text{for ``1''} \\[2mm] 0 & \text{for ``0''} \end{array} \right\} \quad 0 \le t \le T \qquad (3.57)$$

where the amplitude of the lightwave-modulated signal is expressed in terms of the number of photon energy over the time interval, thus the square root of this amount is the amplitude of the field of the lightwave. The characteristic impedance of the medium is set at unity. The distance is then $d = n_p$.

Thus, the BER is given by

$$\text{BER} = \text{erfc}\left(\frac{\sqrt{n_p}}{2N_0}\right). \qquad (3.58)$$

By setting $N_0 = 1$, a BER of 10^{-9} requires $n_p = 4 \times 9 \times 2 = 72$ photon energy for a heterodyne receiver and a 3-dB improvement for a homodyne receiver, thus 36 photon energy under the assumption that no electronic noise is contributed by the electronic amplifier. These are the quantum limits of ASK heterodyne and homodyne detection when the power of the local oscillator is much larger than that of the signal.

3.4.3.2 Coherent Phase and Frequency Shift Keying Systems

$$s_{1,0}(t) = \left\{ \begin{array}{ll} \sqrt{\dfrac{2n_p}{T}}\,\cos\omega_1 t & \text{for ``1''} \\[2mm] \sqrt{\dfrac{2n_p}{T}}\,\cos\omega_0 t & \text{for ``0''} \end{array} \right\} \quad 0 \le t \le T. \qquad (3.59)$$

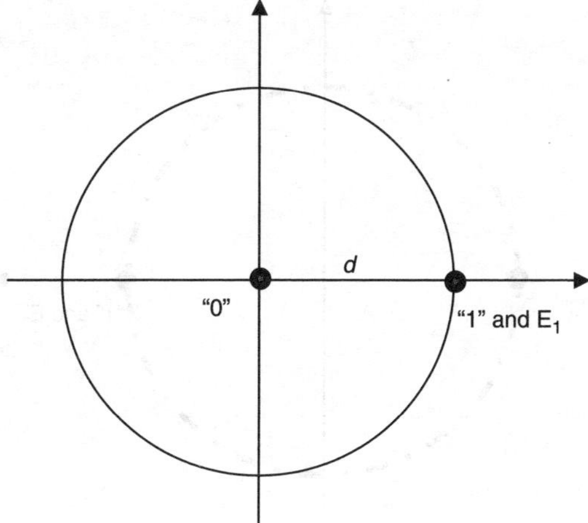

FIGURE 3.5 Signal constellation and energy level and the geometrical distance between "1" and "0" of the ASK system.

The FSK modulation scheme with two distinct frequencies f_1 and f_2 can be represented with a constant envelope and variation of the carrier frequency, or continuous phase between the two states as shown in Figure 3.6. The modulation index can be defined as

$$m = \frac{|\omega_1 - \omega_0|}{2\pi} T. \tag{3.60}$$

And the signal correlation coefficient is given by , under the condition that the two frequencies are large enough, and that the second and higher harmonics are outside the detection region.

$$\rho = \frac{2}{T} \int_0^T s_1(t)s_0(t)\,dt = \frac{2}{T} \int_0^T \cos(\omega_1 t)\cos(\omega_0 t)\,dt \tag{3.61}$$

$$\approx \frac{\sin(2\pi m)}{2\pi m}.$$

Thus, the BER is given as

$$\mathrm{BER} = \frac{1}{2}\,\mathrm{erfc}\left(\sqrt{\frac{n_p}{2}\left(1 - \frac{\sin(2\pi m)}{2\pi m}\right)}\right) \tag{3.62}$$

and, for BER of 1e^−9, we have

$$\mathrm{BER} = 10^{-9} \longrightarrow \sqrt{\frac{n_p}{2}\left(1 - \frac{\sin(2\pi m)}{2\pi m}\right)} = 3\sqrt{2}$$

$$\text{then } n_p = \frac{36}{1 - \dfrac{\sin(2\pi m)}{2\pi m}}. \tag{3.63}$$

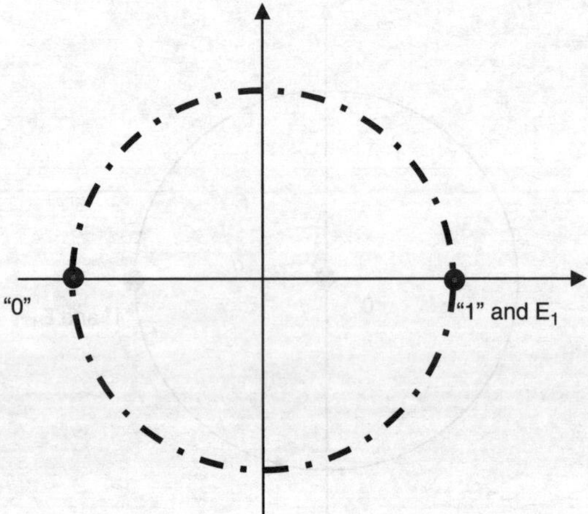

FIGURE 3.6 Signal constellation and energy level evolution of the signal envelope and the continuous phase between the "1" and "0" of the FSK system.

For MSK, the modulation index is 0.25 leading to the required number of photon energy per bit of 60–70, much higher than that of 0.8 at which only 30 photons energy is required per bit. It has however been shown that the MSK can be optimum for the transmission over dispersive medium due to the optimum bandwidth of the modulation scheme, and hence minimum dispersive effects on the phase of the carrier.

The FSK can be implemented with CPFSK, i.e., the phase of the carrier is continuously chirped. Assume that the phase is linearly chirped such that the phase variation for the "1" and "0" are given by

$$
s_{1,0}(t) = \left\{
\begin{array}{l}
\sqrt{\dfrac{2n_\mathrm{p}}{T}}\,\cos\theta_1(t) \quad \text{for "1"} \\[3ex]
\sqrt{\dfrac{2n_\mathrm{p}}{T}}\,\cos\theta_0(t) \quad \text{for "0"}
\end{array}
\right\}
$$

where

$$
\theta_{1,0}(t) = \left\{
\begin{array}{l}
\omega_{\mathrm{IF}}t + \dfrac{m\pi}{T}t \quad \text{for "1"} \\[3ex]
\omega_{\mathrm{IF}}t - \dfrac{m\pi}{T}t \quad \text{for "0"}
\end{array}
\right\} \qquad 0 \le t \le \dfrac{T}{2m}
\tag{3.64}
$$

$$
\theta_{1,0}(t) = \left\{
\begin{array}{l}
\omega_{\mathrm{IF}}t + \dfrac{\pi}{2} \quad \text{for "1"} \\[3ex]
\omega_{\mathrm{IF}}t - \dfrac{\pi}{2} \quad \text{for "0"}
\end{array}
\right\} \qquad \dfrac{T}{2m} \le t \le T.
$$

The correlation coefficient is then given by

$$\text{BER} = 10^{-9} = \frac{1}{2}\,\text{erfc}\left(\sqrt{n_p\left(1-\frac{1}{4m}\right)}\right) \longrightarrow \sqrt{n_p\left(1-\frac{1}{4m}\right)} = 3\sqrt{2}$$

$$\text{then } n_p = \frac{18}{1-\dfrac{1}{4m}}.$$

(3.65)

The variation of the photon energy versus the modulation index for this linear CPFSK is shown in Figure 3.7.

The FSK can be modified with the control of the relative phase of the two carriers between the two bits, thus the data bits can be written as

$$s_{1,0}(t) = \left\{ \begin{array}{l} \sqrt{\dfrac{2n_p}{T}}\,\cos(\omega_1 t + \theta) \ \ \text{for ``1''} \\[2ex] \sqrt{\dfrac{2n_p}{T}}\,\cos\omega_0 t \ \ \text{for ``0''} \end{array} \right\} \qquad 0 \le t \le T$$

(3.66)

where the phase angle t can be chosen as

$$\theta_{1,0}(t) = \left\{ \begin{array}{l} \pi - m\pi \ \ \text{for ``}p \le m \le 2p+1\text{''} \ \ p=0,1,2 \\[1ex] -m\pi \ \ \text{for ``}2p+1 \le m \le 2(p+1)\text{''} \end{array} \right\}$$

(3.67)

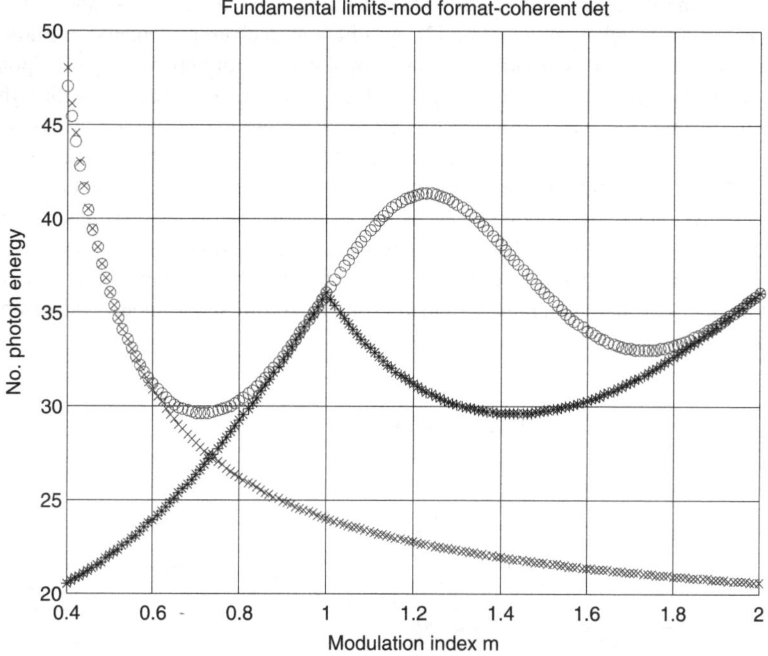

FIGURE 3.7 Fundamental limits of the coherent Phase and frequency shift keying detection for (a) red o– CPFSK (b) 'blue +' FSK (ω_1 and (ω_2) with optimized correlation coefficient with phase control (c) 'black *' linear DMPSK. For a BER=10^{-9}.

with the correlation coefficient

$$\rho = \frac{2}{T}\int_0^T s_1(t)s_0(t)\mathrm{d}t = \frac{2}{T}\int_0^T \cos(\omega_1 t + \theta)\cos(\omega_0 t)\mathrm{d}t \tag{3.68}$$

$$\approx -\frac{|\sin(\pi m)|}{\pi m}.$$

Thus, giving a BER for a phase-controlled FSK as

$$\mathrm{BER} = \frac{1}{2}\mathrm{erfc}\left(\sqrt{\frac{n_\mathrm{p}}{2}\left(1 + \frac{|\sin(\pi m)|}{\pi m}\right)}\right) \tag{3.69}$$

$$\rightarrow n_\mathrm{p} = \frac{36}{1 + \dfrac{|\sin(\pi m)|}{\pi m}} \; for \; a \; \mathrm{BER} = 10^{-9}.$$

The required number of photons is plotted against the modulation index as shown in Figure 3.7b "back *".

3.5   BINARY COHERENT OPTICAL RECEIVER

Another way of amplifying the signal while boosting the accompanying noise above the electronics noise floor is known as coherent detection [3]. A coherent receiver, as depicted in Figure 3.8, combines the signal with a local oscillator (LO) laser by means of an optical coupler. Upon detection, the two fields beat against each other during the electronic generation process within the photodetector, and the average electrical signal results as a mixing product between these time-dependent signals and the local CW. Electronic signal are generated at an intermediate frequency (IF) between the signal and LO. The splitting ratio of the optical coupler must be chosen as high as possible so as not to waste too much signal power, and as low as acceptable to let sufficient LO power reach the detector to achieve shot-noise limited performance (see below). The heterodyne efficiency η accounts for the degree of spatial overlap, as well as for the polarization match between LO field and signal field. If LO and signal are provided co-polarized in single-mode optical fibers, then η approaches unity.

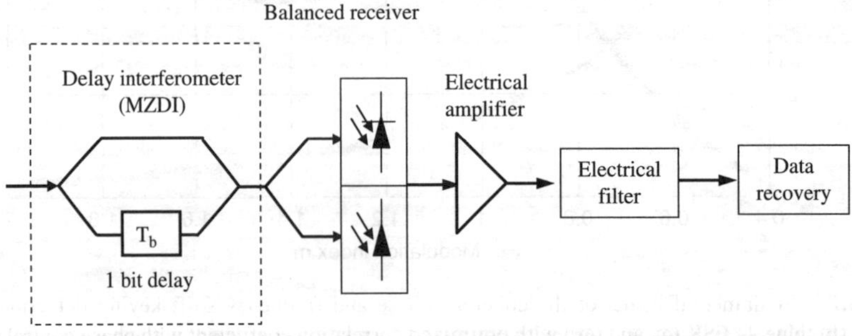

FIGURE 3.8   MZIM optical balanced receiver for optical DPSK detection.

If the frequency of the LO differs from the signal frequency, a heterodyne receiver is present. If LO and signal have the same frequency, such that IF = 0, then we achieve a homodyne receiver. Homodyne detection strictly requires optical phase locking between the LO and signal optical fields, which is a significant technological development. In the desired range of operation, the LO power is chosen much stronger than the signal power, the beating between the local oscillator and that of the signal would dominate. We are then left with an exact replica of the received optical field's amplitude $(P_s)^{1/2}$ and phase $\phi_s(t)$ (as compared with the optical power accessible in direct-detection receivers). Thus, any amplitude or phase modulation scheme can directly be employed in combination with coherent receivers.

Due to the high LO power reaching the detector, the main noise contribution in a coherent receiver in the absence of optical noise at its input is the shot noise produced by the LO, $\sigma^2_{LO-s} = 2qS(1-\varepsilon)P_{LO}B_e$. If this noise term dominates electronics noise $\sigma^2_{LO-s} >> i^2_{Neq}$, optimum receiver performance should be derived depending on the parameters of the receiver. This limit is known as the "shot noise limit" in the context of coherent receivers. The highest receiver sensitivity with the potential of practical implementation can be achieved using homodyne detection of phase shift keying (PSK), where the data bits are directly mapped onto the phase ϕ. $s(t)$ of the optical signal, $\{0, 1\} \rightarrow \{0, \pi\}$. Without going into the derivations, the quantum limit for homodyne PSK can be shown to equal only 9 photons/bit, with a reported receiver sensitivity record of 20 photons/bit at 565 Mbit/s. Using OOK instead of PSK, the sensitivity degrades by 3 dB. If heterodyne detection is employed, then an additional loss of 3 dB in terms of receiver sensitivity would result.

An alternative implementation of coherent balanced receivers is shown in Figure 3.8 It makes use of balanced detection with a pair of photodetectors connected back-to-back so that push–pull signal generations can be generated due to the phase and anti-phase of the combined signals. While a balanced coherent receiver has exactly the same quantum-limited sensitivity as its single-ended equivalent, it offers the advantage of utilizing the complete mixing of the polarization of the fields of the arrived lightwave signal and LO power, and of being more robust to LO relative intensity noise (RIN).

Furthermore, a new type of receiver is the self-heterodyne balanced receiver (Figure 3.3 e.g.,) in which the lightwave by itself is split in two equal paths and one is delayed with respect to the other so as to interferometrically resolve the phases of the lightwaves contained within the two consecutive bit periods. This is then followed by a balanced detection.

The main advantages and disadvantages of direct single detection, balanced receiving and self heterodyne structure of coherent receivers are noted below.

Receiver sensitivities of coherent receivers offer significant enhancement of the sensitivity as compared with those achieved with pin-receivers and APDs, thus allowing for increased transmission distances in unamplified optical links. However, optically pre-amplified receivers exhibit similar receiver sensitivities to coherent receivers, are polarization-insensitive, and take less serious hits in performance if inline amplification is presented.

The possibility of correcting for chromatic dispersion in the microwave regime is offered in coherent detection because both amplitude and phase of the optical field are converted to an electronic signal. However, efficient and adaptive broadband phase corrections can be performed only on RF bandpass signals, asking for heterodyne detection. Because the IF components have to be chosen about at about three times the data rate, thence unrealistically high receiver front-end bandwidths would be needed for the high data rates to be employed. Conversely, all-optical dispersion compensators and adaptive optical filters are attracting extensive development, allowing for efficient phase corrections in the optical regime.

Coherent receivers allow for the separation of closely spaced WDM channels by means of RF bandpass filters with sharp roll-offs. Optical filter technology has advanced dramatically, enabling channel spacing of the order of 10 GHz with sharp optical filter roll-offs that makes possible optical channel filtering even for ultra-dense WDM applications.

Self-heterodyne detection can offer significant improvement in receiver sensitivity and ease the difficulties in detection because this is a non-coherent detection and requires only the coherence within one bit period.

3.6 NON-COHERENT DETECTION FOR OPTICAL DPSK AND MSK

All receiver configurations presented here are based on non-coherent detection, i.e., the detection of the power envelope of the optical lightwaves signals at input of the photodiode(s) is a direct detection with no mixing of the received signals with a local lightwave source. Non-coherent detection mitigates the issues in the coherent receiver schemes induced from phase noise and non-zero linewidth of the coherent laser source.

3.6.1 Photonic Balanced Receiver

The receiver employs an electro-optic Mach–Zehnder delay interferometer (MZDI) with introduction of a delay in one arm which is equal to one bit period. The MZDI acts as a phase comparator. Unlike in the receiver for wireless communications where the phase comparison is performed after the detection process, photonic balance receiver performs the photonic phase differential demodulation prior to opto-electronic detection.

The detected optical signals are converted to the electrical domain by the photodetector that follows square law. That means that the electronic current is proportional to the optical power, hence this is a nonlinear process. In order to overcome this nonlinearity, a coherent detection technique can offer the recovery of the phase of the signals or, alternatively, the optical signals can be preprocessed in the photonic domain before the optoelectronic detection. The photonic balanced receiver employs a differential phase comparator at the front end, a Mach Zehnder delay interferometer, which is an asymmetric interferometer with one arm longer than the other by a physical length equivalent to one-bit period.

Balanced receivers are essential to recover the differentially coded phases of the optical signals. This detection scheme offers a 3-dB improvement over the direct detection using a single photodetector due to its push-pull back-to-back connection of the photodetectors. The delay balanced receiver is shown in Figure 3.8, in which the phase information carrier of the received optical field received and its consecutive pulse are compared by an interferometer in which the lightwaves are split into two arms (one delayed by a symbol period and one without delay) and then recombined via an optical directional coupler.

The received signal at the output of the balanced receiver is given as [9]:

$$P_{\mathrm{D}}(t) = \left| E_{\mathrm{D}}(t) + E_{\mathrm{D}}(t-\tau) \right|^2 - \left| E_{\mathrm{D}}(t) - E_{\mathrm{D}}(t-\tau) \right|^2$$

$$= 4\Re\left\{ E_{\mathrm{D}}(t) \right\} E_{\mathrm{D}}^{*}(t-\tau) = \cos(\Delta\phi + \varsigma) \tag{3.70}$$

where $E_{\mathrm{D}}(t)$ and $E_{\mathrm{D}}(t-\tau)$ are the present and the 1-bit delay version of the optical signals at the delay interferometer, respectively. $\Delta\phi$ and ς represent the differential phase and the phase noise caused by nonlinear phase noise or the MZ delay interferometer imperfections (MZDI). The latter issue is not a severe degradation factor due to the high stability of planar lightwave circuit MZDI using a thin-film heater for tuning any waveguide path mismatch. MZDI-based optical balanced receiver can be used to detect both DPSK and MSK modulated optical signals. However, there is a significant difference in the configuration of the receiver of the two formats which is an additional pi/2 phase shift is introduced in the detection of optical MSK signals (Figure 3.9). This constant phase shift can be located at either arm without affecting the performance of the receiver

3.6.2 Optical Frequency Discrimination Receiver (OFDR)

Detection of optical MSK modulated signals using frequency discrimination principles was studied in the early 1990s. The continuous phase frequency shift keying (CPFSK) and MSK optical lightwave signals could be easily generated with this technique [3]. The detection at this time was mainly based coherent detection. The dual filters used for discriminating two modulating principle

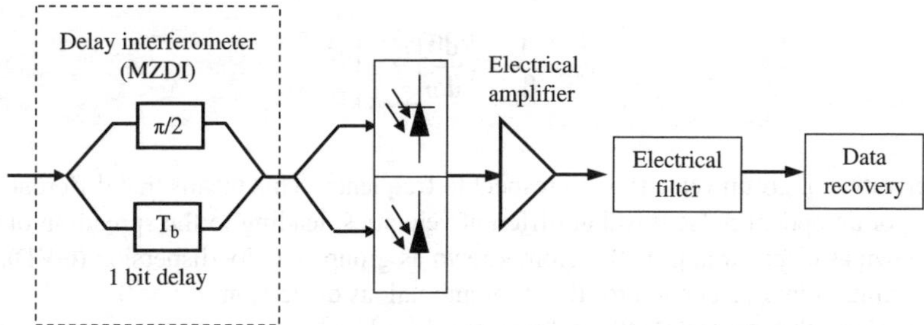

FIGURE 3.9 MZIM optical balanced receiver configuration for optical MSK detection.

frequencies are electrical filters. This is due to the unavailability of optical filters with necessarily narrow bandwidth for use in non-coherent detection of optical CPFSK and MSK signals.

With the advance of technology in filter design and fabrication, especially the micro-ring resonator filters, the 3-dB bandwidth of contemporary optical filters can be narrowed down to about <2 GHz [10]. The availability of such the optical filters enables the frequency discrimination in optical domain for non-coherent detection of optical MSK signals. This scheme has not yet been investigated. The proposed scheme has been reported in [4]. It is expected that the detection technique-based frequency discrimination is less sensitive to the effects of quadratic phase profile of the single mode optical fiber. The detailed configuration of the OFDR, together with the analysis on optimization for the receiver design, is presented in Chapter 6.

3.7 TRANSMISSION IMPAIRMENTS

3.7.1 CHROMATIC DISPERSION (CD)

This section briefly presents the key theoretical concepts describing the properties of CD in a single-mode fiber. Another aim of this section is to introduce the key parameters which will be commonly mentioned in the rest of the chapter.

The initial point when mentioning CD is the expansion of the mode propagation constant or "wave number" parameter, β, which can be approximated around the optical operating frequency using the Taylor series as

$$\beta(\omega) = \frac{\omega n(\omega)}{c} = \beta_0 + \beta_1 \Delta\omega + \frac{1}{2}\beta_2\Delta\omega^2 + \frac{1}{6}\beta_3\Delta\omega^3 \tag{3.71}$$

where β_0 is the propagation constant of the fiber evaluated at the operating frequency, ω is the angular optical frequency, $n(\omega)$ is the frequency-dependent refractive index of the fiber. The higher order parameters

$$\beta_n = \left(\frac{d^n\beta}{d\omega^n}\right)\Bigg|_{\omega=\omega_0}. \tag{3.72}$$

These parameters have different physical meanings. β_0 is involved in the phase velocity of the optical carrier, which is defined as

$$v_p = \frac{\omega_0}{\beta_0} = \frac{c}{n(\omega_0)} \tag{3.73}$$

β_1 determines the group delay that relates to the velocity v_g and the guided mode propagation constant β of the guided mode by [11–12]

$$v_g = \frac{1}{\beta_1} = \left(\left. \frac{d\beta}{d\omega} \right|_{\omega=\omega_0} \right)^{-1} \tag{3.74}$$

β_2 is the derivative of group velocity with respect to frequency. This means that different frequency components of an optical pulse travel at different velocities, leading to the spreading of the pulse, which is known as dispersion. β_2 is therefore known as group velocity dispersion (GVD). The fiber is said to exhibit normal dispersion for $\beta_2 > 0$ or anomalous dispersion if $\beta_2 < 0$.

A pulse having the spectral width of $\Delta\omega$ is broadened by

$$\Delta\tau = \beta_2 L \Delta\omega. \tag{3.75}$$

In practice, a more commonly used factor to represent the chromatic dispersion of a single mode optical fiber is known as D (ps/nm.km). The dispersion factor is closely related to the GVD β_2 and given by

$$D = -\left(\frac{2\pi c}{\lambda^2} \right) \beta_2 \quad \text{at the operating wavelength } \lambda \tag{3.76}$$

β_3 *is* defined as $\beta_3 = d\,\beta_2/d\omega$, which contributes to the dispersion slope, $S(\lambda)$, which is an essential dispersion factor for high-speed multi-channel DWDM transmission. $S(\lambda)$ can be obtained from the higher order derivatives of the propagation constant as

$$S = \frac{dD}{d\lambda} = \left(\frac{2\pi c}{\lambda^2} \right) \beta_3 + \left(\frac{4\pi c}{\lambda^3} \right) \beta_2. \tag{3.77}$$

3.7.2 Chromatic Linear Dispersion

From the view of fiber design [5–6], this dispersion factor is actually a total of two dispersion components, D_M and D_W, the material and waveguide dispersion parameters respectively and given by

$$D = -\left(\frac{2\pi c}{\lambda^2} \right) \beta_2 \equiv D_M + D_W. \tag{3.78}$$

A step-index optical fiber with core radius notation, a, is considered. The refractive indices of the core and cladding of the SSMF are noted to be n_1 and n_2 respectively. The significant transverse propagation constants of guided lightwaves u and v in the core and cladding regions are formulated as

$$u = a\sqrt{k^2 n_1^2 - \beta^2} \quad v = a\sqrt{\beta^2 - k^2 n_2^2} \tag{3.79}$$

where $k^2 n_1^2$ and $k^2 n_2^2$ are the plane-wave propagation constants in the core and second cladding layers, respectively. β is the longitudinal propagation constant of the guided waves along the optical fiber, which can be expressed as

$$\beta = \sqrt{k^2 (b(n_1^2 - n_2^2) + n_2^2)} \tag{3.80}$$

where b denotes the critical normalized propagation constant whose value for guided modes falling within the range of [0,1] is formulated as

$$b = \frac{\beta/k - n_2}{n_1 - n_2}.$$ (3.81)

The normalized frequency is then mathematically expressed as

$$V = ak\sqrt{n_1^2 - n_2^2}.$$ (3.82)

The material dispersion in an optical fiber is due to the wavelength dependence of the refractive index (RI) of the core and cladding. The RI $n(\lambda)$ is estimated by Sellmeier's equation

$$n^2(\lambda) = 1 + \sum_{i=1}^{M} \frac{B_i \lambda^2}{\left(\lambda^2 - \lambda_i^2\right)}$$ (3.83)

where λ_j indicates the ith resonance wavelength and B_j is its corresponding oscillator strength. n stands for n_1 or n_2 for core or cladding regions. These constants of different doping concentration of Geo2 in SiO_2 pure silica are tabulated in Table 3.1 and given in the appendix f [13]. The first three Sellmeier terms, G_1, G_2 and G_3, are normally used. These tabulated coefficients are then used to find material dispersion factor, D_M can be obtained by

$$D_M = -\frac{\lambda}{c}\left(\frac{d^2 n(\lambda)}{d\lambda^2}\right)$$ (3.84)

where c is the velocity of light in vacuum. For pure silica and over the spectral range of 1.25 μm to 1.66 μm, D_M can also be approximated by an empirical relation [11]

$$D_M = 122\left(1 - \frac{\lambda_{ZD}}{\lambda}\right)$$ (3.85)

TABLE 3.1

Sellmeier's Coefficients for Silica-based Material and Doping Concentration of GeO$_2$ for the Design

Type	Doping Conc (%)	SiO$_2$ (%)	B1	B2	B3	λ1	λ2	λ3
A(1)	0	100	0.6961663	0.4079426	0.8974794	0.0684043	0.1162414	9.896161
B(2)	3.1	96.9	0.7028554	0.4146307	0.8974540	0.0727723	0.1143085	9.896161
C(3)	5.8	94.2	0.7088876	0.4206803	0.8956551	0.0609053	0.1254514	9.896162
D(4)	7.9	92.1	0.7136824	0.4254807	0.8964226	0.0617167	0.1270814	9.896161
E(5)	0	pure	0.696750	0.408218	0.890815	0.069066	0.115662	9.900559
F(6)	13.5	86.5	0.711040	0.408218	0.704048	0.064270	0.129408	9.425478
G(7)	9.1	90.9	0.695790	0.452497	0.712513	0.061568	0.119921	8.656641
H(8)	13.3	86.7	0.690618	0.401996	0.898817	0.061900	0.123662	9.098960
I(9)	1	99	0.691116	0.399166	0.890423	0.068227	0.116460	9.993707
J(10)	48.7	51.3	0.796468	0.497614	0.358924	0.094359	0.093386	5.999652

where λ_{ZD} is the zero material dispersion wavelength [7]. For instance, $\lambda_{ZD} = 1.276$ μm for pure silica. λ_{ZD} can vary according to various doping concentrations in the core and cladding of different materials such as Germanium (Ge) or Fluorine (F).

The waveguide dispersion D_W can be calculated as [11,12]

$$D_W = -\left(\frac{n_1 - n_2}{c\lambda}\right) V \frac{d^2(Vb)}{dV^2} \tag{3.86}$$

where b and V are the normalized propagation constant and normalized frequency as defined in Equation 3.50 and Equation 3.51, respectively. $Vd^2(Vb)/dV^2$ is defined as the normalized waveguide dispersion parameter.

An effective approximation based on polynomial interpolation has been developed to calculate the waveguide dispersion not only in a simple SI fiber, but also in a multi-cladding dispersion compensating fiber (MC-DCF) [8].

A well-known parameter to govern the effects of chromatic dispersion imposing on the transmission length of an optical system is known as the dispersion length L_D. Conventionally, the dispersion length L_D corresponds to the distance after which a pulse has broadened by one bit interval. [11]. For high capacity long-haul transmission employing external modulation, the dispersion limit can be estimated in the following Equation [11] (ref Kaminow, Gnauck 1997).

$$L_D = \frac{10^5}{DB^2} \tag{3.87}$$

where B is the bit rate (Gb/s), D is the dispersion factor (ps/nm km) and L_D is in km. Equation 3.87 provides a reasonable approximation, even though the accurate computation of this limit depends the modulation format, pulse shaping and optical receiver design. It can be seen clearly from this equation that the severity of the effects caused by the fiber chromatic dispersion on externally modulated optical signals is inversely proportional to the square of the bit rate. Thus, for 10 Gb/s OC-192 optical transmission on a standard single mode fiber (SSMF) medium with a dispersion of about ±17 ps/nm.km as shown in Figure 3.10, the dispersion length L_D is approximately 60 km,

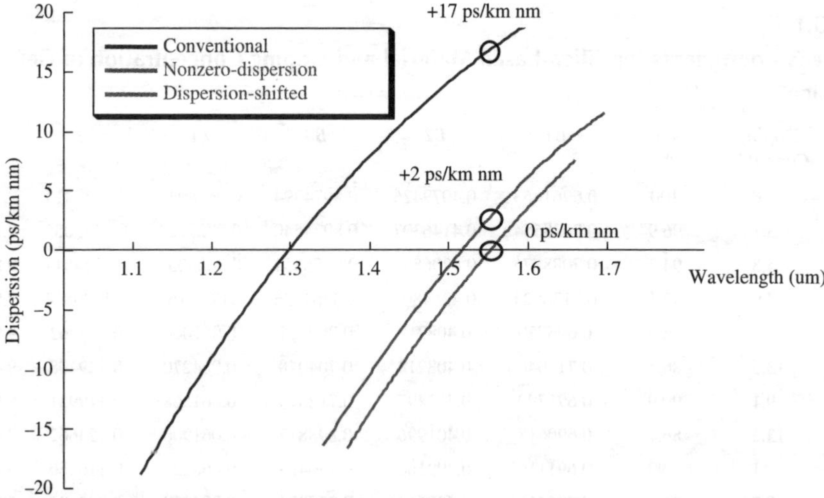

FIGURE 3.10   Dispersion characteristics of different transmission and compensating fibers.

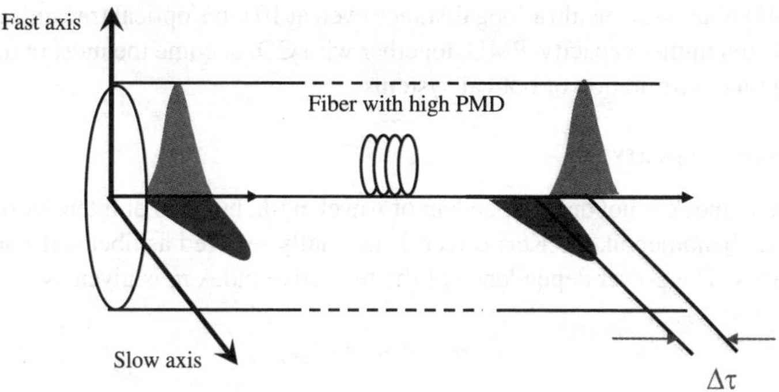

FIGURE 3.11 PMD effects.

i.e., corresponding to a residual dispersion of about ±1000 ps/nm and <4 km (or equivalent to about ±60 ps/nm in the case of 40 Gb/s OC-768 optical systems). These lengths are a great deal smaller than the length limited by ASE noise accumulation. The chromatic dispersion therefore, becomes the one of the most critical constraints for the modern high-capacity and ultra long-haul transmission optical systems.

3.7.3 POLARIZATION MODE DISPERSION (PMD)

PMD represents another type of pulse spreading as shown in Figure 3.11. PMD is caused by the differential group delay (DGD) between two principal orthogonal states of polarization (PSP) of the propagating optical field.

One of the intrinsic causes of PMD is due to the asymmetry of the fiber core. The other causes are derived from fiber deformation, including stress applied on the fiber, fiber aging, variation of temperature over time, or effects from a vibration source. These processes random result in the dynamic of PMD.

The imperfection of the core or deformation of the fiber may be inherent from the manufacturing process or as a result of mechanical stress on the deployed fiber resulting in a dynamic aspect of PMD. The delay between these two PSP is normally negligibly small in 10 Gb/s optical transmission systems. However, at a high transmission bit rate for long-haul and ultra long-haul optical systems, the PMD effect becomes much more severe and degrades system performance [14–17]. The DGD value varies along the fiber following a stochastic process. It is proven that these DGD values comply with a Maxwellian distribution [15,18,19] given by the following expression

$$f\left(\Delta\tau\right)=\frac{32\left(\Delta\tau\right)^{2}}{\pi^{2}\left\langle\Delta\tau\right\rangle^{3}}\exp\left\{-\frac{4\left(\Delta\tau\right)^{2}}{\pi\left\langle\Delta\tau\right\rangle^{2}}\right\}\quad \Delta\tau\geq0 \tag{3.88}$$

where $\Delta\tau$ is differential group delay over a segment of the optical fiber δz. The mean DGD value $\left\langle\Delta\tau\right\rangle$ is commonly termed as the "fiber PMD" and is normally given by the fiber manufacturer. An estimate of the transmission limit due to PMD effect is given as [9]

$$L_{\max}=\frac{0.02}{\left\langle\Delta\tau\right\rangle^{2}R^{2}} \tag{3.89}$$

with R being the transmission bit rate. Therefore, we have (i) $\left\langle\Delta\tau\right\rangle=1$ ps/km (older fiber types) we have Bit rate $=40$ Gbit/s; $L_{\max}=12.5$ Km and bit rate $=10$ Gb/s ; $L_{\max}=200$ Km (ii) $\left\langle\Delta\tau\right\rangle=0.1$ ps/km (contemporary fiber for modern optical systems) and thus for bit rate $=40$ Gb/s ; $L_{\max}=1250$ Km and bit rate $=10$ Gb/s; $L_{max}=20.000$ Km.

In short, PMD is an issue on ultra-long distance even at 10 Gb/s optical transmission. Upgrading to higher bit rate and higher capacity, PMD, together with CD, become the most two critical impairments imposing on the limitation of optical systems.

3.7.4 Fiber Non-linearity

The fiber refractive index is not only dependent of wavelength, but also of intensity of the lightwave. This well-known phenomenon, the Kerr effect, is normally referred as fiber self-phase-modulation (SPM) nonlinearity. The power dependence of the refractive index n_r is given as

$$n'_r = n_r + \bar{n}_2(P/A_{\text{eff}}) \tag{3.90}$$

where P is the average optical intensity inside the fiber, $\bar{n}_2$ is the non-linear-index coefficient, and A_{eff} is the effective area of the fiber.

There are several non-linearity phenomena induced from Kerr effects, including intra-channel self-phase modulation (SPM), cross-phase modulation between inter-channels (XPM), four-wave mixing (FWM), stimulated Raman scattering (SRS) and stimulated Brillouin scattering (SBS). SRS and SBS are not main degrading factors compared with the others. FWM effect degrades performance of an optical system severely if the local phase of the propagating channels is matched with the introduction of the ghost pulse. However, with high local dispersion parameters such as in SSMF or even in NZ-DSF, effect of the FWM becomes negligible. While XPM is strongly dependent on the channel spacing between the channels and also on local dispersion factor of the optical fiber Ref. [10].

In this chapter, only the SPM non-linearity is considered. This is the main degradation factor for a high bit rate transmission system where the signal spectrum is broadened. The effect of SPM is normally coupled with the nonlinear phase shift, which is defined as

$$\phi_{\text{NL}} = \int_0^L \gamma P(z)\mathrm{d}z = \gamma L_{\text{eff}} P$$

$$\gamma = \omega_c \bar{n}_2 / (A_{\text{eff}} c) \tag{3.91}$$

$$L_{\text{eff}} = (1 - e^{-\alpha L})/\alpha$$

where ω_c is the lightwave carrier, L_{eff} is the effective transmission length, and α is the attenuation factor of a SSMF, is normally 0.17–0.2 dB/km for the currently operating wavelengths within the 1550-nm window.

The temporal variation of the non-linear phase is ϕ_{NL}, while the optical pulses propagating along the fiber results in the generation of new spectral components far apart from the lightwave carrier is ω_c, implying broadening of the signal spectrum. The frequency difference $\delta\omega$ is given by

$$\delta\omega = -\frac{\partial\phi_{\text{NL}}}{\partial T}. \tag{3.92}$$

The time dependence of $\delta\omega$ is known as frequency chirping.

3.8 MATLAB® SIMULINK® SIMULATOR FOR OPTICAL COMMUNICATIONS SYSTEMS

A simulation package based on MATLAB® Simulink® platform for optical transmission systems has been developed. The simulator integrates all the mentioned-above transmission impairments over a single optical transmission link. The DWDM optical system can be developed further from this platform with relatively easy prediction.

The simulator provides toolboxes and blocksets for setting up any complicated system configurations under test. Spectral characteristics of new proposed modulation formats and their transmission performance with focus on tolerance to CD and PMD can be easily achieved with various techniques of system evaluation.

Among non-linearity impairments, SPM is considered to be the major shortfalls in the system. Furthermore, XPM can be considered to be negligible in a DWDM system in the following scenarios: (i) highly locally dispersive system (e.g., SSMF and DCF deployed systems); (ii) large channel spacing; and (iii) high spectral efficiency [5, 7, 20–22]. For a transmission system operating at 40 Gb/s, using SSMF 17 ps/(nm.km) at 1550 nm and 0.8 nm spacing between two adjacent DWDM wavelengths, a walk-off length Lw of about 2.95 km is obtained. However, the XPM should be taken into account for the systems deploying non-zero dispersion shifted fiber (NZ-DSF) where the local dispersion factor is low. The values of the NZ-DSF dispersion factors can be obtained.

The other critical degradation factors to the DPSK system are the non-linear phase noise due to the fluctuation of the optical intensity caused by ASE noise via the Gordon–Mollenauer effect [23]. Receiver and electrical amplifier noises are also taken into account.

3.8.1 FIBER PROPAGATION MODEL

3.8.1.1 Non-linear Schroedinger Equation (NLSE)

Evolution of the slow varying complex envelope $A(z,t)$ of the optical pulses along a single mode optical fiber is governed by the non-linear Schroedinger equation (NLSE) [11]:

$$\frac{\partial A(z,t)}{\partial z} + \frac{\alpha}{2} A(z,t) + \beta_1 \frac{\partial A(z,t)}{\partial t} + \frac{j}{2}\beta_2 \frac{\partial^2 A(z,t)}{\partial t^2} - \frac{1}{6}\beta_3 \frac{\partial^3 A(z,t)}{\partial t^3} = -j\gamma \left|A(z,t)\right|^2 A(z,t) \quad (3.93)$$

where z is the spatial longitudinal coordinate, α accounts for fiber attenuation, β_1 indicates the differential group delay (DGD), β_2 and β_3 represent second- and third-order dispersion factors of the group velocity dispersion (GVD) and γ is the nonlinear coefficient. Equation 3.62 involves the following effects in a single-channel transmission fiber: (i) attenuation; (ii) CD; (iii) third-order dispersion factor, the dispersion slope; and (iv) self-phase-modulation (SPM) nonlinearity. Other nonlinear effects such as four-wave mixing (FWM), stimulated Raman scattering (SRS) can be inserted into the NLSE if necessary.

3.8.1.2 Symmetrical Split-Step Fourier Method

In this chapter, solutions of the NLSE and hence the model of pulse propagation in a single-mode optical fiber is numerically solved by using the popular approach of the split-step Fourier method (SSFM) in which the fiber length is divided into a large number of segments of small step size δz.

In practice, dispersion and non-linearity are mutually interactive while the optical pulses propagate through the fiber. However, the SSFM assumes that over a small length δz, the effects of dispersion and the nonlinearity on the propagating optical field are independent. Thus, in SSFM, the linear operator representing the effects of fiber dispersion and attenuation and the non-linearity operator taking into account fiber non-linearities are defined separately as

$$\hat{D} = \frac{i\beta_2}{2}\frac{\partial^2}{\partial T^2} + \frac{\beta_3}{6}\frac{\partial^3}{\partial T^3} - \frac{\alpha}{2} \quad (3.94)$$

$$\hat{N} = i\gamma \, |A|^2$$

where A replaces $A(z, t)$ for simpler notation and $T = t - z/v_g$ is the reference time frame moving at the group velocity.

The NLSE in Equation 3.62 can be rewritten as

$$\frac{\partial A}{\partial z} = (\hat{D} + \hat{N})A \qquad (3.95)$$

and the complex amplitudes of optical pulses propagating from z to $z + \delta z$ are calculated using the approximation as given

$$A(z + h, T) \approx \exp(h\hat{D})\exp(h\hat{N})A(z, T). \qquad (3.96)$$

Equation 3.33 is accurate to second order in the step size δz [11]. The accuracy of SSFM can be improved by including the effect of the non-linearity in the middle of the segment rather than at the segment boundary, as illustrated in Figure 3.12.

Equation 3.65 can now be modified as

$$A(z + \delta z, T) \approx \exp\left(\frac{\delta z}{2}\hat{D}\right)\exp\left(\int_{z}^{z+\delta z} \hat{N}(z')dz'\right)\exp\left(\frac{\delta z}{2}\hat{D}\right)A(z, T). \qquad (3.97)$$

This method is accurate to third order in the step size δz. The optical pulse is propagated down segment from segment in two stages at each step. First, the optical pulse propagates through the first linear operator (step of $\delta z/2$) with dispersion effects taken into account only. The non-linearity is calculated in the middle of the segment. Non-linearity effects are considered over the whole segment. At $z + \delta z/2$, the pulse propagates through the remaining $\delta z/2$ distance of the linear operator. The process continues repetitively in executive segments δz until the end of the fiber. This method requires an appropriate selection of step sizes δz to reserve the required accuracy.

3.8.1.3 Modeling of Polarization Mode Dispersion (PMD)

The first-order PMD effect can be implemented by splitting the optical field into two distinct paths representing two states of polarizations with different propagating delay time $\Delta\tau$, then implementing SSFM over the segment δz before superimposing the outputs of these two paths for the output optical field. The transfer function for first-order PMD is given by [24].

$$H_f(f) = H_{f+}(f) + H_{f-}(f) \qquad (3.98)$$

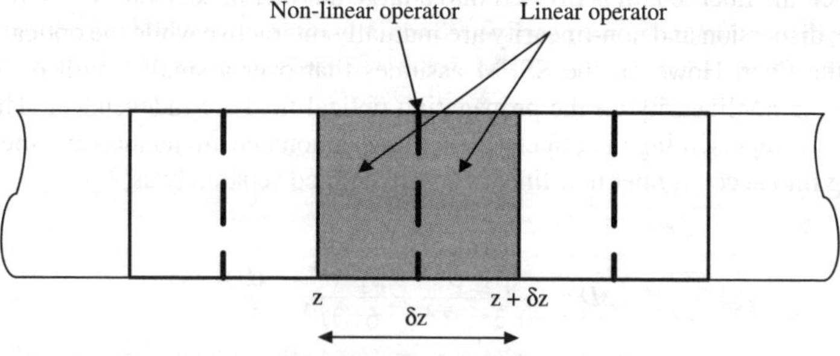

FIGURE 3.12 Schematic illustration of the split-step Fourier method.

$$H_{f+}(f) = \sqrt{\gamma}\,\exp\left[j2\pi f\left(-\frac{\Delta\tau}{2}\right)\right] \quad\text{and}\quad H_{f-}(f) = \sqrt{\gamma}\,\exp\left[j2\pi f\left(-\frac{\Delta\tau}{2}\right)\right] \tag{3.99}$$

γ is the splitting ratio, which normally has a value of 1/2.

3.8.1.4 Optimizing the Symmetrical SSFM

The symmetrical SSFM can be optimized by: (i) split-step length that satisfies the constraint for the convergence; and (ii) evaluation of incident peak power. If it is lower than the threshold of nonlinearities of the transmission fiber, the SPM effect can be neglected and SSFM will be switched to the low pass transfer function (LPTF) method that involves only the fiber dispersion and attenuation effects [12]. The LPTF of the fiber is given as

$$H_f(f) = \exp\left\{-j\left[(1/2)\beta_2'\varpi^2 + (1/6)\beta_3'\varpi^3\right]L\right\}. \tag{3.100}$$

3.8.1.5 Fiber Propagation in Linear Domain

The low-pass equivalent frequency response of the optical fiber, noted as $H(f)$, has a parabolic phase profile, and can be modeled by the following Equation 3.25

$$H_c(f) = e^{-j\alpha_D f^2}, \tag{3.101}$$

where $\alpha_D = \pi^2\beta_2 L, \beta_2$ represents the group velocity distortion (GVD) parameter of the fiber, and L is the fiber length. The parabolic phase profile is the result of the CD of the optical fiber [26]. The third-order dispersion factor β_3 is not considered in this transfer function of the fiber due to negligible effects on 40-Gb/s transmission systems. However, if the transmission bit rate is >40 Gb/s, the spectrum is wider, and the effects of β_3 must be taken into account.

In the model of the optical fiber, it is assumed that the signal is propagating in the linear domain, implying that fiber non-linearities are not included in the model. These non-linear effects are investigated numerically. It is also assumed that the optical carrier has a line spectrum. This is a valid assumption considering the state-of-the-art laser sources have very narrow linewidth and use external modulators in signal transmission.

A pure sinusoidal signal of frequency f, propagating through the optical fiber, experiences a delay of $|2\pi f_D\beta_2 L|$. The standard fibers used in optical communications have a negative β_2 and thus, in low-pass equivalent representation, sinusoids with positive frequencies (i.e., frequencies higher than the carrier) have negative delays, i.e., that early arrival compared with the carrier and the ones with negative frequencies (i.e., frequencies lower than the carrier) have positive delays. Dispersion-compensating fibers have positive β_2 and so have reverse effects. The low-pass equivalent channel impulse response of the optical fiber, $b_c(t)$ would also have a parabolic phase profile, given as

$$h_c(t) = \sqrt{\frac{\pi}{j\alpha_D}}\,e^{-j\alpha^2 t^2/\alpha_D}. \tag{3.102}$$

3.8.2 Non-linear Effects via Fiber Propagation Model

3.8.2.1 SPM Effects

If a single channel has been transmitted and only the SPM effect due to the change of the refractive index of the guided medium is a function of the intensity, then the SPM effects can be represented in the NLSE as the term in the RHS of the Equation 3.99 of the ith channel

$$\frac{\partial A_i(z,t)}{\partial z} + \frac{\alpha}{2} A_i(z,t) + \beta_1 \frac{\partial A_i(z,t)}{\partial t} + \frac{j}{2}\beta_2 \frac{\partial^2 A_i(z,t)}{\partial t^2} - \frac{1}{6}\beta_3 \frac{\partial^3 A_i(z,t)}{\partial t^3} = -j\gamma \left|A(z,t)\right|^2 A(z,t) \quad (3.103)$$

with the nonlinear coefficient given as previously stated as $\gamma = 2\pi n_2/\lambda A_{\text{eff}}$ with n_2 is the nonlinear refractive index and A_{eff} is the effective area covered by the area of the fundamental mode. The instantaneous magnitude of the envelope of the optical field plays an important part in the SPM effect, and thus there is a strong interaction between the SPM and the linear second- and third-order dispersion effects. The nonlinear dispersion length and the linear and nonlinear dispersion lengths are given by

$$L_{\text{D}} = \frac{T_e^2}{\left|\beta^2\right|} \qquad L_{NL} = \frac{1}{\gamma P_p} \qquad\qquad (3.104)$$

where P_p is the peak power of the transmitted pulse, and T_e is the e^{-1} pulse width. In the regime where L_{NL} is much longer than the dispersion length, then the SPM dominates the pulse broadening, and is limited by SPM. This usually happens in dispersion-compensating fibers whose effective area is very small, hence the enhancement effects of the SPM. In such cases, the NLSE reduces to

$$\frac{\partial A_i(z,t)}{\partial z} + \frac{\alpha}{2} A_i(z,t) = -j\gamma \left|A(z,t)\right|^2 A(z,t). \qquad\qquad (3.105)$$

That is, all the terms that relate to the phase variation due to the propagation constant are neglected, and only the phase due to the nonlinear SPM is included and the attenuation of the fiber. This equation can be solved analytically to give a possible solution of

$$A(L,t) = A(0,t)e^{\frac{-\alpha}{2}L} e^{-j\phi_{\text{SPM}}(L,t)} \qquad\qquad (3.106)$$

with the nonlinear phase ϕ_{SPM} given by

$$\phi_{\text{SPM}}(L,t) = \gamma \left|A(0,t)\right|^2 L_{\text{eff}}. \qquad\qquad (3.107)$$

Equation 3.101 reveals that the nonlinear refractive index varies with the evolution of its own intensity along the effective propagation distance. This phase modulation would then turn into frequency modulation in the form of broadening of the spectral characteristics of the signals, and hence pulse compression or dispersion depending on the sign of the nonlinear phase variation.

In the regime in which the effective nonlinear length is very close to that of the dispersion, interplay between linear and nonlinear dispersion effects ors, and the above equations do not represent the overall dispersive effects and only numerical solutions can be reached. This SSFM has already been described in the Section 3.8.1.

If the optical signals propagate in the anomalous regime ($\beta_2 < 0$) then the nonlinear SPM phase change is in the opposite of the linear dispersion, and these two effects can compensate each other. Under some specific power of the pulses, the pulse may preserve its shape over a very long distance. This is called "solitary wave propagation" and is described in Chapter 10. If the change of the linear dispersion is in the same as that of the nonlinear-induced dispersion, then the signals broaden further and significant broadening of the pulse sequence occurs. Optimization or management of the dispersion effects can be used to minimize SPM broadening.

3.8.2.2 XPM Effects

If DWM transmission is employed to extend the total transmission capacity of the system and networks, then the non-linear refractive index generated by the SPM would create inter-modulation effects, usually called the "cross phase modulation" (XPM). This XPM can be represented in the NLSE by coupled equations between different wavelength channels. In the NLSE, the XPM is shown in the second term of the RHS last term

$$\frac{\partial A_i(z,t)}{\partial z} + \frac{\alpha}{2} A_i(z,t) = -\beta_1 \frac{\partial A_i(z,t)}{\partial t} + \frac{j}{2}\beta_2 \frac{\partial^2 A_i(z,t)}{\partial t^2} - \frac{1}{2}\beta_3 \frac{\partial^3 A_i(z,t)}{\partial t^3}$$

$$= -j\gamma \left\{ |A_i(z,t)|^2 + 2\sum_{k,k\neq i} |A_k(z,t)|^2 \right\} A_i(z,t). \tag{3.108}$$

We now assume that only the XPM effect generated by two wavelength channels, then the envelopes of the two wavelength channels are coupled via a coupled differential equations

$$\frac{\partial A_1(z,t)}{\partial z} + \frac{\alpha}{2} A_1(z,t) = -j\gamma \left\{ |A_1(z,t)|^2 + 2|A_2(z,t)|^2 \right\} A_1(z,t)$$

$$\frac{\partial A_2(z,t)}{\partial z} + \frac{\alpha}{2} A_2(z,t) + d_{12}\frac{\partial A_2(z,t)}{\partial z} = -j\gamma \left\{ |A_2(z,t)|^2 + 2|A_1(z,t)|^2 \right\} A_2(z,t). \tag{3.109}$$

With an extra term due to the propagation delay between the two wavelength channels coupled via the propagation constant of the guide waves of different wavelengths given by

$$d_{12}z = D\left[\lambda_1 - \lambda_2)z\right]. \tag{3.110}$$

The solution of the coupled differential equations can be written as

$$\phi_1(L,t) = \phi_{1SPM}(L,t) + \phi_{1XPM}(L,t)$$

$$= \gamma |A(0,t)|^2 L_{eff} + 2\gamma_1 \int_0^L |A_2(0,t) + d_{12}z| e^{-\alpha z} dz \tag{3.111}$$

where the first term of the RHS represents the SPM effects on channel one, and the second term is the inter-modulation XPM effect term which can be expressed in the frequency domain by taking its Fourier transform and given as:

$$|A_2(0,t) + d_{12}z|^2 = P_2(0,t + d_{12}z) = \int_{-\infty}^{+\infty} P_2(0,f)e^{j2\pi f(t+d_{12}z)} df. \tag{3.112}$$

Thus, the term $\phi_{1SPM}(L, t)$ can be written in the frequency domain as

$$\phi_{1XPM}(L,f) = 2\gamma_1 \int_{\infty}^{+\infty} P_2(0,f)e^{j2\pi fd_{12}z} e^{-\alpha z} df = 2\gamma_1 P_2(0,f) \int_{-\infty}^{+\infty} e^{j2\pi fd_{12}z} e^{-\alpha z} df \tag{3.113}$$

$$= P_2(0,f)H_{12}(f)$$

where $H_{12}(f)$ represents the contribution of the XPM effects on the optical passband spectrum of the optical signals given as

$$H_{12}(f) = 2\gamma_1 \int\limits_{-\infty}^{+\infty} e^{j2\pi f d_{12}z - \alpha z} df = 2\gamma_1 \frac{1 - e^{j2\pi f d_{12}L - \alpha L}}{\alpha - j2\pi f d_{12}} \approx \frac{2\gamma_1}{\alpha - j2\pi f d_{12}}. \qquad (3.114)$$

Thus, the effect of the XPM is filtering the spectrum of the optical signals with a low-pass filter whose transfer function in the frequency domain is given by Equation 3.110. For example, consider the transmission fiber SSMF over which the XPM is taking place, with the attenuation coefficient and the dispersion factor D known to be $D = 17$ ps/nm/km and $\alpha = 0.2$ dB/km, and the corner frequency of the XPM low pass filter situated at 540 MHz. As a result, the XPM induced effect would alter the spectrum of the signal channel and would significantly affect the received signals if the bit rate is slow enough, e.g., 10 Gb/s with channel spacing of the order of 100 GHz. This low-frequency filtering effect would not be significant for bit rate of 40 Gb/s and 100 Gb/s.

The XPM effect would be significant under the coherent detection because the phase is preserved under such a scheme, but not for direct detection. Under the latter, this XPM phase is converted into the amplitude modulation, the PM to AM inter-modulation conversion and then through square law detection that would generate the distortion and penalty of the eye opening. Furthermore, this XPM creates propagation delay difference between the channels in DWDM systems and thus walk-off effects between channels.

If intra-channel dispersion is taken into account [13], the XPM increases proportionally with respect to the square of the channel spacing ($\alpha\, P/\Delta f^2$).

3.8.2.3 FWM Effects

FWM is the phenomenon that involves the mixing of the three lightwaves and then generation of the fourth wave which may fall into the spectrum of one of the transmitted channels. This phenomenon occurs when there is matching between the phases of the lightwave channels. This phase matching between the channels occurs when their group velocities are identical. This happens when the transmission fiber has its zero dispersion in the region of these lightwave channels. Thus, the FWM can be distinguished from other non-linear effects affected by the strong intensity of the guided waves in that the phase of the optical fields and the satisfaction of the phase matching condition for the degenerate four-wave mixing are much more critical. The generation of unwanted optical signals shown in Figure 3.13 is due to the mixing of three lightwave channels. The FWM effects can be included in the NLSE as indicated by the last term of the RHS below [14]

$$\frac{\partial A_i(z,t)}{\partial z} + \frac{\alpha}{2} A_i(z,t) + \beta_1 \frac{\partial A_i(z,t)}{\partial t} + \frac{j}{2}\beta_2 \frac{\partial^2 A_i(z,t)}{\partial t^2} - \frac{1}{6}\beta_3 \frac{\partial^3 A_i(z,t)}{\partial t^3} =$$

$$- j\gamma\left\{\left|A_i(z,t)\right|^2 + 2\sum_{k,k\neq i}\left|A_k(z,t)\right|^2\right\} A_i(z,t) + j\gamma \sum_{i=p+q-r,\,pq\neq r} \frac{d}{3}\left\{A_p A_q A_r^* e^{j\Delta\beta_{ipqr}z}\right\} \qquad (3.115)$$

where

$$\Delta\beta_{ipqr} = \beta_i - \beta_p - \beta_q + \beta_r = (\omega_p - \omega_r)(\omega_q - \omega_r) + \frac{1}{2}\beta_{3i}(\omega_p + \omega_q - 2\omega_r)$$

$$d = \begin{cases} 3 & \text{for } q = p \\ 6 & \text{for } q \neq p. \end{cases}$$

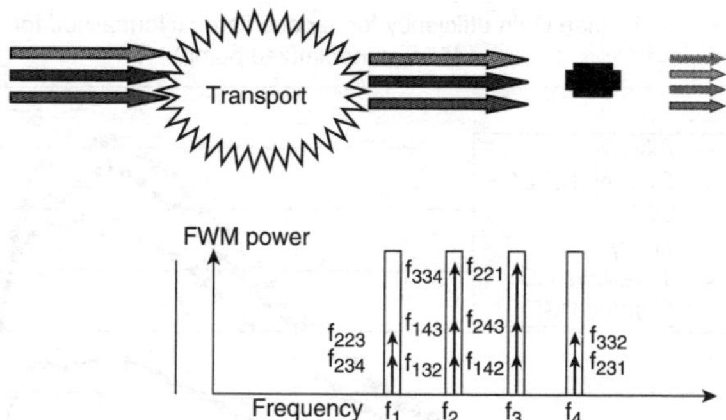

FIGURE 3.13 FWM with the four sources and their mixing to generate other lightwave signals.

Thus, the number of generated waves is

$$M = \frac{1}{2}(N^3 - N^2) \tag{3.116}$$

The effect of FWM can be explained by the non-linear interaction between the lightwaves of frequencies f_1, f_2 and f_3, which can also be represented by the Volterra series transfer function with $(r \neq p,q)$ that would generate new lightwave component that would interfere with other DWDM channels that have the same frequency as that of the generated ones.

For continuous wave channels, the efficiency of the degenerate FWM is given by

$$P_{ipqr} = \left(\frac{d}{3}\gamma L_{\text{eff}}\right)^2 P_p(0)P_q(0)P_r(0)e^{-\alpha L}\eta_{ipqr}. \tag{3.117}$$

The efficiency of the FWM is given by

$$\eta_{ipqr} = \frac{\alpha^2}{\alpha^2 + \Delta\beta_{ipqr}^2}\left[1 + \frac{4e^{-\alpha L}\sin^2(\Delta\beta_{ipqr}L/2)}{(1-e^{-\alpha L})^2}\right]. \tag{3.118}$$

If the fiber is a dispersion-shifted type, the phase is matched with $\beta_2 = 0$, $\Delta\beta_{ipqr} = 0$, thus the efficiency reaches unity and the FWM is maximal under this condition. This is the reason why the NZDSF is designed to reduce the linear dispersion, but must avoid the four-wave mixing effects.

3.8.2.4　SRS Effects

SRS and SBS are the scattering generated from the generation of the phonons in the fiber under high-energy pumping at some spectral distance away from the signal spectrum. The phonons would then be phase-matched to the pump lightwaves to create a stimulated emission condition at the wavelength regions to enable to creation of new photons, the Raman scattering and Brillouin scattering. The difference between SRS and SRS is the vibration of the electronic domain and the elasticity of the crystal structures of silica and doped impurities, respectively.

Various optical fibers already installed in the transmission systems and networks throughout global telecommunication backbone networks are measured for SRS gain efficiency (Figure 3.14). Spacing of the pump to the Stoke wavelength would be of the order of 100 nm or 13 THz.

The SRS effects can be represented in the NLSE as the last term of the RHS of the equation shown below

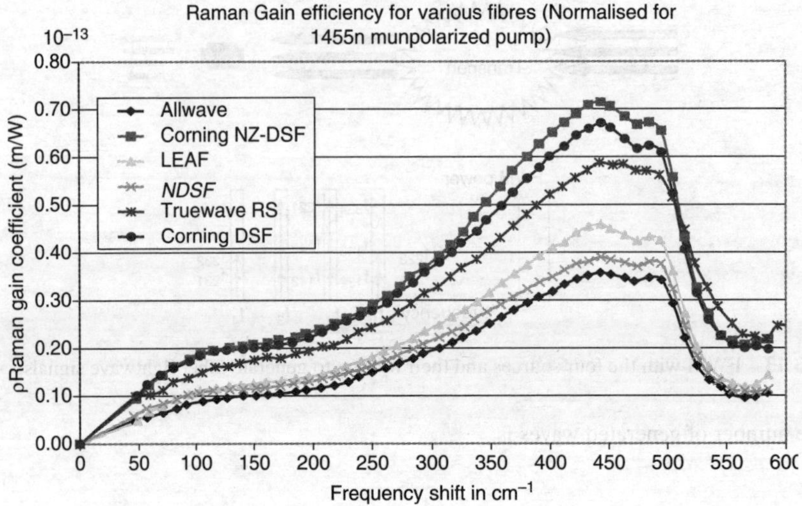

FIGURE 3.14 Raman gain efficiency of various single mode optical fibers.

$$\frac{\partial A_i(z,t)}{\partial z} + \frac{\alpha}{2} A_i(z,t) + \beta_1 \frac{\partial A_i(z,t)}{\partial t} + \frac{j}{2}\beta_2 \frac{\partial^2 A_i(z,t)}{\partial t^2} - \frac{1}{6}\beta_3 \frac{\partial^3 A_i(z,t)}{\partial t^3} =$$

$$- j\gamma\left\{|A_i(z,t)|^2 + 2\sum_{k,k\neq i}|A_k(z,t)|^2\right\}A_i(z,t) + j\gamma\sum_{i=p+q-r, pq\neq r}\frac{d}{3}\left\{A_p A_q A_r^* e^{j\Delta\beta_{ipqr}z}\right\}A_i(z,t) \quad (3.119)$$

$$+ \left\{\sum_{k=1}^{i-1}\frac{\omega_i}{\omega_k}\frac{g_R}{2A_{\text{eff}}}|A_k|^2 + \sum_{k=i+1}^{N}\frac{g_R}{2A_{\text{eff}}}|A_k|^2\right\}A_i(z,t).$$

If we now consider the effect of the SRS only in the NLSE, then we can rewrite the equation as

$$\frac{\partial P_i(z,t)}{\partial z} = -\alpha P_i - \left\{\sum_{k=1}^{i-1}\frac{f_i - f_k}{A_{\text{eff}}}g_R P_k + \sum_{k=i+1}^{N}\frac{f_k - f_i}{A_{\text{eff}}}g_R P_k\right\} \quad (3.120)$$

where the gain coefficient of Raman scattering is represented as $g_R(f_i - f_k)$ for a given frequency separation. This equation can be solved analytically to give a solution of

$$P_i(L,t) = P_i(0)\exp\left\{-\alpha L - \sum_{k=1}^{i-1}\frac{f_i - f_k}{A_{\text{eff}}}g_R P_k + \sum_{k=i+1}^{N}\frac{f_k - f_i}{A_{\text{eff}}}g_R P_k\right\} \quad (3.121)$$

in which the second and third parts of coefficient of the exponential represent the loss and gain of the SRS given as

$$\alpha_{\text{SRS}} = \left\{\sum_{k=1}^{i-1}\frac{f_i - f_k}{A_{\text{eff}}}g_R P_k + \sum_{k=i+1}^{N}\frac{f_k - f_i}{A_{\text{eff}}}g_R P_k\right\}. \quad (3.122)$$

If the gain spectrum of Figure 3.14 can be approximated as a linear distribution, then the gain tilt of the SRS can be written as

$$\alpha_{SRS} = 10\,\mathrm{Log}_e\left(\frac{1}{2}\frac{L_{\mathrm{eff}}}{A_{\mathrm{eff}}}\frac{dg_R}{df}P_{\mathrm{channel}}(N-1)N\Delta f\right)\,\mathrm{dBm}. \tag{3.123}$$

The gain tilt would result in different optical signal levels of different wavelength channels, thus gain equalization should be recommended. This can be implemented by pumping at different pump wavelength as shown in Figure 3.15.

3.8.2.5 SBS Effects

SBS comes from the nonlinear interaction of the photons and the phonons. SBS is to be considered if the optical power within a narrow bandwidth exceeds a certain threshold. The maximum spectral density of the power that can be launched into a single-mode fiber is given by

$$\mathrm{PSD}_{SBS,\mathrm{th}} = \frac{P_{SBS,\mathrm{th}}}{B_{\mathrm{sig}}} \approx \frac{21A_{\mathrm{eff}}}{g_{SBS}L_{\mathrm{eff}}B_{SBS}} \tag{3.124}$$

where g_{SBS} is the Brillouin gain coefficient, and B_{sig} and B_{SBS} represent signal bandwidth and SBS gain bandwidth, respectively. For SSMF with $A_{\mathrm{eff}}=50$ $\mu\mathrm{m}^2$, $L_{\mathrm{eff}}=20$ km $B_{SBS}=20$ MHz, and $g_{SBS}=4.10^{-11}$, the threshold value of the power spectral density is about 6e-11 W/Hz.

Under transmission and detection, the SBS effect can be observed with the low-frequency modulation of the high-frequency signal envelope or effectively the jitter of the eye diagram, and hence the reduction of the sampling time interval of the eye diagram or reduction of the time Q factor. We can observe in later chapters that the ASK format would exhibit a peak power of the optical carrier and none for phase shift keying formats. Thus, ASK would be expected to have a lower SBS threshold and hence low optical signal power available for transmission.

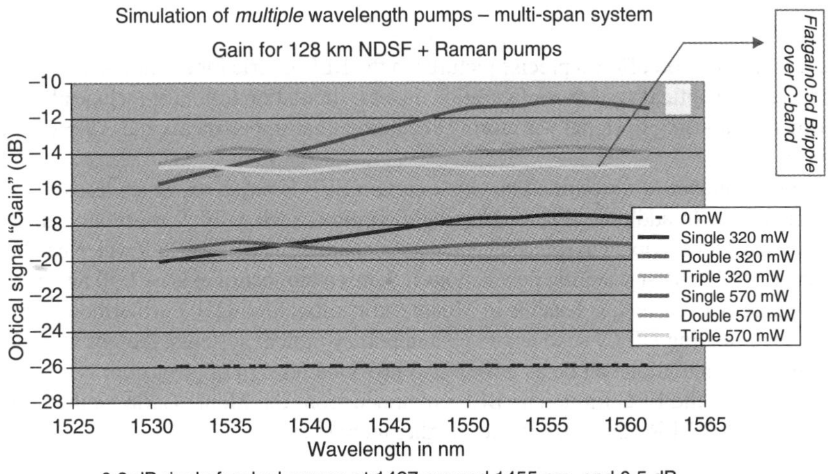

FIGURE 3.15 Simulation of Raman gain flattening using multiple pump sources and total pump power. Vertical axis = gain in dB.

3.9   PERFORMANCE EVALUATION

Performance evaluation of an optical transmission system via the quality of electrically detected signals is an essential aspect in simulation and experiment scenarios. The key metrics reflecting the signal quality include optical signal-to-noise ratio (OSNR) and OSNR penalty, eye opening (EO) and eye opening penalty (EOP) where as bit error rate (BER) is the ultimate indicator for the performance of a system.

In an experimental setup and practical optical systems, BER and the quality factor (Q-factor) can be obtained directly from the BERT set, and data can be exported to a portable memory for post-processing. However, these experimental systems must be run within at least a few hours so that the results are stable and accurate. For investigation of performance of an optical transmission system by simulation, several effective statistical methods have been developed and are outlined in this section.

3.9.1   BER FROM MONTE CARLO METHOD

Conventional methods to calculate Q-factor, Q dB and hence BER are based on assumption of Gaussian distribution of noise. However, new methods based on statistical processes taking into account the distortion dynamics of the optical fibers are necessary to include the common patterning effects and the ISI induced by CD and PMD, especially when the phase or amplitudes of the carrier are under different modulation formats.

The first statistical technique implements the expected maximization theory in which the pdf of the obtained electrical detected signal is approximated as a mixture of multiple Gaussian distributions.

The second technique is based on the generalized extreme values (GEV) theorem [16–18]. Although this theorem [19] is well-known in other fields (e.g., financial forecasting, meteorology, material engineering) to predict the probability of occurrence of extreme values, it has not been applied in optical communications.

Exactly as the BERT set used in experimental transmission, the bit error rate (BER) in a simulation of a particular system configuration is counted. The BER is the ratio of the occurrence of errors (N_{error}) to the total number of transmitted bits N_{total} and given as

$$\text{BER} = \frac{N_{\text{error}}}{N_{\text{total}}}. \tag{3.125}$$

The Monte Carlo method offers a precise picture via the BER metric for all modulation formats and receiver types. The optical system configuration under a simulation test must include all the sources of impairments imposing to signal waveforms, including fiber impairments and ASE (optical)/electronic noise.

A sufficient number of transmitted bits for a certain BER is required, which leads to exhaustive computational time. In addition, time-consuming algorithms such as FFT, especially carried out in symmetrical SSFM, contribute to this long computational time. A BER of 1e-9 which can be considered as "error free" in most scientific publications requires a number of at least 1e10 bits transmitted.

However, 1e-6, even 1e-7, is feasible in Monte Carlo simulation [21]. Furthermore, with the use of forward error coding (FEC) schemes in contemporary optical systems, the reference for BERs to be obtained in simulation can be as low as 1e-3 provided no sign of error floor is shown. This is normally known as the FEC limit. The BERs obtained from the Monte Carlo method are a good benchmark for other BER values estimated in other techniques.

3.9.2   BER AND Q-FACTOR FROM PROBABILITY DISTRIBUTION FUNCTIONS

This method implements a statistical process before calculating values of BER and Q-factor to determine the normalized probability distribution functions (pdf) of received electrical signals (for both "1" and "0" and at a particular sampling instance). The electrical signal is normally in voltage because the detected current after a photo-diode is usually amplified by a trans-impedance electrical amplifier. The pdfs can be determined statistically by the histogram approach. A particular voltage value as a reference for the distinction between "1" and "0" is known as the threshold voltage (V_{th})

The BER in case of transmitting bit "1" (receiving as "0" instead) is calculated from the well-known principle [27]. The integral of the overlap of normalized pdf of "1"exceeds the threshold. Similar calculation for bit "0" can be applied. The shape of the pdf is thus very critical to obtain an accurate BER. If the exact shape of the pdf is known, the BER can be calculated precisely as

$$\text{BER} = P(`1') \, P(`0' \mid `1') + P(`0') \, P(`1' \mid `0') \tag{3.126}$$

where $P(`1')$ is the probability that a "1" is sent; $P(``0" \mid ``1")$ is the probability of error due to receiving "0" where actually an "1" is sent; $P(``0")$ is the probability that a "0" is sent; $P(``1" \mid ``0")$ is the probability of error due to receiving "1" where actually a "0" is sent As commonly used, the probability of transmitting a "1" and "0' is equal, i.e., $P(``1") = P(``0") = 1/2$.

A popular approach in simulation and commercial BERT test-sets is the assumption of pdf of "1" and "0" following Gaussian/normal distributions, i.e., noise sources can be approximated by Gaussian distributions. If the assumption is valid, high accuracy is achieved. This method enables a fast estimation of the BER by using the complementary error functions [27]

$$\text{BER} = \frac{1}{2}\left[\text{erfc}\left(\frac{|\mu_1 - V_{th}|}{\sqrt{2}\sigma_1} \right) + \text{erfc}\left(\frac{|\mu_0 - V_{th}|}{\sqrt{2}\sigma_0} \right) \right] \tag{3.127}$$

where μ_1 and μ_0 are the mean values for *pdf* of "1" and "0", respectively, whereas σ_1 and σ_0 are the variance of the pdfs. The quality factor, Q- or δ-factor, which can be in linear scale or in logarithmic scale, can be calculated from the obtained BER through the expression

$$Q \triangleq \delta = \sqrt{2}\,\text{erfc}^{-1}(2\text{BER}) \tag{3.128}$$

$$Q_{dB} = 20\text{Log}_{10}\left(\sqrt{2}\,\text{erfc}^{-1}(2\text{BER}) \right).$$

3.9.3 Histogram Approximation

The common question is "what are the proper values for number of bins and bin-width to be used in the approximation of the histogram so that the bias and the variance of the estimator are negligible?" According to [28], with a sufficiently large number of transmitted bits (N_0), a good estimate for the width (W_{bin}) of each equally spaced histogram bin is given by

$$W_{bin} = \sqrt{N_0}. \tag{3.129}$$

3.9.4 Optical Signal-to-Noise Ratio (OSNR)

The optical signal-to-noise ratio (OSNR) is a popular benchmark indicator for assessment of the performance of optical transmission systems, especially those limited by the ASE noise from the optical amplifiers (EDFAs). The OSNR is defined as the ratio of optical signal power to optical noise power. For a single EDFA with output power, Pout, the OSNR is given by [22]

$$\text{OSNR} = \frac{P_{out}}{N_{ASE}} = \frac{P_{out}}{(NFG_{op} - 1)h\nu B_o} \qquad (3.130)$$

where NF is the amplifier noise figure, G_{op} is the amplifier gain, $h\nu$ is the photon energy, and B_o is the optical bandwidth by measurement. However, OSNR does not provide good estimation to the system performance when the main degrading sources involve the dynamic propagation effects such as dispersion (including CD and PMD) and Kerr non-linearity effects (e.g., SPM). In these cases, the degradation of the performance is mainly due to waveform distortions rather than the corruption of the ASE or electronic noise.

When addressing a value of an OSNR, it is important to define the optical measurement bandwidth over which the OSNR is calculated. The signal power and noise power is obtained by integrating all the frequency components across the bandwidth leading to the value of OSNR.

In practice, signal and noise power values are usually measured directly from the optical spectrum analyzer (OSA), which does the mathematics for the users and displays the resultant OSNR versus wavelength or frequency over a fixed resolution bandwidth. A value of $\Delta\lambda = 0.1$ nm or $B_o = 12.5$ GHz at 1550 nm region, is widely used as the typical value for calculation of the OSNR.

OSNR penalty is determined at a particular BER when varying value of a system parameter under test. For example, OSNR penalty at BER $= 1e - 4$ for a particular optical phase modulation format when varying length of an optical link in a long-haul transmission system configuration.

3.9.5 Eye Opening Penalty (EOP)

The OSNR is a time-averaged indicator for the system performance where the ratio of average power of optical carriers to noise is considered. When optical lightwaves propagate through a dispersive and nonlinear optical fiber channel, the fiber impairments, including ISI induced from CD, PMD and the spectral effects induced from nonlinearities, cause the distortion of the waveforms. Another dynamic cause of the waveform distortion comes from the ISI effects as the result of optical or electrical filtering. In a conventional OOK system, bandwidth of an optical filter is normally larger than the spectral width of the signal by several times.

The eye-opening penalty (EOP) is a performance measure defined as the penalty of the "eye" caused by the distortion of the electrically detected waveforms to a referenced eye-opening (EO). The latter is the difference between the amplitudes of the lowest mark and the highest space. The benchmark EO is usually obtained from a back-to-back measurement when the waveform is not distorted by any of the above impairments.

The EOP at a particular sampling instance is normally calculated in log scale (dB unit) and given by:

$$\text{EOP}(t_{samp}) = \frac{\text{EO}_{ref}}{\text{EO}_{received}}. \qquad (3.131)$$

The EOP is useful for noise-free system evaluations as a good estimate of deterministic pulse distortion effects. The accuracy of EOP indicator depends on the sampling instance in a bit slot. Usually, the detected pulses are sampled at the instance giving the maximum EO. If noise is present, calculation of the EOP become less precise because of the ambiguity of the signal levels, which are corrupted by noise.

3.9.6 Statistical Evaluation Techniques

The method using a Gaussian-based single distribution involves only the effects of noise corruption on the detected signals and ignores the dynamic distortion effects such as ISI and non-linearity. These dynamic distortions result in a multi-peak pdf as demonstrated in Figure 3.16, which is clearly overlooked by the conventional single distribution technique. As a result, the pdf of the

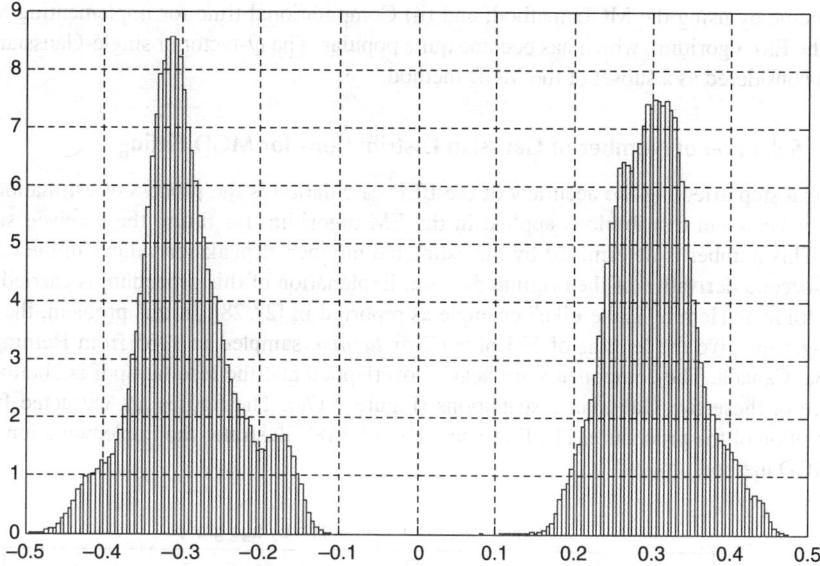

FIGURE 3.16 Demonstration of multi-peak/non-Gaussian distribution of the received electrical signal.

electrical signal cannot be approximated accurately. The addressed issues are resolved with the proposal of two new statistical methods.

Two new techniques proposed to accurately obtain the pdf of the detected electrical signal in optical communications include the mixture of multi-Gaussian distributions (MGD) by implementing the expectation maximization theory (EM) and the generalized Pareto distribution (GPD) of the generalized extreme values (GEV) theorem. These two techniques are well-known in statistics, banking, finance, and meteorology. The required algorithms can be implemented without MATLAB® functions. Thus, these novel statistical methods offer significant flexibility, convenience, and fast-processing while maintaining the errors in obtaining the BER within small and acceptable limits.

3.9.6.1 Multi-Gaussian Distributions (MGD) via Expectation Maximization (EM) Theorem

The mixture density parameter estimation problem is probably one of the most widely used applications of the expectation maximization (EM) algorithm. It comes from the fact that most of deterministic distributions can be seen as the result of superposition of different multi distributions. Given a probability distribution function $p(x\,|\,\Theta)$ for a set of received data, $p(x\,|\,\Theta)$ can be expressed as the mixture of M different distributions

$$p(x|\Theta) = \sum_{i=1}^{M} w_i p_i(x|\theta_i) \tag{3.132}$$

where the parameters are $\Theta = (w_1, \ldots, w_M, \theta_1, \ldots, \theta_M)$ such that $\sum_{i=1}^{M} w_i = 1$ and each p_i is a PDF by θ_i and each pdf assigned a weight w_i, that is the probable event of that pdf.

As a particular case adopted for optical communications, the EM algorithm is implemented with a mixture of multi-Gaussian distributions (MGD). This method offers great potential solutions for evaluation of performance of an optical transmission system for the following reasons: (i) In a linear optical system (low input power into fiber), the conventional single Gaussian distribution fails to take into account the waveform distortion caused by either the ISI due to fiber CD and PMD dispersion or patterning effects. Hence, the obtained BER is no longer accurate. These issues, however, are overcome by using the MGD method; and (ii) Computational time for implementing MGD is fast via the EM algorithm, which has become quite popular. The Q-factor or single-Gaussian distribution is considered as a subset of the MGD method.

3.9.6.2 Selection of Number of Gaussian Distributions for MGD Fitting

The critical step affecting the accuracy of the BER calculation is the process of estimation of the number of Gaussian distributions applied in the EM algorithm for fitting the received signal of the pdf. This number is determined by the estimated number of peaks or valleys in the curves of first and second derivative of the original data set. Explanation of this procedure is carried out via the well-known "Heming Lake Pike" example as reported in [27, 28]. In this problem, the data of five age-groups give the lengths of 523 pike (*Esox lucius*), sampled in 1965 from Heming Lake, Manitoba, Canada. The components are heavily overlapped and the resultant pdf is obtained with a mixture of these five Gaussian distributions (Figure 3.17a). The figures are extracted from for demonstration of the procedure. The flowchart (Figure 3.18) illustrates the implementation process of the MGD method.

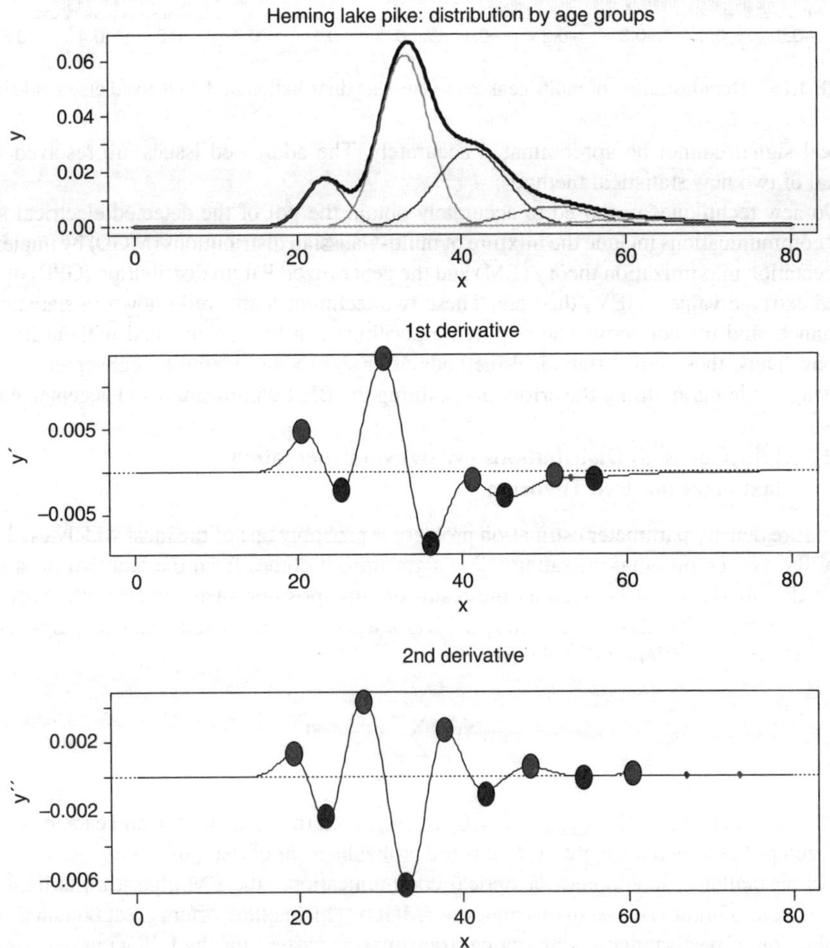

FIGURE 3.17 Estimation of number of Gaussian distributions in the mixed pdf based on first and second derivatives of the original data set.

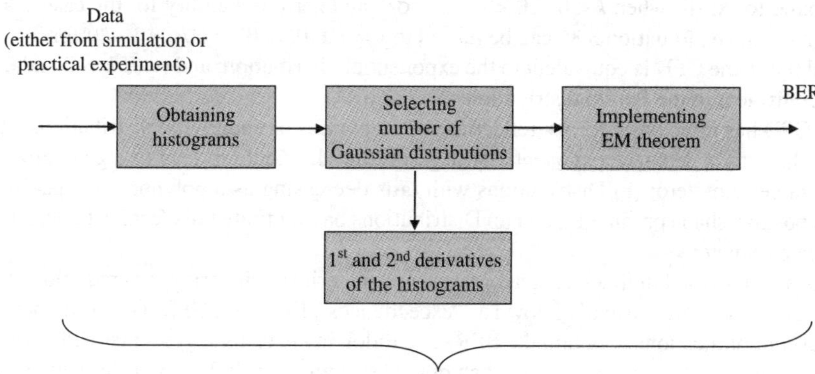

FIGURE 3.18 Implementation process of MGD method.

Estimation of the number of Gaussian distributions in the mixed pdf based on first and second derivatives of the original data set (courtesy of [29]). As seen from Figure 3.17, the first derivative of the resultant pdf shows clearly four pairs of peaks (red) and valleys (blue), suggesting that there should be at least four component Gaussian distributions contributing to the original pdf. However, by taking the second derivative, it is realized that there is actually up to five contributed Gaussian distributions (Figure 3.17).

In summary, the steps for implementing the MGD technique to obtain the BER value is: (i) Obtaining the pdf from the normalized histogram of the received electrical levels; (ii) Estimating the number of Gaussian distributions (N_{Gaus}) to be used for fitting the pdf of the original data set; (iii) Applying EM algorithm with the mixture of N_{Gaus} Gaussian distributions and obtaining the values of mean, variance and weight for each distribution; and (iv) Calculating the BER value based on the integrals of the overlaps of the Gaussian distributions when the tails of these distributions cross the threshold.

3.9.6.2.1 Generalized Pareto Distribution (GPD)

The GEV theorem is used to estimate the distribution of a data set from a function in which the possibility of extreme data lengthen the tail of the distribution. Due to the mechanism of estimation for the pdf of the extreme data set, GEV distributions can be classified into two classes: GEV distribution and the generalized Pareto distribution (GPD).

Currently, there are only a few research reports on the application of this theorem in optical communications systems [23–25]. They report only on the GEV distributions that involve only the effects of noise, and neglect the effects of dynamic distortion factors. Unlike the Gaussian-based techniques but rather similar to the exponential distribution, the generalized Pareto distribution is used to model the tails of distribution [26]. This section provides an overview of the GPD. The probability density function for the GPD is defined as follows

$$y = f(x|k, \sigma, \theta) = \left(\frac{1}{\sigma}\right)\left(1 + k\frac{(x-\theta)}{\sigma}\right)^{-1-\frac{1}{k}} \tag{3.133}$$

for $\theta < x$ when $k > 0$ or for $\theta < x < \dfrac{-\sigma}{k}$ $k < 0$ where k is shape parameter $k \neq 0$, σ is scale parameter and the threshold parameter θ.

Equation 3.80 is subject to some significant constraints (i) when $k > 0$: $\theta < x$ i.e., there is no upper bound for x; (ii) when $k < 0$: $\theta < x < -\sigma/k$ and zero probability for the case $x > -\sigma/k$; (iii) when $k = 0$, i.e., Equation 3.80 can be turned to $y = f(x|0, \sigma, \theta) = (1/\sigma)e^{-(x-\theta)/\sigma}$ for $\theta < x$; (iv) If $k = 0$ and $\theta = 0$, the GPD is equivalent to the exponential distribution; and (v) If $k > 0$ and $\theta = \sigma$, the GPD is equivalent to the Pareto distribution.

The GPD has three basic forms, reflecting different class of underlying distributions. (a) Distributions whose tails decrease exponentially (e.g., normal distribution) lead to a generalized Pareto shape parameter of zero; (b) Distributions with tails decreasing as a polynomial (e.g., Student's t) lead to a positive shape parameter; and (c) Distributions having finite tails (e.g., beta) lead to a negative shape parameter.

GPD is widely used in finance, meteorology, and material engineering for prediction of extreme or rare events which are normally known as "exceedances". However, GPD has not yet been applied in optical communications to obtain the BER. The following reasons suggest that GPD may become a potential quick method for evaluation of an optical system, especially if non-linearity is the dominant degrading factor to the system performance.

The normal distribution has a fast roll-off or short tail distribution. Thus, it is not a good fit to a set of data involving exeedances i.e., rarely happening data located in the tails of the distribution. With a certain threshold value, the GPD can be used to provide a good fit to extremes of this complicated data set.

- When non-linearity is the dominating impairment degrading the performance of an optical system, the sampled received signals usually introduce a long tail distribution. For example, in the DPSK optical system, the distribution of nonlinearity phase noise differs from the Gaussian counterpart due to its slow roll-off of the tail. As the result, the conventional BER obtained from assumption of Gaussian-based noise is no longer valid and it often underestimates the BER.
- A wide range of analytical techniques have recently been studied, such as importance sampling, and multi-canonical method. Although these techniques provide solutions to obtain a precise BER, they are usually very complicated. Whereas, calculation of GPD has become a standard and is available in the recent MATLAB® version (MATLAB 7.1). GPD may therefore provide a very quick and convenient solution for monitoring and evaluating the system performance. Necessary preliminary steps which are fast in implementation must be carried out to find the proper threshold.
- Evaluation of contemporary optical systems requires BER as low as 1e-15. Therefore, GPD can be quite suitable for optical communications.

3.9.6.3 Selection of Threshold for GPD Fitting

Using statistical methods, the accuracy of the obtained BER strongly depend on the threshold value (V_{thres}) used in the GPD fitting algorithm, i.e., the decision where the tail of the GPD curve starts. There have been several suggested techniques for guidelines aiding the decision of the threshold value for the GPD fitting. However, they are not absolute techniques and are quite complicated. In this chapter, a simple technique to determine the threshold value is proposed. The technique is based on the observation that the GPD tail with exceedances normally obeys a slow exponential distribution compared with the faster decaying slope of the distribution close to the peak values. The inflection region between these two slopes gives a good estimation of the threshold value for GPD fitting (Figure 3.19).

Whether the selection of the V_{thres} value leads to an adequately accurate BER or not is evaluated by using the cumulative density function (cdf) and the quantile-quantile plot (Q-Q plot). If there is a high correlation between the pdf of the tail of the original data set (with a particular V_{thres}) and pdf of

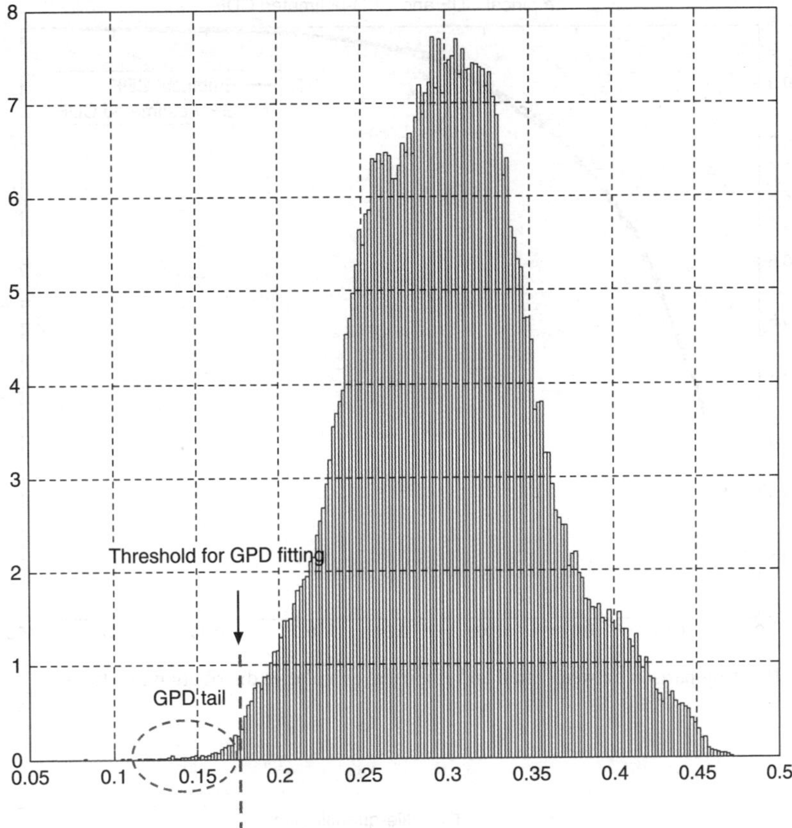

FIGURE 3.19   Selection of threshold for GPD fitting.

the GPD, there would be a good fit between the empirical cdf of the data set with the GPD-estimated cdf with focus at the most right region of the two curves. In the QQ-plot, a linear trend would be observed. These guidelines are illustrated in Figure 3.20 . In this particular case, the value of 0.163 shown in Figure 3.20 is selected to be V_{thres}.

Furthermore, as a demonstration of improper selection of V_{thres}, the value of 0.2 is selected. Figure 3.21 and Figure 3.22 show the incompliance of the fitted curve with the GPD, which is reflected via the discrepancy in the two cdfs and the non-linear trend of the quantile-quantile plot shown in Figure 3.23.

The following flowchart (Figure 3.24) illustrates the implementation process of the GPD method

3.9.6.4   Validation of Novel Statistical Methods

A simulation test-bed of an optical DPSK transmission system over 880 km SSMF dispersion-managed optical link (8 spans) is setup. Each span consists of 100 km SSMF and 10 km of DCF whose dispersion values are +17 ps/nm km and –170 ps/nm km at 1550 nm wavelength, respectively, and fully compensated, i.e., zero residual dispersion. The average optical input power into each span can be set to be higher than the non-linear threshold of the optical fiber. The degradation of the system performance is therefore dominated by the non-linear effects, which are of interest because it is a random process creating non-deterministic errors in the long-tail region of the pdf of the received electrical signals.

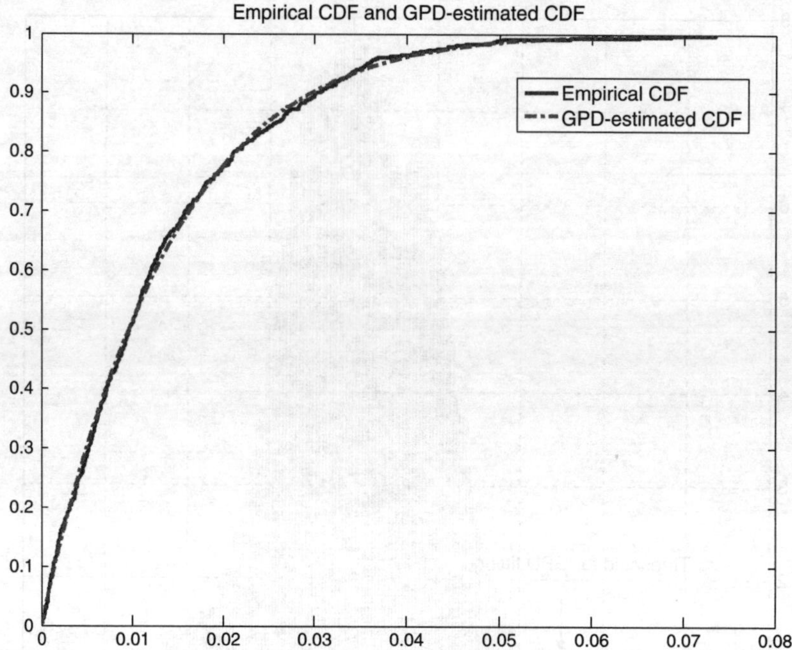

FIGURE 3.20   Comparison between fitted and empirical cumulative distribution functions.

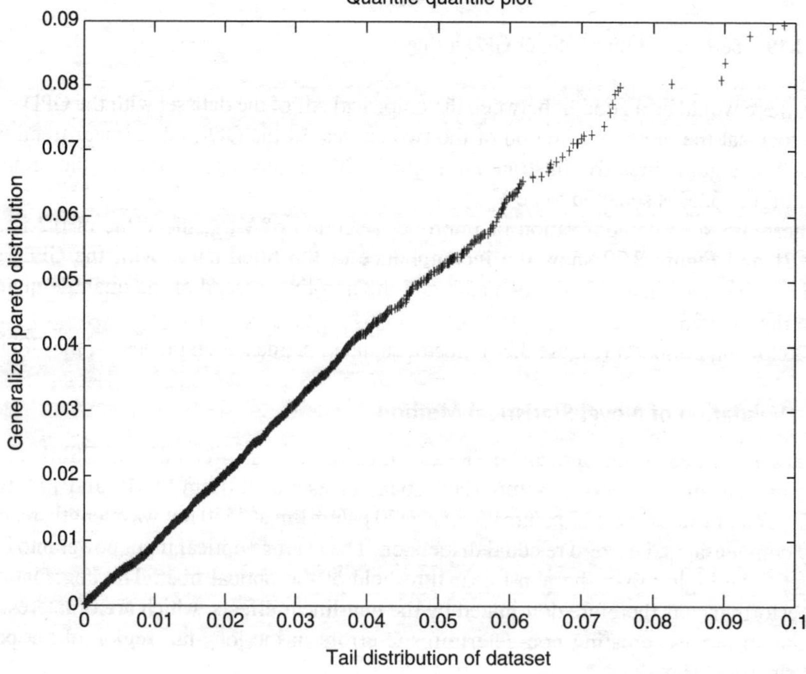

FIGURE 3.21   Quantile-quantile plot.

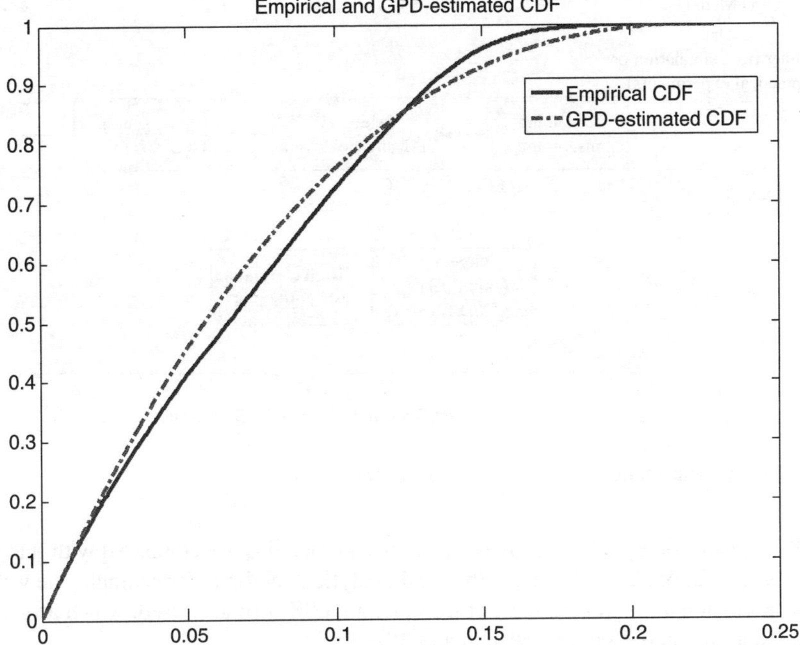

FIGURE 3.22 Comparison between fitted and empirical cumulative distribution functions.

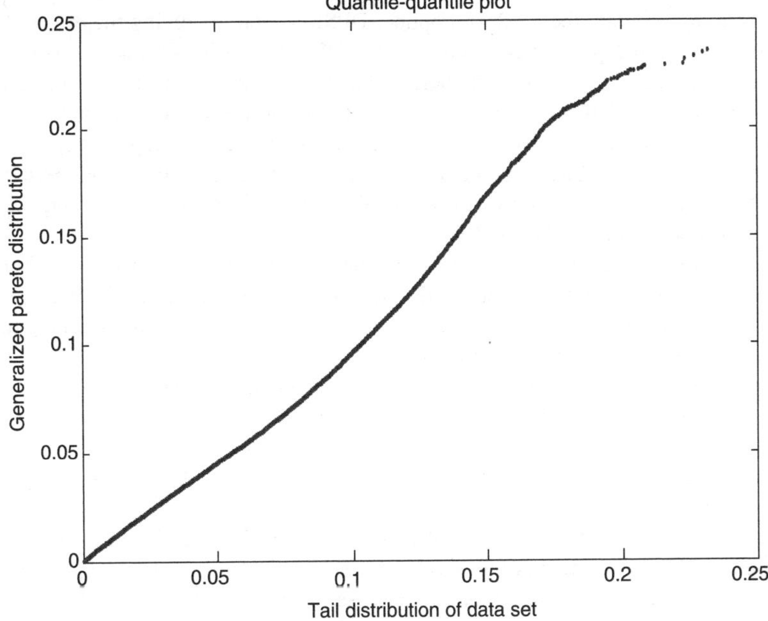

FIGURE 3.23 Quantile-quantile plot.

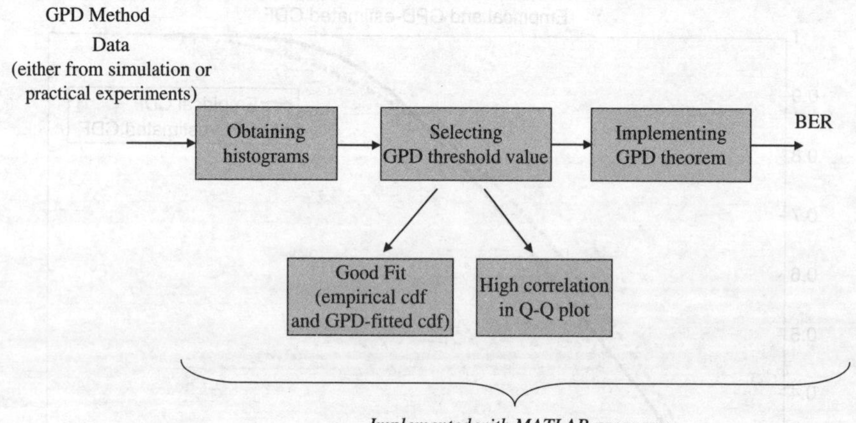

FIGURE 3.24   Implementation process of the MGD method.

The BER results obtained from the novel statistical methods are compared with that from the Monte Carlo simulation, as well as from the semi-analytical method. For example, the well-known analytical expression to obtain the BER of the optical DPSK format is used, which as given in the following equation (Nicholson [29], Blachman [30])

$$\text{BER} = \frac{1}{2} - \frac{\rho e^{-\rho}}{2} \sum_{k=0}^{\infty} \frac{(-1)^k}{2k+1}\left[I_k\left(\frac{\rho}{2}\right) + I_{k+1}\left(\frac{\rho}{2}\right)\right]^2 e^{-\frac{1}{2}(2k+1)^2 \sigma_{\text{NLP}}^2} \tag{3.134}$$

where ρ is the obtained OSNR and σ_{NLP}^2 is the variance of nonlinear phase noise.

In this case, to calculate the BER of a optical DPSK system involving the effect of non-linear phase noise, the required parameters including the OSNR and the variance of non-linear phase noise, are obtained from the simulation numerical data which is stored and processed in MATLAB®. The fitting curves are implemented with the MGD method for the pdf of bit 0 and bit 1 (input power of 10 dBm) as shown in Figure 3.25 and Figure 3.26 for bit 0 and bit 1 respectively.

The selection of optimal threshold for GPD fitting follows the guideline as addressed in detail in the previous section. The BER from various evaluation methods are shown in Table 3.2. The input powers are controlled to be 10 dBm and 11 dBm.

Table 3.2 validates the adequate accuracy of the proposed novel statistical methods with the discrepancies compared with the Monte Carlo method and semi-analytical BER to be within one decade. In short, these methods offer a great deal of fast processing while maintaining the accuracy of the obtained BER within acceptable limits.

Discussions: It is found that the MGD and GPD statistical methods provide adequately accurate BERs. The discrepancies of the BER values obtained by MGD and GPD methods compared with the Monte Carlo method and semi-analytical BER are within one decade.

The BER using the Monte Carlo method at fiber input power of 11 dBm is not provided in the table because it is quite time-consuming to obtain the BER in the range of 1e-8 (may take weeks). To obtain a BER in 1e-8 range, there are at least 1e9 binary bits generated from the random bit pattern generator. Moreover, this is not feasible for simulation to investigate optical transmission systems operating in the non-linear region (in this case, with the existence of non-linear phase noise (NLPN)). In addition, BERs at an input power of ≤9 dBm are not provided because the NLPN does not significantly affect system performance.

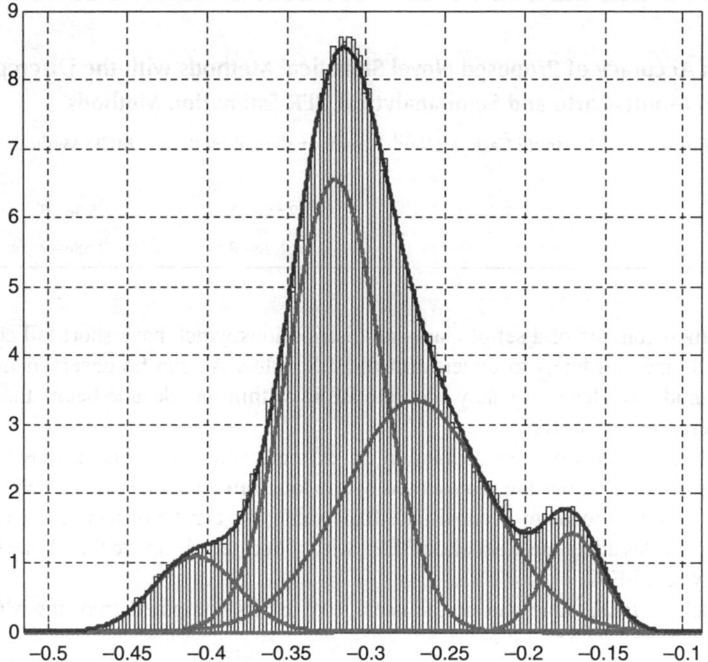

FIGURE 3.25   Fitting curves for bit "0" with MGD method.

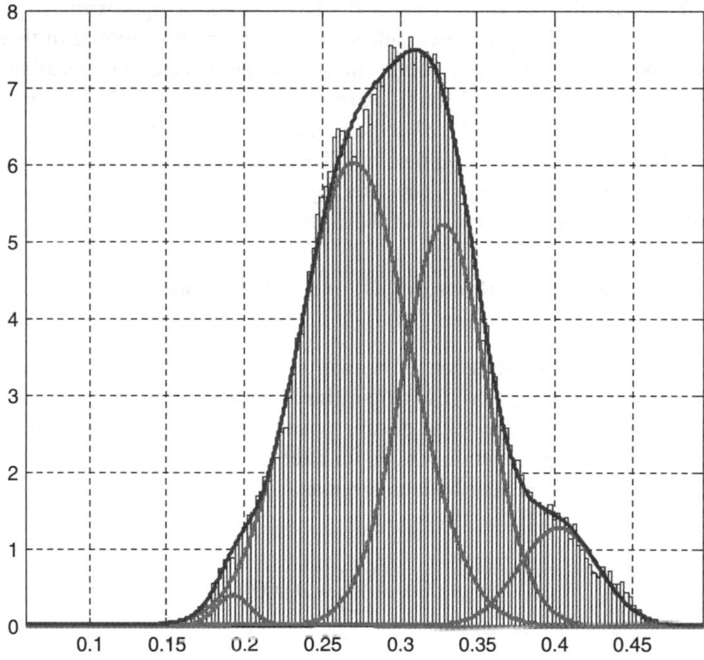

FIGURE 3.26   Fitting curves for bit "0" with MGD method.

TABLE 3.2

Validation of Accuracy of Proposed Novel Statistical Methods with the Discrepancies Compared to Monte-Carlo and Semi-analytical BER Estimation Methods

Evaluation Method-Input power	Monte Carlo Method	Semi-analytical Method	MGD Method	GPD Method
10 dBm	1.7e−5	2.58e−5	5.3e−6	3.56e−4
11 dBm		1.7e−8	2.58e−9	4.28e−8

MGD method consists of a set of Gaussian distributions which have short-tail characteristics. Thus, the MGD method tends to underestimate BER value. As can be observed from the table, the MGD method provides adequately accurate BERs within one decade below the Monte Carlo method and semi-analytical BERs.

GPD method provides precise estimation on extreme values (in this case, errors caused by NLPN). These extreme values are represented by the long-tail characteristic of GPD distributions. Thus, the GPD method tends to overestimate BER value. As can be observed from the table, the GPD method provides adequately accurate BERs within one decade above the Monte Carlo method and semi-analytical BERs.

As a guideline, BER value can be the average of BERs computed from the MGD and GPD methods.

3.9.7 Novel BER Statistical Techniques

The Gaussian-based single distribution involves only the effects of noise corruption on detected signals and ignores the dynamic distortion effects (e.g., ISI and non-linearity). These dynamic distortions result in a multi-peak pdf that is clearly overlooked by the conventional single distribution technique. As the result, the pdf of the electrical signal cannot be approximated accurately. The addressed issues can be resolved with two new statistical methods proposed in this section. Two new techniques proposed to accurately obtain the pdf of the detected electrical signal in optical communications include a mixture of multi-Gaussian distributions (MGD) by implementing the expectation maximization theory (EM) [27–29] and the generalized Pareto distribution (GPD) [29] of the generalized extreme values (GEV) theorem. These two techniques are well-known in statistics, banking, finance, and meteorology. They offer a great deal of flexibility, convenience, and rapid processing while maintaining the accuracy of the BER obtained.

3.9.7.1 Multi-Gaussian Distributions (MGD) and Expectation Maximization (EM) Theorem [30]

The mixture density parameter estimation problem is probably one of the most widely used applications of the expectation maximization (EM) algorithm. It comes from the fact that most deterministic distributions can be seen as the result of superposition of different multi distributions [31,32]. Given a probability distribution function $p(x \mid \Theta)$ for a set of received data, $p(x \mid \Theta)$ can be expressed as a superposition of M different distributions given as

$$p(x|\Theta) = \sum_{i=1}^{M} w_i p_i(x|\theta_i) \tag{3.135}$$

where the parameter are $\Theta = (w_1, \ldots, w_M, \theta_1, \ldots, \theta_M)$ such that $\sum_{i=1}^{M} w_i = 1$ and each p_i is a pdf by θ_i and each pdf having a weight w_i, i.e., the probability of that pdf.

As a particular case adopted for optical communications, the EM algorithm is implemented with a mixture of multi-Gaussian distributions (MGD). This method potentially offers accurate solutions for the evaluation of performance of optical transmission systems due to the following reasons: (i) In a linear optical system (low input power into fiber), the conventional single Gaussian distribution fails to take into account the waveform distortion caused by the ISI due to fiber CD and PMD dispersion, or patterning effects. Hence, the obtained BER is no longer accurate. These issues however are overcome by using the MGD method; and (ii) Computational time for implementing MGD via the EM algorithm is fast.

3.10 EFFECTS OF SOURCE LINEWIDTH

Direct detection differential quadrature phase-shift keying (DD/DQPSK) and other discrete and continuous phase modulation formats such as BDPSK, CPFSK, and MSK have been analyzed and studied in this book as possible modulation techniques for optical systems. In most literature, it is generally assumed that the impact of laser linewidth may be ignored. The aim of this section is to determine when we may neglect the effect of oscillator phase noise. In a DD/DQPSK system, a pair of Mach–Zehnder interferometers and balanced receivers is used to recover the in-phase and quadrature bits. This is illustrated schematically in Chapter 2. In an idealized system, the phase difference in the arms of the Mach–Zehnder is used to recover the phase difference of the two consecutive bits. However, phase noise from the finite linewidth laser will result in this being perturbed. Therefore, this effect is examined to determine when the linewidth of the laser can be ignored.

A 10-GBaud system (i.e., 2×10 Gb/s) is assumed in this section. However, the results can be easily scaled to other line rates. This section is a shortened version of the materials presented in ref [43] in which the bit-error-rate (BER) performance of DQPSK is analyzed using a quadratic decision variable [4]. In this section, a similar approach may be used for optical DQPSK systems. To simplify the analysis, the electrical low-pass filter can be replaced by an integrate-and-dump receiver, giving the decision variable d to within a constant factor of the received optical field as

$$d = \frac{1}{T_s} \int_0^{T_s} e^{j\theta} E_o^H(t - T_s) E_o^H(t) dt + \text{c.c.} \tag{3.136}$$

The superscript H denotes the Hermitian, and θ the phase offset is nominally $\pi/4$; however, it may be perturbed by laser phase noise such that $\theta = \pi/4 + \phi$. ϕ is the phase perturbation, and c.c. represents the complex conjugation. Over the symbol period T_s, the noise and signal may be expressed as a Fourier series, allowing the integral to be evaluated analytically [5]. If the preceding optical filter which has a bandwidth $B = 1/T_s$ can be assumed, then only the first Fourier component need be considered for each polarization. Consider receiving the symbol $(I_k, Q_k) = (1,1)$, such that within a constant factor, can be expressed as

$$d = \sum_{k=1}^{2} e^{j\theta} X_k X_k^* + \text{c.c.} \tag{3.137}$$

with X_k given as

$$X_1 = \sqrt{P} + n_1 + jn_2$$

$$X_2 = \sqrt{P} + n_3 + jn_4$$ (3.138)

$$X_2 = n_5 + jn_6$$

$$X_1 = n_7 + jn_8$$

where n_1–n_4 are the independent identically distributed components of filtered amplified spontaneous emission (ASE) in-phase and quadrature components in both polarizations, assumed to be Gaussian with zero mean. If the optical signal-to-noise ratio (OSNR) can be defined as the ratio of the mean power to the total ASE power measured in an optical bandwidth, then it follows that

$$\langle n_1^2 \rangle = \frac{P}{4\text{OSNR}}.$$ (3.139)

This definition for OSNR allows the use of published results for the quadratic decision variable [4] with the substitution that $\text{OSNR} = \gamma$ by following some algebra as detailed in ref [43], the conditional probability can be obtained as

$$\Pr ob(d < 0|\theta) = \left(\frac{\tan\theta}{2} + \frac{\sec\theta}{2} + \frac{\cot\theta}{2} \right) \frac{e^{-2\gamma(1-\sin\theta)}}{\sqrt{4\pi\gamma\sin\theta}}.$$ (3.140)

A conditional probability that combines with the probability density function of the laser phase noise for both the "1" and "0" allows the determination of the overall BER by evaluating the integral

$$\text{BER} = \frac{1}{2} \int_{-\infty}^{+\infty} [\Pr ob(d < 0|\phi) + \Pr ob(d < 0|-\phi)] p(\phi) d\phi.$$ (3.141)

For a laser with linewidth Δv, the phase noise may be modeled as a Gaussian process, with zero mean and variance. Using an asymptotic technique detailed in the Appendix of ref [38], a relatively simple equation for the BER in the presence of phase noise can be given as

$$\text{BER} = \frac{e^{-2\gamma(\sqrt{2}-2)}}{\sqrt{4\pi\gamma\sqrt{2}}} \left(\frac{3\sqrt{2}}{8} + \frac{1}{2} \right)(1 + \gamma^2\sigma_\phi^2).$$ (3.142)

To validate the approximations made, the analytical solution given by Equation 3.90 can be compared with a numerical evaluation of the integral Equation 3.89 using the condition probability given by Equation 3.88. The agreement is excellent for error rates lower than $1e - 3$, and this region is nearly linear (allowing easily determined for a given BER). By expanding as a Taylor series about $\gamma = 20$, a linear fit can be used to obtain the total BER as

$$\log_{10} \text{BER}(\gamma, \sigma_\phi) \approx -6.05 - 0.265(\gamma - 20)$$ (3.143)

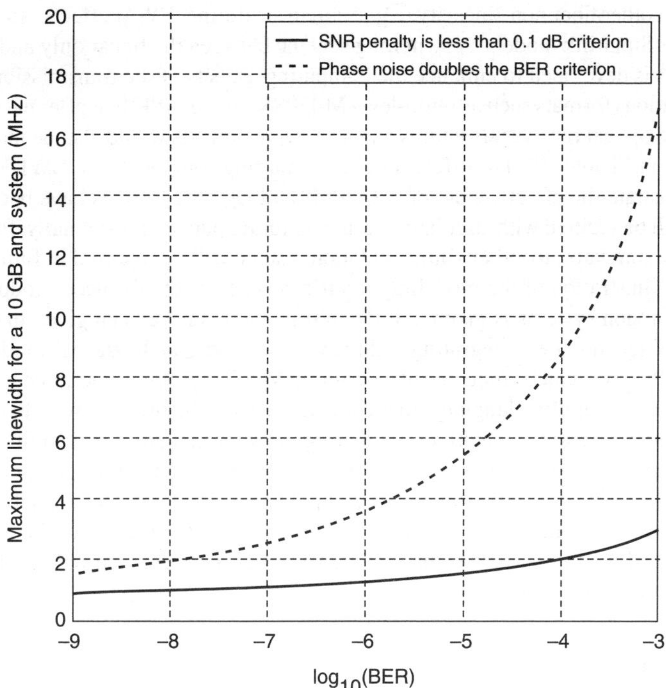

FIGURE 3.27 Criteria for neglecting linewidth in a 10-GBd system.

It is thus possible to consider the scenario under which the effect of the linewidth can be ignored. Two alternating criteria for neglecting the linewidth can be considered: first when the phase noise causes the BER to double, and second a 0.1 dB signal-to-noise-ratio (SNR) penalty. Figure 3.27 illustrates the required linewidth for both criteria for 10-Gb/s systems.

By considering Equation 3.91, it is observed that the BER is doubled when $\sigma_\phi^2 = 1/\gamma^2$. For a 10-GBd DQPSK system operating at a BER of (in the absence of phase noise), the laser linewidth should be <16 MHz. We note from Equation 3.93 that the doubling of BER from to implies an SNR penalty of 0.5 dB. To keep the penalty below 0.1 dB, a similar condition can be obtained. For a 10-GBd DQPSK system operating at a BER of 1e-3 under the Monte Carlo simulation environment, the laser linewidth should be <3 MHz so that the penalty is <0.1 dB. These linewidths are much wider than those of current practical DFB laser sources, and thus direct detection of different modulation formats can be under no significant influence of the laser linewidth.

3.11 REMARKS

An overview of the optical receiver and noise processes in the optical signal detection has been presented. We have also demonstrated the Simulink® modeling of amplitude and phase modulation formats at 40-Gb/s optical fiber transmission. A novel modified fiber propagation algorithm has been used to minimize the simulation processing time and optimize its accuracy. The principles of amplitude and phase modulation, encoding and photonic-opto-electronic balanced detection and receiving modules have been demonstrated via Simulink modules, and can be corroborated with experimental receiver sensitivities. Experimental results on DPSK formats are illustrated in detail in Chapter 5.

The XPM and other fiber non-linearity (e.g., Raman scattering, FWM effects) are not integrated in the MATLAB® Simulink models. A switching scheme between the linear only and the linear and non-linear models is developed to enhance the computing aspects of the transmission model.

Other modulations formats such as multi-level M-DPSK and M-ASK that offer narrower effective bandwidth, simple optical receiver structures and no chirping effects would also be integrated. These systems are given in Chapter 10. The effects of optical filtering components in DWDM transmission systems to demonstrate the effectiveness of the DPSK and DQPSK formats have been measured in this paper, and will be verified with simulation results in future publications. Finally, further development stages of the simulator, together with simulation results, will be reported in future works.

Furthermore, illustration of the modeling of various schemes of advanced modulation formats for optical transmission systems is given. Although transmitters have been given in Chapter 2, the models of transmitter modules integrating lightwaves, pre-coder and external modulators can be included to show that they can be modeled at ease under MATLAB Simulink. Because MATLAB is becoming a standard computing language for academic research institutions worldwide, the models reported herein contribute to the wealth of computing tools for modeling optical fiber transmission systems and teaching undergraduates at a senior level and postgraduate researchers. The models can integrate photonic filters or other photonic components using blocksets available in Simulink. Furthermore, we have used the developed models to assess the effectiveness of the models by evaluating the simulated results and experimental transmission performance of long-haul advanced modulation format transmission systems.

3.12 APPENDICES

3.12.1 SELLMEIR'S COEFFICIENTS FOR DIFFERENT CORE MATERIALS

3.12.2 Total Equivalent Electronic Noises

The principal goal of this section is to obtain an analytical expression of the noise spectral density equivalent to a source looking into the electronic amplifier, including the quantum shot noises of the photodetector. A general method for deriving the equivalent noise current at the input is by representing the electronic device by a Y-equivalent linear network (Figure 3.6). The two current noise sources $d(i_n')^2$ and $d(i_n'')^2$ represent the summation of all noise currents at the input and at the output of the Y-network. This can be transformed into a Y-circuit with the noise current at the input as shown below.

The output voltages V_0 of the two Figure 3.28 can be written as

$$V_0 = \frac{i_N'(Y_f - Y_m) + i_N''(Y_i + Y_f)}{Y_f(Y_m + Y_i + Y_o + Y_L) + Y_i(Y_o + Y_L)} \tag{3.144}$$

and for Figure 3.28b

$$V_0 = \frac{(i_N')_{eq}(Y_f - Y_m)}{Y_f(Y_m + Y_i + Y_o + Y_L) + Y_i(Y_o + Y_L)} \tag{3.145}$$

thus, comparing these two equations, we can deduce that the equivalent noise current at the input of the detector is

$$i_{Neq} = i_N' + i_N'' \frac{Y_i + Y_f}{Y_f - Y_m} \tag{3.146}$$

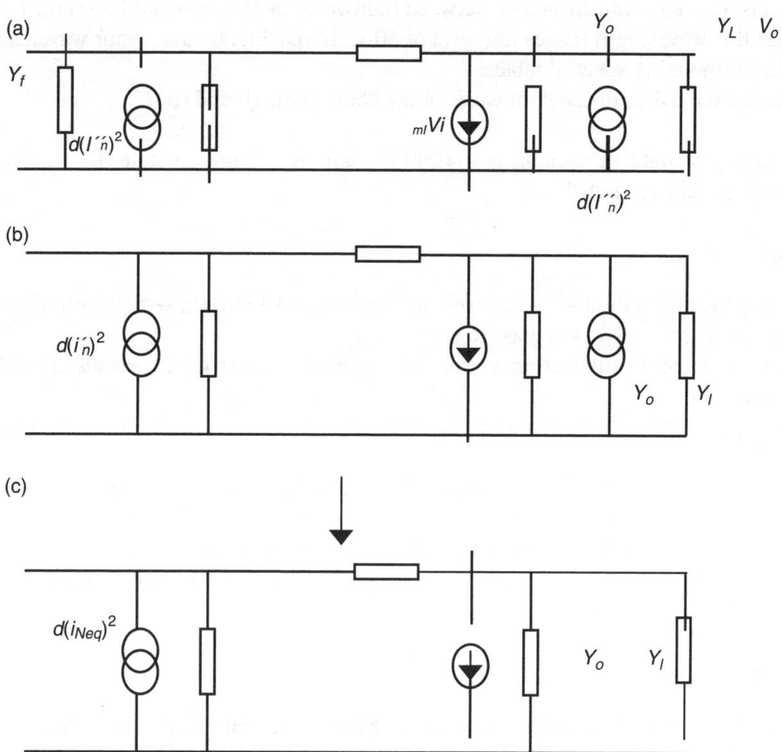

FIGURE 3.28   Small equivalent circuits including noise sources (a) Y-parameter model representing the ideal current model and all current noise sources at the input and output ports (b) with noise sources at the input and output ports and (c) with a total equivalent noise source at the input.

reverting to mean square generators for a noise source, we have

$$d\left(i_{Neq}\right)^2 = d\left(i_N'\right)^2 + d\left(i_N''\right)^2 \left|\frac{Y_i + Y_f}{Y_f - Y_m}\right|^2. \tag{3.147}$$

It is therefore expected that if the Y-matrix of the front-end low-noise amplifier is known, the equivalent noise at the input of the amplifier can be obtained using Equation 3.144.

3.13   PROBLEMS CHAPTER 3

QUESTION 1

(i)   What are the frequencies of lightwaves with wavelengths of 1300 nm and 1550 nm in vacuum?

(ii)   What are the frequencies and wavelengths of these lightwaves if they are propagating in optical fibres with a refractive index of 1.448 at 1330 nm and 1.446 at 1550 nm? You may assume single-mode operation of the lightwaves at these wavelengths.

(iii)   What is the frequency difference between these two lightwaves?

(iv) What is the frequency difference between lightwaves of 1530 nm and 1560 nm? If we now divide this wavelength ranges into grid of 50-GHz spacing, tabulate your wavelength grid and corresponding spectral table.

(v) What are the colors of the lightwaves of the above parts (i) and (iv)?

NOTE: Velocity of light in vacuum is $c = 299{,}792{,}458$ m/s. Can you state the reasons why this velocity must be very accurate?

Question 2

(i) What is the attenuation of lightwaves of 1300-nm and 1550-nm wavelength when propagating through silica-based glass fibers?

(ii) What is the peak attenuation within this spectral range? Give reasons for this peak attenuation.

(iii) Could you obtain a relationship between the loss or attenuation as a function of wavelength?

(iv) Examine Corning single-mode fibers SMF-28 and SMF-28e (available on email request) and spot the differences between them? Give reasons why Corning are spending time and money to develop SMF-28e for metropolitan optical networks.

(v) Write down the frequency responses of a SSMF optical fiber, including the phase and amplitude frequency responses.

Question 3

(i) Examine the standard single-mode fibers SMF-28 Corning type, and write down the principal parameters related to the:
   a. geometrical aspects of the fiber
   b. refractive index distribution
   c. operation of the fibers when modulated lightwaves are propagating
   d. impairments that may cause signal distortion.

(ii) Repeat (i) for Corning LEAF fibers

(iii) Repeat (i) for Corning dispersion-compensating fibres which are available on several websites (e.g., Avenex.com, JDSU.com, or Corning.com (registration required))

(iv) Examine the dispersion factors given in these fibers. Why are there differences between them? Calculate the temporal broadening of lightwave pulse sequence of 10 Gb/s bit rate with lightwave carrier wavelength at 1550 nm, and similarly at 1560 nm.

Question 4

The wavelength channels in Question 2 part (iv) are used as optical carriers for optical transmission systems which are modulated with data rate of 10 Gb/s using on-off-keying non-return-to-zero format.

(i) Sketch the frequency spectrum of a single optical channel whose carrier wavelength is in the region of 1550 nm.

(ii) Thence sketch the spectrum of all possible consecutive multiplexed optical channels in the spectral region 1530–1565 nm.

(iii) Repeat (ii) for data rate of 100 Gb/s. Comment on the spectrum of this part.

NOTE: You should refer to the materials on Digital Communications delivered in the last semester. You may also refer to the table of Fourier Transform accompany this problem set.

QUESTION 5: OPTICAL FIBERS

Refer to the technical data of the following fibers:

(i) Corning SMF-28, SMF-28e, and LEAF, and extract the following information:
   - Geometrical parameters: core diameter, cladding diameter, relative refractive index difference, attenuation versus wavelength, core eccentricity, ITU standard of the fibers.
   - Operational parameters: mode field diameter, effective area, cut-off wavelength or wavelength range (why range?), dispersion factor and its variation as a function of wavelength, dispersion slope thence deriving the values for beta 2 and beta3, non-linear coefficient, polarization mode dispersion factor.

(ii) Compare the parameters of these different fibers and give some applications in optical transmission systems, stating reasons. Why is a large effective area LEAF fiber is preferred for long-haul transmission system?

(iii) Consider the dispersion factor of LEAF fiber. Explain why the fiber is designed so that its zero dispersion is outside the operational spectrum of C- or S-band.

Transform Pairs

FIGURE QUESTION 6: Frequency responses of a FBG filter: Bild (a) Frequency response- power versus wavelength. Bild (b) Group delay response.

OPTICAL FILTERS, MUXES AND DEMUXES

QUESTION 6

Wavelength division multiplexing technique is a common method for increasing the total transmission capacity over a single stand of optical fiber. It is now widely employed in long-haul high-speed optical transmission systems worldwide.

(i) Briefly explain the WDM techniques. State the typical wavelength or frequency spacing between the channels so that it can be named dense wavelength division multiplexing (DWDM) optical transmission systems.

(ii) What are the frequencies and wavelengths of these lightwaves so that they can be transmitted with minimum loss? You may consider the spectral regions called S- C- and L- bands of a silica-based single-mode optical fibers.

(iii) Sketch the structure of a long-haul optical transmission system using multiple spans with optical fibers, optical amplifiers, dispersion compensating fibers, optical filters as multiplexers and demultiplexers, optical transmitters and receivers.

(iv) *Could you search the Internet and find at least one array waveguide multiplexer that would satisfy your DWDM optical transmission system?

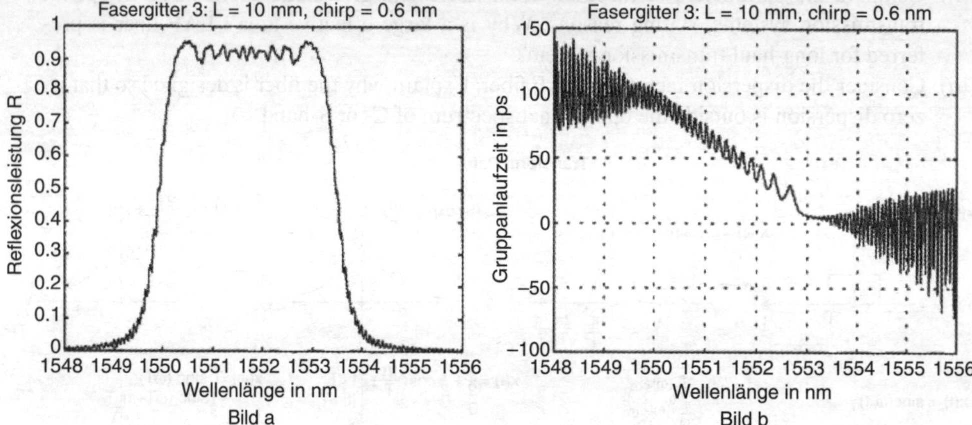

NOTE: Velocity of light in vacuum is $c = 299{,}792{,}458$ m/s. Can you state the reasons why this velocity must be very accurate?

QUESTION 7: OPTICAL FILTER USING MACH–ZEHNDER INTERFEROMETERS

(i) Sketch the basic structure of a symmetric and an asymmetric Mach–Zehnder interferometer.

(ii) Obtain the optical intensity transmittance between an input port to the other two output ports, e.g., T_{11} and T_{12}. Show that a complete power transfer can be achieved with a total phase difference of pi. You may select to state the relationship of the transmittance or derive the equations.

(iii) You could use the z-transform to obtain the transmittance of the filter and thence the continuous transmittance with an appropriate substitution of the z-variable. Determine the zeroes and poles of the transmittance. Sketch the poles and zeroes on the z-plane. How close to the unit circle should be the zeroes of the transfer function so that the filter passband can provide the highest filtering property. Determine the frequency of the cut-off band (−20 dB from the 3-dB roll-off). Sketch the filter passband. Explain the physical meaning of the zeroes and poles of the transfer function.

(iv) With and without using the equations, can you explain the mechanism of the coupling and filtering of optical waves?

(v) Now consider silica on silicon interferometric filters of the above parts. The effective refractive index of the single-mode waveguide of the filter is assumed to be 1.4465 at 1550 nm. Estimate the propagation constant β of the guided wave. Hence estimate the difference length of an asymmetric interferometer so that a complete filtered channel can be achieved at the cross output port.

(vi) If there are some fabrication errors of the path difference, could you suggest a method so that the filter can be tuned to the exact filtering property?

(vii) If the delay difference path is now set to be one-bit long for a 10-Gb/s phase modulation system, suggest a function of this interferometer. Now suppose two consecutive bits of a data sequence of 10 Gb/s whose optical carrier are "0" and "pi" phase or "pi" and "pi" or "0" and "0" phase. Work out the intensity of the outputs of the interferometer under these receiving conditions.

(viii) Determine the length of the delay path of the interferometer so that a one-bit delay can be achieved for part (i).

QUESTION 8: RING RESONATOR FILTER

(i) Refer to the ring resonator given in the lecture set 6. State the power transfer function $H_{11}(z)$ or $H_{11}(\omega)$, including the frequency responses of its power amplitude, its phase, its group delay and dispersion. Use MATLAB® to plot these frequency responses. What is the resonant frequency and the 3-dB bandwidth of the filter, hence its Q-factor? If the resonant frequency is to be set at the equivalent vacuum frequency of the wavelength of 1550 nm, what is the length of the ring resonator? The effective refractive index of a guided wave at 1550 nm of the ring resonator is 1.448.

(ii) Thence by cascading at least two ring resonators, obtain the frequency responses for the power amplitude and phase. This part is best achieved by using the $H_{11}(z)$, i.e., the transfer function would have multiple identical poles. Hence obtain its corresponding group delay and dispersion at the resonant wavelength.

(iii) *How close to the unit circle should be the poles of the transfer function so that the filter passband can provide the highest filtering property? Determine the frequency of the cut-off band (–20dB from the 3-dB roll-off). Sketch the filter passband. Explain the physical meaning of the poles of the transfer function.

(iv) * If there are losses of lightwaves propagating in the ring, how you would correct this loss so that the poles of the resonator can be set close to the unit circle?

(v) Now the frequency of the input lightwaves can be shifted to the right or to the left of the resonant peak frequency. Obtain the dispersion factor at these edges of the filter passband.

(vi) If the ring resonator is used as a dispersion compensator of 100 km optical fiber of standard SMF (about +17 ps/nm/km depending on the wavelength), determine the wavelength at which the resonator should be operating so that a complete compensation can be achieved.

(vii) Comment on the stability of the optical ring resonator and that of the interferometric filter of Question 2.

QUESTION 9

Referring to the lecture notes on optical filters (Section 3.6), the frequency response of the power transmittance of a FabryPerot optical filter is given by:

$$G_{PF} = \cfrac{1}{1 + \left[\cfrac{2\sqrt{R}}{1-R} \sin\left(\cfrac{2\pi nL}{c} f \right) \right]^2}.$$

(i) Sketch the structure of the Fabry Perot filter and indicate all the parameters of the above equation in this diagram.

(ii) State the assumptions and approximations for the derivation of this transfer function.

(iii) Determine the frequencies or wavelengths of the resonant peaks, hence the free-spectral range as a function of n, L and R. Sketch the frequency response of the power transfer between the input and output ports of the filter.

(iv) Determine the 3-dB bandwidth of the filter as a function of n, L and R, thence the Q-factor of the filter and its finesse. Can you determine the cut-off frequency of the filter passband at which the power is −20 dB from the 3-dB roll-off point?

(v) Tunable filter. The center frequency of the FP filter can be tuned so that it can be matched with the centre wavelength of a received data channel. Which parameter can be changed so that the center passband of the filter can be tuned?

(vi) If the FP filter is used for filtering noises of the optical amplifier, the ASE noise, determine the finesse of the filter for ASE noise reduction at a center wavelength of 1565 nm and a bandwidth of 0.1 nm.

QUESTION 10: ARRAY WAVEGUIDE GRATINGS (AWG)

(i) Based on information provided in the lecture notes and manufacturers website (e.g., www. avenex.com, www.jdsu.com) sketch the structure of an integrated optic AWG. State the physical mechanism of the filtering property of the AWG. You may note the path length difference between waveguide channels, hence the interference effects.

(ii) Provide the operation parameters such as center wavelength or frequency spacing, number of channels, insertion loss, and passband cut-off wavelength for an AWG which can be used as a multiplexer or demultiplexer in a 50-GHz spacing DWDM optical transmission system.

(iii) If the average power at the output pig-tail of an optical modulator of an optical transmitter is −10 dBm is launched into the AWG multiplexer of (ii) and a total average power into the transmission fiber is 10 dBm for 44 optical channels is required, find out whether there needs of optical amplification. If it does then what is the gain of the optical amplifier is required?

QUESTION 11: FIBER BRAGG GRATINGS AS OPTICAL FILTERS AND DISPERSION-COMPENSATING DEVICE

Consider the passband reflectance and its group delay characteristics of a FBG as shown in Figure Question 6 below:

(i) What is the center wavelength of the passband and the 3 dB bandwidth (passband) of the filter? Note that the transfer characteristic is plotted in linear scale.

(ii) The group delay is measured using a tunable laser source launched into one end of the FBG. Determine the average dispersion factor of the FBG in ps/nm. Note the sign of the dispersion factor. If you want to reverse the sign of the dispersion, suggest what could you do to the FBG to achieve this. Give your reasons.

(iii) If this FBG is used for compensating a length of standard SMF, determine the length of the fiber.

(iv) Comment on the ripple of the group delay and its effects on the total dispersion characteristics when it is used for dispersion compensation.

(v) Obtain the transmittance of the FBG.

(vi) Show how you can connect the FBG to an inline optical system to compensate for any residual dispersion of the fiber transmission line. You are suggested to use the FBG together with an optical circulator.

QUESTION 12: OPTICAL COUPLER

A 2×2 coupler has an intensity coupling coefficient ε of 0.5. This is called a 3-dB coupler.

(i) write down the transmittance matrix of the coupler, thence

(ii) write down the relationship between the output optical fields and the input fields for the coupler

(iii) sketch the signal flow model of the coupler

QUESTION 13

Consider a Mach–Zehnder interferometric filter as shown below.

(i) Write down the transmittance matrices for the two couplers and the middle section.
(ii) Hence write down its relationship with the input and output fields at the input ports 1,2 and the output ports 1,2.
(iii) Sketch the signal flow graph of the filter, and thence show that for a delay length ΔL which is equivalent to the traveling time of a bit period the output field E_{01} and E_{02} and its input field E_{i1} can be written as:

$$E_{01}(Z) = H_{11}(Z)E_{i1}(Z) = (1 + Z^{-1})E_{i1}(Z) = Z^{-\frac{1}{2}}\left[Z^{+\frac{1}{2}} + Z^{-\frac{1}{2}}\right]E_{i1}(Z)$$

where $z = \exp(j\beta\Delta L) = \exp(j\omega t)$

(iv) Plot the zeroes and poles of the transfer function on the z-plane and hence the transmittance frequency responses of the filter.

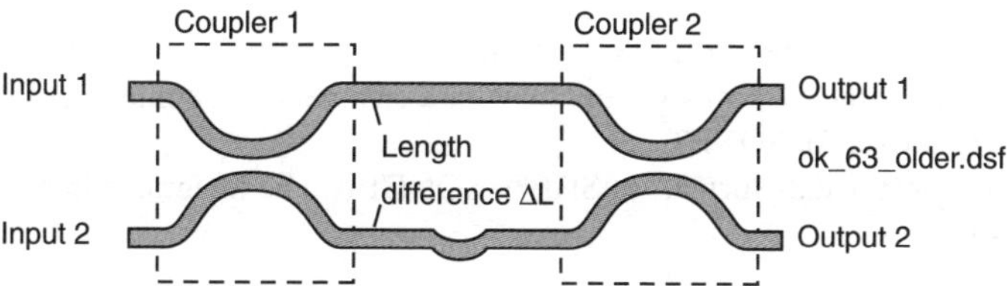

OPTICAL ADD/DROP MUXES

QUESTION 14

Wavelength division multiplexing (WDM) and dense WDM networks are now installed throughout global communication backbone networks.

(i) Define the wavelength or frequency spacing between channels for dense and coarse WDM systems and networks.
(ii) Give reasons why Corning fibers SMF-28e and SMF-28 (ITU G.652 standard) are usually selected for CWDM and DWDM in metropolitan networks. What is the main difference between these types of fibers?
(iii) What are the typical wavelengths that can be used for CWDM? Give reasons. Hint: consider down loading and uploading speed and refer to lecture notes page 7–14.
(iv) Refer to the Problem Set 2 and lecture notes (Section 3.7) and specify the spectral regions for the S-, C- and L- bands for silica-based optical fibers.
(v) For the transpacific transmission systems with a total capacity of 1,806 Gb/s in 2006 and with a 10 Gb/s wavelength channel, how many wavelength carriers are required to carry this capacity? If they are to be transmitted in the C-band, then determine the frequency spacing between the channels.
(vi) Sketch a DWDM optical transmission system with optical transmitters, multiplexers, amplifiers, dispersion compensators and demultiplexers and receivers. Indicate in your system if the signals are in the optical or electronic domain.

Question 15

(i) What are DXC, ADM and LTE in an optical network? You may explain with a sketch of the structure of an optical network.

(ii) Sketch the structure of cascade Mach–Zehnder interferometers that would form 2-channel and 4-channel wavelength demultiplxers. If the wavelength channels are separated by 50 GHz, determine the delay length ΔL.

NOTE: the reference BH. Verbeeck et al., "Integrated four-channel Mach–Zehnder Multi/Demulti-plexer fabricated with Phosphorous Doped SiO_2 Waveguide on silicon", IEEE J. Lightwwave Tech. The condition for splitting the wavelength channels 1 and 2 to the two output ports are given as

$$2\pi n_{\text{eff}} \Delta L \left(\frac{1}{\lambda_1} - \frac{1}{\lambda_2} \right) = \pi.$$

With n_{eff} is the guided refractive index of guided mode and λ_1 and λ_2 are the wavelength of two consecutive channels.

Question 16

(i) Sketch a possible structure of the OSI layer model

(ii) Give a structure of the SHF model, and how data channel can be integrated from STM-1 to STM-16. Likewise for SONET.

(iii) Show clearly how the data formats of SHF or SONET can be integrated into optical transmission system.

Question 17: Optical Switching

(i) Sketch a structure of an 8×8 switching matrix for Benes, strict sense nonblocking and rearrangeably nonblocking architectures. Comment on the advantages and disadvantages of these structures.

(ii) What are the typical switching speed of the optical switches of part (i)? Compare their speed with that of electromechanical switches.

Question 18

A wavelength router can be designed using a wavelength converter and an optical demultiplexer. A wavelength converter is usually formed in a semiconductor optical amplifier (SOA).

(i) Sketch the structure of this wavelength router

(ii) Sketch the structure of a wavelength ADD/DROP multiplexer.

(iii) Can you use the wavelength router of part (i) in the design of the wavelength add/drop of part (II)?

Question 18(b): Inter-channel and Intra-Channel Crosstalk Penalty

(i) Consider an optical add/drop multiplexer as shown in the figure shown below
If –20 dBm of optical power of the channel of wavelength next to the channel λ_2 is present in the channel, estimate the penalty of the eye diagram due o this inter-channel cross talk. The average power of the channel λ_2 is –10 dBm.

If the same amount of optical power of the optical channel on the left of the operating channel is residual in this channel, re-estimate to eye penalty.

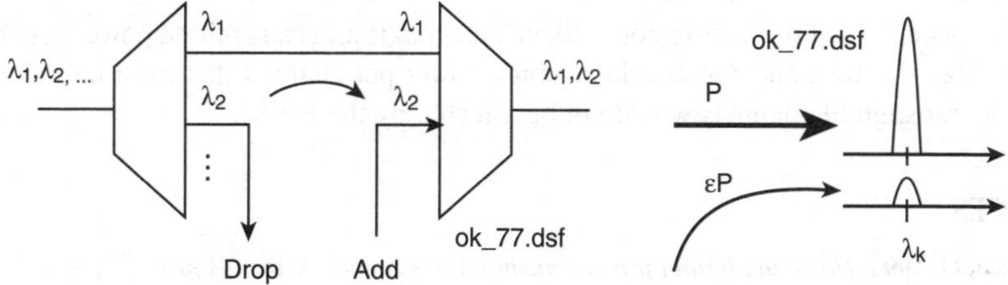

(iii) A DWDM transmission system with an inter-channel spacing of 50 GHz and an average optical power of each channel is –10dBm. The C-band has been filled with all channels. Estimate the number of channels that are contained in the C-band 1530 nm to 1565 nm. The data bit rate is 40 Gb/s and the modulation is return-to-zero ASK.

(iv) An optical filter is now used to extract the channel of wavelength 1552.93 nm which has the following characteristics: center wavelength of 1552.93, optical bandwidth of 40 GHz (3 dB), a roll-off of –12 dB per decade.

   (a) Sketch the spectral distribution of all data channels.

   (b) Sketch the filter frequency response.

   (c) Estimate the power residual of the other wavelength channel on to this channel and hence estimate the eye penalty due to this cross talk. What kind of cross talk is this, inter-channel or intra-channel? You may assume the pulse shape of the data bits being perfect rectangular or Gaussian.

   (d) What can you propose to reduce to power penalty to the eye opening?

      HINT: You may refer to the theoretical analyses given on pages 7-6 of the lecture notes or page 303 of the book by R. Ramaswami, K. N. Sivarajan, "Optical Networks".

QUESTION 19: SELF PHASE MODULATION EFFECTS

The non-linear coefficient due to the intensity of lightwave propagating in a silica fiber is given by $n_2 = 2.5 \times 10^{-20}$ m²/W. A Gaussian-guided mode lightwave is propagating through a length L of standard single-mode optical fiber (SSMF). Its average optical power is 10 dBm.

(i) What is the effective area of the SSMF, i.e., the ITU Standard G.652?

(ii) Thence what is the intensity imposed on the cross section of the SSMF?

(iii) The change of the refractive index of the silica fiber can be given by $\Delta n_{NL} = n_2 I$. Estimate this change of the refractive index of the fiber due to the propagation of this guided mode.

(iv) Thence obtain the change of the phase of the lightwave carrier at 1550 nm after propagating through a length L (km). Hence estimate the fiber length so that the nonlinear phase change is less than or equal to 0.1 radians.

(v) (44) Refer to the lecture notes, write down the Schrödinger wave equation in which the nonlinear effects such as the self-phase modulation is included. Ensure that you state clearly the meaning of all variables and notation of this equation. Refer to lecture notes page 7–9. By using a diagram, explain the split-step Fourier method that can be used to represent and model the propagation of the complex envelope of the lightwave through an optical guided fiber/waveguide.

QUESTION 20: NON-LINEAR FOUR-WAVE MIXING

(i) Give a brief account of four-wave mixing. Indicate clearly the mixing of the waves and the generated wavelength channel.

(ii) Give a condition of the fiber dispersion so that there are significant impacts of FWM on transmission systems. Sketch the generated waves that interfere with other DWDM channels.

(iii) Refer to page 7–13 of the lecture notes, Sketch the maximum transmitted power per channel versus fiber distance and specify the average power per channel that must be satisfied so that the transmitted channels would not be affected by the FWM.

REFERENCES

1. Jacobsen, G. 1994. *Noise in digital optical transmission systems.* Artech House.
2. ITU-REcommendationG.692Optical interface for multichannel systems with optical amplifiers. Technical report.
3. Shimada, H. 1989. Coherent lightwave communications technology.
4. Huynh, T. L., T. Sivahumaran, L. N. Binh, and K. Pang. 2007. Sensitivity improvement with offset filtering in optical MSK narrowband frequency discrimination receiver. *ACOFT – COIN 2007*, Melbourne, Australia.
5. Agrawal, G. P. 2002. *Fiber optic communications systems.* 2nd ed. New York: Wiley.
6. Binh, L. N., and S. V. Chung. 1996. Generalized approach to single-mode dispersion-modified optical fiber design. *Optical Engineering* 35 (08): 2250–61.
7. λ_{ZD} is defined as the wavelength at which the material dispersion factor $D_M(\lambda) = 0$.
8. Binh, L. N., T. L. H. Huynh, K.-Y. Chin, and D. Sharma. 2004. Design of dispersion flattened and compensating fibers for dispersion-managed optical communications systems. *International Journal of Wireless and Optical Communications* 2 (1): 63–81.
9. Kaminov, I., and T. Koch. Vol. IIIa of *Optical fiber communications.* San Diego: Academic Press.
10. Agrawal, G. P. 2002. *Nonlinear fiber optics.* Boston, MA: Academic Press.
11. Agrawal, G. P. 2001. *Nonlinear fiber optics.* New Jersey, USA: Wiley.
12. Elrefaire, A. F., R. E. Wagner, D. A. Atlas, and D. G. Laut. 1988. Chromatic dispersion limitations in coherent lightwave transmission systems. *IEEE Journal of Lightwave Technology* 6 (5): 704–9.
13. Leibrich, J., C. Wree, and W. Rosenkranz. 2002. CF_RZ-DPSK for suppression of XPM on dispersion managed long haul optical WDM transmission on standard optical fiber. *IEEE Photonic Technology Letters* 14: 155–57.
14. Agrawal, G. P. 2002. *Nonlinear fiber optics.* New York: Academic Press.
15. Chang, K. F., and L. N. Binh. 2002. A platform for multi-channel transmission systems frequency domain using Volterra series transfer function. *Proceedings of the First International Conference on Optical Networks*, Singapore, November 2002.
16. Bierlaire, M., D. Bolduc, D. McFadden. "Characteristics of generalized extreme value distributions. Technical report, http://elsa.berkeley.edu/wp/mcfadden0403.pdf, accessed 10 October 2007.
17. http://www.mathworks.com/products/statistics/demos.html?file=/products/demos/shipping/stats/gevdemo.html, accessed 10 October 2007.
18. Markose, S., and A. Alentorn. The generalized extreme value (GEV) distribution, implied tail index and option pricing. http://www.essex.ac.uk/economics/discussion-papers/papers-text/dp594.pdf, accessed 10 October 2007.
19. Kotz, S., and S. Nadarajah. 2000. *Extreme value distributions: theory and applications.* London: ICP Imperial College Press.
20. Ultra-high speed optical transmitters and receivers, www.shf.de
21. Newman, M. E. J., and G. T. Barkema. 1999. *Monte Carlo methods in statistical physics.* Oxford, UK: Oxford Univ. Press.
22. Jacobsen. 2004. *Noise in digital optical communications systems.* Norwood, MA, USA: Artech House.
23. Davidson, A. C. 1990. Models for exceedances over high thresholds. *Journal of Royal Statistics Society* B52: 393–442.
24. Smith, R. L., and I. Weissman. 1994. Estimating the extremal index. *Journal of Royal Statistics Society* B56: 515–28.
25. Smith, R. L. 1989. *Handbook of applicable methematics; extreme value theory.* Hoboken, NJ, USA: Wiley.
26. MATLAB Helpdesk, *Statistical toolbox, generalized pareto distribution,* http://www.mathworks.com/access/helpdesk/help/toolbox/stats, accessed 15 September 2007.

27. Sean Borman. The expectation maximization algorithm: A short tutorial, http://www.seanborman.com/publications/EM_algorithm.pdf, accessed December 2006.

28. Fayyad, U., P. S. Bradley, and C. Reina. Scalable system for expectation maximization clustering of large databases. United States Patent 6263337 accessed December 2007.

29. Jun Sakuma and Shigenobu Kobayashi. Non-parametric expectation-maximization for gaussian mixtures. In *Proceedings of the 9th International Conference on Neural Information Processing (ICONIP'OZ)*, vol. I, 2002, 517–22.

30. Mak M. W., S. Y. Kung, and S. H. Lin., Expectation-maximization theory, sample chapter, Prentice Hall professional technical reference, Jan 3, 2005.

31. Emery, X., and J. M. Ortiz. 2005. Histogram and variogram inference in the multi-Gaussian model. *Journal of Stochastic Environmental Research and Risk Assessment (SERRA)*, 19 (1): 48–58.

32. Wang, Y., F. J. Doyle, III. 2004. Reachability of particle size distribution in semibatch emulsion. *Journal of Polymerization Particle Technology and Fluidization, AIChE Journal* 50 (12): 3049–59.

33. Statistics of weather and climate extremes at http://www.isse.ucar.edu/extremevalues/extreme.html, accessed December 2007.

34. Bierlaire, M. Generalized extreme value model at http://roso.epfl.ch/mbi/papers/discretechoice/node16.html, accessed December 2007.

35. Holmes, J. 2003. Discussion on generalized extreme gust wind speeds distributions. *Journal of Wind Engineering and Industrial Aerodynamics* 91 (7): 965–67.

36. Brabson, B. B., and J. P. Palutikof. 2000. Test of the generalized Pareto distribution for predicting extreme wind speed. *Journal of Applied Meteorology* 39: 1627–40.

37. Glynn, E. F. 2007. *Mixtures of Gaussians.* Stowers Institute for Medical Research, http://research.stowers-institute.org/efg/R/Statistics/MixturesOfDistributions/index.htm, accessed December 2007.

38. Schneider, T. 2000. Analysis of incomplete climate data: Estimation of mean values and covariance matrices and imputation of missing values. *Journal of Climate* 14: 853–71.

39. Bomhoff, E. J. 1995. *Financial forecasting for business and economics.* Spiral ed., San Diego, CA: Academic Press.

40. Statistics Toolbox 6.0, Modeling Tail Data with the Generalized Pareto Distribution, http://www.mathworks.com/products/statistics/demos.html?file=/products/demos/shipping/stats/gparetodemo.html, accessed December 2007.

41. Bhattacharya, K. K., and J. P. Sethna. 1998. Multicanonical methods, molecular dynamics, and Monte Carlo Methods: Comparison for Lennard–Jones glasses. *Physics Review E* 57 (3): 2553–62.

42. Mathworks, 2007. Matrix House, Cambridge Business Park, Cambridge, CB4 0HH, United Kingdom.

43. Savory, S., and A. Hadjifotiou. 2004. Laser linewidth requirements for optical DQPSK systems. *IEEE Photonics Technology Letters* 16 (3): 930–33.

44. This part is for high distinction level students.

4 Coherent Photonic Transmission

4.1 COHERENT DETECTION

Optical coherent detection can be distinguished by the "demodulation" scheme in communications techniques in association with the following definitions: (i) coherent detection is the mixing between two lightwaves or optical carriers, one is information bearing-lightwaves and the other a local oscillator with an average energy much larger than that of the signals; and (ii) demodulation refers to the recovery of baseband signals from the electrical signals.

A typical schematic diagram of a coherent optical communications employing guided wave medium and components is shown in Figure 4.1, in which a narrow-band laser incorporating an optical isolator cascaded with an external modulator is usually the optical transmitter. Information is fed via a microwave power amplifier to an integrated optic modulator, commonly $LiNbO_3$ or EA types. The coherent detection is the principal feature of coherent optical communications, which can be further distinguished by heterodyne and homodyne techniques depending if there is a difference between the frequencies of the local oscillator (LO) and that of the carrier of the signals. A LO is a laser source whose frequency can be tuned and is approximately equivalent to a monochromatic source. A polarization controller would also be used to match its polarization with that of the information-bearing carrier. The LO and the transmitted signal are mixed via a polarization-maintaining coupler and then detected by a coherent optical receiver. Most of previous coherent detection schemes are implemented in a mixture of photonic domain and electronic/microwave domain.

Coherent optical transmission has become the focus of research again. One significant advantage is the preservation of all the information of the optical field during detection, leading to enhanced possibilities for optical multi-level modulation. This article investigates the generation of optical multi-level modulation signals. Several possible structures of optical M-PSK and M-QAM transmitters are theoretically analyzed. Differences in the optical transmitter configuration and the electrical driving lead to different properties of the optical multi-level modulation signals. This is shown by deriving general expressions applicable to every M-ary-PSK and M-ary-QAM modulation format, and exemplarily clarified for square-16-QAM modulation.

Synchronous detection requires an optical phase lock loop (OPLL) which recovers the phase and frequency of the received signals to lock the LO to that of the signal, so as to measure the absolute phase and frequency of the signals relative to that of the LO. Thus, synchronous receivers allow direct mixing of the bandpass signals and the base band; this technique is termed "homodyne reception". For asynchronous receivers, the frequency of the LO is approximately the same as that of the receiving signals, and no OPLL is required. In general, the optical signals are first mixed with an intermediate frequency (IF) oscillator which is about 2–3 times that of the 3-dB passband (or bit rate). The electronic signals can then be recovered using electrical PLL at lower carrier frequency in the electrical domain. The mixing of the signals and a LO of an IF frequency is termed "heterodyne reception".

If no LO is used for demodulating the digital optical signals, then differential or self-homodyne reception may be utilized, which is classically termed an "autocorrelation reception process".

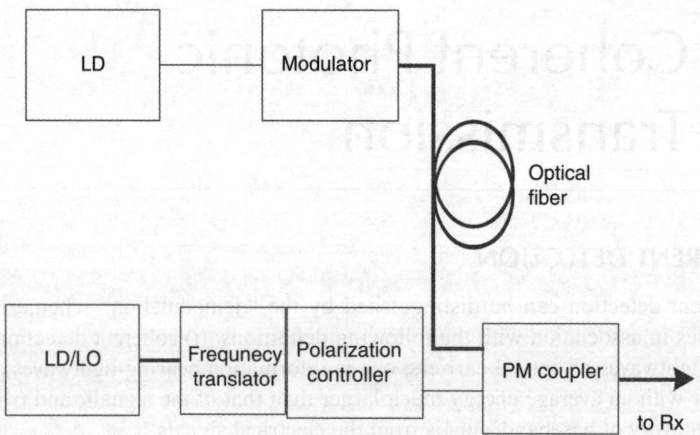

FIGURE 4.1 Typical arrangement of coherent optical communications systems [3, 24] LD/LC is a very narrow linewidth laser diode as a local oscillator. PM is polarization-maintaining fiber coupler.

Coherent systems were important techniques in the late 1980s and the early 1990s, but research was interrupted with the advent of optical amplifiers in the late 1990s that offered up to 20 dB gain without difficulty. Coherent systems have had a resurgence of interest due to the availability of digital signal processing and low-priced components, and partly relaxed receiver requirements at high data rates. Preservation of the temporal phase of coherent detection enables new methods for adaptive electronic compensation of chromatic dispersion (CD). With regard to WDM systems, coherent receivers offer tunability and allow channel separation via steep electrical filtering. Furthermore, only coherent detection permits convergence to the ultimate limits of spectral efficiency. To reach higher spectral efficiencies, multi-level modulation is required. Coherent systems are also beneficial, because all the information of the optical field is available in the electrical domain. Hence, complex optical demodulation with interferometric detection—which must be used in direct detection systems—can be avoided and the complexity transferred from the optical to the electrical domain. Several different modulation formats based on the modulation of all four quadratures of the optical field were proposed in the early 1990s, describing the possible transmitter and receiver structures and calculating the theoretical BER performance. However, a more detailed and practical investigation of multi-level modulation coherent optical systems for present networks and data rates is missing.

Coherent reception has attracted significant interest due to the following: (i) The received signals of coherent optical receivers are in the electrical domain, which are proportional to that in the optical domain. This, in contrast with direct detection receivers, allows exact electrical equalization or exact phase estimation of the optical signals. (ii) Using heterodyne receivers, DWDM channels can be separated in the electrical domain by using electrical filters with sharp roll-off of the passband to the cut-off band. Present availability of ultra-high sampling rate digital signal processors (DSP) allows users to conduct filtering in the DSP in which the filtering can be easily changed.

However, coherent receivers have disadvantages: (i) Coherent receivers are polarization-sensitive, necessitating polarization tracking at the front-end of the receiver; (ii) Homodyne receivers require OPLL and electrical PLL for heterodyne detection that would require control and feedback circuitry (optical or electrical), which may be complicated; and (iii) For differential detection the compensation may be complicated due to the nature of differentiation receiving.

In later chapters, when advanced modulation formats are presented for optically amplified transmission systems, photonic components are extensively exploited for advanced technology of integrated optics. Modulation formats of signals depend on whether the amplitude, phase or frequency of the carrier is manipulated, as mentioned in Chapter 2. In this chapter, the detection is coherently converted to the IF range in the electrical domain and the signal envelope. The down-converted carrier signals are detected and recovered. Binary level and multilevel modulations schemes employing amplitude, phase and frequency shift keying modulation are described in this chapter.

4.1.1 Optical Heterodyne Detection

The basic configuration of optical heterodyne detection is shown in Figure 4.2. The local oscillator, whose frequency can be higher or lower than that of the carrier, is mixed with the information-bearing carrier, thereby allowing down- or up-conversion of the information signals to the IF range. The down-converted electrical carrier and signal envelope is received by the photodetector. This combined lightwave is converted by the photodiode into electronic current signals, which are filtered by an electrical bandpass filter (BPF) and then demodulated by a demodulator. A low-pass filter is also used to remove higher-order harmonics of the non-linear detection photodetection process ("square-law detection"). In an envelope detector, the process is asynchronous, hence the term "asynchronous detection". If the down-converted carrier is recovered and then mixed with IF signals, then this is synchronous detection. Detection is conducted at the IF range in the electrical domain, hence there need for controlling the stability of the frequency spacing between the signal carrier and that of the LO. That means that the mixing of these carriers would result into an IF carrier in the electrical domain before the mixing process or envelope detection to recover the signals.

Coherent detection thus relies on the electric field component of the signal and the LO. The polarization alignment of these fields is critical for the optimal detection. The electric field of the optical signals and the LO can be expressed as

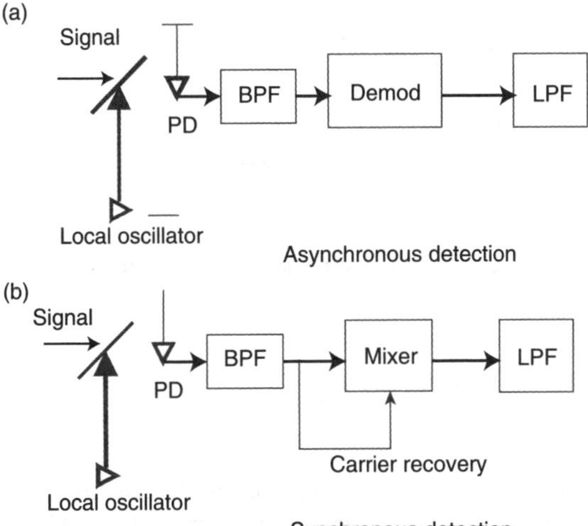

FIGURE 4.2 Schematic diagram of optical heterodyne detection. (a) asynchronous and (b) synchronous, receiver structures. LPF is a low-pass filter. BPF is a bandpass filter. PD is a photodiode.

$$E_{s}(t) = \sqrt{2P_{s}(t)}\cos\left\{\omega_{s}t + \varphi_{s} + \varphi(t)\right\}, \tag{4.1}$$

$$E_{LO} = \sqrt{2P_{L}}\cos\left\{\omega_{LO}t + \varphi_{LO}\right\}, \tag{4.2}$$

where $P_{s}(t)$ and P_{LO} are the instantaneous signal power and average power of the signals and local oscillator respectively, $\omega_{s}(t)$ and ω_{LO} are the signal and LO angular frequencies, ω_{s} and ω_{LO} are the phase (including any phase noise of the signal and the LO), and $\varphi(t)$ is the modulation phase. The modulation can be amplitude with the switching on and off (amplitude shift keying (ASK)) of the optical power or phase or frequency with the discrete or continuous variation of the time-dependent phase term. For the discrete phase, it can be phase shift keying (PSK), differential PSK (DPSK) or differential quadrature PSK-DQPSK and when the variation of the phase is continuous, we have frequency shift keying if the rate of variation is different for the bit "1" and bit "0" (as given in Chapter 2).

Under an ideal alignment of the two fields, the photodetection current can be expressed by

$$i(t) = \frac{\eta q}{h\nu}\left[P_{s} + P_{LO} + 2\sqrt{P_{s}P_{LO}}\cos\left\{(\omega_{s} - \omega_{LO})t + \varphi_{s} - \varphi_{LO} + \varphi(t)\right\}\right], \tag{4.3}$$

where the higher frequency term (the sum) is eliminated by the photodetector frequency response, η is the quantum efficiency, q is the electronic charge and h is Planck's constant and ν is the optical frequency.

Thus, the power of the LO dominates the shot noise process and simultaneously boosts the signal level, enhancing the SNR. The oscillating term is the beating product between the local oscillator. This term is signal proportional with the amplitude and equals to the square root of the product of the power of the local oscillator and the signal.

The electronic signal power S and shot noise N_{s} can be expressed as

$$S = 2\Re^{2}P_{s}P_{LO}$$

$$N_{s} = 2q\Re(P_{s} + P_{LO})B \tag{4.4}$$

$$\Re = \frac{\eta q}{h\nu} = \text{responsivity}$$

with B being the 3 dB bandwidth of the electronic receiver. Thus, the optical signal to noise ratio (OSNR) is given by

$$\text{OSNR} = \frac{2\Re^{2}P_{s}P_{LO}}{2q\Re(P_{s} + P_{LO})B + N_{eq}}, \tag{4.5}$$

where N_{eq} is the total electronic noise equivalent power as seen from the input to the electronic preamplifier of the receiver. From this equation, if the power of the LO is significantly increased so that the shot noise dominates over the equivalent noise, while simultaneously increasing the SNR, the sensitivity of the coherent receiver can be limited only by the quantum noise inherent in the photodetection process. Under this quantum limit, the OSNR_{QL} is given by

$$\text{OSNR}_{\text{QL}} = \frac{\Re P_s}{qB} \tag{4.6}$$

4.1.1.1 ASK Coherent System

Under the ASK modulation scheme, the demodulator of Figure 4.2 is an envelope detector (*in lieu* of the demodulator) followed by decision circuitry. That is, the eye diagram is obtained and a sampling instant is established with a clock recovery circuit, while the synchronous detection would require a locking between the carrier frequency and the local oscillator be obtained. It thus demands tuning to the local oscillator frequency pending on the tracking of the frequency component of the signal. The amplitude-demodulated envelope can be expressed as

$$r(t) = 2\Re\sqrt{P_s P_{\text{LO}}}\,\cos(\omega_{\text{IF}})t + n_x\cos(\omega_{\text{IF}})t + n_y\sin(\omega_{\text{IF}})t, \tag{4.7}$$

$$\omega_{\text{IF}} = \omega_s - \omega_{\text{LO}}.$$

The IF frequency ω_{IF} is the difference between the LO and the signal carrier; n_x and n_y are the expected values of the orthogonal noise power components, which are random variables.

$$r(t) = \sqrt{[2\Re P_s P_{\text{LO}} + n_x]^2 + n_y^2}\,\cos(\omega_{\text{IF}}t + \Phi)t$$

$$\text{with } \Phi = \tan^{-1}\frac{n_y}{2\Re P_s P_{\text{LO}} + n_x}. \tag{4.8}$$

4.1.1.2 Envelop Detection

The noise power terms can be assumed to follow a Gaussian probability distribution and are independent of each other with a zero mean and a variance σ. The probability density function (PDF) can thus be given as

$$p(n_x,n_y) = \frac{1}{2\pi\sigma^2}e^{\frac{-(n_x^2+n_y^2)}{2\sigma^2}} \tag{4.9}$$

with respect to the phase and amplitude, this equation can be written as [4]

$$p(\rho,\varphi) = \frac{\rho}{2\pi\sigma^2}e^{\frac{-(\rho^2+A^2-2A\rho\cos\varphi)}{2\sigma^2}}, \tag{4.10}$$

where

$$\rho = \sqrt{[2\Re\sqrt{P_s(t)P_{\text{LO}}} + n_s(t)]^2 + n_y^2(t)},$$

$$A = 2\Re\sqrt{P_s(t)P_{\text{LO}}}. \tag{4.11}$$

The pdf of the amplitude can be obtained by integrating the phase amplitude pdf over the range of 0 to 2π, and given as

$$p(\rho) = \frac{\rho}{\sigma^2} e^{\frac{-(\rho^2 + A^2)}{2\sigma^2}} I_0 \left\{ \frac{A\rho}{\sigma^2} \right\}, \tag{4.12}$$

where I_0 is the modified Bessel's function. If a decision level is set to determine the "1" and "0" level, then the probability of error and then the bit error rate (BER) can be obtained if assuming an equal probability of error between the "1s" and "0s" is equal, as:

$$\text{BER} = \frac{1}{2} P_e^1 + \frac{1}{2} P_e^0 = \frac{1}{2} \left[1 - Q\left(\sqrt{2\delta}, d\right) + e^{-\frac{d^2}{2}} \right], \tag{4.13}$$

where Q is the Magnum function and δ is given by

$$\delta = \frac{A^2}{2\sigma^2} = \frac{2\Re^2 P_s P_{LO}}{2q\Re\left(P_s + P_{LO}\right)B + i_{N_{eq}}^2}. \tag{4.14}$$

When the power of the LO is much larger than that of the signal and the equivalent noise current power, then this SNR becomes

$$\delta = \frac{\Re P_s}{qB}. \tag{4.15}$$

Physical representation of the detected current and the noises current due to the quantum shot noise and noise equivalent of the electronic pre-amplification are shown in Figure 4.3, in which the signal current can be general and derived from the output of the detection scheme, that from a photodetector or a back-to-back pair of photodetectors of a balanced receiver of the DPSK or DQPSK or CPFSK.

The BER is optimal when setting its differentiation with respect to the decision level δ, an approximate value of the decision level can be obtained as:

$$d_{opt} \cong \sqrt{2 + \frac{\delta}{2}} \Rightarrow \text{BER}_{ASK-e} \cong \frac{1}{2} e^{-\frac{\delta}{4}}. \tag{4.16}$$

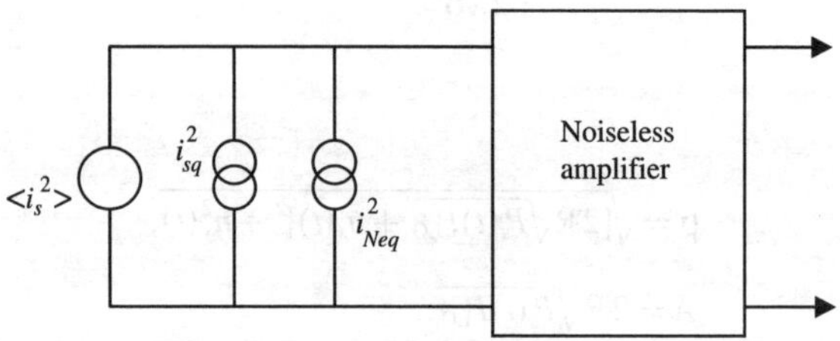

FIGURE 4.3 Equivalent current model at the input of the optical receiver.

4.1.1.3 Synchronous Detection

ASK can be detected using synchronous detection*, and the BER is given by

$$\mathrm{BER}_{\mathrm{ASK-S}} \cong \frac{1}{2}\mathrm{erfc}\,\frac{\sqrt{\delta}}{2}. \tag{4.17}$$

4.1.2 PSK Coherent System

Under the PSK modulation format, detection is similar to that of Figure 4.4 for heterodyne detection, but after the BPF an electrical mixer is used to track the phase of the detected signal. The received signal is given by

$$r(t) = 2\Re\sqrt{P_s P_{\mathrm{LO}}}\,\cos[(\omega_{\mathrm{IF}})t + \varphi(t)] + n_x \cos(\omega_{\mathrm{IF}})t + n_y \sin(\omega_{\mathrm{IF}})t. \tag{4.18}$$

Information is contained in the time-dependent phase term $\varphi(t)$.

When the phase and frequency of the voltage control oscillator (VCO) are matched with those of the signals, then the received signal can be simplified to

$$r(t) = 2\Re\sqrt{P_s P_{\mathrm{LO}}}\,a_n(t) + n_x \tag{4.19}$$

$$a_n(t) = \pm 1.$$

Under Gaussian statistical assumption, the probability of the received signal of a "1" is given by

$$p(r) = \frac{1}{\sqrt{2\pi\sigma^2}}\,e^{\frac{(r-u)^2}{2\sigma^2}}. \tag{4.20}$$

Furthermore, the probability of the "0" and "1" are assumed to be equal. We can obtain the BER as the total probability of the received "1" and "0" as

$$\mathrm{BER}_{\mathrm{PSK}} = \frac{1}{2}\mathrm{erfc}(\delta) \tag{4.21}$$

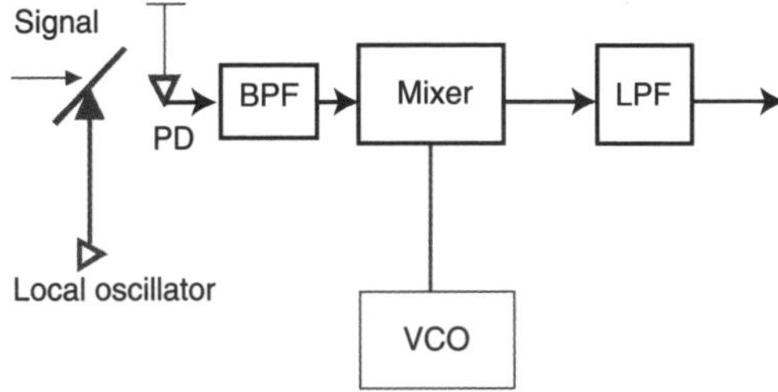

FIGURE 4.4 Optical heterodyne detection for PSK format (schematic).

---

* Synchronous detection is implemented by mixing the signals and a strong local oscillator in association with the phase locking of the local oscillator to that of the carrier.

4.1.2.1　Differential Detection

As observed in synchronous detection, carrier recovery circuitry (usually implemented using a phase lock loop (PLL)) is needed that complicates the overall receiver structure. It is possible to detect the signal by a self-homodyne process by beating the carrier of one bit period to that of the next consecutive bit; this is called "differential detection". The detection process can be modified as shown in Figure 4.5, in which the phase of the IF carrier of one bit is compared with that of the next bit, and a difference is recovered to represent the bit "1" or "0". This requires differential coding at the transmitter and an additional phase comparator for the recovery process. In later chapters on differential PSK, differential decoding is conducted in the photonic domain via a photonic phase comparator by a MZ interferometer incorporating a thermal tuning optical delay line. The BER can be expressed as

$$\mathrm{BER}_{\mathrm{DPSK-e}} \cong \frac{1}{2}e^{-\delta}. \tag{4.22}$$

4.1.3　Optical Homodyne Detection

Optical homodyne detection requires phase matching between the signal carrier and that of the LO (Figure 4.6). Under a perfect phase matching, we would have

$$r(t) = 2\sqrt{P_s P_{\mathrm{LO}}}\,\cos[\pi A_k s(t)] \tag{4.23}$$

with $s(t)$ being the modulating waveform and A_k representing the bit "1" or "0". This is equivalent to the baseband signal and the ultimate limit is the BER of the baseband signal.

The noise is dominated by the quantum shot noise of the LO with its square noise current is given by

$$i^2_{\mathrm{N-sh}} = 2q\Re(P_s + P_{\mathrm{LO}})\int_0^\infty |H(j\omega)|^2 d\omega \tag{4.24}$$

where $H(j\omega)$ is the transfer function of the receiver system, usually a trans-impedance of the electronic preamp and that of a matched filter. Because the power of the LO is much larger than the signal and integrating over the dB bandwidth of the transfer function, this current can be approximated by

$$i^2_{\mathrm{N-sh}} = 2q\Re P_{\mathrm{LO}} B. \tag{4.25}$$

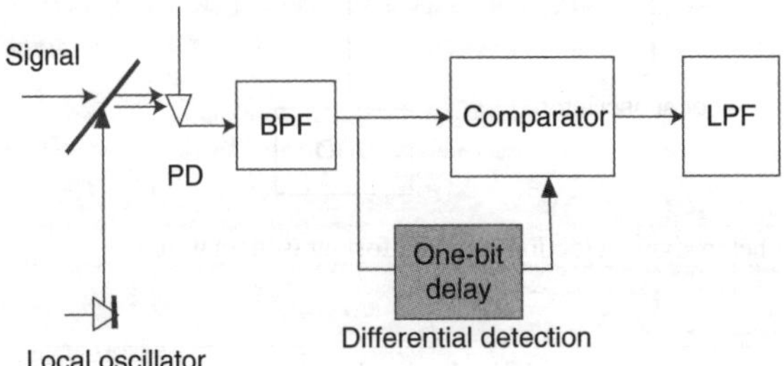

FIGURE 4.5　Optical heterodyne and differential detection for PSK format (schematic).

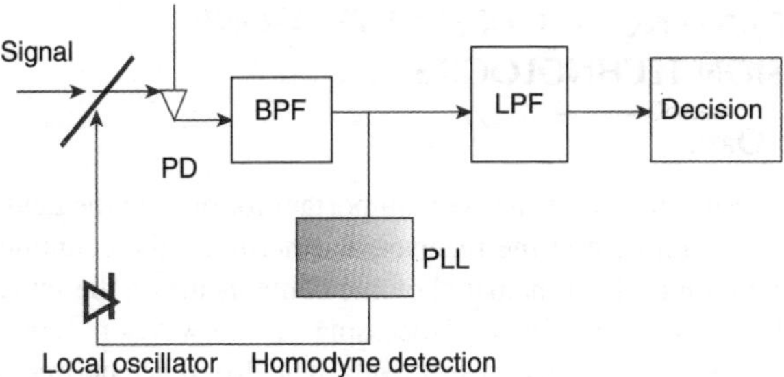

FIGURE 4.6 Optical homodyne detection (schematic).

Hence, the SNR is given by

$$\mathrm{SNR} \equiv \delta \simeq \frac{2\Re P_s}{qB}. \tag{4.26}$$

The BER is the same as that of a synchronous detection, and is given by

$$\mathrm{BER}_{\mathrm{homodyne}} \cong \frac{1}{2}\mathrm{erfc}\sqrt{\delta}. \tag{4.27}$$

The sensitivity of the homodyne process is 3-dB better than that of the heterodyne process, and the bandwidth of the detection is half of its counter part due to the double sideband nature of the heterodyne detection.

4.1.4 FSK COHERENT SYSTEM

The nature of FSK is based on the two frequency components that determine the bits "1" and "0". There are a number of formats related to FSK depending on whether the change of the frequencies representing the bits is continuous or non-continuous: the FSK or CPFSK modulation formats. For non-continuous FSK, the detection is usually performed by a structure of dual frequency discrimination as shown in Figure 4.7 in which two narrow-band filters are used to extract signals. For CPFSK, the frequency discriminator and balanced receiver for PSK detection can be used. Frequency discrimination is preferred compared with balanced receiving structures because it eliminates the phase contribution by the LO or optical amplifiers, which may be used as an optical preamp. We will highlight the performance of these detection schemes for modulation format using non-coherent techniques in the following chapters.

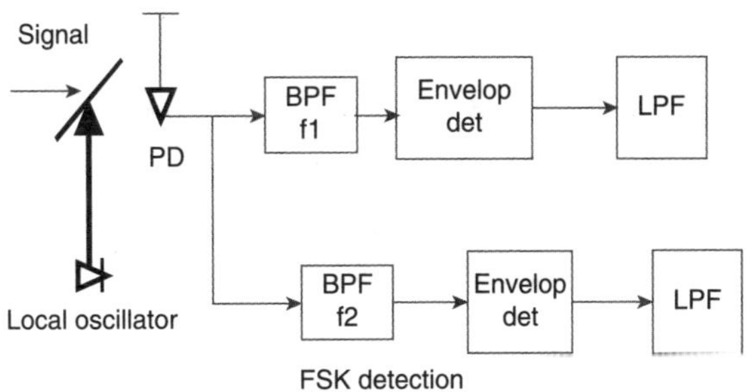

FIGURE 4.7 Optical homodyne detection of FSK format (schematic).

4.2 COHERENT AND NON-COHERENT PHOTONIC TRANSMISSION TECHNOLOGIES

4.2.1 INTEGRATED OPTICS

The polarization of signals and the LO are very important for mixing the fields of these lightwaves to achieve effective resultant field at the receiver. Furthermore, the coupling loss must be minimized. Thus, an integration of the principal photonic components on the same substrate is critical. A typical integrated optic structure for the processing of lightwaves is shown in Figure 4.8a and Figure 4.8b for Z-cut and X-cut LiNbO$_3$, respectively. The difference between these two cuts is the placement of the electrode to maximize the electro-optic effects.

The LO lightwaves and the signals are coupled to the diffused waveguides and then guided to frequency translator and polarization rotator, respectively. The frequency translator tunes the frequency of the LO to the IF region via an interferometric structure, while the polarization rotator is controlled by an interdigital transducer for phase matching of TE mode to TM mode of the diffused waveguide, hence the conversion of polarization. An acousto-optic mode converter can be used [24]. Lightwaves are mixed via an electro-optic directional coupler and its outputs are coupled to photo detectors or to the fiber for further processing.

4.2.2 OPTICAL SOURCE AND MODULATION

In coherent optical communications, the coherence of the lightwave carrier and the LO and its phase noise are critical for achieving maximum demodulation efficiency. Hence, the linewidth of the lasers are critical, as well as the stabilization of the laser frequency.

4.2.2.1 Demands on Laser Linewidth

If envelop detection is used, then it measures the power level of the IF amplitude. If the lightwave source has a finite linewidth, then the beating between the LO and the signals results in some uncertainty on the IF range, i.e., the signal phase would contribute to a phase fluctuation of received signals. The error of the detection would thus be conditioned on the frequency deviation $\Delta\omega$. For the detection of ASK (as shown in Figure 4.9) or FSK, the PDF of this detection process is given as [25]

$$p_{IF} = \frac{1}{\sqrt{\Delta v BT}} e^{-\frac{\Delta\omega^2}{4\pi\Delta vB}} \tag{4.28}$$

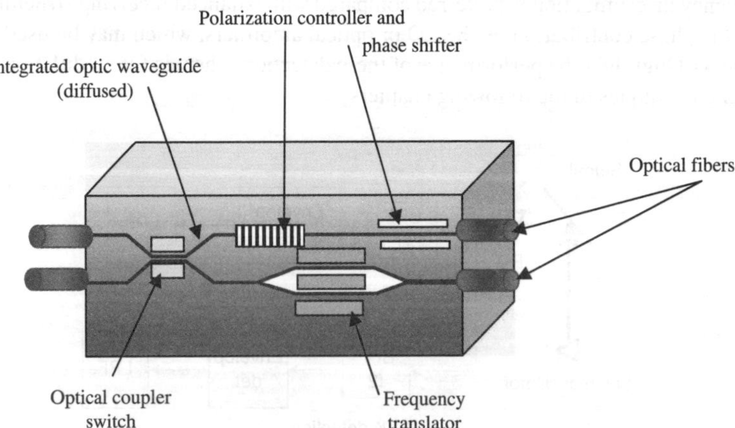

FIGURE 4.8 Integrated optic coherent receiver device Z-cut LiNbO$_3$. (From Stallard, W. A., A. R. Beaumont, and R. C. Booth, *IEEE Journal of Lightwave Technology* LT-4(7), 1986. With permission.)

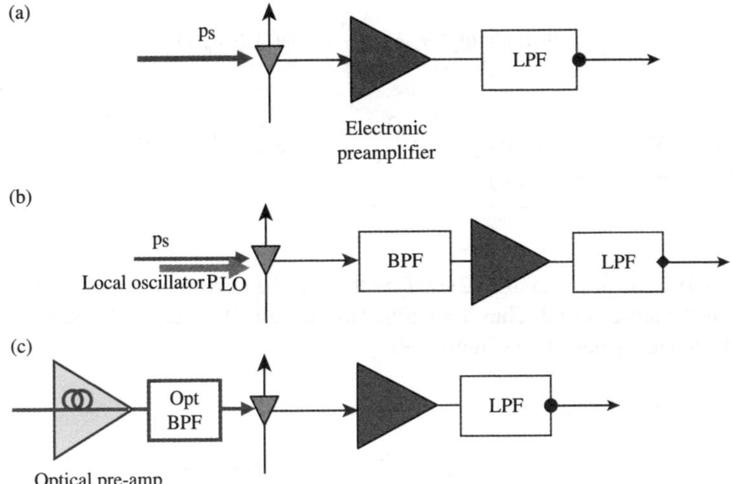

FIGURE 4.9 Optical receiver structures for ASK optical signals (a) direct detection, (b) coherent detection and (c) optically pre-amplification direct detection.

where $\Delta \nu$ is the FWHM IF linewidth of the power spectral density and T is the bit period, and B is the IF average bandwidth.

4.2.2.2 Differential Detection

We can consider discrete differential scheme, DPSK and continuous-phase CPFSK with differential detection. For DPSK detection, as shown in Figure 4.7, the received signal after the IF stage can be written as

$$r(t) = 2\Re \sqrt{P_s P_{LO}} \cos(\omega_{IF} t + \Phi_s(t)) + n_x(t) \cos \omega_{IF} t + n_y \sin \omega_{IF} t \tag{4.29}$$

with $\quad \Phi_s(t) = \varphi_s(t) + \{\varphi_n(t) - \varphi_n(t+T)\} - \{\varphi_{pS}(t) - \varphi_{pS}(t+T)\} - \{\varphi_{pLO}(t) - \varphi_{pLO}(t+T)\}$

where the first term represents the data received (0 or π). The second term is the shot noise generated phase noise mainly contributed by the LO power, and the third and fourth terms are the quantum phase noise due to the beating between the signal and the LO lightwaves.

The PDF of the phase noises would follow a chi-square distribution and one can, by integrating over the full 2π and over the range of the frequency spectrum, obtain the BER as [26]

$$\text{BER} = \frac{1}{2} \frac{\rho e^{-\rho}}{2} \sum_{n=0}^{\infty} \frac{(-1)^n}{2n+1} \left[I_n\left(\frac{\rho}{2}\right) + I_n\left(\frac{\rho}{2}\right) \right]^2 e^{-(2n+1)^2 \pi \Delta \nu_T T} \tag{4.30}$$

where $\Delta \nu_T$ is the total FWHM linewidth of the LO of the transmitter. I_n is the modified Bessel's function.

4.2.2.3 CPFSK Differential Detection

For differentially detected signals, the error probability is dependent on the delay time of the differential detector. It is the phase comparator for detecting the differential phase of the carrier in the two consecutive bits, the frequency deviation, and the phase noise. The detected signal phase at the shot noise limit in the output of the low-pass filter can be expressed as

$$\Phi_s(t) = \omega_c \tau + A_n \frac{\Delta\omega}{2}\tau + \varphi(t) + \varphi_n(t) \tag{4.31}$$

$$\text{BER} = \frac{1}{2}\frac{\rho e^{-\rho}}{2}\sum_{n=0}^{\infty}\frac{(-1)^n}{2n+1}\left[I_n\left(\frac{\rho}{2}\right)+I_n\left(\frac{\rho}{2}\right)\right]^2 e^{-(2n+1)^2\pi\Delta\upsilon_T T}e^{\left|-(2n+1)^2\pi\Delta\upsilon\tau\right|}\cos\left\{(2n+1)\alpha\right\} \tag{4.32}$$

with $\alpha = \pi(1-\beta)/2$ and $\beta = \Delta\omega/\omega_m = 2m\tau/T_0$ with ω_m is the maximum deviation, m is the modulation index, T_0 is the pulse period. Thus β is defined as the ratio of the actual frequency deviation and the maximum frequency deviation (Figure 4.9).

4.2.2.4 Balanced Receiver

A coherent balanced receiver is operating in a push-pull mode with two photodetectors connected back-to-back and receiving the mixed signals from the two ports of a coupler, which are constructive and destructive interferences of the signals and the LO at the input ports. The optical LO is used to mix both branches of the signals. For incoherent differential detection, a one-bit delay line is used for comparison of the phases of the carrier of the two consecutive bits. Thus, the same structure would also be used for DPSK and CPFSK or MASK, with the only difference being that the two input ports are delayed by one bit period with each other. That is, a one-bit self homodyne detection scheme as shown in Figure 4.10.

The fields of the input signal and the LO can be represented as

$$E_s(t) = \sqrt{2P_s}\,\cos\omega_s t$$
$$E_{LO}(t) = \sqrt{2P_{LO}+\Delta P_{LO}}\,\cos\omega_{LO}t. \tag{4.33}$$

The fields at the two output ports are given by

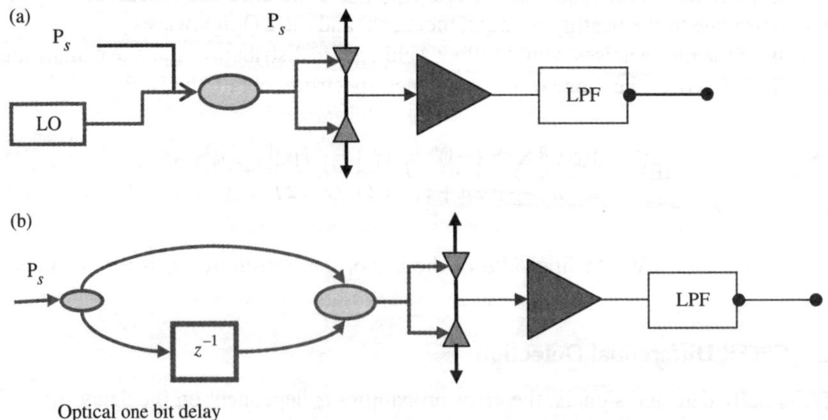

FIGURE 4.10 Optical balanced receiver (a) coherent detection (b) non-coherent differential detection. z^{-1} is a one bit optical delay line.

$$E_1(t) = \frac{P_s}{2} + \frac{P_{LO}}{2} + \sqrt{2P_s P_{LO}}\,[\cos(\omega_s - \omega_{LO})t + \varphi_s - \varphi_{LO} + \varphi(t)]$$

$$E_2(t) = \frac{P_s}{2} + \frac{P_{LO}}{2} - \sqrt{2P_s P_{LO}}\,[\cos(\omega_s - \omega_{LO})t + \varphi_s - \varphi_{LO} + \varphi(t)].$$

$$(4.34)$$

The field E_1 represents the constructive port output and likewise E_2 for the destructive port. Heterodyne detection would require larger bandwidth of the electronic amplifier and the photo-detector responses, while it requires much less for homodyne and self homodyne detection. The balanced receiver has become very popular in non-coherent DPSK detection due to its simplicity and higher sensitivity.

4.2.2.5 Phase Diversity Receiver

Figure 4.11 shows the structure of a phase diversity receiver and a push-pull receiver structure. Phase comparison is made via several directional couplers, which can be structured using a number of 2 × 2 or 3 × 3 couplers and beam splitter or polarization coupler/splitter. Balanced receivers are also used to distinguished the in-phase or out of phase of the carrier in the photonic domain. Although only a 2 × 2 and PM couplers are shown in this section, a higher degree of phase diversity can be performed by cascading or cascading these basic structures.

The in-phase (I) and quadrature (Q) components are received and would be decoded by an electronic decoder. The current outputs can be represented by

$$I_1(t) = \frac{q}{2\Re}\sqrt{P_s P_{LO}}\,\cos(\omega_{LO} - \omega_s)t$$

$$(4.35)$$

$$I_Q(t) = \frac{q}{2\Re}\sqrt{P_s P_{LO}}\,\sin(\omega_{LO} - \omega_s)t.$$

4.2.2.6 Channel-Selective Receiver

If transmitted channels are frequency division multiplexed (FDM), then one of the main advantages of the coherent receiver is that individual channels can be selectively extracted by tuning the LO to a channel so that the RF filter of the heterodyne receiver can be matched to the frequency difference of the signal center frequency and that of the local oscillator. To increase the total capacity of the transmission system, the channel separation is the critical. This depends on the signal spectra, which are affected by the bit rate ratio for various modulation and demodulation schemes.

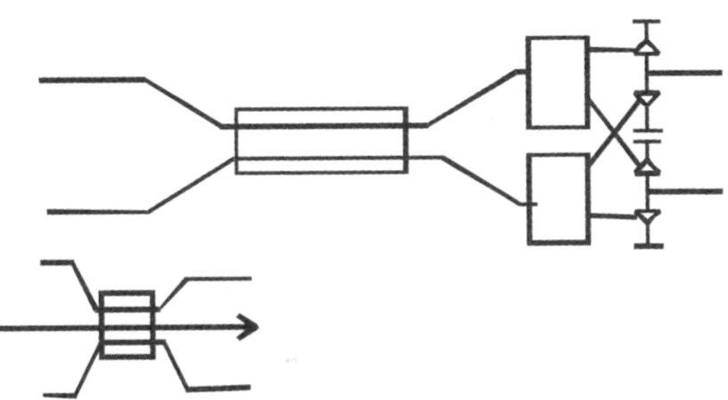

FIGURE 4.11 Integrated optic coherent phase diversity balanced receiver using an optical π/4 hybrid coupler, including an insert of a 3 × 3 coupler.

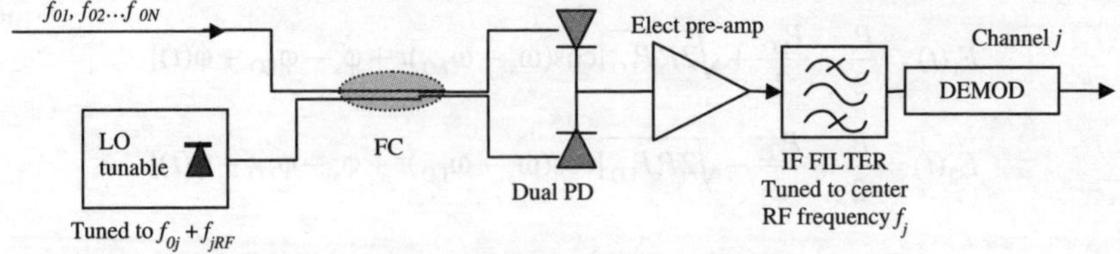

FIGURE 4.12   Channel-selective balanced coherent receiver (schematic).

The schematic diagram of a channel selective receiver is shown in Figure 4.12, in which the center frequencies of multiple channels of $f_{01}, f_{02}....f_{0N}$ are entering an optical mixer with the optical tunable LO. The output ports of the optical coupler are fed into two back-to-back photodetectors. The current generated in this photodetection is electronically pre-amplified, filtered with a band-pass filter and then demodulated to the binary sequence. The output currents of the photodetectors are given by [27]

$$i_1 = \frac{1}{2}\Re P_{LO} + \frac{1}{2}\Re \sum_{k=1}^{N}\sum_{l=1}^{N} E_k E_l^* + \Re\sqrt{P_{LO}P_s}\, m_k \cos(\omega_{IF} t + \phi_N)$$

$$+ \Re\sqrt{P_{LO}P_s}\sum_{k=1, k\neq K}^{N} m_k \cos(\omega_{IF} t + 2\pi(k-N)f_s t + \varphi_K) + i_{Neq1}(t)$$

$$(4.36)$$

$$i_2 = \frac{1}{2}\Re P_{LO} + \frac{1}{2}\Re \sum_{k=1}^{N}\sum_{l=1}^{N} E_k E_l^* + \Re\sqrt{P_{LO}P_s}\, m_k \cos(\omega_{IF} t + \varphi N)$$

$$- \Re\sqrt{P_{LO}P_s}\sum_{k=1, k\neq K}^{N} m_k \cos(\omega_{IF} t + 2\pi(k-N)f_s t + \phi_K) + i_{Neq2}(t)$$

where P_{LO} is the local laser power available at the output of the fiber pigtail of the CW laser, p_s is the peak power of the optical signal, f_k represents the possible amplitude and phase angle modulation of the kth channel, ω_{IF} is the radial frequency of the channel, K is the subscript of the desired channel and N is the total number of the channel; and $n_{eq1,2}$ is the equivalent noise currents as seen by the photodiodes 1 and 2 of the balanced detector pair. The first term of Equation 4.35 is the DC current generated by the LO power, the second term indicates the beating between the fields of the kth and any l^{th} channels of the spectrum of the FDM channels. The third term indicates the beating between the signal and the LO optical fields. The plus and minus signs of the two currents indicate the push-pull effect of the back-to-back connected photodetector pair. Thus, the total current entering the electronic preamplifier can be written as

$$i_T = i_1 - i_2 = 2\Re\sqrt{P_{LO}P_s}\left\{ \begin{array}{l} m_k \cos(\omega_{IF} t + \varphi_N) \\ \\ + \sum_{k=1, k\neq K}^{N} m_k \cos(\omega_{IF} t + 2\pi(k-N)f_s t + \varphi_K) \end{array} \right\} + i_{Neq}(t) \qquad (4.37)$$

with $i^2_{Neq}(t) = i^2_{Neq1}(t) + i^2_{Neq2}(t).$

The first term is the desired signal and the second term is the LO and channel interference currents. The IF bandpass filter eliminates the DC currents generated by the LO as well as that of the inter-channel interference currents, thus they do not deteriorate receiver sensitivity. The power spectral density of i_T can be written as

$$p_{\mathrm{ET}}(f) = 4\Re^2 P_{LO} P_s \{p_{\mathrm{Es}}(f) + p_{\mathrm{Cs}}(f)\} + \xi(f) \tag{4.38}$$

where P_E represents the spectral power density of the detection process, subscript s stands for the signal, C stands for LO and channel interference, and the last term involves the noise power spectral density.

By assuming that the pdfs are Gaussian-like, then the BER of a balanced receiver FDM system can be estimated by

$$\mathrm{BER}_{\mathrm{FDM-BalRx}} = Q(\delta_T) \tag{4.39}$$

where Q is the magnum function and δ_T is given as

$$\delta_T^2 = \frac{1}{\dfrac{1}{\delta^2} + \dfrac{1}{\delta_{\mathrm{CT}}^2}} \tag{4.40}$$

where δ_T is the total SNR, including all the contributed noises, electronic equivalent noise as seen from the input of the electronic preamp, the LO shot noise current, photodetector dark current noises; δ_{CT} is the total SNR with the noise currents resulting from the cross talks between the channels. These terms can be written as

$$\delta = \frac{i_{s"1"} - i_{s"0"}}{2\sqrt{\sigma_{\mathrm{eq}}^2 + \sigma_s^2}}$$

$$\delta_T = \frac{i_{s"1"} - i_{s"0"}}{2\sigma_c}. \tag{4.41}$$

The signals currents generated due to the "1" and "0" received are indicated in the subscripts of Equation 4.40 and the variances of the noise currents due to the signal shot noise current, equivalent electronic current noises and the channel interference noise denoted respectively as σ_s, σ_{eq}, and σ_c.

The channel interference noise should be written as

$$\sigma_C^2 = \int_{-\infty}^{+\infty} p_{\mathrm{Es}}(f) |H_{\mathrm{D}}(f)|^2 \, df \tag{4.42}$$

where $H_{\mathrm{D}}(f)$ is the transfer function of the demodulator.

Thus, observing Equation 4.39, we can see that the BER penalty of FDM system for a BER of 1e-9 can be derived as

$$\mathrm{Penalty}_{\mathrm{FDM}}, \mathrm{dB} = -10 \mathrm{Log}_{10}\left(1 - \frac{36}{\sigma_{\mathrm{CT}}^2}\right). \tag{4.43}$$

4.3 PHOTONIC FILTERS AND PHASE COMPARATORS

Optical filters for single channel or multiplexers and demultiplexers are indispensable in coherent and incoherent ultra-high speed optical fiber communications employing modulation formats, especially the continuous phase (CPFSK, MSK or M-ary MSK) or discrete-phase modulation (DPSK, DQPSK, M-ary DPSK). This section describes the principles of optical filters and phase comparators, and their operational characteristics for modulation format receiving systems.

Table 4.1 tabulates the structures and types of different photonic filters for optical communications employing advanced modulation formats. For example, the inteferometric filter is normally formed using the silica-silicon planar lightwave circuit technology, and it is very common for phase comparison between two consecutive bits in 40 Gb/s optical half band filter transmission systems. At this bit rate, if phase comparison is conducted in the electrical domain it would face difficulty in phase scattering and stray capacitance that would reduce the speed of operation. While using a channel waveguide structure, the delay of the lightwaves can be thermally controlled without difficulty. The MZ stages of different delay length can be cascaded to form all pass (or all zeros) filters [28].

The resonance type filters would normally formed with resonant cavity of either a single pass or multiple pass structures. Multi-layer thin film filters are well known for its sharp roll-off and flat band. However they are available in bulk structure rather than integrated optic type. Hence, some additional loss would occur due to connection. Integrated optic type would consist of a single ring or multiple ring [29] or micro-rings [30] which offer Gaussian like passband and possible narrow band down to even a few nanometers. These reflection-based cavity filters can also be implemented using gratings written in fiber. The refractive index change in these gratings is not large and thus

TABLE 4.1

Integrated Optic and Fiber-based Optical Filters

Filter principle	Filter Type	Structure
Two beam interference	• Directional coupler	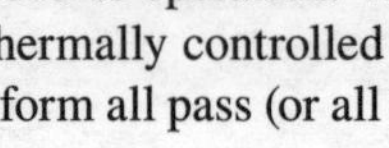
	• MZ interferometric	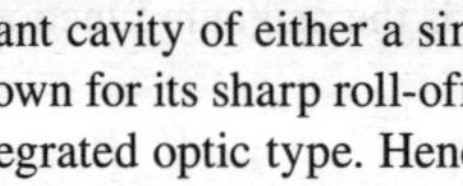
	• Half-band filters (all pass)	
Resonance	• Interference thin-film multilayer	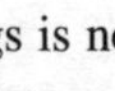
	• Fabry–Perot filter	
	• Ring resonator, micro ring resonator	
	• Fiber Bragg gratings (FBG)	
Multi-beam interference	• Diffraction gratings	
	• Array waveguide gratings	
	• Transversal filter	

it requires a long length of gratings in order to obtain sharp roll off. They would normally exhibit parabolic-like passsband characteristics.

4.3.1 TRANSMISSION CHARACTERISTICS

The transmittance of an interferometric filter for the input port and the through T_{th} and cross T_{cr} coupling ports are given by

$$T_{th} = \cos^2 \frac{\varphi}{2} \qquad T_{cr} = \sin^2 \frac{\varphi}{2}$$

$$\varphi = \frac{2\pi n \Delta L f}{c}.$$

(4.44)

For the ring resonator, the transmittance can be given by

$$T_{0-1,2} = \frac{T_r}{(1 - R_r) + 4R_r \sin^2(\pi\beta r)}$$

(4.45)

$$\text{where} \quad T_r = k e^{-\alpha \pi r} \qquad R_r = (1 - k^2) e^{-2\alpha \pi r}$$

with r is the ring radius, a 1.0 mm corresponds to a free spectral range of 25 GHz. The quality factor of the resonator can be given as

$$Q = \frac{\pi \sqrt{R_r}}{1 - R_r}.$$

(4.46)

An interferometer MZ can incorporate a ring resonator in one path of the interferometer as shown in Figure 4.13c to give a transmittance transfer function from one of the input port to the output ports 3 and 4 of the filter as

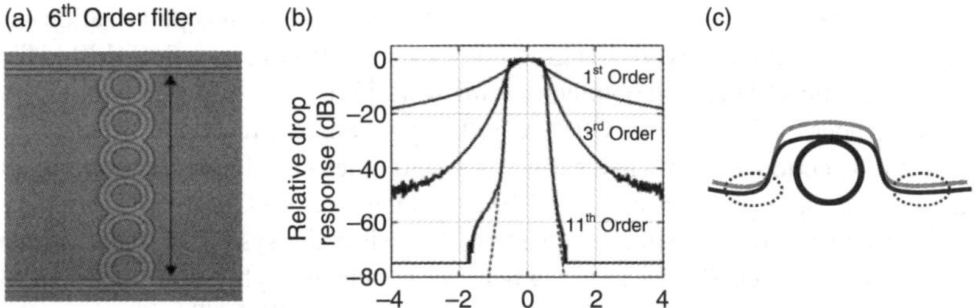

FIGURE 4.13 (a) A sixth-order micro-ring resonator filter fabricated with 17% index contrast high-index material on low index substrate. (b) Measured response for first-, third-, and eleventh-order micro-ring resonator filters similar in fabrication to the device shown in (b). The responses have been normalized to their 3-dB bandwidths. The dashed curve is the theoretical fit to the eleventh-order filter (extracted from [10]). (c) Integration of a ring resonator and a MZ interferometer. Dotted ellipses are optical directional couplers.

$$T_{1-3} = \cos^2\left(\frac{\varphi - \theta}{2}\right)$$

$$T_{1-4} = \sin^2\left(\frac{\varphi - \theta}{2}\right) \tag{4.47}$$

$$\text{where} \quad \varphi = \tan^{-1}\left(\frac{[r^2 - 1]\sin\theta}{2r - (r^2 + 1)\cos\theta}\right) \quad \theta = \beta L_0$$

where ϕ is the phase shift of the ring resonator and θ is the round trip phase delay of the ring resonator, r is the through coupling coefficient and given as $r = (1 - k)^{1/2}$, and k is the intensity coupling coefficient of the coupler. L_0 is the round trip length of the ring resonator.

This shows that the output ports exhibit an orthogonal or 90° shift of the optical filter, which is very useful in the filtering of orthogonal carrier phase signals such as the minimum shift keying (MSK) modulation formats. Thus, the interferometric filter acts as filtering and interference to give constructive and destructive interference signals, or phase comparison, at the two output ports which are very important for push pull operation of balanced receiver.

Recently, multiple resonant ring resonators that offer very narrow passband have been designed [31]. A sixth order and its spectral responses are shown in indicating that a flat band and sharp rolloff can be achieved using integrated optic technology. It is also cascading several of single microring resonators to form a narrow band and closely spaced wavelength channels.

4.4 COHERENT PHASE AND DIFFERENTIAL PHASE SHIFT KEYING

Significant performance improvements have been demonstrated by applying incoherent (i.e., incoherently demodulated) differential binary-phase-shift-keying (DBPSK) modulation to long-haul transmission, as described in Chapter 2. Theoretically, a coherent binary-phase-shift-keying (BPSK) modulation can achieve even higher receiver sensitivity than the incoherent DBPSK format, and thus should further improve system performance. Research on coherent optical fiber communication systems was actively pursued two decades' ago, but practical applications were limited due to complexity and expense. Recently, the coherent detection technique has been revisited, and more cost-effective implementations employing high-speed digital signal processing (DSP) technology have been proposed [32, 33].

Using the homodyne phase-diversity receiver followed by digital signal processing (DSP), it has been demonstrated that the transmission of 16-GHz-spaced 40-Gbit/s polarization-multiplexed QPSK signals, whose symbol rates are 10 GSymbol/s, over a 200-km dispersion-shifted fiber (DSF), the record SE of 2.5 bit/s/Hz can be obtained in this work [34].

For the remainder of this section, CBPSK will imply coherent reception, whereas DPSK will imply incoherent detection, except when explicitly stated otherwise. In this section, we study theoretically the transmission performance of CBPSK in long-haul 40 Gb/s WDM propagation. The experimental demonstration and simulations show that when the system is linear or quasi-linear, CBPSK may outperform DPSK. However, when the system is highly non-linear, BPSK suffers severe performance degradation and performs much worse than DPSK. To investigate the cause of the performance degradation, a series of simulations [35] have been performed and reported, each focusing on a different impairment, including ability to track system phase and polarization variations, cross-phase modulation (XPM), and self-phase modulation (SPM). The SPM effect turns out to be the dominant impairment that degrades the BPSK performance. Simulations also show that the SPM- and XPM-induced penalty on the CBPSK system can be reduced by changing the system into a coherent DBPSK system. The simulation of the system can be seen in Figure 4.14, in which a 500-KHz linewidth laser is assumed as the lightwave source. It is then modulated with a RZ optical modulator and then a data optical modulator for switching the data optical clock pulse

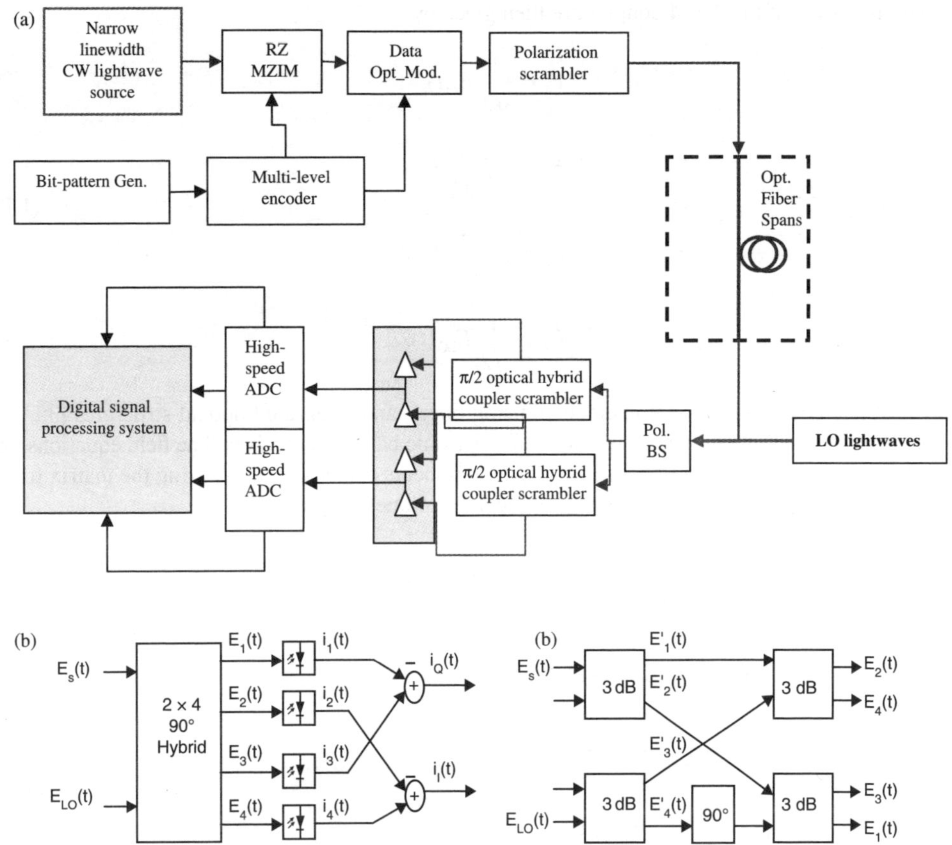

FIGURE 4.14   (a) Structure diagram of the modeled transmission system: transmitter side and coherent receiver for multi-level for coherent BPSK optical transmission, (b) details of the coherent receiver, and (c) structure of 2 × 4 90° hybrid coupler for beating the signals and the local oscillator.

sequence on or off. A polarization scrambler is used to ensure that the lightwave polarization can be estimated and matched with that of the LO after the long-haul transmission. The transmission path can be optically amplified and dispersion compensated both of the linear dispersion and the matched dispersion slope.

After the transmission, the lightwave signals are polarization-split via a polarization beam splitter and then detected coherently with a balanced receiver in combination with pi/2 optical hybrid coupler and a high-power LO. Detected electronic signals are then digitalized and fed into a DSP for processing in the electrical domain. The absolute phase of the detected signals is to be determined in this processor. The coherent receiver front-end shown in Figure 4.14 is where the signals and the LO are beating each other via the opto-electronic conversion, and thus the phase of the product of the total field is preserved. As it can be seen from the receiver front-end, the optical fields of the signals and the LO lightwaves $E_s(t)$ and $E_{LO}(t)$ are combined via the 2 × 4 hybrid coupler, as shown in Figure 4.14c. The field coupling of the 3 dB coupler is given as

$$K_{3dB} = \frac{1}{\sqrt{2}} \begin{pmatrix} 1 & -j \\ -j & 1 \end{pmatrix}. \tag{4.48}$$

The output fields of the 2×4 coupler are then given by

$$E_1^{'} = \frac{1}{\sqrt{2}} E_s(t)$$

$$E_2^{'} = -j\frac{1}{\sqrt{2}} E_s(t)$$

$$E_3^{'} = -j\frac{1}{\sqrt{2}} E_{LO}(t)$$

$$E_4^{'} = \frac{1}{\sqrt{2}} E_{LO}(t). \tag{4.49}$$

These hybrid couplers can be implemented using fibers, and integrated optical structures [35, 36]. The output fields are then detected by an opto-electronic balanced receiver. The field equations and the electronic currents generated for I- and Q- components can be obtained, using the matrix transfer, coupling technique and signal flow graphical techniques [37], as

$$E_k(t) = \frac{1}{\sqrt{4}}\left(E_s(t)e^{jk\frac{\pi}{4}} + E_{LO}(t) \right) \quad k = 1,2,3,4.$$

$$E_1(t) = \frac{1}{2}\left[jE_s(t) + E_{LO}(t) \right]$$

$$E_2(t) = \frac{1}{2}\left[-E_s(t) + E_{LO}(t) \right]$$

$$E_3(t) = \frac{1}{2}\left[-jE_s(t) + E_{LO}(t) \right] \tag{4.50}$$

$$E_1(t) = \frac{1}{2}\left[E_s(t) + E_{LO}(t) \right]$$

$$\text{with} \quad E_s(t) = \sqrt{P_s(t)}e^{jx} \quad \text{where} \quad x = \omega_s(t) + \varphi_{mod}(t) + \varphi(t)$$

$$\text{and} \quad E_s(t) = \sqrt{P_{LO}(t)}e^{jy} \quad \text{where} \quad x = \omega_{LO}(t) + \varphi_{LO}(t).$$

These optical fields are impinged on the photodetectors, and the generated electronic currents for I- and Q-phase components are given by

$$i_1(t) = \Re|E_1(t)| = \frac{\Re}{4}\left[\sqrt{P_s(t)}e^{j(x+\frac{\pi}{4})} + \sqrt{P_{LO}(t)}e^{jy} \right] \cdot \left[-j\sqrt{P_s(t)}e^{-j(x+\frac{\pi}{4})} + \sqrt{P_{LO}(t)}e^{-jy} \right]$$

$$= \frac{\Re}{4}\left[P_s(t) + P_{LO} + 2\Re\sqrt{P_s(t)P_{LO}} \, \cos(x - y + \frac{\pi}{2}) \right] \tag{4.51}$$

$$= \frac{\Re}{4}\left[P_s(t) + P_{LO} - 2\Re\sqrt{P_s(t)P_{LO}} \, \sin(x - y) \right].$$

Thus, the kth current received generated for the kth component is given by

$$i_k(t) = \frac{\Re}{4}\left[P_s(t) + P_{LO} + 2\sqrt{P_s(t)P_{LO}} \, \cos(x + y + k\frac{2\pi}{4}) \right] \tag{4.52}$$

$$i_1(t) = i_4(t) - i_2(t) = \frac{\Re^2}{2}\left[\sqrt{P_s(t)P_{LO}}\,\cos(x-y)\right]$$

$$i_Q(t) = i_3(t) - i_1(t) = \frac{\Re^2}{2}\left[\sqrt{P_s(t)P_{LO}}\,\sin(x-y)\right].$$

(4.53)

It is obvious from the above equations that the phase of the signals can be recovered via the resultant signals from the beating of the LO and the lightwaves whether the demodulation process is homodyne or heterodyne. The phase of the electronic currents are then processed using DSP as indicated in the diagram of the transmission system shown in Figure 4.14a.

Simulations were performed to compare the performance of BPSK and DBPSK in the linear and non-linear regimes. In the simulations, the propagation of a 128-bit pattern was first calculated using the split-step method to capture the signal distortion induced by SPM, XPM, and FWM effects. Then, the distorted 128-bit signal waveform was repeatedly used in a Monte Carlo simulation of the coherent receiver to capture the relatively slow 1-MHz phase and polarization drifting effects. Because a slope-matched system was modeled, all channels had similar performance. The following discussions are focused on the 1544 nm channel as an example. Figure 4.15 compares the simulated BPSK and DBPSK performance at three power levels, corresponding to linear, quasi-linear, and non-linear regimes, respectively. To capture the intra- and inter-channel effects, the powers of the studied channel and its two immediate neighbors were increased and decreased simultaneously. When the system is linear, CBPSK outperforms DPSK by about 2 dB, whereas when the system is highly non-linear, at the –3.5-dBm power level, CBPSK suffers severe performance degradation, and performs about 2 dB worse than DPSK. The result shown in Figure 4.15 combines the effects of all the modeled system impairments.

Therefore it has been shown that in the linear and quasi-linear regimes, the CBPSK system outperforms the incoherent DPSK system but, in the highly non-linear regime, the CBPSK suffers severe performance degradation and performs much worse than the incoherent DPSK. In 40-Gb/s transmission systems, the SPM effect is the dominant impairment that degrades the coherent BPSK performance as observed in Figure 4.16a and Figure 4.16b. The performance impairments on the CBPSK system induced by the intra- and inter-channel non-linear effects can be significantly reduced by changing the system into a CDBPSK system. Figure 4.17b and Figure 4.17a shows the evolution of the phases of the DPSK and CBPSK signals, respectively after the optical transmission as affected by non-linear XPM, SPM and FWM. In that case, however, most of the coherent benefit is lost [15].

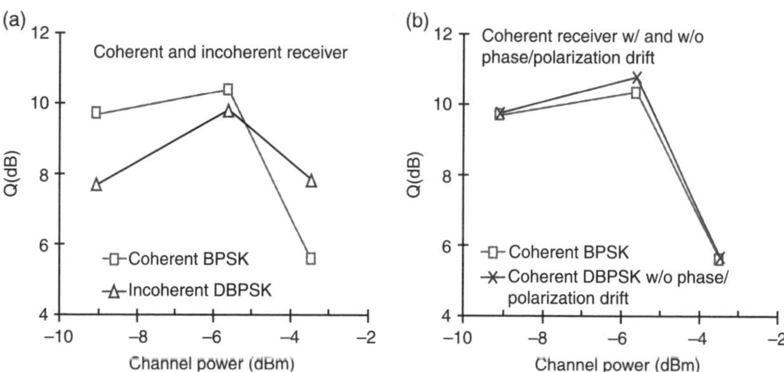

FIGURE 4.15 Simulated results of the system shown in Figure. 4.14 with 9000-km dispersion-compensated optically amplified spans [38].

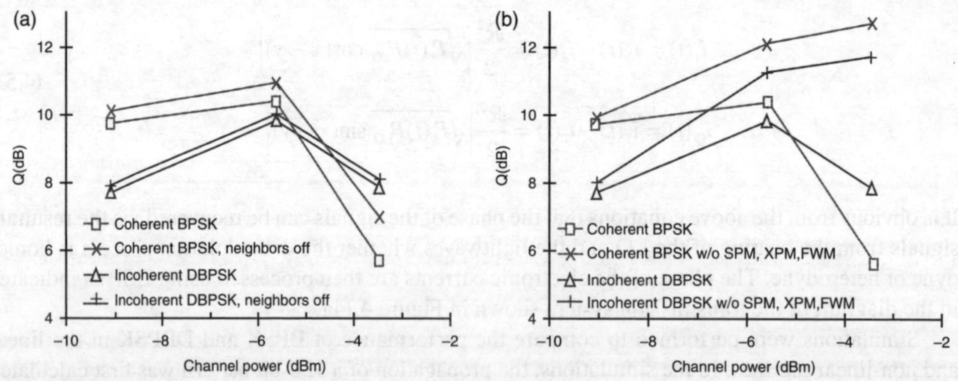

FIGURE 4.16 (a) Effect of XPM on system performance (b) effect of SPM and XPM on system performance under 9000-km dispersion-compensated optically amplified spans.

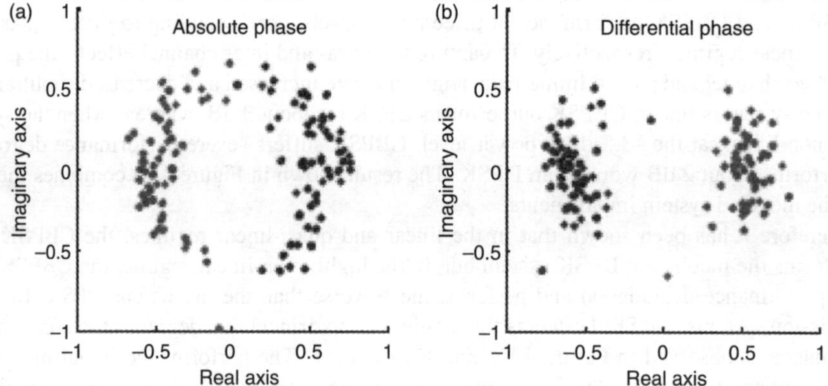

FIGURE 4.17 Scatter plots of the evolution of the phases of the DPSK (b) and CBPSK (a) signals after the optical transmission as affected by non-linear XPM, SPM and FWM.

4.5 FIBER PROPAGATION AND NON-LINEARITIES

4.5.1 Linear Dispersion: Analytical Approach

4.5.1.1 Signal Propagation and the Governing NLSE Equation

The propagation of an optical carrier-modulated signal $u(z,t) = A(z,t).e^{j\omega t}$, whose complex amplitude $A(z,t)$, can be represented by the non-linear Schroedinger wave equation (NLSE)

$$\frac{\partial A}{\partial z} = -\alpha A - j\frac{\beta_2}{2}\frac{\partial^2 A}{\partial t^2} + \frac{\beta_3}{6}\frac{\partial^3 A}{\partial t^3} - j\gamma|A|^2 A \tag{4.54}$$

where the amplitude $A = A(z,t)$ is the complex envelope carried by the lightwaves of wavelength λ, along the propagation z along the axial axis of the optical fiber and t is the time variable; $j = (-1)^{1/2}$. The carrier component $\exp(j\omega_0 t)$ can be removed without loss of generality.

The lightwave pulse propagates along the fiber a guided propagation constant β, i.e., the effective wave number along the z axis of the propagation constant of the guided optical waves, k. This propagation constant varies as a function of the optical frequency and can be represented by a Taylor's series approximation around the operating radial frequency ω_0 as

$$\beta = \beta_0 + (\omega - \omega_0)\beta_1 + \frac{1}{2}(\omega - \omega_0)^2\beta_2 + \frac{1}{6}(\omega - \omega_0)^3\beta_3 + \dots \tag{4.55}$$

where β_2 and β_3 is the group velocity dispersion and dispersion slope, i.e.,

$$\beta_n = \frac{\partial^n\beta}{\partial\omega^n}, \quad \text{that is} \quad \beta_3 = \frac{\partial^3\beta}{\partial\omega^3} \quad \text{and} \quad \beta_2 = \frac{\partial^2\beta}{\partial\omega^2} \tag{4.56}$$

γ is the non-linear coefficient of an optical fiber with an effective area A_{eff} corresponding to a mode spot size r_0, given by:

$$A_{\text{eff}} = \pi r_0^2 \quad \gamma = \frac{n_2\omega}{cA_{\text{eff}}} \tag{4.57}$$

where c is the velocity of light in vacuum. n_2 is the non-linear refractive index that assumes a typical value for silica-based glass of 2.6e-20 m²/W. The pure delay term involving the propagation constant β is eliminated in Equation 4.1.

4.5.1.2 Fiber Transfer Function

A single-mode optical fiber can be modeled as an optical bandpass filter with a flat amplitude response and linear group delay within the bandwidth of the data sequence that modulates the light-waves. The data bandwidth is much smaller than that of the optical carrier frequency. The variation of the group velocity v_g is determined from the chromatic dispersion of the fiber according to the following equation

$$\frac{\partial v_g}{\partial\omega} = \frac{\partial v_g}{\partial\lambda}\frac{\partial\lambda}{\partial\omega} = -\frac{1}{L}\frac{\partial v_g}{\partial\lambda}\frac{\lambda^2}{c}L. \tag{4.58}$$

Using the definition of the dispersion factor $D(\lambda)$

$$D(\lambda) = -\frac{1}{\lambda}\frac{\partial v_g}{\partial\lambda}. \tag{4.59}$$

The group delay v_g of the fiber can be expanded around the center frequency v_c as

$$v_g \simeq v_g(v_c) + (v - v_c)\frac{\partial v_g}{\partial v_c} \tag{4.60}$$

where v_c is the center frequency corresponding to the centered wavelength λ_c. The corresponding phase variation of the each optical frequency component is given by

$$\varphi(\nu) = 2\pi \int_0^\nu \nu_g(\nu')d\nu'$$

(4.61)

$$\varphi(\nu) = 2\pi\nu\nu_g + 2\pi \left[\frac{(\nu - \nu_c)^2}{2} - \frac{\nu_c^2}{2} \right] \frac{\partial\nu_g}{\partial\nu}.$$

Hence, a low-pass equivalent model for a single-mode optical fiber operating in the linear dispersion regime, a purely phase frequency response, can be represented around the center optical frequency wavelength as [33]

$$H(f) = e^{-j\phi(f)} = e^{-j\alpha f^2}$$

(4.62)

$$\text{where} \quad \alpha = \pi D(\lambda)\frac{\lambda^2}{c}L \quad f = \nu - \nu_c.$$

Equation 4.56 indicates that the transfer function is a purely quadratic phase response and a flat amplitude which is affected only by the transmission loss or attenuation of the fiber. This quadratic phase response is similar to the quadratic phase variation in the spatial domain of a light beam passing through a slit [14]. The duality between the temporal and spatial domain is given in a later section of this report. This quadratic phase property generates a chirping effect on optical carrier underlying the pulse envelope.

4.5.1.3 Pulse Shapes

Consider the envelope of a pulse of an arbitrary shape propagating along the z-axis of a single-mode optical fiber $u(z,t)$ as shown in Figure 4.18. The frequency-domain representation of this pulse envelope [37] can be written as

$$U(z,\omega) = U(0,\omega)e^{(\frac{j}{2}\beta_2\omega^2 z - \frac{j}{6}\beta_3\omega^3 z....)}$$

(4.63)

where the $U(0, \omega)$, $U(z, \omega)$ is the Fourier transform of the input pulse at the input and at an arbitrary distance z along the optical fiber. Thus, the time-domain can be found by taking the inverse Fourier response of Equation 4.10.

$$u(z,t) = \frac{1}{2\pi} \int_{-\infty}^{+\infty} U(0,\omega)e^{(\frac{j}{2}\beta_2\omega^2 z - \frac{j}{6}\beta_3\omega^3 z....)} d\omega$$

(4.64)

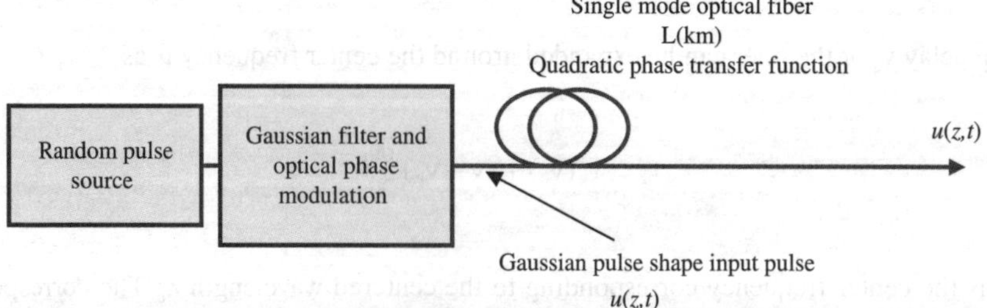

FIGURE 4.18 Propagation of lightwave modulated pulse sequence (schematic).

Thus, given a known pulse shape, one can obtain its frequency domain and time domain representation by integrating the fiber quadratic phase characteristics and performing the Fourier transformation if the pulse intensity is falls within the linear region of the fiber. Otherwise, if the optical power reaches the linear threshold, then one could represent the transfer function of the fiber by a Volterra-series with first-, third- and fifth-order components.

4.5.1.3.1 Unchirped Gaussian Pulse

Now consider that the input pulse to a single-mode optical fiber has a Gaussian shape of the form

$$u(0,1) = e^{-\frac{t^2}{2T_0^2}} \tag{4.65}$$

where T_0 is the pulse width of the pulse at an intensity of e^{-1} of the maximum intensity at $t=0$. It is customary to express the e^{-1} pulse width and the full width half mark (FWHM) as $T_{\text{FWHM}} = 2\sqrt{\ln 2}\,T_0 \simeq 1.665 T_0$. The frequency domain and the time domain can be obtained using Equation 4.10 through Equation 4.11 to give the time domain of the output pulse as

$$u(z,t) = \frac{T_0}{(T_0^2 - j\beta_2 z)} e^{-\frac{t^2}{2(T_0^2 - j\beta_2 z)}}. \tag{4.66}$$

Thus, a Gaussian pulse maintains its shape on the propagation, but its width is increased with z as

$$T(z) = T_0 \sqrt{1 + \frac{z}{L_D}} \quad \text{with} \quad L_D = \frac{T_0^2}{\beta_2} \tag{4.67}$$

and L_D is called the GVD dispersion length. Similarly, the third-order dispersion length is defined as

$$L_{DS} = \frac{T_0^2}{\beta_3}. \tag{4.68}$$

We can now observe that the output pulse is chirped, that is its phase varies within the envelope of the pulse. Under the Gaussian pulse shape, the phase variation can be expressed as

$$u(z,t) = |u(z,t)| e^{j\phi(z,t)} \tag{4.69}$$

where the phase is given by [30]

$$\varphi(z,t) = -\frac{\mathrm{s_i n}(\beta_2)\dfrac{z}{L_D}}{1 + \left(\dfrac{z}{L_D}\right)^2} \frac{t^2}{T_0^2} + \frac{1}{2}\tan^{-1}\left(\frac{z}{L_D}\right). \tag{4.70}$$

The time dependence of the phase of the pulse envelope indicates that there is an instantaneous frequency shift $\delta\omega$ from the central frequency ω_0 given as

$$\partial\omega = -\frac{\partial\varphi}{\partial t} = \frac{\mathrm{sgn}(\beta_2)\dfrac{2z}{L_D}}{\left(1+\left(\dfrac{z}{L_D}\right)^2\right)}\frac{t}{T_0^2}. \tag{4.71}$$

That is, the frequency varies linearly across the pulse shape. This chirp is up or down depending on the sign of the dispersion whether it is anomalous ($\beta_2 < 0$) or normal ($\beta_2 > 0$).

4.5.1.3.2 Chirped Gaussian Pulse

Considering a chirped Gaussian pulse represented as

$$u(0,t) = e^{-\frac{1+jC}{2}\frac{t^2}{T_0^2}} \tag{4.72}$$

where C is the chirp parameter. The chirp is up chirp if $C > 0$ and down chirp if $C < 0$. This chirp factor C can be estimated from the spectral width of the Gaussian pulse given by taking the Fourier transform of the pulse as

$$U(0,\omega) = \left(\frac{2\pi T_0^2}{1+jC}\right)^2 e^{-\frac{\omega^2 T_0^2}{2(1+jC)}} \tag{4.73}$$

and the spectral width at e^{-1} is given as

$$\Delta\omega = \frac{(1+C^2)^{1/2}}{T_0}. \tag{4.74}$$

In the absence of the chirp factor ($C = 0$), the pulse is said to be "transform limited" and we have the relation between the spectral width and the pulse width T_0 as $\Delta\omega T_0 = 1$.

The chirped Gaussian pulse after propagating through the of the fiber of length z is given as

$$u(z,t) = \frac{T_0}{\sqrt{T_0^2 - j\beta_2 z\left(1+jC\right)}} e^{\frac{(1+jC)T^2}{2[T_0^2 - j\beta_2 z(1+jC)]}}. \tag{4.75}$$

Thus, we note that even a chirped Gaussian pulse would maintain its shape when propagating through a single mode optical fiber. After propagating a length z, the width of a chirped Gaussian pulse, $T(z)$ is given as

$$\frac{T(z)}{T_0} = \sqrt{\left(1+\frac{C\beta_2 z}{T_0^2}\right)^2 + \left(\frac{\beta_2 z}{T_0^2}\right)^2}. \tag{4.76}$$

This shows that it is possible to shorten the pulse width by using an appropriate chirp factor C. The distance at which the pulse width reaches its minimum with a chirp factor C and the GVD parameter β_2 with $\beta_2 C < 0$ is given by taking the differentiation of Equation 4.61 as

$$z_{\min} = \frac{|C|}{1+C^2} L_{\mathrm{D}}.$$

(4.77)

This shows that the pulse would be shortened when the chirp parameter $C = -2$.

4.5.2 Non-linear Impairments

The advent of optical amplifiers for long-haul optical communications overcomes the transmission loss, but simultaneously raises concerns about its impacts due to the non-linear phase distortion. In particular, the detrimental feature of fiber non-linearities in multi-carrier or WDM transmission systems is inter-modulation between channels. Besides the self-phase modulation effects, the stimulated Raman scattering effects would induce significant cross-talks across channels of the frequency division multiplexing (FDM) or WDM transmission.

Fiber non-linearities can be classified into two types: one is due the stimulated scattering processes such as the Raman scattering (SRS) and Brillouin scattering (SBS), the other is due the third-order susceptibility of the fiber core glass giving the self phase modulation and cross phase modulation effects.

The threshold power at a critical signal power for the SRS and SBS non-linear effects are given by

$$P_{\mathrm{c-SRS}} = \frac{16 A_{\mathrm{eff}} \gamma}{g_R L_{\mathrm{eff}}}$$

$$L_{\mathrm{eff}} = \frac{1 - e^{-\alpha L}}{\alpha}$$

(4.78)

$\alpha \equiv$ linear attenuation coefficient

$$P_{\mathrm{c-SBS}} = \frac{16 A_{\mathrm{eff}} \gamma}{g_B L_{\mathrm{eff}}}$$

(4.79)

$$P_{\mathrm{c-SPM}} = \frac{0.1\alpha}{\gamma N_A}.$$

(4.80)

Figure 4.19 shows variation of the maximum allowable power per channel versus number of channels that ensures SRS degradation with a power penalty below 1 dB for all channels in the NZ-DSF transmission medium.

4.5.3 Non-linear Phase Noise

The ASE noise of cascaded in-line optical amplifiers can, through the non-linear Kerr effect, alter the non-linear refractive index of the transmission and dispersion-compensating fiber. This has the effect of translating the random intensity fluctuations of optical signals in the fiber due to ASE noise into the random modulation of the phase of the carrier under the enveloped of the signals[†,‡,§]. This kind of phase noise is very critical in long-haul transmission systems whether it is coherent

---

† Gordon, J.P., and L.F. Mollenauer. 1999. Phase noise in photonic communications systems using linear amplifiers. *Opt Lett.* 23: 1351–53.

‡ Saito, S., A. Naka, and M. Murakami. 1992. Kerr non-linearity limits on transmission distance and data rate of inline optical amplifier systems. Proc. Topical Meeting on Optical Amplifiers, Santa Fe, CNM, Paper WC3.

§ Ryu, S. 1991. Signal line width broadening due to fiber non-linearity in long-haul coherent optical uncertainty. *IEE Elect Lett.* 27: 1527–29.

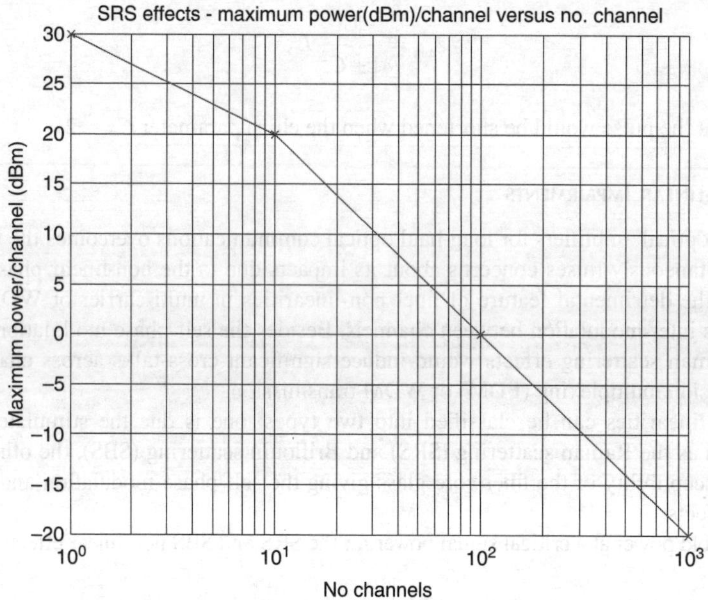

FIGURE 4.19 Maximum power per channel versus number of channels that ensures SRS degradation below 1 dB for all channels in NZ-DSF transmission medium.

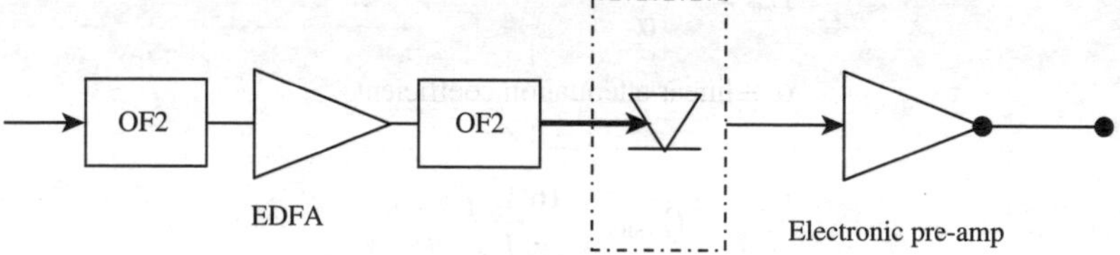

FIGURE 4.20 Schematic diagram of optical pre-amplified receiver. OF is optical filter, EDFA is erbium-doped fiber amplifier.

or non-coherent detection, and has been summarized in Ref[1]. This noise is termed as "non-linear phase noise" (NLPN), which becomes very important in long-haul transmission systems, especially under the coherent detection with phase estimation using DSP at the optical receiver, and direct detection under various modulation formats.

To appreciate the impact of the NLPN in optical transmission systems, let us assume that the transmitter laser phase noise is negligible and the total equivalent noise as seen from the input of electronic pre-amplifier is also very small and not taken into account. The schematic diagram of the optical pre-amplified receiver is shown in Figure 4.20, which consists of an optical amplifier inserted between two optical filters (as mentioned previously) and then a photodetector and an electronic preamplifier. If balanced receiver is used, then the input power is split, and the noise can be considered accordingly without much difficulty. The NLPN is strongly polarization-dependent and thus influences the performance of modulation formats and detection techniques that require accurate measurements of the absolute phase. Similarly, for intensity-direct detection, the phase noise is transferred to amplitude noise and then interfering with the eye opening at the receiver.

The current generated at the output of the photodetector or equivalently the total current at the input of the electronic pre amplifier is given by[**]:

---

[1] Jacobsen, G. *Noise in digital optical transmission systems*. Boston, USA: Artech House.
[**] Extracted Jacobsen, G. 1994. Noise in digital optical communications. *Artech House* 308–10.

For a "ONE" received we have

$$\langle i_s(T_B)\rangle_1 = \sum_{i=1}^{M_2}\left(\left|1+n_{i,x}+jn_{i,y}\right|^2+\left|n'_{i,x}+jn'_{i,y}\right|^2\right)-\frac{M_1}{M_2}\left(\frac{1}{\tau_1}\int_0^{T_B}\varphi^2(t)dt-\sum_{i=1}^{M_1}\left(\frac{1}{\tau_1}\int_{(i-1)\tau_1}^{iT_B}\varphi(t)dt\right)^2\right) \tag{4.81}$$

With $M_1 = T/\tau_1$, $M_2 = T/\tau_2$, this is obtained by taking the summation of all the inphase and quadrature components of the ASE noises impinged on the photodetector and the integration of the phase noises at the output of the optical filters as influenced by the time constant of the filters. The square law detection is applied as can be obviously observed.

For a "ZERO" received, we have

$$\langle i_s(T_B)\rangle_0 = \sum_{i=1}^{M_2}\left(\left|1+n_{i,x}+jn_{i,y}\right|^2+\left|n'_{i,x}+jn'_{i,y}\right|^2\right). \tag{4.82}$$

The inphase and quadrature ASE noise components follow a mutually independent zero-mean Gaussian distribution with variances

$$\sigma = \frac{n_{sp}^e M_2}{2n_{ps}(1-\sigma_1^2)^N} \tag{4.83}$$

where n_{ps} is the number of photons generated by the received signal power within the bit interval. The term $(1-\sigma_1^2)$ represents the degradation of the SNR caused by the coupling of the signal into the NLPN noise power at each of the span length with σ_1 given by

$$\sigma_1 = \frac{2n'_{sp}L^2k_2^2(G_{il}-1)^3 h\nu P}{T_B(G_{il}\ln G_{il})^2}B_{ASE} \tag{4.84}$$

where L is the span length or distance between two discrete optical amplifiers, $h\nu$ is the photon energy, k_2 is the non-linear phase modulation coefficient, P is the inline amplifier output power, and B_{ASE} is the amplifier noise bandwidth which is extremely wide (several nm).

The non-linear phase noise is given by $\phi(t)$, has a white noise power spectral density pdf as compared with the phase noise of the laser source. The pdf of the optical amplifier follows a chi-square distribution with non-central parameter M_2 for a ONE and for ZERO bits, with variance σ_1. The pdf of NLPN can be found within the band limited bandwidth M_1/T_B. The phase $\phi(t)$ is represented by an orthogonal series representation of Karhunen-Loe`ve expansion[††].

The phase noise due to NLPN can be written as:

$$\sigma_\varphi^2 = \frac{2n'_{sp}N^2L^3\alpha k_2^2 G_{il}(G_{il}-1)^3 h\nu P}{3T_B(G_{il}\ln G_{il})^3}M_1 \tag{4.85}$$

with α being the attenuation factor. This noise express allows us to see that the noise is proportional to the output power of the optical amplifier. When booster amplifiers are used at the output of each span, this would create the phase noise because its power is essentially high to increase the optical power available for transmission.

Figure 4.21 shows the eye opening penalty due to the effects of ASE noise and NLPN and ASE noise as a function of the bandwidth of the optical filter. The contribution of NLPF is quite significant and reaches 1 dB for an optical filter bandwidth of 1 nm.

†† Lepian, D. S., and H.O. Pollack. 1961. Polate spheroidal wave functions, Fourier analysis and uncertainty. *Bell Syst. Tech. J* 40: 43–63.

Jacobsen, K. *Noise in digital optical communications.* Artech House, 309–10.

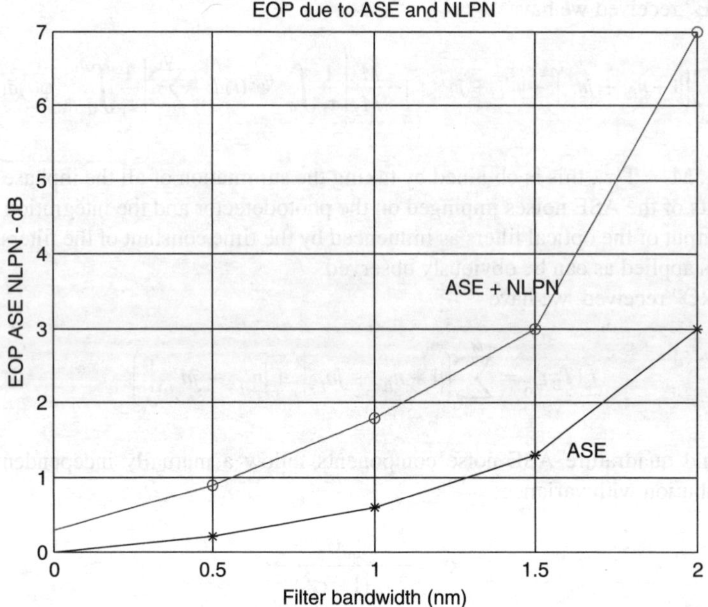

FIGURE 4.21 Simulated results of eye-opening penalty at BER of 1e-12 versus first stage filter bandwidth – circle contribution by ASE + NLPN and star (*) ASE only. Adapted from [8].

4.6 OPTICAL MULTI-LEVEL MODULATION: GENERAL THEORY

Coherent optical transmission returns into the focus of research. One significant advantage is the preservation of all the information of the optical field during detection, leading to enhanced possibilities for optical multi-level modulation. This paper investigates the generation of optical multi-level modulation signals. Several possible structures of optical M-PSK and M-QAM transmitters are shown and theoretically analyzed. Differences in the optical transmitter configuration and the electrical driving lead to different properties of the optical multi-level modulation signals. This is shown by deriving general expressions applicable to every M-PSK and M-QAM modulation format and exemplarily clarified for Square-16-QAM modulation.

Coherent systems were an important topic of investigation in the late eighties and the early nineties, but then research came to interruption with the advent of optical amplifiers that offer up to 20 dB gain without difficulty. Nowadays however, coherent systems come again into the focus of interest, due to the availability of digital signal processing and low-priced components, the partly relaxed receiver requirements at high data rates and several advantages that coherent detection provides. The preservation of the temporal phase of the coherent detection makes possible new emerging methods for adaptive electronic compensation of chromatic dispersion. When concerning WDM systems coherent receivers offer tunability and allow channel separation via steep electrical filtering. Furthermore, only the use of coherent detection permits to converge to the ultimate limits of spectral efficiency. To reach higher spectral efficiencies the use of multi-level modulation is required. Concerning this matter coherent systems are also beneficial, because all the information of the optical field is available in the electrical domain. That way complex optical demodulation with interferometric detection – which has to be used in direct detection systems – can be avoided and the complexity is transferred from the optical to the electrical domain. Several different modulation formats based on the modulation of all four quadratures of the optical field were proposed in the early nineties, describing the possible transmitter and receiver structures and calculating the theoretical BER performance. However, a more detailed and practical investigation of multi-level modulation coherent optical systems for today's networks and data rates is missing so far.

This section investigates the generation of optical multi-level modulation signals. Several optical transmitters for M-ary Phase Shift Keying (M-PSK) and M-ary Quadrature Amplitude Modulation (M-ary-QAM) are theoretically analyzed and compared. Chapter 2 has described the basics of optical multi-level modulation in which expressions for the electrical driving signals are setup, and the envelope and phase characteristics of the optical multi-level modulation signals applicable to every modulation format are calculated for different transmitter types. We will see that differences in the transmitter configuration and the electrical driving lead to different properties of the multi-level modulation signals.

4.6.1 OPTICAL MULTILEVEL MODULATION: GENERAL THEORY

In digital transmission with multi-level modulation, m bits are collected and mapped to a complex symbol chosen from an alphabet

$$d_k = d_{rk} + jd_{ik} \in \left\{ d_0, d_1, d_2, \ldots d_{M-1} \right\} \quad M = 2^m.$$ (4.86)

To each symbol interval (denoted by the integer k) of length $T_S = mT_B$, where $1/T_B$ is the bit rate, one of the symbols of this alphabet is assigned, depending on the respective bit combination and defined in a so called constellation diagram. Different from direct detection systems, in coherent optical systems, the constellation points can be equally spaced as for electrical systems, because the noise does not depend on the power of the detected symbols. For further calculations, it is useful to scale d_{rk} and d_{ik} to unity.

$$s(k) = i_k + jq_k \quad \text{with} \quad i_k = \frac{d_{rk}}{\max\{d_{rk}\}} \quad \text{and} \quad q_k = \frac{d_{ik}}{\max\{d_{ik}\}}.$$ (4.87)

The complex envelope and its components are defined as

$$A(t) = a(t)e^{j\psi(t)} = \frac{1}{\sqrt{2}}\sum_k s(k)p(t - kT_s) = \frac{1}{\sqrt{2}}I(t) + j\frac{1}{\sqrt{2}}Q(t)$$

$$\text{where} \quad I(t) = \sum_k [i_k p(t - kT_s)]; \quad Q(t) = \sum_k [q_k p(t - kT_s)];$$ (4.88)

$$a(t) = \frac{1}{\sqrt{2}}\sqrt{I^2(t) + Q^2(t)} \quad \psi(t) = \tan^{-1}\frac{Q(t)}{I(t)}.$$

The amplitude and phase of the complex envelope are represented by $a(t)$ and $\psi(t)$, respectively. The arctan-function is defined here for all four quadrants by case differentiation. $I(t)$ and $Q(t)$ describe the in-phase component and the quadrature component of the complex envelope, respectively, and $p(t)$ is the pulse shape of the electrical driving signals. For NRZ transmission, the optical multi-level modulation signal is created by modulating the optical carrier (emitted from continuous wave (CW) laser) with the complex envelope given by Equation 4.88, resulting in Equation 4.89. For RZ modulation the CW carrier is formed additionally by an intensity modulation using a Mach–Zehnder modulator (MZM) driven with a sinusoidal electrical RF signal at a DC bias point at $-\pi/2$, resulting in the additional cosine-term in Equation 4.89. $E_I(t)$ and $E_Q(t)$ are the in-phase and quadrature components of the modulated optical carrier.

$$E_s(t) = E_I(t) + jE_Q(t) = a(t)e^{j\psi(t)}\tilde{E}_{CW}(t) = \frac{1}{\sqrt{2}}I(t)\tilde{E}_s e^{j\omega_s(t)} + \frac{1}{\sqrt{2}}Q(t)E_s e^{\tilde{j}\omega_s(t)} \quad \text{for } NRZ$$

$$\text{(4.89)}$$

$$E_s(t) = E_I(t) + jE_Q(t) = a(t)e^{j\psi(t)}E_{CW}(t)\cos\left[0.25\pi\sin\left(2\pi\frac{1}{T_s} - \pi/2\right) - \pi/4\right] \quad \text{for } RZ$$

4.6.2 Multi-phase and Amplitude Format Optical Transmitters

In this section, various structures of multilevel-modulation transmitters are shown and theoretically analyzed. The optical part of the transmitters is common for almost every M-PSK and M-QAM modulation format. That is why they are called multiformat transmitters. Only the electrical driving has to be adapted to realize a special modulation format. It is possible that an optical multilevel-modulation signal can be generated from an intensity modulator followed by a phase modulator (Figure 2.2a) or from an I-Q modulator composed of two arms with two orthogonal carriers, where the I component of the complex envelope modulates the optical carrier in the I arm and the Q component modulates the $90°$ phase-shifted optical carrier in the Q arm (Figure 2.2b). In the latter case, only an amplitude modulation (which can be realized with a single-drive MZM driven by bipolar driving signals at a DC bias point at $-\pi$) has to be performed in each arm and the electrical driving signals have a smaller number of states. A further reduction of the number of states can be reached by replacing the amplitude modulation in each arm by separate intensity and phase modulations (Figure 2.1). Then, the intensity modulator is driven by a unipolar RF driving signal at a DC bias point at $-\pi$ and the phase modulator changes the phase between 0 and π for positive and negative values of i_k and q_k, respectively. Furthermore, it is possible to save two modulators compared to (c) and to realize the intensity and phase modulation with only one component, using a dual-drive MZM (Figure 2.2). The latter can be simultaneously driven in the push-pull mode for intensity modulation, and in the push-push mode for phase modulation, and the RF driving signals for intensity and phase modulation have to be electrically combined before being injected into the MZM inputs.

For the I-Q-configurations, it is assumed that the signals in the I-arm and the Q-arm have the same polarization, so that the in-phase and quadrature components optimally combine in the second 3dB-coupler. From a practical viewpoint, an integrated design seems to be more realistic than a hybrid built-up, in particularly when regarding stabilization of the optical IQ-balance. Every transmitter structure has its own advantages and drawbacks. The serial configuration (a) features a simple optical part, but the electrical driving signals have a high number of states (e.g., 12-ary signals are required for phase modulation for Square-16-QAM). The IQ-transmitters are composed of two arms, leading to a bigger optical complexity and necessitating integration, but the electrical driving signals have fewer number of states. For example, for Square-16-QAM, 4-ary electrical signals are required for the single-drive MZM configuration (b), whereas only 2-ary signals for the configuration with separate intensity and phase modulation (c). For the dual-drive MZM configuration (d) two modulators can be saved compared to configuration (c), but electrical combining of the driving signals is necessary. In the following it is shown that the differences in the transmitter built-ups and different characteristics of the driving signals lead to different properties of the multi-level modulation signals. For the case of using the modulators shown in Figure 4.22, the expressions for the driving signals can be obtained using the phasor development, as shown in Chapter 2. For the IQ-transmitters, the driving signals of the Q-arm can be obtained substituting ik with qk. For amplitude and intensity modulation, the driving signals are classified as "ideal" and "non-ideal". The ideal driving signals compensate completely the MZM characteristic, but are not very practical to generate (possibly by DSP), whereas the so-called non-ideal driving signals are easier to generate and yield to the correct optical constellation diagrams, but lead to small differences during the symbol transitions because they do not fully compensate the cosine MZM characteristic. Figure 4.23 shows detailed arrangement of an optical transmitter for multi-level modulation using two optical modulators, one dual-drive MZIM and electronic pre-coder for data and the other for generating RZ format.

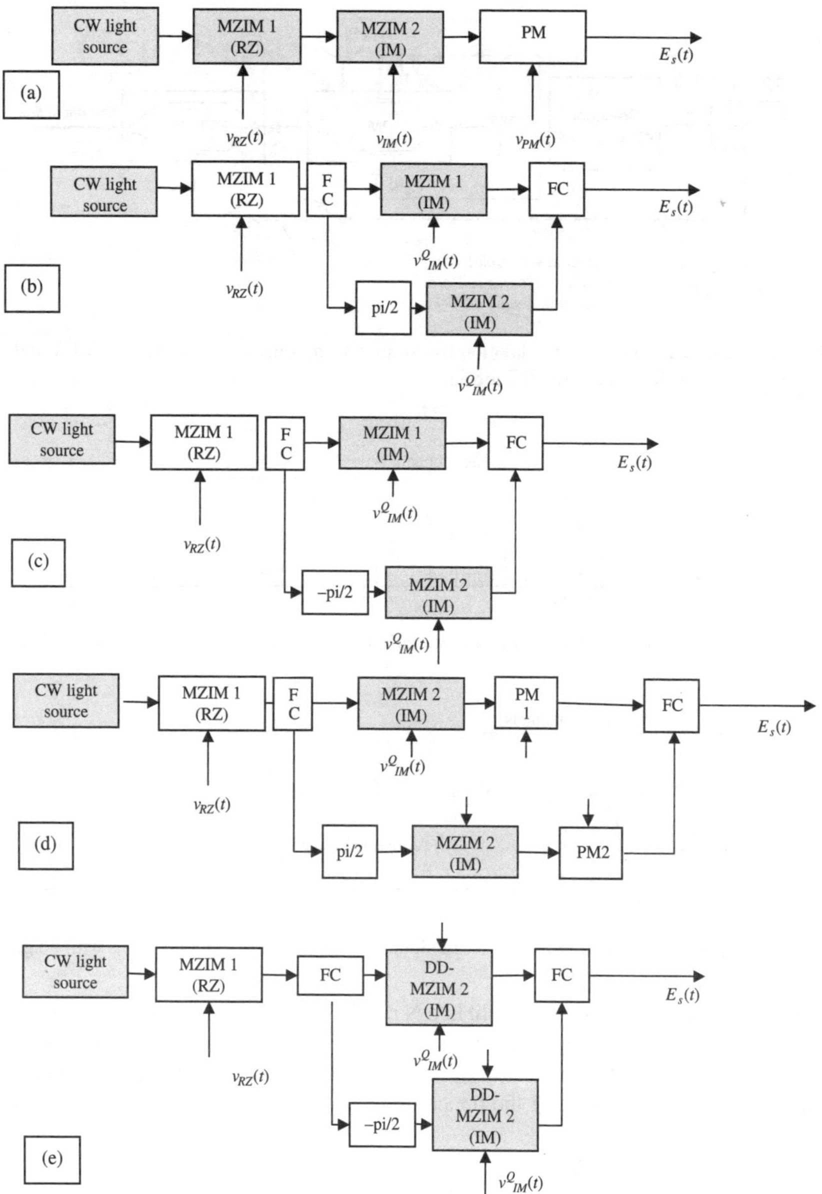

FIGURE 4.22 Optical multi-level modulation transmitters. (a) serial configuration, (b) IQ-configuration with single-drive MZM, (c) IQ-configuration with MZM and PM, (d) I-Q-configuration with dual-drive DD-MZM [38]. The modulated output is represented by lightwave field.

4.7. COHERENT OFDM

4.7.1 PRINCIPLES

Orthogonal FDM is a multi-carrier transmission technique that uses multiple frequencies to simultaneously transmit multiple signals in parallel; their spectra are shown in Figure 4.24, and its counter parts using the orthogonality between channels are shown in Figure 4.25. The efficiency in the bandwidth of the channels is very clear.

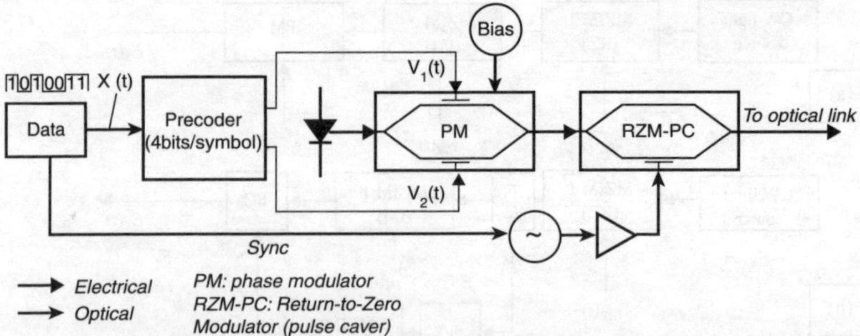

FIGURE 4.23 Optical multi-level modulation transmitter using only one dual-drive MZIM and electronic pre-coder; the second is for generating RZ format.

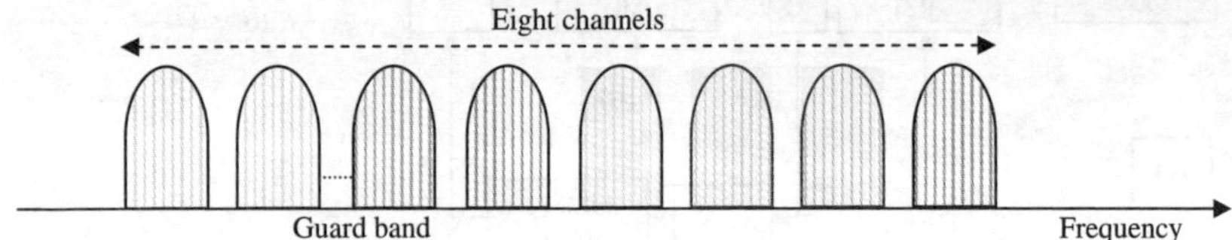

FIGURE 4.24 Spectra of FDM multi-carrier modulation schemes with eight channels with guard bands.

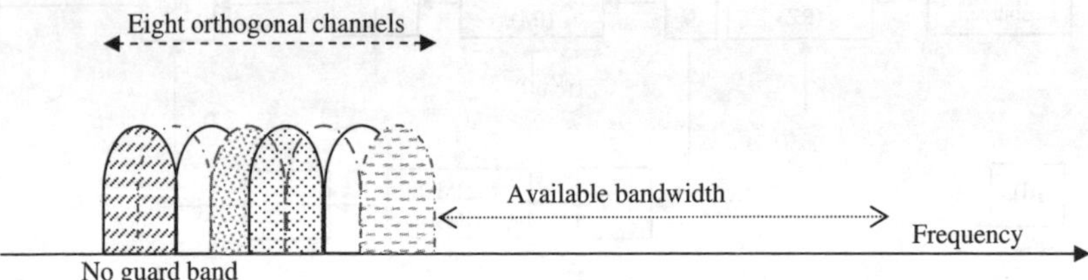

FIGURE 4.25 Spectra of OFDM multi-carrier modulation schemes with eight channels with no guard bands.

In OFDM, the serial data stream is split into N parallel sub-streams (i.e., from the least significant to the most significant bits) via a serial to parallel converter. Each line sub-channel would then be multiplied with individual RF carriers (in the electrical domain) and then are combined to generate time-domain signals for modulating the optical lightwaves. Refer to Figure 4.26, the time domain signal $s(t)$ can be written as

$$s(t) = \sum_k \sum_{i=0}^{N} a_i[k] g_i(t - kT) \tag{4.90}$$

where k is the time index, N is the number of sub-carriers, $T = NT_s$ is the stretched OFDM symbol period (due to the serial-to-parallel conversion), $a_i[k]$ is the kth data symbol of the ith sub-carrier ($i = 0,1,...N$) and $g_i(t)$ is the baseband pulse given by

$$g_i(t) = e^{j\omega_i t} g(t) \tag{4.91}$$

with $g(t)$ being the pulse shaping function. The orthoginality condition requires that the base band pulses satisfying

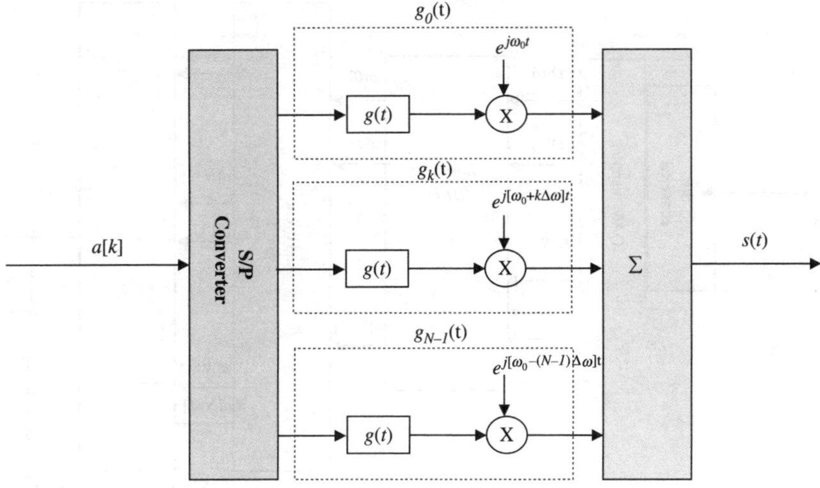

FIGURE 4.26 Time-domain generation of OFDM multi-sub-carrier modulation schemes. Sub-carriers can be generated automatically via inverse discrete Fourier transform g(t) is impulse response of shape function filter.

$$\int_{-\infty}^{+\infty} g_i(t)g_j^*(t)\mathrm{d}t = \frac{1}{T}\int_{-T/2}^{+T/2} g_i(t)g_j^*(t)\mathrm{d}t = \delta_0(i-j). \qquad (4.92)$$

For orthogonality, each sub-carrier should have been an integer number of cycles over a symbol period T, and the number of cycles between adjacent sub-carriers differs by exactly one unit. Thus, they are the multiple harmonics of the fundamental frequency, which is the inverse of the symbol period. Typically, the shaping function follow a sinc or raised cosine function, i.e., the transmission filter function that requires the function crosses the zero position at each symbol period. This property is also translated to the frequency domain and the frequency spectra of each sub-channel would be at maximum when the others fall to zero values.

The requirement of orthogonality and discrete spectra of the sub-channels can be fully supplied by the discrete Fourier transform (DFT) techniques. Thus, a block diagram of the discrete model for OFDM using N-point FFT/IFFT can be seen in Figure 4.27. Details of OFDM operations can be found in Chapter 13. The electronic processing parts can be implemented using fast high-sampling rate equipment such as the Tektronix arbitrary signal generator‡‡.

4.7.2 Experimental Implementation of Coherent Detection

Figure 4.4 shows the schematic functional blocks and an experimental setup for a coherent optical OFDM system. OFDM signals in the electrical domain can be generated using a Tektronix arbitrary waveform generator (Tektronix AWG710B). Time domain waveform is first generated with a MATLAB® program including mapping $2^{15}-1$ PRBS into corresponding 128 QPSK encoded sub-carriers within multiple OFDM symbols, which are subsequently converted into time domain using IFFT, and inserted with guard interval (GI).

The digital waveform is then uploaded into the Tektronix arbitrary waveform generator (AWG) operated at 4 GSymbs/s to produce a real-time analog RF OFDM signal. The AWG performs all the necessary FFT and IFFT of the block diagram shown in Figure 4.4a, which generates the

---

‡‡ AWG7000 Series, http://www.tek.com/site/ps/0,,76-19779-INTRO_EN,00.html

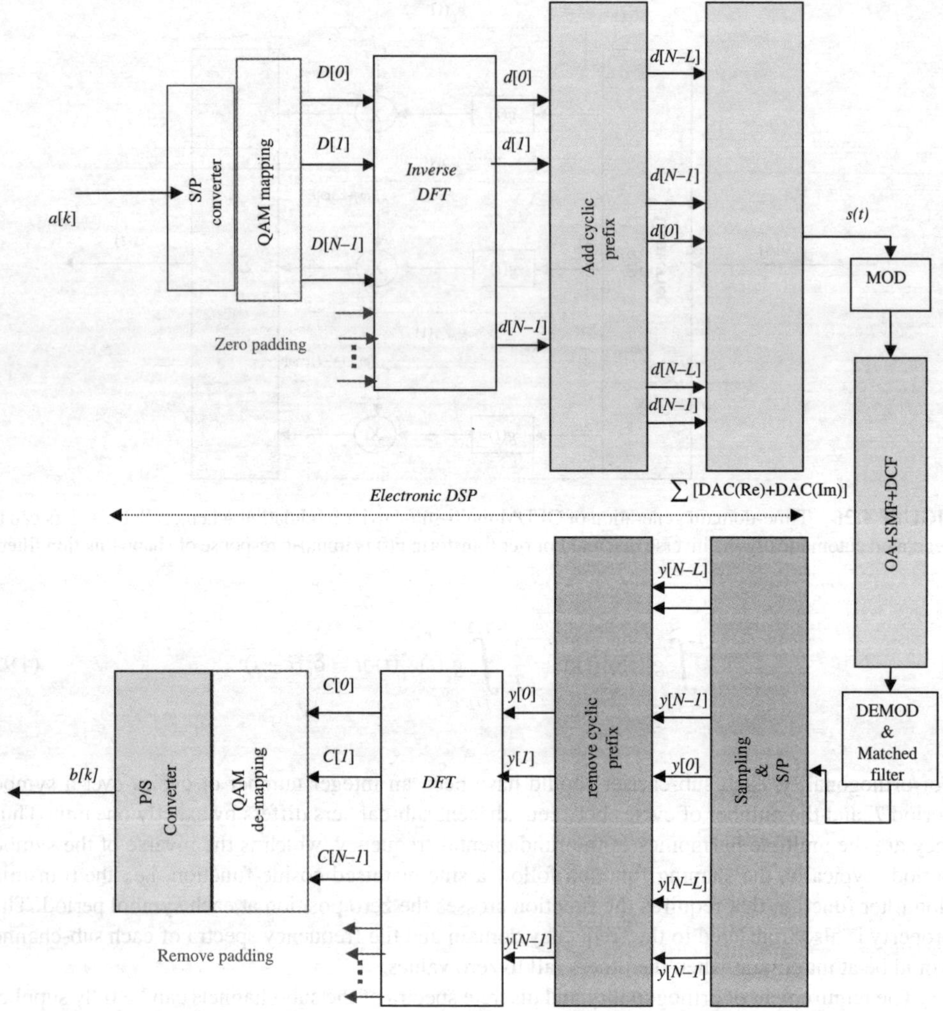

FIGURE 4.27 Block diagram of the discrete model or OFDM system using N-point FFT/IDFT. Adapted from [4.41, 42].

analog signals from the multi-channel OFDM signals and then feeds them to the optical modulator – DQPSK (I and Q) or other modulation formats, which can then be used at this stage to produce required lightwave signals for transmission. The IDFT block generates orthogonal signals, hence OFDM in the electrical domain and the modulation of optical domain signals is straight forward as described in Chapter 2.

An optical I-Q modulator should be used for direct up-conversion to the optical domain, with the real and imaginary part of the OFDM signal feeding into the real and imaginary arm of the optical modulator. However, the Tektronix AWG has only one output, equivalent of only the real part of the OFDM signal. Thence, a single-drive MZIM that can be biased at zero output is used for RF-to-optical OFDM up-conversion. Thus, equivalently only one modulation is supplied in the optical I-Q modulator (see Figure 4.28).

Such waveform generation still produces 128 OFDM sub-carriers, but with the constraint expressed as

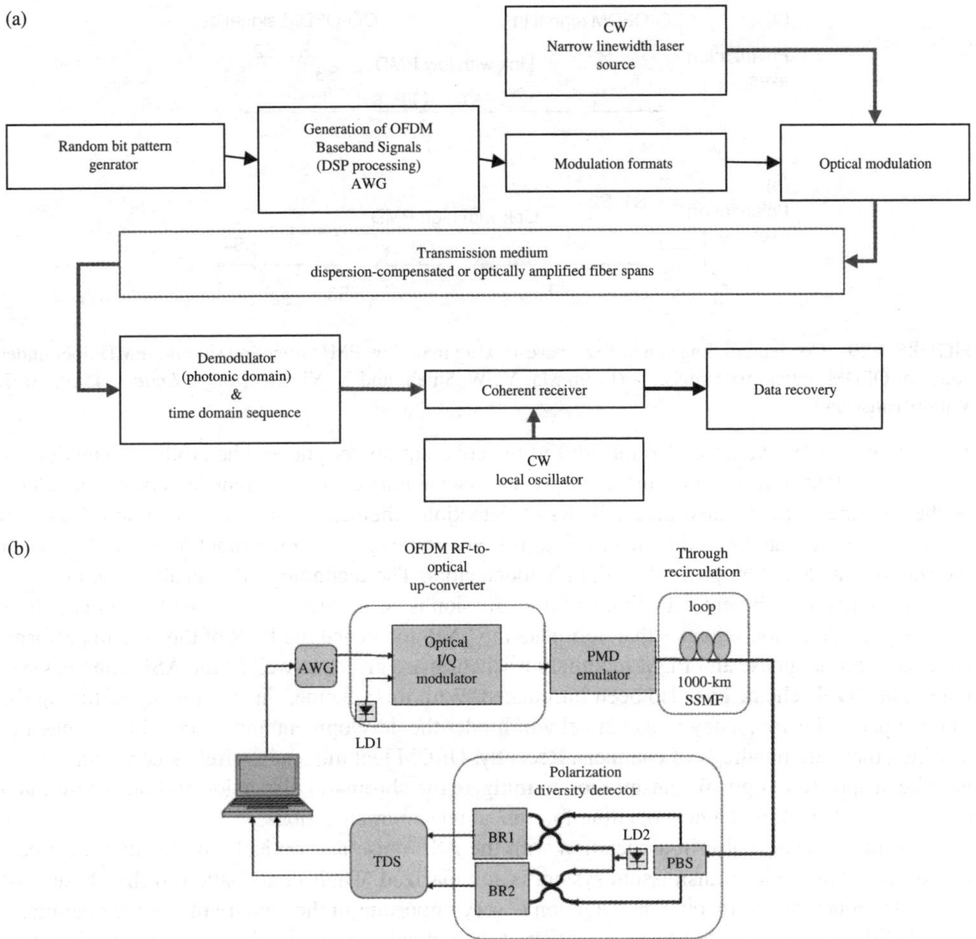

FIGURE 4.28 Schematic diagram of (a) generic blocks (b) experimental test-bed of OFDM coherent transmission system. PBS: polarization beam splitter, TDS: time domain sampling scope, AWG: arbitrary waveform generator, BR1, BR2: Balanced receiver [extracted from [4.43]]. LD1 is the lightwave carrier, LD2 is the local oscillator source. (From Shieh, W., X. Yi, and Y. Tang, *Electronics Letters*, 43(3), 2007. With permission.)

$$a_k + a_{N-k} = [00] \quad a_k \in \{00,11\}$$
$$a_k + a_{N-k} = [11] \quad a_k \in \{01,10\} \qquad \text{with} \quad N = 128. \qquad (4.93)$$

A coherent OFDM receiver could be arranged similar to other modulation formats with the LO combined with the lightwave OFDM channels at the balanced receiver, as shown in Figure 4.29a and Figure 4.29b. Any effects in the transmission of real optical transmission medium can be emulated, such as the PMD emulator shown in Figure 4.29b [23,24]. Non-linearity can be implemented in simulation using the NLSE split step and the power of the signals, while in the practical system it is simply an increase of the power level of the carrier.

4.8 REMARKS

This chapter has introduced the concept of coherent optical communications systems with the transmitters and receiver concepts for achieving the true modulation formats and improvement of

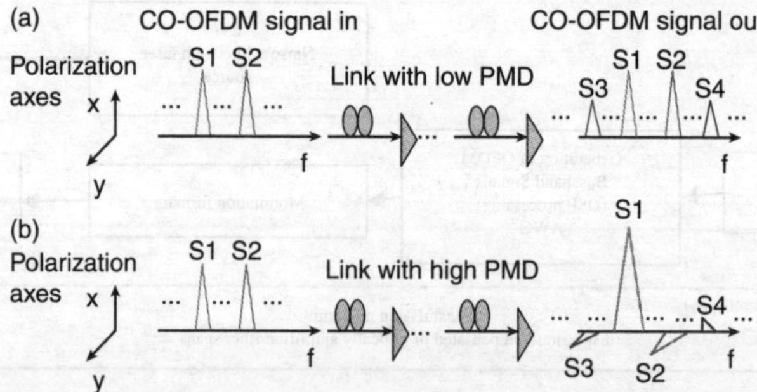

FIGURE 4.29 Conceptual diagram of four-wave-mixing in (a) low PMD fiber and (b) high PMD fiber under coherent OFDM optical transmission. (From Ma, Y., W. Shieh, and X. Yi, *Electronics Letters*, 43(3), 2007. With permission.)

receiver sensitivity. We have also introduced the coherent homodyne and heterodyne detection, as well as the balanced receiver to further improve the sensitivity of the coherent detection. The effects of the LO linewidth are also given. BERs of detection schemes are assumed to follow Gaussian pdfs, and are estimated based on the SNR of the electronic signals at the input of the receiver after the square law detection process of the photodetectors. The technology that enables the practical implementation of coherent detection and transmission is described, in particular the optical filters and the photodetection schemes that dominate the SNR and hence the BER of the system performance. Differential phase and FDM modulation schemes are given, as well as the ASK coherent systems. The FDM scheme has also been introduced. With this scheme, the complexity of tuning the LO to a particular frequency of a channel will hinder the development and practicality of schemes in which there are hundreds of channels. Recently, OFDM techniques in wireless communications have been applied to optical transmission to mitigate the chromatic dispersion and other unwanted effects. This has offered a new opening for optical telecommunications.

Non-linear phase noise in association with the ASE noise and optical amplification over cascaded optical amplifier transmission system is summarized. Their contribution to the phase shift keying for coherent and incoherent detection is very important in the penalty of the eye opening.

OFDM has attracted significant attention of the optical communications community due to its potentials in the reduction of channel bandwidth of the sub-channels, hence long-haul transmission without using dispersion compensating fibers and associated optical amplifiers. Furthermore, the orthogonality of the channels can be generated using ultra-fast DSP and the nature of the orthogonality of the FFT and IFFT and multicarrier feature make this scheme very attractive.

REFERENCES

1. Yamamoto, Y. 1980. Receiver performance evaluation of various digital optical modulation-demodulation systems in the 0.5–1.0 Mn wavelength region. *IEEE Journal of Quantum Electronics* QE-16: 1251– 59.
2. Alferness, R. C. 1981. Guided wave devices for optical communication. *IEEE Journal of Quantum Electronics* QE-17: 946–59.
3. Stallard, W. A., A. R. Beaumont, and R. C. Booth. 1986. Integrated optic devices for coherent transmission. *IEEE Journal of Lightwave Technology* LT-4 (7).
4. Gee, C. M., and G. D. Thurmond. 1984. High-speed integrated-optics traveling-wave modulator. In *Proceedings 2nd Euro. Conf. Integrated Opt.*, October 17–18, 1983, Firenze, Italy, 118–21.
5. Shimmada, S., ed. 1992. *Coherent lightwave communications technology*. London: Chapman & Hall.
6. Binh, L. N., J. Livingstone, and D. H. Stevens. 1980. Optimization of a collinear acousto-optic TEm–TMn mode converter in LiNbO3', *Proc. IEE Microwave, Optics and Acoustics* 127, Pt. H, No. (6): 323—29.
7. Shimada, S. 1995. *Coherent lightwave communications technology*. London: Chapman & Hall.

8. Shimada,S.,ed. 1995. *Coherent lightwave communications technology*, Chapter 2 of *Theory of optical coherent detection*, 27. London: Chapman & Hall.

9. Kazovsky, L. G. 1987. Multi-channel coherent optical communications systems. *IEEE Journal of Lightwave Technology* LT-5: 1095–02.

10. Mikroulis, S., H. Simos, E. Roditi, A. Chipouras, and D. Syvridis. 2006. 40-Gb/s NRZ and RZ operation of an all-optical AND logic gate based on a passive InGaAsP/InP microring resonator. *Journal of Lightwave Technology* 24 (3): 1159–64.

11. Salamon, A., et al. 2003. *MILCOM'2003*, Boston, MA.

12. Malyon, D. J. 1983. Digital transmission over 30-km monomode fiber link using optical homodyne detection. In Proceedings 9th Euro. *Conf. Opt. Commun.* 555–58.

13. Taylor, M. 2004. *IEEE Photonics Technology Letters* 16 (2): 674–76.

14. Satoshi Tsukamoto, Dany-Sebastien Ly-Gagnon, Kazuhiro Katoh and Kazuro Kikuchi. 2006. Coherent demodulation of 40-Gbit/s polarization-multiplexed QPSK signals with 16-GHz spacing after 200-km transmission. PDP29, *Proceedings of OFC*.

15. Cai, Y., L. Liu, A. N. Pilipetskii, M. Nissov, and N. S. Bergano. 2006. On performance of coherent phase-shift-keying modulation in 40 Gb/s long-haul optical fiber transmission systems, OFC, Anaheim CA USA, Paper JThB11.

16. Hoffmann, D., H. Heidrich, G. Wenke, G. Langenhorst, and E. Dietrich. 1989. Integrated optic eight port90o-hybrid on LiNbO3. *IEEE Journal of Lightwave Technology* 7 (5): 794–98.

17. Paiam, M. R., and R. I. MacDonald. 1997. Design of phased array wavelength division multiplexers using multimode interference couplers. *Applied Optics* 36 (21): 5097–5108.

18. Le Nguyen Binh. 2007. *Photonic signal processing*, Chapter 2. Orlando, FL: CRC Press, Taylor & Francis.

19. Seimetz, M. 2005. Multi-format transmitters for coherent optical M-PSK and M-QAM transmission. *ICTON 2005,225*, Th.B1.5.

20. Heatley, D. J. T., R. A. Lobbett, T. K. White, A. J. Farthing, K. I. Gilbert, W. A. Stallard, and A. R. Beaumont. Analogue video transmission over cabled monomode fiber using optical frequency modulation with heterodyne detection. *Electronics Letters* 21.

21. Emura, K., M. Shikada, S. Yamazaki, K. Komatsu, I. Mito, and K. Minemura. 1985. 400-Mbitis optical DPSK heterodyne detection experiments using DBR laser diodes with external optical feedback. In *Proceedings 5th Int. Conf. Integrated Opt. Optical Commun.*, vol. 1, October 1–4, Venice, Italy, 401–4.

22. Shieh, W., X. Yi, Y. Tang.2007. Transmission experiment of multi-gigabit coherent optical OFDM systems over 1000 km SSMF fiber. *Electronics Letters* 43(3).

23. Ma, Y., W. Shieh, X. Yi. 2007. Characterization of nonlinearity performance for coherent optical OFDM signals under the influence of PMD. *Electronics. Letters.*

24. Ma, Y., W. Shieh, X. Yi. 2007. Characterization of nonlinearity performance for coherent optical OFDM signals under the influence of PMD. *Electronics Letters.*

25. Smith, D. W., T. G. Hodgkinson, D. J. Malyon, and P. Healey. 1981. Coherent communication schemes. *Journal of Optical Communications* 2: 89–96.

26. Alferness, R.C. 1982. Waveguide electrooptic modulators. *IEEE Journal of Microwave Theory and Techniques* MTT-30: 1121–37.

27. Booth, R. C. 1985. LiNb0,-integrated optic devices for coherent optical fiber systems. *Thin Solid Films* 126: 167–76.

28. Ritchie, S., and A. G. Steventon. The potential of semiconductors for optical-integrated circuits. In *Proceedings Conf. Digital Optical Circuit Tech.*, vol. 362, NATO, France: 1111–120.

29. Favre, F., and D. Le Guen. 1982. Effect of semiconductor laser phase noise on BER performance in an optical DPSK heterodyne-type experiment. *Electronics Letters* 18: 964–65.

30. Malyon, D. J., T. G. Hodgkinson, D. W. Smith, R. C. Booth, and B. E. Daymond-John. 1983. PSK homodyne receiver sensitivity measurements at 1.5 pm. *Electronics Letters* 19: 144–45.

31. Wyatt, R., T. G. Hodgkinson, and D. W. Smith. 1983. 1.52-pm PSK heterodyne experiment featuring external cavity diode laser local oscillator. *Electronics Letters* 550–52.

32. Heatley, D. J. T., and T. G. Hodgkinson. 1984. Video transmission over cabled monomode fiber at 1.523 pm using PFM with 2-PSK heterodyne detection. *Electronics Letters* 20: 110–12.

33. Malyon, D. 1984. Digital fiber transmission using optical homodyne detection. *Electronics Letters* 20: 281–83.

34. Shikada, M., E. Emura, S. Fujita, M. Kitamura, M. Arai, M. Kondo, and K. Minemura. 100-Mb/s ASK heterodyne detection experiment using 1.3-pm DFB laser diodes. *Electronics Letters* 20: 1171.

35. Linke, R. A., B. L. Kasper, N. A. Olsson, R. C. Alferness, L. L. Buhl, and A. R. McCormick. 1985. Coherent lightwave transmission over 150 km fiber lengths at 400 Mbit/s and 1 Gbit/s data rates using DPSK modulation. In *Proceedings 5th Int. Conf. Integrated Opt. Optical Fiber Commun.*, vol. 3, October 1–4, Venice, Italy, 3, 35–38.

36. Stallard, W. A., T. G. Hodgkinson, K. R. Preston, and R. C. Booth. 1985. A novel $LiNbO_3$-integrated optic component for coherent optical heterodyne detection. *Electronics Letters* 21: 1077–79.

37. McCaughan, L., and E. J. Murphy. 1983. Influence of temperature and initial titanium dimensions on fiber-Ti:$LiNbO_3$ waveguide insertion loss at A = 1.3 pm. *IEEE Journal of Quantum Electronics* QE-19: 131–36.

38. Riviere, L., A. Carenco, A. Yi-Yan, and R. Guglielmi. 1985. Normalized diagrams for diffused wave-guides optical properties: Application to Ti: $LiNbO_3$, electro-optical directional coupler design. In *Proceedings 3rd Euro. Conf. Integrated Optics*, May 6–8, Berlin, Germany, 53–57.

39. Eisenstein, G., S. K. Korotky, L. W. Stulz, J. J. Veselka, R. M. Hopson, and K. L. Hall. 1985. Antire-flection coatings on lithium niobate waveguide devices using electron beam evaporated yttrium oxide. *Electronics Letters* 21: 363–64.

40. Alferness, R. C., V. R. Ramaswamy, S. K. Korotky, M. D. Divino, and L. L. Buhl. 1982. Efficient single-mode fiber to titanium diffused lithium niobate waveguide coupling for λ=1.32 µm. *IEEE Journal of Quantum Electronics* QE-18: 1807–12.

41. Alferness, R. C., and M. D. Divino. 1984. Efficient fiber to X-cut Ti:LiNbO3, waveguide coupling for λ=1.32 µm. *Electronics Letters* 20: 465–66.

42. Lowery, A. J., L. Du, and J. Armstrong. 2006. Orthogonal-frequency division multiplexing for disper-sion compensation of long haul optical systems. *Proceedings of OFC*, Anaheim, USA, PDP39.

43. Ali, A. 2006. Orthogonal frequency division multiplexing in optical transmission systems with high spectral efficiency. Master Thesis Dissertation, Chair of Communications, Germany: CA University-Kiel.

5 Differential Phase Shift Keying Photonic Systems

5.1 INTRODUCTION

Owing to tremendous growing demand on high-capacity transmission over the Internet, high data rate of 40 Gb/s per channel has appeared to be an attractive feature in the next generation of lightwave communications systems. Under the current 10-Gb/s DWDM optical system, overlaying 40 Gb/s on the existing network can be considered to be the most cost-effective method for upgrading. However, there are a various technical difficulties confronted by communications engineers involving interoperability that requires a 40-Gb/s line system to have signal optical bandwidth, tolerance to chromatic dispersion, resistance to non-linear crosstalk, and susceptibility to accumulated noise over the multi-span of optical amplifier to be similar to the 10-Gb/s system.

In view of this, advanced modulation formats have been demonstrated as an effective scheme to overcome 40-Gb/s system impairments. Differential phase shift keying (DPSK) modulation format has elicited extensive studies due to its benefit over conventional on–off keying (OOK) signaling format. This includes 3-dB lower optical signal-to-noise ratio (OSNR) [1] at a given bit-error rate (BER), more robust to narrow-band optical filtering, and more resilience to some non-linear effects (e.g., cross-phase modulation and self-phase modulation). Moreover, spectral efficiencies can be improved by using multi-level signaling. Also, coherent detection is not critical because DPSK detection requires comparison of two consecutive pulses, hence the source coherence is required over only one bit period.

Nevertheless, the DPSK format involves rapid phase change, causing intensity ripples due to chromatic dispersion that induces pattern dependent SPM-GVD (group velocity dispersion) effect [2]. Therefore, return-to-zero (RZ) pulse can be employed in conjunction with DPSK to generate more tolerance to the data pattern-dependent SPM-GVD effect. In addition, RZ improves dispersion tolerance and non-linear effects, particularly in long-haul networks at high data rate. Specific RZ formats such as carrier-suppressed RZ (CSRZ) help reduce the inherent wide spectral bandwidth.

At 40 Gb/s, generation of RZ pulse is not feasible because it is at 10 Gb/s due to the large bandwidth requirement. Thus, 40-Gb/s RZ signals are produced optically in "pulse carver" by driving the modulator with a 20-GHz RF signal. With the remarkable advancement in external modular, especially Mach–Zehnder interferometic (intensity) modulator (MZIM), this is easily achieved by utilizing the microwave optical transfer characteristic of MZIM.

Despite the telecommunications boom and subsequent bust, Internet traffic has been steadily growing. This growth requires new investment in telecommunications infrastructure to provide long-haul communications capacity between major population centers in the world. The transmission route between Melbourne to Sydney and *vice versa* is one of the most intensive, demanding upgrading to higher capacity, especially from the current 10 Gb/s to 40 Gb/s. One of the most cost-effective ways to provide such upgrades is to use existing fiber infrastructure, and upgrade the transmitters and receivers at either end of the long-haul-links. That means the transmission rate of 40 Gb/s over the existing 10G RZ format dense multi-wavelength optical communications systems.

Because of the properties of the installed fibre (which is older and also degraded than state-of-the-art fibers used in laboratory "hero" experiments), the transmission methods must be highly tolerant to chromatic dispersion (CD) and polarization-mode dispersion (PMD). This favors advanced modulation formats (such as variants of phase-shift-keying) rather than ultra-high-rate time division-multiplexed (TDM) schemes, because the effects of CD and second-order PMD are proportional to bit-rate squared. However, there are various optical filters installed throughout the DWM optical transmission systems such as optical multiplexers (mux), demultiplexers (demux) and add-drop muxes that would affect the spectral properties of the multiplexed channels, particularly with a hybrid transmission of 10 Gb/s and 40 Gb/s channels.

Advanced modulation formats are considered to play a significant part in enhancing the effectiveness of bandwidth reduction and effective transmission over long distance [1,2]. These advanced modulation formats are new to optical communications, but very well known in wire and wireless communications systems. Accordingly, the modulation can be done by manipulating the amplitude, frequency or phase of the carriers corresponding to the coding of the input data sequence. Coherent or incoherent transmission and detection techniques have been possible over the years for optical systems. However, incoherent detection is preferred so as to minimize the linewidth obstacles of the lasers at the transmitter and the local oscillators required for homodyne or heterodyne coherent detection systems. Furthermore, the phase comparison in differential detection can be easily implemented using the delay interferometric photonic component. This is extremely difficult if implemented in the electrical domain at very high bit rate. Therefore, differential discrete or continuous phase modulation formats may be preferred.

This chapter investigates the transmission of 40-Gb/s channels over a 10-Gb/s DWDM optical system in which "standard" single optical filters (SSMF) usually employed for 10 Gb/s systems are used. Differential phase shift keying (DPSK) is transmitted and compared with those employing RZ and NRZ amplitude shift keying (ASK). We demonstrate that with the passband of optical filters of the order of 0.5 nm, the transmitted 40-Gb/s channels are not penalized and *vice versa* for 10 Gb/s DWDM channels. Transmission BER and receiver sensitivities are reported for different transmission scenario.

Due to the high demand of data transmission capacity, and with the advent of wavelength division multiplexing (WDM) technology, data transmission has moved into a revolutionary stage for increasing higher system capacity. Higher bit rates up to 40 Gb/s and growing numbers of WDM channels are challenging the limit of amplitude shift keying (ASK) format, numerous modulation formats using phase shift keying, frequency shift keying and multi-subcarrier emerge a potential techniques to stretch the distance of optical communications.

5.2 OPTICAL DPSK MODULATION AND FORMATS

Figure 5.1 shows the structure of a 40-Gb/s DPSK transmitter in which two external LiNbO$_3$ optical interferometric modulators (MZIM) are used [3]. Operational principles of on MZIM were presented in Chapter 2. The MZIM used in Figure 5.1 can be a single- or dual-drive type. As previously described, the optical DPSK transmitter consists of a narrow linewidth laser source to generate a lightwave of wavelength conformed to the ITU grid which is then modulated via two cascade MZIMs: one to generate a periodic (clock-like) return to zero or CSRZ formats before feeding through the data modulator that would switch on and off for amplitude modulation of exerting a phase shift to the lightwave carrier for discrete phase modulation. This shift can be continuous between the states, as in the case of continuous-phase FSK or minimum shift keying modulation formats, which are presented in later chapters.

5.2.1 GENERATION OF RETURN-TO-ZERO PULSES

The first MZIM, commonly known as the "pulse carver", is used to generate the periodic pulse trains with a required return-to-zero format (RZ) format. The suppression of the lightwave carrier can also be carried out at this stage if necessary and known as carrier-suppressed RZ (CSRZ). CSRZ pulse shape is found to have attractive attributes in long-haul WDM transmissions compared

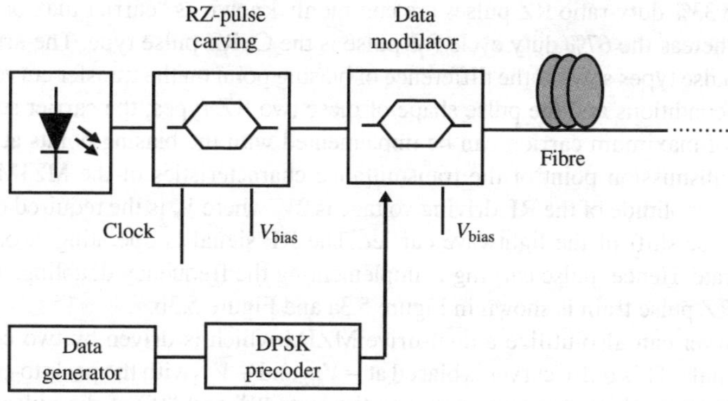

FIGURE 5.1 Schematic diagram of a discrete PSK optical transmitter using cascaded MZI modulators.

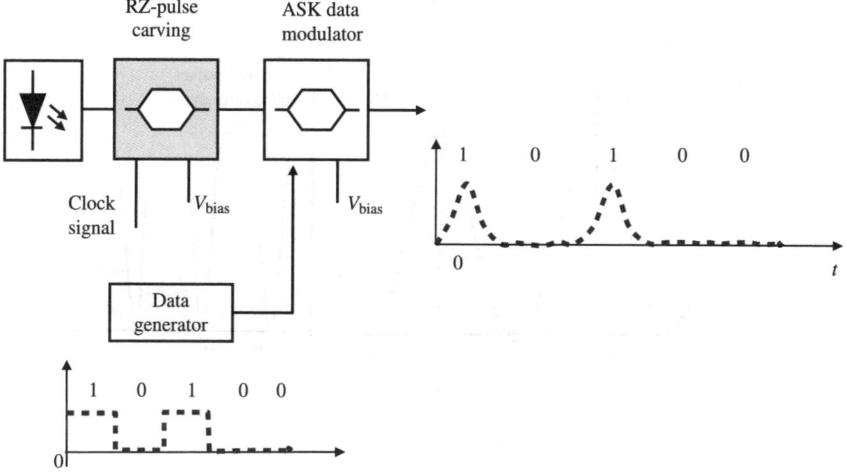

FIGURE 5.2 Cascaded MZIM for generation of RZ or CSRZ DPSK lightwave signals.

with other RZ types, including optical phase difference of π in adjacent bits, suppression of the optical carrier component in optical spectrum and smaller spectral width. Figure 5.2 shows the structure of a 40-Gb/s ASK transmitter in which two external LiNbO$_3$ optical interferometric modulators (MZIM) [4] are used. Operational principles of MZIM were presented in Chapter 2. The MZIM used in Figure 5.2 can be a single- or dual-drive type. The optical ASK transmitter would consist of a narrow linewidth laser source to generate a lightwave of wavelength conformed to the ITU grid.

Different types of RZ pulses can be generated depending on the driving amplitude of the RF voltage and the biasing schemes of the MZIM. The equations governing the RZ pulses electric field waveforms are [1]

$$E(t) = \begin{cases} \dfrac{1}{\sqrt{E_b}} \sin\left[\dfrac{\pi}{2}\cos\left(\dfrac{\pi t}{T_b}\right)\right] & \text{67\% duty-ratio RZ pulses} \\[4mm] \dfrac{1}{\sqrt{E_b}} \sin\left[\dfrac{\pi}{2}\left(1+\sin\left(\dfrac{\pi t}{T_b}\right)\right)\right] & \text{33\% duty-ratio RZ pulses} \end{cases} \tag{5.1}$$

where E_b is the pulse energy per transmitted bit.

The first type 33% duty-ratio RZ pulses are commonly known as "carrier max or conventional" (RZ33) pulses, whereas the 67% duty cycle RZ pulse is the CSRZ pulse type. The art in generation of these two RZ pulse types stays at the difference of biasing point on the transfer curve of an MZIM. The bias voltage conditions and the pulse shape of these two RZ types, the carrier suppression and non-suppression of maximum carrier, can be implemented with the biasing points at the minimum and maximum transmission point of the transmittance characteristics of the MZIM, respectively. The peak-to-peak amplitude of the RF driving voltage is $2V_\pi$ where V_π is the required driving voltage to achieve a π phase shift of the lightwave carrier. The RF signal is operating at only half of the transmission bit rate. Hence, pulse carving is implementing the frequency doubling. The generation of RZ33 and CSRZ pulse train is shown in Figure 5.3a and Figure 5.3b.

The pulse carver can also utilize a dual-drive MZIM which is driven by two complementary sinusoidal RF signals. This pulse carver is biased at $-V_{\pi/2}$ and $+V_{\pi/2}$ with the peak-to-peak amplitude of $V_{\pi/2}$. Thus, a π phase shift is created between the state "1" and "0" of the pulse sequence and hence the RZ with alternating phase 0 and π. If the carrier suppression is required, then the two electrodes are applied with voltages V_π and swing voltage amplitude of V_π.

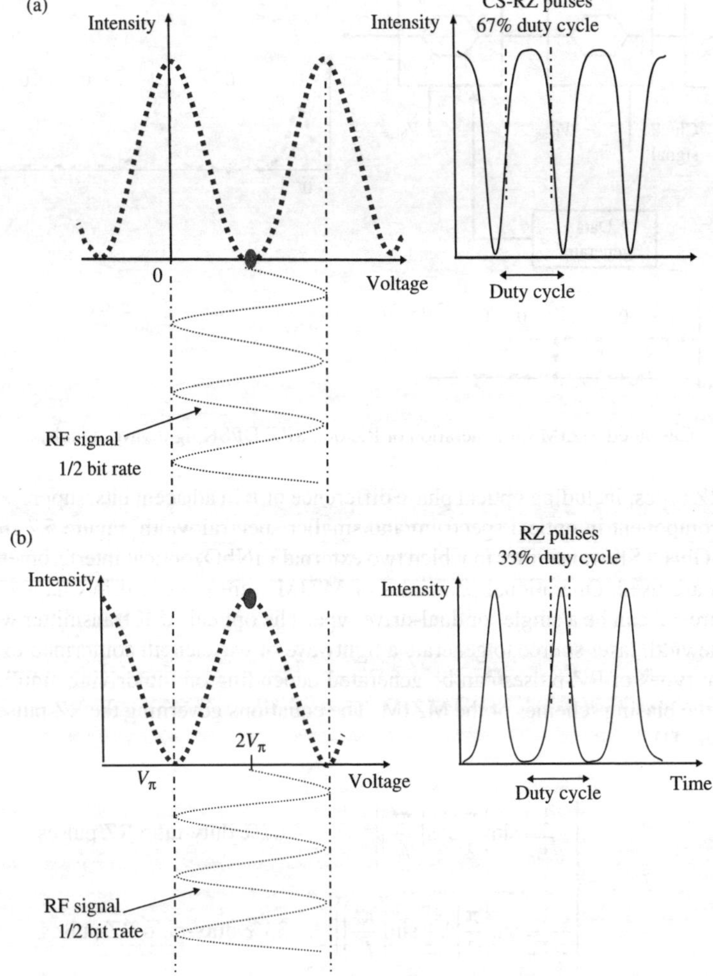

FIGURE 5.3 Biasing and driving sinusoidal electrical waveform to generate (a) CS-RZ and (b) RZ-33%.

RZ optical systems have been proven to be more robust against impairments, especially self-phase modulation and polarization mode dispersion. Although RZ modulation offers improved performance, RZ optical systems usually require more complex transmitters than those in NRZ optical systems. Compared with only one stage for modulating data on the NRZ optical signals, two modulation stages are required for generation of RZ optical pulses.

5.2.2 Phasor Representation

Recalling the derivation of a phase modulated interferometer as given in Chapter 2, we have the output field of the MZIM given by

$$E_{o} = \frac{E_{i}}{2}\left[e^{j\varphi_1(t)} + e^{j\varphi_2(t)}\right] = \frac{E_{i}}{2}\left[e^{j\pi v_1(t)/V_\pi} + e^{j\pi v_2(t)/V_\pi}\right]. \tag{5.2}$$

The modulating process for generation of RZ pulses can be represented by a phasor diagram, as shown in Figure 5.4. This technique gives a clear understanding of the superposition of the fields at the coupling output of two arms of the MZIM. Here, a dual-drive MZIM is used, i.e., the data $[V_1(t)]$ and the complementary data ($\overline{\text{data}}$: $V_2(t) = -V_1(t)$) are applied into each arm of the MZIM, respectively, and the RF voltages swing in inverse directions. Applying the phasor representation, vector addition and simple trigonometric calculus, the process of generation RZ33 and CSRZ is explained in detail and verified.

The width of these pulses are commonly measured at the position of full-width half maximum (FWHM). The measured pulses are intensity pulses, whereas we are considering the addition of the fields in the MZIM. Thus, the normalized E_0 field vector has the value of $\pm 1/\sqrt{2}$ the FWHM intensity pulse positions, and the time interval between these FWHM points gives the FWHM values.

5.2.3 Phasor Representation of CS-RZ Pulses

Key parameters including the V_{bias}, the amplitude of driving voltage and its correspondent initialized phasor representation are shown in Figure 5.5a and Figure 5.5b.

Swing voltage of driving RF signal on each arm has the amplitude of $V_\pi/2$ (i.e., $V_{\text{p-p}} = V_\pi$). RF signal operates at half of bit rate ($B_R/2$). At the FWHM position of the optical pulse, the $E_{\text{output}} = \pm 1/\sqrt{2}$ the component vectors V_1 and V_2 form with vertical axis a phase of $\pi/4$, as shown in Figure 5.6.

Considering the scenario for generation of 40-Gb/s CSRZ optical signals (25 ps pulse width), the modulating frequency is $f_{\text{m}} = 20\,\text{GHz} = B_R/2$. At the FWHM positions of the optical pulse, the phase is given by the following expressions

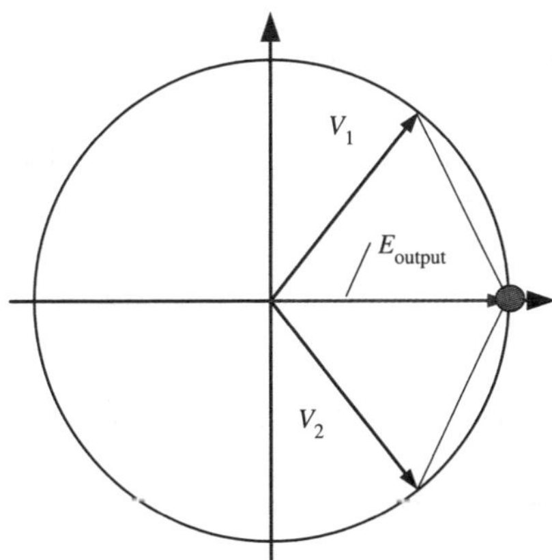

FIGURE 5.4   Phasor representation for generation of output field in dual-drive MZIM.

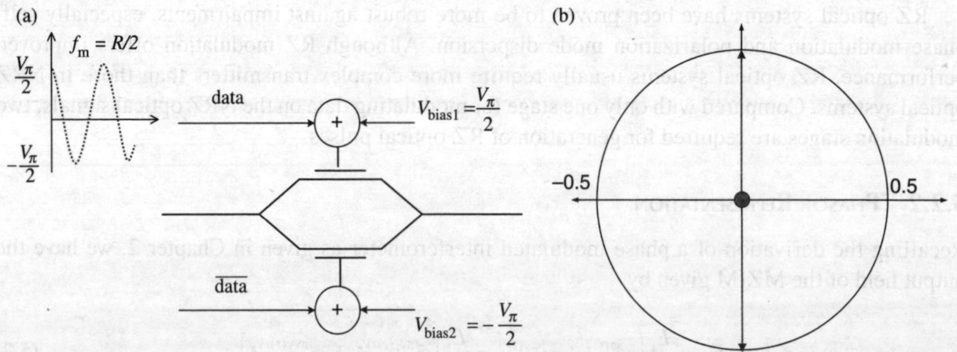

FIGURE 5.5 Biasing voltages of (a) MZIM structure and (b) phasor representation for RZ pulse generation. V_{bias} is $\pm V_\pi/2$.

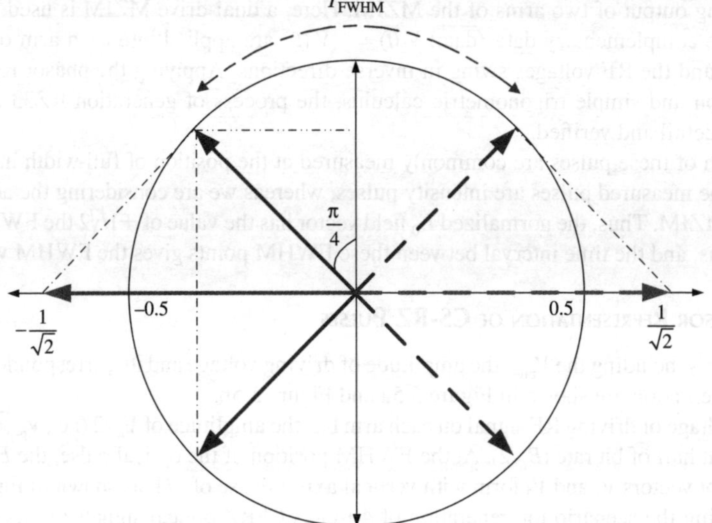

FIGURE 5.6 Phasor representation of evolution of CSRZ pulse generation using dual-drive MZIM.

$$\frac{\pi}{2}\sin(2f_m) = \frac{\pi}{4} \Rightarrow \sin 2f_m = \frac{1}{2} \Rightarrow 2f_m = \left(\frac{\pi}{6}, \frac{5\pi}{6}\right) + 2n\pi. \tag{5.3}$$

Thus, the calculation of TFWHM can be carried out and hence, the duty cycle of the RZ optical pulse can be obtained, as given in the following expressions

$$T_{FWHM} = \left(\frac{5\pi}{6} - \frac{\pi}{6}\right)\frac{1}{B_R\,2\pi} = \frac{1}{3}\times\frac{1}{B_R} \Rightarrow \frac{T_{FWHM}}{T_{BIT}} = \frac{1.66\times10^{-11}}{2.5\times10^{-11}} = 66.67\%. \tag{5.4}$$

The result clearly verifies the generation of CSRZ optical pulses.

5.2.4 Phasor Representation of RZ33 Pulses

Key parameters, including the V_{bias}, the amplitude of driving voltage and its correspondent initialized phasor representation are shown in Figure 5.7a and Figure 5.7b.

At the FWHM position of the optical pulse, the $E_o = \pm 1/\sqrt{2}$ and the component vectors V_1 and V_2 form with horizontal axis a phase of $\pi/4$, as shown in Figure 5.8.

Considering the scenario for generation of 40-Gb/s CSRZ optical signal (25 ps pulses), the modulating frequency is $f_m = 20\,\text{GHz} = B_R/2$. At the FWHM positions of the optical pulse, the phase is given by the following expressions

$$\frac{\pi}{2}\cos(2f_m t) = \frac{\pi}{4} \Rightarrow t_1 = \frac{1}{6f_m} \tag{5.5}$$

$$\frac{\pi}{2}\cos(2f_m t) = -\frac{\pi}{4} \Rightarrow t_2 = \frac{1}{3f_m}. \tag{5.6}$$

Thus, the calculation of TFWHM can be carried out and hence, the duty cycle of the RZ optical pulse can be obtained as given in the following expressions

$$T_{\text{FWHM}} = \frac{1}{3f_m} - \frac{1}{6f_m} = \frac{1}{6f_m} \quad \therefore \quad \frac{T_{\text{FWHM}}}{T_b} = \frac{1/6f_m}{1/2f_m} = 33\%. \tag{5.7}$$

Therefore, the result clearly verifies the generation of RZ33 optical pulses.

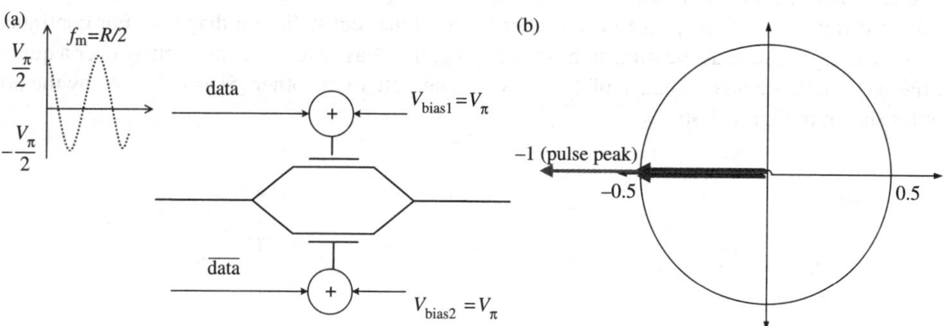

FIGURE 5.7 Generation of 33% RZ lightwave pulse sequence. V_{bias} is V_π for both arms, Swing voltage of driving RF signal on each arm has the amplitude of $V_\pi/2$ (i.e., $V_{\text{p-p}} = V_\pi$) and RF signal operates at half of bit rate ($R/2$).

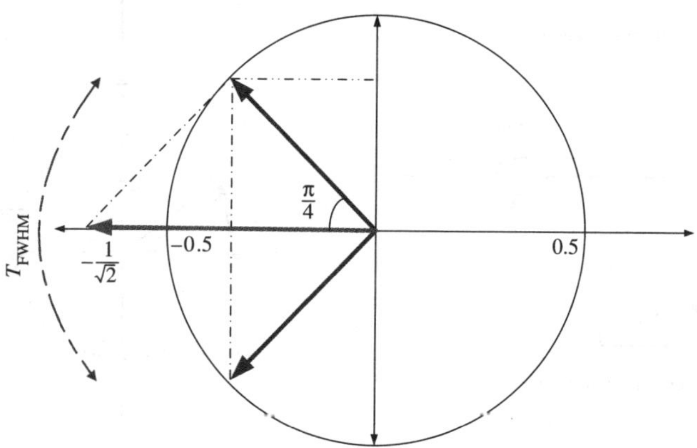

FIGURE 5.8 Phasor representation of RZ33 pulse generation using dual-drive MZIM.

5.2.5 Discrete Phase Modulation (DPSK)

5.2.5.1 Principles of DPSK and Theoretical Treatment of DPSK and DQPSK Transmission

For PSK signals, the data information is contained in the discrete phase shifts versus the phase of the unmodulated carrier. The required absolute phase at the receiver is to be provided by the recovered carrier such as the OPLL or an electrical PLL. The later technique is more complicated. Thus, it is more effective to process the phase difference of the two consecutive symbols.

Information encoded in the phase of an optical carrier is commonly referred to "optical phase shift keying". In the early days, PSK required precise alignment of the transmitter and demodulator center frequencies. Hence, PSK system was not widely deployed. When the DPSK scheme was introduced, coherent detection was not critical because DPSK detection requires only source coherence over a one bit period by comparison of two consecutive pulses.

A binary "1" is encoded if the present input bit and the past encoded bit are of opposite logic and a binary "0" is encoded if the logic is similar. This operation is equivalent to an XOR logic operation. Hence, an XOR gate is usually employed in a differential encoder. NOR can also be used to replace the XOR operation in differential encoding, as shown in Figure 5.9a. The decoding logic circuit is also shown in this figure.

In optical applications, electrical data "1" is represented by a π phase change between the consecutive data bits in the optical carrier, while state "0" is encoded with no phase change between the consecutive data bits. Hence, this encoding scheme gives rise to two points located exactly at π phase difference with respect to each other in the signal constellation diagram. For continuous phase shift keying such as the minimum shift keying, the phase evolves continuously over a quarter of the section, thus a phase change of $\pi/2$ between one state to the other [5] as indicated by the bold circles shown in Figure 5.9b.

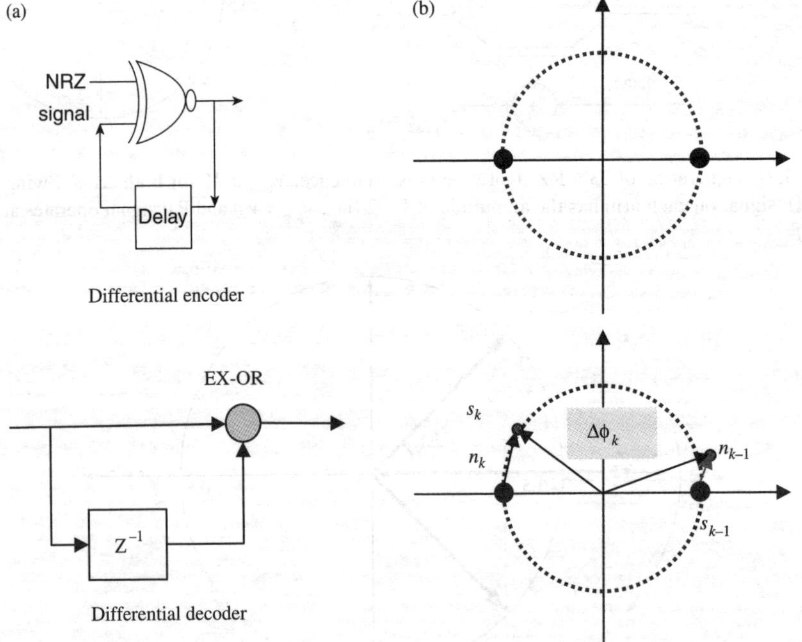

FIGURE 5.9 (a) DQPSK pre-coder logic diagram and modeling blocks (b) Signal constellation under ideal (no noise) and under noisy conditions.

For differential quadrature phase shift keying (DQPSK), another quadrature component or subsystem would be superimposed on the DPSK (the in-phase) as that they are orthogonal to each other. The phase constellation of Figure 5.9b would have another imaginary ($\pi/2$) axis with the states of $+\pi/2$ and $-\pi/2$.

DPSK under no noise and with phase noise.

5.2.5.2 Optical DPSK Transmitter

Figure 5.10 shows the structure of a 40-Gb/s DPSK transmitter in which two external LiNbO$_3$ optical interferometric modulators [3] (MZIM) are used. Operational principles of on MZIM were presented in Chapter 2. The MZIM used in Figure 5.10 can be either a single- or dual-drive type. The optical DPSK transmitter would consist of a narrow linewidth laser source to generate a lightwave of wavelength conformed to the ITU grid.

As shown in Figure 5.11, the optical RZ pulses are then fed into the second MZIM, through which the RZ pulses are modulated by the pre-coded NRZ binary data to generate RZ-DPSK optical signals. Electrical data pulses are differentially pre-coded in the pre-coder using the XOR coding scheme. Without pulse carver and sinusoidal RF signal, the system becomes a NRZ-DPSK transmitter.

In DPSK modulation, the MZIM is biased at the minimum transmission point. The pre-coded electrical data has the peak-to-peak amplitude equal to $2V_\pi$ and operates at the transmission bit rate. The modulation principles are demonstrated in Figure 5.11.

Electro-optic phase modulator might also be used for generation of DPSK signals but Mach–Zehnder wave guide structure has several advantages over the phase modulator described in [1]. Using an optical phase modulator, the transmitted optical signal is chirped whereas, using electro-optic intensity modulator (e.g., MZIM), especially X-cut versions, chirp-free signals are produced. In practice, a small amount of chirp might be useful for transmission [6].

5.2.6 DPSK BALANCED RECEIVER

Consider two consecutive symbols whose signals can be represented by

$$s_{k-1} = A\sin(2\pi f_c t + \phi_{k-1}) \quad \text{for} \quad (k-1)T_b \leq t \leq kT_b$$
$$s_k = A\sin(2\pi f_c t + \phi_k) \quad \text{for} \quad kT_b \leq t \leq (k+1)T_b$$

$$(5.8)$$

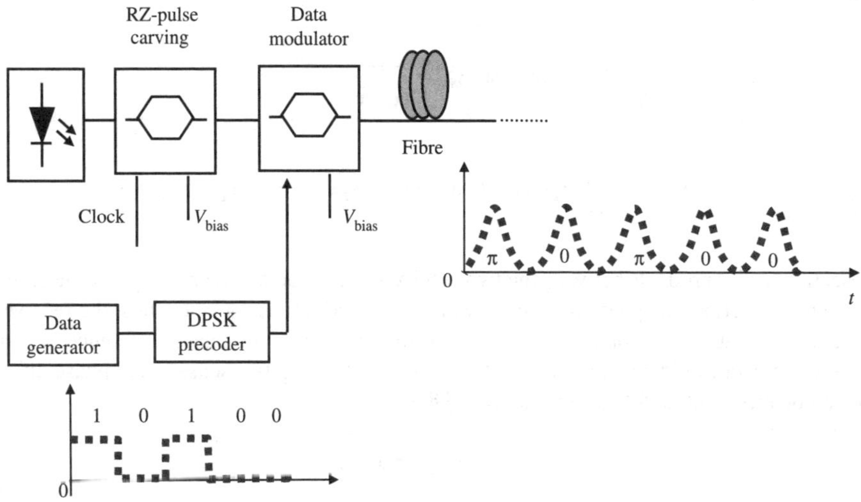

FIGURE 5.10 Generation of RZ formats DPSK modulation.

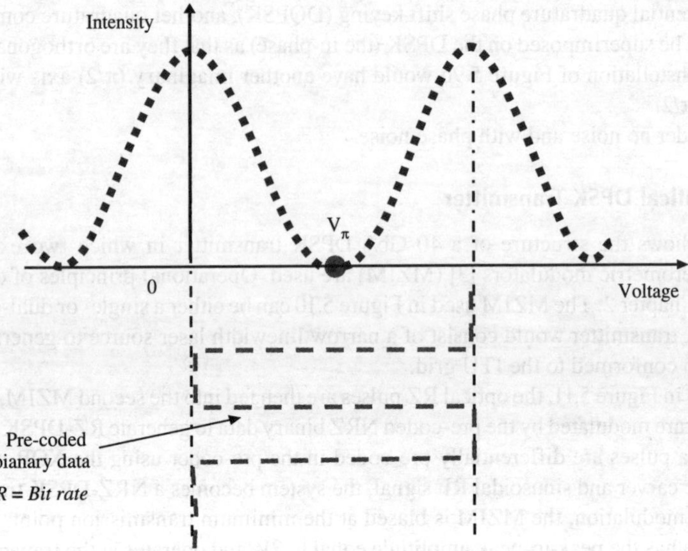

FIGURE 5.11   Voltage bias and amplitude swing levels for NRZ DPSK modulation formats.

where A is an arbitrary amplitude and f_c the carrier frequency. At the receiver side, the phase difference $\Delta\phi_k = \phi_k - \phi_{k-1}$ must be determined. If there is no distortion, then the phase difference is that of the original phase difference generated at the transmitter. However, in general, the signals are distorted when reaching the receiver, thus we have

$$x_{k-1}(t) = s_{k-1}(t) + n_{k-1}(t)$$
$$x_k(t) = s_k(t) + n_k(t).$$

$$(5.9)$$

Figure 5.9b shows the position of the data signals at the receiver side, including the additional contribution of the noise vectors of the two consecutive intervals. The phase angle between the two vectors can then be written as

$$\cos\Delta\phi_k = \frac{x_k x_{k-1}}{|x_k||x_{k-1}|} \quad \text{and} \quad \sin\Delta\phi_k = \frac{x_k x_{k-1}^*}{|x_k||x_{k-1}^*|}$$

$$(5.10)$$

$$\text{with} \quad x_k x_{k-1} = \int_0^T x_k(t)x_{k-1}(t)\,dt \quad \text{and} \quad x_k x_{k-1}^* = \int_0^T x_k(t)x_{k-1}^*(t)\,dt.$$

Representing scalar products between the two received signal vectors x_k and x_{k-1}. For the case when there are no distortion, the phase difference would be 0 and π thus $\Delta\phi_k = \pi$. Thus, the received signals are in push-pull states, bipolar nature due to the fact that if the phase difference is zero, the received waveform would be maximum and minimum in the negative sense if a pi phase difference which can be represented, following Equation 5.8 as

$$b_k = \mathrm{sgn}(x_k x_{k-1})$$
$$\rightarrow b_k = \mathrm{sgn}(\cos\Delta\phi_k)$$

$$(5.11)$$

The receiver for determining of the differential phase can be implemented using a balanced receiver operating in the push–pull mode. The delay balanced receiver is shown at the left side in Figure 5.12 in which the received optical field is split into two arms, one delayed in the optical domain by a symbol period and one which has no delay. The delay time must be matched to the propagation time of the lightwave carrier over the guided region of the optical fiber or integrated waveguide. A thermal tuning section is normally used to adjust to any mismatched time delay section. Two photo-detectors are connected in balanced configuration, hence producing a push-pull eye diagram (Equation 5.10) as shown in Figure 5.12b.

Figure 5.12 shows the structure of a MZDI-based DPSK balanced receiver with an insert of a typical eye diagram at the output of the MZDI balanced receiver for 50 ps, i.e., 20-Gb/s DPSK optical pulses. The diagram is of a RZ format so there is no continuous base line. The received signal at the output of the balanced receiver is given as [7]

$$P_{\mathrm{D}}(t) = \left|E_{\mathrm{D}}(t) + E_{\mathrm{D}}(t - \tau)\right|^2 - \left|E_{\mathrm{D}}(t) - E_{\mathrm{D}}(t - \tau)\right|^2$$

$$= 4\Re\left\{E_{\mathrm{D}}(t)\right\} E_{\mathrm{D}}^*(t - \tau) = \cos\left(\Delta\phi + \varsigma\right)$$

(5.12)

where $E_{\mathrm{D}}(t)$ and $E_{\mathrm{D}}(t - \tau)$ are the current and the one-bit delay version of the optical DPSK pulses, respectively. $\Delta\phi$ and ς represent the differential phase and the phase noise caused by the MZ delay interferometer imperfections (MZDI). The latter issue is not a severe degradation factor in modern MZDI-based DPSK receiver due to the use of a thin-film heater for tuning any waveguide path mismatch. The tuning has high stability with the implementation of the electronic feedback control circuit.

5.3 DPSK TRANSMISSION EXPERIMENT

5.3.1 COMPONENTS AND OPERATIONAL CHARACTERISTICS

The components and operating remarks of the optical transmission system shown in Figure 5.13 for ASK and DPSK modulation and related formats of RZ, NRZ and CSRZ are given in Table 5.1.

5.3.2 SPECTRA OF MODULATION FORMATS

The center wavelength is set at 1551.72 nm. The experimental spectra of NRZ, RZ and CS-RZ of ASK and DPSK formats are shown in Figure 5.14. For DPSK spectra, the carrier at the center is at the same level as that of the signal. The RZ spectra are wider than that of the NRZ as expected due to the shorter pulse width. The influence of the spectra of the optical signals on the dispersion and

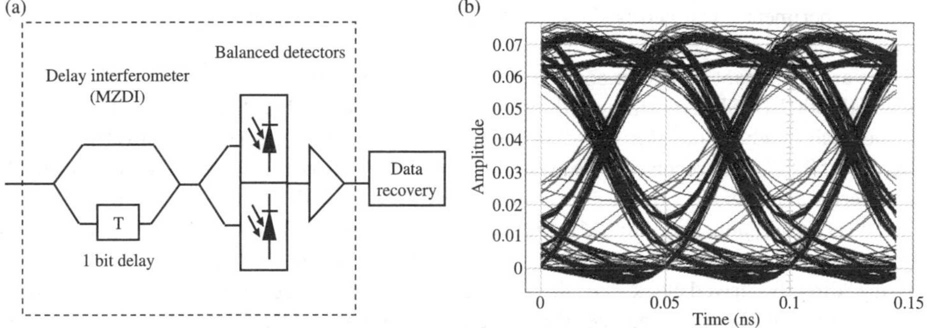

FIGURE 5.12 Schematic diagram of the balanced receiver for detection of the phase difference of two consecutive bits (a) structure (b) a sample of phase detected eye diagram. Bit rate = 20 Gb/s format RZ.

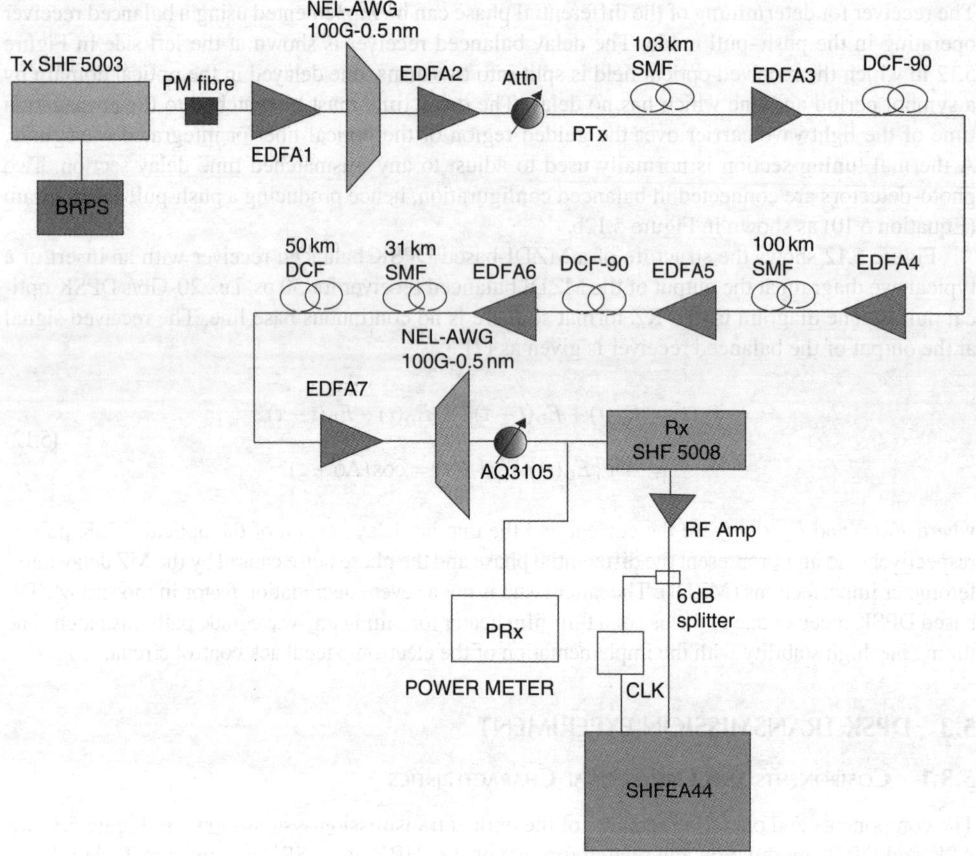

FIGURE 5.13 Experimental setup for evaluation of modulation formats of ASK and DPSK modulation, 230-km three-optically amplified spans transmission, 234 km SMF plus 230 km DCF. (From Le Nguyen Binh, *Radio Science Bulletin*, 321, 2007. With permission.)

nonlinear impairments will be discussed in a later session. For CSRZ, the carrier is clearly suppressed. Due to the resolution of the spectrum analyzer, deep suppression is not observed.

5.3.3 DISPERSION TOLERANCE OF OPTICAL DPSK FORMATS

A typical experimental test-bed is shown in Figure 5.15. The eye diagrams obtained for different dispersion length are shown in Figure 5.16.

There are significant points for consideration in the setup of the test bed as follows: (i) 40-Gbps Tx is a SHF 5003 DPSK Transmitter which can generate ASK and DPSK data; (ii) In the case of ASK, the "Lab Buddy" Optical Receiver (45 GHz with built-in amplifier) is used; (iii) In the case of DPSK, SHF 5008 DPSK receiver is used (MZ – 1 bit delay). Also, the electrical amp is utilized to drive the error analyzer; (iv) JDS 1.2-nm filter was utilized after EDFA to decrease the ASE noise level; and (v) Power launched into SMF (right after the transmitter Tx) is low, hence nonlinearities do not have an impact. The launched power is recorded as shown in Table 5.2.

Using "directly derived clock", i.e., directly from the bit-pattern generator and hence without the clock recovery unit, the error analyzer directly using clock pulse of the PRBS. Laser source center wavelength is set at 1551.72 nm and the pulse pattern of PRBS is $2^{31} -1$. Figure 5.18 shows the performance of CSRZ/RZ/NRZ-DPSK formats over lengths of 0 to 4 km (from top down to bottom) of SSMF using the eye diagrams of 40-Gb/s transmission. Hence, the dispersion tolerance to the transmission fiber SSMF with a dispersion factor of D = + 17 ps/nm.km for DPSK format versus the

TABLE 5.1
Photonic Components and Operating Characteristics

Photonic Sub-systems	Description	Remarks
Laser source	Anritsu tunable laser set at 1551.72 nm for center wavelength (lamda 5); 1548.51 nm for lamda 1 and 1560 nm for lamda 16.	Max output power of 8 dBm
Multiplexer and demultiplexer	NEL-AWG = frequency spacing 100G 3 dB BW of 0.5 nm. Circulating property so spectrum would appear cyclic (note the spectrum), note of the channel input and out put – e.g., input at port 5 of Lamda 5 then output at port 8. Input Lamda 1 at port 1 then output at port 8 (Lamda 1). This demux is a NEL array waveguide product. The AWG has a circulant property and thus can be used as cascade filters.	
Photonic transmitters (Tx) for different modulation format generation	The Tx can be set at CSRZ-DPSK or RZ- DPSK, no DQPSK format facility is available.	A pair of external modulators of MZIM single drive. SHF model 5003
Clock Recovery Module	Clock recovery using 6-dB splitter	SHF module – SHF 1120A
Optical receiver	Balanced receiver used when DPSK format is used. Otherwise Discovery Semiconductor Lab Buddy DS-10H is employed. A MZD interferometer with thermal tuning of the delay path included.	Tuning of MZ phase decoder at the Rx is necessary when wavelength (lamda) channels are changed
In-line optical amplifiers	EDFA 1 driven at 178 mW pump power (saturation mode). EDFA2 driven at 197 mW (saturation mode).	All EDFAs are driven in the saturation mode
Optical filters	NEL AWG Mux/Demmux 0.45 nm bandwidth[1] AWG JDS 8 channel Mux/Demux 0.35 nm bandwidth JDS – 1.2 nm bandwidth Santec – 0.5 nm bandwidth with slow roll-off.	
PRBS	Bit pattern generator SHF-BPG 44	
Error analyzer	SHF model EA-44	
"Directly derived clock"	Direct synchronization of the sync signals to the error analyzer. If no "directly derived clock" is indicated, then a clock recovery module is used for synchronization and sampling of the received data sequence.	
Signal monitoring	Oscilloscope: Tektronix oscilloscope with 70 GHz optical plug-in and Agilent oscilloscope with 50 GHz optical plug-in Spectrum Analyzer – Photonetics OSA.	
Other components	Optical attenuator is used to vary the optical power input at the receiver for setting the operating condition in the linear or nonlinear region. About 5 dBm is required to set the onset level of nonlinear operation. Optical attenuator is inserted in front of the receiver to ensure no damage of the photodetector.	

[1] The bandwidth here is considered as the 3 db bandwidth.

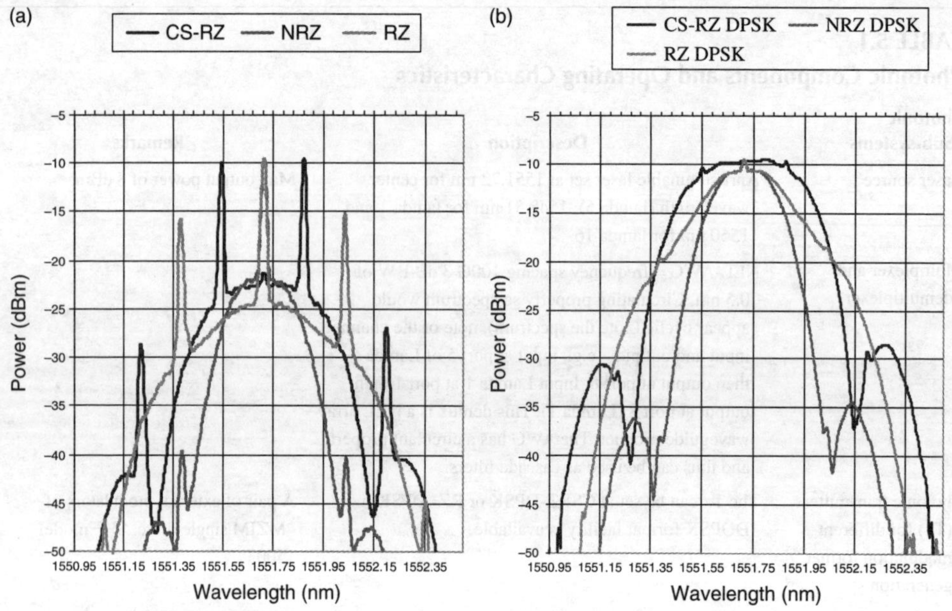

FIGURE 5.14 Optical spectra of transmitted optical signals of formats NRZ, RZ and CSRZ for (a) ASK (b) DPSK.

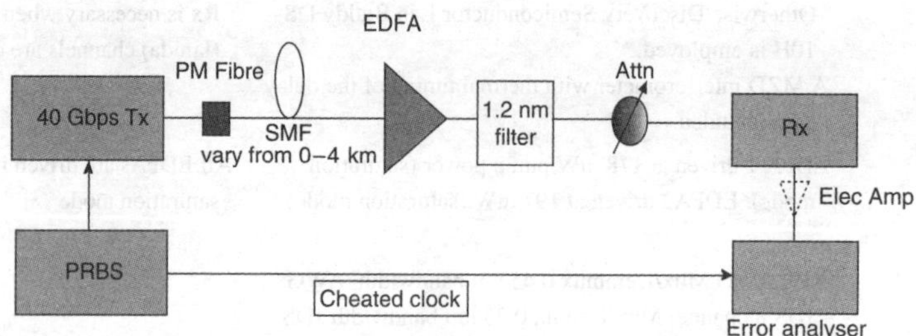

FIGURE 5.15 Experimental test bed for investigation of dispersion tolerance of transmission system.

TABLE 5.2

Maximum Power Pounced into Input of the Optical Fiber Transmission for Different Modulation ASK and DPSK with RXZ, NRZ and CSRZ Formats

	NRZ-ASK	RZ-ASK	CSRZ-ASK	NRZ-DPSK	RZ-DPSK	CSRZ-DPSK
Launched power (dBm)	−4.2	−7.7	−6.2	−3.1	−6.2	−4.8

conventional ASK can be obtained. Figure 5.19a shows the experimental results on power penalties-induced chromatic dispersion limits of ASK and DPSK formats and Figure 5.19b shows simulation results on power penalties induced chromatic dispersion limits of ASK and DPSK formats.

The DPSK system offers approximately 3-dB better receiver sensitivity compared with the same ASK system. The obtained results confirm the theoretical analysis.

5.3.4 Optical Filtering Effects

Initially, the impacts of the optical filtering characteristics of the mux are evaluated using a back-to-back setup as shown in Figure 5.20. We set the sampling clock directly from the auxiliary clock

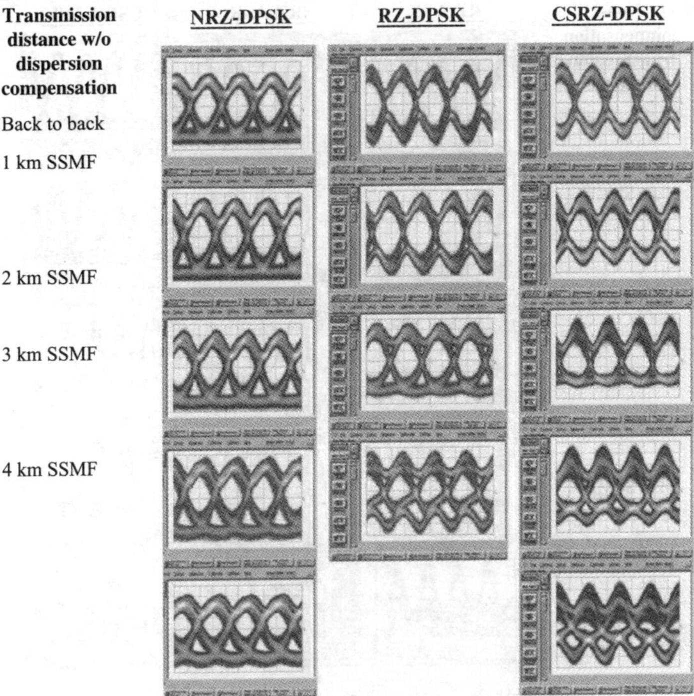

FIGURE 5.16 Eye diagrams of DPSK transmission after 0, 1, 2, 3 and 4 km SSMF for evaluation of the dispersion tolerance.

output of the $2^{31}-1$ pattern generator. Although two AWGs can be used as muxes and demuxes, we use only one AWG at the transmitter or receiver sites. The other filter can be substituted with a multilayer thin-film optical filter. Two optical filters acting as mux and demux at the transmitting and receiving ends are used to evaluate their impacts on 40-Gb/s channels. We observe insignificant degradation of the BER, as shown in Figure 5.21.

Figure 5.22 shows the BER versus the receiver sensitivity curves obtained for ASK and DPSK and DQPSK with RZ, NRZ or CSRZ formats (Figure 5.21). The sensitivities do not change significantly under the influence of the 0.5-nm optical filter on 40-Gb/s channels operating under different modulation formats. We can now observe the optical spectra of ASK and PSK formats as shown in Figure 5.1a and Figure 5.1b. The Gaussian-like or $\cos^2$ profile of the pulses generated at the output of both optical modulators and the parabolic passband properties of the AWG can tolerate a wide signal spectrum. We do not observe any degradation of the BER versus the sensitivity for the cases of wideband optical filters (1.2 nm) and 0.5-nm optical mux filtering. The detected eye diagrams under RZ and CS RZ DPSK transmission are shown in Figure 5.16. The MZDI acting as a phase comparator is thermally tuned so as to obtain a maximum eye opening, thence optimum BER. Similarly obtained eye diagrams for other ASK modulation formats, RZ-ASK, NRZ-ASK and CSRZ_ASK are shown in Figure 5.17.

The characteristics of multiplexer/de-multiplexers which are currently utilized for 10-Gb/s systems still offer sufficiently good performance for 40-Gb/s transmission systems. The experimental results suggest the feasibility of upgrading the currently deployed system to higher transmission rate: 40 Gb/s. Higher bit rate, such as 80 Gb/s or 160 Gb/s, need to be further investigated.

5.3.5 PERFORMANCE OF CSRZ-DPSK OVER A DISPERSION MANAGED OPTICAL TRANSMISSION LINK

A dispersion-managed optical transmission link over 328 km length is setup as shown in Figure 5.13 with three optically amplified spans. The dispersion factor of the transmission fiber is matched with

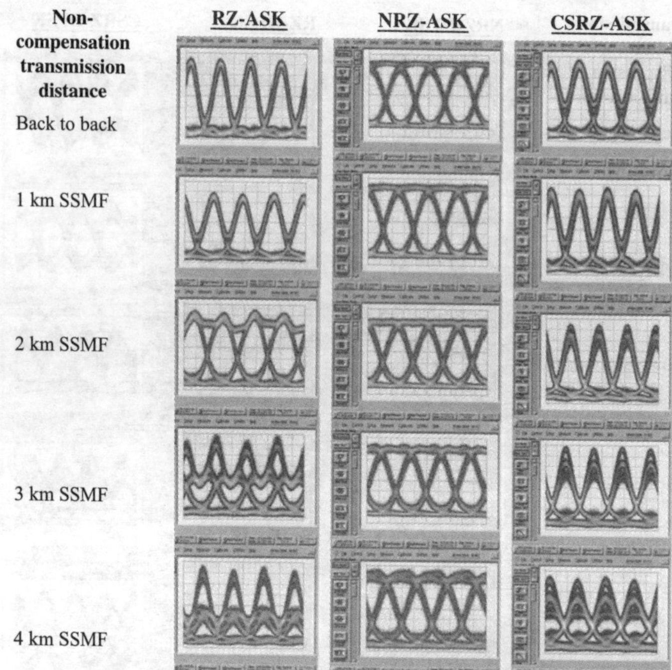

FIGURE 5.17 Eye diagrams of ASK transmission after 0, 1, 2, 3 and 4 km SSMF for evaluation of the dispersion tolerance of the transmission system.

that of a length of dispersion compensation fiber. The bit pattern generator (BPBG) is used to drive the modulators of the transmitter. RZ and CSRZ can also be generated with an insert of sinusoidal wave generated from a signal synthesizer. The nonlinear effect is explored by varying the optical power launched into the SMF from 0 dBm to 15 dBm, as shown in Figure 5.23. The power launched into the DCF is kept unchanged at 0 dBm (non-linearities are generated in SMF, not DCF).

Mopping-up the residual dispersion using a tunable dispersion compensator is used, and setting the wavelength at 1555.75 nm can be shown to be optimal as compared with the BER performance curve obtained at $\lambda_5 = 1551.72$ nm. Hence, laser source is tuned to this wavelength.

5.3.6 Mutual Impact of Adjacent 10 G and 40 G DWDM Channels

The 320-km transmission is also conducted to assess the performance of 10 G NRZ ASK and a CS-RZ DPSK 40 G channels. The transmission of adjacent and non-adjacent channels is demonstrated and evaluated with a 100-GHz AWG mux and a 1.2-nm tunable filter at the input of the Rx. Insignificant power penalty is observed when adjacent 40-G channel is co-transmitted. The setup of the transmission system is shown in Figure 5.24 with 320-km SSMF and dispersion compensating module with two optical filters and splitters, the 100 GHz AWG, which are used as muxes and demuxes. An optical filter is inserted at the front-end to measure the impact of adjacent CS-RZ DPSK 40-G channel on 10-G NRZ-ASK performance. No significant impact was noted when 40-G channel, adjacent or non-adjacent, is co-transmitted, as evaluated in Figure 5.25.

5.4 DQPSK MODULATION FORMAT

When digitizing data for transmission across many hundreds of kilometres, many digital modulation formats have been proposed and investigated. This section investigates the DQPSK modulation format. This modulation scheme, although having been in existence for quite some time, has been implemented only in the electrical domain. Its application to optical systems proved difficult in the past as constant

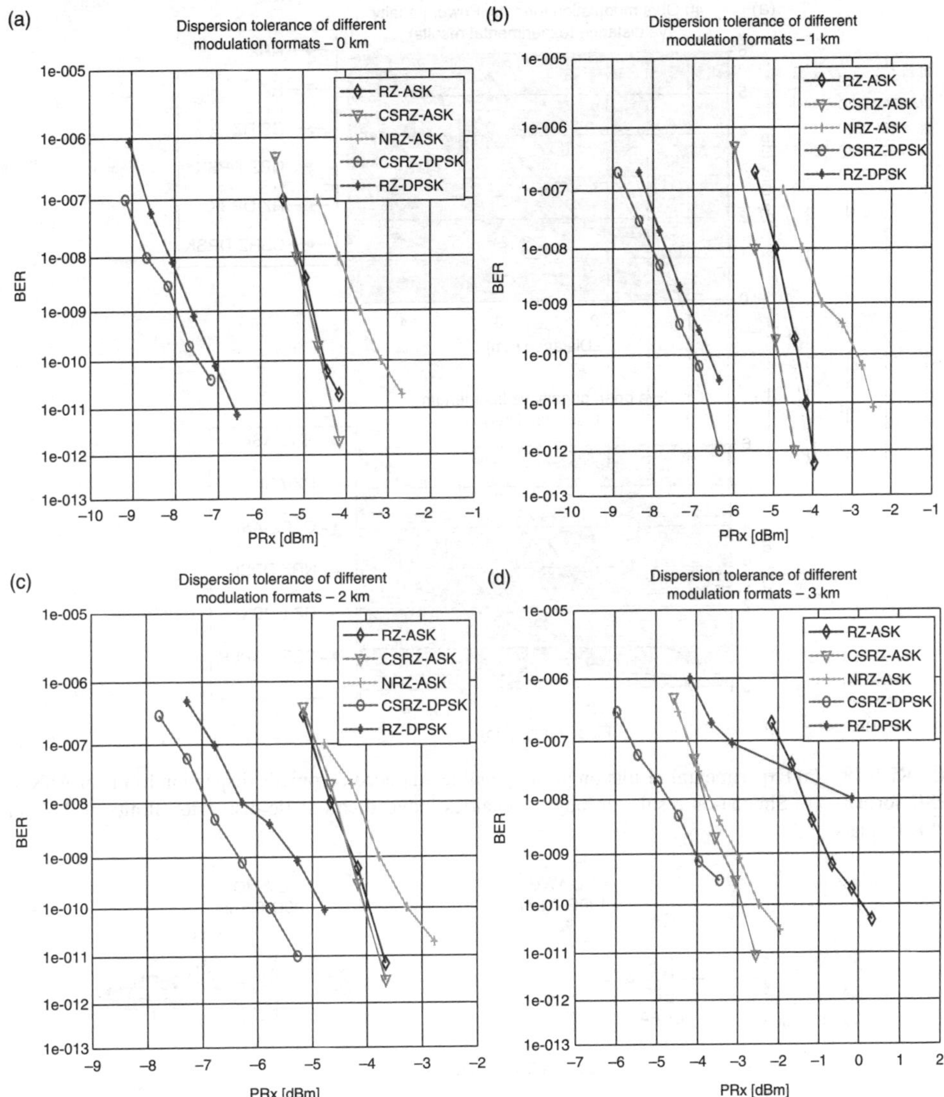

FIGURE 5.18 Performance of dispersion tolerance of advanced modulation formats in the case of 1-km SSMF.

phase shifts in optical carriers (OC) required maintenance. However, with the improvement of optical technology and alternative transmitter design setup, these difficulties are eliminated.

5.4.1 DQPSK

One of the main attractive features of the DQPSK modulation format is that it offers twice as much bandwidth and increased spectral efficiency compared with OOK. For example, comparisons between 8×80 Gb/s DQPSK systems and 8×40 Gb/s OOK systems show that DQPSK modulation offers more superior performance, i.e., spectral efficiency. The very nature of the signaling process also allows non-coherent detection at a receiver to be possible, thus reducing the overall cost of the system design.

The idea behind DQPSK modulation format is to apply a differential form of phase shift modulation to the optical carrier which encodes the data. DQPSK is an extension of the simpler DPSK (differential phase shift keying) format. Rather than having two possible symbol phase states (0 or π phase shift) between adjacent symbols, DQPSK is a four-symbol equivalent $\{0, \pi/2, \pi \text{ or } 3\pi/2\}$. Depending

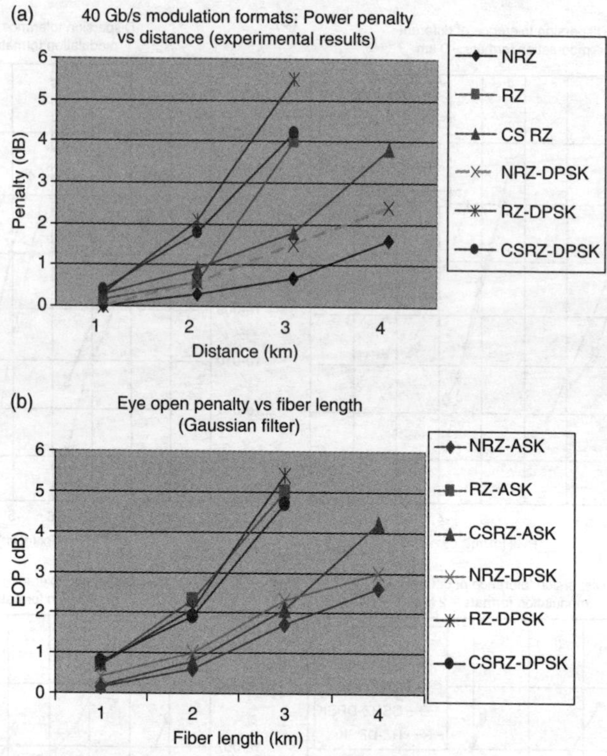

FIGURE 5.19 (a) Experimental results on power penalties-induced chromatic dispersion limits of ASK and DPSK formats. (b) Simulation results on power penalties induced chromatic dispersion limits of ASK and DPSK formats.

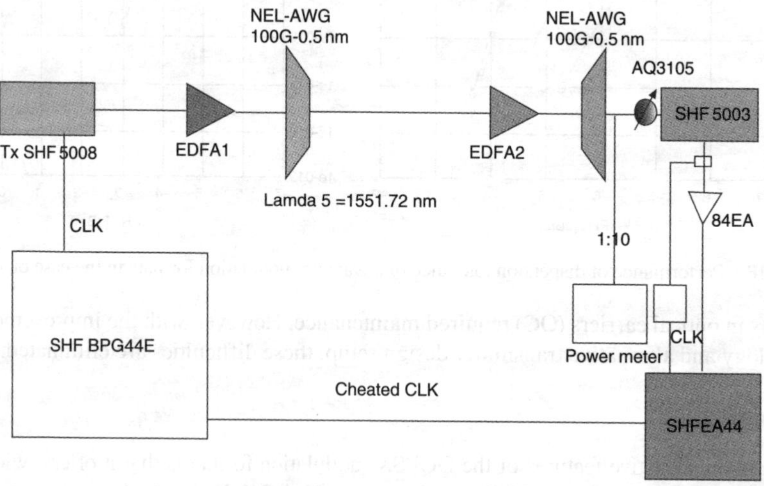

FIGURE 5.20 Back-to-back experimental system for investigation of the optical AWG filtering impacts on modulation formats.

on the desired di-bit combination to be encoded, the difference in phase between the two adjacent symbols (optical carrier pulses) is varied systematically. The Table 3.3 outlines this behavior.

DQPSK transmitter design is based on a previously proposed experimental design [8]. First, we implement a RZ pulse carving MZM which generates the desired RZ pulse shape as described in

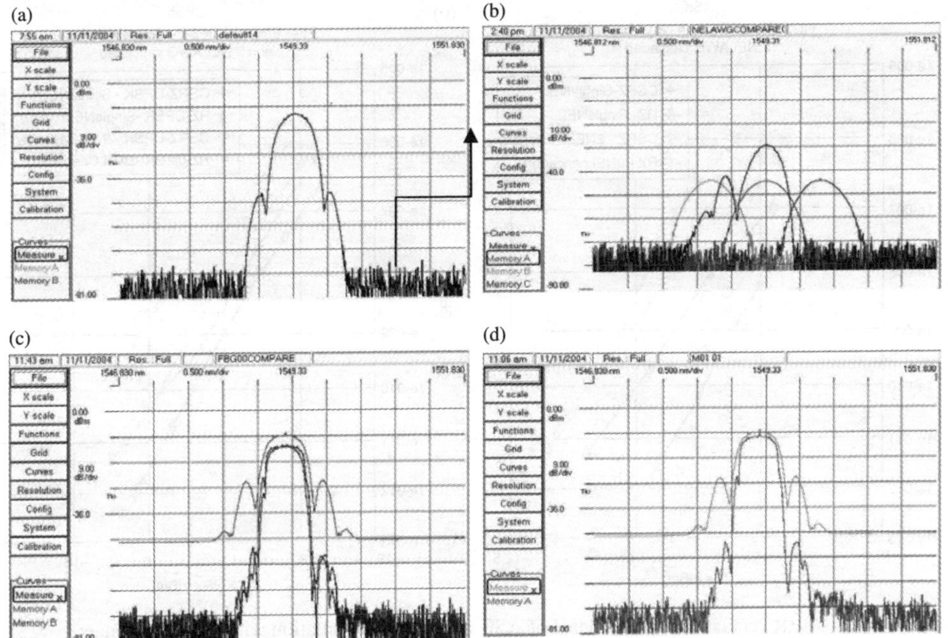

FIGURE 5.21 Optical passbands of AWG filters (a) top left corner: signal spectrum of a channel (b) right top: Optical passbands of the multiplexed output of the AWG – note the parabolic passband characteristics and the "black" curve of the output spectra of a wavelength channel (c) Signal spectrum and its out put of the AWG (d) same as (c) but different wavelength region.

Chapter 2. Next, an MZIM (generating 0 or π phase shift of OC) is to be coupled along with a phase modulator (PM) which induces a 0 or $\pi/2$ phase shift of the optical carrier. When placed in this configuration, the four phase states of the OC required by the DQPSK modulation format, {0, $\pi/2$, π or $3\pi/2$}, can be achieved. The MZM and PM are to be driven by random binary generators operating at 40 Gb/s. Figure 5.27 is a scheme of the system transmitters implementing the RZ-DQPSK modulation. We originally started with the NRZ-DQPSK design, but the RZ format is used more readily in practice and has proved to be more robust to system non-linearities [9]. The pulse sequence generated after the pulse carver can be observed in Figure 5.28.

For DQPSK, the signal constellation can be considered to be two DPSK orthogonal to each other. That is a phase shift of $\pi/2$ between the constellation points. Thus, the transmitter and receiver can be implemented with the following phase difference

$$\Delta\phi_k = \left\{0, \frac{\pi}{2}, \pi, -\frac{\pi}{2}\right\} \quad \text{or} \quad \Delta\phi_k = \left\{\frac{\pi}{4}, \frac{3\pi}{4}, -\frac{\pi}{4}, -\frac{3\pi}{4}\right\}. \tag{5.13}$$

In the latter case, the receiver simply gives the outputs of the binary bits of the imaginary and real parts of the DQPSK signals.

$$b_{k\_I} = \cos\Delta\phi_k = \frac{\mathbf{x}_k \mathbf{x}_{k-1}}{|\mathbf{x}_k||\mathbf{x}_{k-1}|} \quad \text{and} \quad b_{k\_Q} = \sin\Delta\phi_k = \frac{\mathbf{x}_k \mathbf{x}^*_{k-1}}{|\mathbf{x}_k||\mathbf{x}^*_{k-1}|}$$

$$\tag{5.14}$$

$$\text{with} \quad \mathbf{x}_k \mathbf{x}_{k-1} = \int_0^T x_k(t) x_{k-1}(t)\mathrm{d}t \quad \text{and} \quad \mathbf{x}_k \mathbf{x}^*_{k-1} = \int_0^T x_k(t) x^*_{k-1}(t)\mathrm{d}t.$$

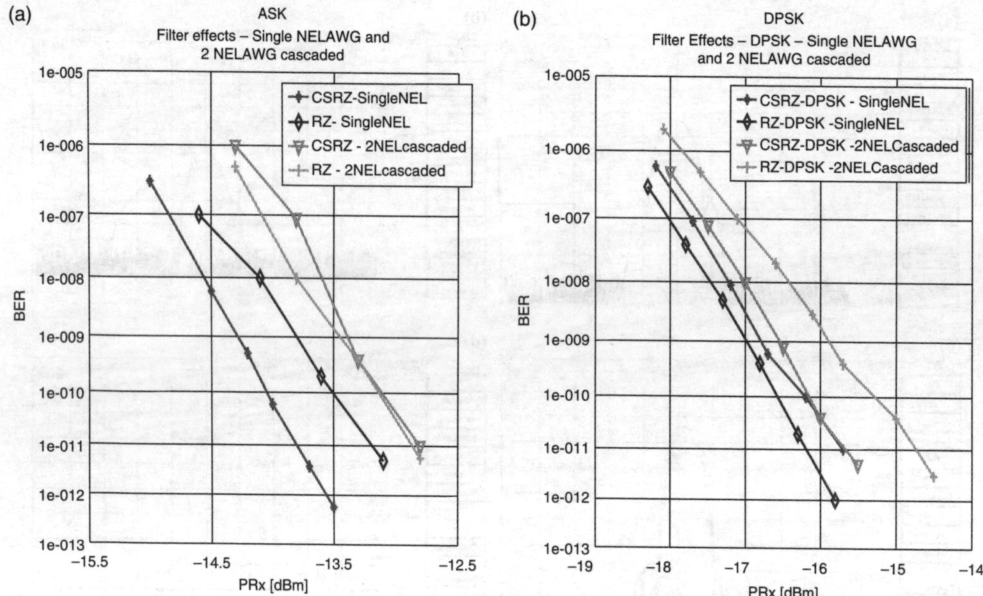

FIGURE 5.22 BER versus input power level of ASK and DPSK modulation with various formats.

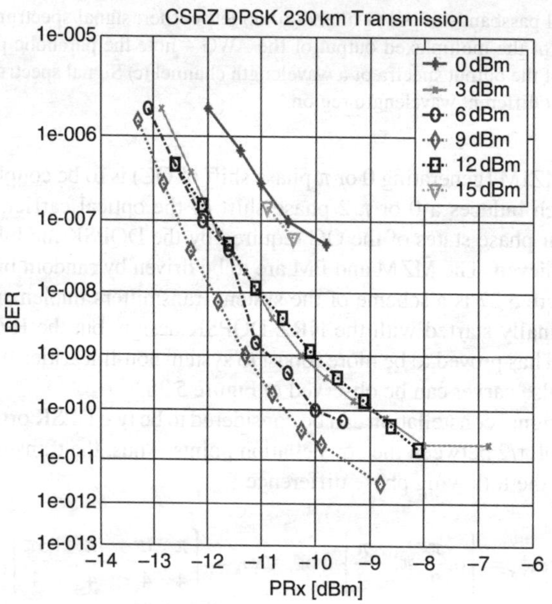

FIGURE 5.23 Transmission performance of CSRZ DPSK under different power launched levels.

This can be explained as follows. The data information transmitted during the time interval $[(k-1)T_s, kT_s]$ with T_s as the symbol period, is carried by the in-phase and quadrature components of the signal and thus given by the sign of the terms sin and cos as given in Equation 5.13. An offset of the signal constellation by $\pi/4$ would assist the simplification of the transmitter and receiver due to the fact that

$$\Delta\phi_k = \left\{0, \frac{\pi}{2}, \pi, -\frac{\pi}{2}\right\} \to \sin\left\{0, \frac{\pi}{2}, \pi, -\frac{\pi}{2}\right\} \to \{0, \pm 1\} \qquad (5.15)$$

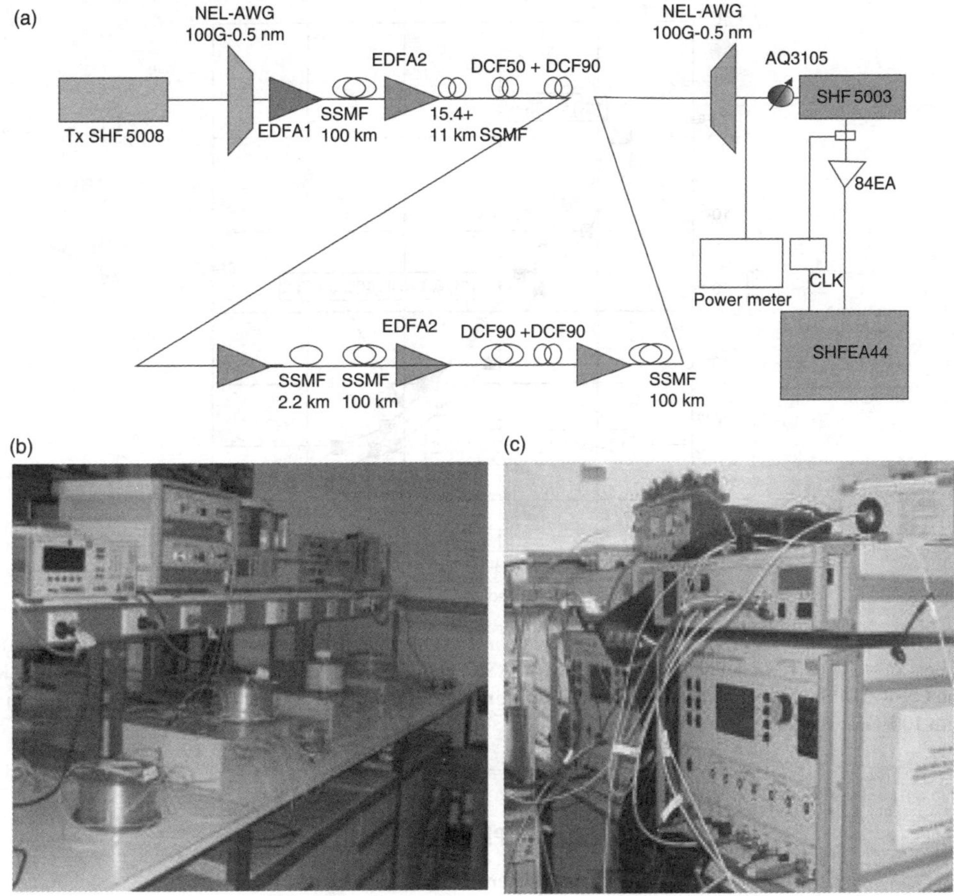

FIGURE 5.24   (a) Setup of the optically amplified long-haul optical transmission system – the span length can be varied for 50 km to 320 km (b) Optically amplified and dispersion compensated fiber transmission line (left) – transmitter and receiver plus (c) 40 Gb/s bit pattern generator and error analyzer (right).

Or effectively there would be three output levels while the $\pi/4$ shift would give binary levels.

For a noiseless case, the correlation between the two bits are given as

$$s(t)s(t - T_{\mathrm{b}}) = A\sin(2f_{\mathrm{c}}t + \phi_k)\, A\sin(2f_{\mathrm{c}}t - 2\pi f_{\mathrm{c}}T_{\mathrm{b}} + \phi_{k-1})$$

$$= \frac{A^2}{2}\{\cos(\phi_k - \phi_{k-1} + \alpha) - \cos(4f_{\mathrm{c}} + \phi_k + \phi_{k-1} - \alpha)\} \tag{5.16}$$

detected signals   filtered by lowpass filter.

With the phase $\Delta\phi_k = \phi_k - \phi_{k-1} \in \{0, \pi\}$ leading to $\cos(\Delta\phi_k + k\pi) \in \{1, -1\}$ so that it ensures a maximum distance between the two difference symbols.

For the rest of this section, some of the key principles allowing for the successful decoding of the DQPSK modulated signal are described. We describe an exact receiver model before fiber propagation to allow comparisons between pre- and post-fiber effects of the received eye diagram. The comparison of the phase of the lightwave carrier contained within the two consecutive bits is the optical delay interferometer is shown in Figure 5.29. Due to the differential nature of the modulation

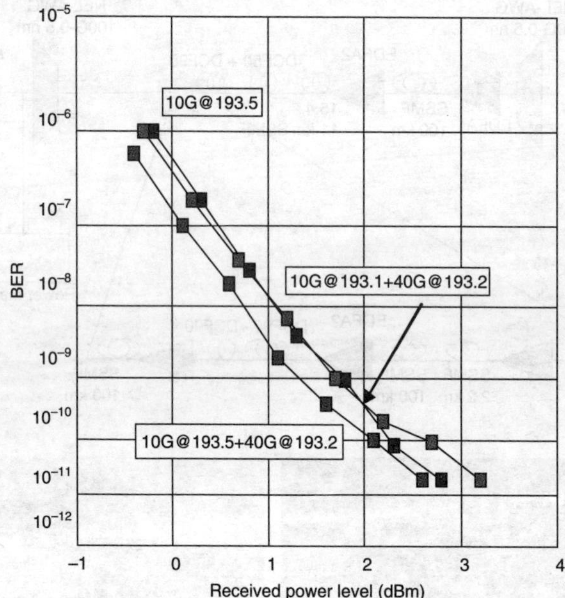

FIGURE 5.25 320-km transmission 40 G impact on 10 G channel: BER versus receiver sensitivity (dBm)-effects of 40 G (CS_RZ DPSK) with 10 G (NRZ-ASK) channel simultaneously transmitted for NRZ-ASK and CS-RZ DPSK formats – blue dots for 1.2-nm thin-film filter and red dots for 0.5-nm AWG filter (demux with 100-GHz spacing).

TABLE 5.3
DQPSK Modulation Phase Shifts

Dibit	Phase Difference $\Delta\phi = \phi_2 - \phi_1$ (degrees)
00	0
01	90
10	180
11	270

The parameters ϕ_1 and ϕ_2 as indicated in are the phase of adjacent symbols. The above table can also be represented in the form commonly known as a "constellation diagram" (see Figure 5.26), this graphically explains the signals state in both amplitude and phase.

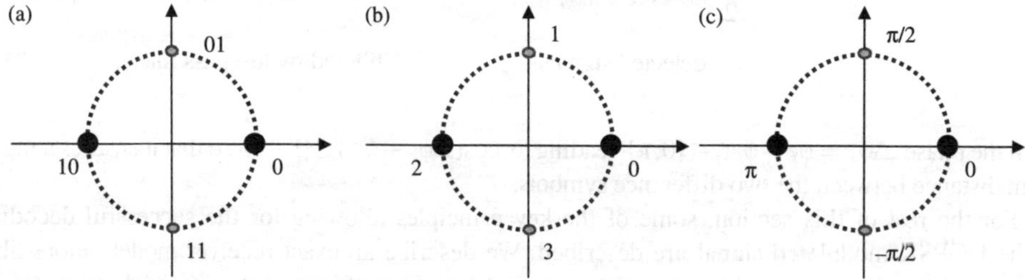

FIGURE 5.26 DQPSK signal constellation and assignment of phase symbols (a) quarternary digits (b) two-digit binary word and (c) phase complex plane.

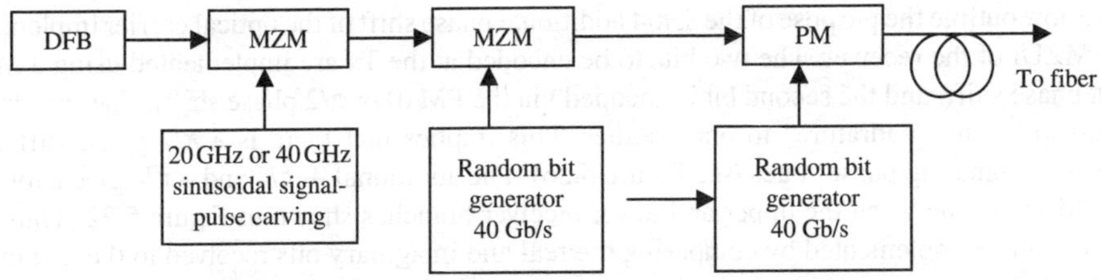

FIGURE 5.27 Channel-implementing RZ-DQPSK modulation (schematic).

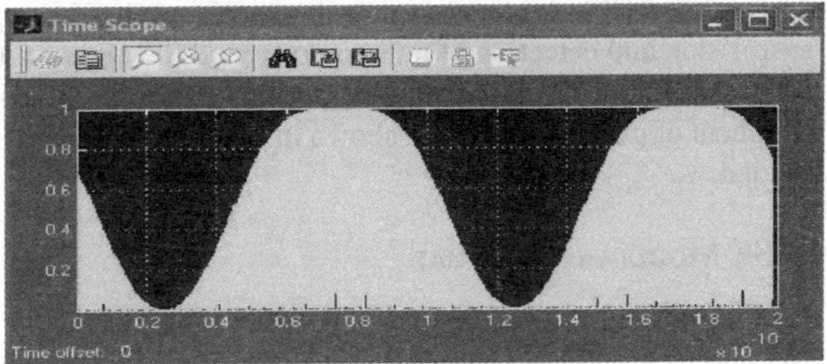

FIGURE 5.28 The optical carrier after RZ pulse carving using an MZM driven by a 10-GHz electrical signal.

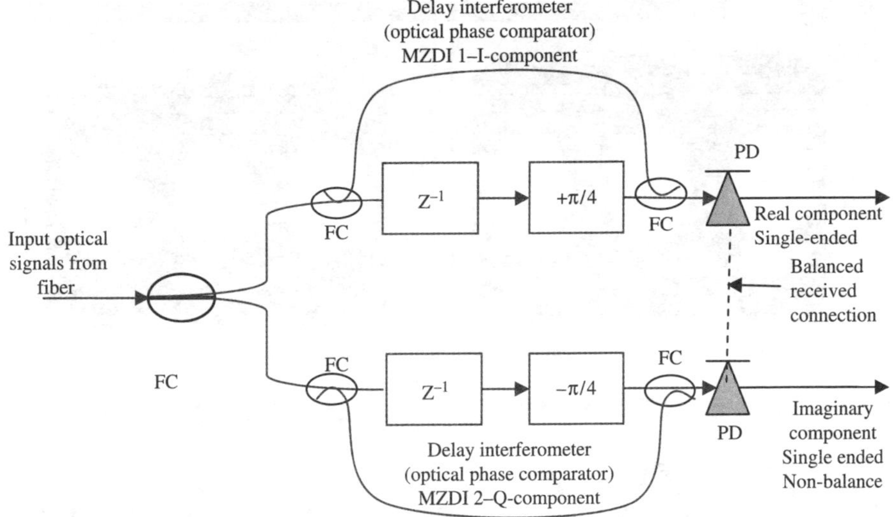

FIGURE 5.29 DQPSK receiver configuration using non-coherent direct detection. FC = fiber coupler, MZDI = Mach–Zehnder delay inteferometer.

process, the demodulation and detection stage can be considered as a "non-coherent" or "direct detection" or effectively this can be known as a delay autocorrelation or self heterodyne detection scheme. The absence of a local oscillator (LO) and any other extra hardware required in conventional detectors makes this demodulation technique attractive. The receiver configuration shown in Figure 5.28 is capable of demodulating the signal transmitted along the 96-km total fiber span. Since the modulation format used in this section typically encodes two bits of data per symbol, it is necessary to extract both bits (termed "real" and "imaginary" bits [8]) from the one received symbol.

We now outline the purpose of the $\pm\,\pi/4$ additional phase shift of the optical carrier implemented in the MZDI of the receiver. The two bits to be encoded at the Tx are implemented using a MZIM (0 or π phase shift) and the second bit is encoded via the PM (0 or $\pi/2$ phase shift). The two devices are said to be in "quadrature" to one another. This implies that there is a $\pi/2$ phase difference between all signaling phase states (see Figure 5.26). The additional $+\,\pi/4$ and $-\,\pi/4$ give a total $\pi/2$ phase difference between the upper and lower receiver branches shown in Figure 5.22. Thus, data recovery can be implemented by comparing the real and imaginary bits received to those transmitted. However, one can consider only the "real component" of the received signal and assess the overall performance of the system via eye diagram analysis (using Q factor method and BER). In practice, the demodulation of the DQPSK signal using the two branch configuration in Figure 5.29 can be successful demonstrated [10]. A balanced detection structure has been proved to be more sensitive. Optical transmission and detection of non-compensated SSMF are obtained as shown in Figure 5.30, with 10 km and 5 km for the Q- and I-components, respectively. The ODQPSK signal constellation and assignment of phase symbols are shown in Figure 5.32 in which the constellations of DPSK are also included.

5.4.2　Offset DQPSK Modulation Format

In this section, we introduce offset-DQPSK (O-DQPSK) transmission. ODQPSK have been applied in the transmission over nonlinear satellite channel that offers a smother transition between the phase states and hence avoids pi phase jumps. In contrast with DQPSK, which requires detection at the symbol rate, the demodulation of ODQPSK can however be achieved on the bit rate, thus allowing

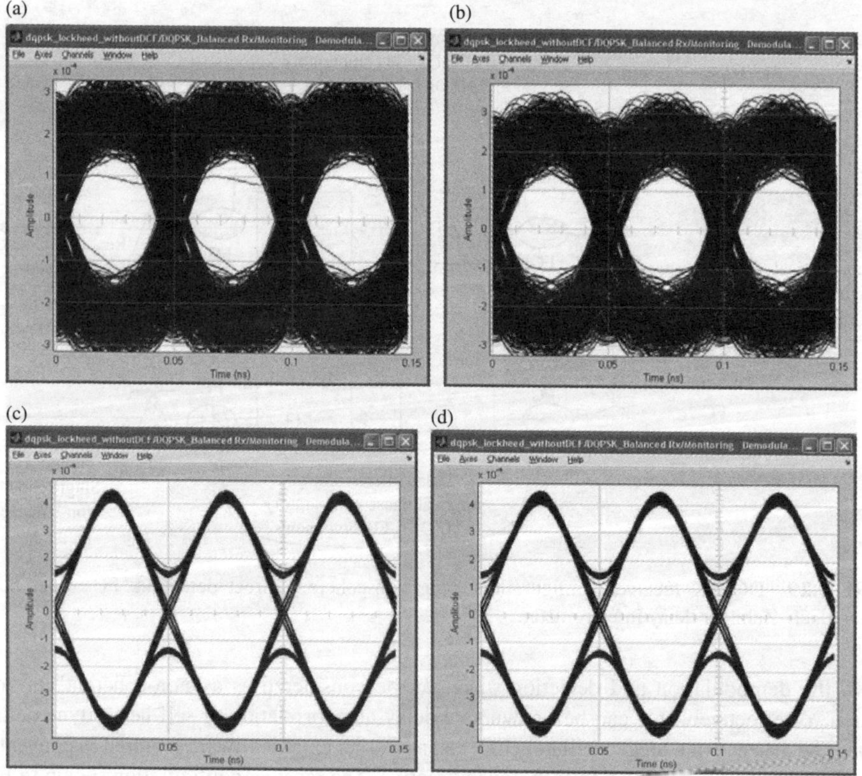

FIGURE 5.30 Eye diagrams achieved for I-component (a) 10 km-SSMF (c) 5-km SSMF; and for Q-components (b) 10-km SSMF (d) 5-km SSMF mismatch of dispersion over two 200-km span dispersion compensated and optically amplified spans – RZ-DQPSK 50% RZ pulse shape.

detection of the ODQPSK with only one set of balanced receiver and one set of MZDI. The optical ODQPSK can be generated by two binary RZ-DPSK signals operated at half the bit rate, which can be merged optically by a 3-dB coupler with a one-bit delay in one path, as shown in Figure 5.31.

Similar to the case of DQPSK, a carrier phase difference of $\pi/2$ must be guaranteed between two parallel signals to ensure the $\pi/2$ phase shift between the states of the constellation points similar to the phase plane shown in Figure 5.32. The receiver for DQPSK modulation format can be implemented using only one MZDI and balanced receiver, but only one MZDI is required and the a $p/2$ phase shifted must be inserted into one path of the MZDI to ensure the demodulation of the $\pi/2$ phase difference between consecutive bits, and hence the inphase and quadrature components. If there is a cosine continuity of the phase shift between the constellation points, then the scheme ODQPSK becomes a MSK scheme that will be presented in Chapter 6 and Chapter 7.

The receiver sensitivity of DQPSK, ODPSK, and DPSK have also been measured of the transmission over a 400-km span with optical amplifiers and DCF as shown in Figure 5.32c and Figure 5.32d. The bit rate for DPSK is 40 Gb/s and 20 Gb/s for DQPSK and the input power to the fiber is set at 0 dBm to ensure the systems operate under the linear regime. There are no difference between the DQPSK and offset DQPSK. A 6-dB difference between DQPSK and DPSK between their receiver sensitivity using a balanced receiver, and a 3.5-dB difference between single detector receiver and balanced receiver for ODQPSK systems that is expected as the distance between the "1" and "0" is double that of a single detector, as shown by the detected eye diagrams in Figure 5.16. Figure 5.33 shows the receiver sensitivity of these differential phase shift keying modulation formats.

There are two principal differences between the receiver sensitivity of RZ-DPSK and RZ-DQPSK (i) Using the same total average power (i.e., the same radius on the phasor diagram in Figure 5.32) the binary level to the quaternary level would require an increase of a factor of $\sqrt{2}$ as we can observe from the signal constellations of the DPSK and DQPSK. Thus, a 3-dB increase in the power required for DPSK as compared with DQPSK. (ii) For the MZDI self-heterodyne detection, the detection seems

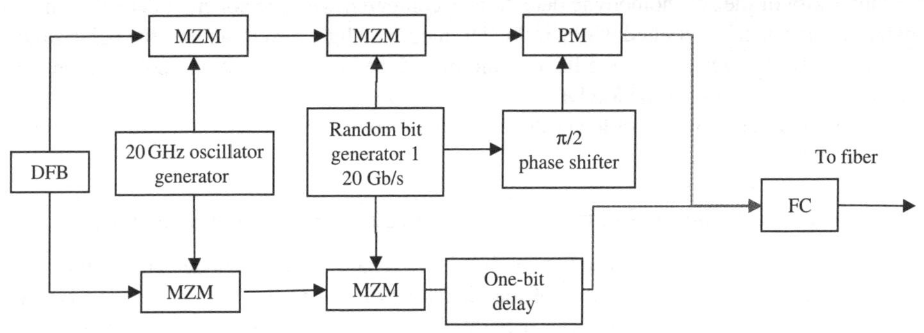

FIGURE 5.31 Optical offset DQPSK transmitter (schematic).

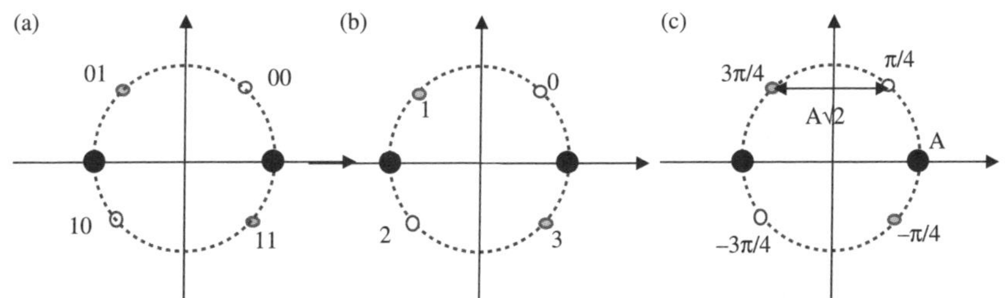

FIGURE 5.32 ODQPSK signal constellation and assignment of phase symbols (a) quaternary digits (b) two-digit binary word and (c) phase complex plane with two state (horizontal line) circles are for DPSK constellation points.

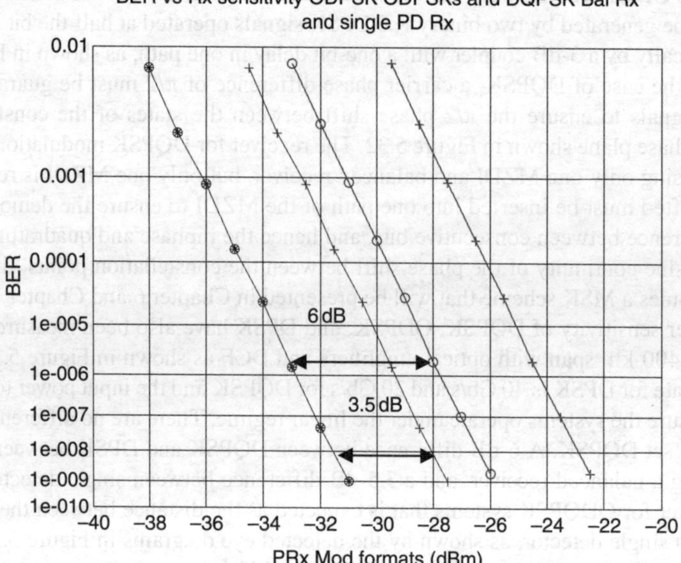

FIGURE 5.33 Optical DPSK and ODPSK transmission over 4×100 km SSMF of DWDM 8×40 Gb/s transmission. Simulated results of BER versus receiver sensitivity of modulation formats: ($-$ red o) Offset DQPSK_Balanced Rx, ($-$ red +) Offset DQPSK_single detector, (blue *) DQPSK_Balanced Rx, (blue +) DQPSK_single PD. Black *----offset DQPSK.

to add an additional power penalty in the splitting and combining of the received optical fields. This is the complexity of the self-homodyne detection as compared with the coherent detection in which the polarization can be diversified while, in MZDI integrated lightwave circuitry, the polarization of the input lightwave coupled from the fiber would be reduced due to the strong polarization dependence of the rib waveguide of the MZDI.

These original mechanisms of the reduction of the receiver sensitivity can be further explained as discussed below.

5.4.2.1 Influence of the Minimum Symbol Distance on the Receiver Sensitivity

Assuming the noise pdf distribution at the receiver is Gaussian, the coherent detection with matched filter would follow the well-known rule of the bit error probability that depends on the Euclidean distance between the symbols and the variances at the levels of the symbols. In general, noise power would have a mean of zero and variance σ^2. With a minimum distance d between the symbols, the bit error probability P_e is given by:

$$P_e = \mathrm{erfc}\left(\frac{d/2}{\sigma}\right). \tag{5.17}$$

If the total average power of the signals is A^2, then a binary DPSK has a P_e of

$$P_e = \mathrm{erfc}\left(\frac{A}{\sigma}\right). \tag{5.18}$$

While the DQPSK would have a P_e of

$$P_e = \mathrm{erfc}\left(\frac{A/\sqrt{2}}{\sigma}\right). \tag{5.19}$$

The DQPSK allows the receiving of four bits in a symbol period while only two for the binary DPSK. Thus, the capacity of the DQPSK is double that of the DPSK without increasing the noise contribution at the receiver output.

5.4.2.2　Influence of Self Homodyne Detection on the Receiver Sensitivity

The complexity of coherent detection with the use of a local laser and the synchronization of the phase of both the signals and the LO has been given in Chapter 3. With the advancement of integrated lightwave technology, the MZDI can be implemented without difficult and the self-heterodyne detection can be assisted with the phase comparison. Thus, self-heterodyne detection has attracted attention in current DPSK and DQPSK receivers or any receiver that would require the detection of the I and Q components using a balanced receiver. Under a matched filter condition of binary DPSK, the BER is approximately given by

$$P_e = \frac{1}{2}\exp\left(-\frac{A^2}{\sigma^2}\right). \tag{5.20}$$

For QPSK signals using MZDI balanced receiver as self-homodyne detection, then the BER is given by [11]

$$P_e = Q(\vartheta,\mu) - \frac{1}{2}\exp\left(-\frac{\vartheta^2}{\sigma^2}\right)I_0\left(\frac{\vartheta^2\sqrt{2}}{\sigma^2}\right) \tag{5.21}$$

$$\text{with}\quad \mu = A'\sqrt{2}\cos\frac{\pi}{8}\quad \vartheta = A'\sqrt{2}\sin\frac{\pi}{8}.$$

With Q representing the Marcum function, I_0 the modified Bessel function of zeroth order, and A' the amplitude of the signal at the input of the PD. The BER versus the SNR given in Equation 5.18 to Equation 5.21 indicate a 3 dB gain and wider eye opening for the coherent detection scheme using balanced receiver [13]. However, a 2-dB improvement for DQPSK for the case of coherent over self-heterodyne detection is obtained. These results are considered without considerations of the noise contribution in the positive and negative electrical signal levels.

5.4.3　MATLAB® Simulink® Model

5.4.3.1　The Simulink® Model

MATLAB® Simulink models of the transmitter and receivers for DQPSK modulation format are shown in Figure 5.34 and Figure 5.35, respectively, with a π/4 phase offset of the signal constellation. A RZ-DQPSK optical transmission system is modeled with a pulse carver inserted in front of the data modulator MZIM DQPSK precoder of 2-bit per symbol providing the two output signals (in electrical domain) to drive the two arms of the optical MZIM. A π/2 phase modulator is also used to assign the inphase or quadrature phase components of the DQPSK format.

In the receiver, side-differential detection or self-homodyne receiving structure are used with a phase offset of π/4 to obtain directly the amplitude and phase of the receiving signals as discussed above.

5.4.3.2　Eye Diagrams

The transmission of the DQPSK with a bit rate of 40 Gb/s or 20 Giga symbols/sec. is conducted over SSMF length of 10 km and 5 km and is shown in Figure 5.36.

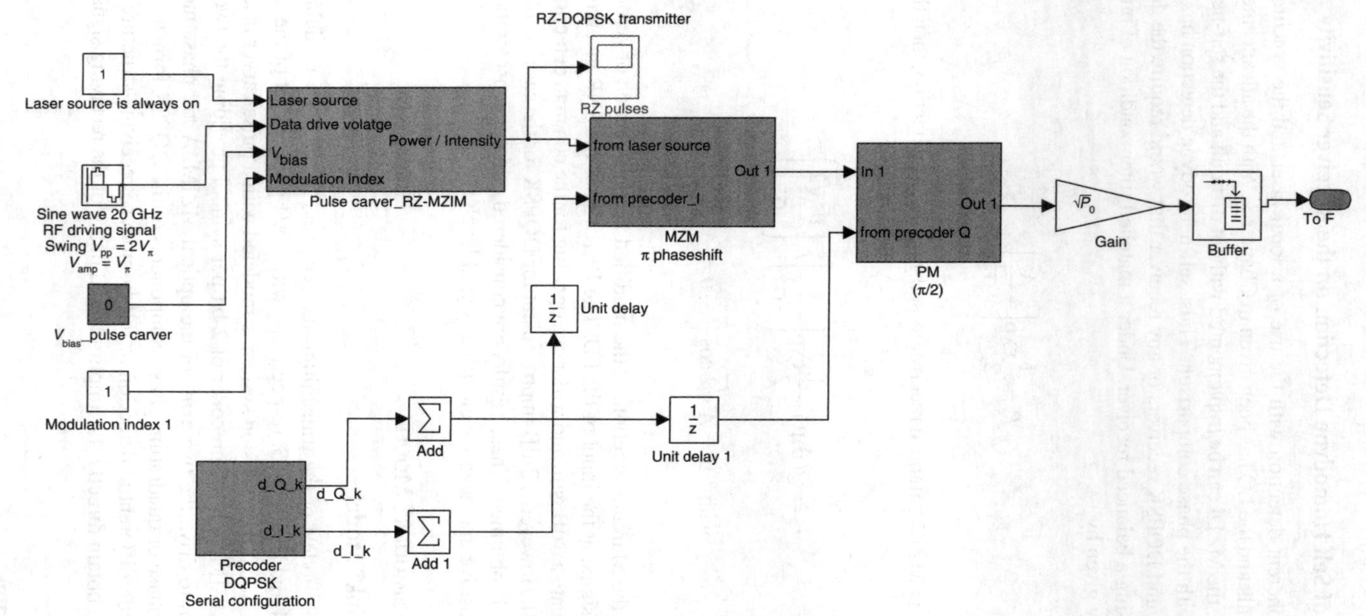

FIGURE 5.34   Transmitter MATLAB® Simulink® model for DQPSK optical transmission.

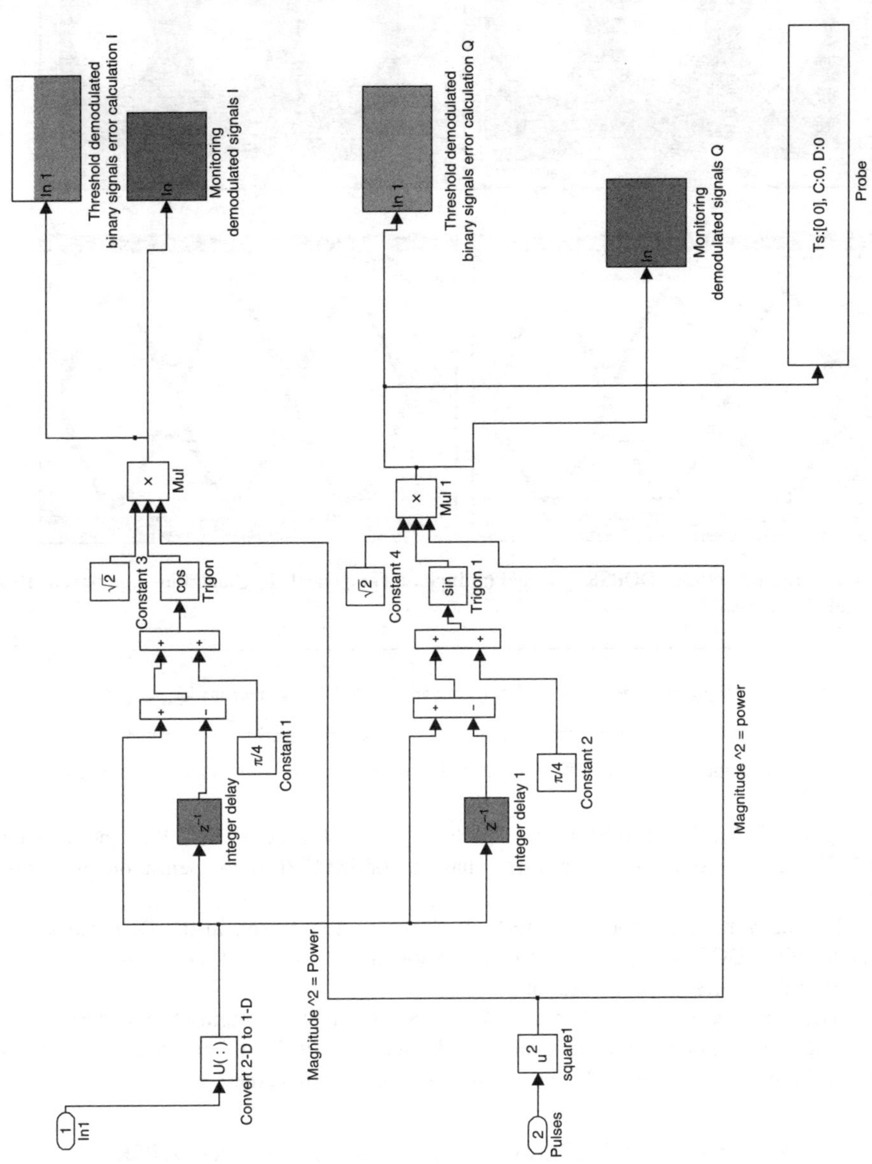

FIGURE 5.35 Receiver Simulink model for DQPSK optical transmission.

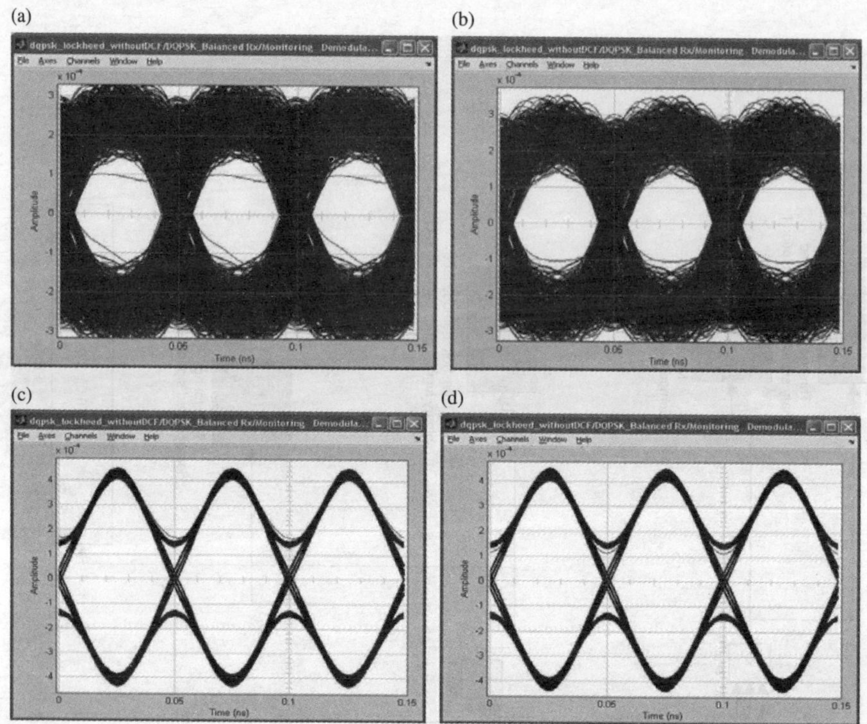

FIGURE 5.36   Eye diagram of DQPSK detected (a) Inphase component (b) Quadrature component 10-km SSMF (c) and (d) 5-km SSMF.

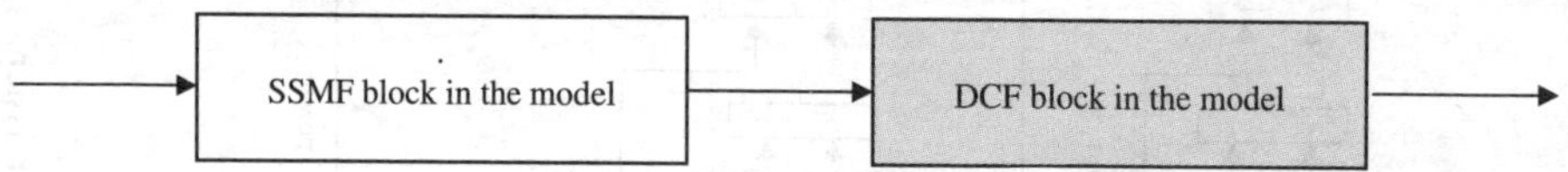

FIGURE 5.37   Arrangement of dispersion compensated spans for transmission of DQPSK signals.

The length of SSMF is set to be 10 km, the length of DCF is set to be 2 km, and dispersion factor of DCF is to be five times more negative than that of SSMF (full compensation) as shown in Figure 5.37.

The following Simulink® model Figure 5.38 illustrates how to set, in the DCF block in the model, the length of DCF to be 2 km and the dispersion factor to be five times more negative than that of SSMF (full dispersion compensation).

The modeled and measured eye diagrams of DQPSK are shown in Figure 5.39 and Figure 5.40 respectively at a total bit rate of 108 Gb/s and 110 Gb/s (or about 50 Gsymbols/sec.) when the I- and Q-components are detected separately and when both channels are turned on.

5.5   COMPARISONS OF DIFFERENT FORMATS AND ASK AND DPSK

5.5.1   BER AND RECEIVER SENSITIVITY

5.5.1.1   RZ ASK and NRZ ASK

If the NRZ and RZ (50% duty) amplitude shift keying pulse sequences whose average power is the same, then the FWHM of the RZ is half that of NRZ, thus its peak power is higher. This is proportional for 33% and 67% duty cycle RZ pulses. The higher the peak power of the RZ pulses, the

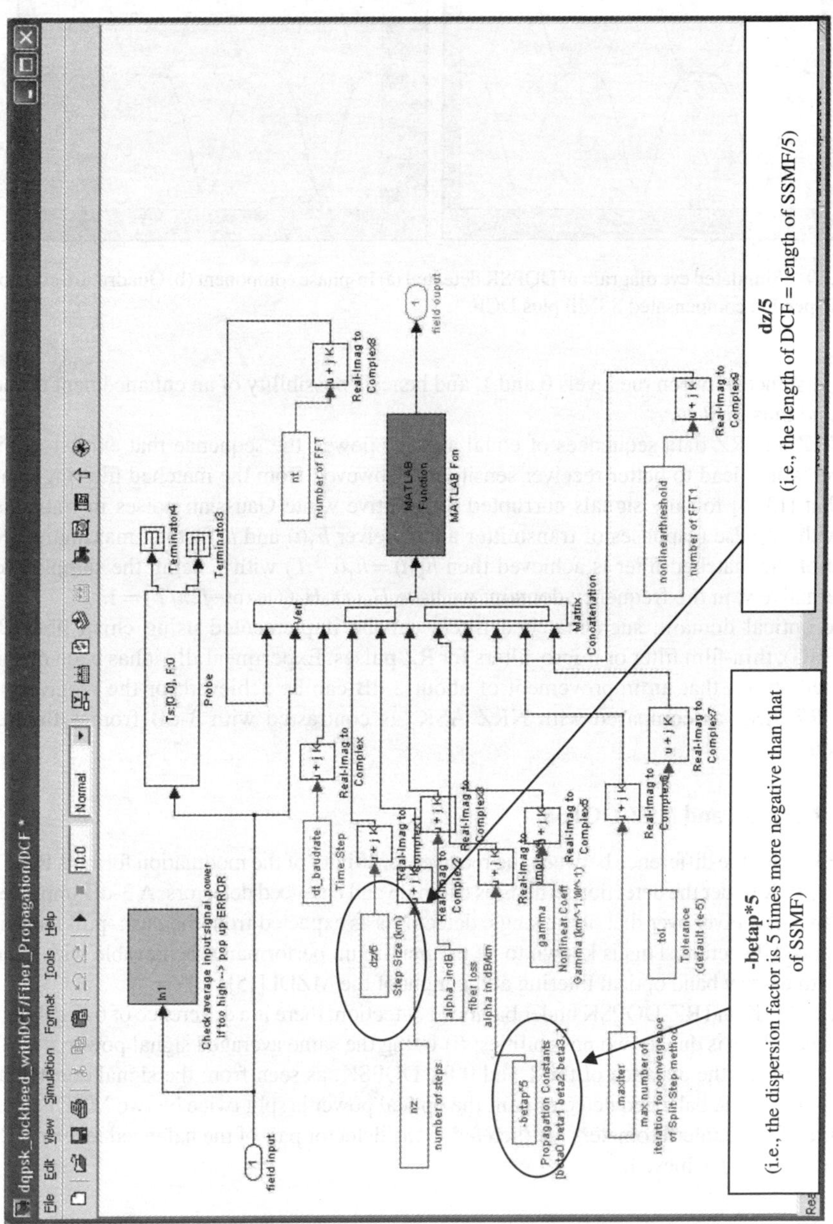

FIGURE 5.38 Simulink model of the fiber sections of dispersion compensating fiber for fully compensated span.

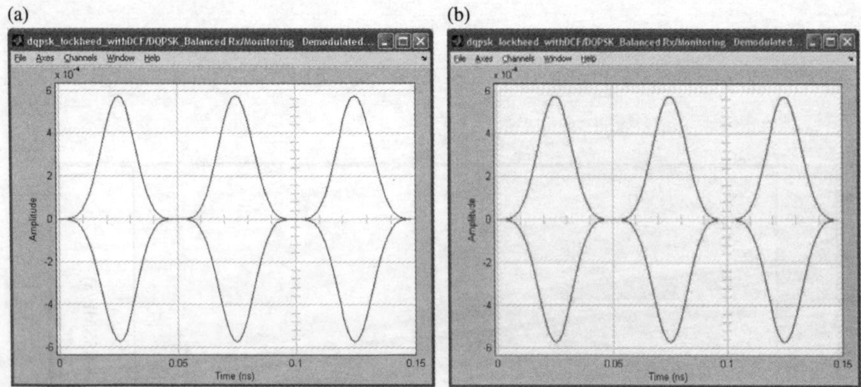

FIGURE 5.39   Simulated eye diagram of DQPSK detected (a) In-phase component (b) Quadrature component, of a fully dispersion compensated SSMF plus DCF.

longer the distance between the levels 0 and 1, and hence a possibility of an enhancement of the eye opening, and thus BER.

For NRZ and RZ data sequences of equal average power, the sequence that exhibits a higher peak power would lead to better receiver sensitivity. However, from the matched filtering concept, it states that [13,14] for any signals corrupted by additive white Gaussian noises in transmission systems with impulse responses of transmitter and receiver $h_T(t)$ and $h_R(t)$ then maximum SNR at the output of the matched filter is achieved then $h_R(t) = h_T(t - T)$ with T being the sampling delay time. Alternatively, in the frequency domain, we have $H_R(f) . H_T(f) \exp(-j2\pi fT) = 1$.

In the optical domain, such matched filters can be implemented using chirp fiber Bragg grating (FBG), thin-film filter or micro-filters for RZ pulses. Experimentally it has been observed, as described above, that an improvement of about 2 dB can be achieved for the receiver sensitivity for RZ ASK as compared with NRZ ASK, as contrasted with 3-dB from a theoretical expectation.

5.5.1.2   RZ DPSK and NRZ DQPSK

Figure 5.33 shows the difference between the receiver sensitivity of the modulation formats RZ-DPSK and RZ- DQPSK under the detection structures of single and balanced detectors. A 3-dB improvement of the balanced receiver over that of the single detector is as expected from the push–pull mechanism of the balanced detection. This is known to be the maximum performance achievable with balanced receiver with narrow band optical filtering at the input of the MZDI [15].

For RZ-DPSK and RZ-DQPSK under balanced detection, there is a difference of 6–7 dB between these schemes. This is due to two possibilities: (i) Using the same averaged signal power, a $\sqrt{2}$ factor is reduced between the distance of the 1 and 0 for DQPSK, as seen from the signal constellations; (ii) At the input of the balanced detector pair, the optical power is split twice by two MZDIs, the four output ports of these interferometers are then fed to the detector pair of the balanced receivers. So no gain in power can be achieved.

5.5.1.3   RZ ASK and NRZ DQPSK

The comparison between ASK, DPSK and DQPSK is important for the case where differential phase with balanced receiving is used in the upgrading for 10 G to 40 G rates as mentioned in Section 5.3.3. RZ-DPSK is expected, theoretically and confirmed experimentally as shown in Figure 5.22, of 3-dB enhancement over the RZ-ASK, and about 6 dB with NRZ ASK. Thus, the

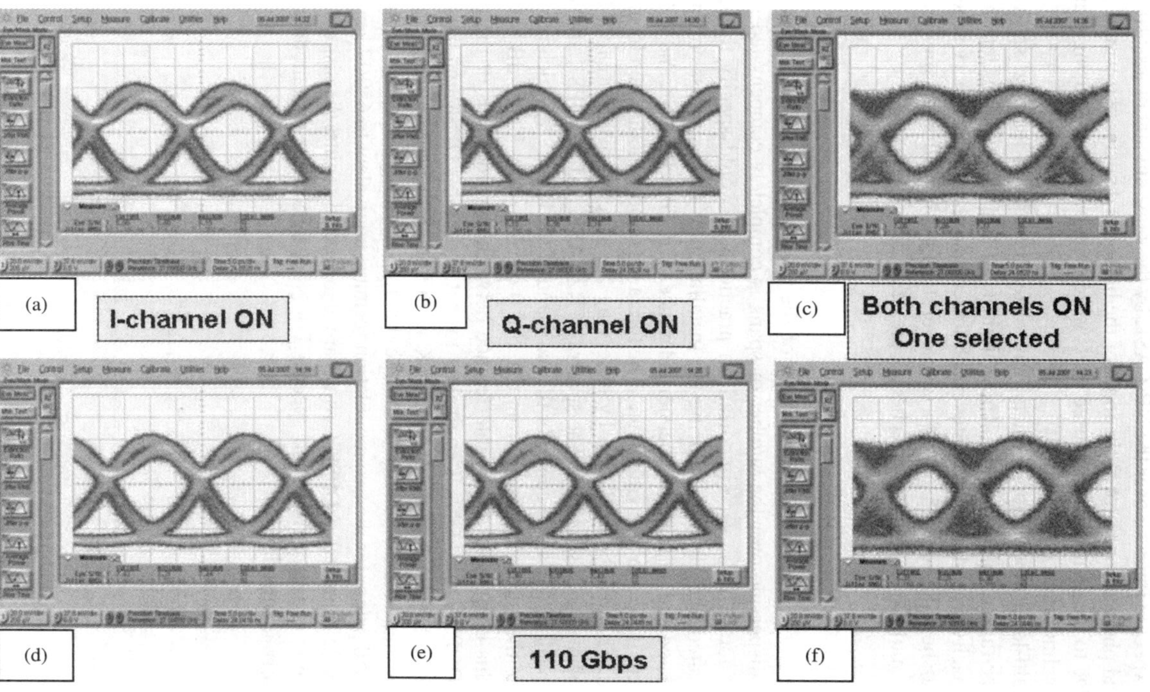

FIGURE 5.40 Measured eye diagrams of 108-Gb/s RZ-DQPSK detected (50 Gsymbols/sec.) (a) In-phase component (b) Quadrature component (c) both channels on; and for 110 Gb/s (d) In-phase component (e) Quadrature component (f) both channels transmitted and one component is selected by tuning the MZDI. (Courtesy of SHF Ltb, Berlin, Germany).

performance of RZ-DQPSK would nearly be the same as RZ-ASK, but with a symbol rate twice that of the NRZ or RZ-ASK. The Simulink® MATLAB® model for RZ DQPSK is given in the Appendix.

This can be explained as follows: (i) The distance between the "1" and "0" of the ASK is nearly the same as that for DQPSK signals of the same average power, as seen from the signal constellation shown in Figure 5.41; (ii) RZ format outperforms the NRZ format by 3 dB (maximum). Thus, a RZ-DQPSK would expect to improve over the NRZ ASK by the same amount or less; (iii) The use of MZDI as an optical phase comparator at the input of the balanced detector pair would introduce optical power loss due to the splitting and unused ports which can be improved using planar light-wave circuit (PLC) technology.

In summary, multi-level modulation formats may suffer the optical loss and splitting when MZDI balanced receiving technique is used. These multi-level schemes would offer an effective high information capacity with the symbol rates much lower than that of the binary schemes. For example, 100 Gb/s can be reduced to 25 Gb/s if a two-level star-QAM (quadrature amplitude modulation) is used. These multi-level modulation schemes are described in Chapter 7.

5.5.2 Dispersion Tolerance

Dispersion tolerance is the measure of the penalty of the eye opening at the receiver after the transmission whose dispersion over the number of spans is completely compensated with some extra residual dispersion. This is very important in practice and would specify the maximum length of uncompensated fiber allowable during the installation of the transmission link. Figure 5.42 shows the simulated and measured dispersion tolerance of 10 Gb/s RZ ASK and RZ DPSK at 20 Gb/s, effectively equivalent in the symbol rate. The tolerance of both modulation formats seems very much the same. This can be explained by observing the spectral distribution of these two modulation formats shown in Figure 5.14. The spectra of RZ_ASK and RZ-DPSK are very much the same except the peak power at the centre of the passband of the RZ-ASK, which is understood as the contribution of the amplitude rising from the 1 to 0 or *vice versa*. The 3-dB bandwidth is thus the same and this contributes to the interference between the sidebands of the modulated lightwave. The RZ-DQPSK at 20 Gb/s is equivalent to that of 10 Gb/s DPSK and thus we could state that the dispersion tolerance of the RZ-DQPSK would be equivalent to that of the 10-Gb/s DPSK modulation format. A 1-dB penalty of the eye diagram or the eye opening penalty (EOP) is at a dispersion of 650 ps/nm for the two modulation formats, equivalent to 650/17 km of SSMF. The EOP of the NRZ ASK and NRZ DPSK formats are also plotted against the dispersion to give a very similar pattern as expected from their spectra. For NRZ DPSK, the data modulator is dual modulator and no chirp is introduced.

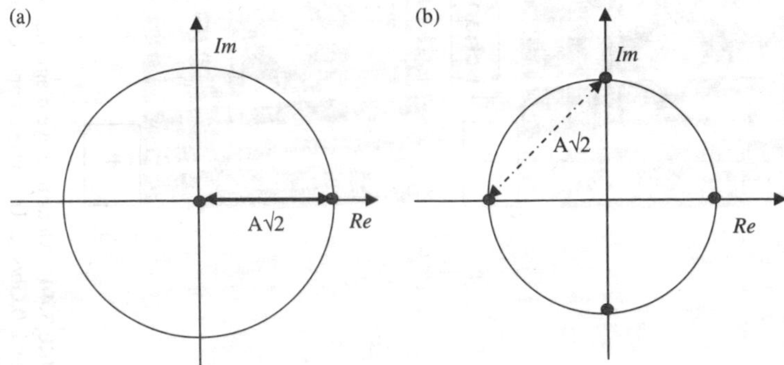

FIGURE 5.41   Signal constellation of (a) ASK and (b) DQPSK for the same average power.

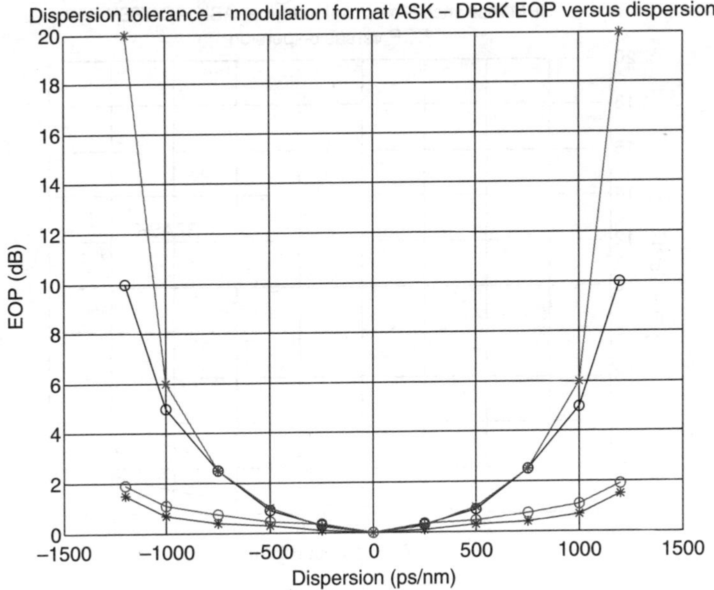

FIGURE 5.42 Measured and simulated dispersion tolerance of ASK and DQPSK (* red – RZ-DQPSK, o black RZ ASK) – both RZ format of 50% duty cycle and NRZ_ASK (blue *), NRZ_DPSK (red circle).

5.5.3 POLARIZATION MODE DISPERSION TOLERANCE

Refs [16–20] have investigated the impacts of PMD on DPSK and DQPSK direct detection formats. In systems with impairments dominated by PMD first order, smaller duty cycle RZ formats would improve the resilience to the PMD by the ASK and DPSK systems because the ISI is smaller for narrower pulses. RZ-DQPSK is much more resilient to the first-order PMD. DQPSK allows a DGD of at least two times higher for the same EOP due to PMD as compared with that suffered by ASK and DPSK.

In systems in which the PMD is compensated, then the second-order PMD would be minimized for modulation formats that offer narrower spectra due to the fact that the second order is wavelength-dependent, and hence narrower spectrum would offer more resilience due to second order PMD effects.

5.5.4 ROBUSTNESS TOWARDS NON-LINEAR EFFECTS

5.5.4.1 Robustness Towards SPM

The setup for investigation, both experimental and simulation, consists of 4 spans with 100 km SSMF and dispersion compensating modules (DCM) and associated EDF amplifiers at the end of SSMF and at the output of DCM with a booster EDFA and attenuator to adjust the launched input power to the transmission link. Four wavelength channels in the middle of the C-band are used. The EOP of a wavelength channel at 1552.95 nm is monitored and plotted against the launched power, as shown in Figure 5.43 for RZ-ASK and RZ-DQPSK modulation formats. It is observed that the nonlinear SPM threshold for 1-dB EOP is 3 dBm, which is consistent with a 0.1π phase shift due to the nonlinear phase effect of a fiber with a non-linear coefficient $n^2 = 2.3 \times 10^{-20}$ m^{-1}. The nonlinear threshold for RZ-DQPSK is observed at around 7 dBm. By inspecting the spectra of the ASK and DPSK, we note that due to the amplitude switched on and off, there is transitional time or equivalently the sampling time of the sequence and the Fourier transform of the sequence

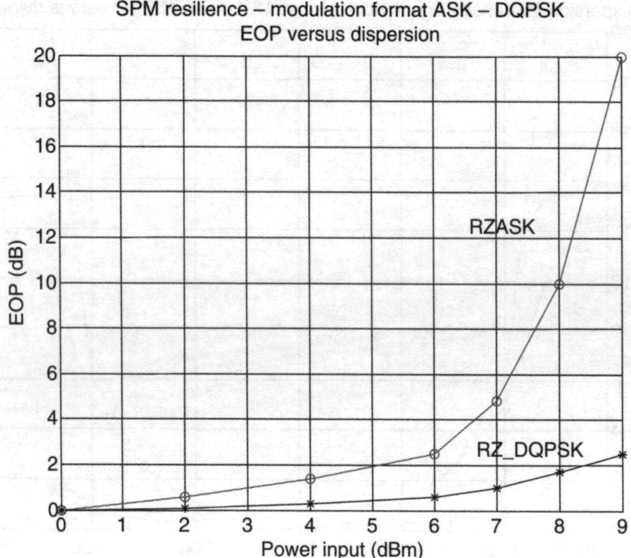

FIGURE 5.43 EOP induced by SPM versus input power at the beginning of the fiber transmission links of 4×100 km fully compensated SSMF spans at a bit rate of 40 Gb/s. RZ_ASK (red circle) RZ-DQPSK (star black).

would have spikes of the sampling frequency. This peak power level and hence average power of the ASK sequence is mostly contained in the carrier peaks. Unlike DPSK or DQPSK modulation formats whose envelops are constant and the energy is contained mostly in the signals. Thus we could observe no peaks in their spectra shown in Figure 5.14b. Thus, RZ-DQPSK and RZ-DPSK data sequence would be more tolerable to the SPM effects.

5.5.4.2 Robustness Towards XPM

In 40-Gb/s systems, the nonlinear SPM effects dominate the impairments due to fiber non-linearity. However, at 10-Gb/s DWDM systems, the situation is different, especially for 100-GHz spacing [21]. The XPM is caused by the non-linear phase effects of a channel in the DWDM system, especially when its intensity is fluctuating, which would cause a phase disturbance in other channels. This PM disturbance would then be transferred to other channels with amplitude modulation. The XPM effects can be considered to contribute to the frequency spectrum via a transfer function [22] given by

$$H_{12} \approx \frac{2\gamma_1}{\alpha - j2\pi f d_{12}} \tag{5.22}$$

where γ_1 is the nonlinear factor due to the intensity fluctuation of wavelength λ_1 channel, α is the attenuation factor, and d_{12} $(d_{12}z = D(\lambda_1 - \lambda_2)z)$ is the group delay difference between the two channels λ_1 and λ_2 over a distance z. Thus, for 100-GHz spacing and over SSMF of $D = 17$ ps/nm/km, the 3-dB passband of this low-pass filter transfer function is about 540 MHz, which would fall into the signal bands of a 10-Gb/s bit rate system.

Shown in Figure 5.44 is the EOP due to XPM of 10-Gb/s 100-GHz spacing. DWDM transmission of RZ-ASK and RZ-DPSK over multi-spans link in which dispersion and attenuation effects are equalized with DCM and EDFAs. Clearly the DPSK format is much more tolerable to the XPM effects as compared with that of RZ-ASK. XPM effect could be reduced by using an electrical filter

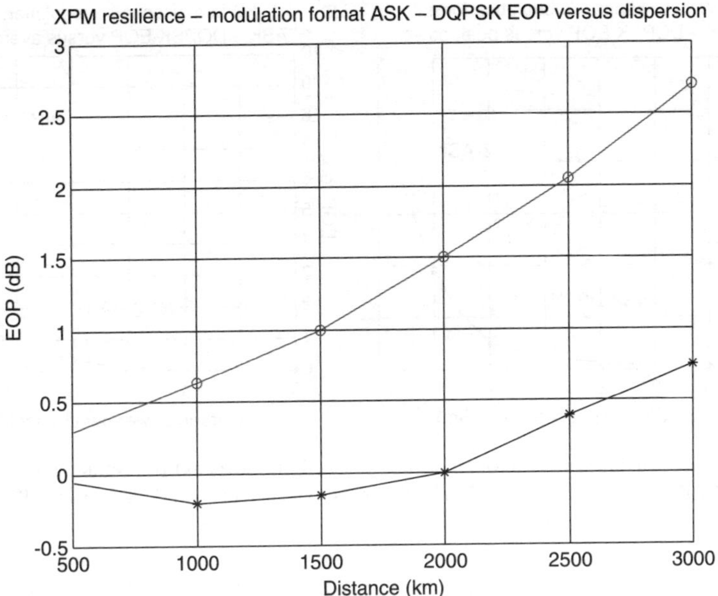

FIGURE 5.44 EOP induced by XPM versus total transmission distance (dispersion- compensated and attenuation equalized spans).

of a corner frequency of about 540 MHz with a sharp roll-off factor placed at the output of the electronic preamplifier.

5.5.4.3 Robustness Towards FWM

Unlike the impairments due to SPM and XPM, which are mainly caused by the optical power of the channels, FWM effects come from the relative phase difference between the channels. We can summarize the influence of FWM on DPSK and DQPSK signals as shown below.

For dispersion-shifted fibers, the phase velocity of different channels around the zero dispersion wavelength region is the same, and thus phase matching would occur and enhance the generation of the fourth wave that would fall within the active band of equally spaced DWDM system. Figure 5.45a and Figure 5.45b shows simulated results of the EOP of the fourth channel versus the peak and average power of three NRZ-ASK and NRZ-DPSK channels with 100-GHz spacing in the region of zero dispersion wavelength of DSF.

5.5.4.4 Robustness Towards SRS

The impact of SRS on different modulation formats has been reported [23], including the formats DPSK and DQPSK, which have been shown to improve significantly the resilience to SRS crosstalk. Especially when the RZ formats are employed, the SRS effects are negligible. The principal reasons are due to the periodic variation of the optical power or field propagating through the fiber, hence no low frequency components exist to enhance the SRS effects. The effect is very much like that of the XPM, but at lower frequency. Thus, RZ DPSK would exhibit no low frequency generated by the SRS. Simulated results of the effects of SRS on the broadening of NRZ pulse ASK modulation over the number of spans of SSMF and NZDSF, completely optically amplified and dispersion compensated, are shown in Figure 5.46. The SRS effects on the pulse broadening of RZ_DPSK are also shown to be significantly undisturbed by the periodic variation of the phase of the carrier, and hence strong resilience of the modulation format

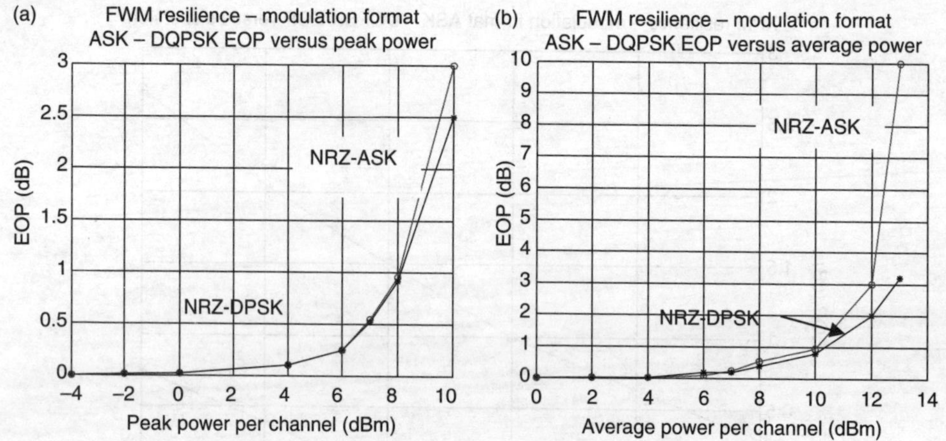

FIGURE 5.45 EOP induced by FWM versus (a) launched power (peak) (b) average power of three NRZ ASK and NRZ DPSK channels after 100-km DSF with zero dispersion wavelength at 1552.93 nm.

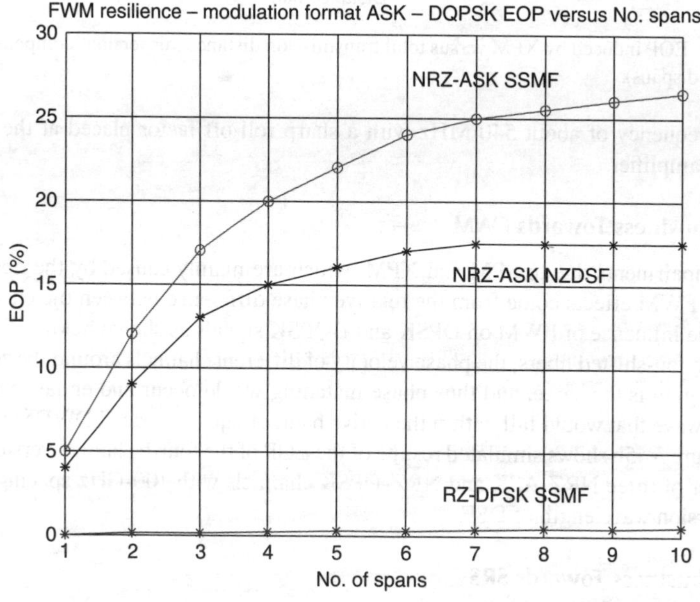

FIGURE 5.46 Percentage broadening of the "1" level (% of pulse period) due to SRS crosstalk for 11 channels 4 nm spacing 10 Gb/s bit rate under SSMF and NZDSF fiber spans. Modulation format NRZ-ASK. Span length = 100 km SSMF fully dispersion compensated.

to SRS. The standard deviation of the pulse amplitude over the average signal level is measured as the percentage.

5.5.4.5 Robustness Towards SBS

All DPSK and DQPSK do not exhibit the carrier component at the centre of the passband of the spectrum. Thus the SBS would have no role for phase shift keying modulation formats.

FIGURE 5.47 Siemens TranXpress Multi-wavelength Transport Systems.

5.6 REMARKS

This chapter gives the fundamental aspects of the optical transmission of discrete-phase modulation formats (DPSK, DQPSK and ASK) as well as the structures of the transmitters and receivers. Experimental setup for transmission of modulation formatted signals, NRZ, RZ, CS-RZ, NRZ-DPSK, RZ-DPSK, and CS-RZ DPSK over the optically amplified dispersion compensating system have been described. The experimental results have been accomplished to investigate the filtering effects on performance of the transmission systems, the dispersion tolerance and receiver sensitivity performance of the RZ-ASK and RZ/CS-RZ DPSK. The filtering properties of the muxes and demuxes do not affect significantly the transmission performance in term of BER and receiver sensitivity of the 40-Gb/s RZ or CSRZ DPSK modulation formats.

We have also reported and simulated the nonlinear effects including SRS, SPM, XPM and SBS on the eye opening at the receiver of the phase shift keying modulation, in particular the DPSK and DQPSK with RZ formats. Phase shift keying modulation is much more resilient to the non-linear effects as compared with amplitude shift keying.

The negligible mutual effects of adjacent 40-Gb/s DPSK channels on 10-Gb/s ASK-modulated transmission channel enable the feasibility of the upgradeability of the 10-Gb/s current system to 40-Gb/s optical DPSK modulated transmission. We have also measured the transmission quality of 40-Gb/s and 10-Gb/s channels, and observe no significant degradation of one channel by the other. Our next step is to launch the 40-Gb/s modulation format over a commercial system, the Siemens multi-wavelength transport Tranxpress system shown in Figure 5.47, and then over installed multi-wavelength 10-Gb/s transmission link between cities, e.g., Melbourne and Sydney in Australia.

REFERENCES

1. Winzer, P. J. 2002. Optical transmitters, receivers, and noise. In *Wiley encyclopedia of telecommunications*, ed. J.G. Proakis, 1824–40. New York: Wiley.
2. Winzer, P. J., and R.-J. Essiambre. 2003. Advanced optical modulation formats. In *Proceedings of ECOC 2003, invited paper Th2.6.1,* Rimini, Italy, 1002–3.
3. Binh, L. N. 2005. Lithium niobate optical modulators. In *International conference on material and technology symposium M: Material and devices*, Singapore, July.
4. Linke, R. A., and A. H. Gnauck. 1988. High-capacity coherent lightwave systems. *Journal of Lightwave Technology* 6(11): 1750–69.

5. Cai, J.-X., D. G. Foursa, C. R. Davidson, Y. Cai, G. Domagala, H. Li, L. Liu, et al. 2003. A DWDM demonstration of 3.73 Tb/s over 11 000 km using 373 RZ-DPSK channels at 10 Gb/s. In *Proceedings of OFC 2003, Postdeadline paper PD22,* Atlanta, GA.

6. Livas, J. C., E. A. Swanson, S. R. Chinn, E. S. Kintzer, R. S. Bondurant, and D. J. DiGiovanni. 1995. A one-watt, 10-Gbps high-sensitivity optical communication system. *IEEE Photonics Technology Letters* 7: 579–81.

7. Gnauck, A. H., X. Liu, X. Wei, D. M. Gill, and E. C. Burrows. 2004. Comparison of modulation formats for 42.7-Gb/s single-channel transmission through 1980 km of SSMF. *IEEE Photonics Technology Letters* 16: 909–11.

8. Wree, C. 2006. RZ-DQPSK format with high spectral efficiency and high robustness towards fiber nonlinearities. Technical report, University of Kiel, Germany.

9. Winzer, P. J., and H. Kim. 2003. Degradations in balanced DPSK receivers. *IEEE Photonics Technology Letters* 15: 1282–84.

10. Wree, C. 2003. Experimental Investigation of receiver sensitivity of RZ-DQPSK modulation format using balanced detection. In *Proceedings of OFC2003Paper ThE5,* vol. 2, p. 456.

11. Okunev, Y. 1997. *Phase and phase difference in digital communications.* Artec House.

12. Wree, C. 2004. Differential phase shift keying for long haul fiber-optic transmission based on direct detection. Dr. Ing. Dissertation, CAU University zu Kiel, Kiel Germany, Figure 4.6, p. 75.

13. Benedetto and Biglieri. 1999. *Principles of digital communications with wireless applications.* Norwell, MA, USA: Kluwer Academics.

14. Pang, K. K. 2002. *Digital communications.* Lecture notes: Monash University.

15. Gnauck, A. H., and P. J. Winzer. 2005. Optical Phase-Shift-Keyed Transmission. *IEEE Journal of Lightwave Technology* 23(1): 115–30.

16. Xie, C., L. Moeller, H. Haustein, and S. Hunsche. 2003. Comparison of system tolerance to polarization mode dispersion between different modulation formats. *IEEE Photonics Technology Letters* 15(8): 1168–70.

17. Wang, J., and J. M. Kahn. 2004. Impact of chromatic and polarization mode dispersion on DPSK systems using interferometric demodulation and direct detection. *IEEE Journal of Lightwave Technology* 10: 96–100.

18. Kim, H., C. R. Doer, P. Pafchek, L. W. Stulz, and P. Bernasconi. 2002. Polarization mode dispersion impairments in direct detection differential phase-shift keying systems. *IEE Electronics Letters* 38: 1047–48.

19. Griffin, R. A., R. I. Johnstone, R. G. Walker, S. D. Wadsworth, K. Kerry, A. C. Carter, M. J. Wale, J. Hughes, P. A. Jerram, and N. J. Parsons. 2002. 10 Gb/s optical differential quadrature phase shift keying (DQPSK) transmission using GaAs/AlGaAs integration. In *Proceedings of optical fiber communication OFC2002, Postdeadline paper FD-6,* 1–3.

20. Griffin, R. A., and A. C. Carter. 2002. Optical differential quadrature phase shift keying (oDPSK) for high capacity optical transmission. In *Proceedings of optical fiber communication conference OFC2002, paper WX-6,* Optical Fiber Conference 2002, Annaheim, CA, USA, Paper WX-6, 1–3.

21. Mitra, P., and J. Stark. 2001. Nonlinear limits to the information capacity of optical fiber communications. *Nature* 411: 1027–70.

22. Wree, C. 2002. Differential phase shift keying for long haul transmission based on direct detection. Dr Ing Dissertation, CAU University zu Kiel, p. 24.

23. Schoemann, S., J. Leibrich, C. Wree, and W. Rosenkranz. 2004. Impact of SRS-induced crosstalk for different modulation formats. In *Proceedings of optical fiber conference, OFC 2004, paper FA5,* Optical Fiber Conference 2005, Annaheim, CA, USA, Paper FA5, 1–3.

APPENDIX: MATLAB® SIMULINK® MODEL FOR DQPSK OPTICAL SYSTEM

A number of models in MATLAB® Simulink® are given in Figure 5.48, Figure 5.49 (same as Figure 5.34), Figure 5.50 and Figure 5.51 (same as Figure 5.35). Figure 5.48 gives the block diagram of the whole transmission system, generally an optical transmitter, the optically amplified fiber spans, and then the receiver including sub-systems for evaluation of BER versus parameters such as receiver sensitivity, and eyeopening penalty. Figure 5.49 gives the model of optical transmitters in association with Figure 5.50 as a pre-coder in electrical domain to generate signals for driving the electrodes of the MZIM. Figure 5.51 gives the model of balanced receiver incorporating MZDI as phase comparator for DPSK.

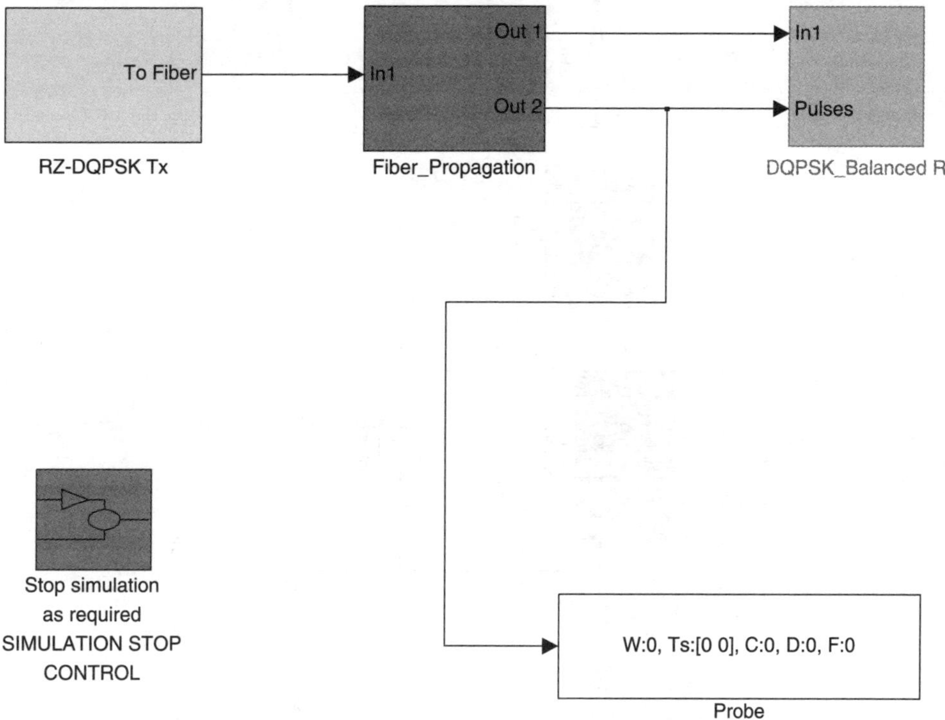

FIGURE 5.48 DQPSK system (schematic).

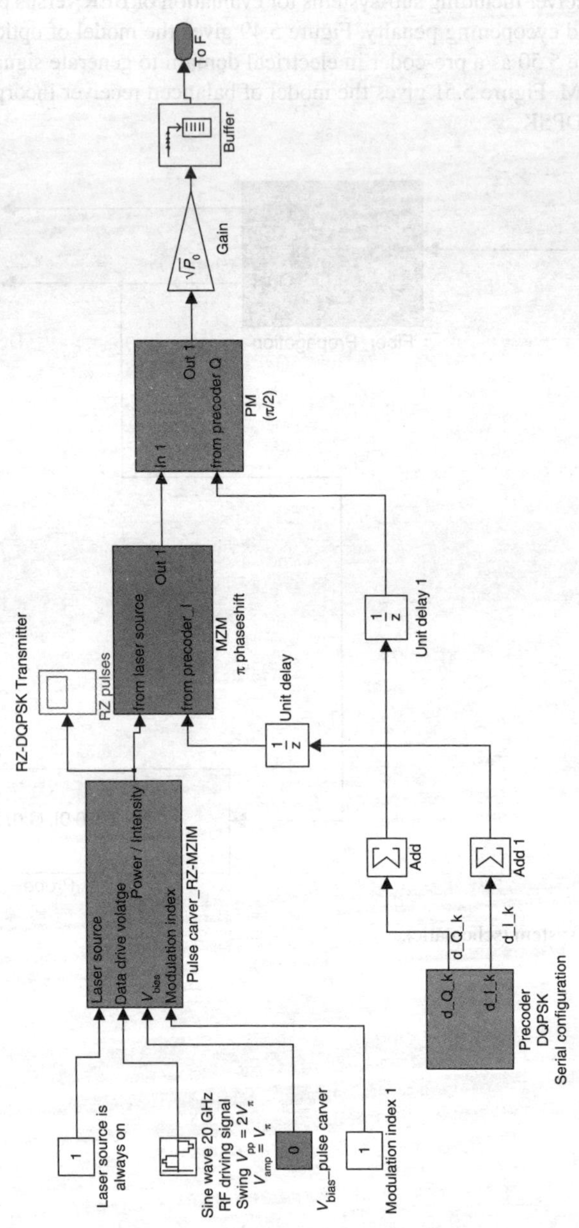

FIGURE 5.49   Simulink model of the optical transmitter for RZ DQPSK.

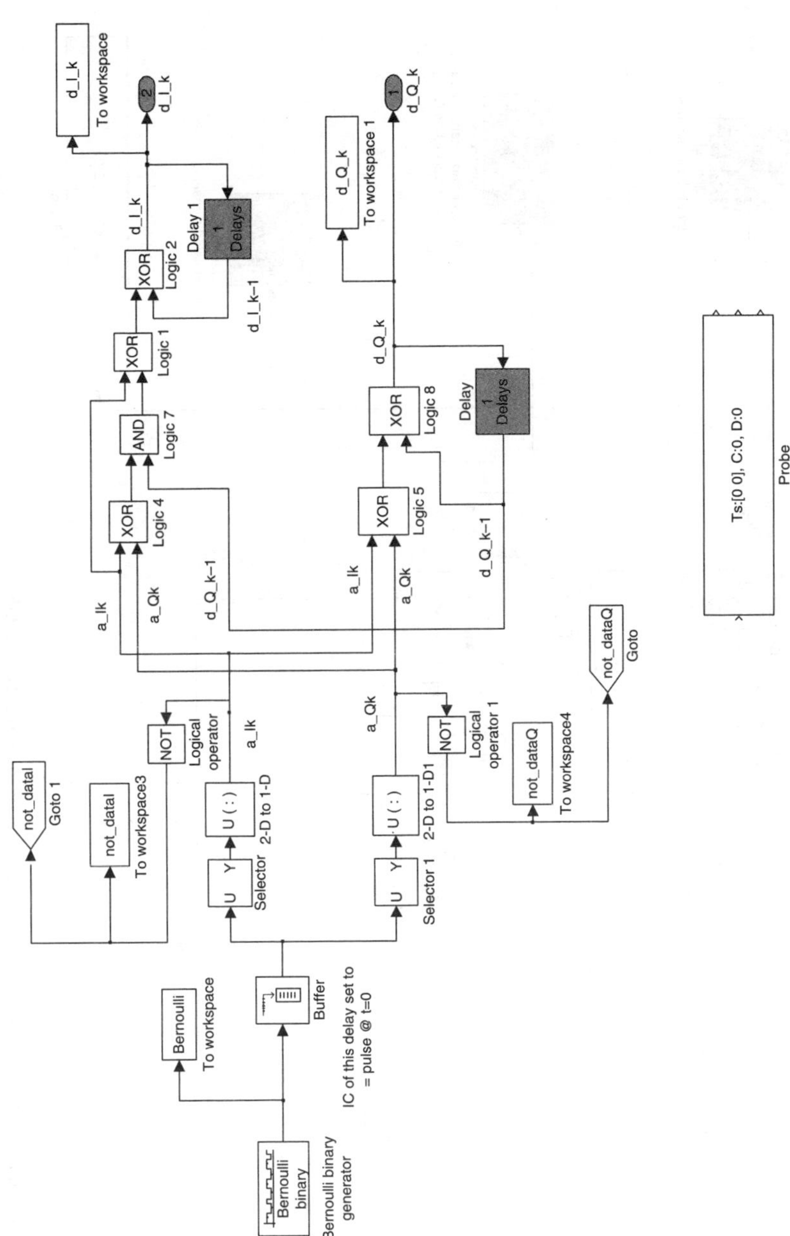

FIGURE 5.50 Simulink model of the electrical pre-coder for optical transmitter for RZ DQPSK. The block "Probe" is used to monitor the signal values for later processing.

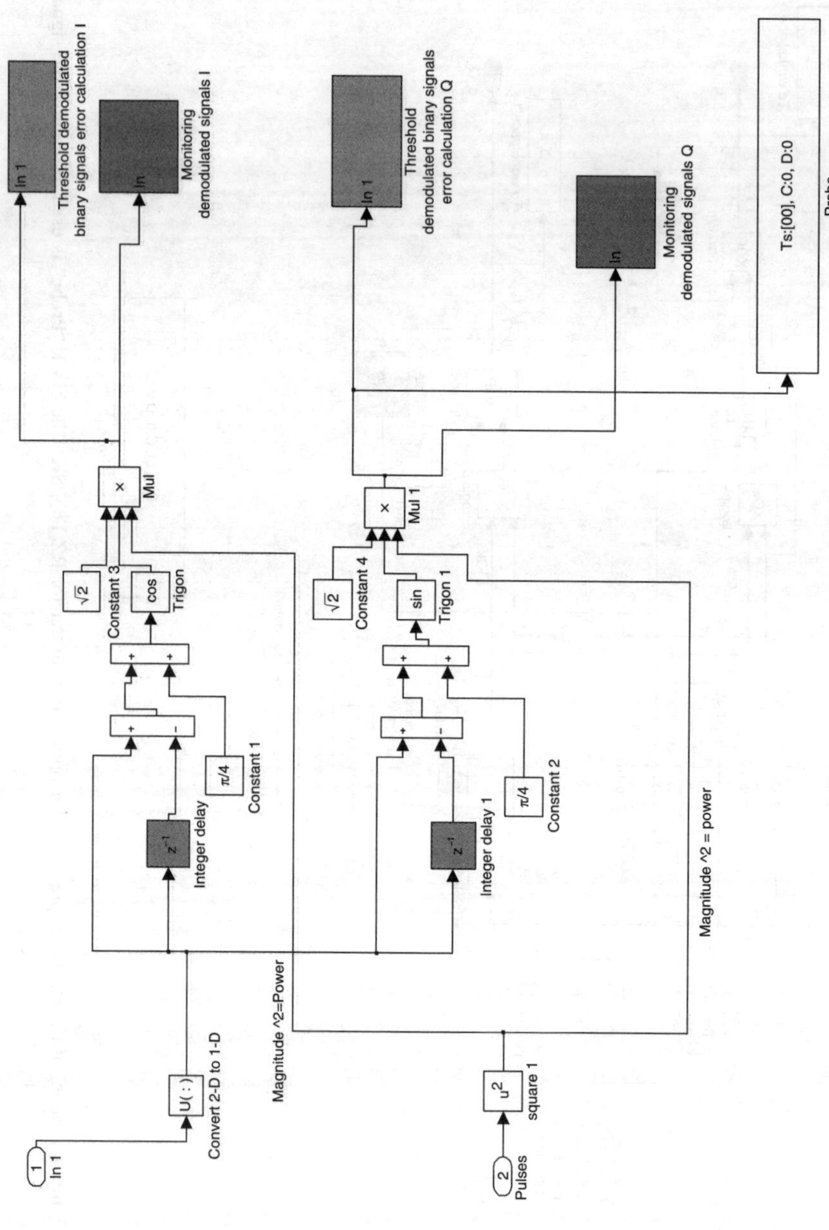

FIGURE 5.51 Simulink model of the MZDI optical balanced receiver for RZ DQPSK.

6 Continuous Phase Modulation Format Optical Systems

6.1 INTRODUCTION

The generation of MSK requires a linear variation of the phase, hence a constant frequency of the optical carrier. However, the generation of the optical phase may be preferred by driving an optical modulator using a sinusoidal signal for practical implementation. Thus a non-linear variation of the carrier phase, hence some distortion effects are produced. In this article, we investigate linear and non-linear phase-shaping filtering and their impacts on MSK-modulated optical signals transmission over optically amplified long-haul communications system. The evolution of the phasor of the in-phase and quadrature components is illustrated for lightwave-modulated signal transmission. The distinct features of three different MSK modulation formats (linear MSK, weakly non-linear MSK and strongly non-linear MSK) and their transmission are simulated. Transmission performance obtained indicates the resilience of the MSK signals in transmission over multi-optically amplified multi-spans.

Recent years have witnessed intensive interest in the employment of advanced modulation formats to explore their advantages and performance in high-density and long-haul transmission systems. Return-to-zero (RZ) and NRZ without or with carrier suppression (CS-RZ) formats are associated with shift keying (SK) modulation schemes such as amplitude (ASK), phase (PSK) and differential phase (DPSK, DQPSK) [1, 2]. Differential detection offers the best technological implementation due to the non-requirement of coherent sources and avoidance of polarization control of the mixing of the signals and a local oscillator at multi-THz frequency range. Continuous phase modulation (CPM) is also another form of phase shift keying in which the phase of the optical carrier evolves continuously from one phase state to the other. For minimum phase shift keying (MSK), the phase change is limited to $\pi/2$. Although MSK is a well-known modulation format in radio frequency digital communications, it has only been attracting interest in optical system research in the last few years [3–5]. The phase-continuous evolution of the MSK has many interesting features: the main lobe of the power spectrum is wider than that of quadrature phase-shift keying (QPSK) and DPSK, and the side lobes of the MSK signal spectrum are much lower, allowing the ease of optical filtering and hence less distortion due to dispersion effects. In addition, higher signal energy is concentrated in the main lobe of the MSK spectrum than its side lobes, leading to better signal-to-noise ratio (SNR) at the receiver.

Advanced modulation formats for 40 Gb/s and higher bit rate long-haul optical transmission systems have recently attracted intensive research, including various amplitude and discrete differential phase modulation and pulse shape formats (ASK-NRZ/RZ/CSRZ, DPSK and DQPSK-NRZ/RZ/CSRZ). However, there are only a few reports on optical continuous phase modulation (CPM) schemes using external electro-optical modulators [1–4]. Compared to discrete phase modulation, CPM signals have very interesting and attractive characteristics, including high spectral efficiency, higher energy concentration in the signal bands, and more robustness to inter-channel crosstalk in DWDM due to greater suppression of the side lobes, which has recently risen as a critical issue in dense wavelength division multiplexed (DWDM) optical systems [5, 6].

In bandwidth-limited digital communication including wireless and satellite digital transmission, MSK has been proved to be a very efficient modulation format due to its prominent spectral efficiency, high sideband suppression, constant envelope and high energy concentration in the main spectral lobe. In modern high-capacity and high-performance optical systems, we have witnessed the acceleration of transmission bit rate approaching 40 Gb/s and even to 160 Gb/s. At these ultra-high bit transmission speeds, the essence of efficient modulation formats for long-haul transmission is critical. There are several technical works on advanced modulation techniques for optical transmission, which mostly focus on discrete phase modulation, including binary DPSK and DQPSK [1–3] with various pulse carving formats including non-return-to-zero (NRZ) and CS-RZ [2, 4]. However, there is only limited number of works on MSK modulation format [5–8]. The features of MSK, compared with other modulation formats, have prompted researchers to investigate its suitability for high-capacity long-haul transmission. The side lobes of MSK power spectrum are greatly suppressed, giving it good dispersion tolerance and avoiding inter-channel crosstalks. Thus, this modulation is of interest for further investigations.

Two proposals of transmitter configurations for generation of linear and non-linear phase optical MSK modulation are reported in another paper in this conference [9]. The brief operation descriptions of these two transmitter configurations are presented in Section 6.2, whereas the direct detection techniques for linear and non-linear optical MSK signals are stated in Section 6.3. Our simulation models for the modulation and system transmission are based on MATLAB® Simulink® platform [10] with detailed discussions in Section 6.4. In this article, we have also proved that the optical MSK signal can propagate over an optically amplified and fully dispersion compensated system. Not only the performance of conventional MSK format, but also that of the weakly and strongly non-linear optical MSK sequences is evaluated.

The MSK format exhibits dual alternating frequencies and offers orthogonal property of the two consecutive bit periods. More interestingly, MSK can be considered as a special case of continuous phase frequency shift keying (CPFSK) or a staggered/offset QPSK in which I and Q components are interleaved with each other [7]. These characteristics are greatly advantageous to optically amplified long-haul transmissions because such a compact spectrum potentially gives a good dispersion tolerance, making MSK a suitable candidate for DWDM systems.

This chapter thus investigates a number of novel structures of photonic transmitters for generation of optical MSK signals. The theoretical background of MSK modulation formats discussed in two different approaches is presented in Section 6.2. Section 6.3 proposes two configurations of optical MSK transmitters that employ (i) two cascaded electro-optic phase modulators (E-OPMs) and (ii) parallel dual-drive Mach–Zehnder modulators (MZMs). Different types of linear and non-linear phase-shaped optical MSK sequences can be generated. The pre-coder for I-Q optical MSK structure is also derived in this section. We present in Section 6.4 a simple non-coherent configuration for detection of the MSK- and non-linear MSK-modulated sequences. The optical detection of MSK and non-linear MSK signals employs a pi/2 phase shift in one arm of the Mach–Zehnder interferometric delay (MZIM)-balanced receiver. New techniques in the measurement of the BERs with the probability density function (pdf) of received signals after decision sampling to be computed with superposition of a number of weighted Gaussian pdfs. The following performance results and observations are obtained in Section 6.5: (i) spectral characteristics of linear and non-linear MSK-modulated signals; (ii) improvement on dispersion tolerance of MSK and non-linear MSK over ASK and DPSK counterparts.

Recently, advanced modulation formats have attracted intensive research for long-haul optical transmission systems, including various amplitude and discrete differential phase modulation and pulse shape formats (ASK-NRZ/RZ/CSRZ, DPSK and DQPSK-NRZ/RZ/CSRZ). For phase modulation, the phases of the optical carrier are discretely coded with "0", "π" (DPSK) or "0", "π" and "$\pi/2$, $-\pi/2$" (DQPSK). Although the differential phase modulation techniques offer better spectral properties, higher energy concentration in the signal bands, and more robustness to combat the non-linearity impairments as compared with amplitude modulation formats, phase continuity would offer even better spectral efficiency and at least 20-dB better in the suppression of the side lobes. MSK exhibits a dual alternating frequency between the two consecutive bit periods that is considered to offer the best scheme because this offers orthogonal property of the two-frequency modulation of the carrier lightwaves embedded within the consecutive bits. Furthermore, it offers the most simplicity in the implementation of the modulation in the photonic domain.

This chapter thus investigates various novel structures of photonic transmitters and differential non-coherent balanced receivers for generation and detection of MSK optical signals, as described below.

Cascaded electro-optic phase modulators MSK transmitter: This structure employs two cascaded optical phase modulators (OPM). The first OPM has the role of modulating the binary data logic into two carrier frequencies deviating from the optical carrier of the laser source by a quarter of the bit rate. The second OPM enforces the phase continuity of the lightwave carrier at every bit transition. The driving voltage of this second OPM is pre-coded in such as a way that the phase discrepancy due to frequency modulation of the first OPM will be compensated, hence preserving the phase continuity characteristic of the MSK signal. Utilizing this double-phase modulation configuration, different types of linear and non-linear CPM phase-shaping signals including MSK, weakly-MSK and linear-sinusoidal MSK, can be generated. The optical spectra of the modulation scheme obtained confirm the bandwidth efficiency of this novel optical MSK transmitter.

Parallel dual-drive Mach–Zehnder modulators optical MSK transmitter: In this second configuration, MSK signals with small ripple of approximately 5% signal amplitude level can be generated. The optical spectrum is demonstrated equivalent to the configuration of (1). This configuration can be implemented using commercially available dual-drive intensity interferometric electro-optic modulators.

We also present a simple non-coherent configuration for detection of the MSK and non-linear MSK modulated sequences. The optical detection of MSK and non-linear MSK signals employs a $\pi/2$ OPM followed by the Mach–Zehnder interferometric delay (MZIM)-balanced receiver.

The following performance results and observations are obtained: (a) modeling platform based on MATLAB® Simulink® developed for the transmission systems; (b) spectral characteristics of linear and non-linear MSK-modulated signals; (c) improvement on dispersion tolerance of MSK and non-linear MSK over ASK and DPSK; and (d) novel techniques in the measurement of the BERs with the probability density function (pdf) of received signals after decision sampling to be computed with superposition of a number of weighted Gaussian pdfs. This technique has been proven to be effective, convenient but accurate for estimating the non-Gaussian pdf for continuous phase modulation and its transmission over long-haul optically amplified systems. The MSK format has been demonstrated via modeling as a promising candidate in the selection of advanced modulation formats for dense wavelength division multiplexed (DWDM) long-haul optical transmission systems.

6.2 GENERATION OF OPTICAL MSK MODULATED SIGNALS

The two optical MSK transmitter configurations can be referred in more detailed in [6]. The descriptions of these configurations are briefly addressed in Section 6.2.1 and Section 6.2.2.

6.2.1 Optical MSK Transmitter Using Two Cascaded Electro-Optic Phase Modulators

Electro-optic phase modulators and interferometers operating at high frequency using resonant-type electrodes have been proposed in [7, 8] In addition, high-speed electronic driving circuits evolved with the ASIC technology using 0.1-µm GaAs P-HEMT or InP HEMTs [9] enable the feasibility in realization of the proposed optical MSK transmitter structure. The base-band equivalent optical MSK signal is represented in Equation 6.1

The first E-OPM enables the frequency modulation of data logics into upper side bands (USB) and lower side bands (LSB) of the optical carrier with frequency deviation of f_d. Differential phase pre-coding is not necessary in this configuration due to the nature of the continuity of the differential phase trellis. By alternating the driving sources $V_d(t)$ to sinusoidal waveforms for simple implementation or combination of sinusoidal and periodic ramp signals, which was first proposed by Amoroso in 1976 [10], different schemes of linear and non-linear phase-shaping MSK transmitted sequences can be generated.

The second E-OPM enforces the phase continuity of the light wave carrier at every bit transition. The delay control between the E-OPMs is usually implemented by the phase shifter shown in Figure 6.1. The driving voltage of the second E-OPM is pre-coded to fully compensate the transitional phase jump at the output $E_{01}(t)$ of the first E-OPM. Phase continuity characteristic of the optical MSK signals is determined by the algorithm in Equation 6.8.

$$\Phi(t,k) = \frac{\pi}{2}\left(\sum_{j=0}^{k-1} a_j - a_k I_k \sum_{j=0}^{k-1} I_j\right) \tag{6.1}$$

where $a_k = \pm 1$ are the logic levels; $I_k = \pm 1$ is a clock pulse whose duty cycle is equal to the period of the driving signal $V_d(t)$, f_d is the frequency deviation from the optical carrier frequency, and $h = 2f_d T$ is previously defined as the frequency modulation index. In optical MSK, $h = 1/2$ or $f_d = 1/(4T)$. The phase evolution of the continuous phase optical MSK signals are shown with detailed discussion in Figure 6.2. To mitigate the effects of unstable stages of rising and falling edges of the electronic circuits, the clock pulse $V_c(t)$ is offset with the driving voltages $V_d(t)$.

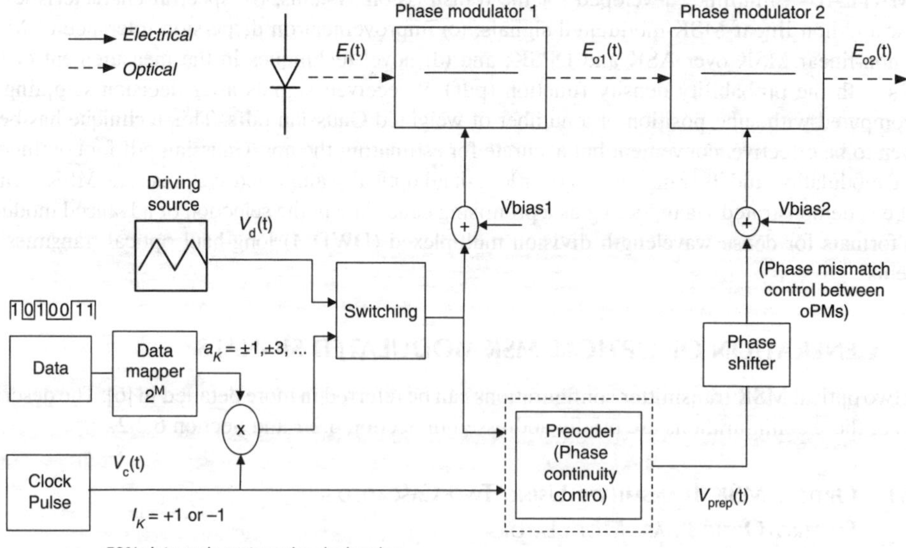

FIGURE 6.1 Block diagram of optical MSK transmitter employing two cascaded optical phase modulators. (From Huynh, T. L., L. N. Binh, K. K. Pang, and L. Chan, Paper 6353–85, *presented at APOC*, 3–7, 2006. With permission.)

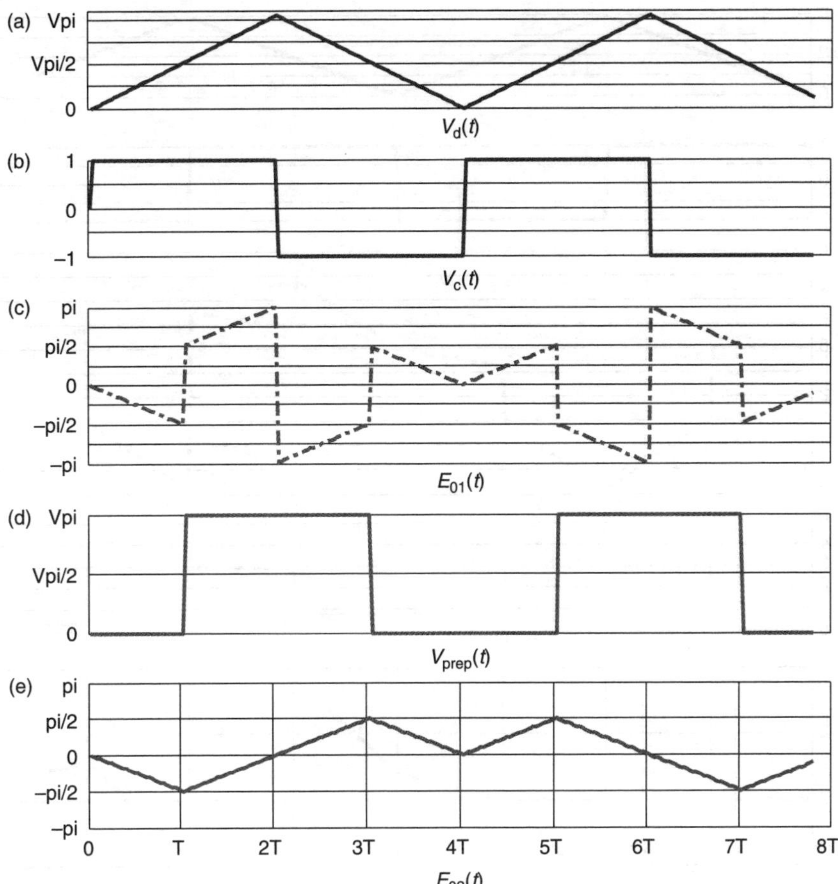

FIGURE 6.2 Evolution of time-domain phase trellis of optical MSK signal sequence [−1 1 1 −1 1 −1 1 1] as inputs and outputs at different stages of the optical MSK transmitter The notation is denoted in Figure 6.1 accordingly; (a) $V_d(t)$: periodic triangular driving signal for optical MSK signals with duty cycle of 4 bit period, (b) $V_c(t)$: the clock pulse with duty cycle of $4T$, (c) $E_{01}(t)$: phase output of oPM1 (d) $V_{prep}(t)$: pre-computed phase compensation driving voltage of oPM2 and (e) $E_{02}(t)$: phase trellis of a transmitted optical MSK sequences at output of oPM2. (From Huynh, T. L., L. N. Binh, K. K. Pang, and L. Chan, Paper 6353–85, *presented at APOC*, 3–7, 2006. With permission.)

The generated optical signal envelope can be written as

$$\tilde{s}(t) = A \exp\left\{ j \left[a_k I_k f_d \frac{2\pi t}{T} \Phi(t,k) \right] \right\}, \quad kT \leq t \leq (k+1)T \quad \text{with} \quad a_k = \pm, \pm 3, \ldots \quad (6.2)$$

6.3 DETECTION OF M-ARY CPFSK MODULATED OPTICAL SIGNAL

The detection of linear and non-linear optical M-ary CPFSK utilizes the well-known structure Mach–Zehnder delay inteferometer (MZDI)-balanced receiver. The addition of $\pi/2$ phase on one arm of MZDI is also introduced. The time delay being a fraction of bit period enables the phase trellis detection of optical M-ary CPFSK. The detected phase trellis using the proposed technique is shown in Figure 6.3 An optimized ratio of switching frequencies results in the maximum

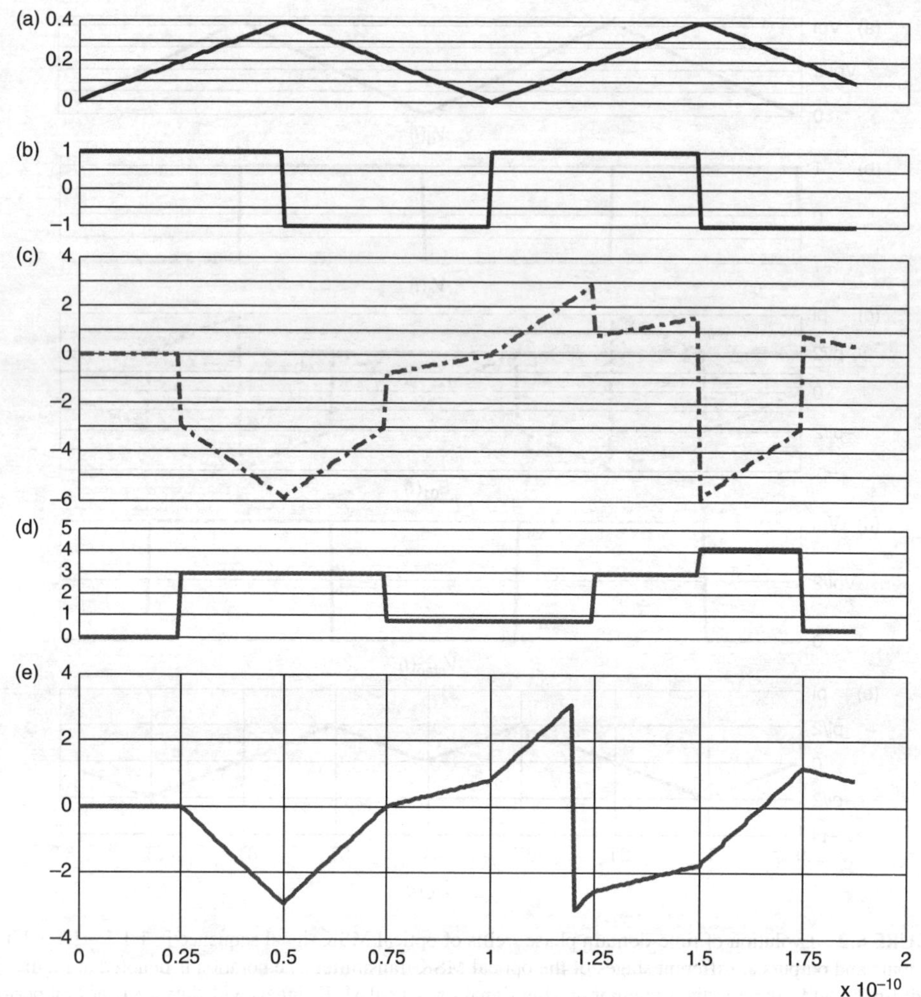

FIGURE 6.3 Demonstration of phase compensation for enforcement of phase continuity at bit transitions. (a) shows the periodic triangular driving signal whose peak voltage to be $V_{\pi/8}$ and duty cycle of 4 bit period. (b) shows the clock pulse corresponding to the driving signal. (c) shows frequency switching with $h = \pm 1/8$, $\pm 7/8$. (d) shows the computed phase compensation. (e) shows the phase trellis of an optical quaternary CPFSK. (From Huynh, T. L., L. N. Binh, K. K. Pang, and L. Chan, Paper 6353–85, *presented at APOC*, 3–7, 2006. With permission.)

eye open. The detection of optical M-ary CPFSK with delay of $T_b/2$ at $t = (k+1)T_b/2$ is expressed in Equation 6.3.

$$\sin(\Delta\Phi) = \sin\left[a_{k+1} \frac{2\pi f_d (k+1)^{T_b}/_2}{T_b} - a_k \frac{2\pi f_d\, 2\pi f_d k^{T_b}/_2}{T_b} \right]. \tag{6.3}$$

The detected eye diagram of a M-ary CPFSK is shown in Figure 6.4. On the same slope of phase in the phase trellis, the differential phase and hence, the modulated frequency levels can be mapped to detected amplitude levels via (4).

$$\sin(\Delta\Phi) = \sin(a_{k+1}\pi f_d) \tag{6.4}$$

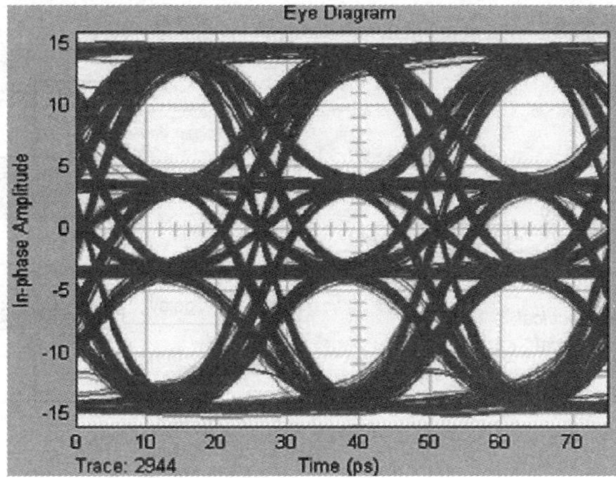

FIGURE 6.4   Eye phase trellis detection of optical M-ary CPFSK-modulated signal.

6.3.1   OPTICAL MSK TRANSMITTER USING PARALLEL I-Q MACH–ZEHNDER INTENSITY MODULATORS

The conceptual block diagram of optical MSK transmitter is shown in Figure 6.5a. The transmitter consists of two dual-drive electro-optic MZM modulators generating chirpless I and Q components of MSK-modulated signals, and is considered as a special case of staggered or offset QPSK. The binary logic data is pre-coded and de-interleaved into even and odd bit streams which are interleaved with each other by one bit-duration offset. Figure 6.5b shows the block diagram configuration of band-limited phase shaped optical MSK. Two arms of the dual-drive MZM modulator are biased at $V_{\pi/2}$ and $-V_{\pi/2}$ and driven with data and data complement. Phase-shaping driving sources can be a periodic triangular voltage source in linear MSK generation or simply a sinusoidal source for generating a non-linear MSK-like signal which also obtain linear phase trellis property, but with small ripples introduced in the magnitude. The magnitude fluctuation level depends on the magnitude of the phase-shaping driving source. High spectral efficiency can be achieved with tight filtering of the driving signals before modulating the electro-optic MZMs. Three types of pulse shaping filters are investigated, including Gaussian, raised cosine and squared-root raised cosine filters. The optical carrier phase trellis of linear and non-linear optical MSK signals are shown in Figure 6.6.

The generation of linear and non-linear optical MSK sequences can be briefly discussed as detailed below.

6.3.1.1   Linear MSK

The pulse-shaping waveform for linear MSK is triangular waveform with duty cycle of 4 Tb. The triangular waveform for quadrature path is delayed by one bit period with respect to the in-phase path, hence they are interleaved with each other. The optical MSK signal is the superposition of even and odd waveforms from the MZIMs. The amplitude of the signal is perfectly constant, clearly displaying the constant amplitude characteristic of continuous phase modulation. The phase trellis is perfectly linear and the phase transition is continuous as shown in Figure 6.6a. The signal constellation is a perfect circle.

6.3.1.2   Weakly Non-linear MSK

The pulse-shaping waveform for weakly non-linear MSK is a sinusoidal waveform with amplitude of $1/4V_\pi$, and its symbol period is equal to two bit period (symbol rate of 20 Gbits/s). The in-phase pulse shaper is a cosine waveform, whereas the quadrature pulse shaper is a sine

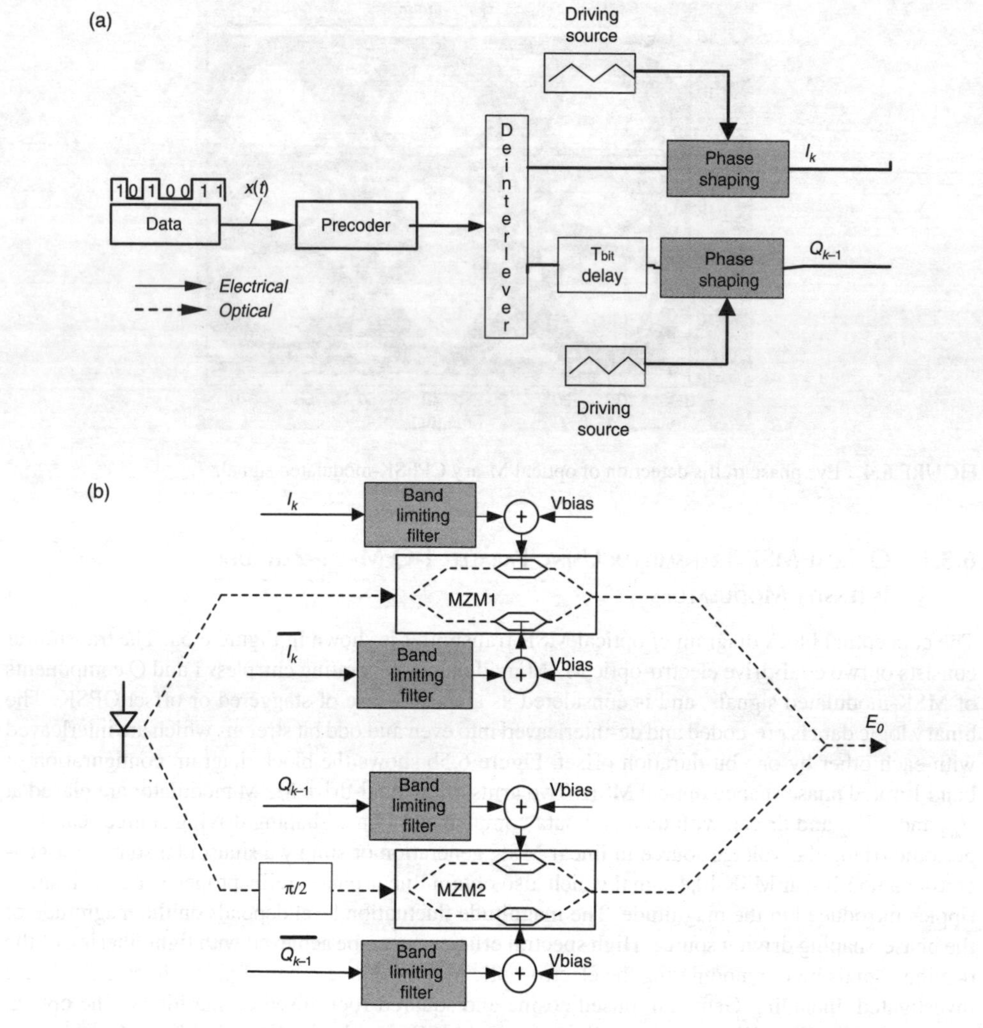

(a)

(b)

FIGURE 6.5 Block diagram configuration of band-limited phase-shaped optical MSK. (From Huynh, T. L., L. N. Binh, K. K. Pang, and L. Chan, Paper 6353–85, *presented at APOC*, 3–7, 2006. With permission.)

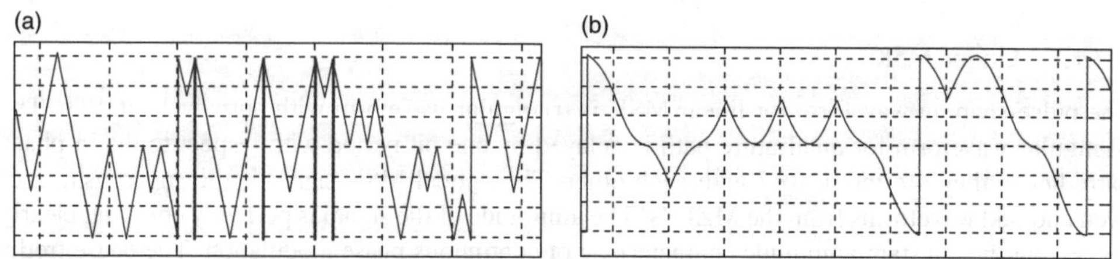

(a)

(b)

FIGURE 6.6 Phase trellis of linear and non-linear MSK transmitted signals.

waveform. There is ripple of approximately 5%. Due to the sinusoidal pulse shaper, the variation of phase with time, which is represented by the phase trellis, is non-linear. Therefore, the rate of frequency change is not constant. This causes mismatch of MZIM when the modulated waveforms are added up, resulting in ripple (Figure 6.6b).

6.3.1.3 Strongly Non-linear MSK

The pulse-shaping waveform for strongly non-linear MSK is sinusoidal waveform, the same as for weakly non-linear. However, the amplitude of pulse shaper is $V_{\pi/2}$. The waveforms are interleaved with each other. The optical MSK signal has ripple of approximately 26%. This ripple is also caused by the mismatch of MZIM because the modulating waveform is strongly non-linear. The effect of non-linearity is obvious in the phase trellis in Figure 6.4b.

The signal state constellations and eye diagrams of optical MSK sequences are shown in Figure 6.7a–c of linear and non-linear schemes, respectively. The magnitude ripples can be clearly observed in Figure 6.7c. In non-linear configuration, MSK signals with small ripple of approximately 5% signal amplitude level can be generated. This configuration can be implemented using commercially available dual-drive intensity interferometric electro-optic modulators.

The conceptual block diagram of optical MSK transmitter is shown in Figure 6.1 and Figure 6.5. The transmitter consists of two dual-drive electro-optical MZMs modulators generating chirpless I and Q components of MSK modulated signals which is considered as a special case of staggered or offset QPSK. The binary logic data is pre-coded and de-interleaved into even and odd bit streams which are interleaved with each other by one bit duration offset.

Two arms of the dual-drive MZM modulator are biased at $V_{\pi/2}$ and $-V_{\pi/2}$ and driven with data and inverted data. Phase-shaping driving sources can be a periodic triangular voltage source in linear MSK generation or simply a sinusoidal source for generating a non-linear MSK-like signal which also obtain a linear phase trellis property but with small ripples introduced in the magnitude.

(a)　　　　　　　　　　(b)　　　　　　　　　　(c)

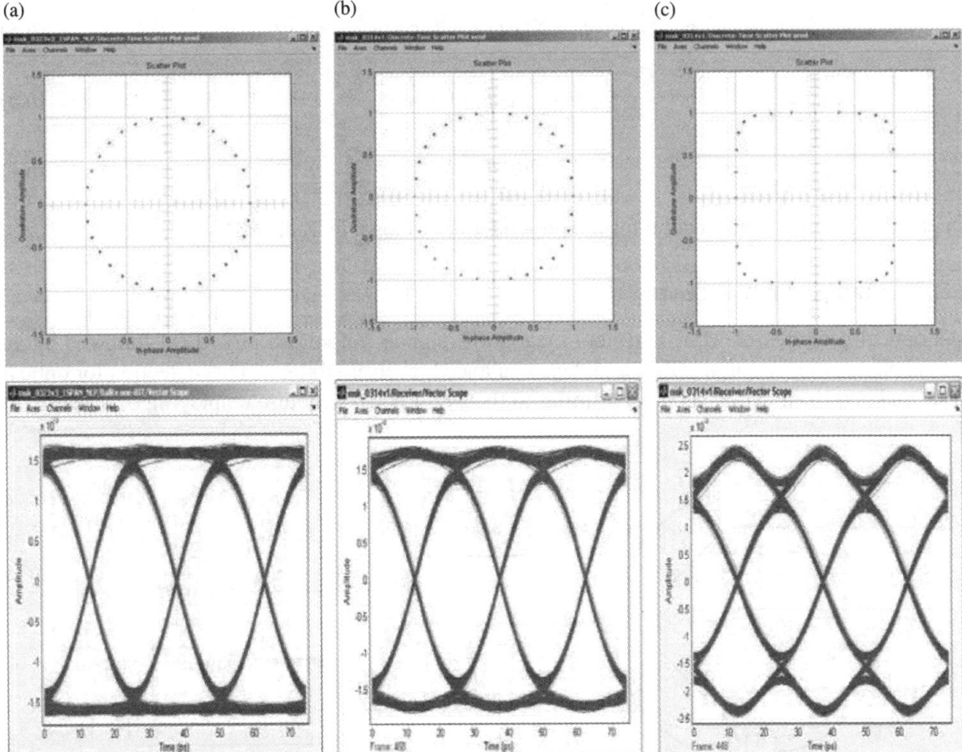

FIGURE 6.7 Constellation diagrams and eye diagrams of optical MSK transmitted signals: (a) linear; (b) weakly non-linear and (c) strongly non-linear transmission.

The magnitude fluctuation level depends on the magnitude of the phase-shaping driving source. High spectral efficiency can be achieved with tight filtering of the driving signals before modulating the electro-optic MZMs. Three types of pulse shaping filters are investigated, including Gaussian, raised cosine and squared-root raised cosine filters. Narrow spectral width and high suppression of the side lobes can be achieved.

The logic gates in the precoder are constructed based on the state diagram because this approach eases the implementation of the pre-coder. As seen from the state diagram, the current state of the signal is dependent on the previous state because the state of the signal advances corresponding to the binary data from the previous state. Therefore, memory is needed to store the previous state. The state diagram in Figure 6.8a is developed into a logic state diagram in Figure 6.8b to enable the construction of truth table. $S_0 S_1 = 00$ or $S_0' S_1' = 00$ corresponds to state 0, $S_0 S_1 = 01$ or $S_0' S_1' = 01$ corresponds to state $\pi/2$, $S_0 S_1 = 10$ or $S_0' S_1' = 10$ corresponds to state π, while $S_0 S_1 = 11$ or $S_0' S_1' = 11$ corresponds to state $-\pi/2$, with $S_0 S_1$ as current state and $S_0' S_1'$ as previous state.

Two delay units in Figure 6.8b function as memory by delaying the current state and feedback into the pre-coding logic block as the previous state. The pre-coding logic block which consists of logic gates would compute the current state and output based on the feedback state (previous state) and binary data from the Bernoulli binary generator.

The truth table is constructed based on the logic state diagram and combinational logic diagram above. For positive half cycle cosine wave and positive half cycle sine wave, the output is 1; for negative half cycle cosine wave and negative half cycle sine wave, the output is 0. Then, Karnaugh maps are constructed to derive the logic gates within the pre-coding logic block, based on the truth table. The following three pre-coding logic equations are derived as

$$S_0 = \overline{b_n S_0'} \; \overline{S_1'} + b_n \overline{S_0'} \, S_1' + \overline{b_n} S_0' \, \overline{S_1'} + b_n S_0' \, \overline{S_1'} \tag{6.5}$$

$$S_1 = \overline{S_1'} = \overline{b_n S_1'} + b_n \overline{S_1'} \tag{6.6}$$

$$output = \overline{S_0}. \tag{6.7}$$

The final logic gates construction for the pre-coder is as shown in Table 6.1

6.3.2 Optical MSK Receivers

The optical detection of MSK and non-linear MSK signals employs a pi/2 OPM followed by an optical phase comparator, a MZDI and then a balanced receiver (BalRx). This detection for optical MSK and non-linear MSK signals is similar to the well-known structure employing on one arm of

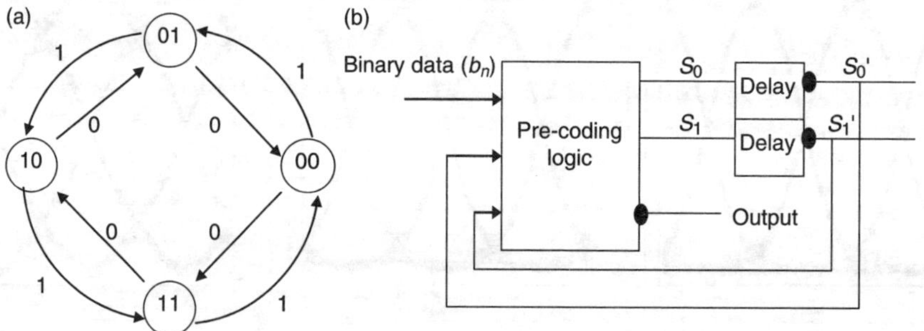

FIGURE 6.8 (a) State diagram for MSK. The arrows indicate continuous increment or decrement of the phase of the carrier. (b) Combinational logic, the basis of the logic for constructing the truth table of the pre-coder.

TABLE 6.1
Truth Table Based on MSK State Diagram

$b_n S_0' S_1'$	$S_0 S_1$	Output
100	01	1
001	00	1
010	01	1
101	10	0
110	11	0
111	00	1
000	11	0
011	10	0

MZDI is introduced to detect $\pm\pi/2$ phase difference of two adjacent bits due to the nature of MSK modulation.

A new technique for evaluation of the BERs is implemented. The probability density functions (pdf) of noise-corrupted received signals after decision sampling are computed with superposition of a number of weighted Gaussian pdfs. The technique implements the expected maximization theorem and has shown its effectiveness in determining arbitrary distributions [11, 12].

6.4 OPTICAL BINARY-AMPLITUDE MSK FORMAT

6.4.1 GENERATION

The optical MSK transmitters from [13–15] can be integrated in the proposed generation of optical MAMSK signals. However, in this article, we propose a new simple-in-implementation optical MSK transmitter configuration employing high-speed cascaded electro-optic phase modulators (E-OPMs) as shown in Figure 6.2. Electro-optic phase modulators and interferometers operating at high frequency using resonant-type electrodes have been studied and proposed in [7, 8]. In addition, high-speed electronic driving circuits evolved with the ASIC technology using 0.1-μm GaAs P-HEMT or InP HEMTs [9] enable the feasibility in realization of the proposed optical MSK transmitter structure. The base-band equivalent optical MSK signal is represented in (6).

$$\tilde{s}(t) = A\exp\left\{j\left[a_k I_k 2\pi f_{\mathrm{d}} t + \Phi(t,k)\right]\right\}, \quad kT \le t \le (k+1)T$$

$$= A\exp\left\{j\left[a_k I_k \frac{\pi t}{2T} + \Phi(t,k)\right]\right\}$$

$$(6.8)$$

where $a_k = \pm 1$ are the logic levels; $I_k = \pm 1$ is a clock pulse whose duty cycle is equal to the period of the driving signal $V_d(t)$, f_d is the frequency deviation from the optical carrier frequency and $h = 2f_d T$ is defined in Equation 6.2 and Equation 6.3 as the frequency modulation index. In optical MSK, $h = 1/2$ or $f_d = 1/(4T)$.

The first E-OPM enables the frequency modulation of data logics into upper side bands (USB) and lower side bands (LSB) of the optical carrier with frequency deviation of f_d. Differential phase pre-coding is not necessary in this configuration due to the nature of the continuity of the differential phase trellis. By alternating the driving sources $V_d(t)$ to sinusoidal waveforms or combination of sinusoidal and periodic ramp signals, different schemes of linear and non-linear phase-shaping MSK transmitted sequences can be generated [16]. The second E-OPM enforces the phase continuity of the light wave carrier at every bit transition. The delay control between the E-OPMs is usually implemented by the phase shifter shown in Figure 6.1. The driving voltage of the second E-OPM is pre-coded to fully compensate the transitional phase jump at the output $E_{01}(t)$ of the first E-OPM. Phase continuity characteristic of the optical MSK signals is determined by the algorithm in Equation 6.2. To mitigate the effects of unstable stages of rising and falling edges of the electronic circuits, the clock pulse $V_c(t)$ is offset with the driving voltages $V_d(t)$.

$$\Phi(t,k) = \frac{\pi}{2}\left[\sum_{j=0}^{k-1} a_j - a_k I_k \sum_{j=0}^{k-1} I_j\right]. \tag{6.9}$$

Binary-amplitude minimum shift keying (BAMSK) modulation format has been proposed for optical communications. We report numerical results of 80-Gb/s 2-bit-per-symbol BAMSK optical system on spectral characteristics and residual dispersion tolerance to different types of fibers. BER of 1e-23 is obtained for 80-Gb/s optical BAMSK transmission over 900-km Vascade-fiber systems enabling the feasibility of long-haul transmission for the proposed format.

6.5 OPTICAL MINIMUM SHIFT KEYING

MSK, which is well-known in radio frequency and wireless communications, has recently been adapted to optical communications. A few optical MSK transmitter configurations have recently been reported [17, 18]. For optically amplified communications systems, if multi-level concepts can be incorporated in those reported schemes, the symbol rate would be reduced and hence bandwidth efficiency can be achieved. This is the principal motivation for the proposed modulation scheme.

BAMSK is a special case of M-ary continuous phase modulation (CPM) format which enables binary-level (PAM- or QAM-like) transmission scheme while the bandwidth efficiency due to transitional phase continuity properties between two consecutive symbols (CPM-like signals) are preserved. The generation of M-ary CPM sequences can be expressed in Equation 6.1 [19]

$$s(t) = A_n \cos(\omega_c t + \phi_n(t,a)) + \sum_{m=1}^{N-1} B_m \cos(\omega_c t + \phi_m(t,b_m)) \tag{6.10}$$

where

$$\phi_N(t,a) = \pi h a_n q(t-nT) + \pi h \sum_{k=-\infty}^{n-1} a_k \quad nT \leq t \leq (n+1)T \tag{6.11}$$

$$\phi_m(t,b_m) = \pi a_n\left(h + \frac{b_{mn}+1}{2}\right) q(t-nT) + \pi \sum_{k=-\infty}^{n-1} a_k\left(h + \frac{b_{mk}+1}{2}\right) \quad mT \leq t \leq (m+1)T. \tag{6.12}$$

In a generalized M-ary CPM transmitter, values of a_n and b_{mn} are statistically independent and taken from the set of $\{\pm 1, \pm 3, \ldots\}$. A_n and B_m are the signal state amplitude levels, which are in phase or pi phase shift with the largest level component at the end of nth symbol interval; $q(t)$ is the pulse shaping function, and h is the frequency modulation index. In BAMSK, Equation 6.2 and Equation 6.3 which show the constraints of ϕ_m to maintain the phase continuity characteristic of CPM sequences, are simplified to Equation 6.4 and Equation 6.5, respectively, where $h = 1/2$ and the phase-shaping function $q(t - nT)$ is a periodic ramp signal with duty cycle of $4T$.

$$\phi_n(t,a) = \pi h a_n \frac{t - nT}{T} + \pi h \sum_{k=-\infty}^{n-1} a_k \qquad nT \leq t \leq (n+1)T \tag{6.13}$$

$$\phi_m(t,b_m) = \pi I_n \left(h + \frac{b_{mn}+1}{2} \right) \frac{t - nT}{T} + \pi \sum_{k=-\infty}^{n-1} b_k \left(h + \frac{b_{mk}+1}{2} \right) \qquad mT \leq t \leq (m+1)T. \tag{6.14}$$

Any configuration of the reported optical MSK transmitters in [15, 17, 18, 20] can be utilized in the proposed generation scheme of optical BAMSK signals. Figure 6.9a shows the block structure of the optical BAMSK transmitter in which two optical MSK transmitters are integrated in parallel configuration. The amplitude levels are determined from Equation 6.1, Equation 6.4, and Equation 6.5 by the splitting ratio at the output of a high precision power splitter. The logic sequences $\{\pm 1,..\}$ of a_n and b_n are pre-coded from the binary logic $\{0,1\}$ of d_n as $a_n = d_n - 1$ and $b_n = a_n (1 - d_n - 1/h)$ [19]. The signal-space trajectories of BAMSK signals are shown in Figure 6.9b.

A simple non-coherent configuration for detection of the optical BAMSK sequences consists of phase and amplitude detections. In BAMSK, i.e., $n = 2$, the system effectively implements 2-bits per symbol with two amplitude levels. Phase detection is enabled with employment of the well-known integrated optic phase comparator MZDI-balanced receiver with one-bit time delay on one arm of MZDI [21]. An additional $\pi/2$ phase shift is introduced. Figure 6.10a and Figure 6.10b show the eye diagrams of the amplitudes and phases of the optical BAMSK signals, respectively. In phase detection, the decision threshold is plotted in broken-line style is at zero level, whereas amplitude levels are determined by different thresholds. A new technique in calculation of BER for dispersive and noise corrupted received signals, which exploits the expected maximization (EM) theorem, is implemented with superposition of a number of weighted Gaussian probability distribution functions. [11, 22].

A simple non-coherent configuration for detection of a linear and non-linear optical M-ary MSK sequences consists of phase and amplitude detections, which are very well-known in the discrete phase shift keying schemes such as differential PSK (DPSK) or quadrature DPSK [23, 24]. Phase detection is enabled with employment of the well-known integrated optic phase comparator

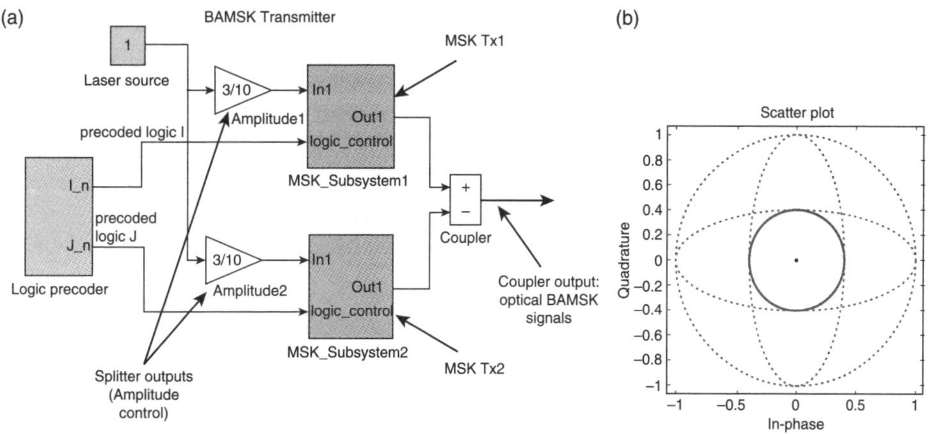

FIGURE 6.9 (a) Block diagram of optical BAMSK transmitter. (b) Signal trajectories of optical BAMSK transmitted signals of multi-amplitude MSK.

MZDI-balanced receiver with one-bit time delay on one arm of MZDI. An additional pi/2 phase shift is introduced to detect the differential $\pi/2$ phase shift difference of two adjacent optical MSK pulses. Figure 6.10a and Figure 6.10b show the eye diagrams of the amplitudes and phases of the optical B-AMSK-detected signals, respectively. In case of $N=2$ and $N=3$, the system effectively implements 2-bits-per-symbol scheme and 3 bits-per-symbol scheme with two and four amplitude levels, respectively. In phase detection, the decision threshold is plotted in broken-line style is at zero level because only in-phase and π differential phase are of interest.

6.6 NUMERICAL RESULTS AND DISCUSSIONS

The state diagram for MSK is shown in Figure 6.11. The continuous wave carrier source is modulated by two cascaded MZIMs, which are driven by a voltage level conditioning (broadband microwave amplifiers) fed by the output level of the pre-coder. The arrows indicate continuous increment or decrement of the phase of the carrier. These information-bearing lightwave signals then propagate along the fiber spans and are detected via a photonic phase comparator, the MZDI, and then detected via a balanced receiver. The obtained eye diagram is then statistically analyzed. More efficient detection schemes using frequency discrimination will be presented in Chapter 8.

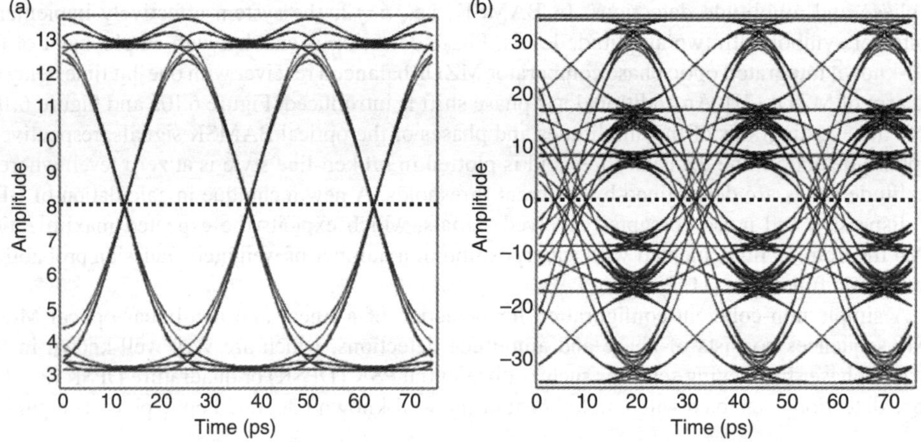

FIGURE 6.10 Eye diagrams of amplitude (a) and phase (b) detection of optical B-AMSK received signals with normalized amplitude ratio of 0.285/0.715. The decision threshold is shown in broken-line style.

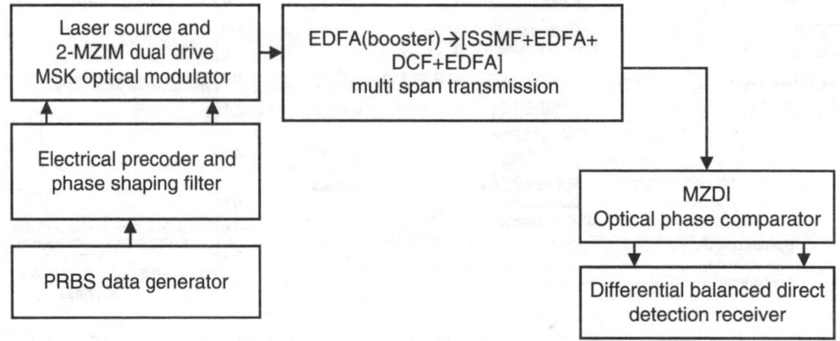

FIGURE 6.11 Optically amplified optical transmission system (schematic).

6.6.1 Transmission Performance of Linear and Non-linear Optical MSK Systems

The block diagram of simulation setup is shown in Figure 6.11. The dispersion tolerance of linear, weakly non-linear and strongly non-linear optical MSK signals is numerically investigated and the results shown in Figure 6.12. Among the three types, linear MSK is most tolerant to residual dispersion with 1-dB eye open penalty at 72 ps/nm/km. Strongly non-linear MSK suffers a severe penalty when residual dispersion exceeds 85 ps/nm/km or equivalent of 5-km SSMF. Figure 6.13 shows the performance of three types of optical MSK modulated signals versus optical signal-to-noise ratio (OSNR) in transmission over 540 km Vascade fibers of optically amplified and fully compensated multi-span links (6 spans × 90 km/span). The receiver sensitivity of the differential phase comparison balanced receiver is −24.6 dBm. In Vascade fibers, the dispersion factors of the dispersion-compensating fiber is negatively opposite to that of the transmission fiber, a standard single mode

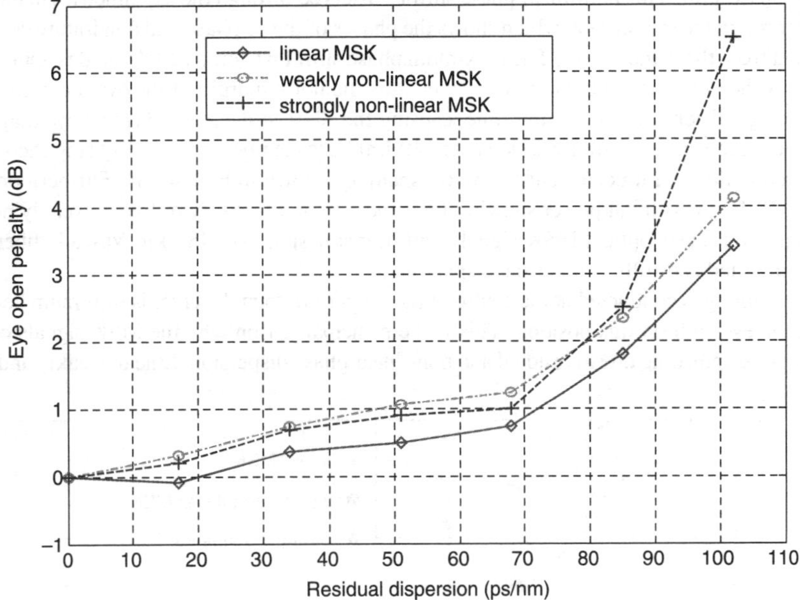

FIGURE 6.12 Dispersion tolerance of 40-Gb/s linear MSK, weakly non-linear MSK and strongly non-linear MSK optical signals.

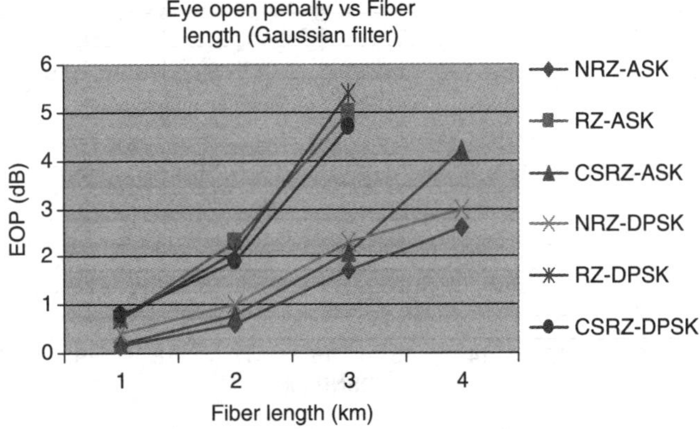

FIGURE 6.13 Simulation of EOP versus transmission distance 1–4 km of SSMF.

fiber, of $+17.5$ ps/nm/km at 1550-nm wavelength. In addition, the dispersion slopes of these fibers are also matched. Optical amplifiers of EDFA types are placed one at the end of the transmission fiber and one after the DCF so that it boosts the optical power to the appropriate level, which is equal to that of the launched power. The EDFA optical gain of 19-dB and 5-dB ASE noise figure is used. The noise margin reduces severely after the propagation over 6×90 km spans. Effects of positive and negative dispersion mismatch and mid-link non-linearity on phase evolution are shown in Figure 6.12.

The tolerance of this MSK modulation using transmission models to non-linear effects is also studied, and simulation results are shown in Figure 6.12 and Figure 6.13. The input power into the fiber span is increased, whereas the length of transmission link is constant at 180 km. At BER = 1e-9, linear MSK could tolerate an increase of the input launched power up to 10.5 dBm, weakly non-linear could tolerate up to 10.2 dBm compared with 9.2 dBm in strongly non-linear MSK. The non-linear phase shift is proportional to the input power, therefore increasing the input power would also increase non-linear phase shift. This non-linear phase shift is observed through the asymmetries in the eye diagram and through the scatter plot, which shows the phases of the in-phase and quadrature components have shifted from the x and y axes. The maximum phase shift that could be tolerated is approximately 15° of arc. Although increasing input power increases the noise margin of the eye diagram, it is paid off by the large distortion at sampling time, causing the SNR to decrease. Typical eye diagrams for fully compensation and after transmission over 540-km Vascade fibers of optically amplified and fully compensated multi-span links (6 spans $\times$ 90 km/span) are shown in Figure 6.14. Furthermore, Figure 6.15 shows BER versus input power showing robustness to non-linearity of linear, weakly non-linear and strongly non-linear optical MSK signals with transmission over 180-km Vascade fibers of two optically amplified and fully compensated span links.

If the sampling is conducted at the centre of the bit period, then the error is minimum because the ripples of the eyes fall on this position. This is the principal reason why the MSK signals can suffer minimum pulse spreading due to residual and non-linear phase dispersion. Linear, weakly and strongly

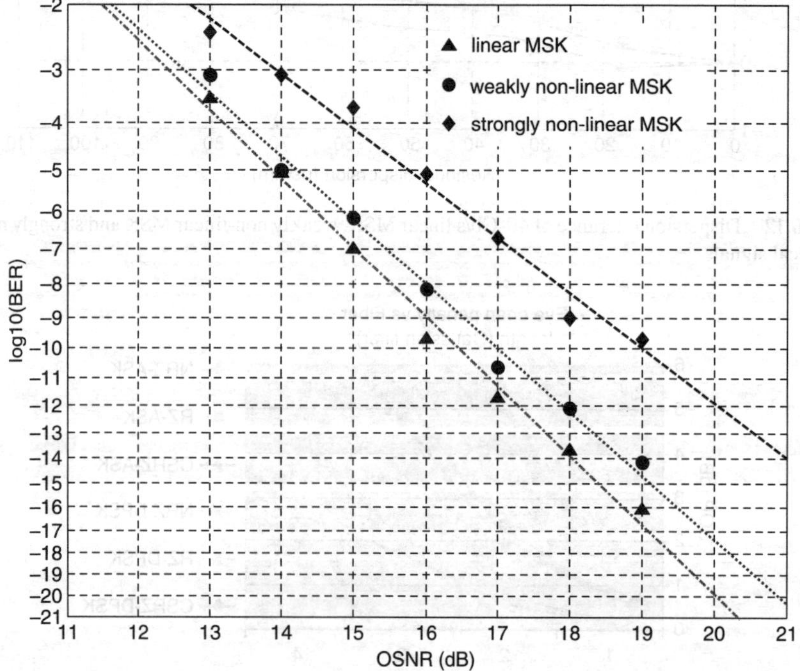

FIGURE 6.14 BER versus optical SNR for transmission of three types of modulated optical MSK signals over 540-km Vascade fibers of optically amplified and fully compensated multi-span links (6 spans $\times$ 90 km/span).

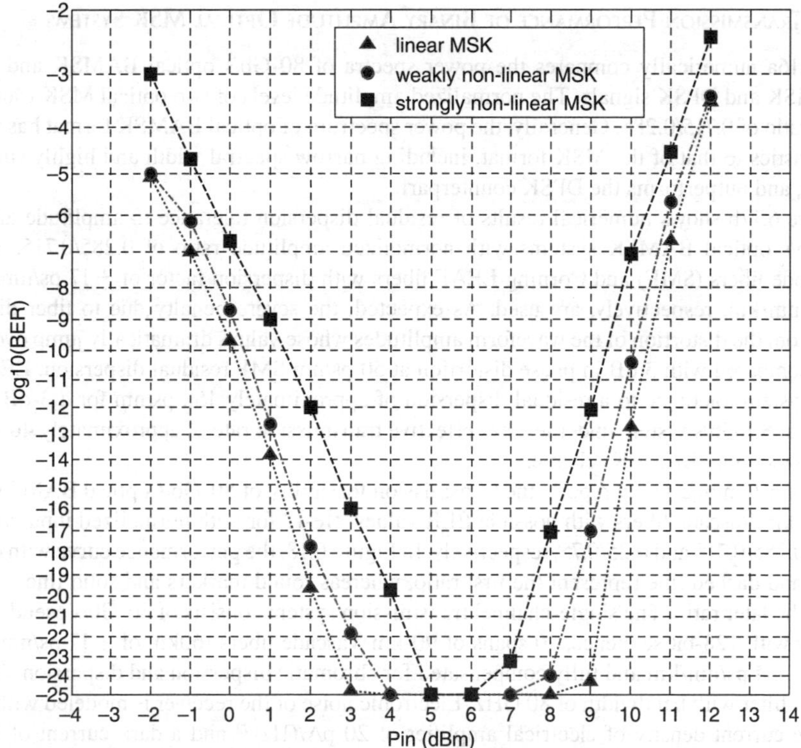

FIGURE 6.15 BER versus input power showing robustness to non-linearity of linear, weakly non-linear and strongly non-linear optical MSK signals with transmission over 180-km Vascade fibers of two optically amplified and fully compensated span links.

non-linear MSK phase shaping functions are investigated. It has been proved that optical MSK is a very efficient modulation that offers excellent performance. With OSNR of about 17 dB, BER is obtained to be 1e-9 and reaches 1e-17 for an optical SNR of 19 dB under linear MSK modulation. The modulation formats of linear and non-linear phase-shaping MSK is also highly resilient to non-linear effects. Non-linear distortion appears when the total average power reaches about 9 dBm, i.e., about 3–4 dB above that of NRZ ASK format over a 50-μm diameter SSMF fiber.

The non-linear phase-shaping filters offer better implementation structures in the electronic domain for driving the dual-drive MZIMs than the linear type, but suffer some power penalty, but they are still better than those candidates of other amplitude or phase or differential phase modulation formats.

Weakly non-linear MSK offers much lower power penalty and ease of implementation for pre-coders and phase-shaping filters. Thus, it would be the preferred MSK format for long-haul transmission over optically amplified multi-span systems. At a BER of 1e-12, linear MSK is 0.3-dB and 1.2-dB more resilient to non-linear phase effects than weakly non-linear MSK and strongly non-linear MSK, respectively.

Compared with the DPSK counterpart in 40-Gb/s transmission, various types of filtered MSK modulated signals are more tolerant to residual dispersion. The eye open penalty of 3 dB is obtained at 4 km of SSMF in NRZ-DPSK/ASK or 68 ps/nm accordingly, whereas linear MSK can tolerate up to 98 ps/nm or 6 km accordingly.

Pulse-shaping using raised cosine filter offers better dispersion tolerance and lower penalty for RZ-DPSK and CSRZ-DPSK after transmitting 3 km. Approximately 2-dB improvement is observed. Thus, pulse shaping compromises the deficits of RZ-pulses due to broader spectrum compared to NRZ pulse shapes. We observe no significant improvement on NRZ pulses for DPSK and ASK signals.

6.6.2 Transmission Performance of Binary Amplitude Optical MSK Systems

Figure 6.16a numerically compares the power spectra of 80-Gb/s optical BAMSK and 40-Gb/s optical MSK and DPSK signals. The normalized amplitude levels of two optical MSK transmitters take the ratio of 0.715/0.285. Generally, the power spectrum of optical BAMSK format has identical characteristics to that of the MSK format, including narrow spectral width and highly suppressed side-lobe, and outperforms the DPSK counterpart.

Figure 6.17b shows numerical results of residual dispersion tolerance in amplitude and phase of 80-Gb/s optical BAMSK systems with normalized amplitude ratio of 0.285/0.715. Standard single mode fibers (SMF) and Corning-LEAF fibers with dispersion factor of ± 17 ps/nm/km and ±4.5 ps/nm/km, respectively, are used. As expected, the severe penalty due to fiber dispersion derives from the distortion of the waveform amplitudes whose values dramatically jump over 20-dB penalty compared with 3 dB in phase distortion at 50 ps/nm SMF residual dispersion. LEAF fiber enables system tolerance to a residual dispersion of approximately 150 ps/nm for a 3-dB penalty. The optical 80-Gb/s system under test has effective transmission rate of approximately 40 Gb/s due to 2 bit-per-symbol BAMSK scheme.

Figure 6.16 numerically reports the transmission feasibility of 80-Gb/s optical BAMSK signals over 900-km Vascade fibers with possible BER values <1e-12 for both normalized input amplitude ratios of 0.285/0.715 and 0.25/0.75, respectively. In Figure 6.18, the performance curve with diamond markers and dashed line represent the first ratio, whereas round markers and solid line curves are used for the later ratio. The single-channel transmission system consists of 80-Gb/s pseudo random generator with 128-bit sequence, 10 spans of 90-km Vascade fibers (60km of + 17 ps/nm/km and 30 km of −34 ps/nm/km and fully compensated for chromatic dispersion and dispersion slope) and an optical filter with bandwidth of 80 GHz. Electronic noise of the receiver is modeled with equivalent noise current density of electrical amplifier of 20 pA/(Hz)$^{1/2}$ and a dark current of 2*10 nA (two photodiodes in balanced receiver structure). This configuration yields the back-to-back receiver sensitivity at BER = 1e-9 to be approximately −23 dBm. The eye diagrams are obtained

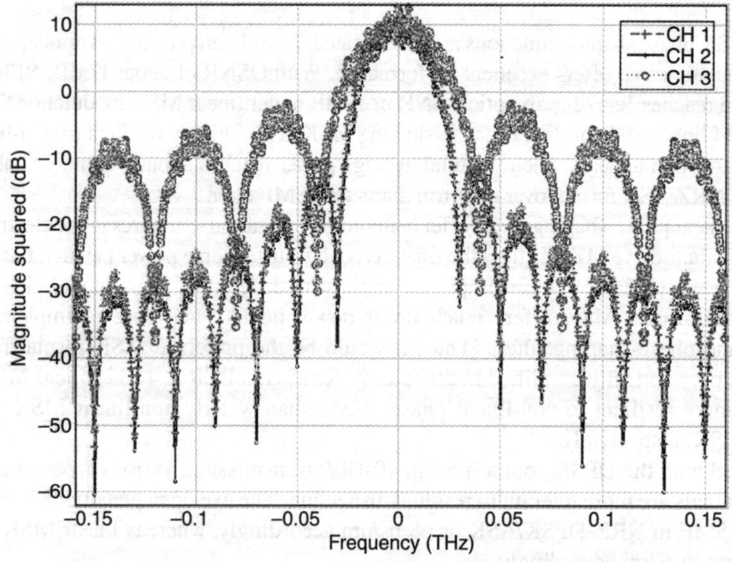

FIGURE 6.16 Comparison of spectra of 80-Gb/s optical BAMSK, 40-Gb/s optical MSK and 40-Gb/s optical binary DPSK signals.

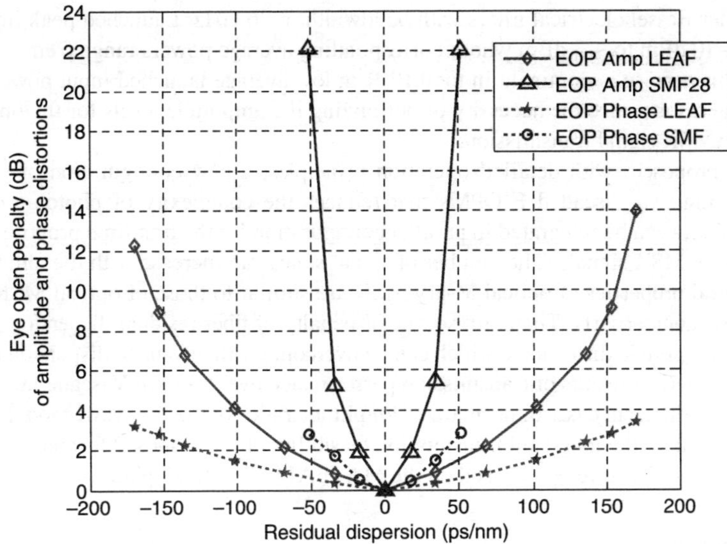

FIGURE 6.17   Numerical results on residual dispersion tolerance of 80-Gb/s optical BAMSK systems (effectively 40-Gb/s symbol rate) with normalized amplitude ratio of 0.285/0.715 in amplitude and phase detection.

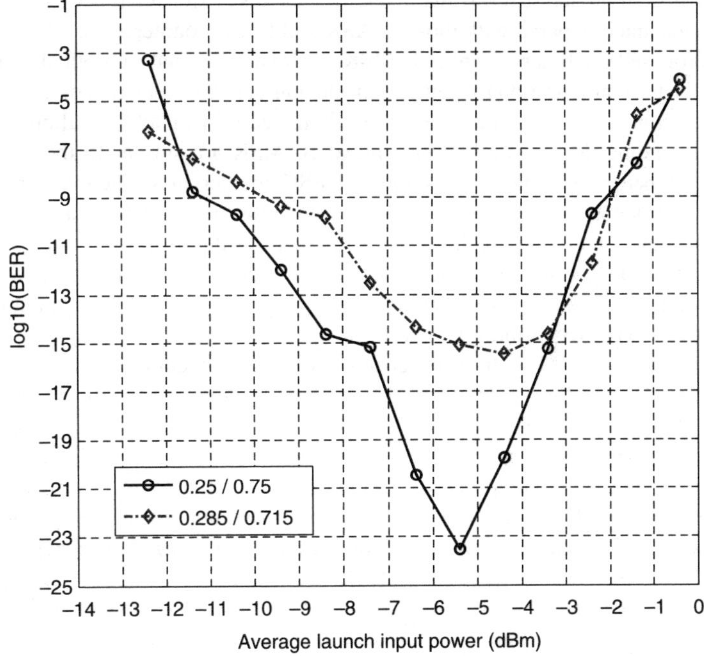

FIGURE 6.18   Transmission performance of 80-Gb/s optical BAMSK over 900-km Vascade fiber systems in two cases of normalized amplitude ratios: 0.25/0.75 (round markers and solid line) and 0.285/0.715 (diamond markers and dashed line), respectively.

after fifth-order Bessel electrical filters with bandwidth of 36 GHz. Launched peak input power is varied from −10 dBm to +3 dBm whose corresponding average powers range from −12.5 dBm to −0.5 dBm. Phase noise is dominant in total BER in low average launched input powers (less than −3 dBm). The results raise the necessity of optimizing the amplitude levels for the optimum BER for optical BAMSK signal transmission.

We have proposed, with detailed operation principles, a new configuration of optical MSK transmitter using two cascaded E-OPMs that reduces the complexity of photonic components. These transmitters can be integrated in parallel configuration for the first-time proposed generation of optical M-ary MSK signals. The number of signal states can increase with the number of transmitters. Spectral properties of optical M-ary MSK are similar to those of optical MSK, and better than the DPSK counterparts. The main source of penalty of fiber residual dispersion is caused by the waveform amplitude distortions, which can be overcome with advanced dispersion equalization techniques. Numerical results of transmission performance over 900-km Vascade fibers have been presented in two different cases of normalized amplitude ratios of the B-AMSK modulation format. BER of 1e-23 enables the long-haul transmission feasibility of the proposed format. The simulation test-bed is successfully developed based on the MATLAB® Simulink® platform.

6.7   REMARKS

We have proposed two new schemes of optical MSK generation and detection. These two optical transmitter configurations can generate linear and various types of non-linear optical MSK modulation formats. The pre-coder for I-Q optical MSK structure is shown. The direct detection of optical lightwaves is utilized with implementation of the well-known differential non-coherent balanced receivers. Simulated spectral characteristics and dispersion tolerance to 40-Gb/s transmission are presented and compared with those of ASK and DPSK counterparts.

We have proposed the binary-amplitude MSK modulation format for optical communications. We have reported the configuration for generation and non-coherent detection of the optical BAMSK sequences. The number of signal states can be easily increased with additional optical MSK transmitters. Spectral properties of the 80-Gb/s optical BAMSK are similar to those of the 40-Gb/s optical MSK, and better than the 40-Gb/s optical DPSK counterparts. The main source of penalty in dispersion tolerance is caused by waveform amplitude distortions, which can be overcome with advanced dispersion equalization techniques. Numerical results of transmission performance for 80-Gb/s optical BAMSK system over 900-km Vascade fibers have been reported. BER of 1e-23 enables the long-haul transmission feasibility of the proposed format. The simulation test-bed is developed on the MATLAB® Simulink® platform.

It has been proved that MSK transmitter models for these modulation formats can be implemented using parallel structure of dual-drive MZIM data modulators. The differences in the implementations of these transmitter models are the pulse-shaping waveforms, in which linear MSK follows a triangular periodic waveform, weakly non-linear and strongly non-linear MSK have sinusoidal waveforms, but differ in amplitudes. The models are simulated under different conditions to investigate the effects of fiber characteristics such as fiber loss, non-linear effects, and dispersion on the performance of the models. The rotation of the scatter plots confirm the behavior of MSK modulated signals due to linear chromatic dispersion and non-linear phase shaping effects. The simulated optically amplified distance is 540-km fully compensated SSMF. The linear and non-linear optical MSK modulation formats can thus be offered as an alternative advanced modulation format for long-haul optically amplified transmission.

The visualization of the evolutions of the MSK signal phasor under self phase modulation is not fully described in this article, but will be reported in a future article. Electronic compensation technique can be implemented by design of the pre-distorted MSK signals at the input of the shaping filters. The optical MSK pre-compensating transmission system will be reported in the near future.

REFERENCES

1. Hoshida, T., O. Vassilieva, K. Yamada, S. Choudhary, R. Pecqueur, and H. Kuwahara. 2002. Optimal 40Gb/s modulation formats for spectrally efficient long haul DWDM systems. *IEEE Journal of Lightwave Technology* 20: 1989–96.

2. Zhu, Y., et al. 2004. 1.6 bit/s/Hz orthogonally polarized CSRZ – DQPSK transmission of 8x40 Gbit/s over 320 km NDSF, *OFC'04* Tu-F1.

3. Sakamoto, T., T. Kawanishi, and M. Izutsu. 2005. Initial phase control method for high-speed external modulation in optical minimum-shift keying format. *ECOC'05* 4: 853–54.

4. Ohm, M., and J. Spiedel. 2004. Optical minimum shift keying with direct detection. In *Proceedings of SPIE on optical transmission, switching and systems* 5281: 150–61.

5. Huynh, T. L., L. N. Binh, and K. K. Pang. 2006. Optical MSK long-haul transmission systems, *to be published in SPIE proceedings of APOC'06*, paper 6353–86, Thu9a, Asia Pacific Optical Fiber Conference, September 2006, Kwangju, Korea.

6. Huynh, T. L., L. N. Binh, and K. K. Pang. 2006. Linear and weakly nonlinear optical continuous phase modulation formats for high performance DWDM long-haul transmission. Presented at the European Conference on Optical Communications, Cannes, France.

7. Kawanishi, T., S. Shinada, T. Sakamoto, S. Oikawa, K. Yoshiara, and M. Izutsu. 2005. Reciprocating optical modulator with resonant modulating electrode. *Electronics Letters* 41(5): 271–72.

8. Krahenbuhl, R., J. H. Cole, R. P. Moeller, and M. M. Howerton. 2006. High-speed optical modulator in LiNbO3 with cascaded resonant-type electrodes. *Journal of Lightwave Technology* 24(5): 2184–89.

9. Kaminow, I. P., and T. Li. 2002. *Optical fiber communications*. Volume IVA, Chapter 16. Burlington, MA, USA: Elsevier Science.

10. Amoroso, F. 1976. Pulse and spectrum manipulation in the minimum frequency shift keying (MSK) format. *IEEE Transactions on Communication* 24: 381–84.

11. Ding, L., W. D. Zhong, C. Lu, and Y. Wang. 2004. New bit-error-rate monitoring technique based on histograms and curve fitting. *Optics Express* 12(11): 2507–11.

12. Shake, I., H. Takara, and S. Kawanishi. 2003. Simple Q factor monitoring for BER estimation using opened eye diagrams captured by high-speed asynchronous. *Photonics Technology Letters* 15: 620–22.

13. Mo, J., D. Yi, Y. Wen, S. Takahashi, Y. Wang, and C. Lu. 2004. Novel modulation scheme for optical contiuous-phase frequency-shift keying. Optical Fiber Communications Conference, OFC'02 Annaheim, CA, USA, Paper OFG2, 1–3.

14. Mo, J., D. Yi, Y. Wen, S. Takahashi, Y. Wang, and C. Lu. 2005. Optical minimum-shift keying modulator for high spectral efficiency WDM systems, *ECOC'05*, 4: 781–82.

15. Ohm, M., and J. Speidel. 2004. Optical minimum-shift keying with direct detection (MSK/DD).In *Proceedings of SPIE* 5281: 150–61.

16. Huynh, T. L., L. N. Binh, K. K. Pang, and L. Chan. 2006. Photonic MSK transmitter models using linear and non-linear phase shaping for non-coherent long-haul optical transmission. Paper 6353-85, presented at the Asia Pacific Optical Communications Conference, APOC 2006, September 3–7.

17. Sakamoto, T., T. Kawanashi, and M. Izutsu. 2005. Optical minimum-shift keying with external modulation scheme. *Optic Express* 13(20): 7741–47.

18. Huynh, T. L., L. N. Binh, K. K. Pang, and L. Chan. 2006 Photonic MSK transmitter models using linear and non-linear phase shaping for non-coherent long-haul optical transmission, In *SPIE Proceedings of of APOC'06*, paper 6353–85, presented at Asia Pacific Optical Communications Conference, Kwangju, Korea, September 3–7.

19. Proakis, J. G. 2001. *Digital communications*. 4th ed. Chapter 4. New York: McGraw-Hill, 199–202.

20. Mo, J., D. Yi, Y. Wen, S. Takahashi, Y. Wang, and C. Lu. 2005. Optical minimum-shift keying modulator for high spectral efficiency WDM systems. *ECOC'05* 4: 781–82.

21. Wree, C., J. Leibrich, J. Eick, W. Rosenkranz, and D. Mohr. 2003. Experimental investigation of receiver sensitivity of RZ-DQPSK modulation using balanced detection. *OFC'03* 2: 456–57.

22. Redner, R., and H. Walker. 1984. Mixture densities, maximum likelihood and the EM algorithm. *SIAM Review* 26(2): 195–239.

23. Sekine, K., N. Kikuchi, S. Sasaki, S. Hayase, C. Hasegawa, and T. Sugawara. 2005. 40Gbit/s, 16-ary (4bit/symbol) optical modulation/demodulation scheme. *Electronics Letters* 41(7): 430–32.

24. Proakis, J. G. 2001. *Digital communications*. 4th ed. New York: McGraw-Hill, 185–213.

7 Multi-level Amplitude and Phase Shift Keying Optical Transmission

7.1 INTRODUCTION

Under the conventional on–off keying (ASK) modulation format, the transmission bit rate beyond 40 Gb/s per optical channel is very expensive because the electronic signal processing technology may have reached its fundamental speed limit. It is expected that advanced photonic modulation formats such as M-ary amplitude and differential phase shift keying will replace ASK in the near future. These advanced formats would offer efficient spectral properties, thus be able to increase transmission rate without placing stringent requirements on high-speed electronics and to use the existing photonic communication infrastructure.

Coherent communications developed in the mid-1980s extensively exploited different modulation techniques to improve the optical signal-to-noise ratio OSNR [1]. However, coherence detection faced considerable difficulties due to the stability of the source spectrum and the laser linewidth for a mere gain in the receiver sensitivity of 3 dB for heterodyne detection and 6 dB for homodyne detection to extend the repeater-less distance 60–80 km of standard single-mode fiber (SSMF).

The invention of the optical amplifier (OA) in the early 1990s overcame the fiber attenuation limit and thus offered a significant improvement in optical transmission technology. Due to this, ultra-long-haul and ultra-high-capacity optical transmission systems have been deployed widely around the world in the last decade. The technology has been matured with ASK modulation reaching 10 Gb/s per optical channel, total channel count of hundreds, and with 100/50 GHz channel spacing [2].

Based on proven efficient spectra and transmission technology, especially the controllable total dispersion of the transmission and compensating fibers, it is much more advantageous that these spectral regions be efficiently used. Therefore, the contribution of advanced modulation techniques and formats would offer higher spectral efficiency for photonic transmission.

Further, digital modulation techniques have been well established over the last half century with amplitude, frequency, or phase modulations [3]. These techniques, especially phase modulation, which rely principally on detection schemes (i.e., whether it is coherent or pseudo-coherent differential detection) have been heavily exploited in wireless communication networks. In the photonic domain, for a long time the technological difficulties associated with manufacturing narrow linewidth lasers have prevented the use of coherent and differential phase modulation. Only over the last several years, due to the maturity of the laser technology, particularly the successful development of distributed feed back (DFB) laser, laser linewidth has reached a level that is much smaller than the modulation bandwidth. The coherence of the sources is now sufficient for differential phase modulating and detecting applications that require the phase of the sources to remain stable over at least two consecutive symbol periods [4].

Recently, advanced modulation techniques have attracted significant interest from the photonic transmission research and system engineering community. Several modulation schemes and formats, such as binary differential phase shift keying (BDPSK), differential-quadrature phase shift keying (DQPSK), duo-binary ASK associated with non-return-to-zero (NRZ), return-to-zero (RZ), and carrier-suppress return-to-zero (CSRZ) formats have been widely studied [5–8]. However, what have not been widely explored are optical multi-level modulation schemes. Although multi-level schemes have been intensively exploited in wireless communications, [6,9,10], there are few works that incorporate the incoherent multi-level optical amplitude-phase shift keying modulation schemes that offer the following advantages: (i) lower symbol rate, hence for the same available spectral region a multi-level modulation scheme would offer a transmission capacity higher than binary modulation counterparts; (ii) Efficient bandwidth utilization, photonic transmission of these multi-level signals could be implemented over the existing optical fiber communications infrastructure without significant alteration of the system architecture, thus saving the cost of capital investment and easing system management; and (iii) The complexity of the coder and demodulation sub-systems falls within the technological capabilities of current microwave and photonic technologies.

The principal objectives of this chapter are: (i) To evaluate different modulation and coding techniques and signal pulse formats for long-haul ultra-high-capacity transmission; thence determine novel modulation schemes, the multi-level amplitude-phase shift keying, and others to be determined, for research studies; (ii) To develop analytical, simulation and experimental test-beds to demonstrate the uniqueness and superiority of our novel schemes; (iii) A comparative study of the modulations formats so as to unveil the principal directions for photonic modulation and transmission technologies for the next transmission generation; and (iv) A novel photonic communication system based on advanced multi-level optical modulation format and implementation of the system on Simulink platform to demonstrate its effectiveness and superiority to its other counterparts can be demonstrated to be useful platform for desk top computer simulation.

Thence, a conceptual photonic transmission system is proposed based on a hybrid technique that combines the phase and amplitude modulation, the multi-level amplitude-differential phase shift keying (MADPSK) format. This technique combines two modulation formats: the well-known M-ary ASK, and the M-ary DPSK to take advantage of high receiver sensitivity and dispersion tolerance (DPSK), and the enhancement of total transmission capacity (M-ASK) as compared to the traditional ASK format.

The models of MADPSK transmitter and receiver have been structured for MADPSK signaling. A simulation model based on the MATLAB® Simulink® platform has been developed for the proof-of-concept. The system performance is evaluated for back-to-back and long-haul transmission. Analytical and simulation results of the transmission configurations are demonstrated. The followings are presented: (i) Noise mechanisms e.g., quantum shot noises, quantum phase noises, optically amplified noises, noise statistics, non-linear phase noises; hence design of an optimum detection and decision-level schemes for MADPSK; (ii) Linear, non-linear and polarization dispersion impairments and their impacts on MADPSK system performance; (iii) Matched filter design for optimum MADPSK signal detection; (iv) Offset MADPSK (O-MADPSK) modulation schemes; (v) Multi-level amplitude-minimum shift keying (MAMSK) modulation; (vi) MADPSK modulation for applications in sub-carrier transmission systems, especially for metropolitan wide area multi-add/drop networks; and (vii) Other issues or additional modulation formats suitable for MADPSK.

This chapter is thus organized as follows: Section 1 gives a brief review of a number of advanced photonic modulation formats. Section 2 reviews and compares different modulator structures used for generating advanced photonic modulation signals and emphasizes the advantages of dual-drive MZIM as modulator for generating MADPSK signal, the main object of research. In Section 4, a novel photonic transmission system with MADPSK modulation format is proposed. Section 5 summarizes the preliminary works and results. Section 3 identifies a number of critical issues and alternative multi-level signaling for optical systems.

7.2 AMPLITUDE AND DIFFERENTIAL PHASE MODULATION

7.2.1 ASK MODULATION

7.2.1.1 NRZ-ASK Modulation

ASK has been the dominant modulation technique from the early days of optical communications. The main advantage of this modulation is that ASK signal is not sensitive to the phase noise. ASK modulation can take two principal formats: the first one is called NRZ-ASK and the "1" optical bit occupies the whole bit period; the second one, RZ-ASK, has the "1" bit presented in only the first half of the bit period.

Figure 7.1 shows the spectrum of a 40-Gb/s NRZ-ASK signal, with the carrier seen at the highest peak and the 3-dB bandwidth reaches the bit rate. The main advantage of NRZ-ASK signal is that its spectrum is generally the most compact compared with that of other formats such as RZ-ASK, CSRZ-ASK. Conversely, it has non-linear effects as compared with its RZ- and CSRZ-ASK counterparts.

7.2.1.2 RZ-ASK Modulation

A RZ-ASK signal (Figure 7.2) is similar to NRZ-ASK except for the "1" bit occupying only the first half of the bit period. This signal can be generated by a transmitter shown in the same figure in which a NRZ-ASK transmitter is followed by a pulse caver driven by a pulse train synchronized with the data source. The pulse train is frequency equal to the data rate. The RZ-ASK pulse width can take the form of 33%, 50%, 66% duty-ratio. Because of its narrower pulse width, the spectrum of RZ-ASK signal (Figure 7.3) is larger than that of NRZ-ASK signal, leading to less spectrum efficiency. In this spectrum, the carrier is seen as highest peak, the two side peaks are RF modulating signals positioned 80-GHz apart.

7.2.1.3 CSRZ-ASK Modulation

CSRZ-ASK modulation format [11] is similar to standard RZ-ASK, except that the neighboring optical pulses have π- phase difference. The carrier in neighboring time slots is cancelled out and

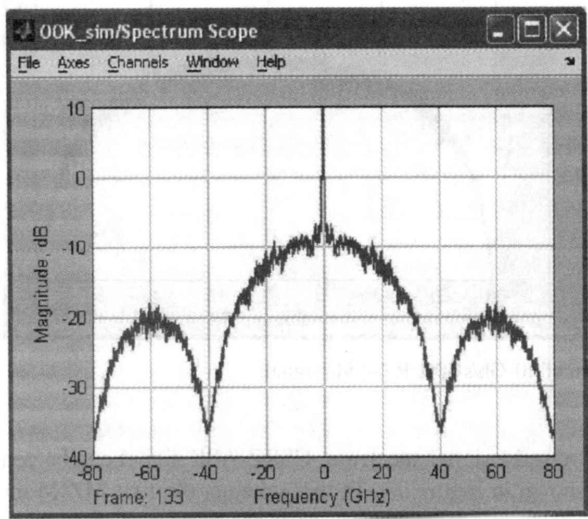

FIGURE 7.1 Spectrum of 40-Gb/s NRZ-ASK signal.

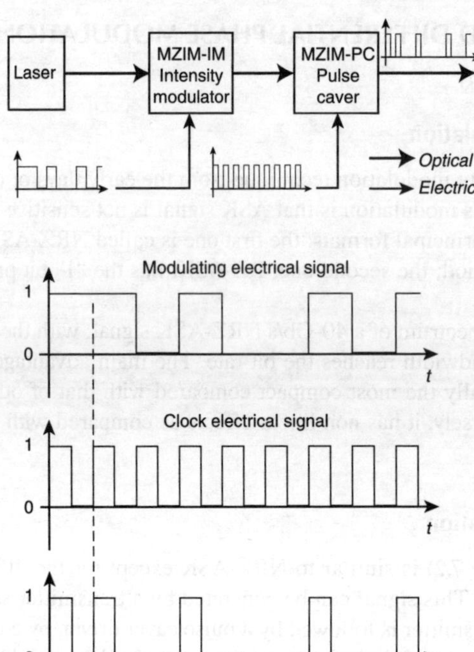

FIGURE 7.2 RZ-ASK transmitter and signal.

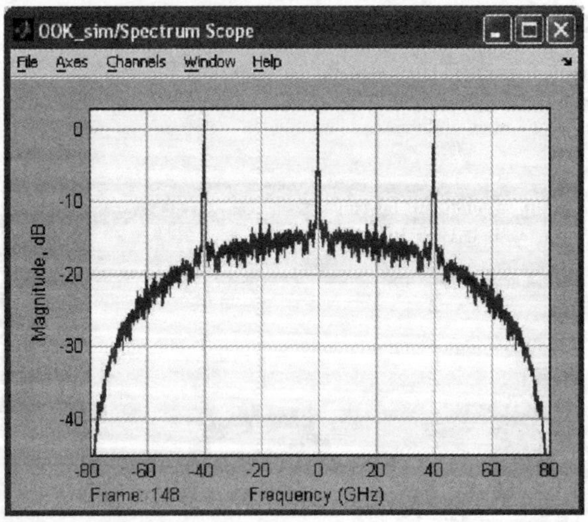

FIGURE 7.3 Spectrum of 40-Gb/s 50% RZ-ASK signal.

effectively excluded from the signal spectrum. CSRZ-ASK signal can be generated by a transmitter with the scheme shown in Figure 7.4. In this scheme, the first MZIM modulates the intensity of optical signal coming from a laser source, while the second MZIM, driven by a clock signal at the haft data rate, carves the NRZ pulses into RZ ones. Because the second MZIM is biased at the minimum-intensity point, it provides a RZ pulse train at the data rate with alternating phase 0 and π

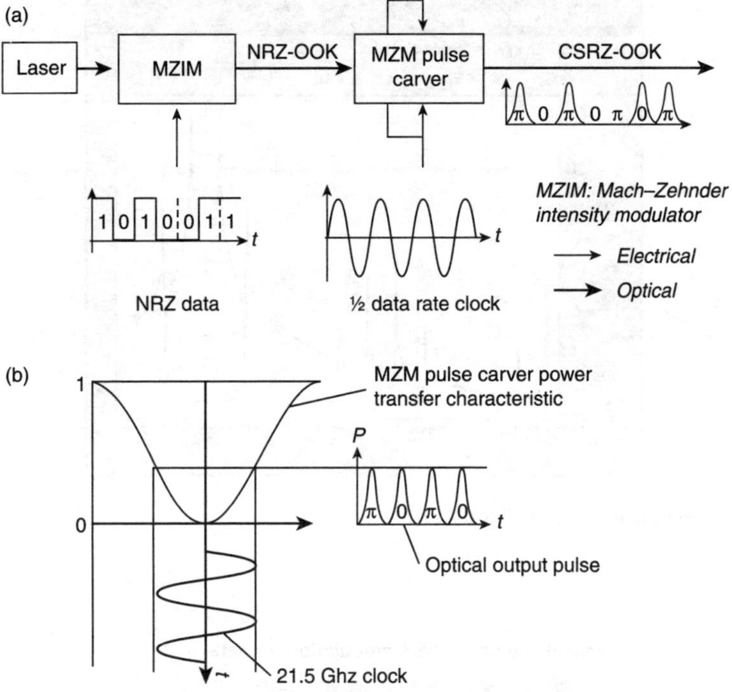

FIGURE 7.4 (a) Block diagrams of CSRZ-ASK transmitter (b) generation of optical pulse with alternative phase using biasing control and amplitude. (From Miyamoto, Y., S. Hayase, N. Kikuchi, K. Sekine, and S. Sasaki, *IEEE Electronics Letters*, 35(23), 2041–42, 1990. With permission.)

for neighboring time slots. CSRZ-ASK signal can be also detected by a direct-detection receiver because it would not be phase-sensitive.

The main advantages of CSRZ-ASK include narrower spectrum, higher tolerance to dispersion, and stronger robustness against fiber non-linear effects as compared with standard RZ-ASK. Because its peak optical power is much lower than that of other formats, it is less affected by self-phase modulation (SPM) and cross-phase modulation (XPM) [7]. Figure 7.5 shows the spectrum of 40-Gb/s CSRZ-ASK signal with very low carrier power level [12].

Amplitude shift keying (ASK) is a modulation technique that generates a signal $s(t)$ by multiplying a digital signal $m(t)$ by a carrier f_c [13]

$$s(t) = Am(t)\cos 2\pi\, f_c t, \quad 0 < t < T \tag{7.1}$$

where A is amplitude envelope; digital signal $m(t)$ may take one of M levels $[b_0, b_1, ..., b_M]$. When $M = 2$, $s(t)$ is a binary ASK signal with ASK is a special case. ASK is also implemented in NRZ, RZ, and CSRZ formats, whose spectra are shown in Figure 7.6 in the same graph for the purpose of comparison. Like their ASK analogs, NRZ-ASK has the most compact spectrum while RZ-ASK has the broadest. In terms of energy, CSRZ-ASK has the lowest peak power because the carrier signal has been effectively removed.

7.2.1.4 Differential Phase Modulation

Under ASK/ASK modulation schemes with the associated NRZ, RZ, and CSRZ formats, the amplitude of optical carrier varies accordingly. Phase modulation modulates carrier phase and thus facilitates

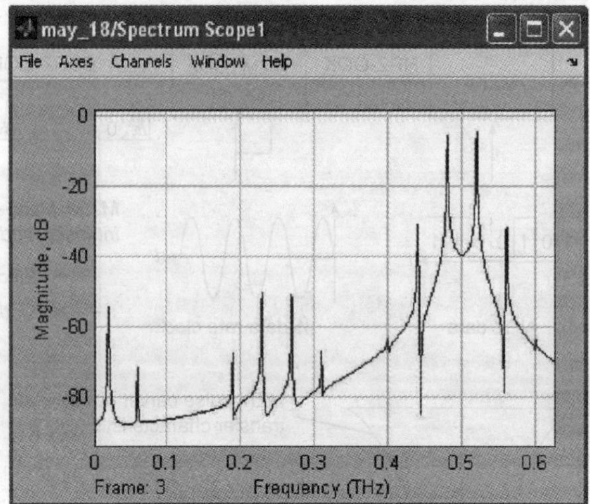

FIGURE 7.5 Spectrum of 40-Gb/s CSRZ-ASK.

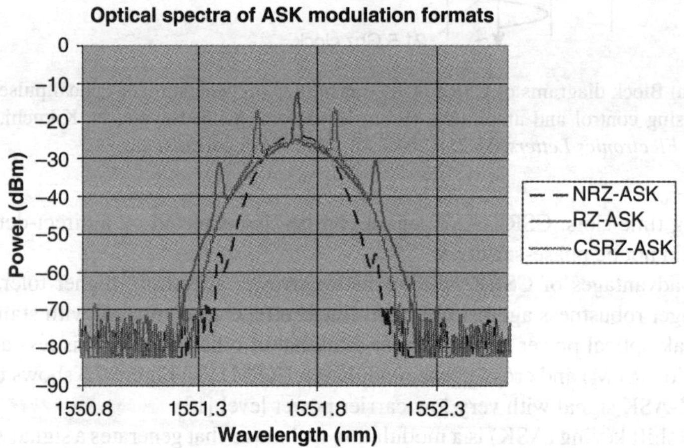

FIGURE 7.6 Spectrum of NRZ-ASK, RZ-ASK, and CSRZ-ASK signals.

the use of bi-polar signals "± 1". This distinguished feature means that phase modulation offers significant improvement in receiver sensitivity as compared with ASK modulation. With the recent advancement in photonic lightwave technology, especially integrated optic delay interferometer, differential phase modulation and demodulation and balanced receiver have become realizable. This section gives a brief overview on the differential modulation techniques and their implementations in photonic domain, especially the MADPSK.

The term NRZ-BPSK, or traditionally NRZ-DPSK, is commonly used for denoting a modulation technique in which optical carrier is always present with a constant power, only its phase is alternated between 0 and π. The modulation rule is as follows: (i) At the transmitter: Initially a reference "0" bit is entered as the present encoded bit. Then the next data bit is compared with the present encoded bit. If they are different, then the next encoded bit is "1" for which a phase change of Π occurs, or else the next encoded bit is "0" which causes no (or 0) phase change. (ii) At the receiver, the phase of the carrier at the present bit slot is compared with that of the previous one. If the phase difference is π, then the data is decoded as "1", otherwise the data is "0" when phase difference is 0.

One of the NRZ-DPSK transmitter structures is shown in Figure 7.7. User data are first encoded by a differential encoder into the driving voltage that then alternates phase of the carrier signal between 0 and π. In detecting a NRZ-DPSK signal, a delay Mach–Zehnder interferometer (MZI), in combination with balanced an optoelectronic receiver, can be used. The interferometer acts as the phase comparator with constructively and destructively interfered outputs. As shown in Figure 7.7, the received optical signal is split into two arms of a MZI, one of which has a one-bit optical delay. The MZI compares the phase of each bit with the phase of the previous bit and the photodetector converts the phase difference to intensity. When there is no phase shift between two bits, they are added constructively and give maximum rise to the output signal, otherwise they cancel out when the phase shift is equal to π. If the differential phase shift is Δϕ then the differential current at the output of the balanced photodetector can be written as:

$$i = A^2 \cos \Delta\phi. \tag{7.2}$$

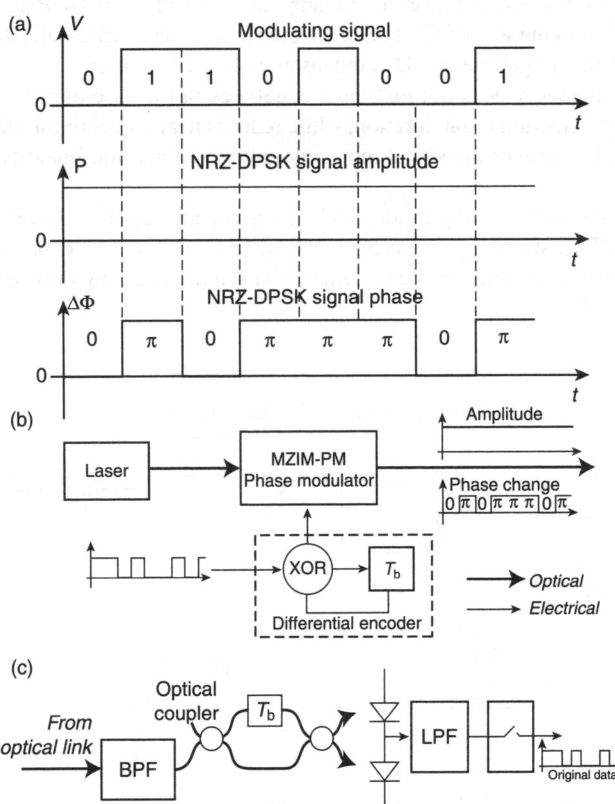

FIGURE 7.7　(a) NRZ-DPSK signal (b) transmitter and (c) receiver. (From Hui, R., and S. Zhang, *Advanced optimal modulation formats and their comparison in fiber-optic systems*, Information and Telecommunication Technology Center, University of Kansas, 2004. With permission.)

Because the balance receiver uses constructive and destructive ports of the MZI, the detected signal level can swing from "1" to "−1". Compared with ASK or with the use of unbalanced receiver where signal amplitude is limited between "1" and "0", DPSK can offer a 3-dB improvement in receiver sensitivity.

Due to its constant envelope, the NRZ-DPSK signal is less sensitive to power modulation-related non-linear effects, such as SPM and XPM, than its NRZ-ASK counterpart [15,16]. Conversely, long-haul DPSK systems, including NRZ and RZ, with optical amplifier are affected by non-linear phase noise. Amplified spontaneous emission (ASE) noise of optical amplifiers is converted into phase noise leading to waveform distortion and, consequently, signal degradation. The spectrum of NRZ-DPSK signal is shown in Figure 7.8, together with other DPSK formats. It can be seen that the NRZ-DPSK signal has the most compact spectrum compared with that of other DPSK formats. This can be explained by the fact that the NRZ-DPSK signal amplitude remains constant, regardless whether bit "1" or bit "0" is transmitted, and thus the energy is distributed more equally when comparing with RZ- and CSRZ-DPSK signals.

RZ-DPSK format is similar to the NRZ-DPSK format, the only difference is that, instead of constant optical power, a pulse narrower than bit period appears in each bit slot as shown in Figure 7.9. The RZ-DPSK transmitter, however, resembles a RZ-ASK transmitter with the phase modulator (PM) replacing the intensity modulator (IM). The RZ-DPSK signal can also be detected by the same receiver used for the NRZ-DPSK signal. Due to it narrow pulse, RZ-DPSK format is expected to minimize the effects of inter-symbol interference and is capable of achieving a longer transmission distance [7]. Narrow pulse, however, spreads the spectrum of RZ-DPSK signal wider than that of NRZ-DPSK, making RZ-DPSK systems more susceptible to chromatic dispersion (CD). To reduce the effect of this impairment, CD compensation devices are used.

RZ-DPSK signal energy does not distribute equally as in the case of NRZ-DPSK. Most of it concentrates in only a fraction of bit duration, while reduces nearly to zero for the rest of time. This large energy fluctuation makes the signal more susceptible to fiber non-linearity and signal detection more difficult.

The carrier suppression technique can also be used in conjunction with RZ-DPSK modulation to produce CSRZ-DPSK signal which has been demonstrated as one of the most attractive modulation formats in high spectral efficiency wavelength division multiplexing (WDM) and dense WDM (DWDM) systems [15].

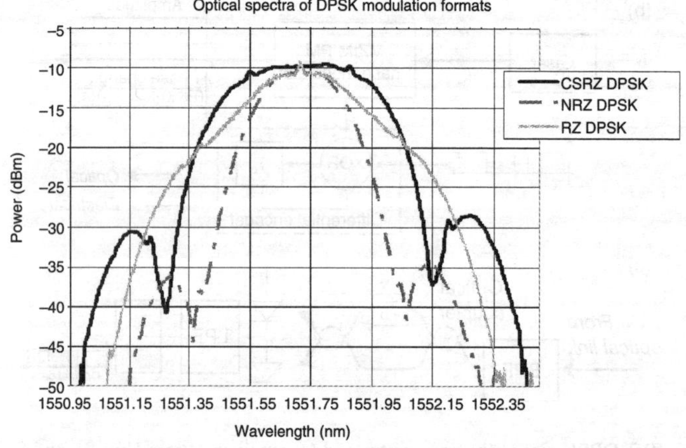

FIGURE 7.8 Experimentally measured spectra of NRZ-DPSK, RZ-DPSK, CSRZ DPSK signals.

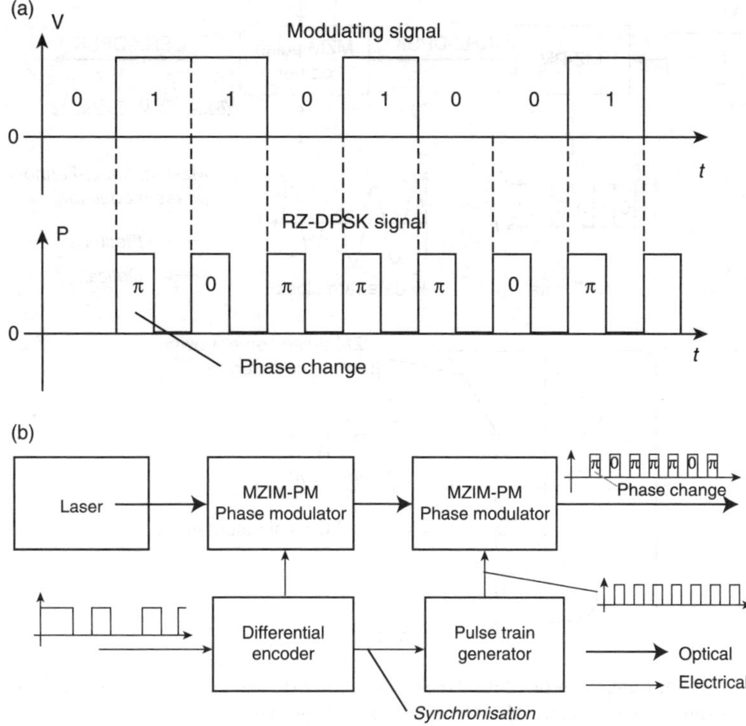

FIGURE 7.9   (a) RZ-DPSK signal (b) and transmitter structure.

Due to higher energy and spectral efficiency, the modulation format CSRZ-DPSK can be more tolerant to fiber non-linearity induced impairments, CD and polarization-mode dispersion (PMD) as compared to its RZ-DPSK counterpart.

The CSRZ-DPSK signal and its constellation can be generated by a transmitter whose scheme, shown in Figure 7.10 and Figure 7.11 respectively. The ASK parts are similar to that of CSRZ-ASK. The main difference is for CSRZ-DPSK transmitter, in which a PM replaces the IM in the CSRZ-ASK transmitter. The receiver for CSRZ-DPSK has the same structure as that of NRZ-DPSK scheme.

To increase transmission bit rate without suffering bandwidth requirement, one can code more than one bit into a data symbol. DQPSK modulation is the first step in the realization of this idea [17–19].

A signal constellation or signal space is the best way to represent a DQPSK signal, in which the points representing phase-modulated signals are located in two orthogonal axes called I and Q (for in-phase and quadrature components, respectively). Each two data bits $[D_1, D_0]$ are first pre-encoded into a symbol, then the symbol is encoded into phase shift, which may take one of four values $(0, \pi/2, \pi, 3\pi/2)$ depending on the bit combination it represents. DQPSK symbol rate is thus equal to only half of bit rate. Intuitively, one can say that, with the same bandwidth available, DQPSK can offers twice the transmission capacity compared with ASK and binary DPSK counterparts (See Table 7.2).

DQPSK signal can be generated by a transmitter shown in Figure 7.12. This structure consists of two MZIMs connected in parallel. A $\pi/2$ phase shift is introduced in one of these MZIM, making optical signals in two paths orthogonal to each other. A pre-coder encodes user data in

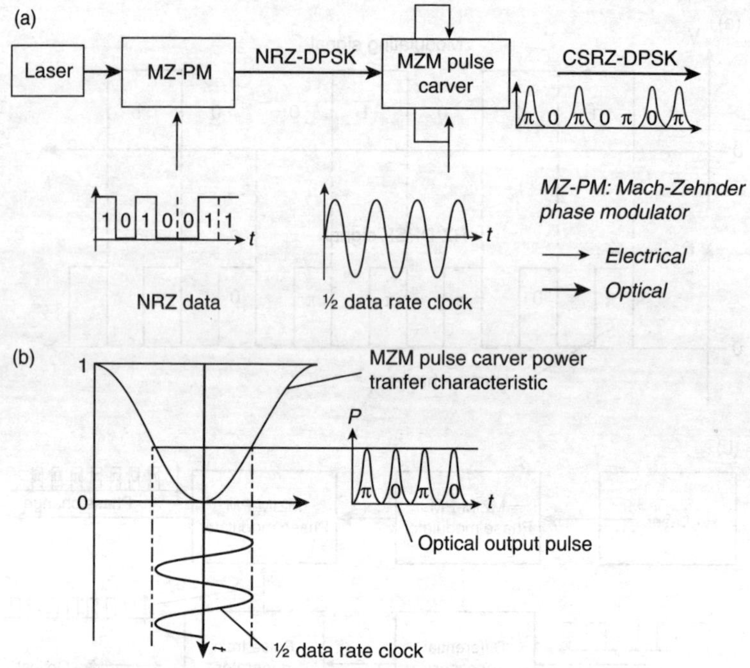

FIGURE 7.10 (a) Block diagrams of CSRZ-DPSK transmitter (b) and generation of optical pulse with alternative phase by driving the dual drive MZIM with a 2 V_π voltage swing.

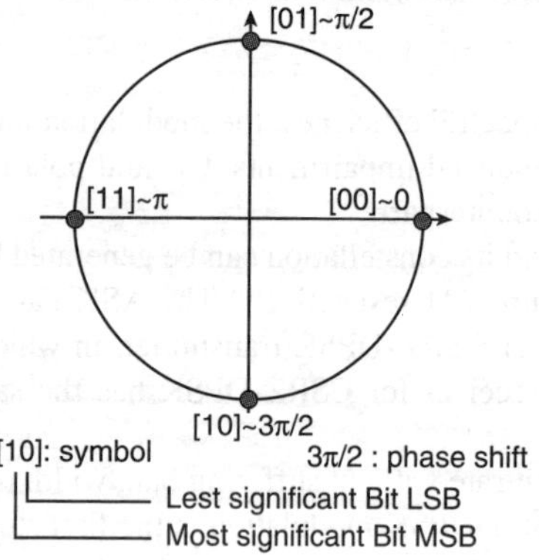

FIGURE 7.11 DQPSK signal constellation.

accordance with the differential rule to generate the I and Q driving voltages, which then modulate the phase of the carrier in two optical paths. Modulated carrier components are then combined at the output of the MZI. If the two normalized driving signals are denoted by I and Q, respectively, then the output signal is [20]

$$E_{\text{output}} = I \cos 2\pi \, f_c t + Q \sin 2\pi \, f_c t \tag{7.3}$$

where f_c is frequency of optical carrier.

DQPSK receiver uses two set of MZ delay interferometers (DI) and balance receivers to detect in-phase (I) and quadrature-phase (Q) components of the received signal, each set is similar to

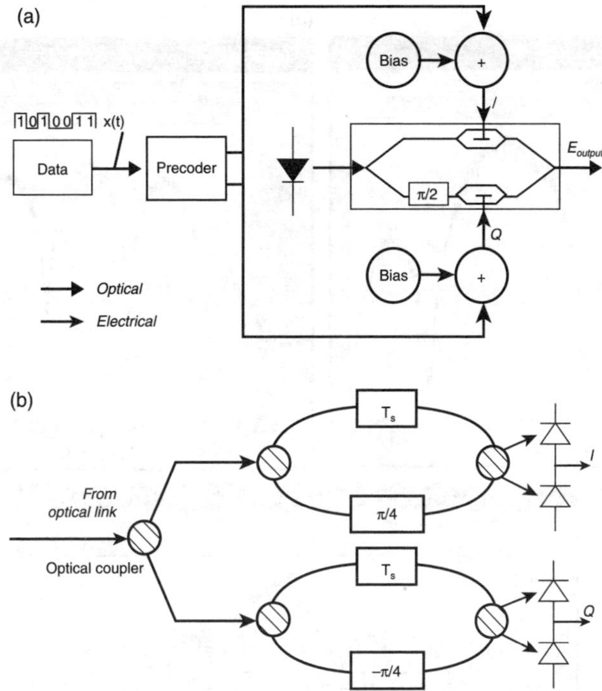

FIGURE 7.12   Parallel structure of DQPSK transmitter, T_s=symbol duration.

the one used in NRZ-DPSK receiver. There are, however, two main differences: first, the delay introduced in the first branches of interferometers is now replaced by the symbol duration T_s; second, the phases of signal in the second branches are shifted by $+\pi/4$ and $-\pi/4$ for I and Q components, respectively. These additional phase shifts are needed to separate two orthogonal phase components I and Q (Table 7.1).

Figure 7.13a shows the spectrum of 40-Gb/s NRZ-DQPSK signal with the single-sided bandwidth of the main lobe equal 20 GHz, which is only half of transmitted bit rate. The spectra of RZ-DQPSK signal, Figure 7.13b, is much broader with strong harmonics beside the main lobe.

Despite of numerous advancements in optical modulation techniques the number of levels encoded in a signal symbol falls far behind 256 or 1024 achieved in microwave modulation schemes [6] The phase noise associated with optical sources and OAs have hindered the use of phase-related modulation schemes to current fluctuations in the photodetection, and hence the degradation of the BER. Differential phase demodulation processes based on the phase comparison of two consecutive symbols requires that the phase should remain stable over two symbol periods. Thus, narrow linewidth lasers are critical for phase-modulated systems. It has been shown that to achieve a power penalty < 1 dB [21], $\Delta v/B < 1\%$ Δv and B, the laser linewidth and system bit rate, respectively. In optical transmission systems where OAs are used, the ASE noises intermingle with the fiber non-linear phase effect, thus enhancing the non-linear phase noise. SPM-induced non-linear phase noise is dominant phase noise in single optical channel systems; XPM-induced phase noise is main phase noise for multi-channel (WDM) systems.

Significant phase noises caused by optical sources and OA have prevented optical DPSK schemes from having many levels in each symbol. To increase the number of levels in the signal space, and thus the number of bits per symbol, the most preferred solution is a combination of DQPSK and ASK.

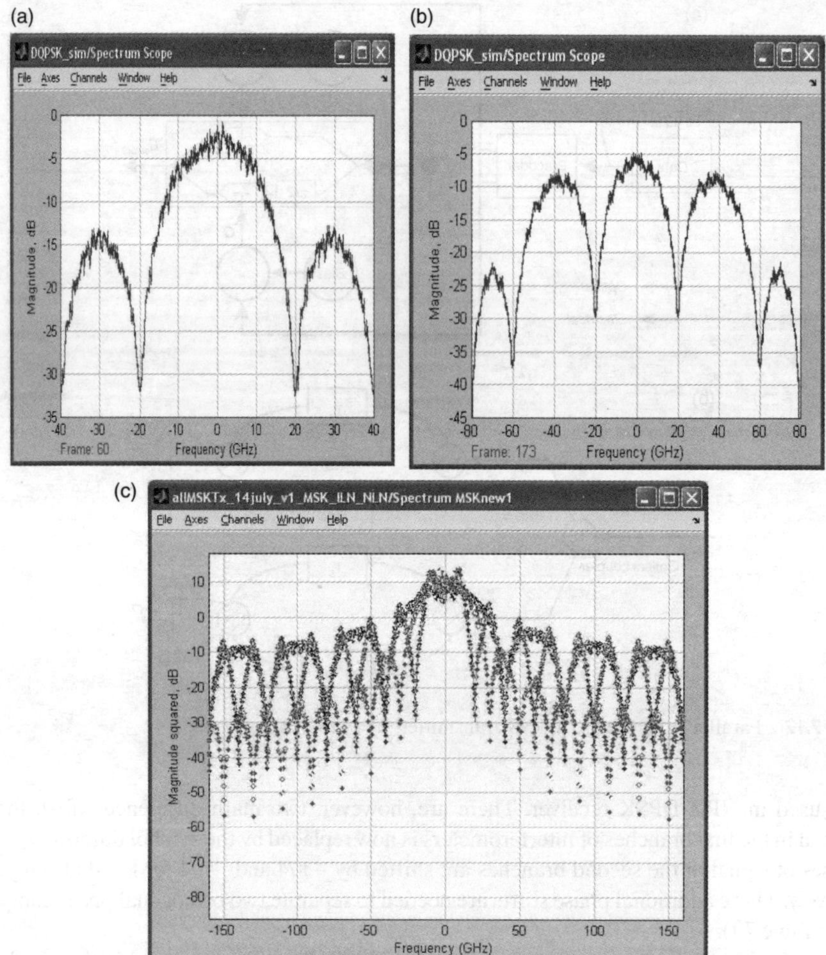

FIGURE 7.13 Optical spectra of 40 Gb/s (a) NRZ-DQPSK (b) and 50% RZ-DQPSK (c) DPSK as compared with MSK (red curve).

Recently, Hayase et al. [9] have demonstrated experimentally a 30-Gb/s eight-states per symbol optical modulation system using a combined ASK and DQPSK modulation scheme, as shown in Figure 7.16, maps three bits into a symbol, thus creating transmission bit rate three times higher than symbol rate. The transmitter consists of two cascaded PMs and an amplitude modulator (AM). The first PM, driven by data bit D_0, creates 0 and π phase shifts, while the second, driven by D_1, forces two further phase shifts 0 and $\pi/2$, the quadrature phase to generate four distinct phases of DQPSK signal. The AM, driven by D_2 bit, shifts the four phases between two amplitudes to create eight signal points in total.

At the receiver side, optical signals are detected in the amplitude and differential phase. An ASK demodulator detects the D_2 bit. The other is a DQPSK demodulator and detected to recover D_1 and D_0 bits. Seikine et al. [6] reported experimentally a similar scheme, but with 4-bits $[D_3, D_2, D_1, D_0]$ mapped into a symbol: $[D_1, D_0]$ bits are used to generate a "normalized" DQPSK signal while $[D_3, D_2]$ bits manipulate the amplitude of this DQPSK signal between four concentric circles. Thus, a 16-ary MADPSK signal can be generated. This would offer 40-Gb/s bit rate with the symbol rate of only 10 Gbauds. Figure 7.14 shows the optical spectra of (i) 40-Gb/s NRZ-DQPSK; (ii) 40 Gb/s 67% CSRZ-DQPSK; and (iii) 100 Gb/s CSRZ 16-ADPSK. This indicates the compatibility of 100 Gb/s 16-ADPSK with 40 Gb/s 67% CSRZ DQPSK. Typical eye diagrams at the output of the receiver are shown in Figure 7.15.

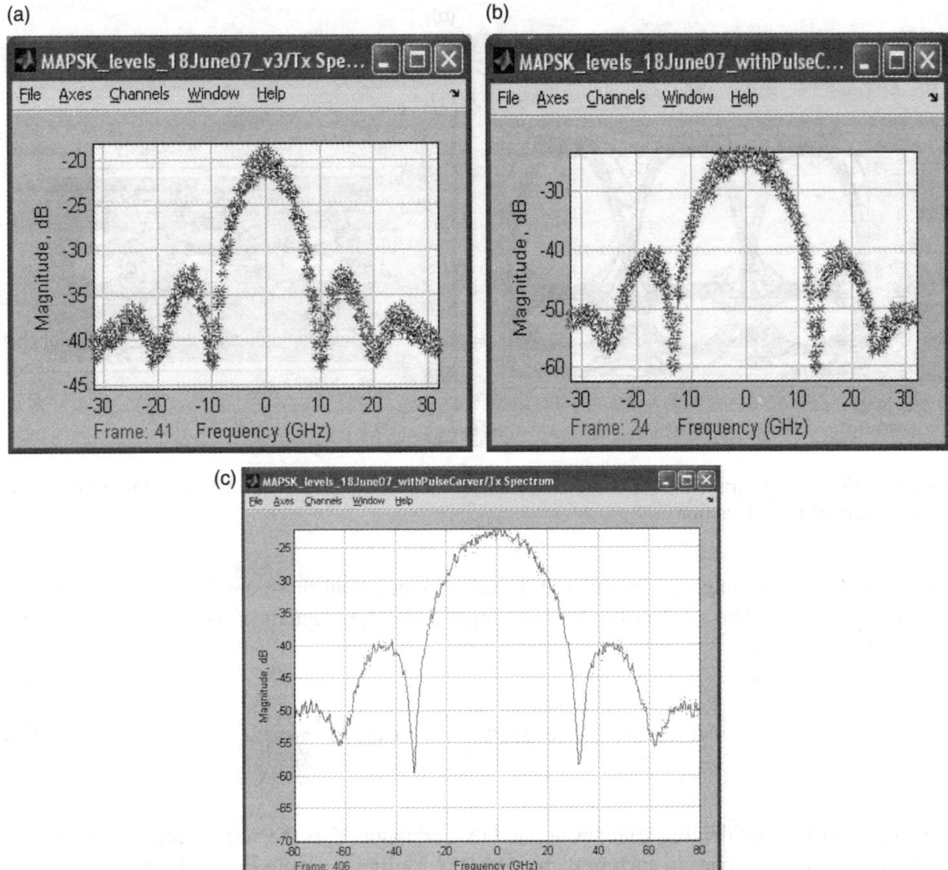

FIGURE 7.14 Optical spectra of 40-Gb/s (a) NRZ-DQPSK (b) 67% CSRZ-DQPSK (c) 100 Gb/s CSRZ 16-ADPSK.

7.2.2 Comparison of Different Optical Modulation Formats

Different amplitude and phase optical modulation formats are summarized in Table 7.2. In most cases, NRZ-ASK parameters are used as references. From the comparison, it can be concluded that MADPSK has an advantage over other modulation formats in terms of spectral efficiency and ability to significantly increase transmission bit rate which are very, if not the most, important parameters for an optical transmission system. It is also expect that MADPSK inherits good properties (and bad ones, if any) from two basic ASK and DPSK modulation formats.

7.2.3 Multilevel Optical Transmitter

In this section several optical transmitter structures used for generating DQPSK signal are described. It is necessary because, based on DQPSK modulation format, a novel optical transmission system will be developed. All these structures have MZIM as their base component, which can have a single- or dual-electrode structure.

Unlike single-drive MZIM, a dual-drive electrode structure with two traveling wave RF electrodes can modulate the phase of optical signals in both of its branches, hence push–pull operation. Interference at the output of dual-drive MZIM will produce a phase-modulated signal. However, when the effects of phase modulation in the two branches are exactly equal but opposite in sign, the

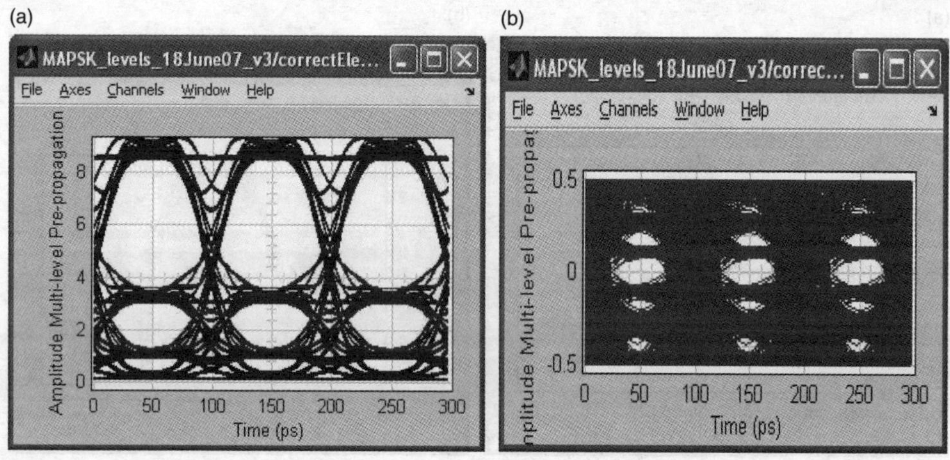

FIGURE 7.15 Eye diagram – amplitude detection section of 40Gb/s – (a) 5-km SSMF transmission (b) Quadrature phase – detection.

output signal becomes intensity modulated. In this manner, dual-drive can be used for phase and intensity modulation. The relationship between input and output signals of a dual-drive MZIM can be described by [22,12]

$$E_{\text{output}} = \frac{E_{\text{input}}}{2} \left[\exp\left(j\pi \frac{V_1(t)}{V_\pi} \right) + \exp\left(j\pi \frac{V_2(t)}{V_\pi} \right) \right] \tag{7.4}$$

where $V_1(t)$ and $V_2(t)$ are driving voltages applied to modulator, V_π is the voltage required to provide a π phase shift of the carrier in each branch of MZIM. Unlike in single-drive MZIMs, chirp effect does not exist in dual-drive MZIMs.

The transmitter structure shown in Figure 7.12, is called parallel type. It is only one of the several structures that can be used for generating DQPSK signals, i.e., parallel structure, serial structure, single PM structure, and dual-drive MZIM structure. These terms are used to indicate the structuring of MZIMs whether they are connected in tandem, parallel or just a pure PM with a single electrical drive port.

In an electro-optic transmitter of the serial type shown in Figure 7.17, a MZIM-generating in-phase component and a PM generating quadrature component are connected in tandem. Pre-encoded data generate two signals: one is used for driving the MZIM, and the other for driving the PM. Usually the square shape of the pre-encoded waveforms are replaced by the raised cosine one before being fed to the modulators [20]. Furthermore, the biasing conditions and the amplitude of the modulators can be used to generate 33% to 67% pulse width RZ formats. The pulse shape would also follow a $\cos^2$ profile due to the property of the intensity modulator. This transmitter would suffer chirping effects due to the rise time of the electrical driving signals and hence would contribute to the distortion of the lightwave signals, in particular when switching between the lowest level to the highest level.

The single PM structure (Figure 7.18) uses only one MZIM as a phase modulator. Pre-encoded data are added up to create a single driving voltage. One of the two pre-encoded data is amplified and, together with the other signal, represents 4 positions of the DQPSK signal [20].

The dual-drive MZIM structure (Figure 7.19) uses two driving voltages for modulating optical carrier phase in two branches of a MZIM. Data are first pre-encoded following the differential rule, and then used to create driving voltages $V_1(t)$ and $V_2(t)$ to the signal constellation points [20].

In the four transmitter structures described above, the parallel and serial structures are the most complex and difficult to implement because they have discrete devices connected together. The dual-drive

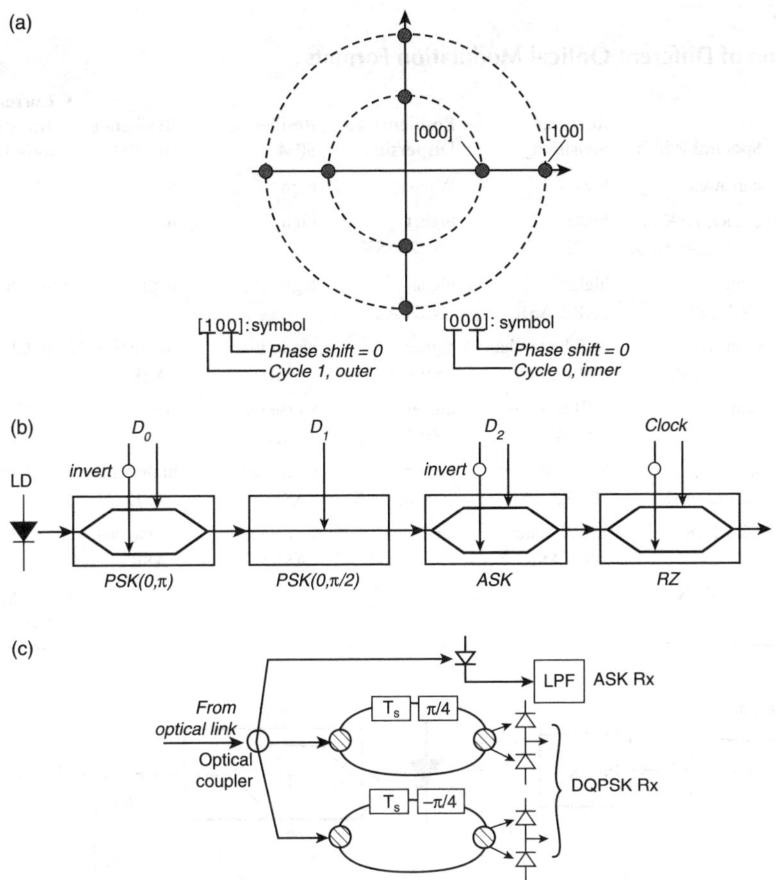

FIGURE 7.16 Eight-ary APSK modulation experimental configuration (extracted from [6]), 10-GHz clock assign synchronization of symbol rate, data modulator and qudrature phase shift in optical domain using the PM, Two balanced receivers for differential phase shift detection and direct detection for amplitude detection. (a) 8-ary ASK-DPSK signal, (b) transmitter configuration, and (c) receiver configuration. (From Hayase, S., N. Kikuchi, K. Sekine, and S. Sesaki, Proposal of 8-state per symbol (binary ASK and QPSK) 30-Gb/s optical modulation demodulation scheme. In *Proceedings of European Conference on Optical Communication, Paper Th2.6.4*, ECOC 2003, 1008–9, September, Rimini, Italy, 2003. With permission.)

TABLE 7.1

DQPSK Signal Bit-phase Mapping D_1, D_0

D_1	D_0	I	Q	Phase Shift
0	0	0	0	0
0	1	0	1	$\pi/2$
1	1	1	1	π
1	0	1	0	$3\pi/2$

MZIM and single PM structures are much simpler because they require fewer discrete devices. Furthermore, as it will be shown in the next section, dual-drive MZIM can be configured to work as phase and amplitude modulators simultaneously, so it can easily generate not only DQPSK, but generally MADPSK signals. Thus, a dual-drive MZIM is the principal part of the MAPSK transmission system.

TABLE 7.2

Comparison of Different Optical Modulation Formats

Mod. Format	Spectral Width	Receiver Sensitivity	Resilience to Dispersion	Resilience to SPM	Resilience to XPM	Current Transmission Bit Rate Limits
NRZ-ASK	narrowest	lowest	Worst	high	high	40 Gb/s
RZ-ASK	$2 \times$ RZ-ASK (at 50% duty ratio)	higher NRZ-ASK	higher NRZ-ASK	high	high	40 Gb/s
CSRZ-ASK	same as RZ-ASK	higher NRZ-ASK	higher NRZ-ASK	high	high	40 Gb/s
NRZ-DPSK	same as NRZ-ASK	3 dB better than NRZ-ASK	higher NRZ-ASK	worse than ASK	worse than ASK	40 Gb/s
RZ-DPSK	same as RZ-ASK	3 dB better than RZ-ASK	higher NRZ-ASK	worse than ASK	worse than ASK	40 Gb/s
CSRZ-DPSK	same as CSRZ-ASK	3 dB better than CSRZ-ASK	higher CSRZ-ASK	higher than ASK	higher than ASK	40 Gb/s
DQPSK	1/2 DPSK	1.5 dB better than ASK	UR*	worse than ASK	worse than ASK	$2 \times$ DPSK
MADPSK	1/M DPSK					$(\log_2 M) \times$ DPSK-expected

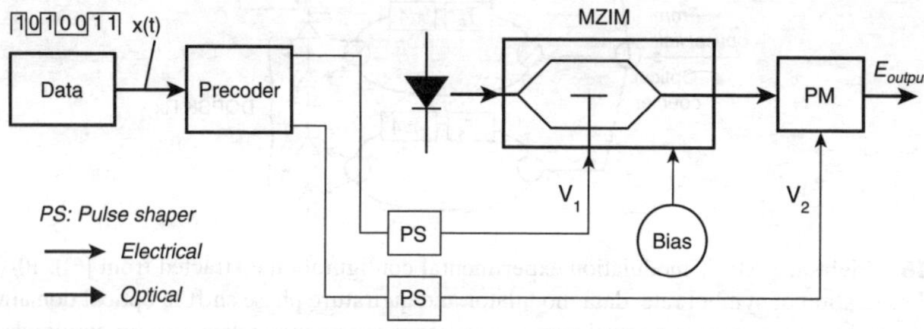

FIGURE 7.17 Cascade PM and MZIM for DQPSK signal generation.

Table 7.2 gives a comparison between transmitter structures, the dual-drive modulator is outstanding as a combined amplitude and phase switching between the states of a multi-circular constellation.

7.2.4 SINGLE DUAL-DRIVE MZIM TRANSMITTER FOR MADPSK

The main reason why the dual-drive MZIM structure has attracted our attention in this chapter is that it can have a role in amplitude and phase modulators simultaneously, which is impossible with other transmitter structures. This means that to generate a MADPSK signal, there is no need to employ separate phase and amplitude modulators, as it has been implemented in the works of Seikine K. et al. [6] and Hayase S. et al. [9]. This section describes a method for generating 16-ary MADPSK signal using this dual-drive MZIM structure.

The 16-ary MADPSK signal constellation of interest is shown in Figure 7.20. It is actually a combination of a 4-ary ASK and a DQPSK signal, where four bits $[D_3, D_2, D_1, D_0]$ are mapped into a symbol, among them two bits $[D_1, D_0]$ are coded into four phases $[0, \pi/2, \pi, 3\pi/2]$ and two bits $[D_3, D_2]$ are coded into four amplitude levels $[I_3, I_2, I_1, I_0]$. As it has been shown in [6], with the use of balanced receiver and DI, which is a solely available practical optical phase demodulator, the MADPSK signal produces clear DQPSK eye patterns centered at zero-voltage decision level only when constellation points are positioned in a radial pattern.

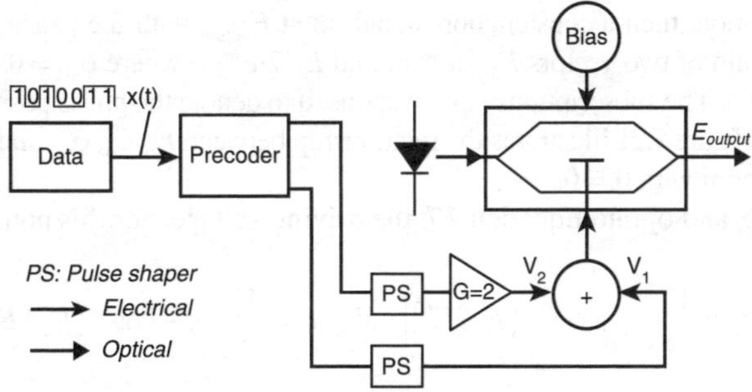

FIGURE 7.18 Structure of single PM.

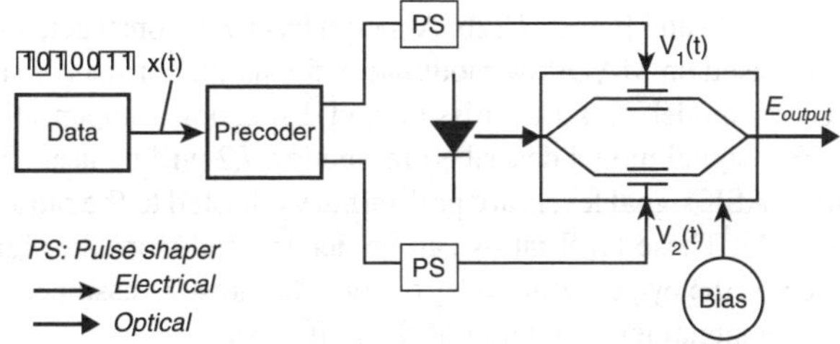

FIGURE 7.19 Structure of dual-drive MZIM.

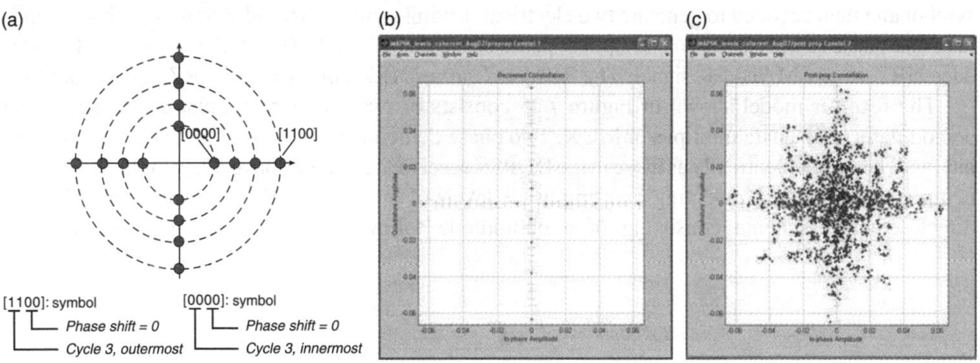

FIGURE 7.20 16-ary MADPSK signal bit-phase mapping (a) design (b) Simulink® scattering plot before transmission (c) after 200-km transmission with 2-km mismatch in dispersion.

Recall that the signal at the output of the dual-drive MZIM can be represented as [22,12]:

$$E_{\mathrm o} = \frac{E_{\mathrm i}}{2} e^{j\phi_1} + \frac{E_{\mathrm i}}{2} e^{j\phi_2}, \tag{7.5}$$

with $\phi_1 = \pi(V_1(t)/V\pi)$, $\phi_2 = \pi(V_2(t)/V\pi)$ where $E_{\mathrm i}$ and $E_{\mathrm o}$ are electrical fields of the input and output optical signal, respectively; $V_1(t)$, $V_2(t)$ are driving voltages applied to the modulator, and V_Π is voltage required to provide a Π phase shift of for the carrier in each MZIM branch.

Equation 7.7 suggests that with properly chosen input signal E_{input} and driving voltages $V_1(t)$, $V_2(t)$ all signal points of the constellation in Figure 7.20 can be constructed from two phasor signals $E_{\mathrm i}/2e^{j\phi_1}$ and $E_{\mathrm i}/2(e^{j\phi_1})$. If $E_{\mathrm i}$ is chosen to be equal to the electrical field corresponding to signal points in the largest circle of the constellation, then a constellation signal point E_{output} with the phase shift θ_i in the circle n can be found as a sum of two vectors $E_i/2(e^{j\phi_{ni1}})$ and $E_i/2(e^{j\phi_{ni1}})$ where $\phi_{ni1} = \theta_i + \arccos(E_n/E_{\text{input}})$, $\phi_{ni2} = \theta_i - \cos^{-1}(E_n/E_i)$. The subscriptions i and n are used to denote the phase position and the order of the circle of interest. Figure 7.21 illustrates the relationship between E_i, E_o, ϕ_{ni1} and ϕ_{ni2}. For simplicity, the signal point is chosen with $\theta_i = 0$.

By substituting ϕ_1 and ϕ_2 into Equation 7.7, the driving voltages for this point can be obtained:

$$V_{ni1}(t) = \frac{V_\pi}{\pi}\left[\theta_i + \cos^{-1}\left(E_n/E_i\right)\right] \quad V_{ni2}(t) = \frac{V_\pi}{\pi}\left[\theta_i - \cos^{-1}\left(E_n/E_i\right)\right]. \tag{7.6}$$

7.3 MADPSK OPTICAL TRANSMISSION

In general, the structures of the photonic transmitter and optical receiver the MADPSK can be given as shown in Figure 7.22a and Figure 7.22b. A model has been constructed for investigating the performance of systems based on MADPSK modulation format. It consists of signal coding model, transmitter model, receiver model, and transmission and dispersion compensation fiber models.

The 16-ary MADPSK signal model described in Section 7.2 will be used. To balance the ASK and DQPSK sensitivities, ASK signal levels are preliminary adjusted to the ratio $I_3/I_2/I_1/I_0 = 3/2/1.5/1$ [6] as shown in Figure 7.23. These level ratios can be determined from the signal to noise ratio at each separation distance of the eye diagram or q-factor. The noise is assumed to be dominated by the beat noise between the signal level and that of the ASE noise.

The transmitter model shown in Figure 7.24 is used to produce the 16-ary MADPSK signal. It consists of a DFB laser source generating CW light (carrier) which is then modulated in phase and amplitude by a dual-drive MZIM. Each four bits of user data $[D_3,D_2,D_1,D_0]$ are first grouped into a symbol and then encoded to generate two electrical driving signals $V_1(t)$ and $V_2(t)$ under which amplitude and phase of the carrier in two optical paths of the dual-drive MZIM will be modulated to produce NRZ 16-ary MADPSK signal. The RZM-PC carves NRZ pulse train into RZ data sequence.

The receiver model shown in Figure 7.25 consists of two phase demodulators, an amplitude demodulator, and a data multiplexer MUX. Two phase demodulators are used for extracting $[D_1,D_0]$ bits, and they work identically as those in the DQPSK receiver described above. The amplitude demodulator (AD) is used for detecting four amplitude levels of the MADPSK signal. It is a well-known direct-detection scheme consisting of a photodiode followed by an electronic receiver. The

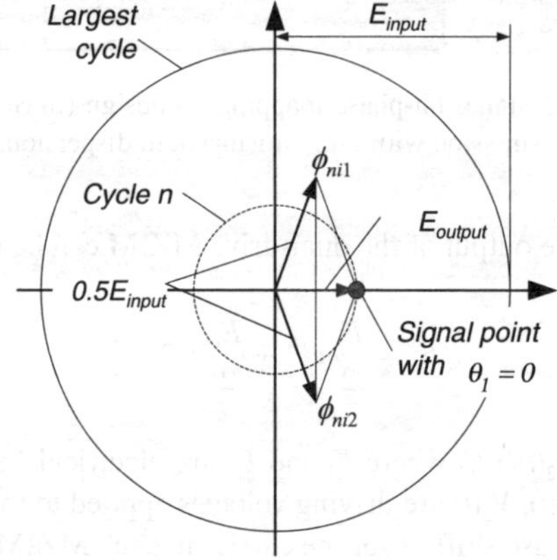

FIGURE 7.21   Relationship between E_i, E_o, ϕ_{ni1} and ϕ_{ni2} using phasor representation.

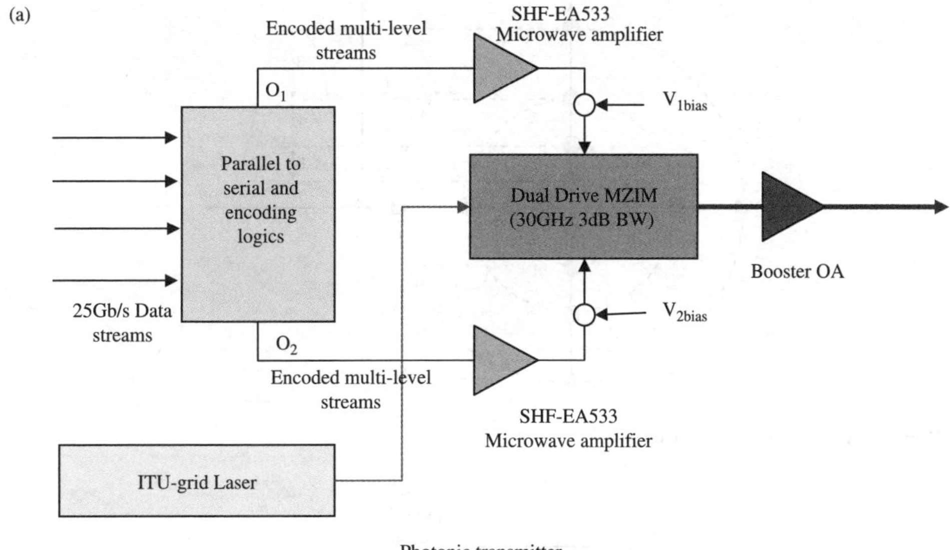

Photonic transmitter

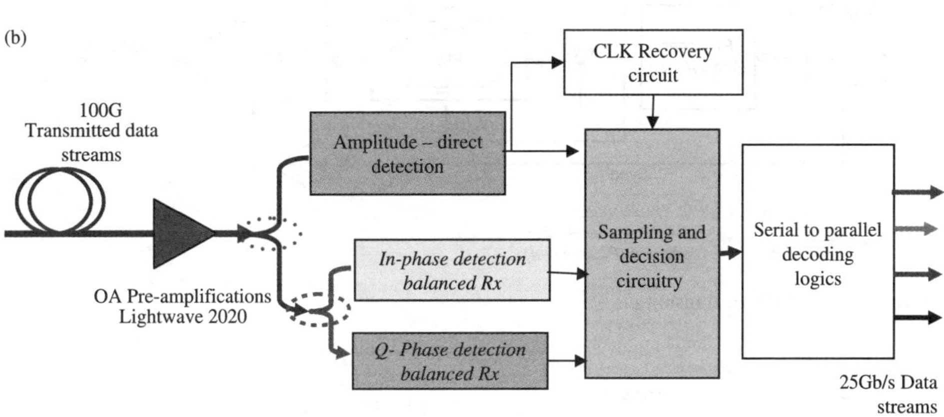

Amplitude-phase detection balanced receiver and decoding.

FIGURE 7.22 Schematic diagram of the photonic transmitters and receivers for the 16 ADQPSK transmission scheme (a) transmitter (b) receivers with branches for detection of amplitude, In-phase and quadrature phase components.

amplitude-modulated signal is then threshold detected in association with a clock recovery circuit to recover two bits $[D_3, D_2]$. Two bits $[D_3, D_2]$ are interleaved with two bits $[D_1, D_0]$ by the MUX to reconstruct the original binary data stream.

7.3.1 Performance Evaluation

Under performance evaluation, the following main parameters are investigated: (i) System bit error rate BER versus SNR: a solution for system BER will be found analytically and BER will be computed against different SNR values and bit rates. System BER versus SNR will also be obtained by system simulation and cross-checked with BER versus obtained analytically. Graphs of BER versus SNR will be plotted; (ii) System BER versus receiver sensitivity: BER versus receiver sensitivity will be obtained analytically and by simulation, and the results will be cross-checked. Graphs of BER versus receiver sensitivity will be plotted; (iii) Dispersion tolerance: transmission over fibers of types

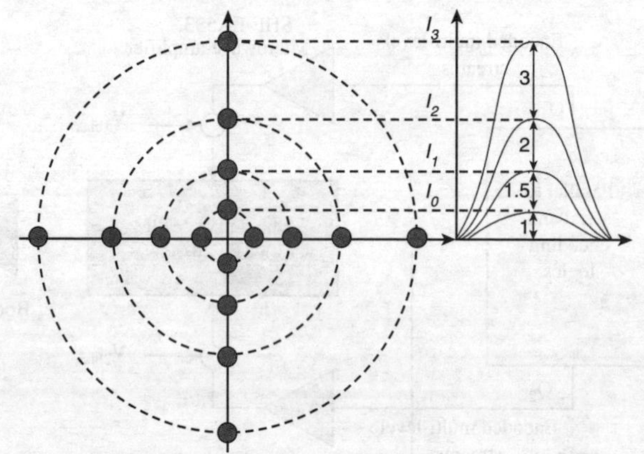

FIGURE 7.23 ASK inter-level spacing.

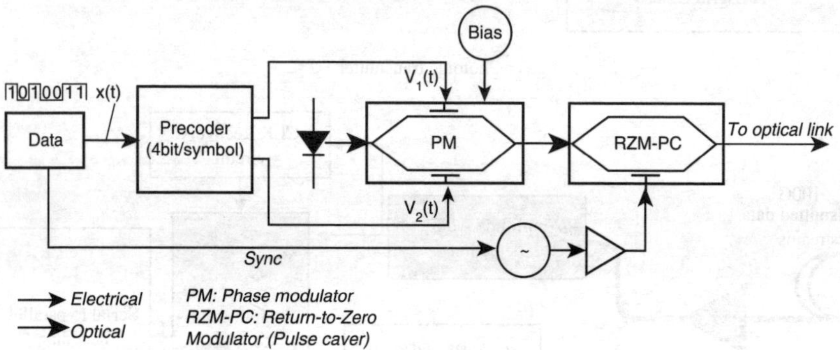

FIGURE 7.24 MADPSK transmitter.

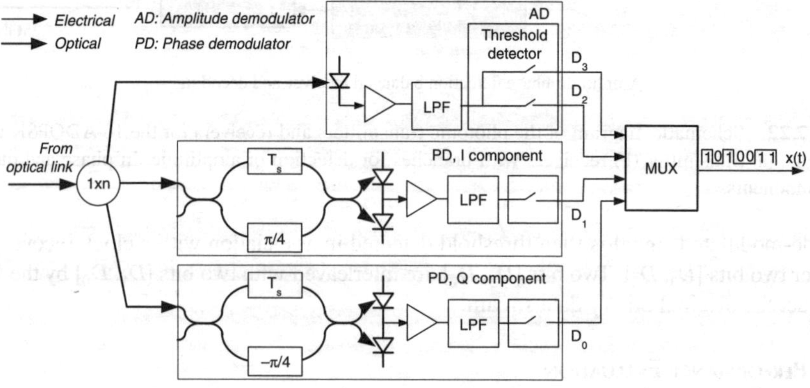

FIGURE 7.25 Amplitude direct detection and photonic phase comparator with balance receiver for MADPSK demodulation.

ITU-G652, ITU-G.655, and LEAF with corresponding dispersion factors will be considered. Graphs of the power penalty due to the dispersion as compared with back-to-back will be plotted against the dispersion factor in ps/nm; (iv) Tolerance to other system impairments: like with dispersion tolerance, power penalty due to other impairments such as laser and OA phase noise, and receiver phase error, will be investigated. Corresponding graphs will be plotted.

Performance evaluation is conducted under the effects of the following conditions or contexts: (i) Different pulse shapes: raised cosine, rectangular, Gaussian; (ii) Modulation formats: NRZ, RZ, CSRZ; (iii) ASE noise of optical amplifiers; (iv) Transmitter impairments: laser noise; (v) Receiver impairments: phase error of DI-based phase demodulators; (vi) Change of ASK inter-level spacing; (vii) Optical and electrical filtering; (viii) Multi-channel environment: system performance in combination with DWDM technology will be reported in future.

7.3.2 IMPLEMENTATION OF MADPSK TRANSMISSION MODELS

7.3.2.1 System Modeling

The following simulation models have been built on the MATLAB® Simulink® platform for proving the working principles and for investigating the performance of systems using optical MADPSK modulation. A transmitter is simulated to generate 16-ary MADPSK signal, and a receiver is used to reconstruct the original binary signal. These models run over a simulated single mode optical fiber. Laser chirp, OA phase noise, non-linearities, CD, PMD, and other impairments will be involved in later stages to evaluate different performance characteristics of the modulation format: system BER, receiver sensitivity, power penalties due to different impairments.

The phases and the driving voltages for creating signal points of the 16-ary MADPSK constellation can be computed and tabulated as shown in Table 7.3.

TABLE 7.3

Phase and Driving Voltages for 16-ary MADPSK Constellation

Positions	θ_i	V_{i1}	V_{i2}
	Circle 3		
1100	0^0	$0.0V_\pi$	$0.0V_\pi$
1101	90^0	$0.5V_\pi$	$0.5V_\pi$
1111	180^0	$1.0\,V_\pi$	$1.0\,V_\pi$
1110	270^0	$1.5V_\pi$	$1.5V_\pi$
	Circle 2		
1000	0^0	$0.2952V_\pi$	$-0.2952V_\pi$
1001	90^0	$0.7949V_\pi$	$0.2046V_\pi$
1011	180^0	$1.2947V_\pi$	$0.7043V_\pi$
1010	270^0	$1.7944V_\pi$	$1.2041V_\pi$
	Circle 1		
0100	0^0	$0.3919V_\pi$	$-0.3919V_\pi$
0101	90^0	$0.8917V_\pi$	$0.1078V_\pi$
0111	180^0	$1.3914V_\pi$	$0.6076V_\pi$
0110	270^0	$1.8912V_\pi$	$1.1073V_\pi$
	Circle 0		
0000	0^0	$0.4575\,V_\pi$	$-0.4575V_\pi$
0001	90^0	$0.9573\,V_\pi$	$0.0422V_\pi$
0011	180^0	$1.4570\,V_\pi$	$0.5420V_\pi$
0010	270^0	$1.9568\,V_\pi$	$1.0417V_\pi$

7.3.3 Transmitter Model

MATLAB® Simulink® model of the system is shown in Figure 7.26. The transmitter model using the dual-drive MZIM structure is shown in Figure 7.27. The purpose of blocks are as follows: (i) "User data and ADPSK pre-coder" block generates a pseudo-random data sequence to simulate user data stream and encodes each group of four data bits into a symbol; (ii) "Voltage driver 1" and "Voltage driver 2" blocks map pre-encoded data into driving voltages for modulating amplitude and phase of the carrier in the dual-drive MZIM; (iii) Two "complex phase shift" blocks simulate two optical paths of the dual-drive MZIM; (iv) "Sum block" simulates the combiner at the output of MZIM; (v) "Gaussian noise generator" block simulates noise source; and (vi) "Amplifier" block simulates optical amplifier. Table 7.4 gives a comparison of a number of structures of optical transmitters for the generation of DQPSK modulation formats that can be integrated in a M-ary ADPSK modulator.

7.3.4 Receiver Model

The receiver structure is shown in Figure 7.28. The functions of the blocks are as follows: (i) Each DI is simulated by a set of two "magnitude-angle" blocks, a "delay block", and a "sum" block. The "delay block" stores the phase of the previous symbol, the "magnitude-angle" blocks extract the phase and amplitude of present and previous symbols, which will be used in the followed different phase demodulation and detection operations; (ii) "constant pi/4" and "constant −pi/4" and next two "sum" blocks simulate extra phase delay in each branch of DI; (iii) Two "cos" blocks and two "product" blocks simulate two balanced receivers; (iv) "amplitude detector D2 and D3" block simulates ASK detector for D_2 and D_3 bits; (v) Three "analog filter design" blocks simulate electrical low pass filters; and (vi) "phase detector D0_I" and "phase detector D1_Q" blocks simulate the threshold detectors for D_0 and D_1 bits (I and Q component of a DQPSK signal), respectively.

7.3.5 Transmission Fiber and Dispersion Compensation Fiber Model

The propagation of optical signal in a fiber medium which is dispersive and non-linear is best described by the non-linear Schrödinger equation (NLSE) [21] as described in Chapter 2. Other parameters are explained below. Transmission fiber model as shown in Figure 7.29 is used to simulate the propagation of optical signal. This fiber model simulates the impairments which have impacts on the system performance.

All characteristic parameters of the fiber medium, together with optical input signal, are taken by the "matrix concatenation" block and then processed by a MATLAB® function that solves the NLSE using the split-step Fourier method [23].

The dispersion compensation fiber model has the same structure of the propagation fiber model, except that the sign of the propagation constant beta2 in the two models are opposite.

7.4 Transmission Performance

7.4.1 Signal Spectrum, Signal Constellation and Eye Diagram

The spectrum of 40-Gb/s 16-ary MDAPSK signal obtained by running transmitter model is given in Figure 7.30. As seen in the graph, the single-sided bandwidth of the main lobe equals 10 GHz. Numerically, it amounts to only one-fourth of the transmission bit rate, and from that it can be concluded that MADPSK is a high bandwidth efficient modulation format.

Figure 7.31 shows the signal constellation recovered at the receiving end. Noise and non-linear property of fiber cause amplitude and phase fluctuations and scatter signal points around some mean value. The MADPSK eye diagram is shown in Figure 7.32 for the I component (the Q component should have the similar diagram). This eye diagram clearly shows four amplitude levels associated with two phase shifts 0 and π.

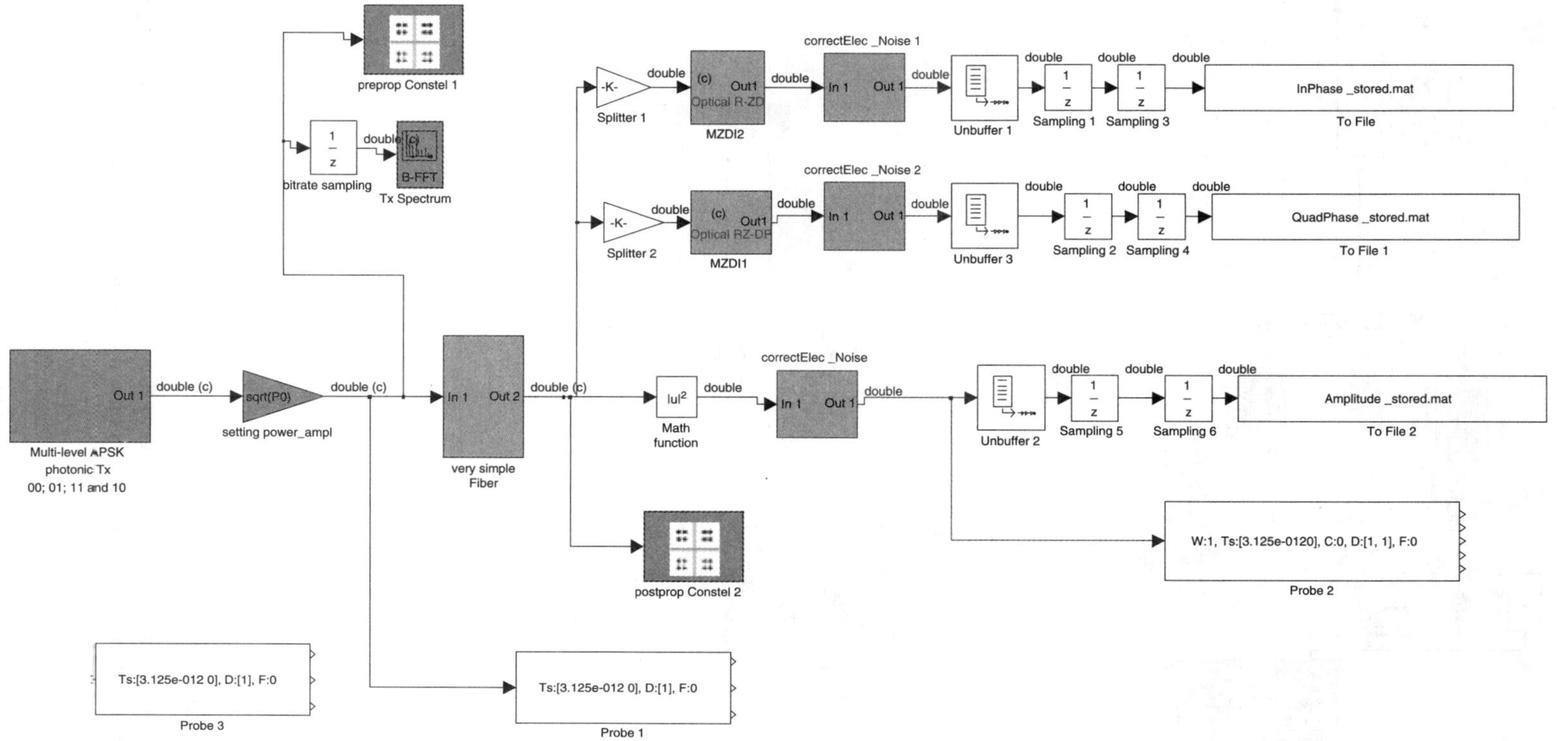

FIGURE 7.26 MATLAB®-simulated system model.

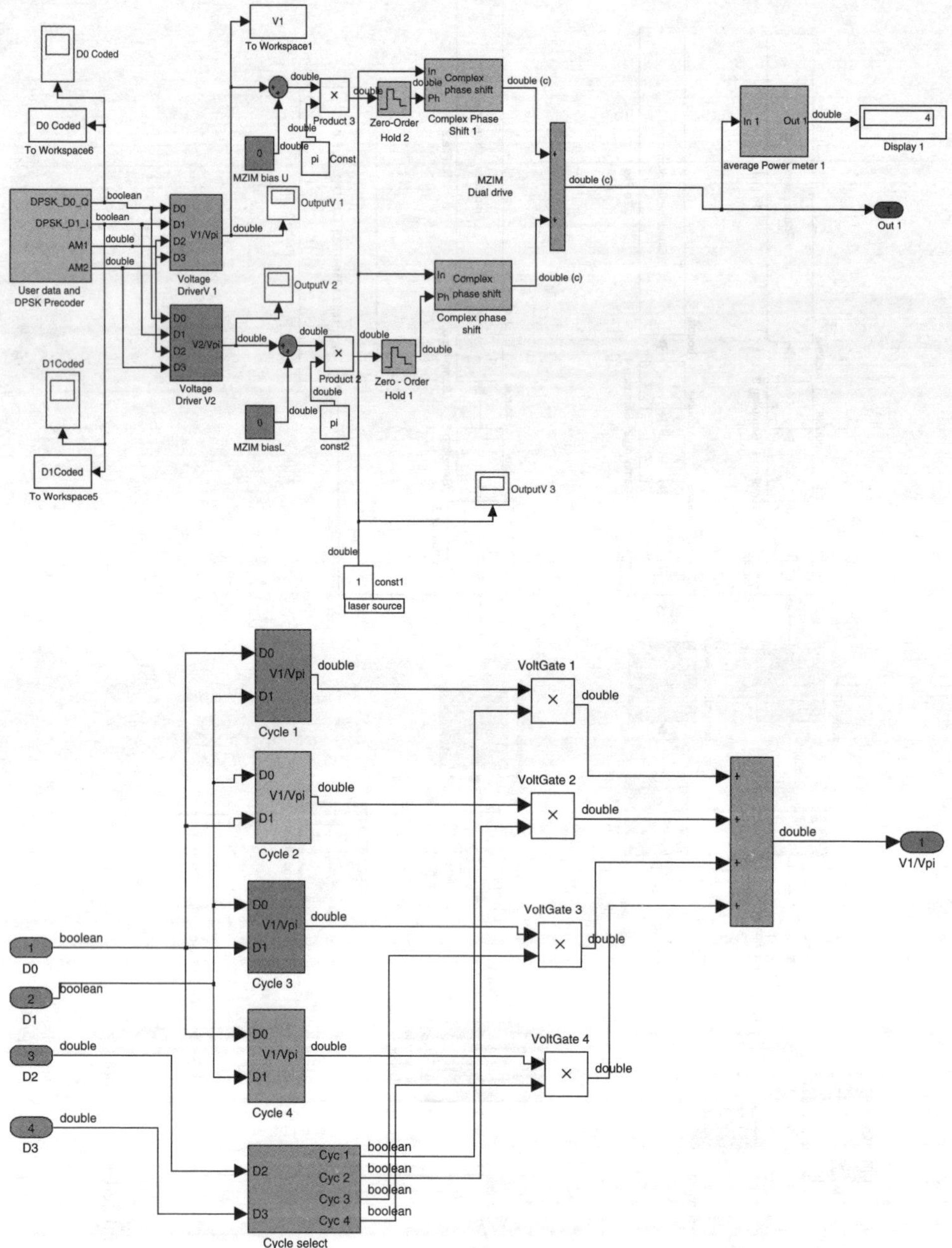

FIGURE 7.27 MATLAB-simulated MADPSK (a) transmitter (b) Logic pre-coder.

7.4.2 BER Evaluation

The MADPSK system can be considered as consisting of two sub-systems, ASK and DQPSK, and its error probability can be evaluated as a joint error probability of the two

$$P_{\text{ADPSK}} = \left[\frac{1}{2} P_{\text{ASK}} + \frac{1}{2} P_{\text{DPSK}} - \frac{1}{2} P_{\text{ASK}} \frac{1}{2} P_{\text{DPSK}} \right] = \frac{1}{2} \left[P_{\text{ASK}} + P_{\text{DPSK}} - P_{\text{ASK}} P_{\text{DPSK}} \right] \tag{7.7}$$

where P_{ASK} and P_{DPSK} are error probabilities of ASK and DQPSK sub-systems, respectively.

TABLE 7.4

Comparison of DQPSK Transmitter Structures

Parameters for Comparison	Parallel MZIM	Serial MZIM&PM	Single PM	Dual-drive MZIM
Complexity of circuit design	Complicated in matching of ultra-high frequency electrical paths; high insertion loss. Flexible in biasing.	Complicated in matching of ultra-high frequency electrical paths; high insertion loss. Flexible in biasing.	Simple in photonic but complicated in realization of ultra-high frequency signal connections.	Simplest but required multi-level voltage switching at symbol rate (microwave speed).
Ability to create MADPSK signal	Not possible. A separate ASK modulator required.	Not possible. A separate ASK modulator required.	Impossible. A separate ASK modulator required.	Dual-drive MZIM acts as ASK and DPSK simultaneously.

7.4.2.1 ASK Sub-System Error Probability

Figure 7.33 shows four ASK signal levels b_0, b_1, b_2, b_3 three decision levels d_1, d_2, d_3 and standard deviation of noise at different signal levels σ_0, σ_1, σ_2, σ_3. The error probability of the ASK sub-system can be evaluated by [24]

$$P_{ASK} = \frac{2}{M+1}\sum_1^M Q\left(\frac{b_i-d_i}{\sigma_i}\right) = \frac{2}{3+1}\left[Q\left(\frac{b_1-d_1}{\sigma_1}\right) + Q\left(\frac{b_2-d_2}{\sigma_2}\right) + Q\left(\frac{b_3-d_3}{\sigma_3}\right)\right]. \tag{7.8}$$

For example, in our system: (i) $b_1=8.08\mathrm{e}-2$, $b_2=1.45\mathrm{e}-1$, $b_3=2.42\mathrm{e}-1$; (ii) $d_1=5.11\mathrm{e}-2$, $d_2=1.08\mathrm{e}-1$, $d_3=1.88\mathrm{e}-1$; (iii) $\sigma_1=5.00\mathrm{e}-3$, $\sigma_2=6.70\mathrm{e}-3$, $\sigma_2=8.65\mathrm{e}-3$ at an OSNR $=20\,\mathrm{dB}$.

The error probability of the ASK sub-system thus equals

$$P_{ASK} = \frac{1}{2}\left[Q\left(\frac{(8.08\mathrm{e}-2)-(5.11\mathrm{e}-2)}{5.00\mathrm{e}-3}\right) + Q\left(\frac{(1.45\mathrm{e}-1)-(1.08\mathrm{e}-1)}{6.70\mathrm{e}-3}\right) + Q\left(\frac{(2.42\mathrm{e}-1)-(1.88\mathrm{e}-1)}{8.65\mathrm{e}-3}\right)\right]$$

$$P_{ASK} = \frac{1}{2}\left[Q(5.94) + Q(5.52) + Q(6.24)\right]$$

$$= \frac{1}{2}\left[(1.47\mathrm{e}-9) + (1.73\mathrm{e}-8) + Q(2.26\mathrm{e}-10)\right]$$

$$= 9.49\mathrm{e}-9.$$

The error probability of ASK sub-system over a range of OSNR from 6 dB to 24 dB is evaluated and shown in Figure 7.34.

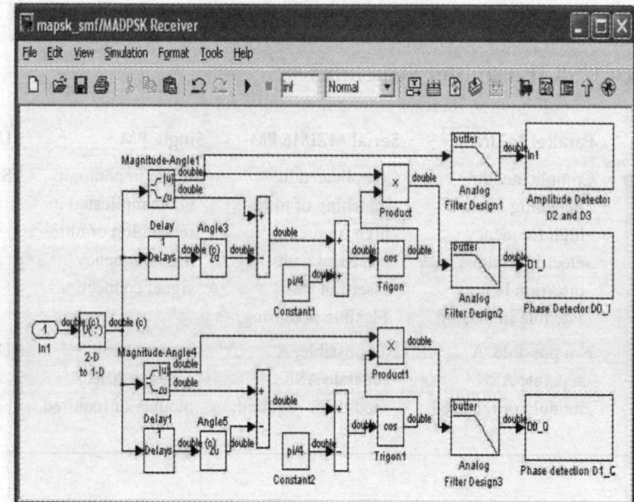

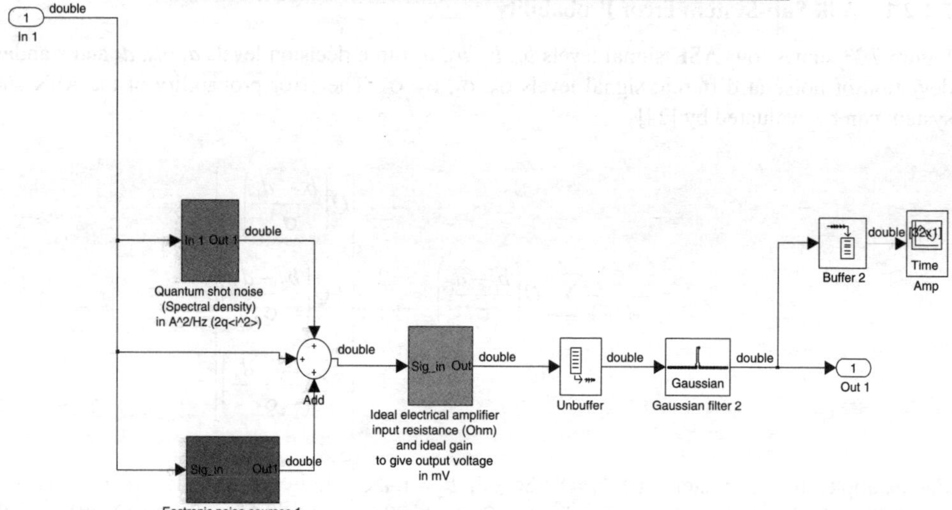

FIGURE 7.28 MATLAB® Simulink® model of MADPSK receiver (a) amplitude direct detection (b) balanced receiver detection – inphase and quadrature.

7.4.3 DQPSK Sub-System Error Probability Evaluation

In terms of differential phase shift keying modulation, the system can be broken up into four independent DQPSK sub-systems corresponding to circle 0, circle 1, circle 2, and circle 3 of the signal constellation. The error probability of each sub-system is evaluated first and then they are averaged to obtain the error probability of the DQPSK sub-system.

Each DQPSK sub-system in turn can be thought of as being made from two 2-ary DPSK sub-systems. The error probability of each 2-ary DPSK sub-system is evaluated and they are averaged to get the error probability of DQPSK sub-system

$$P_{\text{DQPSK}} = 1 - (1 - P_{\text{DPSK\_I}})(1 - P_{\text{DPSK\_Q}}) = P_{\text{DPSK\_I}} + P_{\text{DPSK\_Q}} - P_{\text{DPSK\_I}}P_{\text{DPSK\_Q}} \tag{7.9}$$

where $P_{\text{DPSK\_I}}$ and $P_{\text{DPSK\_Q}}$ is error probability of in-phase (I) and quadrature-phase (Q) components of each DPSK sub-system (circle). Because I is coded by bit D_0,Q is coded by D_1, I and Q are detected in the same way, and D_0 and D_1 are supposed to be equally probable, then (20) becomes

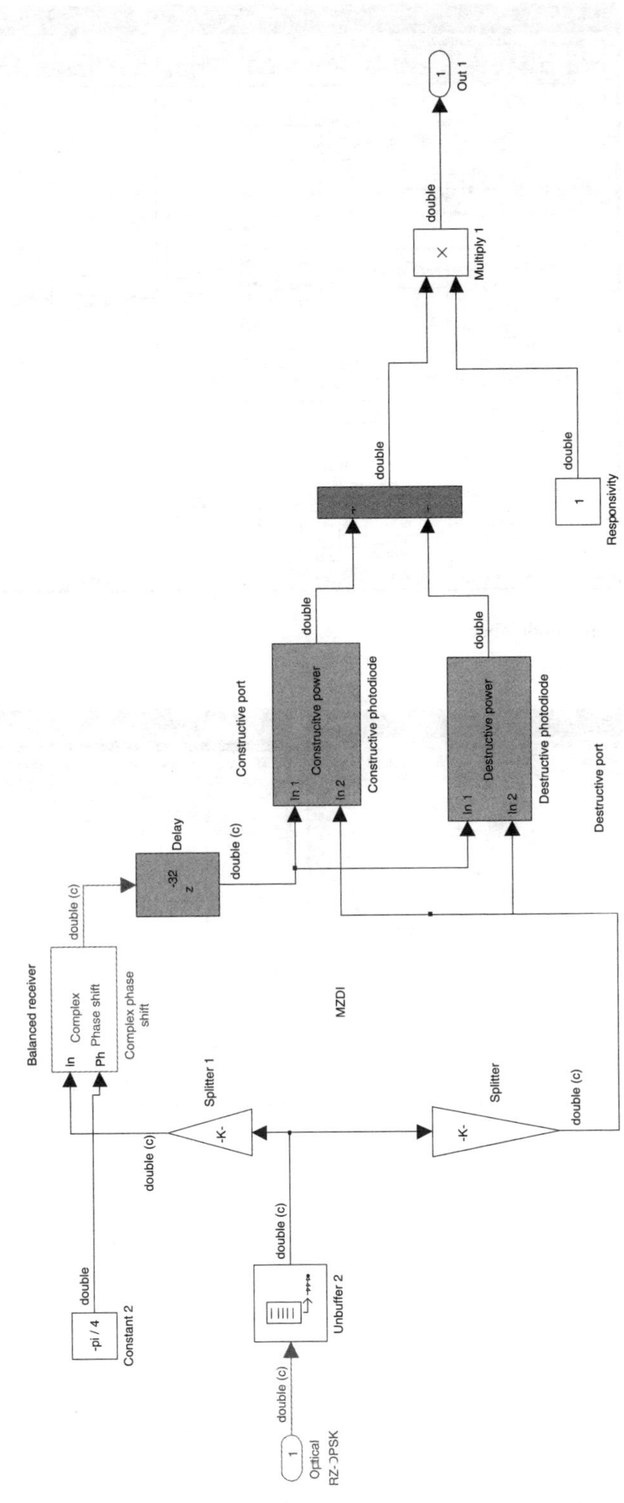

FIGURE 7.28 (*Continued*)

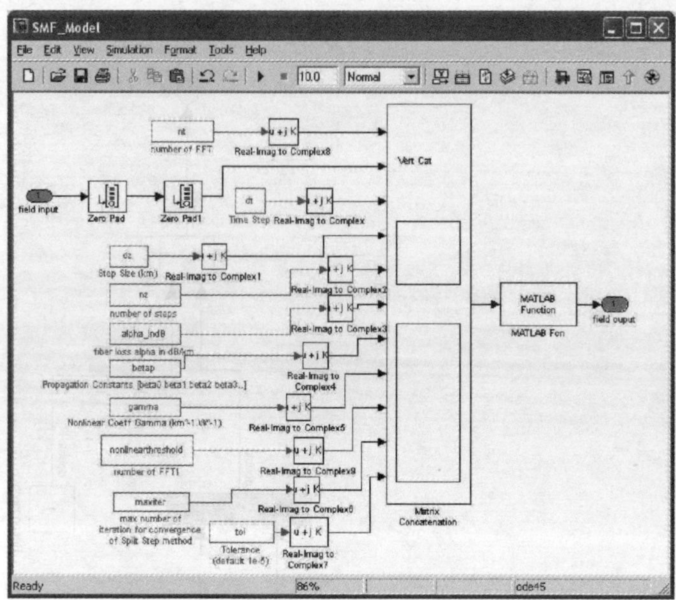

FIGURE 7.29   Single-mode fiber model.

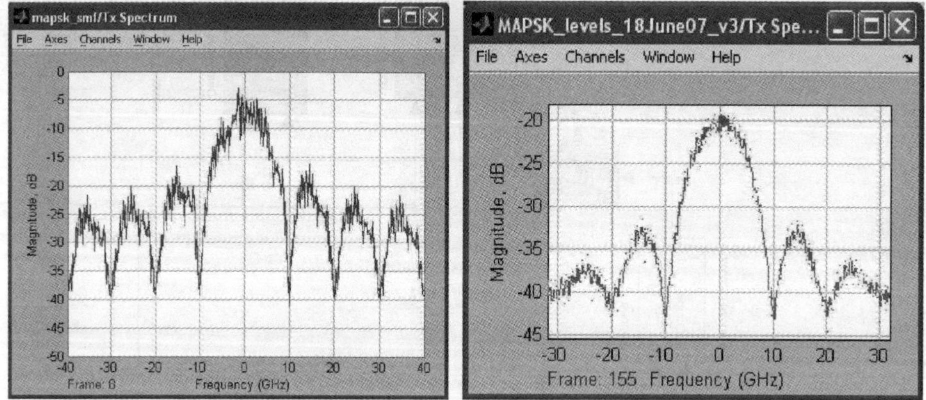

FIGURE 7.30   40-Gb/s MADPSK spectrum.

$$P_{\text{DQPSK}} = 2P_{\text{DPSK\_I}} - P_{\text{DPSK\_I}}^2 = 2P_{\text{DPSK\_Q}} - P_{\text{DPSK\_Q}}^2 \tag{7.10}$$

$P_{\text{DPSK\_I}}$ evaluated based on the δ-factor [21]:

$$P_{\text{DPSK\_I}} = \frac{1}{2}\left(\frac{\delta}{\sqrt{2}}\right) \approx \frac{\exp\left(-\delta^2/2\right)}{\delta\sqrt{2\pi}} \tag{7.11}$$

where $Q = (i_{\text{H}} - i_{\text{L}})/(\sigma_{\text{H}} + \sigma_{\text{L}})$, i_{H}, i_{L} and σ_{H}, σ_{L} are mean value and standard deviation of signal currents at high and low levels at the input of the receiver, respectively. For example, the

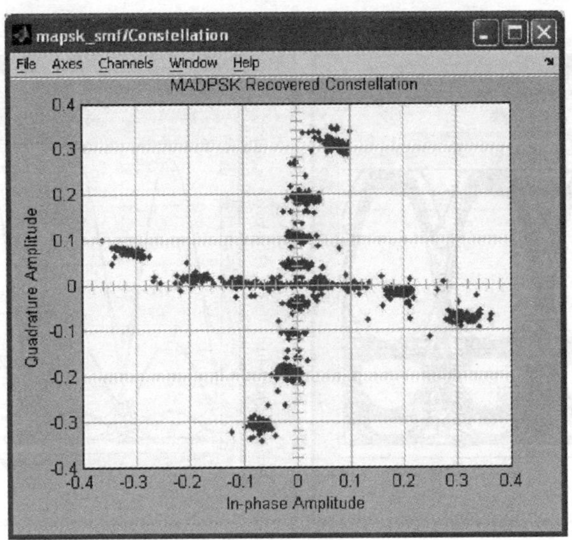

FIGURE 7.31   40-Gb/s MADPSK constellation recovered at the receiver.

transmission parameters can be set as follows: $i_H = 3.23e - 02$, $i_L = (-3.23e - 02)$, at OSNR=20 dB $\sigma_H = \sigma_H = 3.16e - 3$. The δ-factor for a single DQPSK sub-system of circle 0 thus equals, and the corresponding the error probability $P_{DPSK\_1\_CYCLE0} = 1/2 \text{erfc}(10/\sqrt{2}) \approx 7.7e - 24$.

The error probability of circle 0 (innermost circle) is $P_{DQPSK\_CYCLE0} = 2*(7.7e - 24) - (7.7e - 24)^2 = 1.54*10e - 23$. Thus, the error probability of all four circles is:

$$P_{DQPSK} = \frac{1}{4}\left[P_{DQPSK\_CYCLE0} + P_{DQPSK\_CYCLE1} + P_{DQPSK\_CYCLE2} + P_{DQPSK\_CYCLE3}\right]. \tag{7.12}$$

P_{DQPSK} a range of OSNR from 6dB to 24dB is evaluated and shown in Figure 7.35.

7.4.4   MADPSK System BER Evaluation

The MADPSK system error probability is evaluated based on Equation 7.7. Figure 7.36 shows the graphs of error probability for the ASK sub-system, DQPSK sub-system and MADPSK system in the same co-ordinates for comparison. As it can be observed from Figure 7.36, at OSNR = 24 dB the MADPSK data sequence can be received without errors. For the same value of OSNR, especially when it is high, DQPSK sub-system outperforms the ASK counterpart, and the overall performance of MADPSK system is dominated by the ASK sub-system performance. Thus, the spaces between ASK levels could be adjusted for a better balance between BER ASK and BER DQPSK to achieve a better overall MADPSK BER performance. This probably is caused mainly by the inter-symbol interference during the transition of different levels.

Figure 7.36 shows the simulation results of 16 ADPSK at 100-Gb/s transmission (extreme left graph) in comparison with other modulation formats such as duo-binary 50 and duo-binary 67 and experimental results of CSRZ DPSK. The bit rates of these other transmission results are at 40 Gb/s. For 16 MADPSK, the receiver sensitivity is close to the –28-dBm performance standard used in 10-Gb/s NRZ transmission, and performs better at 100 Gb/s than the other modulations operating at the lower rate of 40 Gb/s. However, this superior performance at 100 Gb/s is still at penalty of approximately 3 dB compared with 10-Gb/s transmission systems. Fortunately, this penalty can easily be compensated for using a low-noise optical pre-amplifier at the receiver end. For example,

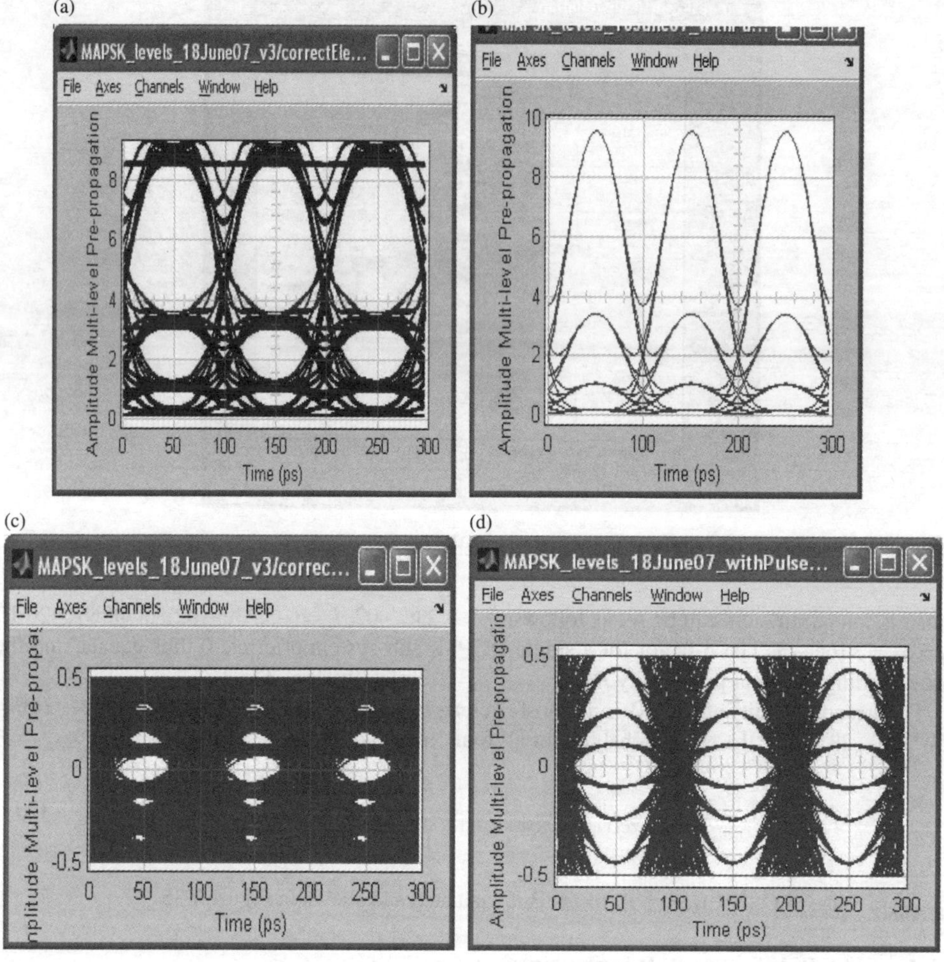

FIGURE 7.32 40-Gb/s MADPSK eye diagram at OSNR=20 dB (a) NRZ amplitude (b) CSRZ amplitude (c) NRZ in phase (d) CSRZ in phase.

a 15-dB gain optical pre-amplifier with a 3-dB noise figure would adequately resolve the issue. The BER versus the receiver sensitivity of 16 ADPSK and duobinary formats and ASK are shown in Figure 7.37. It indicates a 2–3-dB improvement of the MADPSK.

The detection of the lowest level may have been affected by the noise level of the optical preamp when only the amplitude information was used. This can be improved significantly if the phase detection and amplitude detection are used, as we could observe from Figures 7.32c and Figures 7.32d.

7.4.5 Chromatic Dispersion Tolerance

The residual chromatic dispersion of the optical link is characterized by the DL product defined as product of the dispersion coefficient D and the total fiber length L. Figure 7.38 shows the signal phase evolution under the effect of chromatic dispersion. With a predetermined $DL = 50$ ps/nm, all signal points are rotated around the [0,0] origin by the same angle of approximately 0.125 rad. This confirms the parabolic phase shift due to the chromatic dispersion. This phenomenon is called "linear phase distortion" in contrast to the non-linear phase distortion caused by the fiber non-linearity.

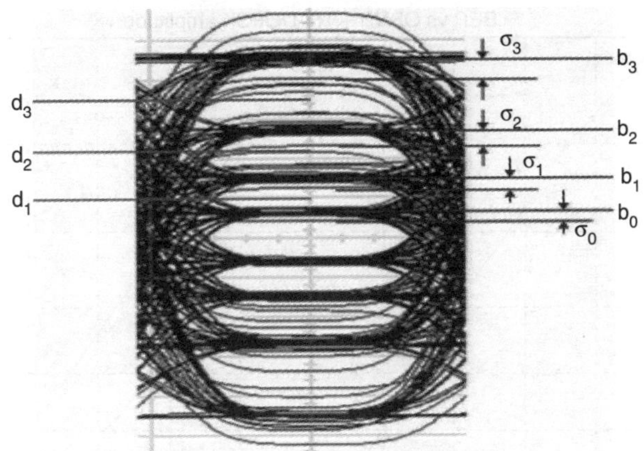

FIGURE 7.33　MADPSK eye diagram: signal levels, decision levels, and standard deviation of noise.

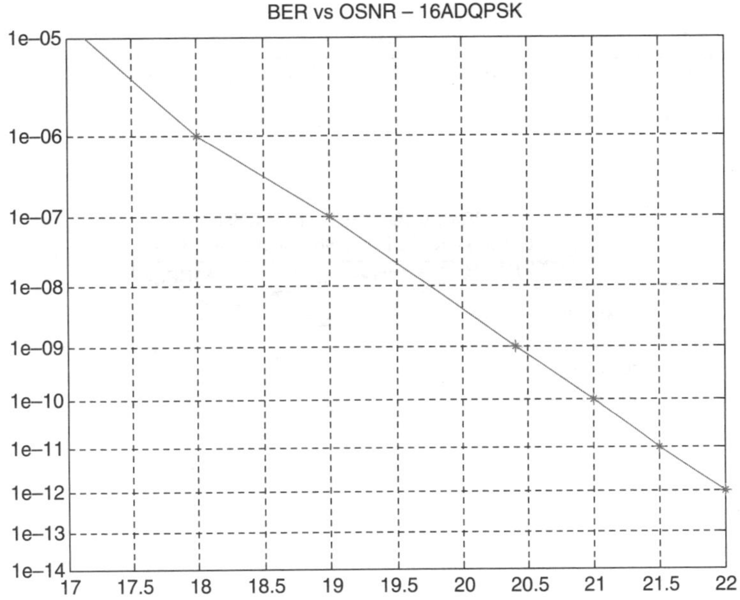

FIGURE 7.34　Error probability of ASK sub-system versus OSNR.

Figure 7.39 shows the BER penalty versus different values of DL product. The BER performance of NRZ format is severely affected by the fiber dispersion. When DL increases from 0 ps/nm (fully CD compensated) to 35 ps/nm, its BER performance is improved by 1.5 dB, but sharply degraded by 28-dB penalty at $DL = 50$ ps/nm, and should be worse for a higher value of dispersion. It is therefore undesirable to use NRZ format in MADPSK systems because the optical link residual dispersion usually can not be compensated to a small amount, and ineffective dispersion management and control plan could lead to a very high BER.

The 66%-RZ format can tolerate a much higher degree of chromatic dispersion. Its BER performance is even slightly improved at $DL = 50$ ps/nm, and the BER penalty is just less than 1 dB at $DL = 100$ ps/nm. That is equivalent to the transmission over a 6-km of uncompensated standard SMF fiber without significant giving up the BER performance.

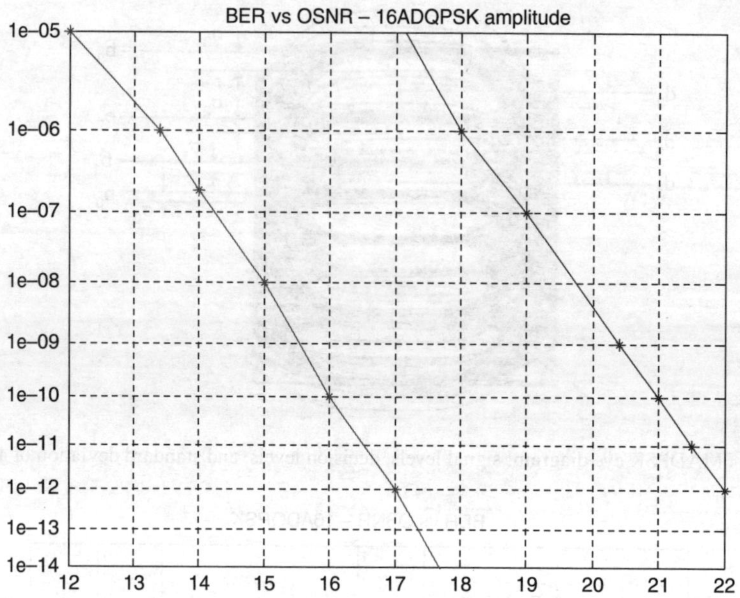

FIGURE 7.35 Error probability of DQPSK sub-system versus OSNR.

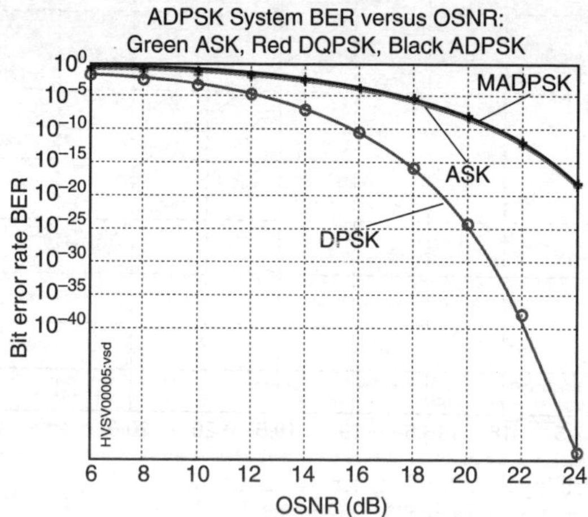

FIGURE 7.36 Error probability of MADPSK (solid) system versus OSNR, logarithmic scale. Error probability of ASK (dotted) and MADPSK are nearly coincided.

The MAPSK offer lower symbol rate and hence higher channel capacity that would allow the upgrading of higher rate merging in a low bit rate optical fiber transmission system without modifying the photonic infrastructure of the optical networks.

7.4.6 Critical Issues

This section outlines the critical issues in the evaluation of the performance of MADPSK systems.

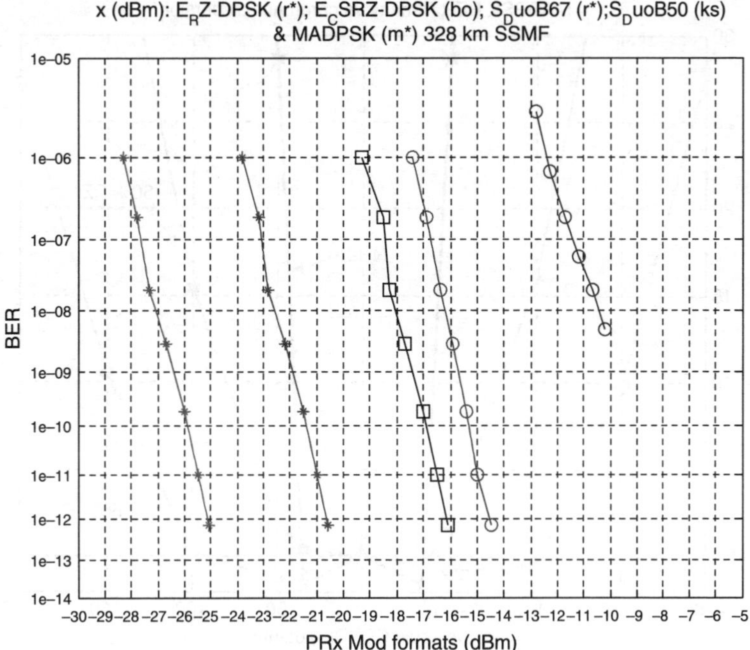

FIGURE 7.37   BER versus receiver sensitivity for MADPSK format and other duo-binary and ASK (simulation) and CSZ and CSRZ DPSK (experimental). Legend: Curve indicated by "*" (far left) is for MADPSK; "*" is for RZ-DPSK; far right "o" is for NRZ_DPSK; "o" Duo-binary RZ-50%; "square" duo-binary RZ-67%.

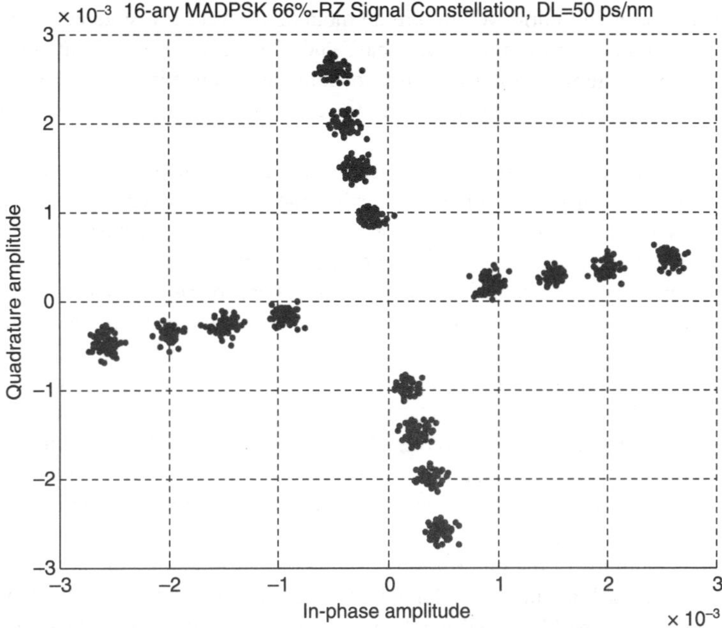

FIGURE 7.38   Evolution of the phase scattering of the MADPSK signal constellation under chromatic dispersion effects.

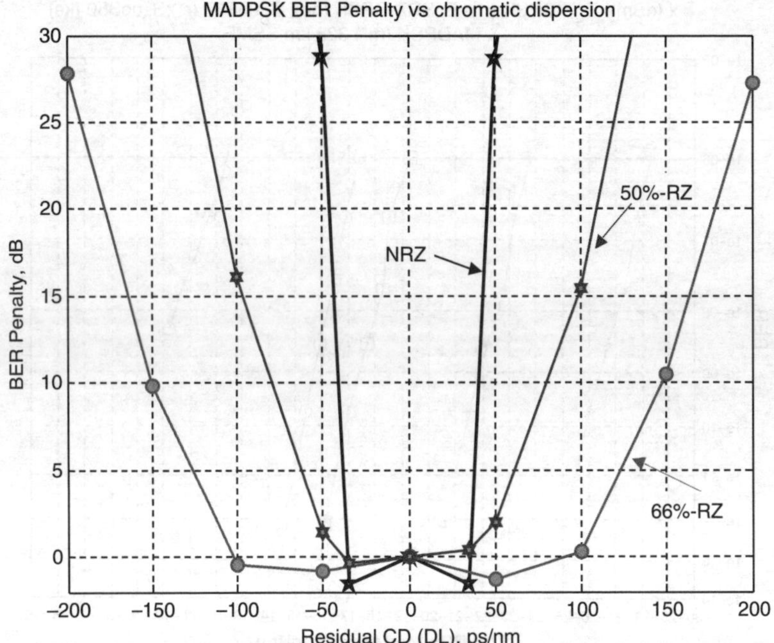

FIGURE 7.39 Error probability of MADPSK system versus the dispersion-length DL product.

7.4.6.1 Noise Mechanism and Noise Effect on MADPSK

Although receiver noises in multi-level amplitude modulation were investigated intensively in the 1980s, little has been reported for multi-level phase and differential phase modulation. One of the principal goals in the system design, especially for long-haul transmission systems, is to achieve high receiver sensitivity. At a given optical power, the error probability depends on the noise power, and hence the receiver sensitivity.

Quantum shot noise is the fundamental noise mechanism in photodiodes, which leads to the fluctuation in the detected electrical current even when the incident optical signal has a constant or varied power. Thus, it is signal dependent. Furthermore, the beating of the currents of the signal and the optical phase noise would generate an amplitude-dependent noise at different level signals of the MADPSK. It is caused by random generation of electrons contributing to the photoelectric current, which is a random variable. All photodiodes generate some current even in the absence of optical signal because of the stray light and/or thermal generation of electron-hole pairs, the dark current.

In MADPSK, the amplitude of the signal of the outermost circle of the constellation would be affected by the quantum shot noises which are strongly signal amplitude-dependent, especially when there is an optical preamplifier. Conversely, it is desirable that the innermost constellation would have the largest magnitude to maximize the optical signal energy for long-haul transmission. Therefore, an optimum receiving scheme must be developed by analysis, modeling and eventually by experimental demonstration.

However, the amplitude of the outermost constellations is limited by the non-linear self-phase modulation effects that will be further explained in the next few sections. Thus, the lower and upper limits of the amplitude of the MADPSK would be extensively investigated in the next phases of the research.

The electronic equivalent noise as seen from the input of the electronic pre-amplifier following the photodetector can be measured and taken into account to the total noise process caused by thermal noises of the input impedance, the biasing current shot noise and the noises at the output of the electronic pre-amplifier. These noises are combined with the signal-dependent quantum shot noises so as to gauge their contribution to the MADPSK receiver. Thus, we may consider new structures of electronic amplifiers or matched filter at the input of the receiver to achieve optimal MADPSK receiver structure.

For long-haul transmission systems, ASE of optical amplifier is probably the most important noise mechanism. In optical amplifiers, even in the absence of input optical signal, spontaneous emission always occurs stochastically when electron-hole pairs recombine and release energy in the form of light. This spontaneous emission is noise, and it is amplified by the optical amplifiers together with the useful optical signal and accumulated along optical transmission link [25].

Noises reduces the SNR, hence system BER and receiver sensitivity. Noise models also affect the design of optimum detection schemes such as decision thresholds. A thorough investigation of the noise mechanism and their impacts on multi-level signaling has never been reported, except some preliminary results for 10-Gb/s 4-ary ASK schemes [10]. Thus, all noise sources and the mechanism by which they affect the system performance must be thoroughly investigated. These noises are used to estimate the optimum decision level of the detection of the amplitude of the multilevel eye diagram.

7.4.6.2 Transmission Fiber Impairments

For optical signals, the transmission medium is an optical fiber with associated OAs and dispersion compensation devices, or a leased wavelength running on top of a DWDM system. Impairments are always part of the transmission medium, among them CD, PMD, and non-linearity, are critical.

When an optical pulse propagates along a fiber, its spectral components disperse due to the differential group delay (DGD) and the output pulse will be broadened. CD is proportional to the fiber length and the laser linewidth, especially the spectrum of the lightwave modulate signals. CD may cause optical pulses to overlap each other, thus leading to inter-symbol interference and increase system BER, especially for ASK systems. DPSK systems are more CD-tolerant. For MADPSK systems, the phase constellation, as shown in Figure 7.9, is rotating when the MADPSK is under the linear chromatic dispersion effect. It is also well known and developed in our model that this CD can be compensated by dispersion-compensating fiber modules. However, the mismatching of the dispersion slopes of the transmission and compensating fibers is very critical for multi-channel multi-level modulation schemes.

Optical pulse is also broadened by PMD, which is the time mismatching between two orthogonal polarizations of the optical pulse when they traverse along a fiber. In the ideal optical fiber having truly homogeneous glass and truly coaxial geometry of the core, the two optical polarizations would propagate with the same velocity. However, it is not the case for a real fiber, so the two polarizations have different speeds and will reach the fiber end at different times.

Similar to the CD effects, PMD can cause pulse overlapping and thus increase system BER. However, unlike CD, which is practically constant over time and can be in a large scale compensated, PMD is a stochastic process and cannot be managed easily. It is well known that PMD has the Maxwellian probability density function with a mean value (PMD) $= K_{\text{PMD}} \sqrt{L}$ where K_{PMD} is defined as PMD coefficient whose measured values vary from fiber to fiber in the, [0.01–1 ps/$\sqrt{\text{km}}$], and L is fiber length. Under the MADPSK, the signal-space of the constellations would be affected in the magnitude or phases by PMD, but expected to be dominated by the phase distortion. It is well known that the PMD first and second effects are critical for ASK modulation. For DPSK, it is expected that the principal axes of the polarization modes propagating through the fiber would be minimally affected. Thus, under hybrid amplitude-phase modulation scheme, there are several issues remained to be resolved. Under the MADPSK scheme, the delay of the polarization modes would generate the phase difference or phase distortion on I and Q components, hence enhancement of the distortion effects of the ISI. The amplitude distortion would then be increased but considered to be secondary effect.

7.4.6.3   Non-linear-Effects on MADPSK

Non-linear effects occur due to the non-linear response of the fiber glass to the applied optical power. Fiber non-linearity can be classified into stimulated scattering and the Kerr effect. Among several stimulated scattering effects, stimulated Raman scattering, caused by interaction between light and acoustical vibration modes in the fiber glass, is the most critical. Under this mechanism, optical signal is reflected back to the transmitter, and in WDM systems its power is also transferred from shorter to longer wavelengths, thus attenuating signal and causing crosstalk. The Kerr effect is the root of intensity-dependent phase shift of the optical field. It is shown in three forms: self-phase modulation (SPM), cross-phase modulation (XPM), and four-wave mixing (FWM) provided the phase matching is satisfied.

SPM is usually the dominant effect in a single-channel DPSK system. The changes in instantaneous power of optical pulses, together with the ASE from associated OAs, lead to intensity-dependent changes, the Kerr effect, in the guided medium refractive index, hence the effective index of the guided mode. These changes are converted to the phase shifts or phase noise of the lightwave carriers. At the receiver, the phase noise is transferred back to intensity noise, which degrades BER [26]. As mentioned above, the contribution of noises into different levels of the MADPSK scheme is very critical to determine optimum decision thresholds. This is further complicated by these additional non-linear effects, especially the non-linear phase noises (NLPN) usually contributed by the SPM due to the outer most constellation. These NLPN effects from the outermost constellation to other inner circle signal spaces have not been investigated.

XPM becomes the most critical non-linearity in WDM systems where the phase shifts (noise) in one channel comes from refractive index fluctuations caused by power changes in other channels. XPM becomes more pronounced when neighboring channels have equal bit rate [26]. FWM is basically a crosstalk phenomenon in WDM systems. When three wavelengths with frequencies, ω_1, ω_2 and ω_3 propagate in a non-linear fiber medium at which the dispersion is zero, they combine and create a degenerate fourth wavelength which would fall on the location of an active wavelength channel. If these parametric wavelengths fall into the spectral region of the other channels, they will cause crosstalk and the performance of the system. FWM, although is expected to reduce the receiver sensitivity, in the MADPSK system the impairment due to fiber non-linearity can be avoided if the maximum average power of optical signals is set not higher than the non-linear threshold. This maximum power dictates the amplitude of signal points in the outermost circle (circle 3), and hence other circles, of the signal space. Thus optimization of the signal amplitude levels for MADPSK is very critical.

7.4.6.4   Offset Detection

The 16-ary MADPSK signal model described in Section 7.2 can be modified. To balance the ASK and DQPSK sensitivities, ASK signal levels are preliminary adjusted to the ratio $I_3/I_2/I_1/I_0 = 3/2/1.5/1$ and rotate by $\pi/4$ [6] as shown in Figure 7.40. These level ratios can be determined from the SNR at each separation distance of the eye diagram or q-factor. The noise is assumed dominated by the beat noise between the signal level and that of the ASE noise. The eye opening is expected to improve significantly as shown in Figure 7.41. The signal constellations after the transmitter and that after propagation through the system are shown in Figure 7.42. Figure 7.43 shows the eye diagram at the output of a balanced receiver under two-level amplitude modulation scheme. A 16-ary RZ ADPSK has also been reported [27]. The measured BER curves of different modulation formats using 2 bits/symbol are shown in Figure 7.44.

Multi-level modulation can be combined with multi-carrier modulation techniques such as OFDM to significantly improve the tranmission performance [27–31]. A system arrangement for OFDM using multilevel modulation can be seen in Figure 7.45. Using different number of sub-carriers for OFDM with different launched power into the fiber would affect the OSNR quality as shown in Figure 7.46.

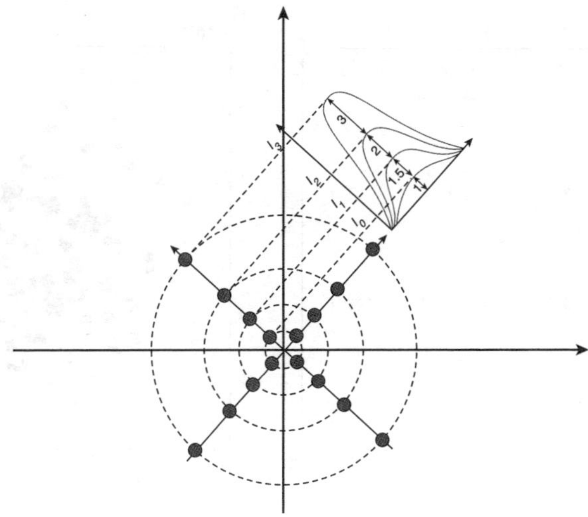

FIGURE 7.40   ASK inter-level spacing and offset modulation and detection line.

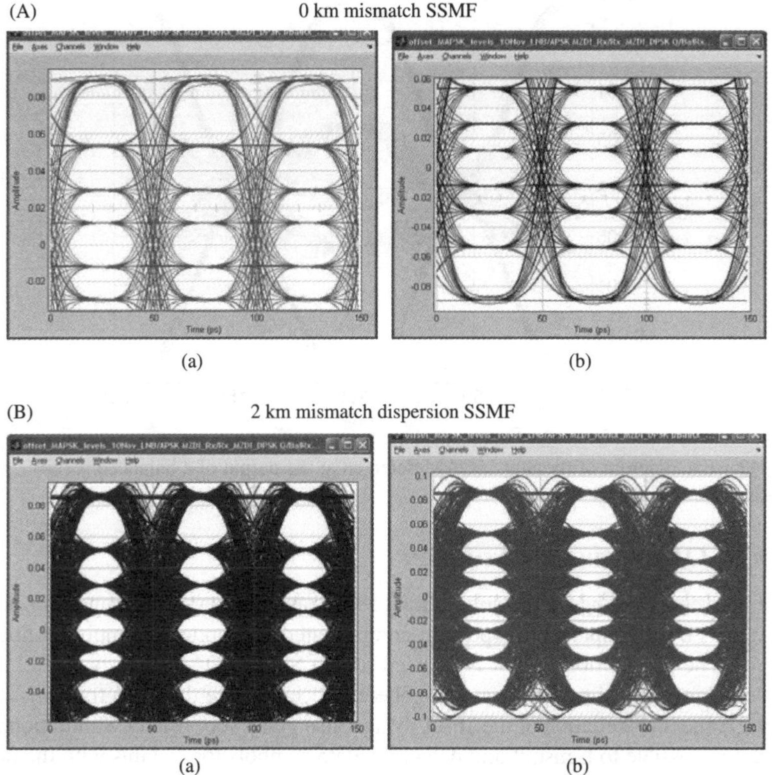

FIGURE 7.41   40-Gb/s MADPSK eye diagrams of the I (a) and Q (b) components (A) 0 km – back to back (B) 2-km SSMF mismatch over three 100-km SSMF transmission spans (dispersion compensated).

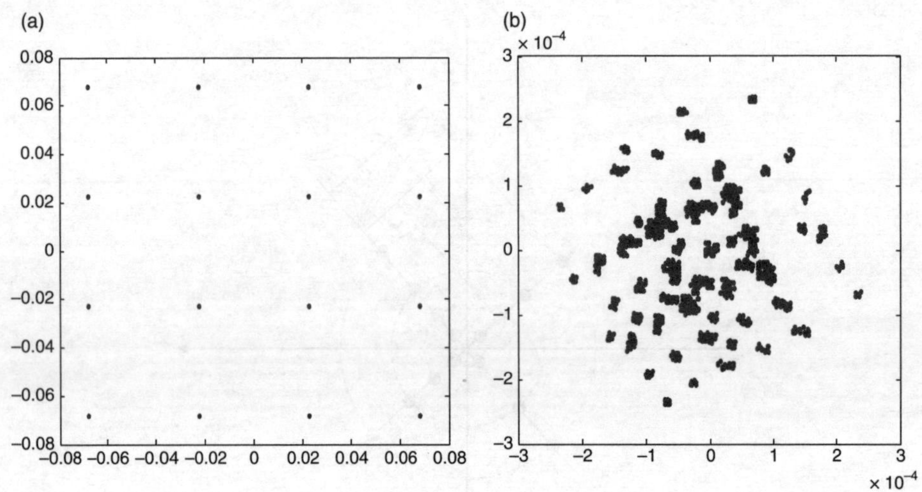

FIGURE 7.42 Constellation of 16-square QAM after two optically amplified spans and 2-km SSMF dispersion mismatch (a) pre-propagation and (b) post- propagation.

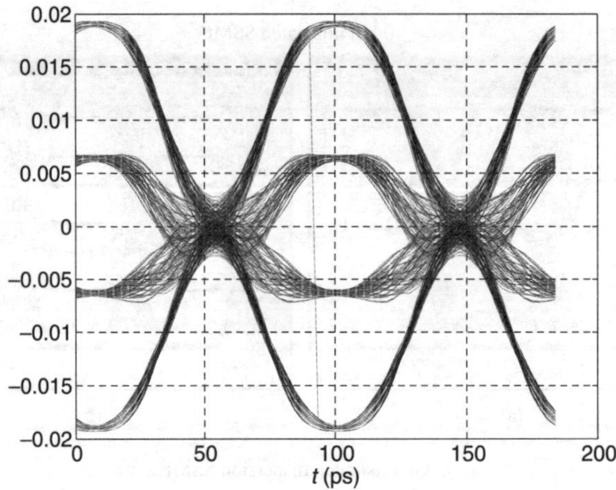

FIGURE 7.43 Eye diagram of the Star-16QAM detected at the output of a balanced receiver. Bit rate of 40 Gb/s and baud rate of 10 Gb/s.

7.5 STAR 16-QAM OPTICAL TRANSMISSION

This section gives a briefing on the simulation of the transmission performances of optical transmission systems over 10 spans of dispersion-compensated and optically amplified fiber transmission systems. The modulation format is focused on the Star 16-QAM with two-level and 8-phase state constellation. Optical transmitters and coherent receivers are the main transmission terminal equipments – other constellation of the 16-QAM are also given very briefly. Simulation results have shown that it is possible to transmit and detect the data symbols for 43 Gb/s with the possibility of scaling to 107 Gb/s without much difficulty. The OSNR with 0.1-nm optical filters is achieved with 18 dB and 23 dB for back-to-back and long-haul transmission cases with a dispersion tolerance of 300 ps/nm.

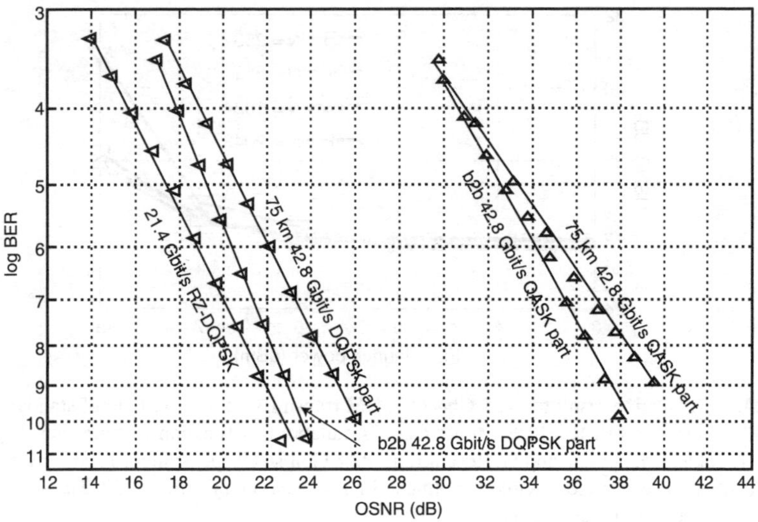

FIGURE 7.44 Measurement BER results for 16-ary inverse RZ ADPSK modulation.

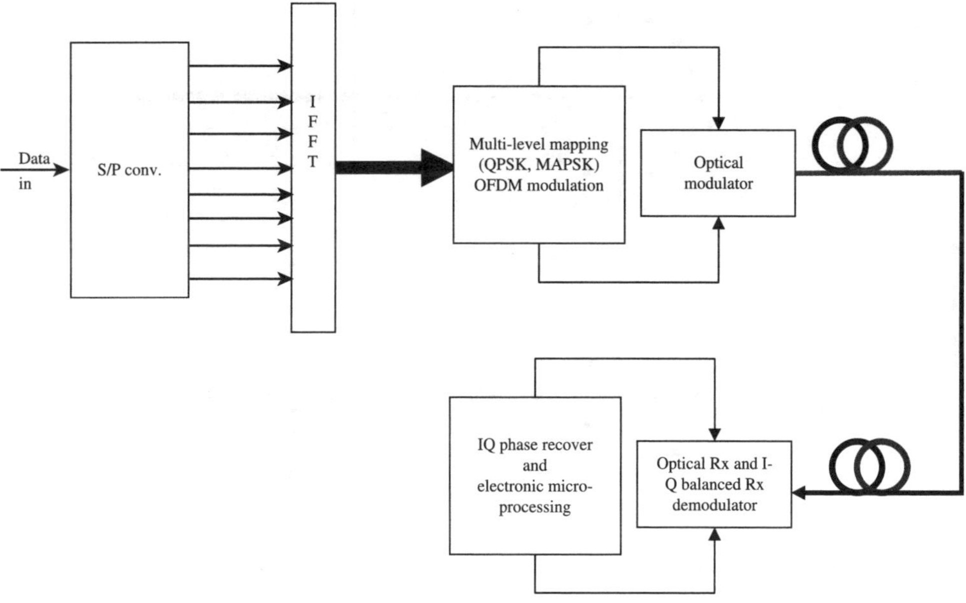

FIGURE 7.45 Schematic diagram of optical OFDM transmitter of a long-haul optical transmission system using multilevel modulation formats.

7.5.1 INTRODUCTION

To increase the channel capacity and bandwidth-efficiency in optical transmission, multi-level modulation formats like QAM formats are of interest [32–42]. In digital transmission with multi level (*M*-levels) modulation, m bits are collected and mapped onto a complex symbol from an alphabet with $M = 2m$ possibilities at the transmitter side.

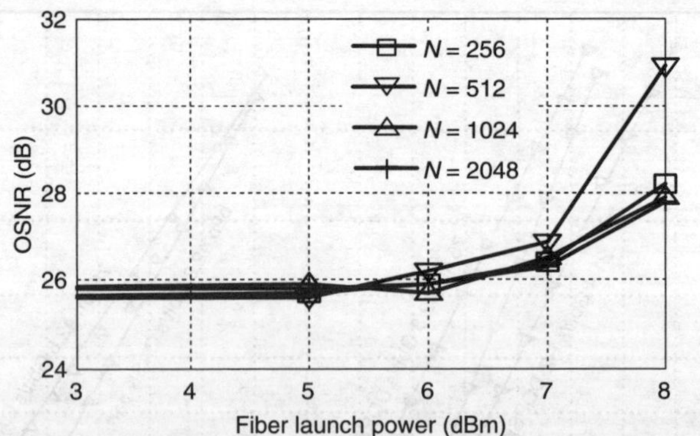

FIGURE 7.46   Simulation result for 42.7 Gb/s OOFDM transmission over 640 km of standard single-mode fiber; OSNR required for BER $= 10^{-3}$ (Monte Carlo simulation) as function of fiber launch power. (From Serbay, M., T. Tokle, P. Jeppesen, and W. Rosenkranz, 42.8 Gbit/s4 bits per symbol 16-ary inverse-RZ-QASK-DQPSK transmission experiment without polmux. In *Proceedings of OFC 2007, Paper OThL2*, March, Anaheim, USA, 2007. With permission.)

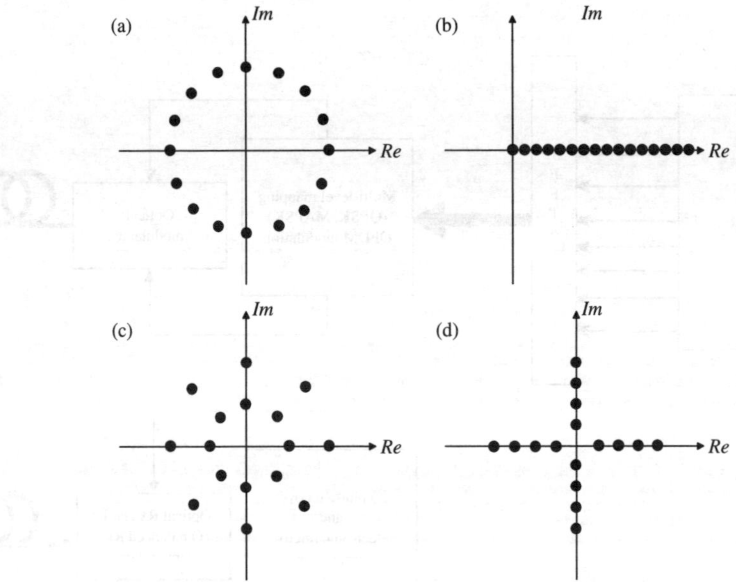

FIGURE 7.47   Constellation of symbols in the complex plane for (a) 16-DPSK; (b) 16-ASK; (c) Star 16-QAM; (d) 16ADPSK.

The symbol duration is $T_s = mT_B$ with T_B as the bit duration and the symbol rate is $f_s = f_B/m$ with $f_B = 1/T_B$ as the bit rate. This shows that for a given bit rate, the symbol rate decreases if the modulation level increases. That means higher bandwidth efficiency can be achieved by a higher-order modulation format. For 16-QAM format, $m = 4$ bits are collected and mapped to one symbol from an alphabet with $M = 16$ possibilities. In comparison to the case of binary modulation format, only $m = 1$ bit is mapped to one symbol from an alphabet with $M = 2$ possibilities. With 16-QAM format and a data source with a bit rate of $f_B = 40$Gb/s, only a symbol rate of $f_s = 10$ Gbaud/s is necessary. From a commercial viewpoint, it means a 40-Gb/s data rate can be transmitted with 10-Gb/s transmission devices. In binary transmission, the transmitter needs a symbol rate of $f_s = 40$ Gbaud/s. It means 16-QAM transmission requires four times slower transmission devices than that for the binary transmission. It is noted that 10.7 Gsymbols are used as the symbol rate so as to compare the simulation results with well-known 10.7 Gb/s modulation schemes such as DPSK, and CSRZ DPSK. For 107 Gb/s bit rate, the transmission performance (i.e., sensitivity and OSNR) can be scaled accordingly without difficulty.

This section gives a general approach regarding to the design and simulation of Star 16-QAM with two amplitude levels and eight phase states forming two star circles. We term this Star 16-QAm as 2A-8P Star 16-QAM, 2 amplitude – level and eight-phase states. The transmission format is discussed with theoretical estimates and simulation results to determine the transmission performance. The optimum Euclidean distance is defined for the design of star 16-QAM. Then, in the second section, the two detection schemes (direct detection and the coherent detection) for star-QAM constellations are discussed.

7.5.2 Design of 16-QAM Signal Constellation

There are many possibilities to design 16-QAM signal constellation. Three most popular constellations can be introduced. For 16-QAM modulation schemes are (i) Star 16-QAM; (ii) Square 16-QAM; and (iii) Shifted-square 16-QAM. The first two of these constellations are implemented. However, only the Start 16-QAM with two amplitudes and eight phases per amplitude level are employed in this section.

7.5.2.1 Signal Constellation

The signal constellation for star 16-QAM with Gray coding is shown in Figure 7.48. The binary presentation of the symbols in the figure is shown in mapping Table A.1 of Table 7.5. As can be seen form the figure, the symbols are evenly distributed on two rings and the phase difference between the neighboring symbols on the same ring are equal (pi/4). In order to detect a received symbol, its phase and amplitude must be determined. In other words, between two amplitude levels of the rings and among eight phase possibilities there are a number of ways to form this constellation.

The ring ratio (RR) for this constellation is defined as: $RR = b/a$ where a and b are the ring radii as shown in Figure 7.48. The RR can be set to different values to optimize the transmission performance.

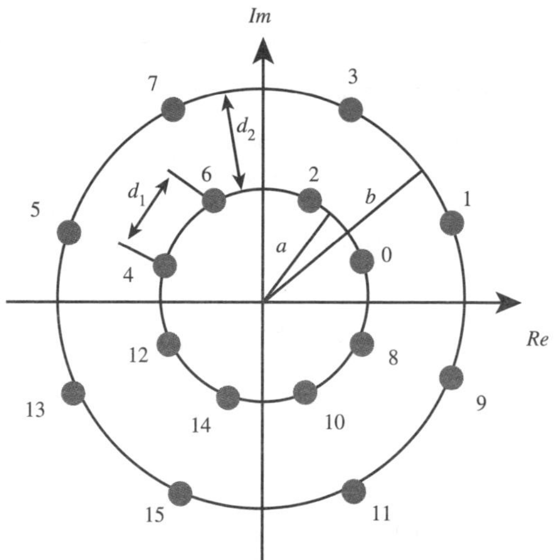

FIGURE 7.48 Theoretical arrangement of the modulation constellation for Star 16-QAM.

TABLE 7.5
Symbol Mapping and Coding for Star 16-QAM

Table A.1: Symbol to Bit Presentation

0→0000	4→0100	8 →1000	12→1100
1→0001	5→0101	9 →1001	13→1101
2→0010	6→0110	10→1010	14→1110
3→0011	7→0111	11→1011	15→1111

Table A.2: Gray Coding for Star 16-QAM

0→1	4→7	8 →15	12→ 9
1→0	5→6	9 →14	13→ 8
2→3	6→5	10→13	14→11
3→2	7→4	11→12	15→10

Table A.3: Mapping for star 16-QAM

0→1	4→ 9	8 →13	12→5
1→0	5→ 8	9 →12	13→4
2→3	6→11	10→15	14→7
3→2	7→10	11→14	15→6

Table A.4: Gray Coding for Square 16-QAM

0→12	4→11	8 →13	12→4
1→10	5→ 2	9 → 5	13→3
2→15	6→ 8	10→14	14→7
3→ 9	7→ 1	11→ 6	15→0

Table A.5: Mapping for Square 16-QAM

0→10	4→11	8 → 4	12→15
1→ 6	5→ 1	9 →14	13→ 3
2→ 5	6→ 2	10→13	14→ 0
3→ 9	7→ 8	11→ 7	15→12

7.5.2.2 Optimum RR for Star Constellation

From Figure 7.44, it can be seen that there are many possibilities to choose the RR for the star 16-QAM constellation. Here the theoretical best RR is defined to minimize the error probability in an AWGN channel by maximizing the minimum distance d_{min} between the neighboring symbols. The results for AWGN channel can be used approximately for optical transmission. For Star 16-QAM, the minimum distance d_{min} is maximized, when

$$d_1 = d_2 = b - a = d_{min}. \tag{7.13}$$

With some geometrical calculations it can be obtained that

$$d_{min} = 2a \sin(22.5°) \tag{7.14}$$

which leads to the optimal ring ratio of

$$RR_{opt} = b/a = (d_{min} + a)/a = (2a \sin 22.5 + a)/a \approx 1.77. \tag{7.15}$$

The average power of the star 16-QAM constellation can be determined as

$$P_0 = (8a^2 + 8b^2)/16 = (a^2 + b^2)/2 . \tag{7.16}$$

Thus, we have the relationship between the average optical power and the minimum distance between the two rings of the two amplitude levels as

$$d_{min} \approx 0.53(P_0)^{1/2}. \tag{7.17}$$

The obtained $RR_{opt} = 1.77$ does not depend on P_0 and is constant for each P_0 value. For an average power of 5 dBm (3.16 mW), $d_{min} = 2.98 \cdot 10^{-2}\sqrt{W}$, $a = 3.89 \cdot 10^{-2}\sqrt{W}$ and $b = 6.87 \cdot 10^{-2}\sqrt{W}$ are obtained.

7.5.2.2.1 Square 16-QAM

The signal constellation of the square 16-QAM with Gray coding is shown in Figure 7.49. The binary presentation of the symbols in the figure is shown in mapping Table A.1 in Table 7.5. In the constellation of the square 16-QAM, the 16 symbols have equal distance with direct neighbors and totally 12 different phases, i.e., three phases per quarter, distributed on three rings. The phase differences between neighboring symbols on the inner- and outer rings are equal ($\pi/2$) but the phase differences between neighboring symbols on the middle ring are different (37° or 53°). If the distance between direct neighbors in the square 16-QAM is rotated as $2d$, the average symbol power (P_0) of the constellation is:

$$P_o = 10 \cdot d_2. \tag{7.18}$$

For an average power of 5 dBm (3.16 mW), it can be computed that $d = 1.77 \cdot 10^{-2}\sqrt{W}$ and from it: $a = 2.5 \ 10^{-2}\sqrt{W}$, $b = 5.6 \cdot 10^{-2}\sqrt{W}$ and $c = 7.5 \cdot 10^{-2}\sqrt{W}$. In comparison with star 16-QAM, here the distances between the middle ring and the outer ring are much smaller. It means, to achieve the same BER, square 16-QAM needs a higher average power than star 16-QAM. The decision method for the square 16-QAM is more complicated than that for star 16-QAM. The signal constellation of the square 16-QAM is shown in Figure 7.50 in which three circles are drawn to include all

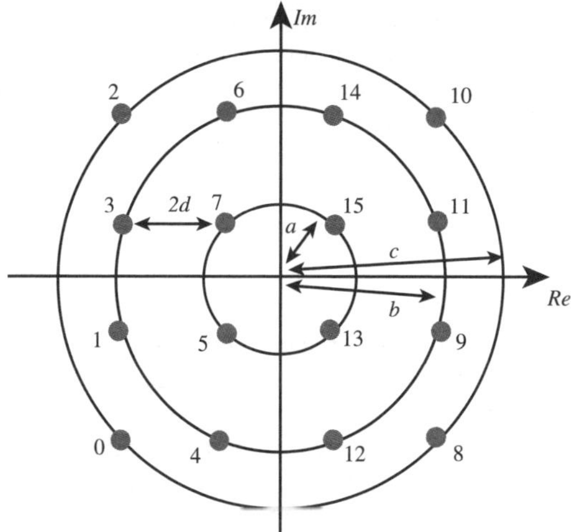

FIGURE 7.49 Square 16-QAM signal constellation.

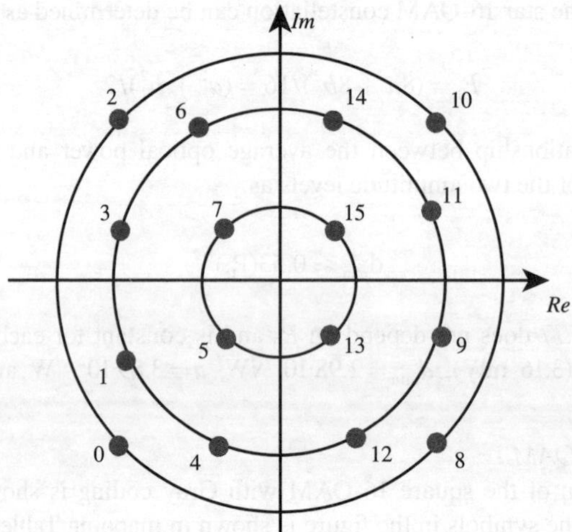

FIGURE 7.50 Square 16-QAM signal constellation.

points. Firstly, the decision between three amplitude possibilities of each ring should be made then, depending on the ring level, the decision is made between four or eight phase possibilities.

7.5.2.2.2 Offset-square 16-QAM

To optimize the phase detection of the middle ring, it is envisaged that the phase differences between neighboring symbols on the middle ring in square 16-QAM should be equal. Thus, the shifted-square 16-QAM is introduced by shifting (rotation) of symbols on the middle ring to obtain equal phase differences between all neighboring symbols, as shown in Figure 7.50. After shifting the symbols on the middle ring, the distances between all direct neighbors are not necessarily equal. In comparison with square 16-QAM, this constellation may offer more robust detection against phase distortions according to our amplitude and phase detection method introduced in later section.

7.5.3 Detection Methods

In differential encoding for the 16-QAM format, there are two detection methods to demodulate and recover the data in the receiver: (i) Direct detection; and (ii) Coherent detection.

In this section, direct detection means detection with MZDI or (2×4) 90°. Hybrid and coherent are similar, except that an local oscillator, a very narrow linewidth laser, is used to mix the signal and its lightwaves to generate the IF or base band signals with preservation of the modulated phase states. Each of these two receiving methods have different implementations which can be introduced as described below.

7.5.3.1 Direct Detection

In contrast to coherent detection, the differential decoding is done for direct detection in the optical domain. This is equivalent to a self-homodyne coherent detection. This has the disadvantage that the transmitted absolute phase is lost after differential decoding. However, the relative phase (phase of differential decoded signal) remains in electrical domain, and it makes the electrical equalization still possible. The equalization with relative phases is more difficult and the results are worse then equalization with absolute phases. The advantage of direct detection, compared with coherent detection, is that the synchronization of local laser with that of the signal light wave is omitted.

There are two methods to implement the direct detection. One is with MZDI and the other is with a (2×4) 90° hybrid coupler.

7.5.3.2 Coherent Detection

In a coherent receiver, a local oscillator (LO) is used to mix its signal with the incoming signal light wave for demodulation. As a result, it is possible to preserve the phase in the electrical domain. This makes electrical equalizing very effective in coherent detectors. For coherent detectors, the differential decoding is done in the electrical domain. Depending on the intermediate frequency (f_{IF}) defined as $f_{IF}=f_s-f_{LO}$ three different coherent methods can be distinguished: (i) Homodyne receiver; (ii) Heterodyne receiver; (iii) Intradyne receiver. Only homodyne receiver is included in this section and the other two are only briefly mentioned.

7.5.3.3 Homodyne Receiver

A receiver is called homodyne when the carrier frequency (f_s) and the LO frequency (f_{LO}) are the same. It means

$$f_{IF} = f_s - f_{LO} = 0. \tag{7.19}$$

In practice, because of the laser linewidth, a carrier synchronization must be implemented to set the center frequency and the phase of LO to the same values as those in the incoming signal. For homodyne receivers, carrier synchronization can be implemented in the optical domain via an optical phase locked loop (OPLL). Carrier synchronization failure causes degradation in receiver performance but, in this document, this effect is not considered and a perfect synchronization in the receiver (a perfect single spectrum line) is assumed. Alternatively, as mentioned below, heterodyne receiver using only one 90° hybrid coupler with associated electronic demodulation circuitry can be used to simplify the receiver configuration for coherent detection. Polarization control is another critical difficulty in coherent receivers, which is not included in this work either. The implementation of homodyne receiver for the star 16-QAM is described in several text books.

7.5.3.4 Heterodyne Receiver

For this kind of receiver, the following applies

$$f_{IF} = f_s - f_{LO} \neq 0 > B_{opt}. \tag{7.20}$$

B_{opt} is the optical bandwidth of the transmitted signal. The IF will be mixed in the electrical domain with a synchronous or asynchronous method in the low-pass domain. In synchronous demodulation, the phase synchronization can be done in the electrical domain. The implementation complexity of heterodyne receivers in optical domain is less than that of homodyne receivers.

7.5.3.5 Intradyne Receiver

Intradyne receiver requires

$$f_{IF} = f_s - f_{LO} \neq 0 < B_{opt}. \tag{7.21}$$

The phase synchronization in the intradyne receiver can be done in the digital domain. This makes the intradyne receiver less complex in the optical domain than the homodyne receiver. Intradyne receiver compared with the heterodyne receiver has the advantage that the processing bandwidth of intradyne receiver is smaller. The disadvantage of the intradyne receiver is the higher requirement on laser linewidth than that of heterodyne receiver.

7.5.4 Star 16-QAM Format

In this section, the transmission performance of star 16-QAM are evaluated by simulations and compared. The implementation of the transmitter for star 16-QAM is introduced in the first section. Direct and coherent homodyne receivers for star 16-QAM are described in the second section. In the case of the homodyne receiver (without laser phase noise and ideal frequency locking), two receiver models, one without phase estimation and the other one with phase estimation, are realized. The simulation results of all three receivers are shown in the third section and the comparison of them is in the fourth section.

7.5.4.1 Transmitter Design

There are many possibilities to implement the transmitter for star 16-QAM described in the previous section. For the simulations in this work, the parallel transmitter shown in Figure 7.51 is implemented. The bit stream enters the differential encoder module after serial-to-parallel converting.

The differential encoder provides the following processes: (i) The four bits that have parallel arrived at the module are mapped (Gray coding) into symbols according to the mapping Table A.2 in Table 7.1; (ii) The pre-coded symbols are differentially encoded (differential coding); and (iii) The differentially encoded symbols are mapped again to other symbols to drive the MZMs according to the mapping Table A.3 of Table 7.5.

Each symbol at the output of the differential encoder module is represented by four bits. The bits are sent to pulse formers. The first two bits drive the first two Mach–Zehnder modulators (MZM), with lightwaves generated from the continuous wave (CW) laser. If the input bit is equal to "1", then the output of MZM is "–1" and in other case the output of MZM is a "1" (after sampling). After combining the output signals of these two MZMs and considering the 90° phase delay in one arm, we obtain the QPSK signal shown in Figure 7.51.

The third bit from the differential encoder output drives a phase modulator (PM) to obtain the 8-PSK signal constellation from the QPSK signal. If this bit is equal to "1", then the QPSK symbol will rotate by pi/4. The 8-PSK signal constellation is shown in Figure 7.52. To achieve the two-level star 16-QAM signal constellation, another MZM is used to generate the second amplitude. If the

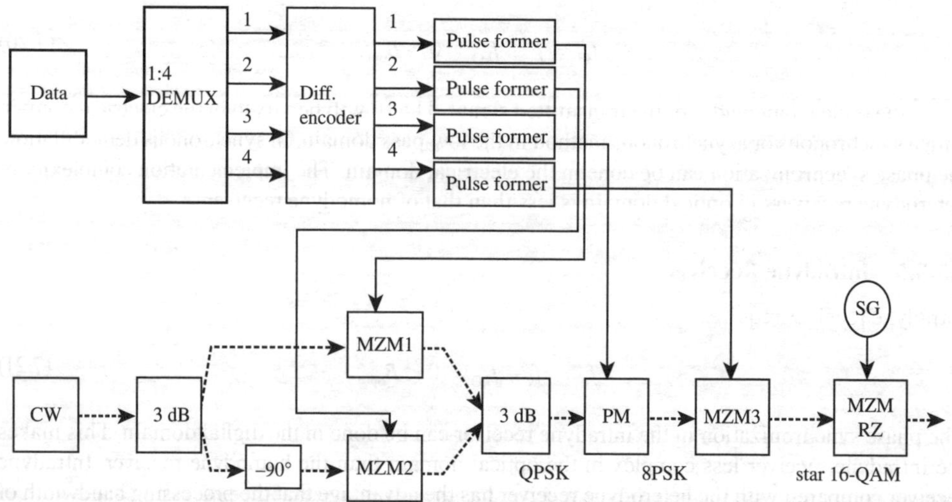

FIGURE 7.51   Optical transmitter for Star16-QAM (schematic).

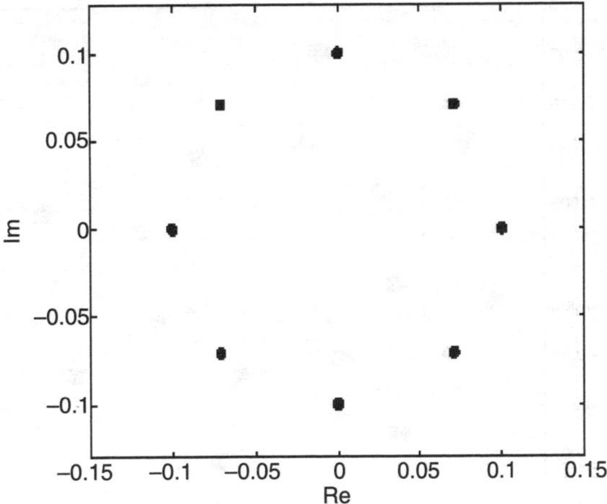

FIGURE 7.52   Constellation of the first amplitude level generated from the optical transmitter for Star 16-QAM.

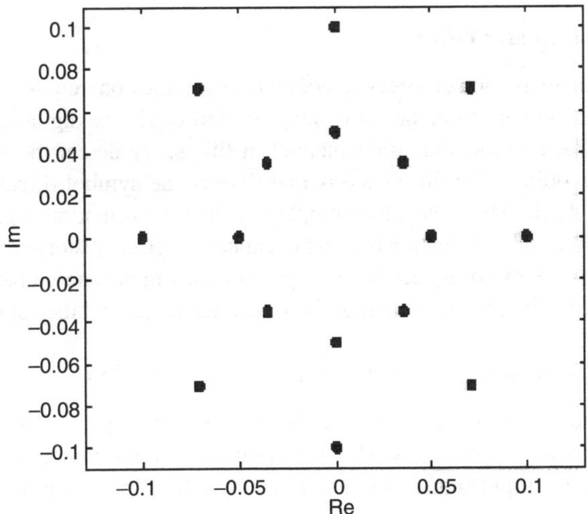

FIGURE 7.53   Constellation of the first and second amplitude level generated from the optical transmitter for Star 16-QAM.

fourth bit of differential encoder output is "1", then this output symbol is set on the outer ring of the constellation, otherwise on the inner ring. This MZM sets the RR of constellation. The signal constellation after MZM3 is shown in Figure 7.53. Furthermore, Figure 7.54 shows the signal constellation of the first and second amplitude levels at the receiver of Star 16-QAM after 10-spans of dispersion compensated standard SMF links.

The signal constellation in Figure 7.53 can be constructed from that the whole constellation in Figure 7.48 with a rotation of pi/6. The advantage of this rotation is that, on the real and imaginary axis of the constellation, only eight different amplitude levels instead of nine levels exist. So another PM can be used between MZM3 and MZM-RZ to rotate the constellation by pi/6°. This additional PM is not shown in Figure 7.51. To increase the receiver sensitivity and reduce the signal chirp, a return to zero (RZ) pulse curving with a duty cycle of 50% should be implemented at the end of the transmitter with a MZM driven by a sinus signal generator (SG). In our simulations, the MZMs in Figure 7.51 work in a push–pull operation, and the PMs are MZMs working as phase modulators.

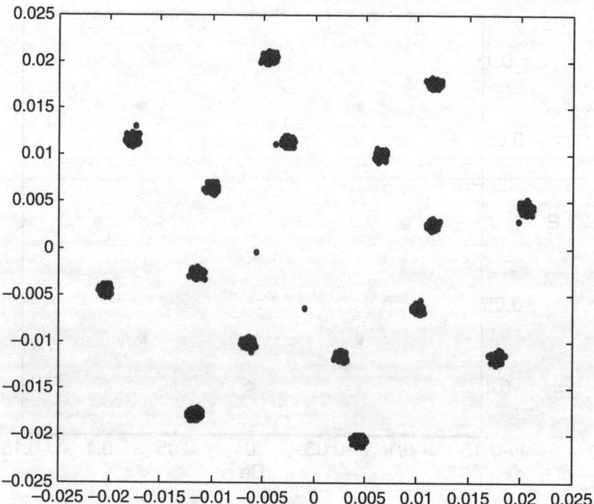

FIGURE 7.54   Constellation of the first and second amplitude level at the receiver of Star 16-QAM after 10-spans of dispersion compensated standard SMF links.

7.5.4.2   Receiver for 16-star QAM

In this section, the implementation of direct detection receiver and coherent receiver for star 16-QAM are explained. For coherent detection, there are many possibilities in the digital domain of the receiver to recover the data. The two methods implemented in this study detect the symbol before realizing the differential decoding. The difference is one detects the symbol directly using the method described in Section 4.2.1, while the other employs a phase estimation algorithm, described in Section 4.2.2, before the symbol detection in order to cancel the phase distortions (phase synchronization between LO and the received signal). Another possibility, which is not implemented in this work, is first carrying out the differential decoding of the incoming signal and then symbol detection.

7.5.4.3   Coherent Detection Receiver without Phase Estimation

The structure of a coherent detection receiver is shown in Figure 7.55a. After transmission over the fiber, the signal is amplified by an EDFA. The input power of EDFA can be changed via an attenuator to set the OSNR. The output signal of EDFA is sent to a band pass (BP) filter in order to reduce the noise bandwidth.

An attenuator is used to set the OSNR as required. The output signal of EDFA is sent to a BP filter in order to reduce the noise bandwidth. The signal from LO and the output of BP filter are sent to a (2×4) $\pi/2$-hybrid and after it to two balanced detectors. The (2×4) 90°-hybrid and the balanced detectors demodulate and separate the received signal into in-phase (I) and quadrature (Q) components. The structure of (2×4) $\pi/2$-hybrid coupler, the balanced detectors and their mathematical description can be found in may published works for coherent optical communication technology. This coherent detector can be simplified further if heterodyne detection is used as shown in Figure 7.55b and commercially available by Discovery Semiconductor*. However, the electrical signals should be demodulated into I and Q components as shown above rather than the balanced receiver of Discovery Semiconductor.

Furthermore an amplitude direct detection of in the electronic digital processor must be able to process the magnitude of the vector formed by I and Q components to decide the amplitude and

---

* C Wree et al. "Coherent Receivers for Phase-Shift Keying Transmission", Discovery Semiconductors Inc.

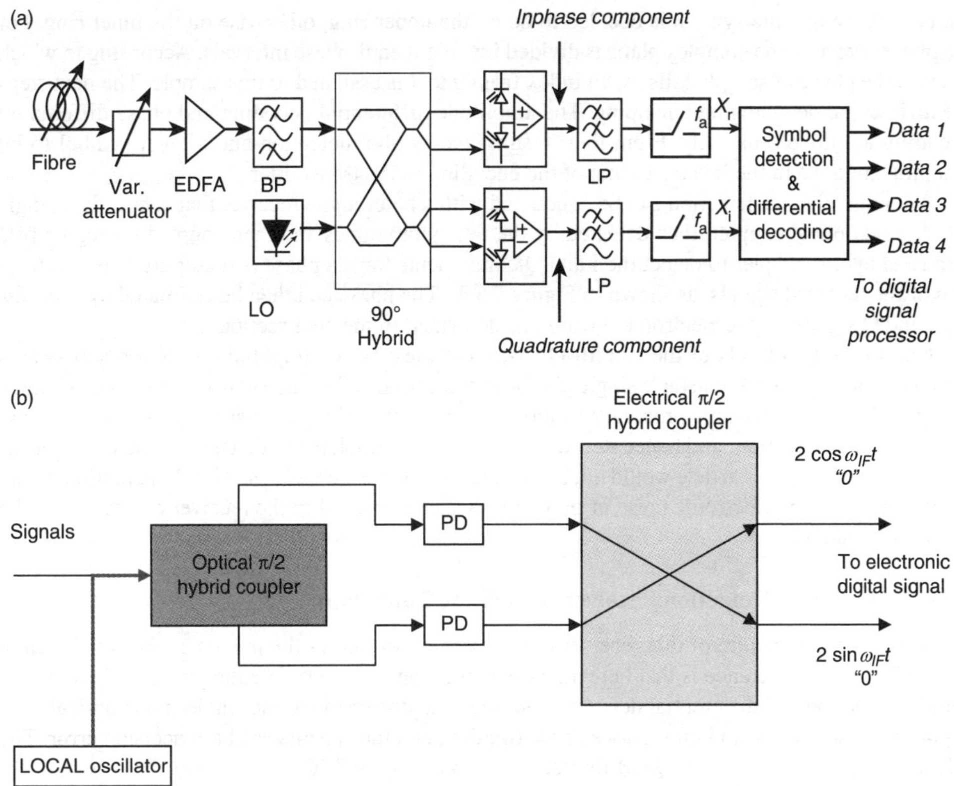

FIGURE 7.55 Coherent receiver for star 16-QAM (a) homodyne and heterodyne I and Q detection model (b) heterodyne model with optical and electrical pi/2 hybrid couplers–electrical detection of I and Q components.

phase of the received signals and hence their corresponding position on the constellation and the decoding of the data symbols.

After low pass (LP) filtering and sampling of I and Q components, the samples are sent to the symbol detection and differential decoding module. The sampling is done in the center of the eye diagram. In the symbol detection and differential decoding module, we firstly recover the symbols from the incoming samples, then make the differential decoding of symbols. In order to recover the symbols, I and Q components are added together to a complex signal. According to the original signal constellation, a decision must be made to which symbol our complex sample must be mapped. This decision has two parts. At first an amplitude decision is made to determine to which ring the sample belongs. After this, a phase decision takes place to determine to which of eight possible symbols on a ring our sample belongs.

For amplitude decision, a known bit sequence for receiver (training sequence) is used and an amplitude threshold a_{th} is defined according to

$$a_{th} = \{\max 1 \le k \le n |s_1(k)| + \min 1 \le k \le m |s_2(k)|\}/2 \qquad (7.22)$$

where $s_1(k)$ is the kth complex received samples of symbols on the inner ring and n is the whole number of symbols on the inner ring. $s_2(k)$ is the kth complex received sample of symbols on the outer ring and m is the whole number of symbols on the outer ring. If the amplitude of one sample is larger than the threshold, the symbol is decided to be on the upper ring, otherwise on the inner ring. For the phase decision, the complex plane is divided into eight equi-phase intervals. According to which interval the phase of sample falls in, an index from 0 to 7 is assigned to this sample. The next steps are differential decoding and mapping. The amplitude differential decoding and phase differential decoding are done separately. From their results, the symbol detection and after it symbol to bit mapping are done in the inverse to that of the encoding in the transmitter.

Alternatively, the detection can be conducted with a heterodyne receiver that uses only a single pi/2 hybrid optical coupler, then detected by the two photodiodes and then coupled through a pi/2 electrical hybrid coupler to detect the I and Q components for the phase and amplitude reconstruction of the received signals, as shown in Figure 7.55b. The phase can then be estimated by processing I and Q signals in the electronic domain, as described in the next section.

Due to the two levels of the star 16-QAM, there must be an amplitude detection sub-system which can be implemented using a single photo-detector followed by an electronic preamp, as shown in Figure 7.57. An electronic processor would be able to determine the position of the received signals on the constellation, and hence decoding can be easily implemented. The transmission performance presented in this article would not be affected. Only the technological implementation would be affected, and the electronic noise or optical noises contribution to the receiver can thus must be taken into account.

7.5.4.4 Coherent Detection Receiver with Phase Estimation

The method and structure of this receiver is almost the same as for the previous receiver shown in Figure 7.55. The difference is that here a phase estimation is done before the phase decision in the symbol detection and differential decoding module. The dispersion of the single-mode optical fiber is purely phase effects, and thus causes phase rotation, and thus results as phase decision error. The effect of dispersion on star 16-QAM format is shown in Figure 7.56.

The left plot (a) in the figure shows the sampled input signal into the fiber, and the plot on the right side shows the sampled output of the fiber with a chromatic dispersion of 300 ps/nm. The fiber is considered as linear in this simulation. Comparison of point A in both figures shows that this point is spread and rotated due to dispersion. The spreading causes phase and amplitude distortion

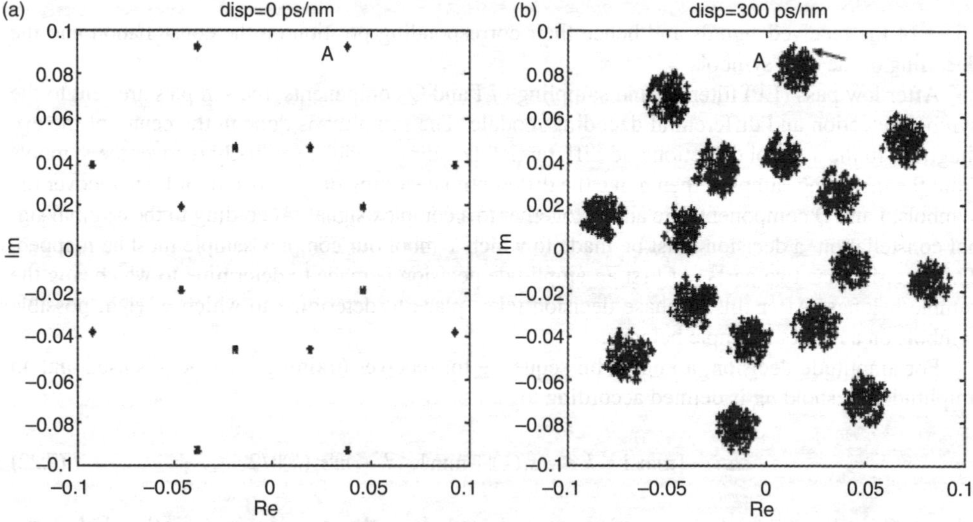

FIGURE 7.56 Signal constellation before (a) and after (b) propagation through the optical fiber link-phase rotation.

while rotation causes only phase distortion. To solve the problem of phase rotation, the following phase estimation method is implemented. Generally, this phase estimation method is for phase synchronization between LO with linewidth and signal to replace the optical phase lock loop.

The incoming signal after sampling can be described as

$$c(k) = Ae^{j(\Phi_{tot}(k)+\phi_{mod}(k))}. \tag{7.23}$$

Where Φ'_{tot} is the phase distortion due to the dispersion and noise and ϕ_{mod} is the signal phase that must be recovered. Now, ϕ_{tot} must be eliminated from $c(k)$. Φ_{mod} values are $\pi/8$, $3\pi/8$, $5\pi/8$, $\pi/8$, $9\pi/8$, $11\pi/8$, $13\pi/8$ and $15\pi/8$. If this phase values are multiplied by 8 then

$$c_8(k) = c^8(k) = A^8 e^{j(8'tot(k)+8\phi_{mod}(k))} = A^8 e^{j(8'tot(k)+\pi)} = -A^8 e^{j(8'tot(k))} \tag{7.24}$$

and from this

$$\phi'_{tot}(k) = 1/8 \arg(-c_8(k)). \tag{7.25}$$

$\Phi'_{tot}(k)$ is the estimated phase for $\phi'_{tot}(k)$. In the simulations in this work $\arg(c_8(k))$ is filtered (the filter makes average between 20 neighbor symbols) to avoid the phase jumps from symbol to symbol. The filter order of 20 is not optimized for each CD.

The signal phase $\phi_{mod}(k)$ can be estimated as

$$\phi_{mod}(k) = \arg(c(k)) - \phi'_{tot}(k) = \Phi_{mod}(k) + \phi'_{tot}(k) - \phi'_{tot}(k). \tag{7.26}$$

After this phase estimation, the signal decision takes place using the same method as for amplitude decision case.

7.5.4.5 Direct Detection Receiver

The block diagram of the direct detection receiver is shown in Figure 7.57. After the optical filter, the signal is split into two branches via a 3-dB coupler. We name these two branches: intensity branch and phase branch. In the latter, the phase differential demodulation is done in the optical domain. The signal and the delayed signal at Ts (symbol duration) are sent into the (2 × 4) 90° hybrid and after

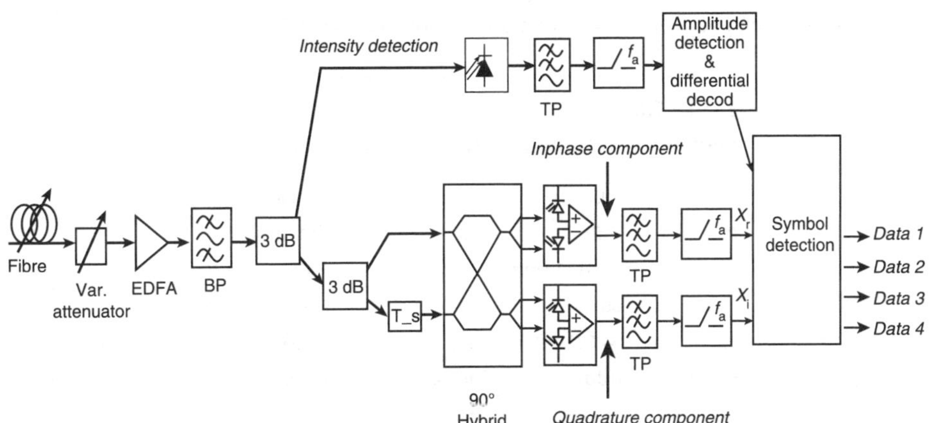

FIGURE 7.57 Direct detection receiver for star 16-QAM.

that again into balanced detectors. At the output of the balanced detectors, the in-phase and quadrature component of the demodulated and differential decoded and received signal can be derived. After electrical filtering, the signal is sampled and then sent into the symbol detection module. In amplitude branch, the amplitude is determined and differentially decoded. After the photodiode, the signal is low-pass filtered and sampled and then fed into the amplitude detection and differential decoding module, a well known optical coherent structure can be used to accomplish the amplitude decision and differential decoding. At the end, the in phase and quadrature component and the amplitude branch are send to the symbol detection module for further processes as symbol detection and symbol-to-bit mapping.

7.6 COHERENT RECEIVER WITHOUT PHASE ESTIMATION

In this section, the required OSNR at 10^{-4} BER (using Monte Carlo simulation) are determined for the coherent and incoherent direct detection receivers. For each detection method, the optimum RR is obtained to minimize the OSNR at $BER = 10^{-4}$ (BER is determined via Monte-Carlo simulations). The dispersion tolerance (DT) at 2-dB OSNR penalty at $BER = 10^{-4}$ is determined. The OSNR penalty is defined as the OSNR difference in dB between the OSNR of back-to-back case and the OSNR of other CD values. DT is the CD interval that can be achieved with a certain OSNR penalty. In practice, DT describes how much dispersion (residual dispersion) a system can tolerate with a OSNR penalty <2 dB. The simulations in this work are done via the simulation tool. The simulations are done for the linear channel and for the non-linear channel. The simulation parameters are given in Table 7.6. The average input power of non-linear fiber in this work is always 5 dBm.

7.6.1 LINEAR CHANNEL

In Figure 7.58b, the optimum RR (RR_{opt}) can be seen for each CD. RR_{opt} is the RR, which minimize the OSNR for the given CD. The optimum RR changes here is nearly linear with CD and can be expressed as

$$RR_{opt} = -0.002 \mid CD \mid + 1.92 \text{ for } 50 \text{ ps/nm} \leq \mid CD \mid \leq 300 \text{ ps/nm}. \tag{7.27}$$

RR_{opt} increases with CD because increasing of CD means increasing of phase rotation due to dispersion. This causes more phase detection errors and thus a higher OSNR requirement. Increasing of RR reduces the phase error probability, but increases the amplitude error probability. RR_{opt} is the best trade-off between phase errors and amplitude errors for each CD.

TABLE 7.6

Simulation Parameters For 16-STAR QAM

f_s	=	10.7 GHz	λ_c	=	1550 nm
P_{laser}	=	7 dBm	α_{att}	=	0.21 dB/km
γ	=	0.00137 1/W/m	F_n	=	5 dB
G_{EDFA}	=	30 dB	fil-opt.	=	Gaussian 1. order
B_{opt}	=	44 GHz	λ_{lo}	=	1550 nm
P_{LO}	=	0 dBm	fil-el.	=	Butterworth 3. degree
B_e	=	11 GHz	$R_{photodiode}$	=	1A/W

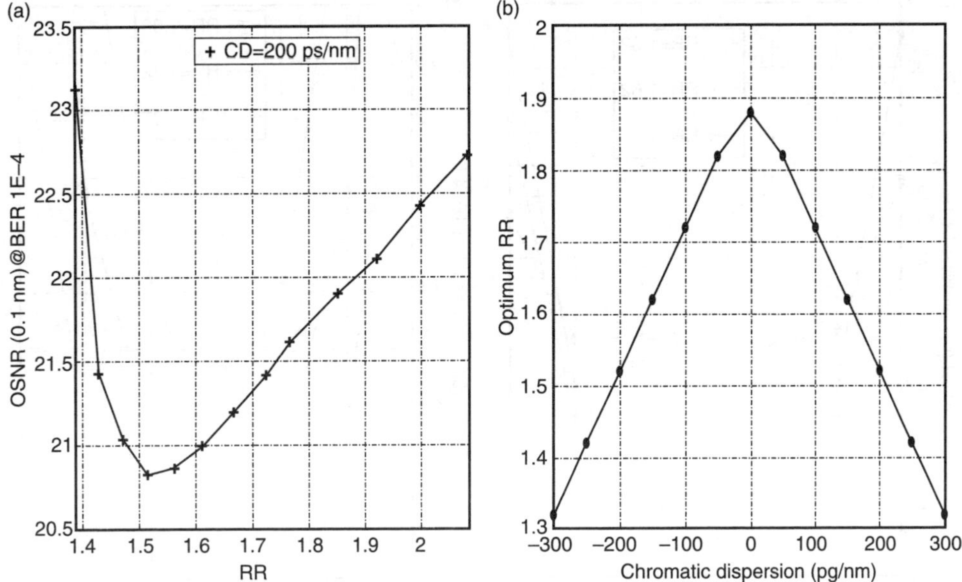

FIGURE 7.58   OSNR vs. RR (left) (a) and optimum RR vs. CD (right) (b) for coherent receiver without phase estimation.

In the back-to-back case, RR_{opt} is about 1.87. The theoretical value for RR_{opt} is 1.77. The difference is because in an introduced coherent receiver, at first the symbol detection is done and then the differential decoding. In differential decoding before symbol detection, the optimum RR is about 1.77, as expected. To determine the RR_{opt} for each CD, RR is changed for each CD. The RR value, which yields the smallest OSNR is RR_{opt}. The RR step (determines RR accuracy) in simulations is 0.05. The characteristic for other CDs is similar to Figure 7.58a. To compare the OSNRs, the reference in this work is the OSNR from the back-to-back case. In case of residual dispersion is of interest, how the system performance changes with the change of RR. The simulation results of three RR can be seen in Figure 7.57. As shown in Figure 7.57a, the OSNR performance for the back-to-back case is decreased if (in simulated interval) the RR from 1.87 decreases. A degradation of 6.5 dB is determined if the RR from 1.87 decreases to 1.32. From the other side, the dispersion tolerance (DT) at 2 dB OSNR penalty increases (Figure 7.9b). The DT for RR = 1.87 is 220 ps/nm, and is 460 ps/nm for RR = 1.32. To understand the reason for this behavior, Figure 7.58a should be considered again. For each CD, if the RR decreases from the RR_{opt} value, the OSNR increases rapidly. This means that for a CD of 300 ps/nm and a RR of 1.32 the required OSNR is lowest for the back-to-back case. This OSNR value is further reduced if phase estimated is used. This effect causes a larger DT at a certain OSNR penalty. In a practical system according to higher requirement for OSNR or DT, the RR can be chosen.

In Figure 7.59a on the left, the required OSNR for |CD| = 300 ps/nm makes a jump compared with other CDs. With |CD| = 350 ps/nm, it is not possible to reach a BER of 10^{-4}. The reason is that |CD| = 350 ps/nm is the limit of the system. For this CD, it is not possible to transmit error-free even without noise due to phase rotations caused by dispersion (phase detection error). The signal constellation of received signal after sampling with CD = 350 ps/nm can be seen in Figure 7.60. For example, some of received symbols of A are over the phase threshold line and they generate the detection errors. A typical eye diagram at the output of the coherent receiver at the limit of the distortion is shown in Figure 7.61. Phase estimation can be implemented in the digital signal processor.

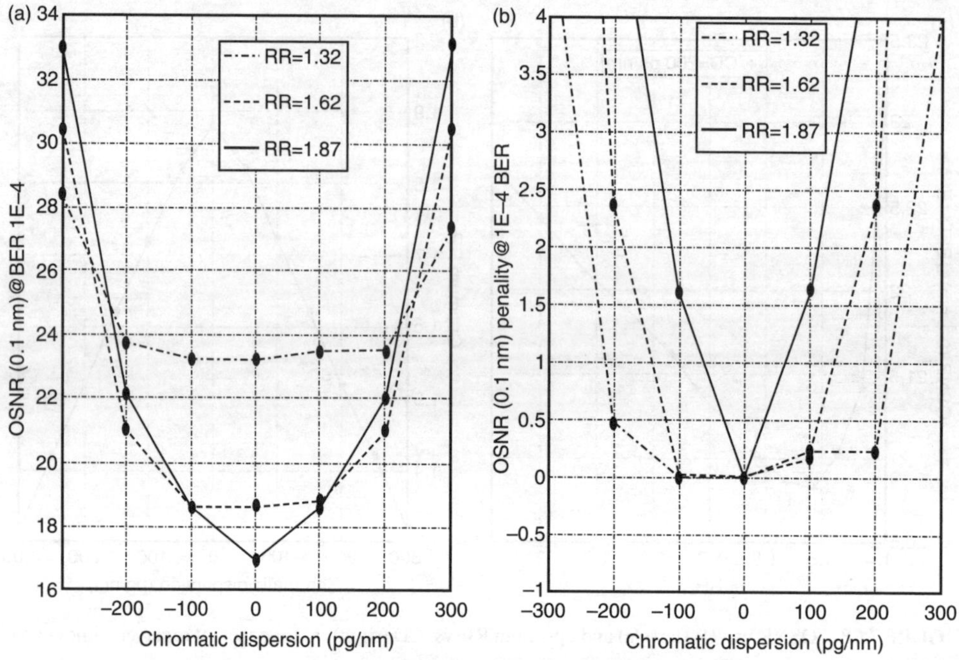

FIGURE 7.59 OSNR vs. CD (a) and OSNR penalty vs. CD (b) for coherent receiver.

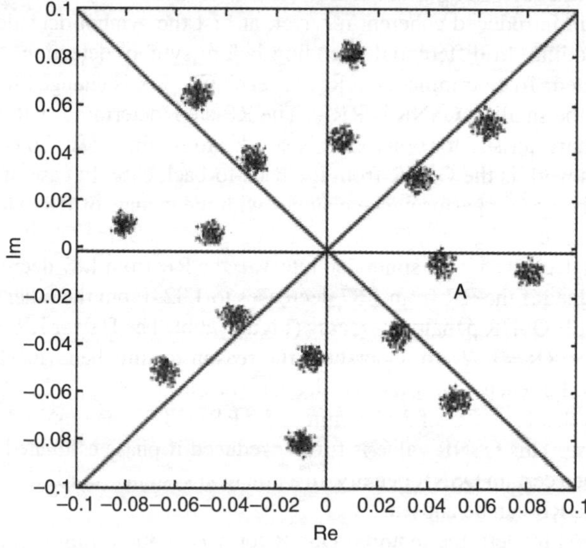

FIGURE 7.60 Received signal constellation with CD= 350 ps/nm without hout noise for coherent receiver without phase estimation in linear channel.

The signal constellation is rotated uniformly due to the property of the single-mode fiber that is purely phase distortion, and thus in the processing of the constellation it is best if the reference frame of the phasor diagram is rotated to align with the constellation, thus it may simplify the phase estimation process at $\pi/8$ and its multiple values.

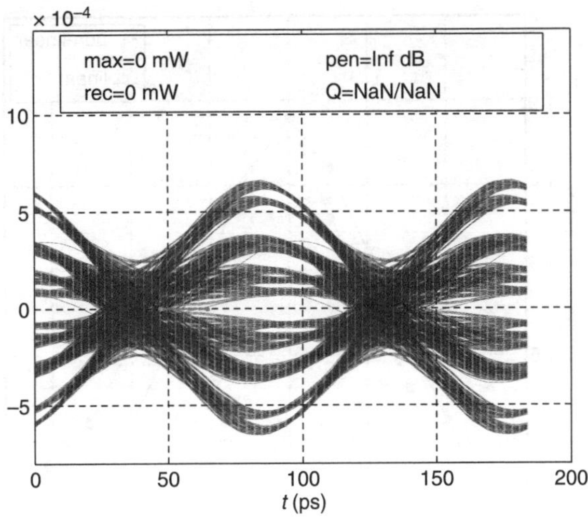

FIGURE 7.61   Typical eye diagram at the output of the coherent receiver under significant distortion limit of the 2A-8P star 16-QAM.

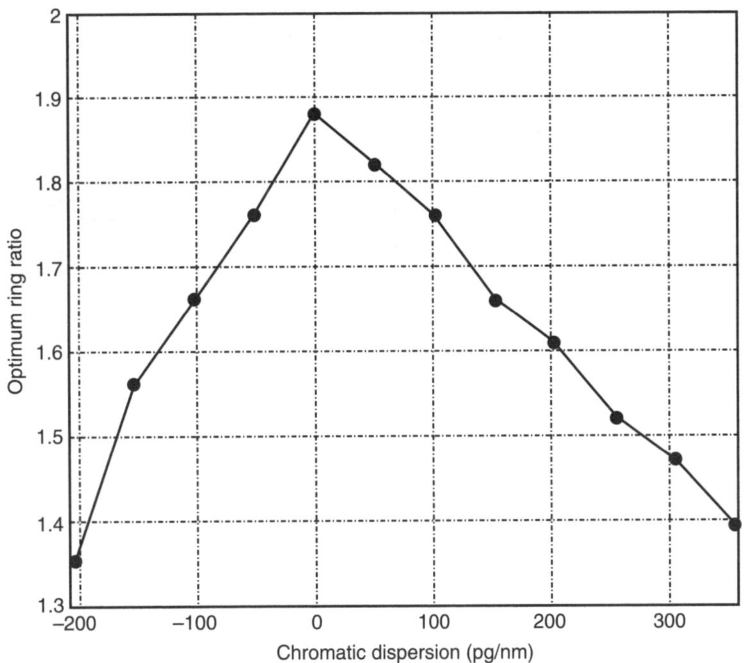

FIGURE 7.62 RR_{opt} vs. CD in non-linear channel for coherent receiver without phase estimtion.

7.6.2 Non-linear Effects

The optimum RR for the non-linear channel for different CDs is shown in Figure 7.62. As mentioned for the linear channel, the RR_{opt} also decreases with increase of CD. The difference with regard to Figure 7.58 is that the diagram is not symmetric. The reason of this is the interaction between dispersion and SPM [43–52]. For a positive CD, the dispersion and SPM have a constructive interaction, which

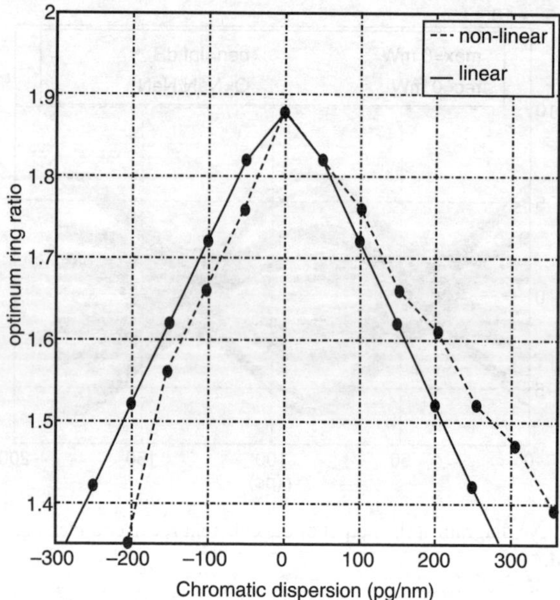

FIGURE 7.63　RR_{opt} comparison in linear and non-linear channel for coherent receiver without phase estimation.

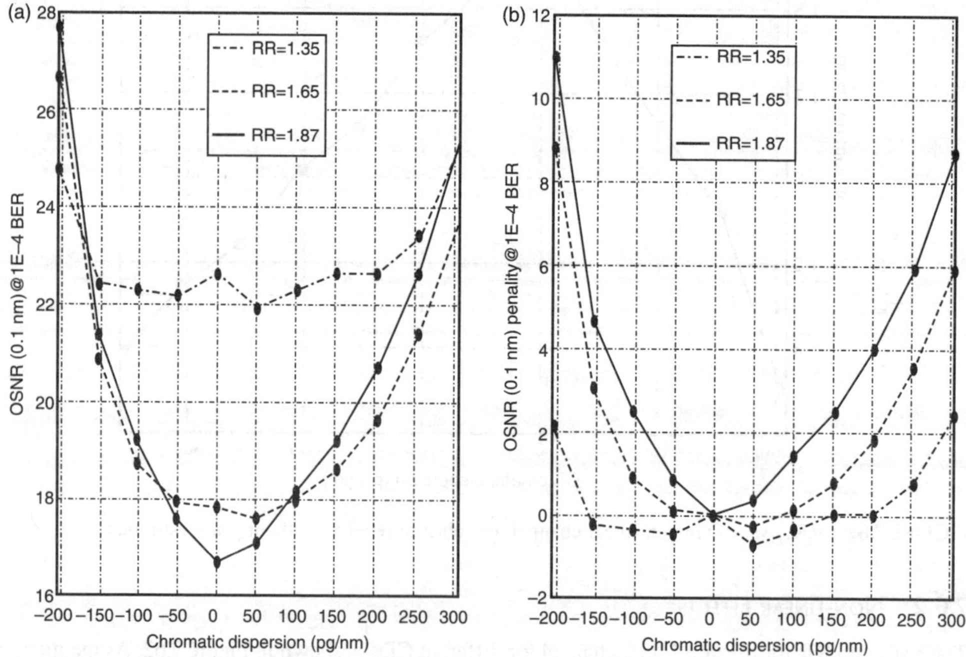

FIGURE 7.64　OSNR vs. CD (a) and OSNR penalty vs. CD (b) for coherent receiver without phase estimation in non-linear channel. Note–no equalization only phase estimation processing of I and Q receives components in electrical domain.

results in a better OSNR performance. For a negative CD, the dispersion and SPM have destructive interaction, and it results in a much worse OSNR performance [53–57].

This effect can be seen if Figure 7.58b (right side) is compared with Figure 7.62 as shown in Figure 7.63. The curve slope for negative CD is higher in non-linear channel. It means the phase distortion is higher in the non-linear channel. For the positive CD, the slope in non-linear channel is lower and it means the phase distortion is less. The simulation results for different RR are shown in Figure 7.64. The required OSNR for the back-to-back case increases with decrease of RR. Similarly with the same reason as for the linear case, the dispersion tolerance at 2dB OSNR penalty increases as well.

7.7 REMARKS

The design of a Star 16-QAM modulation scheme is proposed for coherent detection for ultra-high speed ultra-high capacity optical fiber communications schemes. Two amplitude levels and eight phases (2A-8P 16_QAM) are considered to offer simple transmitter and receiver configurations and at the same time the best receiver sensitivity at the receiver. An optical SNR of about 22 dB is required for the transmission of Star 16-QAM over optically amplified transmission dispersion–compensated link. Dispersion tolerance of 300 ps/nm is possible with 1-dB penalty of the eye-opening at 40 Gb/s bit rate or 10 Giga-symbols/sec. with an OSNR of 18 dB. AN OSNR could be about 22 dB for 107 Gb/s bit rate and a symbol rate of 26.75 G symbols/sec. The transmission link consists of several spans of total 1000 km dispersion compensated optically amplified transmission link. Optical gain of the in-line optical amplifiers are set to compensate for the attenuation of the transmission and compensating fibers with a noise figure of 3 dB.

The optical transmitter and receivers incorporating commercially available coherent receiver are structured and sufficient for engineering of the optical transmission terminal equipment for the bit rate of 107 Gb/s and a symbol rate of 26.3 Gsymbols/sec.

Furthermore, electronic equalization of the receiver phase shift keying signals can be done using blind equalization that would improve the dispersion tolerance much further. For a symbol rate of 10.7 Gb/s, this dispersion tolerance for 1-dB penalty would reach 300 km of standard SMF. This electronic equalization can be implemented without any difficulty at 10.7 Gsymbols/sec. For 107 Gb/s bit rate, similar improvement of the dispersion tolerance can be at 26.5 Gsymbol/sec provided that the electronic sampler can offer more than 50 Gig-sampling rate.

7.8 OTHER MULTILEVEL AND MULTI-SUBCARRIER MODULATION FORMATS FOR 100 GB/S ETHERNET TRANSMISSION

Numerous technologies have been introduced in recent years to cope with the ever-growing demand for transmission capacity in optical communications. Although the optical single-mode fiber offers enormous bandwidth in the order of magnitude of 10 THz, efficient exploitation of bandwidth started to become an issue two years ago. Moreover, limited speed of electronics and electro-optic devices such as modulators and photo receivers are considered as bottleneck for further increase of data rate based on binary modulation.

With the arguments presented above and previous chapters, optical modulation formats using multi-level with a high number of bits per symbol will lower the demands on the high speed electronics. They would offer the essential ingredients for next generation ultra-high speed optical transmission systems and networks.

Based on the demand for transmission technologies offering high ratio of bits per symbol, two promising candidates to achieve a data rate of 100 Gb/s per optical carrier are discussed, namely optical orthogonal frequency division multiplexing and 16-ary multilevel modulation.

In this section, two promising candidates, namely optical orthogonal frequency division multiplexing and 16-ary multilevel modulation, are introduced. Performance is analyzed by means of numerical simulation and by experiment.

7.8.1 Remarks on Multilevel Modulation

Optical modulation formats incorporating four or eight bits per symbol were investigated in numerous contributions in the last two years (e.g., DQPSK [58] and 8-DPSK [59]). However, to carry out the step from 10 Gb/s to 40 Gb/s data rate using devices designed for 10 Gsymb/s, a number of 16 bits are required per symbol. The main challenge is to find the optimal combination of ASK and DPSK formats. Several approaches, which are reviewed in [28], are possible, as discussed below.

The simplest structure can be an extension of a 30-Gb/s 8-DPSK by an additional phase modulator, resulting in 40 Gb/s 16-DPSK. That is, in the complex plane, 16 symbols are placed onto a unit circle, as shown in Figure 7.46a. Depending on the current bit at the data input of the additional phase modulator, the 8-DPSK symbol is shifted by $\pi/8$ in case of a "1", while in case of a "0", the incoming phase of the symbol is preserved. Although it seems simple, experimental implementations have shown that the phase stability of the modulators and corresponding demodulators is very critical. Thus, experimental setup must be stabilized. Moreover, 16-DPSK is sub-optimal regarding exploitation of the full area that the complex plane offers. That is, the ratio of symbol distance and signal power is low, resulting in poor receiver sensitivity.

Similar behavior can be found for the other extreme case of 16-ASK, as shown in Figure 7.46(b). Here, the 16 symbols are placed onto the positive real axis. The symbol distance is extremely narrow, thus also resulting in poor sensitivity.

Much improved performance can be achieved by combining the amplitude and phase shift keying modulation of ASK and DPSK, the M-ary ADPSK as described above. There are a number of combinations of the M-ary ADPSK. One approach can be the extension of a 8-DPSK by two rings of ASK levels. Thus, the 16 symbols appear as two rings in the complex plane with eight symbols per ring, as shown in Figure 7.46c. Alternatively, this topology is known as star-16 QAM. A second structure is given by combining the DQPSK with four-level ASK (or equivalently M-ary ADPSK), resulting in four rings with four symbols each as analyzed in above sections and shown in Figure 7.46d). Both structures utilize effectively the complex plane. However, both structures require that the sensitivity or the magnitude (diameters) of the rings have to be optimized to compromise the sensitivity performance of the DPSK and the ASK geometrical distribution. Especially, the inner ring has to be of sufficient size to enable distinction of the different phases of the symbols on this ring. They are limited by the non-linear effects of the transmission fiber and the noises contributed by the receiver and the in-line optical amplifiers. Hence, the distance between the constellations are limited by these two limits.

A strategy to mitigate this trade-off was introduced in [58] by using a special pulse shaping called inverted RZ. For binary ASK in conjunction with inverse RZ, a "0" is encoded as temporary breakdown of the optical power, while for a "1" the optical power remains at high level. Using this pulse shaping, for the M-ary ADPSK format, the four levels of the QASK part are transmitted by means of four different values for temporary decay of optical power. The DQPSK part, however, is transmitted by modulating the phase of the signal in the time slot between, implying that in the transmitter the phase of the signal can be detected while the signal has maximum power.

Measurement results for this modulation format are depicted in Figure 7.44, where the BER is plotted as a function of the OSNR measured within a 0.1 nm optical filter bandwidth. The main outcome is that the DQPSK-part is insignificantly disturbed by an additional QASK part. Moreover, even after transmission over 75 km of standard single-mode fiber, the DPSK part shows a very low penalty. In contrast, the QASK component inherently shows low performance due to low symbol Euclidean distance. Improvement might be achievable by optimizing the duty cycle of inverse RZ and the ring ratio. A simulated eye diagram of the 16 square QAM is shown in Figure 7.43 with the constellation before and after the optical transmission link of two optically spans with 2-km SSMF mismatch shown in Figure 7.42.

7.8.2   Optical Orthogonal Frequency Division Multiplexing (oOFDM) [27]

OFDM is a transmission technology that is primarily known from wireless communications and wired transmission over copper cables [29]. It is a special case of the widely known frequency division multiplexing (FDM) technique for which digital or analog data is modulated onto a certain number of carriers and transmitted in parallel over the same transmission medium. The main motivation for using FDM is that due to parallel data transmission in frequency domain, each channel occupies only a small frequency band. Signal distortions originating from frequency-selective transmission channels, the fiber chromatic dispersion, can be minimized. The special property of OFDM is characterized by its very high spectral efficiency. While for conventional FDM, the spectral efficiency is limited by the selectivity of the bandpass filters required for demodulation, OFDM is designed such that the different carriers are pair-wise orthogonal. This way, for the sampling point, the inter-carrier interference (ICI) is suppressed although the channels are allowed to overlap spectrally.

Orthogonality is achieved by placing the different RF-carriers onto a fixed frequency grid and assuming rectangular pulse shaping. It can be shown that in this special case the OFDM signal can be described as the output of a discrete inverse Fourier transform with the parallel complex data symbols as input. This property has been one of the main driving aspects for OFDM in the past since modulation and demodulation of a high number of carriers can be realized by simple digital signal processing (DSP) instead of using many local oscillators in transmitter and receiver. Recently, OFDM has become an attractive topic for digital optical communications [28,30,31,58,59]. It is just another example of the current tendency in optical communications to consider technologies that are originally known from classical digital communications. Using OFDM appears to be very attractive since the low bandwidth occupied by a single OFDM channel increases the robustness towards fiber dispersion drastically, allowing the transmission of high data rates of 40 Gb/s and more over hundreds of kilometers without the need for dispersion compensation [32]. In the same way as for modulation formats like DPSK or DQPSK that were introduced in recent years, the challenge for optical system engineers is to adapt a classical technology to the special properties of the optical channel and the requirements of optical transmitters and receivers.

Thus, two approaches have been reported recently. An intuitive approach introduced by Lorente et al. [59] makes use of the fact that the wavelength-division multiplexing (WDM) technique itself already realizes data transmission over a certain number of different carriers. By means of special pulse-shaping and carrier wavelength selection, the orthogonality between the different wavelength channels can be achieved resulting in the so-called orthogonal WDM technique (OWDM). However, this way the option of simple modulation and demodulation by means of discrete Fourier transforms (DFT) cannot be utilized as this kind of digital signal processing is not available in the optical domain.

An alternative and popular method is generation of an electrical OFDM signal by means of electrical signal processing followed by modulation onto a single optical carrier [28,30,31]. This approach is known as optical OFDM (oOFDM). Here, the modulation is a two-step process: first, the electrical OFDM signal already is a broadband bandpass signal which is then modulated onto the optical carrier. Second, to increase data throughput, oOFDM can be combined with WDM, resulting in multi-Tb/s transmission system, as shown in Figure 7.45. Nevertheless, oOFDM itself offers different options for implementation. An important issue is optical demodulation that can be realized by means of direct detection (DD) or coherent detection (CD) using a local oscillator. The DD is preferable due to its simplicity. However, for DD the optical intensity has to be modulated. Due to the fact that the electrical OFDM signal is quasi-analog with zero mean and high peak-to-average ratio, most of the optical power has to be wasted for the optical carrier (i.e., an additional DC-value of the complex baseband signal) resulting in low receiver sensitivity. For CD, in addition the bandwidth efficiency is twice as high as for DD since for pure intensity modulation inherently a double-sideband signal is generated. For CD, a complex optical I-Q modulator composed of two real modulators in parallel followed by superposition with $\pi/2$ phase shift allows for transmission of twice as much data within the same bandwidth. For intensity modulation, the bandwidth efficiency may be increased by suppressing one of the redundant sidebands, resulting in optical OFDM with single-sideband (SSB) transmission. Firstly the input serial data sequence is converted to parallel streams. These parallel data sequences are then mapped to QAM constellation in the frequency domain, then by IFFT converted back to the time domain. The time domain signals are in the I- and Q- components, which are then fed into I and Q optical modulator. This optical modulation can be DQPSK or any other multilevel modulation sub-system. At the end of the optical fiber transmission, I and Q components are detected by direct detection or coherent detection. For coherent detection, a 2×4 90° hybrid coupler is used to mix the polarizations of local oscillator and that of the received signals. The outputs of the couplers are then fed into the balanced optical receivers. The mixing of the local laser source and that of the signals preserves the phase of the signals, which are then processed by a high-speed electronic processor. For direct detection, the I and Q components are detected differentially, the amplitude and phase detection are then compared and processed similarly as for the coherent case.

In order to show the robustness of oOFDM towards fiber dispersion and also fiber non-linearity, numerical simulations are carried out for a data stream of 42.7 Gb/s data rate. The number of OFDM channels can be varied between $N_{min} = 256$ and $N_{max} = 2048$. A guard interval of 12 ns can be inserted, a strategy belonging inherently to OFDM technology that ensures orthogonality of the different channels in case of a transmission channel with memory. For the optical modulation, intensity modulation using a single Mach–Zehnder modulator in conjunction with SSB filtering and direct detection was implemented. The non-linear optical transmission channel consisted of eight 80-km non-DCF spans of SSMF. As a criterion for performance, required OSNR for a BER of 10^{-3} (Monte Carlo) is measured. Using an FEC, the BER can be lower than 1e-9.

Figure 7.47 shows the required OSNR as function of fiber launch power for different values of N. The most important result is that transmission is possible over 640 km over SSMF without any dispersion compensation. It can be explained by the fact that even for the lowest value of $N_{min} = 256$, each sub-channel occupies a bandwidth of approximately 42.7 GHz/256 = 177 MHz, resulting in high robustness towards fiber dispersion.

The principal difficulties of optical OFDM are that the pure delay due to the variation of the refractive index of the fiber with respect to the optical frequency lead to bunching of the sub-channels and hence the increase of the optical power, thus unexpected SPM may occur in a random manner.

7.8.3   100 Gb/s 8-DPSK _ 2-ASK 16-Star QAM

7.8.3.1   Introduction

Multi-level modulation scheme enables the transmission baud rate to be reduced, thus obtaining the spectral efficiency. Another significant advantage of this modulation scheme is to reduce the requirement of high-speed processing electronics. This is of particular interest for high-speed optical transmission systems.

This part of the report investigates a multilevel modulation scheme which has eight phase and two amplitude levels. This scheme, which is named in short as 8-DPSK_2-ASK, effectively utilizes four bits per one symbol for transmission, in which the first three bits are for coding phase information while the coding of the amplitude levels is implemented with the fourth bit. As a result, the transmission baud rate is equivalently a quarter of the bit rate from bit pattern generator.

This section is organized as follows: Sub-section 2 presents detailed description of the optical transmitter for generating 8-DPSK_2-ASK signals. In Sub-section 3, the detailed configuration of the receiver is provided. The configuration of the optical transmitter and receiver is referred from those reported by Ivan, et al. [33]. Sub-section 4 provides study on dispersion tolerance and transmission performance of the 8-DPSK_2-ASK scheme. Finally, a short summary of the report is provided.

7.8.3.2   Configuration of 8-DPSK _ 2-ASK Optical Transmitter

There have been several different configurations of an optical transmitter for generating multi-phase/level optical signals using amplitude or phase modulators arranged in serial or parallel configurations

[33–37]. However, the optical transmitters reported in [33–37] require a pre-coder with high complexity. Conversely, the configuration reported Ivan, et al. [33] utilizes the Gray mapping technique to differentially encode the phase information and this significantly reduce the complexity of the optical transmitter. In addition, as elaborated in more detail in section 3, this pre-coding technique enables the detection scheme using the I-Q demodulation techniques, as equivalently in coherent transmission systems.

The optical transmitter of the 8-DPSK_2-ASK scheme employs the I-Q modulation technique with two Mach–Zehnder Intensity Modulators (MZIMs) in parallel and a $\Pi/2$ optical phase modulator, as shown in Figure 7.65. At each kth instance, the absolute phase of transmitted lightwaves κ_k is expressed as: $\kappa_k = \kappa_{k-1} + \Delta\kappa_k$ where κ_{k-1} is the phase at $(k-1)^{th}$ instance and $\Delta\kappa_k$ is the differentially coded phase information. The encoding of this $\Delta\kappa_k$ for generating 8-DPSK_2-ASK modulated optical signals (four bits per one transmitted symbol) follows the well-known Gray mapping rules. This Gray mapping phasor diagram is shown in Figure 7.66. The phasor is normalized with the maximum energy on each branch, i.e., $E_1/2$.

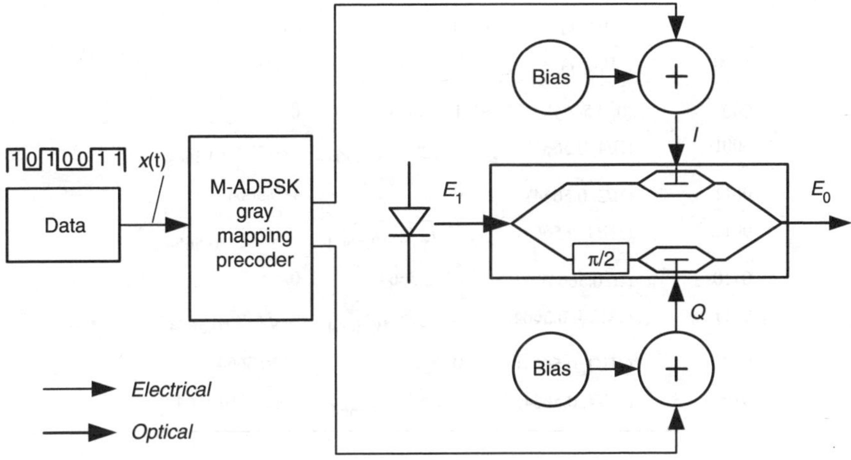

FIGURE 7.65 Optical transmitter configuration of the 8-DPSK_2-ASK modulation scheme.

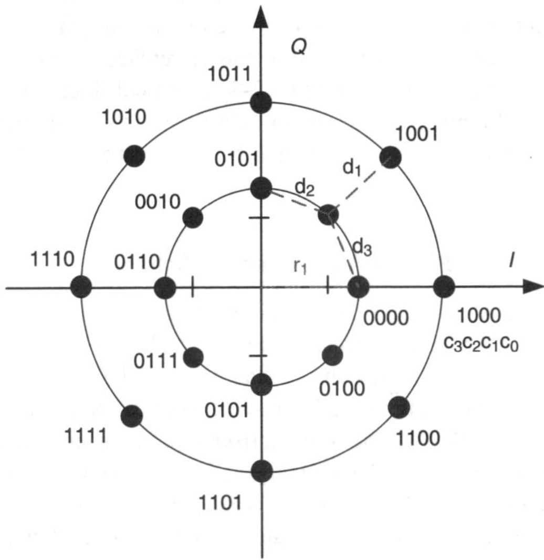

FIGURE 7.66 Gray mapping for optimal 8-DPSK_2-ASK modulation scheme.

TABLE 7.7

I and Q Field Vectors in 8-DPSK_2-ASK Modulation Scheme Using Two MZIMs in Parallel

Binary Sequence	$(\Delta\kappa_k,$ Amplitude)	I_k	Q_k
1000	$(0, 1)$	1	0
1001	$(\Pi/4, 1)$	$-\sqrt{2}/2$	$-\sqrt{2}/2$
1011	$(\Pi/2, 1)$	0	1
1010	$(3\Pi/4, 1)$	$-\sqrt{2}/2$	$-\sqrt{2}/2$
1110	$(\Pi, 1)$	-1	0
1111	$(-3\Pi/4, 1)$	$-\sqrt{2}/2$	$-\sqrt{2}/2$
1101	$(-\Pi/2, 1)$	0	-1
1100	$(-\Pi/4, 1)$	$-\sqrt{2}/2$	$-\sqrt{2}/2$
0000	$(0, 0.5664)$	$1*0.5664$	0
0001	$(\Pi/4, 0.5664)$	$-\sqrt{2}/2\,*0.5664$	$-\sqrt{2}/2\,*0.5664$
0011	$(\Pi/2, 0.5664)$	0	$1*0.5664$
0010	$(3\Pi/4, 0.5664)$	$-\sqrt{2}/2\,*0.5664$	$-\sqrt{2}/2\,*0.5664$
0110	$(\Pi, 0.5664)$	$-1*0.5664$	0
0111	$(-3\Pi/4, 0.5664)$	$-\sqrt{2}/2\,*0.5664$	$-\sqrt{2}/2\,*0.5664$
0101	$(-\Pi/2, 0.5664)$	0	$-1*0.5664$
0100	$(-\Pi/4, 0.5664)$	$-\sqrt{2}/2\,*0.5664$	$-\sqrt{2}/2*0.5664$

The amplitude levels are optimized in order that the Euclidean distances d_1, d_2, and d_3 are equal, i.e., $d_1 = d_2 = d_3$. After derivation, we obtain: $r_1 = 0.5664$. The I and Q field vectors corresponding to Gray mapping rules from the M-ADPSK pre-coder (see Figure 7.65) are provided in Table 7.7.

The above-described transmitter configuration can be replaced with a dual-drive MZIM. The explanation and derivation for generating 8-DPSK_2-ASK optical signals are also based on the phasor diagram of Figure 7.66. In this case, the output field vector is the summation of two component field vectors, each of which is not only determined by the amplitude, but also by initially biased phases.

7.8.3.3 Configuration of 8-DPSK _ 2-ASK Detection Scheme

The detection of 8-DPSK_2-ASK optical signals is implemented with the use of two Mach–Zehnder delay interferometric (MZDI)-balanced receivers (Figure 7.67).

Several key notes in this detection structure is stated as: (i) The MZDI introduces a delay corresponding to the baud rate; (ii) One arm of MZDI has a $\Pi/4$ optical phase shifter while the other arm has an optical phase shift of $-\Pi/4$; (iii) The outputs from two balanced receivers are superimposed positively and negatively, which leads to I and Q detected signals, respectively. The I and Q detected components are expressed as $I = \mathrm{Re}\{E_k E_{k-1}^*\}$ and $Q = \mathrm{Re}\{E_k E_{k-1}^*\}$; and (iv) I-Q detected components are demodulated using the popular I-Q demodulator in the electrical domain. These detected signals are then sampled and represented as shown in the signal constellation.

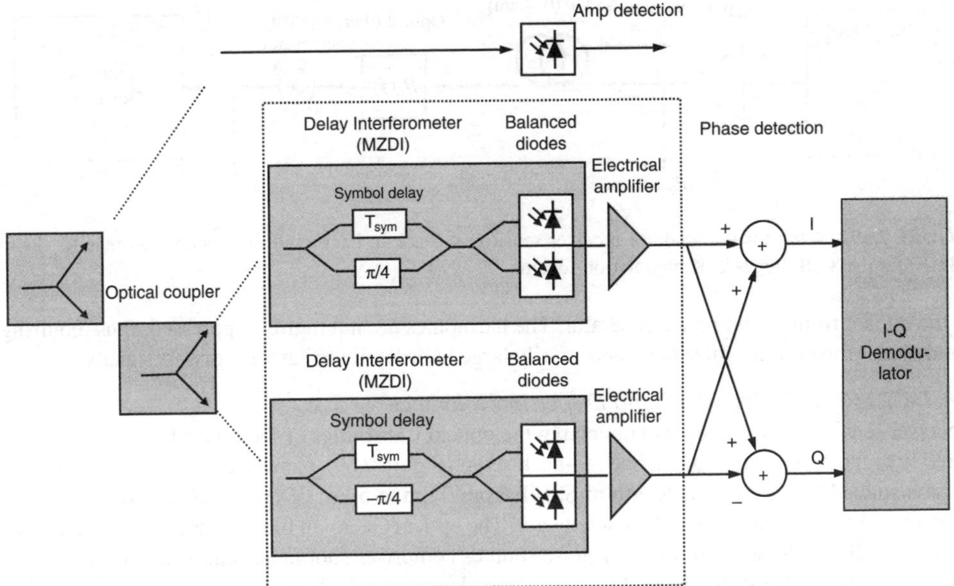

FIGURE 7.67 Detection configuration for the 8-DPSK_2-ASK modulation scheme.

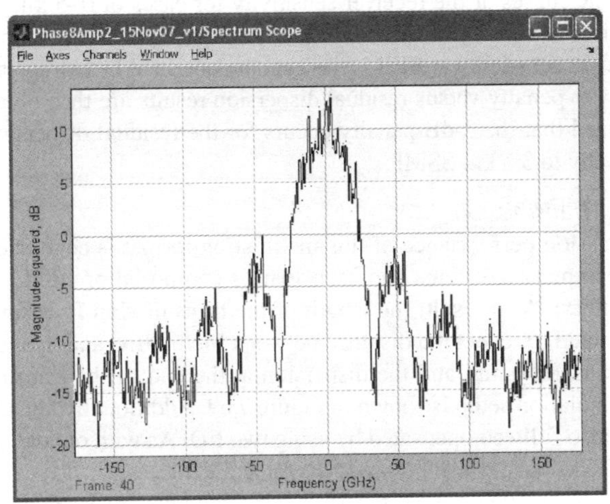

FIGURE 7.68 Power spectrum of 8-DPSK_2-ASK signals.

7.8.3.4 Transmission Performance of 100 Gb/s 8-DPSK_2-ASK Scheme

Performance characteristics of the 8-DPSK_2-ASK scheme operating at 100 Gb/s bit rate are studied in terms of receiver sensitivity, dispersion tolerance and the feasibility for long-haul transmission. Bit error rates (BER) are the pre-Forward Error Correct (FEC) BERs and the pre-FEC limit is conventionally referenced at 2e-3. In addition, the BERs are evaluated by the Monte Carlo method.

7.8.3.4.1 Power Spectrum
The power spectrum of 8-DPSK_2-ASK optical signals is shown in Figure 7.68. The main lobe spectral width is about 25 GHz as the symbol baud rate of this modulation scheme is equal to a quarter

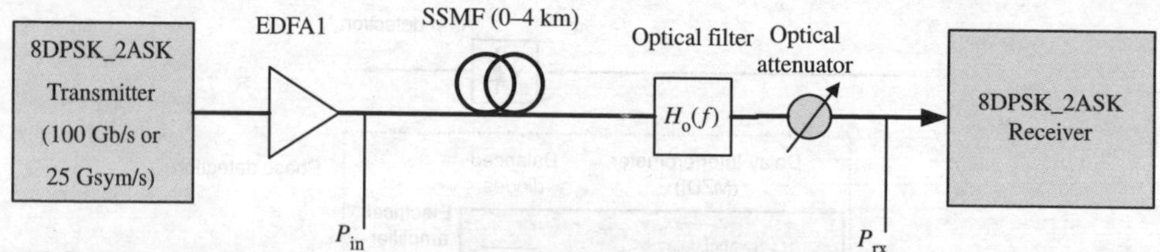

FIGURE 7.69 Setup for the study of receiver sensitivity (back to back) and dispersion tolerance (0–4-km SSMF) for the 8-DPSK_2-ASK modulation scheme.

of the bit rate from the bit patter generator. The harmonics are not highly suppressed, thus requiring bandwidth of the optical filter to be necessarily large in order not to severely distort signals.

7.8.3.4.2 *Receiver Sensitivity and Dispersion Tolerance*

Receiver sensitivity is studied by connecting the optical transmitter of the 8-DPSK_2-ASK scheme directly to the receiver to make a back-to-back setup (Figure 7.69). Conversely, the dispersion tolerance is studied by varying the length of SSMF from 0 km to 5 km ($|D| = 17$ ps/(nm km)). Received powers are varied by using an optical attenuator. The optical Gaussian filter has BT = 3 (B is approximately 75 GHz). Modeling of receiver noise sources comprises shot noise, equivalent noise current density of 20 pA/$\sqrt{Hz}$ at the input of the trans-impedance electrical amplifier and dark current of 10 nA for each of the two photodiodes in balanced structure. A fifth-order Bessel electrical filter with a bandwidth of BT = 0.8 is used.

The numerical BER curves of the receiver sensitivity for cases of 0–5 km SSMF are shown in Figure 7.70. The receiver sensitivity of the 8-DPSK_2-ASK scheme is approximately –18.5 dBm at BER = 1e-4. The receiver sensitivity at BER = 1e-9 can be obtained by extrapolating the BER curve of 0 km case. The power penalty versus residual dispersion results are then obtained and plotted in Figure 7.71. It is realized that the 2-dB penalty occurs for the residual dispersion of approximately 60 ps/ nm or equivalently to 3.5 km SSMF.

7.8.3.4.3 *Long-haul Transmission*

The long-haul transmission performance of this modulation format is conducted over 10 optically amplified and fully compensated spans and each span is composed of 100 km SSMF and 10-km DCF100 (Sumitomo fiber). As a result, the length of the transmission fiber link is 1100 km. This long-haul range is selected to reflect the distance between Melbourne and Sydney in Australia. The wavelength of interest is 1550 nm, and the dispersion at the end of the transmission link is fully compensation. The simulation setup is shown in Figure 7.69. Additionally, the fiber attenuation due to SSMF and DCF is also fully compensated by using two EDFAs with optical gains, as depicted in Figure 7.72. These EDFAs have Noise Figure (NF) set at 5 dB.

Numerical transmission BERs are plotted against received powers in Figure 7.73 and compared to the back-to-back BER curve. The BER curve of 1100 km follows a linear trend and feasibly reaches 1e-9 if extrapolated, as shown in Figure 7.73. This transmission performance can be significantly improved with a high-performance forward error coding (FEC) Scheme.

7.9 CONCLUDING REMARKS

The ever-increasing bandwidth hungry of telecommunication network infrastructures which based mainly on optical fiber communication technology indicates that low-efficient modulation formats such as ASK will no longer satisfy the demands of transmission capacity. Thus, it is expected that new advanced optical modulation schemes would replace the ASK in the near future in long-haul optical transmission systems. Advanced optical modulation schemes, especially the multi-level amplitude and phase schemes presented above, are able to: (i) provide long reach, error-free, and

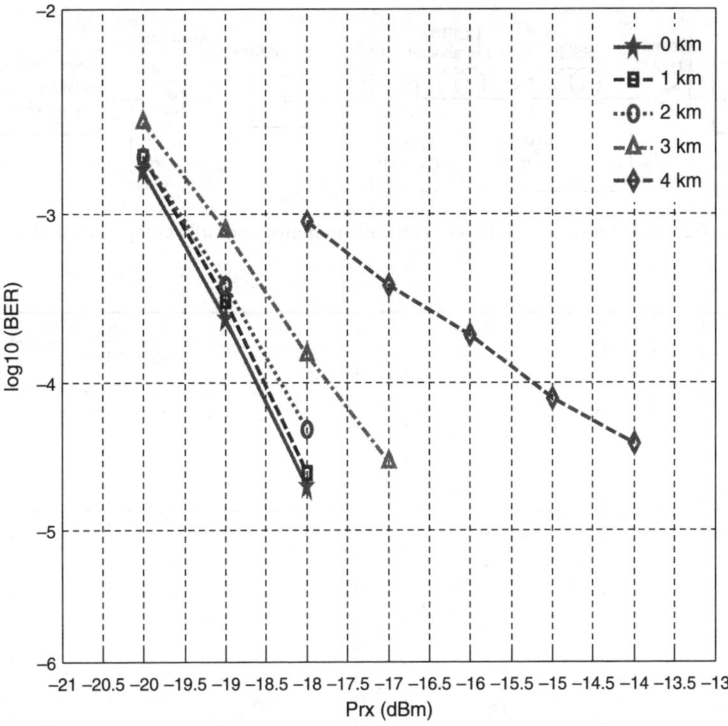

FIGURE 7.70 Receiver sensitivity (back-to-back) and dispersion tolerance (0–4-km SSMF) for the 8-DPSK_2-ASK modulation scheme.

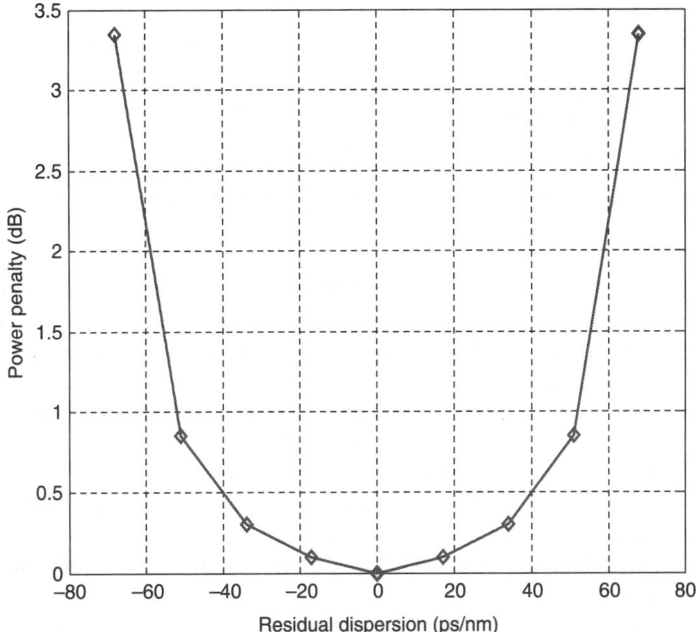

FIGURE 7.71 Power penalty due to residual dispersions for the 8-DPSK_2-ASK modulation scheme.

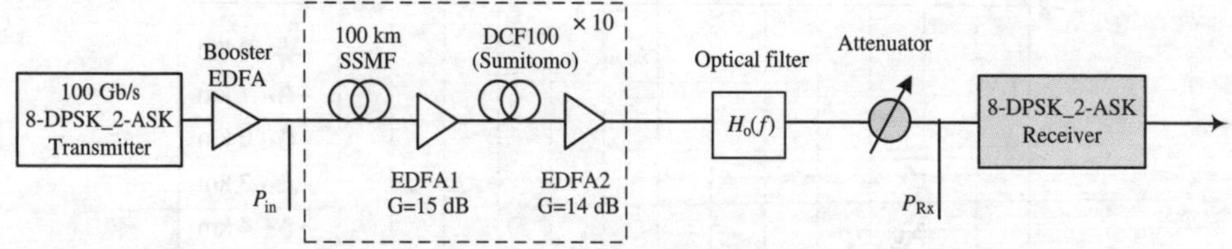

FIGURE 7.72 Transmission setup of 1100 km optically amplified and fully compensated fiber link.

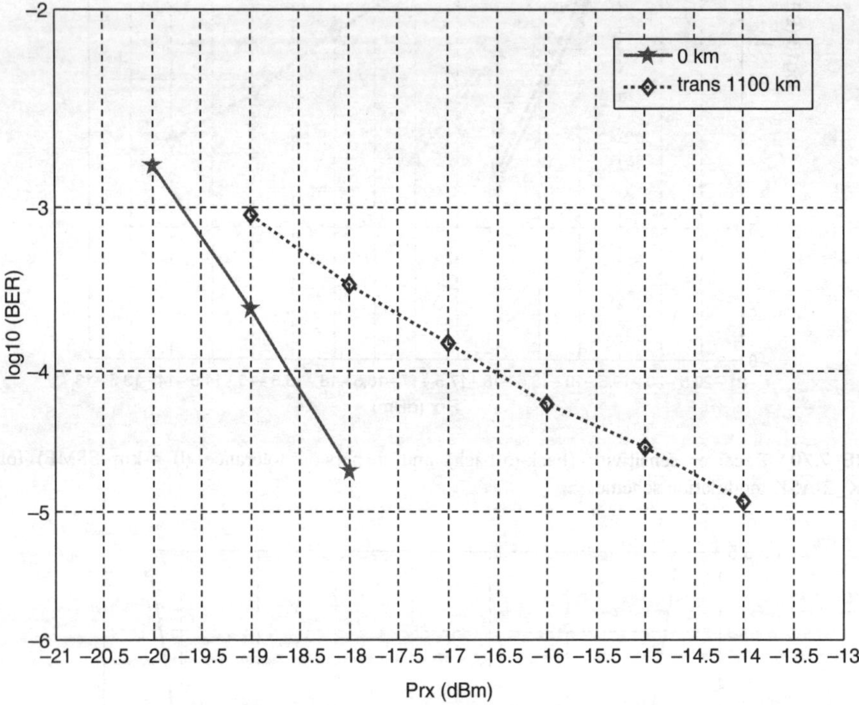

FIGURE 7.73 BER versus receiver sensitivity for 8DPSK 2ASK modulation format transmission.

high-transmission capacity; (ii) have high bandwidth efficiency (no. bits/Hz parameter); (iii) push the bit rate well above that could be offered by electronic technology, e.g., 100 Gb/s with the detection at the symbol rate; (iv) tolerate dispersion and non-linearity; and (v) to maximally utilize the existing optical network infrastructure.

Current developments of photonic technology have enabled differential phase modulation and demodulation in the optical domain. Presently, BDPSK and DQPSK formats have received great attention due to their improvement in the receiver sensitivity compared with ASK. Furthermore, the RZ and CSRZ modulation formats would assist mitigation of impairments due to non-linear effects. However, as long as transmission capacity and bandwidth efficiency are concerned, MADPSK modulation formats would offer better performance with a higher complexity of the receiver structure.

Alternatively, there are other possible multi-level modulation schemes that would offer further improvement of optical transmission performance [10].

7.9.1 OFFSET MADPSK MODULATION

The binary DPSK and quaternary DPSK (DQPSK) is just a special case of a more general class of differential phase modulation formats that maps data bits into phase difference between neighboring symbols. This phase difference $\Delta\phi_i$ can be described as

$$\Delta\phi_i = \theta + \frac{2\pi(i-1)}{M}, \qquad i = 1, 2 \ldots M-1 \tag{7.28}$$

where θ is the initial phase, M is total number of phase levels. This class of formats is called offset DPSK (O-DPSK) and denoted θ-MS-DPSK pecifically, $\theta = 0$, $M = 2$ or $M = 4$ θ-MS-DPSK become θ-2-DPSK or θ-2-DPSK are the conventional binary DPSK and DQPSK mentioned above.

ODPSK has been used to transmit over satellite non-linear channel because its phase transition is smooth and can avoid 180°phase jump [38]. As fiber medium also exhibits non-linearity, this modulation format attracts our attention as a candidate, together with ASK, for creating a new multi-level modulation format, possibly termed as Offset MADPSK.

7.9.2 MULTI-LEVEL AMPLITUDE-MINIMUM SHIFT KEYING (MAMSK) MODULATION

Minimum shift keying (MSK) is a form of OQPSK with sinusoidal pulse weighting [39]. In MSK, data bits are first coded into bipolar signals ± 1 which are then separated into $V_I(t)$ and $V_Q(t)$ streams consisting of even and odd bits, respectively. In the next stage, $V_I(t)$ and $V_Q(t)$ are used to modulate a carrier f_c create a MSK signal $s(t)$ which can be presented as [39]

$$s(t) = V_I(t)\cos\left(\frac{\pi t}{2T}\right)\cos\left(2\pi f_c t\right) + V_Q(t)\cos\left(\frac{\pi t}{2T}\right)\sin\left(2\pi f_c t\right). \tag{7.29}$$

MSK is a well-known modulation format in wireless communications with efficient spectral characteristics due to high compactness of its main lobe as compared with DPSK, and high suppression of the side lobes [40]. These characteristics indicate that MSK signal is highly dispersion tolerant. A combination of MSK and ASK into multi-level amplitude-minimum shift keying (M-AMSK) modulation would even improve the transmission performance without increasing the complexity of the detection scheme.

Multilevel techniques for 100 Gb/s are given, in particular the OFDM with multi-carrier and multilevel amplitude modulation with orthogonality between adjacent channels are proven to be cost effective and appropriate for current electronic technologies. Two interesting approaches to achieve data transmission of 40 Gb/s and beyond (e.g., 100 Gb/s Ethernet) based on low symbol rate are discussed. On the one hand, oOFDM can combine a large number of parallel data streams into one broadband data stream with high spectral efficiency. Simulation results are shown for different values of the number of parallel data streams in non-linear environment. On the other hand, 16-ary modulation formats enable 40 Gb/s transmission with 10 GSym/s (i.e., 100 Gb/s with 25 GSym/s). For a special case, namely inverse RZ 16ADQPSK, measurement results are demonstrated.

7.9.3 STAR QAM COHERENT DETECTION

These last two sections of the chapter describe two optical transmission schemes using coherent and incoherent transmission and detection techniques, the 2A-8P Star QAM and 8DPSK 2ASK 16-Star QAM for 100 Gb/s or 25 G Symbols/sec.

Firstly, the design of a Start 16-QAM modulation scheme is proposed for coherent detection for ultra-high speed ultra-high capacity optical fiber communications schemes. Two amplitude levels and eight phases (2A-8P 16_QAM) are considered to offer significant simple transmitter and receiver configurations and at the same time the best receiver sensitivity at the receiver. An optical SNR of about 22 dB is required for the transmission of Star 16-QAM over optically amplified transmission dispersion–compensated link. Dispersion tolerance of 300 ps/nm is possible with 1-dB penalty of the eye-opening at 40 Gb/s bit rate or 10 Giga-symbols/sec. with an OSNR of 18 dB. AN OSNR could be about 22 dB for 107 Gb/s bit rate and a symbol rate of 26.75 G symbols/sec. The transmission link consists of several spans of total 1000-km dispersion-compensated optically amplified transmission link. Optical gain of the in-line optical amplifiers are set to compensate for the attenuation of the transmission and compensating fibers with a noise figure of 3 dB. The optical transmitter and receivers incorporating commercially available coherent receiver are structured and sufficient for engineering of the optical transmission terminal equipment for the bit rate of 107 Gb/s and a symbol rate of 26.3 Gbauds/sec.

Furthermore, electronic equalization of the receiver phase shift keying signals can be done using blind equalization that would improve the dispersion tolerance much further. For a symbol rate of 10.7 Gb/s, this dispersion tolerance for 1 dB penalty would reach 300 km of standard SMF. This electronic equalization can be implemented without any difficulty at the 10.7 Gbauds/sec. For 107 Gb/s bit rate, similar improvement of the dispersion tolerance can be at 26.5 Gbauds/sec provided that the electronic sampler can offer more than 50 Gig-sampling rate.

Secondly, the transmitter and receiver configurations for generating 8-DPSK_2-ASK optical signals and direct detection are described as well as the transmission performances. In addition, performance characteristics of this modulation format at 100 Gb/s (equivalently 25 Gbauds/s) has also been investigated in terms of the receiver sensitivity, dispersion tolerance the long-haul transmission performance. The simulation results shows that 8-DPSK_2-ASK is a promising modulation for very high-speed (100 Gb/s) and long-haul optical communications.

REFERENCES

1. Linke, R. A., and A. H. Gnauck. 1988. High-capacity coherent lightwave systems. *IEEE Journal of Lightwave Technology* 6 (11): 1750–69.
2. Binh, L. N. 2004a. Monash optical communication system simulator. Part I: Ultra-long ultra-high speed optical fiber communication systems. *Manual for technical development*, ECSE Monash University. Melbourne, Australia:
3. Proakis, G. 2001. *Digital communication*. New York: McGraw-Hill.
4. Binh, L. N., and B. Laville. 2005. *SIMULINK models for advanced optical communications: Part IV-DQPSK modulation format*. Technical Report, ECSE Monash University, Melbourne Australia.
5. Gnauck, A. H., X. Liu, X. Wei, D. M. Gill, and E. C. Burrows. 2004. Comparison of modulation formats for 42.7Gbs single-channel transmission through 1980 km of SSMF. *IEEE Photonics Technology Letters* 16 (3): 909–11.
6. Sekine, K., N. Kikuchi, S. Sasaki, S. Hayase, C. Hasegawa, and T. Sugawara. 2005. 40 Gb/s 16-ary (4 bit/symbol) optical modulation/demodulation scheme. *IEEE Electronics Letters* 41 (7).
7. Mizuochi, T., Ishida, K., T. Kobayashi, J. Abe, K. Kinjo, S. Kuroda, and T. Mizuochi. 2003. A comparative study of DPSK and ASK WDM transmission over transoceanic distances and their performance degradations due to nonlinear phase noise. *IEEE Journal of Lightwave Technology* 21 (9): 1933–43.
8. Hoshida, T., O. Vassilieva, K. Yamada, S. Choudhary, R. Pecqueur, and H. Kuwahara. 2002. Optimal 40 Gb/s modulation formats for spectrally efficient long-haul DWDM systems. *IEEE Journal. of Lightwave Technology* 20 (12): 1989–96.
9. Hayase, S., N. Kikuchi, K. Sekine, and S. Sasaki. 2003. Proposal of 8-state per symbol (binary ASK and QPSK) 30-Gb/ optical modulation demodulation scheme. *In Proceedings of European Conference on Optical Communication, Paper Th2.6.4,* ECOC 2003, 1008–9, September, Rimini, Italy.
10. Walklin, S., and J. Conradi. 1999. Multilevel signaling for increasing the reach of 10Gb/s lightwave systems. *IEEE Journal of Lightwave Technology* 17 (11): 2235–48.
11. Miyamoto, Y., A. Hirano, K. Yonenaga, A. Sano, H. Toba, K. Murata, and O. Mitomi. 1999. 320 Gb/s (8 × 40Gb/s) WDM transmission over 367-km zero-dispersion-flattened line with 120-km repeater spacing using carrier-suppressed return-to-zero pulse format. *IEEE Electronics Letters* 35 (23): 2041–42

12. Binh, L. N., H. S. Tiong, T. L. Huynh, and D. D. Tran. 2006. DPSK RZ modulation formats generated from dual drive interferometric optical modulators. Proceedings of Asia Pacific Optical Communications Conference, September 2006, Kwangju, Korea.
13. Binh, L. N., and Y. L. Cheung. 2005. DWDM advanced optical communication–Simulink models: Part I–optical spectra. Technical Report, ECSE Monash University, Melbourne, Australia.
14. Hui, R., and S. Zhang. 2004. *Advanced optical modulation formats and their comparison in fiber-optic systems*. Technical Report, Information and Telecommunication Technology Center, University of Kansas.
15. Lee, D., M. S. Lee, Y. J. Wen, and A. Nirmalathas. 2004. Electrically band-limited CSRZ-DPSK signal with a simple transmitter configuration and reduced linear crosstalk in high spectral efficiency DWDM systems. *IEEE Photonics Technology Letters* 16 (9): 2135–37.
16. Wree, C. 2002. RZ-DQPSK format with high spectral efficiency and high robustness towards fiber nonlinearities. University of Kiel, Germany.
17. Kim, H., and P. J. Winzer. 2003. Robustness to laser frequency offset in direct detection DPSK and DQPSK systems. *IEEE Journal of Lightwave Technology* 21 (9): 1887–91.
18. Griffin, R. A., P. Boffi, L. Marazzia, L. Paradisoa, P. Parolaria, A. Righettia, D. Settia, R. Sianoa, R. Cigliuttib, D. Mottarellab, P. Francoc, and M. Martinellia. 2002. 10 Gb/s optical differential quadrature phase shift key (DQPSK) transmission using GaAs/AlGaAs integration. *Proceedings of OFC 2002, Postdeadline Paper FD6*, February 2002, Annaheim, CA, USA.
19. Yoshikane, N., and I. Morita. 2004. 1.14 b/s/Hz spectrally-efficient 50x85.4 Gb/s transmission over 300 km using co-polarized CS-RZ DQPSK signals" *Proceedings of OFC 2004, Postdeadline Paper PDP3*, March 2004, Annaheim, CA, USA.
20. Tran, D. D., L. N. Binh, T. L. Huynh, H. S. Tiong. 2005. Geometrical and phasor representation of multi-level amplitude-phase modulation formats and photonic transmitter structures. In *Proceedings of IEEE Tencon'05, November*, Melbourne, Australia.
21. Agrawal, G. P. 1992. *Fiber-optic communication systems*. Wiley.
22. Ho, K-P. 2005. Generation of arbitrary quadrature signals using one dual-drive modulator. *IEEE Journal of Lightwave Technology* 23 (2): 764–70.
23. Agrawal, G. P. 1995. *Nonlinear fiber optics*. 2nd ed. Boston, MA, USA: Academic Press.
24. Muoi, T. V. 1978. Stepped-index optical fiber systems. Ph.D. Thesis, University of Western of Australia.
25. Binh, L. N., K.-Y. Chin, and D. Lam. 1997. Monash optical communication system simulator. Wavelength division multiplexed optical communication systems and networks. *Manual for technical development*. Melbourne, Australia: Monash University.
26. Gumaste, A., and T. Antony. 2002. *DWDM network designs and engineering solutions*. http://www.ciscopress.com, Cisco Press.
27. Leibrich, J., A. Ali, and W. Rosenkranz. 2007. Optical OFDM as a promising technique for bandwidth-efficient high-speed data transmission over optical fiber. In *Proceedings of the 12th International OFDM-Workshop 2007, InOWo 2007*, 29–30 August, Hamburg, Germany.
28. Lowery, A. J., L. Du, and J. Armstrong. 2006 Orthogonal frequency division multiplexing for adaptive dispersion compensation in long haul WDM systems. *In Proceedings of OFC 2006, Paper PDP39*, March, Anaheim, USA.
29. Yoon, H., D. Lee, and N. Park. 2005. Performance comparison of optical 8-ary differential phase-shift keying systems with different electrical decision schemes. *Optics Express* 13 (2): 371.
30. Hanzo, L., M. Münster, B. J. Choi, and T. Keller. 2003. *OFDM and MC-CDMA for broadband multi-user communications, WLANs and broadcasting*. Brisbane, Australia: Wiley.
31. Shieh, W., and C. Athaudage. 2006. Coherent optical orthogonal frequency division multiplexing. *Electronics. Letters* 42: 587–89.
32. Binh, L. N. 2004b. Monash optical communication system simulator. Part II: Ultra-long Ultra-high speed optical fiber communication systems. *Manual for technical development*. Melbourne, Australia: ECSE Monash University.
33. Djordjevic, I. B., and B. Vasic. 2006. Multilevel coding in M-ary DPSK/Differential QAM high-speed optical transmission with direct detection. *IEEE Journal of Lightwave Technology* 24: 420–28.
34. Serbay, M., C. Wree, and W. Rosenkranz. 2004. Implementation of different precoder for high-speed optical DQPSK transmission. *Electronics Letters* 40: 1288–89.
35. Serbay, M., C. Wree, and W. Rosenkranz. 2005. Experimental investigation of RZ-8DPSK at 3 × 10.7Gb/s, LEOS Annual Meeting 2005, Sydney, Australia, 23–27, paper WE3.

36. Seimetz, M., M. Noelle, and E. Patzak. 2007. Optical systems with high-order DPSK and star QAM modulation based on interferometric direct detection. *IEEE Journal of Lightwave Technology* 25: 1515–30.
37. Yoon, H., D. Lee, and N. Park. 2005. Performance comparison of optical 8-ary differential phase-shift keying systems with different electrical decision schemes. *Optics Express* 13: 371–76.
38. Wree, C., M. Serbay, J. Leibrich, W. Rosenkranz. 2004. Offset-DQPSK modulation format for 40Gb/s and comparison to RZ-DQPSK in WDM environment. In *Proceedingsof OFC 2004, Paper MF62*, February, Los Angeles, California, USA.
39. Pasupathy, S. 1979. Minimum shift keying: A spectrally efficient modulation. *IEEE Communications Magazine* 17: 14–22.
40. Sakamoto, T., T. Kawanishi, and M. Izutsu. 2005. Optical minimum-shift keying with external modulation scheme. *Optics Express* 13: 7741–47.
41. Huynh, T. L., L. N. Binh, D. D. Tran, and Q. H. Lam. 2005. Long-haul ASK and DPSK optical fiber transmission systems: Simulink modeling and experimental demonstration test-beds. In *Proceedings of IEEE Tencon'05*, November, Melbourne, Australia.
42. Ohm, M., and J. Speidel. 2003. Quaternary optical ASK-DPSK and receivers with direct detection. *IEEE Photonics Technology Letters* 15 (1):159–61.
43. Boyraz, O., Y. Liu, C. W. Chow, H. K. Tsang, and S. P. Wong. 2004. Self phase modulation induced spectral broadening in silicon waveguides. In *Proceedings of CLEO 2004*, vol. 2: Paper CThj2, May
44. Goeger, G., M. Wrage, and W. Fischler. 2004. Cross-phase modulation in multispan WDM systems with arbitrary modulation formats. *IEEE Photonics Technology Letters* 16 (8): 1858–60.
45. Turin, G. 1960. An introduction to matched filters. *IEEE Transactions on Information Theory* 6 (3): 311–29.
46. Zhu, B. 2001. 3.08 Tbit/s (77 × 42.7 Gb/s) WDM transmission over 1200 km fiber with 100 km repeater spacing using dual C- and L-band hybrid Raman/erbium-doped inline amplifiers. *IEEE Electronics Letters* 37.
47. Hui, R., B. Zhu, R. Huang, C. T. Allen, K. R. Demanrest, and D. Richards. 2005. Subcarrier multiplexing for high-speed optical transmission. *IEEE Journal of Lightwave Technology* 20 (3): 417–27.
48. Gnauck, A. H., and P. J. Winzer. 2005. Optical phase-shift-keyed transmission. *IEEE Journal of Lightwave Technology* 23 (1): 115–30.
49. Ho, K. P. 2004. Error probability of DPSK signals with cross-phase modulation induced nonlinear phase noise. *IEEE Journal of Selected Topics in Quantum Electronics* 10 (2): 421–27.
50. Agrawal, G. P. 1989. Self-phase modulation and spectral broadening of optical pulses in semiconductor laser amplifiers. *IEEE Journal of Quantum Electronics* 25 (11): 2297–2306.
51. Gordon, J. P., and L. F. Mollenauer. 1990. Phase noise in photonic communications systems using linear amplifiers. *Optics Letters* 15 (23): 1351–53.
52. Kikuchi, N. 2005. Amplitude and phase modulated 8-ary and 16-ary multilevel signaling technologies for high-speed optical fiber communication. In *Proceedings of APOC 2005, Paper 602127*, October, Shanghai, China.
53. Michel, C. 1984. Techniques for estimating the bit error rate in the simulation of digital communication systems. *IEEE Journal of Selected Areas in Communications SAC-2* (1): 153–70.
54. Kim, H., and A. H. Gnauck. 2005. Experimental investigation of the performance limitation of DPSK systems due to nonlinear phase noise. *IEEE Journal of Lightwave Technology* 15 (2): 320–22.
55. Wree, C., J. Leibrich, and W. Rosenkranz. 2002. RZ-DQPSK format with high spectral efficiency and high robustness towards fiber nonlinearities. In *Proceedings of ECOC* 2002, Paper 9.6.6, September, Copenhagen, Denmark.
56. Serbay, M., C. Wree, and W. Rosenkranz. 2005. Experimental investigation of RZ-8DPSK at 3x 10.7Gb/s. In *Proceedings of LEOS 2005, Paper WE3*, October, in Sydney, Australia.
57. Serbay, M., T. Tokle, P. Jeppesen, and W. Rosenkranz. 2007. 42.8 Gbit/s4 bits per symbol 16-ary inverse-RZ-QASK-DQPSK transmission experiment without polmux. In *Proceedings of OFC 2007, Paper OThL2*, 25–29 March, Anaheim, USA.
58. Jansen, S. L., I. Morita, N. Takeda, and H. Tanaka. 2007. 20-Gb/s OFDM transmission over 4160-km SSMF enabled by RF-pilot tone phase noise compensation. *In Proceedings of OFC 2007, Paper PDP15*, March, Anaheim, USA.
59. Llorente, R., J. H. Lee, R. Clavero, M. Ibsen, and J. Martí. 2005. Orthogonal wavelength-division-multiplexing technique feasibility evaluation. *Journal of Lightwave Technology* 23: 1145–51.

8 Frequency Discrimination Receiver for Optical MSK

8.1 OPERATIONAL PRINCIPLES OF ONFDR

In MSK modulation format, binary information is modulated in upper side band (USB) frequency (f_1) and lower side band (LSB) frequency (f_2), which are defined as $f_1 = f_c + f_d$ and $f_2 = f_c - f_d$, where f_c is the carrier frequency of the lightwave, and f_d is the frequency deviation from f_c. In optical MSK modulated signals, f_d is equal to a quarter of the transmission bit rate (B_R), i.e., $f_d = B_R/4$. The block diagram of the proposed narrowband frequency discrimination receiver is shown in Figure 8.3.

The operation of the receiver is based on the principles of frequency discrimination. Optical MSK lightwaves at the output of the optical fiber are split into two paths by a 3-dB optical power splitter after being amplified with a low noise optical amplifier (Figure 8.1). Two optical narrowband filters F1 and F2 are used to discriminate the USB and the LSB frequencies, respectively. The center frequencies and the bandwidths of the two filters F1 and F2 are selected so that F1 captures most of the MSK signal only when a "+1" is transmitted, and F2 captures most of the MSK signal only when a "0" is transmitted.

A constant optical delay line (ODL), which is easily implemented in integrated optics, is introduced in one path to compensate the differential group delay $2\pi f_D \beta_2 L$ between f_1 and f_2 where $f_D = |f_1 - f_2| = R/2$, β_2 represents group velocity delay (GVD) parameter of the fiber, and L is the fiber length. If the differential group delay is compensated completely, the optical MSK pulses at the output of the two optical discrimination filters arrive at the photodiodes simultaneously. The selection of the appropriate path is such that to have the ODL depends on the sign of the GVD. In normal case of SMF optical fibers, i.e., $\beta_2 < 0$, ODL is used in the path of the USB frequency.

The receiver sensitivity of the scheme can be controlled by adjusting the gain of the optical amplifier. By using dual narrowband optical filters, ASE noise induced by the optical amplifier is mostly filtered and hence highly suppressed. As shown in Figure 8.1, the output of the filters is then converted to the electrical domain through two photodiodes arranged in a balanced receiver configuration. Trends of ISI effects caused by fiber dispersion and narrowband optical filters are shown in Figure 8.2.

8.2 RECEIVER MODELING

The baseband equivalent description of a MSK modulated signal, $x(t)$ is given by [12],

$$x(t) = \sqrt{\frac{E_b}{T}} e^{j\Phi(t;I)} \tag{8.1}$$

where, $\Phi(t;I)$ is given by

$$\Phi(t;I) = \left[\pi/2 \sum_{k=-\infty}^{n-1} I_k\right] + 2\pi f_d(t + T/2 - nT)I_n$$

$$= \theta_{n-1} + 2\pi f_d(t + T/2 - nT)I_n, \quad nT \quad T/2 \leq t \leq nT + T/2. \tag{8.2}$$

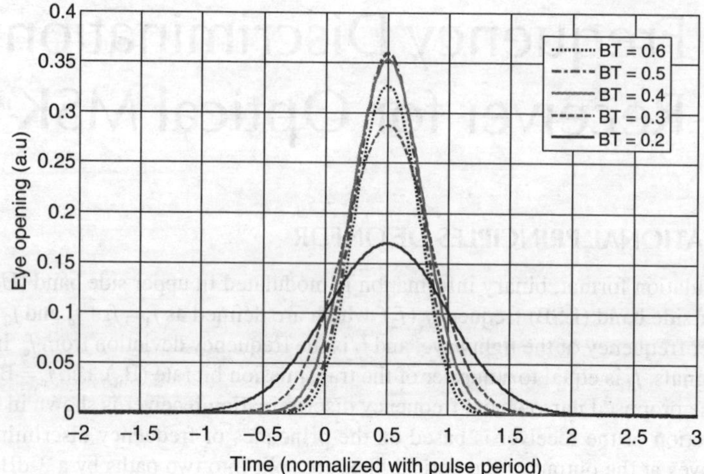

FIGURE 8.1 Eye opening for different BT products to demonstrate effects of power leakage when using dual optical frequency discrimination filters. (From Huynh, T. L., L. N. Binh, and K. K. Pang, Optimal MSK long-haul transmission systems. In *SPIE Proceedings of APOC'06* 6353–86, Thuga, Kwangju, Korea, 2006; and Huynh, T. L., T. Sivahumaran, K. K. Pang, and L. N. Binh, A narrowband filter receiver acheiving 225 ps/nm residual dispersion tolerance for 40 Gb/s optical MSK tranmission systems, *Conference on Optical Communication, OFC'07*, CA, USA, 1–3, 2007. With permission.)

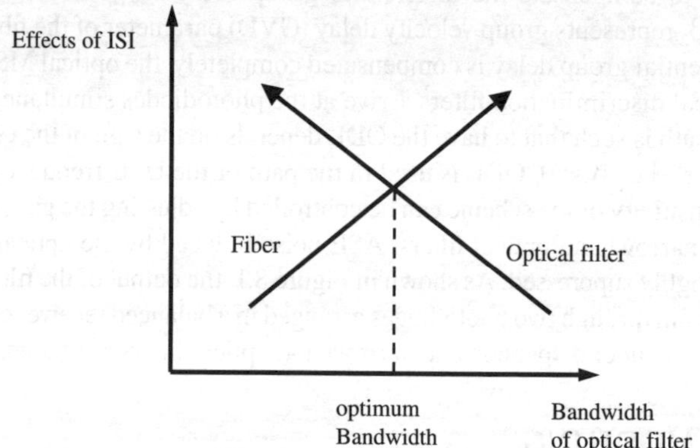

FIGURE 8.2 Trends of ISI effects caused by fiber dispersion and narrowband optical filters. (From Huynh, T. L., L. N. Binh, and K. K. Pang, Optimal MSK long-haul transmission systems. In *SPIE Proceedings of APOC'06* 6353–86, Thuga, Kwangju, Korea, 2006; and Huynh, T. L., T. Sivahumaran, K. K. Pang, and L. N. Binh, A narrowband filter receiver acheiving 225 ps/nm residual dispersion tolerance for 40 Gb/s optical MSK tranmission systems, *Conference on Optical Communication, OFC'07*, CA, USA, 1–3, 2007. With permission.)

In Equation 8.2, $I_n \in \{+1, -1\}$ is the data symbol transmitted at symbol interval n, T is the symbol period and $f_d = 1/4T$ is the frequency deviation. Thus, when $I_n = +1$, the carrier's frequency in the nth symbol period is shifted by $+f_d$ and when $I_n = -1$, the carrier's frequency is shifted by $-f_d$. The switching of frequencies at rate $1/T$ however, results in the spectrum of MSK spilling over a range of frequencies centered around f_d and $-f_d$. For the purpose of the analysis, we express the baseband equivalent MSK signal as a sum of MSK pulses in each symbol period, that is

$$x(t) = \sum_n x_n(t) \tag{8.3}$$

where $x_n(t)$ is defined as

$$x_n(t) = \sqrt{\frac{E_b}{T}} e^{j(2\pi f_d(t+T/2-nT)I_n + \theta_{n-1})} s(t-nT). \tag{8.4}$$

In Equation 8.4, $s(t)$ is a square pulse of duration T, i.e., $s(t) = u(t-T/2) - u(t+T/2)$ where $u(t)$ is the unit step function. Thus, $x_n(t)$ describes the MSK signal in the time interval $nT - T/2 \leq t \leq nT + T/2$ and has zero amplitude elsewhere.

The low-pass equivalent frequency response of the optical fiber, $H(f)$ has a parabolic phase profile and can be modelled by the following equation, [13]

$$H_c(f) = e^{-j\alpha_D f^2}, \tag{8.5}$$

where, $\alpha_D = \pi^2 \beta_2 L$, β_2 represents the GVD parameter of the fiber, and L is the length of the fiber. The parabolic phase profile is the result of the chromatic dispersion of the optical fiber [2]. As in [2], we have omitted the constant phase term and the phase terms linear in f as they do not introduce distortion*. Thus, a pure sinusoidal signal of frequency f, propagating through the optical fiber, experiences a delay of $2\pi f \beta_2 L$. The standard fibers used in optical communications have a negative β_2 and thus, in low-pass equivalent representation, sinusoids with positive frequencies (i.e., frequencies higher than the carrier) have negative delays, i.e., arrive early compared with the carrier and the ones with negative frequencies (i.e., frequencies lower than the carrier) have positive delays and arrive delayed. The dispersion-compensating fibers have positive β_2 and so have the reverse effects. The low-pass equivalent channel impulse response of the optical fiber, $h_c(t)$ has also got a parabolic phase profile and is given as

$$h_c(t) = \sqrt{\frac{\pi}{j\alpha_D}} e^{-j\alpha^2 t^2/\alpha_D}. \tag{8.6}$$

When an optical MSK signal is passed through the optical channel, the output $y(t)$ is given by the following convolution†. When an optical MSK signal is passed through the optical channel, the output $y(t)$ is given by the following convolution‡

$$y(t) = x(t) * h_c(t) = \sum_n x_n(t) * h_c(t) \tag{8.7}$$

The output of each of these filters is then fed into the optical MSK narrowband frequency discrimination receiver shown in Figure 8.3. The outputs of the delay lines $d_{F1}(t)$ and $d_{F2}(t)$ can be modelled as a convolution of the channel output $y(t)$ with each of the filters impulse responses, $h_{F1}(t)$ and $h_{F2}(t)$ respectively and correctly delayed delta pulses, $\delta(t \pm t_d/2)$, that is

$$d_{F1}(t) = \frac{1}{\sqrt{2}} y(t) * h_{F1}(t) * \delta(t + t_d/2)$$

and $\tag{8.8}$

$$d_{F2}(t) = \frac{1}{\sqrt{2}} y(t) * h_{F2}(t) * \delta(t + t_d/2).$$

---

* We also do not include any fiber non-linearities in the model since as discussed above it can be easily avoided in MSK transmission.
† The optical carrier with a line spectrum is assumed. This is a valid assumption considering the fact that the source is always on and external modulators are used.
‡ The optical carrier with a line spectrum is assumed.

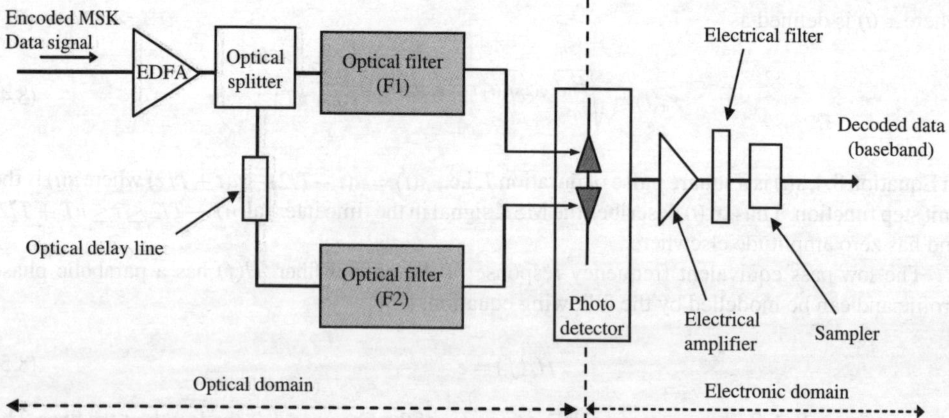

FIGURE 8.3　Narrowband optical filter detector for optical MSK.

To simplify the above convolutions, two template functions $\psi_{F1,I}(t), I \in -1, +1$ for filter F1, and two functions $\psi_{F2,I}(t), I \in -1, +1$ for filter F2, are defined as follows

$$\psi_{F1,I}(t) = \sqrt{\frac{E}{T}}\, e^{j2\pi f_d(t+T/2)I} s(t) * h_c(t) * h_{F1}(t) * \delta(t + t_d/2),\ I \in -1, +1, \tag{8.9}$$

$$\psi_{F2,I}(t) = \sqrt{\frac{E}{T}}\, e^{j2\pi f_d(t+T/2)I} s(t) * h_c(t) * h_{F2}(t) * \delta(t + t_d/2),\ I \in -1, +1, \tag{8.10}$$

These template functions represent the correctly delayed outputs of filters F1 and F2 for each of the +1 and –1 MSK modulated pulse inputs. An example of the template function is shown in Figure 8.4 and discussed in the next section. Using these template functions, and Equation 8.7 and Equation 8.8, the expression for the output of the delay lines can compactly be expressed as follows

$$d_F(t) = \sum_n \psi_{F,I_n}(t + T/2 - nT)e^{j\theta_{n-1}},\ \ F \in F1, +F2, \tag{8.11}$$

The outputs of the delay lines are then converted to the electrical domain using photodiodes in a balanced receiver configuration. The electrical signal is amplified, sampled and sent to the decision device. The decision device selects +1 as the transmitted symbol if the input is positive or –1 otherwise. To express the sampled values going to the decision device, let us express the delay line outputs at the sampling instant $t = kT$ by

$$d_F(kT) = \sum_n \psi_{F,I_n(kT-nT)}e^{j\theta_{n-1}} = \left[\underbrace{\psi_{F,I_k}(0)}_{\text{desired}} + \underbrace{\sum_{n \neq k} \psi_{F,I_n}(kT-nT)e^{j\phi_{n,k}}}_{\text{ISI}}\right]e^{j\theta_{k-1}} \tag{8.12}$$

where, the phase offset ϕ_n, k is given by

$$\phi_n, k = \pi/2 \left[\sum_{n'=k+1}^{n-1} I_{n'}\right]\ \text{if } n > k = -\pi/2 \left[\sum_{n'=n}^{k} I_{n'}\right]\ \text{if } n < k. \tag{8.13}$$

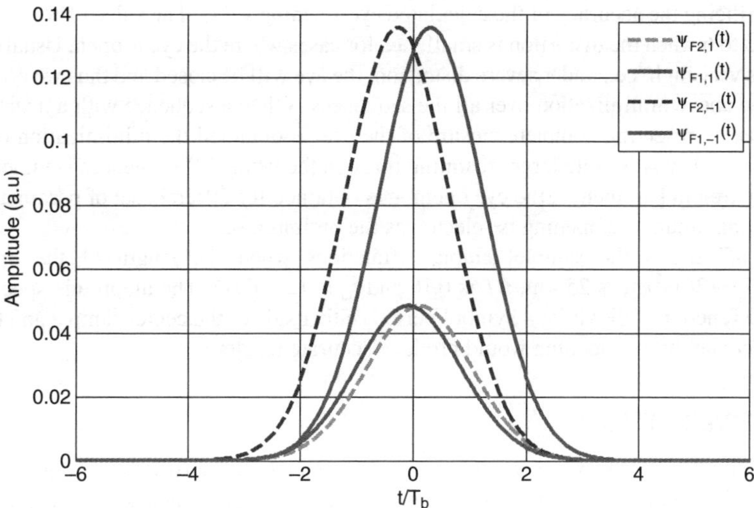

FIGURE 8.4 Eye openings given by the template functions ψ_F of the discrimination filters F1 and F2 for each the symbol "+1" and −"1".

The last term $e^{j\theta_{k-1}}$, in Equation 8.12, need not be considered further because it is lost when the signal is passed through the photodiode. The output of the balanced receiver at sampling instant $t = kT$ is then given by $|d_{F1}(kT)|^2 - |d_{F2}(kT)|^2$.

The photodiode introduces quadratic point non-linearity into the system and complicates the analysis of the receiver design. Therefore, we present a semi-analytical method to evaluate receiver performance. This method shows the mutual interactions among the receiver's key parameters and thus provides the design guidelines for selection of their optimum values.

When system performance is limited by inter-symbol interference, the eye opening gives a good indication of system performance. The proposed method relies on obtaining the eye opening for different values of the filter parameter. The parameters that give maximum eye opening will provide the best performance[§]. To obtain the eye opening, we first consider the transmission of +1 at symbol interval 0, and find out the minimum sample value, $d_{\min}^{+1}$ over all the input train sequences that have +1 as its 0th symbol, i.e.,

$$d_{\min}^{+1} = \max\left\{\ldots,,\ldots|d_{F1}(0)|^2 - |d_{F2}(0)|^2, 0\right\}. \tag{8.14}$$

Similarly we obtain the minimum negative sample value when −1 is transmitted, i.e.,

$$d_{\min}^{-1} = \max\left\{\ldots,,\ldots -|d_{F1}(0)|^2 - |d_{F2}(0)|^2, 0\right\}. \tag{8.15}$$

The eye opening is then given by $d_{\min}^{+1} + d_{\min}^{-1}$. Carrying out the inner minimization over all the possible sequences is prohibitive, and so we limit the length of the sequence to $2N + 1$, i.e., N symbols on either side of the 0th symbol. We select the value for N such that the maximum time span over which the template functions have a significant absolute value is less than $(2N + 1)T$. Thus, we limit the minimization operation in Equation 8.14 and Equation 8.15 to be carried over just 2^{2N} sequences

---

[§] The eye spectrum is necessary to find out exactly the optimum parameters. However, considering the fact that most number of errors occur when the eye is at its minimum, considering only the eye opening is accurate enough for most purposes.

without sacrificing the accuracy of the calculated eye-opening value. The value of N is usually quite small, e.g., 3 to 6 when the distortion is small, i.e., for cases where the eye is open. Usually when the value of N is very high, i.e., under severe distortion, the eye will be closed and thus we do not need to carry out the above minimization over all the sequences. When a sequence with a negative sample value (that is, representing complete closure of the eye) is obtained the minimization operation is stopped. Thus, when N is quite large, scanning through the whole 2^{2N} sequences is unnecessary. To obtain the optimum parameters, the eye opening is obtained for different set of parameters and the set with the maximum eye opening is selected as the optimum set.

Figure 8.7 shows the sample template functions when the length of the optical fiber $(\beta_2 = -2.17e - 26)$ s^2/m is 25 km, BT is 0.16 and f_ω is 12.5 GHz. The amplitude spectrum of all the template functions fall within 4 symbol periods either side of the center sample and thus taking $N = 6$ to calculate the eye opening would produce accurate results.

8.3 RECEIVER DESIGN

The proposed optical MSK receiver is based on the principles of narrowband frequency discrimination and relies on the alignment of the signal arrivals at two photodiodes for a receiver-balanced configuration. Therefore, it is important to optimize the parameters that have effects on receiver performance. Key parameters include the bandwidth and the center frequency of the optical discrimination filters and the ODL. In this section, an intuitive justification for the effects of key parameters on the receiver performance based on the four template functions $\psi_{F,I}(t)$ is provided. Discussions on the receiver design also provide design guidelines for the selection of optimum filter parameters.

8.3.1 BANDWIDTH OF THE OPTICAL FILTER

The receiver performance critically depends on the bandwidth B of the two optical discrimination filters. They have to be sufficiently narrow compared with the transmission bit rate $(R = 1/T)$ to discriminate the USB and LSB frequencies f_d and $-f_d$ effectively (written in low-pass equivalent format). The switching of frequencies at the rate of $1/T$ results in the spectrum of MSK spilling over a range of frequencies centered around f_d and $-f_d$ corresponding to either "+1" or "−1" is transmitted. This "spectral spilling" causes the leakage problem and in the MSK modulation scheme, where the two transmission modulation frequencies are close to each other, the leakage terms are substantial. For example, consider the transmission of a "+1" pulse. The bulk of the energy of a "+1" modulated MSK pulse would be captured by filter F1 and a small amount of the leakage energy will be captured by filter F2. For the system to discriminate the "+1" pulse effectively, the leakage term should be comparatively smaller. The effect would be reversed for a "0" MSK modulated pulse.

The ratio of the signal energy to the leakage energy from F1 and F2 increases with the decrease in the BT product of the filters. However, if the bandwidths of the filters are too narrow, the energy captured by F1 and F2 goes down, reducing the discrimination property. Thus, there should be an optimal BT product that gives the best performance for the detection. Under this circumstance, the ISI is not taken into account. The above justifications are demonstrated in Figure 8.2. The values of BT within the range of 0.4 to 0.5 obtain the maximum EOs and thus, are the optimum values for the optical discrimination filters. Increasing the BT product, i.e., increasing the filter bandwidth over 0.5 does not further open the eye.

When propagating through the fiber, different spectral components of the optical pulse experience different delays and thus arrive at the receiver at different times. This leads to pulse spreading in the time domain, causing the ISI. When optical signals are passed through the discrimination filter, the filter attenuates frequencies away from its center frequency. The lower the bandwidth of the filter, the more frequency components of the signal are suppressed, reducing ISI effects caused by fiber CD. As a result, when the length of the fiber increases, i.e., the dispersion caused by the fiber is getting more severe, the bandwidth of the optical discrimination filter should be reduced equivalently. However, when the bandwidth of the filter is lowered, its impulse response starts to spread more in the time domain, introducing the ISI caused by the filter's impulse response itself. Therefore, there should be an optimum bandwidth which is the intercept point of these two trends of ISI effects, as illustrated in Figure 8.2.

The performance of the receiver depends critically on the bandwidth, B, of the two filters. They must be sufficiently narrow compared with the transmission bit rate ($R = 1/T_b$) to discriminate the two frequencies f_1 and f_2 effectively. The bulk of the energy of a "+1" modulated MSK pulse would be captured by filter F1, and a small amount of leakage energy will be captured by filter F2. The effect would be reversed for a "0" modulated pulse. The ratio of the energies captured by F1 and F2 increases with the decrease in time-bandwidth product, BT, of the filters. However, if the bandwidth of the filters is too narrow, the energy captured by F1 and F2 decreases, resulting in a lower ratio. In addition, the narrower the bandwidth, the higher is the inter-symbol interference (ISI). Thus, there is an optimal BT product that gives the best performance for detection. The two discriminating filters F1 and F2 used in the analysis are modeled as Gaussian filters having the same bandwidth-time product of BT and with center frequencies at $f_1 = f_c + f_d$ and $f_2 = f_c - f_d$ respectively.

When the received signal is passed through a filter, the filter attenuates frequencies away from its center frequency. Thus, the dispersion due to the channel is reduced and the pulse spreading and ISI are reduced. The lower the bandwidth of the filter, the more the frequency components of the signal are suppressed, which leads to lower ISI. However, when the bandwidth of the filter is lowered, its impulse response starts to spread more in the time domain, leading to more ISI. The optimum bandwidth of the filter is when the pulse spreading after filtering is equal to the spread of the filters impulse response.

One consequence of this criterion is that, for optimal performance, with increasing fiber lengths, the bandwidth of the filters should be reduced when the length of the fiber increases the dispersion caused by the channel increases. Thus, to compensate for this effect, the bandwidth of the filter should be reduced equivalently for optimal performance. Although not shown here, this has been validated by analysis.

The switching of frequencies at rate $1/T$ results in the spectrum of MSK spilling over a range of frequencies centered around f_d and $-f_d$ corresponding to whether +1 or −1 is transmitted. When such a signal is passed through the optical channel, the different frequencies travel at different speeds and thus arrive at the receiver at different times. This leads to pulse spreading in the time domain.

8.3.2 Center Frequency of the Optical Filter

The MSK spectrum spilling over a range of frequencies centered around f_d and $-f_d$ causes another problem due to leakage. For example, consider the transmission of a +1 pulse. It has frequency components centered around f_1. The filter F1 with center frequency f_ω will capture most of the signal. However, frequency components of the pulse falling in the bandwidth of F2 called the "leakage" term will be captured by F2. For the system to discriminate the +1 pulse, the leakage term should be comparatively smaller. In the MSK modulation scheme, where the two transmission frequencies are very close to each other, the leakage terms are substantial.

One way to reduce the leakage terms is to select the filters with center frequencies away from the transmission frequencies used by the MSK signal, i.e., $f_\omega > f_d$. Although off-setting the center frequency of the filters away from the transmission frequencies reduces the leakage term, it also reduces the signal captured by the primary filter, leading to reduced discrimination properties. So the optimum value for the center frequency is obtained when the reduction in the leakage term offsets the reduction in the amount of signal passing through the primary filter.

One consequence of the above design criteria is that, for optimal performance, the center frequency of the filters should be offset more when the bandwidth of the filters increases. When the bandwidth of the filters used is large, the leakage term would be high and offsetting the center frequency of the filters offers performance improvements.

The selection of the optimal value of ODL, the optimal bandwidth (B) and the offsets from the nominal center frequency of the optical discrimination filters (F1 and F2) is studied.

8.3.3 Selection of Optimal Values of Optical Delay Line

Figure 8.5 shows the analytical results for eye opening of the receiver for different BT products of F1 and F2, and for different mismatch values of ODL for 25-km SSMF transmission or effectively 425 ps/nm of residual dispersion. Here, and in the rest of the text, length of the fiber refers to a residual distance that is not compensated for chromatic dispersion. The Gaussian-type optical discrimination filters are used in the analysis. This analysis is based on obtaining the expressions for eye-opening taking into account the effects of the ISI of pre-cursor and post-cursor symbols. Zero mismatch represents the value of $2\pi f_D|\beta_2|L$ for the delay, and the mismatches are measured from this value as a percentage of the pulse width T of 25 ps. Larger values for eye opening signifies better system performance, and a zero value indicates complete closure of the eye. The mesh shown in Figure 8.5 presents a guideline for optimizing the performance of the proposed receiver at a particular length of fiber, and also indicates if the system will have an error floor for a particular setting.

8.4 SELECTION OF OPTIMUM BANDWIDTH AND CENTER FREQUENCY OF OPTICAL FREQUENCY DISCRIMINATION FILTERS

Narrowband filtering has a trade-off between the discrimination property and the ISI mitigation. Figure 8.6 shows the analytical results for eye opening against different BT products for various lengths of the fiber.

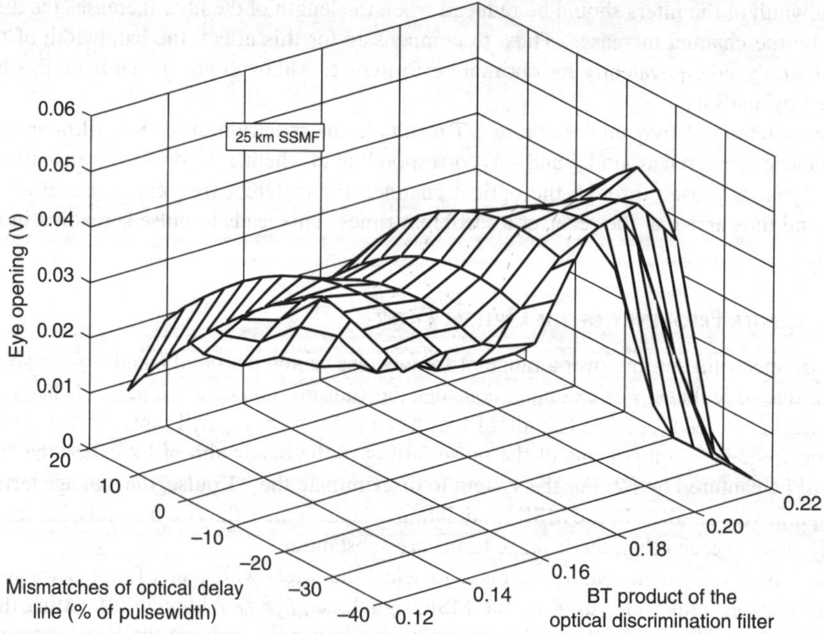

FIGURE 8.5   Analytical results for eye opening of the receiver for different BT products of F1 and F2 and for different mismatch values of ODL for 25-km SSMF. (From Huynh, T. L., L. N. Binh, and K. K. Pang, Optimal MSK long-haul transmission systems. In *SPIE Proceedings of APOC'06* 6353–86, Thuga, Kwangju, Korea, 2006; and Huynh, T. L., T. Sivahumaran, K. K. Pang, and L. N. Binh, A narrowband filter receiver acheiving 225 ps/nm residual dispersion tolerance for 40 Gb/s optical MSK tranmission systems, *Conference on Optical Communication, OFC'07*, CA, USA, 1–3, 2007. With permission.)

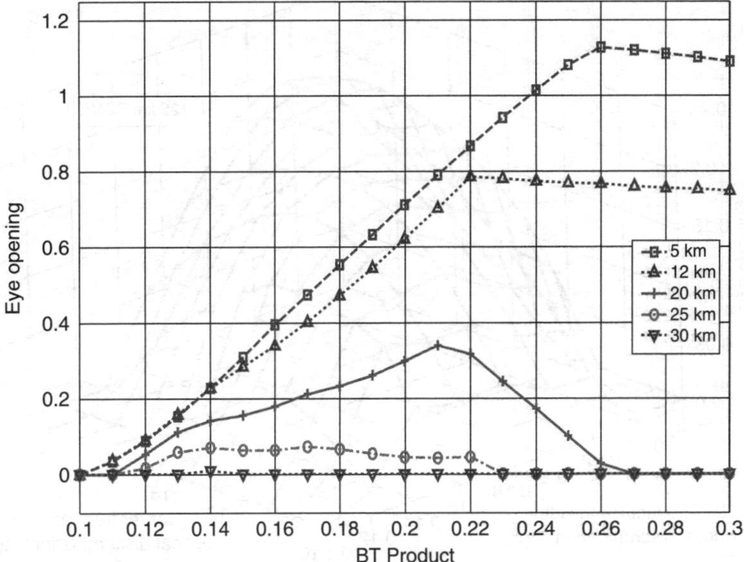

FIGURE 8.6 Eye opening against different BT products under transmission of various lengths of the fiber. (From Huynh, T. L., L. N. Binh, and K. K. Pang, Optimal MSK long-haul transmission systems. In *SPIE Proceedings of APOC'06* 6353–86, Thuga, Kwangju, Korea, 2006; and Huynh, T. L., T. Sivahumaran, K. K. Pang, and L. N. Binh, A narrowband filter receiver acheiving 225 ps/nm residual dispersion tolerance for 40 Gb/s optical MSK tranmission systems, *Conference on Optical Communication, OFC'07, CA*, USA, 1–3, 2007. With permission.)

Here, and in the rest of the text, length of the fiber refers to residual distance that is not compensated for chromatic dispersion. Larger values of eye opening signify better system performance, and a zero value represents failure of the receiver. From Figure 8.6, it can be seen that, for long span of standard single mode fiber (SSMF), i.e., 25 km and 30 km, the optimal BT is obtained at 0.14 (that is, $B = 5.6$ GHz) while it is 0.28, 0.22 and 0.21 for the length of 5 km, 12 km and 20 km SSMF, respectively.

Analytical results of offset filtering for different bandwidth (that is different *BT* products where $T = 25$ ps) of the Gaussian-type filters are shown in Figure 8.7 for 25 km SSMF (effectively 425 ps/nm) of residual dispersion. The analysis is based on obtaining the expressions for eye opening taking into account the effects of the inter-symbol interference (ISI) caused by pre-cursor and post-cursor symbols. Larger values for eye opening signify better system performance, whereas a zero value indicates failure of the proposed receiver. As shown in Figure 8.8, offset filtering can offer significant performance improvements with appropriate bandwidth for the optical discrimination filters. For higher offset frequencies, the optimum bandwidth of the optical filters is increased. The optimum BT is obtained at about 0.22, and the optimum center frequency of the optical filters is shifted from 10 GHz towards 15 GHz, i.e., an offset of 5 GHz.

A mesh for the analysis on transmission of 40 Gb/s optical MSK systems over a length of 35 km SSMF is shown in Figure 8.8. Figure 8.8a shows the eye openings vs BT and center frequencies of optical discrimination filters, while Figure 8.8b shows that there are optimal values of BT and f_c to give the positive value in the eye opening.

The results show that the optical MSK receiver still has eye opening and thus, the narrowband frequency discrimination receiver may offer error-free demodulation of data for an optical

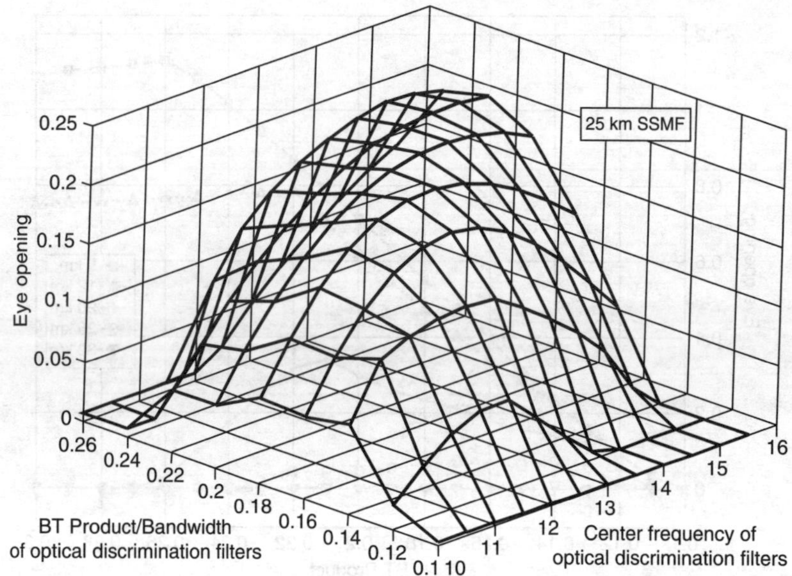

FIGURE 8.7 Mesh diagram for decision of optimum eye opening with respect to the BT product and centre frequency of the discrimination filter. (From Huynh, T. L., L. N. Binh, and K. K. Pang, Optimal MSK long-haul transmission systems. In *SPIE Proceedings of APOC'06* 6353–86, Thuga, Kwangju, Korea, 2006; and Huynh, T. L., T. Sivahumaran, K. K. Pang, and L. N. Binh, A narrowband filter receiver acheiving 225 ps/nm residual dispersion tolerance for 40 Gb/s optical MSK tranmission systems, *Conference on Optical Communication, OFC'07*, CA, USA, 1–3, 2007. With permission.)

link of up to 35-km SSMF non-compensated optical link for 40 Gb/s optical MSK transmission. In terms of 10-Gb/s transmission systems, the above figure corresponds to 560 km SSMF respectively.

8.5 RECEIVER PERFORMANCE

8.5.1 NUMERICAL RESULTS VALIDATING RECEIVER DESIGN

The analysis shown in Figure 8.9 is validated with simulation results. Figure 8.9 shows the simulation results of the receiver sensitivity penalty against different center frequency offsets for various BT products of the optical discrimination filters at a SSMF length of 25 km. For relatively narrow bandwidths (BT = 0.13 − 0.17), the receiver achieves optimum receiver sensitivity with center frequencies in the range of 11 GHz to 12.5 GHz. Within this optimum range, the change in the receiver performance is gradual, whereas the performance severely degrades when the offset frequencies fall outside this range. However, when considering relatively large bandwidths of the optical discriminating filters (BT = 0.2 and 0.25), the optimum center frequency shifts towards 14.5 GHz rather than 12.5 GHz. The simulation results in Figure 8.9 closely agree with the analytical results.

Figure 8.10 shows the numerical results for receiver sensitivity penalty for different optical delay mismatches. Zero mismatch represents the value of $2\pi f_D \beta_2 L$ for the delay, and mismatches are measured from this value as a percentage of the pulse width T ($T = 25$ ps). As shown in Figure 8.10, when the distance increases, the optimum delays shift to smaller values. For 5-km SSMF, the optimum value is at zero, whereas for 25-km SSMF, it is −12.5%. Secondly, the receiver is less sensitive to delay mismatches at short fiber lengths or small residual dispersions. In 5-km SSMF, the tolerance range is about 45% (−20 to +25%) at 1-dB power penalty and about 22% in 25-km SSMF.

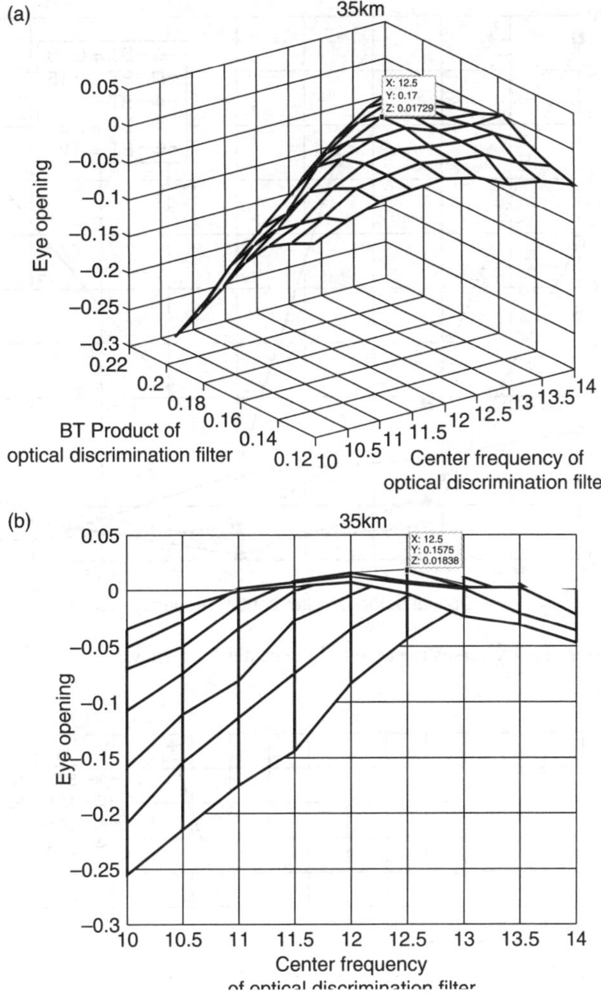

FIGURE 8.8 (a) Eye openings vs BT and center frequencies of optical discrimination filters (b) Optimal values of BT and f_c. (From Huynh, T. L., L. N. Binh, and K. K. Pang, Optimal MSK long-haul transmission systems. In *SPIE Proceedings of APOC'06* 6353–86, Thuga, Kwangju, Korea, 2006; and Huynh, T. L., T. Sivahumaran, K. K. Pang, and L. N. Binh, A narrowband filter receiver acheiving 225 ps/nm residual dispersion tolerance for 40 Gb/s optical MSK tranmission systems, *Conference on Optical Communication, OFC'07*, CA, USA, 1–3, 2007. With permission.)

8.6 ROBUSTNESS TO CHROMATIC DISPERSION OF ONFDR

8.6.1 40-Gb/s Transmission

Figure 8.11 shows the simulation system configuration considered for the investigation of the performance and dispersion tolerance of non-coherent 40 Gb/s optical MSK systems using the proposed narrowband filter receiver.

The input power into fiber (P_0) was kept lower than the non-linear threshold power. The ideal EDFA2 provides the gain to completely compensate for the attenuation of the signals propagating through SSMF. As shown in Figure 8.12, the optical received power (P_{Rx}) is measured at the input of the proposed MSK receiver. In order to obtain the BER curve at a particular length of fiber, P_{Rx} is varied by adjusting the attenuator. The key simulation parameters used in the experiment are

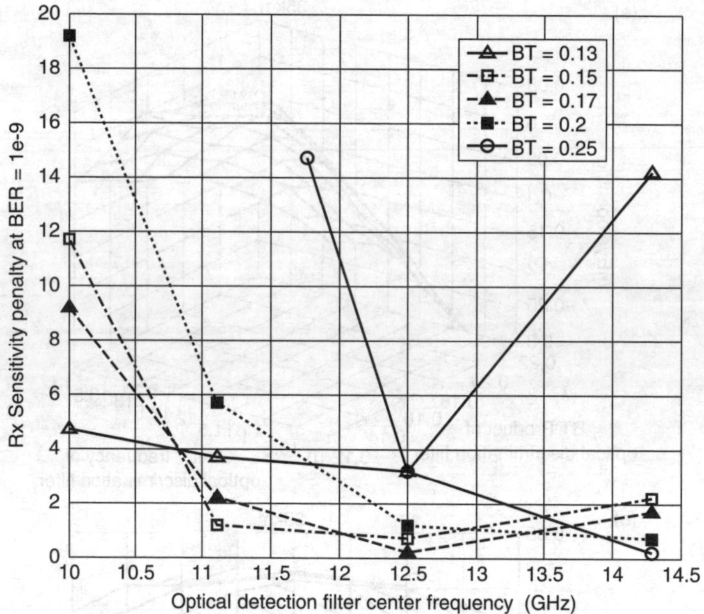

FIGURE 8.9 Receiver penalty versus the offset of the center frequency.

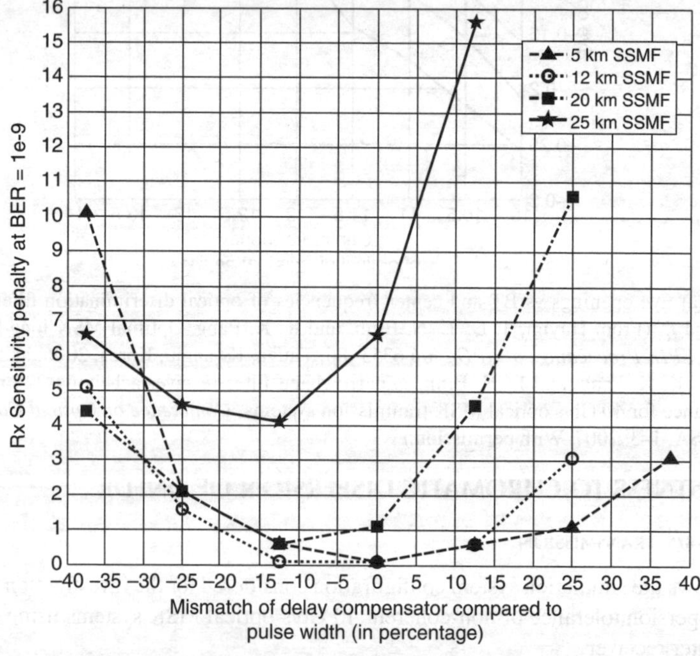

FIGURE 8.10 Receiver penalty versus the mismatch of the delay comparator.

given in Table 8.1. Various residual dispersions are obtained by varying the lengths of SSMF from 0 km to 30 km. The receiver sensitivity of -23.2 dBm at BER $=$ 1e-9 is achieved in back-to-back experiment in case of nominal filtering. The optical delay line value is based on the differential

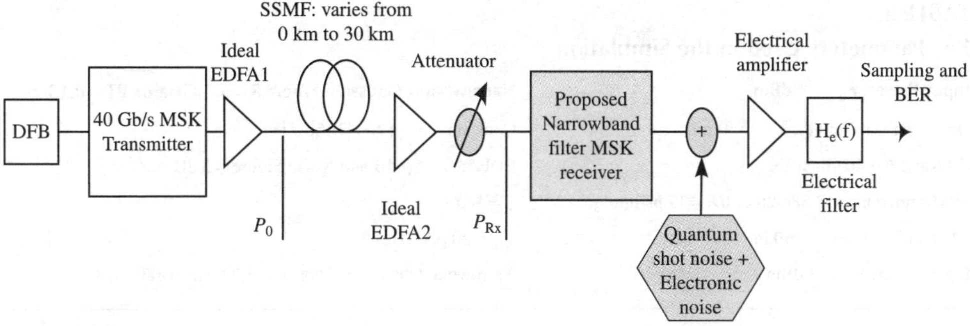

FIGURE 8.11 Simulation setup for investigation on dispersion tolerance of the 40-Gb/s MSK system using narrowband filter receiver.

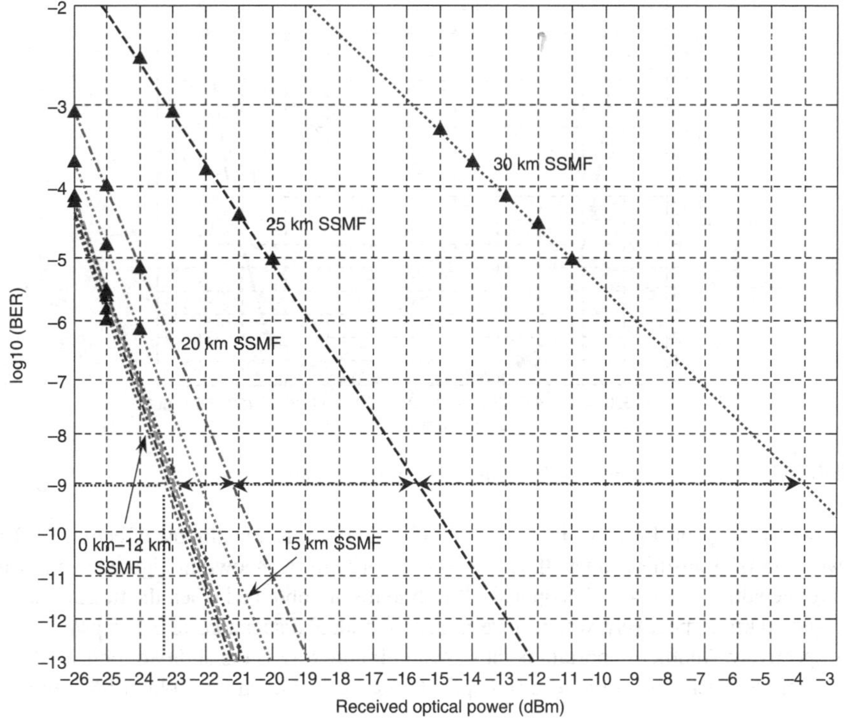

FIGURE 8.12 BER vs received optical received powers for transmissions over 0 km to 30 km SSMF without dispersion compensation.

group delay between f_1 and f_2 calculated as $2\pi f_D \beta_2 L$ where $f_D = |f_1 - f_2| = R/2$, β_2 represents group velocity delay parameter of the fiber and L is the fiber length. The numerical results are obtained via Monte Carlo simulation.

8.6.2 DISPERSION TOLERANCE

Figure 8.12 shows the BER versus optical received power curves corresponding to different lengths of transmission fiber. The power penalty caused by residual dispersion is shown in Figure 8.13. The numerical results are obtained via Monte Carlo simulation (triangular markers as shown in Figure 8.14)

TABLE 8.1
Key Parameters Used in the Simulation

Input Power: $P_0 = -3$ dBm	**Narrowband Gaussian Filter: $B = 5.2$ GHz or BT = 0.13**
Operating wavelength: $\lambda = 1550$ nm	Constant Delay: $t_d = |2\pi f_d \beta_2 L|$
Bit Rate: $R = 40$ Gb/s	EDFA: $G = 15$ dB and Noise Figure $= 5$ dB
SSMF fiber: $|\beta_2| = 2.68\text{e-}26$ or $|D| = 17$ ps/nm/km.	$i_d = 10$ nA
Attenuation: $\alpha = 0.2$ dB/km	$N_{eq} = 20$ pA/ (Hz)$^{1/2}$
Input power: $P_0 = -3$ dBm	Narrowband Gaussian filter: $B = 5.2$ GHz or BT = 0.13

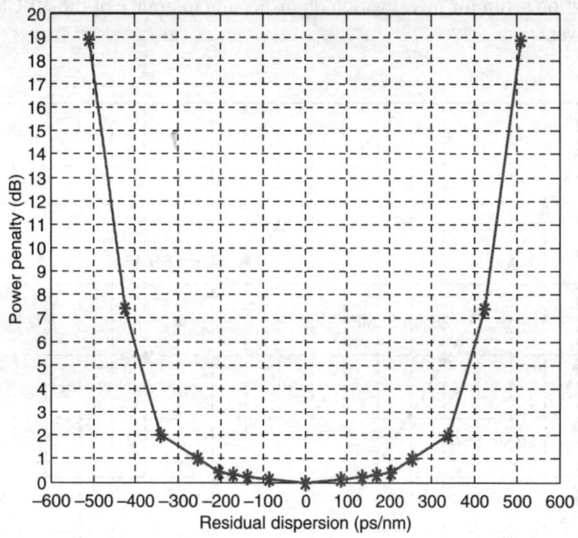

FIGURE 8.13   Dispersion tolerance with optimum values of key parameters.

with the low BER tail of the curve linearly extrapolated. There is a negligible effect on BER performance when SSMF transmission length is from 0 km to 12 km. As shown in Figure 8.4, the 1-dB and 2-dB power penalties of the 40-Gb/s optical MSK systems are obtained when the transmission length is 15 km and 20 km respectively, which corresponds to residual dispersions of ± 225 ps/nm and ± 340 ps/nm, respectively. Another important point shown in Figure 8.12 is the ability of optical MSK transmission system to reach 1e-9 at the received optical power of -4 dBm for 30-km SSMF transmission length.

Comparison of dispersion tolerance of the proposed 40-Gb/s optical MSK narrowband filter receiver in both cases of optimum offset filtering (dashed line and triangular markers) and nominal filtering with center frequencies at ± 10 GHz (solid line and square markers) are shown in Figure 8.14. The filter bandwidth in this case is optimal at BT = 0.14.

There is no advantage in using offset filtering at low residual dispersion less than ± 340 ps/nm or equivalently 20-km SSMF in 40-Gb/s transmission. However, offset filtering offers significant improvement in receiver sensitivity at high residual dispersion.

Approximately 2-dB and 9-dB gain of receiver sensitivity are achieved at residual dispersions of ± 425 ps/nm and ± 510 ps/nm or equivalently to 25 km and 30 km of SSMF, respectively. Also, the system can operate up to ± 595 ps/nm or effectively 35 km SSMF, whereas nominal filtering fails when residual dispersion exceeds ± 510 ps/nm. In terms of 10-Gb/s transmission systems, the above figures may effectively correspond to ± 10115 ps/nm or 560-km SSMF, respectively.

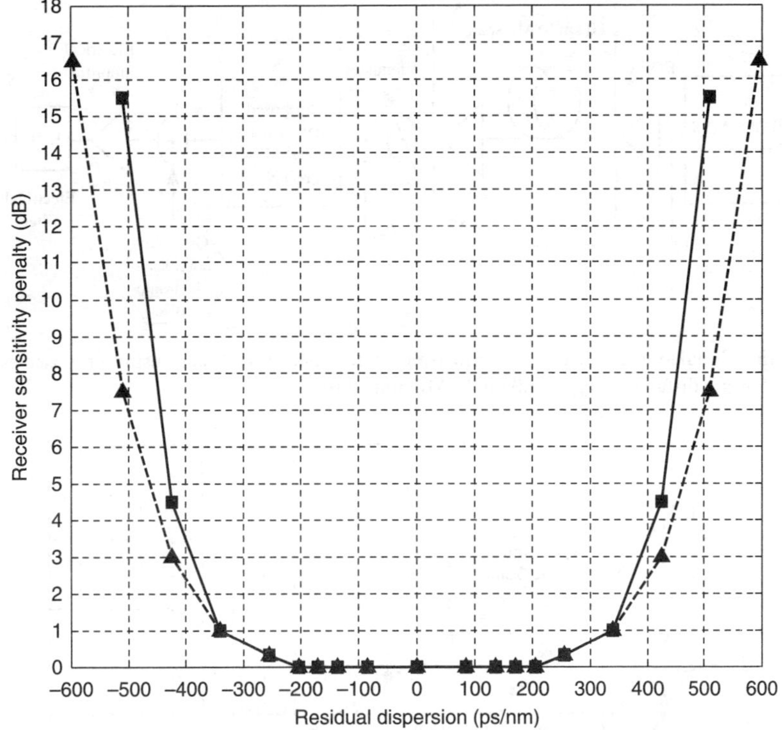

FIGURE 8.14 Dispersion tolerance of 40-Gb/s optical MSK systems with ONFDR in two cases: nominal values (square markers and solid line) and optimum values (triangle markers and dashed line) of key parameters of ONFDR.

8.6.3 10-Gb/s Transmission

Figure 8.15 shows the simulation test-bed for the 10-Gb/s optical MSK transmission system over 560 km SSMF link. Figure 8.16 shows the BER versus optical received power corresponding to different lengths of transmission fiber. The numerical results are obtained via Monte Carlo simulation. As shown in Figure 8.16, the 1-dB and 2-dB power penalties of the 10-Gb/s optical MSK systems are obtained when the transmission length is approximately 320 km and 390 km SSMF, respectively, which corresponds to residual dispersions of ± 5440 ps/nm and ± 6630 ps/nm, respectively. For transmission over 560-km SSMF uncompensated optical link, the performance of 10-Gb/s optical MSK narrowband frequency discrimination receiver is able to reach the BER of 1e-9 at $P_{Rx} = -7$ dBm even though its suffers about 15-dB penalty in receiver sensitivity. Equivalently, a residual dispersion tolerance of up to ± 9520 ps/nm is achieved.

8.6.4 Robustness to Polarization Mode Dispersion (PMD) of ONFDR

Figure 8.17 shows the tolerance to PMD of 40-Gb/s optical MSK signals with use of the narrowband optical filter receiver. At normalized ratio of $<\Delta\tau>/T_o = 0.35$ or $<\Delta\tau> = 8.75$ ps, the receiver performance degrades by 1-dB power penalty [14,15].

8.6.5 Resilience to Non-Linearity (SPM) of ONFDR

The BER versus average optical input power P_o are shown in Figure 8.18. The results are obtained via Monte Carlo simulation. The optical MSK is more sensitive to SPM than CSRZ- and RZ-DPSK,

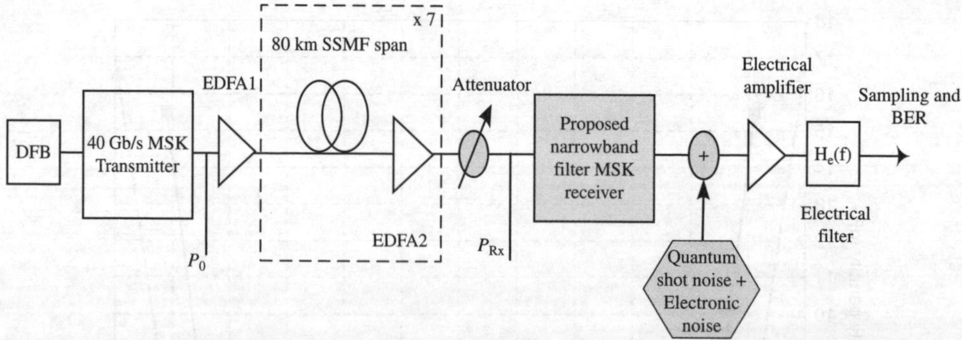

FIGURE 8.15 Simulation test-bed for transmission of 10-Gb/s MSK system using proposed narrowband filter receiver over a distance of up to 560 km SSMF optical link.

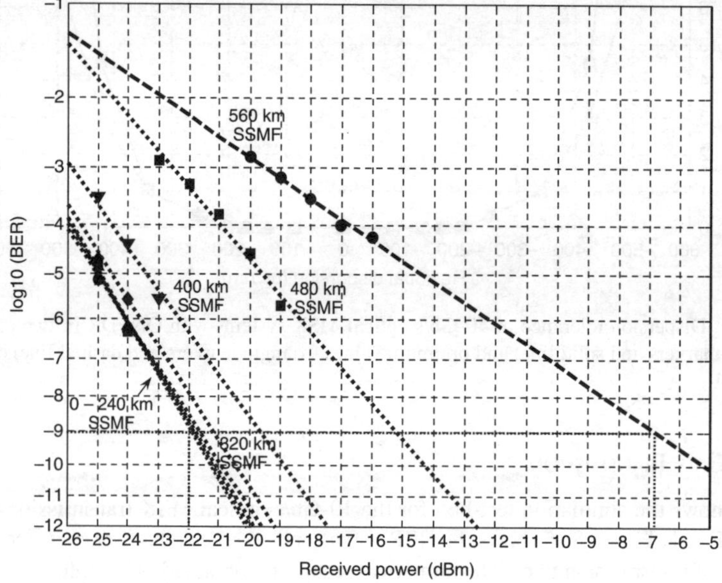

FIGURE 8.16 BER versus received optical powers for 10-Gb/s optical MSK transmission over uncompensated optical links from back-to-back 560-km SSMF with detection based on optical narrowband frequency discrimination receiver.

with the power penalties approximately 3 dB and 4 dB, respectively, at 1e-4. However, optical MSK signal is more robust to SPM than the CSRZ-ASK counterpart.

8.6.6 Transmission Limits of OFDR-based Optical MSK Systems

One common question (and also a requirement for a new modulation format or new detection scheme in optical communications) relates to the transmission limit of an non-regenerated optical link compared to the well-known ones. Thus, in this section, the transmission limit of optical MSK transmission systems with the OFDR for detection is numerically investigated. The simulation results also compare OFDR detection performance to the MZDI-counterpart, and also to the CSRZ-DPSK modulation format. The setting of the input power level and PMD are $P_{in} = -3$ dBm, $<\Delta\tau> = 0.5$ ps/$\sqrt{km}$ for contemporary and new SMF fiber. Figure 8.19 shows the simulated results of the transmission performance of different modulation formats and MSK signals over non-optical regeneration 1760-km and 2640-km

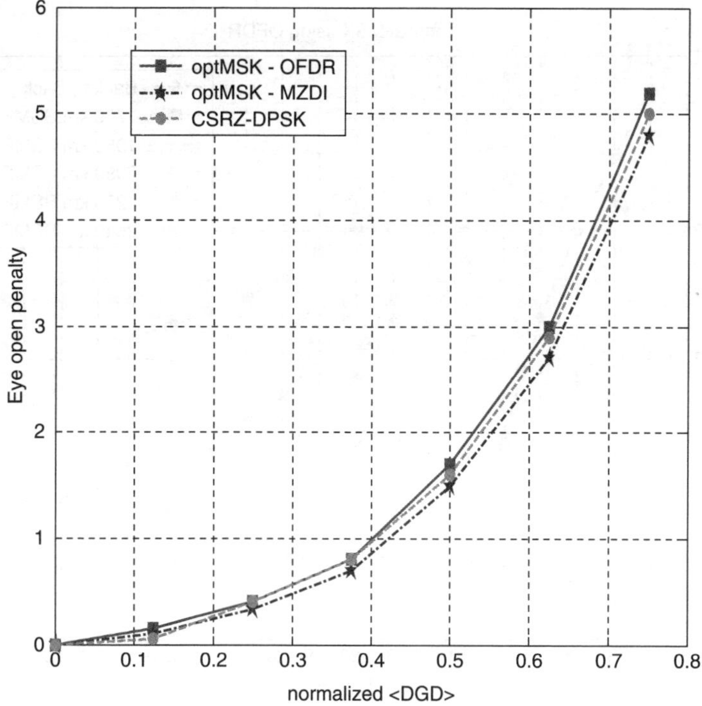

FIGURE 8.17   PMD tolerance of 40-Gb/s optical MSK signals with optical narrowband frequency receiver and comparison with MZDI-based receiver and CSRZ-DPSK.

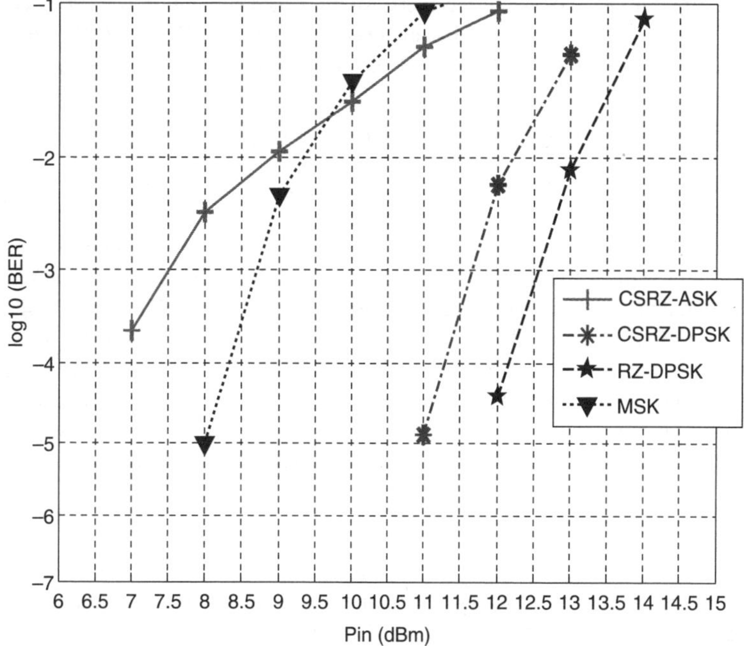

FIGURE 8.18   Comparison of robustness to SPM non-linearity of 40-Gb/s optical MSK signals with narrowband optical filter receiver to CSRZ/RZ-DPSK and ASK.

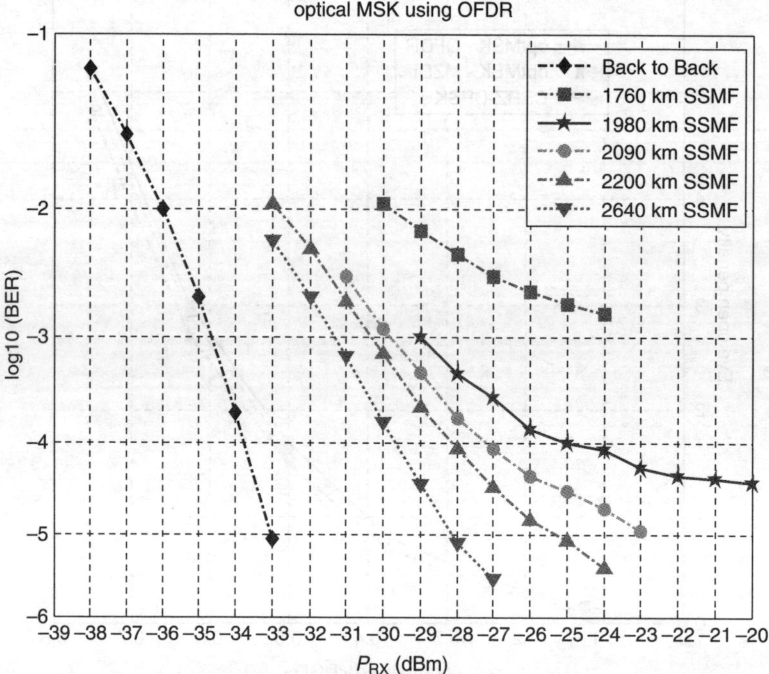

FIGURE 8.19   Transmission performance over non-optical regeneration 1760-km and 2640-km SSMF of different modulation formats and MSK OFDR and MZDI receivers.

SSMF incorporating OFDR and MZDI receivers. Monte-Carlo simulation is used. A comparsion of BER against receiver sensitivity for transmission over 1980-km SSMF under MSK OFDR detection and MZDI receiver as compared with CSRZ DPSK is also illustrated in Figure 8.20.

8.7   DUAL-LEVEL OPTICAL MINIMUM SHIFT KEYING (MSK)

The migration of externally modulated MSK format into optics has recently been reported in [1–6] and in Chapters 9 and 10. If a multi-level scheme can be incorporated into the MSK format, the symbol rate would be reduced, which is of particular interest in optically amplified long-haul communications systems. This is the principal feature of the dual-level MSK modulation scheme, whose generation and detection configurations are proposed for the first time. Dual-level MSK can be seen as a superposition of two optical MSK signals of different amplitudes [7,16–19]. This modulation scheme enables the transmission bandwidth efficiency due to the utilization of two bits per a modulated symbol and at the same time, exploiting the advantage of the narrow spectral characteristic of the MSK format. The constellation of the dual-level MSK modulated signals is shown in Figure 8.21. The mapping scheme of the data information for the dual-level MSK format is shown in Table 8.2.

8.7.1   GENERATION SCHEME

Any configuration of the reported optical MSK transmitters can be utilized for the proposed generation of the optical dual-level MSK signals. Figure 8.22 shows the block structure of the optical dual-level MSK transmitter in which two optical MSK transmitters are integrated in a parallel configuration.

The ratio of the lightwave intensity levels input into the optical MSK transmitters is critical bcause it is used to manipulate the amplitude level of the transmitter signal. This ratio can be obtained with a

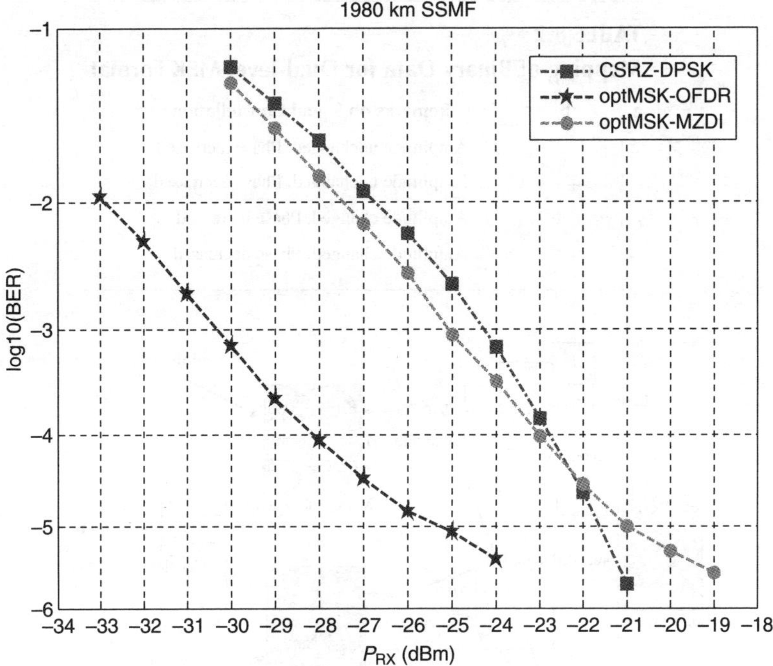

FIGURE 8.20 Comparison of BER against receiver sensitivity for transmission over 1980-km SSMF under MSK OFDR detection and MZDI receiver and compared with CSRZ DPSK.

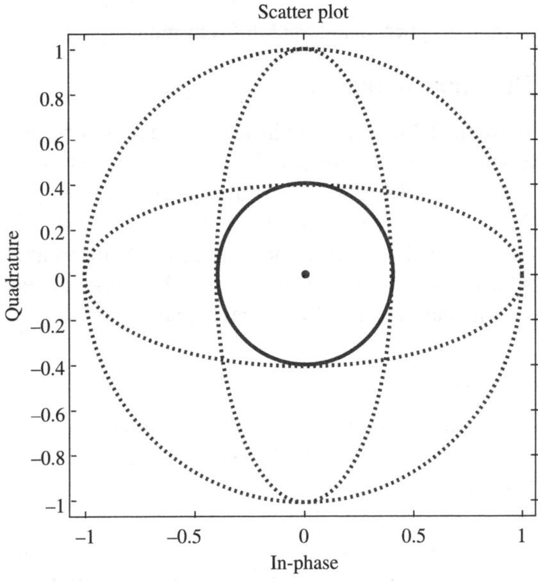

FIGURE 8.21 Signal trajectory of optical dual-level MSK.

high-precision power splitter, or simply by using a 3-dB coupler followed by the optical attenuator on each arm for the intensity manipulation. At an nth instance, the logic sequence of a_n and b_n of $\{\pm 1\}$ are pre-coded from the binary logics d_n of $\{0,1\}$ such that $a_n = 2d_n - 1$ and $b_n = a_n(1 - d_n - 1/h)$ where $h = 1/2$ in the case of MSK.

TABLE 8.2

Mapping of Binary Data for Dual-level MSK Format

a	b	Remarks on Signal Constellation
1	1	Amplitude unchanged, Phase increased
1	−1	Amplitude unchanged, Phase decreased
−1	1	Amplitude changed, Phase increased
−1	−1	Amplitude changed, Phase decreased

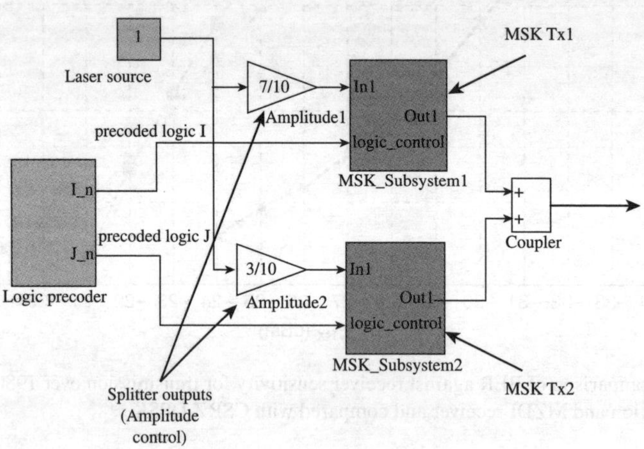

FIGURE 8.22　Block diagram of optical dual-level MSK transmitter.

8.7.2 INCOHERENT DETECTION TECHNIQUE

As a marriage between optical MSK and multi-level modulation formats, the demodulation of dual-level MSK optical signals requires amplitude and phase detection. Similar to the incoherent detection technique for optical MSK, a MZDI-balanced receiver is implemented for phase detection. An additional $\pi/2$ phase shift is introduced in one arm of the MZDI. The amplitude component of the lightwaves is detected by a single photo-detector. A differential decoder is used so that the amplitude changes between two consecutive bits can be demodulated. Figure 8.23 shows the detected eye diagrams of the dual-level MSK optical signals.

8.7.3 OPTICAL POWER SPECTRUM

Figure 8.24 compares the optical power spectra of three modulation formats: 80-Gb/s dual-level MSK, 40-Gb/s MSK and 40-Gb/s NRZ-DPSK. The intensity-splitting ratio for the optical dual-level MSK format is set at "0.8/0.2".

The power spectrum of the optical dual-level MSK format has identical characteristics to that of the MSK format. The main lobe of the spectral width is narrower than that of the NRZ DPSK. The base-width takes a value of approximately ± 32 GHz on either side, compared with ± 40 GHz in the DPSK format.

8.7.4 RECEIVER SENSITIVITY

Three values of the intensity-splitting ratio are studied and compared: "0.7/0.3", "0.8/0.2" and "0.9/0.1". Their amplitude and phase receiver sensitivities are shown in Figure 8.25.

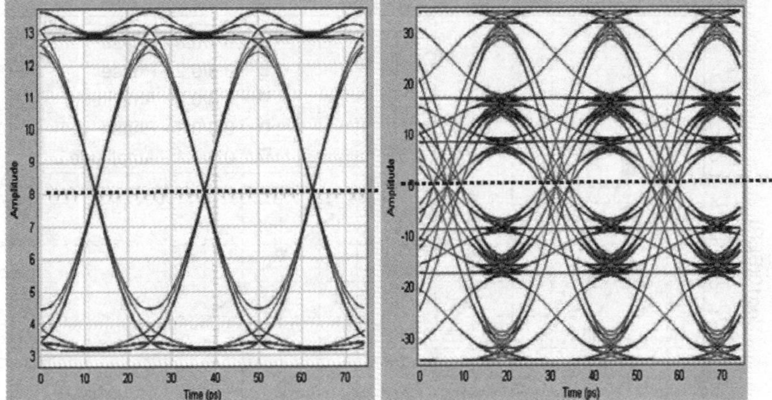

FIGURE 8.23 Eye diagrams of amplitude and phase detection of optical dual-level MSK signals.

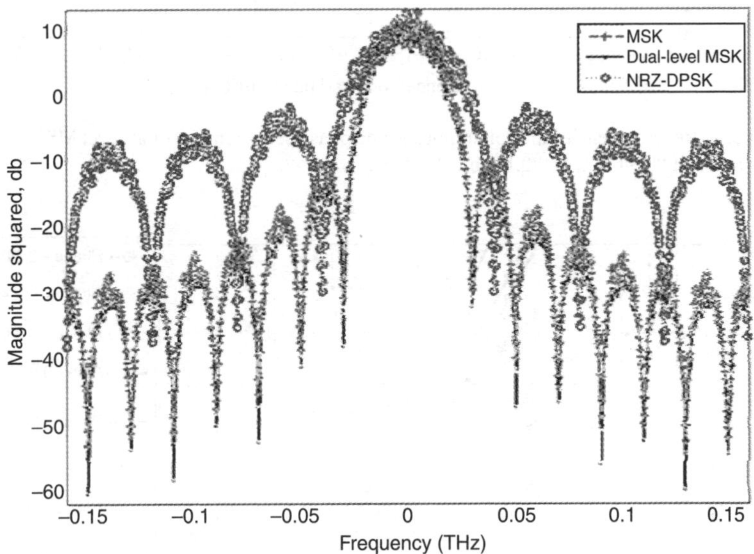

FIGURE 8.24 Spectral properties of 40-Gb/s MSK (dash), 80Gb/s dual-level MSK (solid) and 40Gb/s NRZ DPSK (dot).

The average received optical powers are measured after the 3-dB optical power splitter. At BER = 1e-9, the receiver sensitivies for amplitude and phase detection of "0.8/0.2" are −21.5 and −17.5 dBm, whereas they are −20.2 and −21 dBm for "0.9/0.1". The performance of 80-Gb/s optical dual-level MSK systems is then evaluated over 540 km Vascade fiber comprising 6 optically amplified spans (90 km per span). The Vascade fiber has a complete compensation for chromatic dispersion and dispersion slope. The simulation results are shown in Figure 8.26. The BERs are obtained with the Monte-Carlo method and linearly interpolated. "0.8/0.2" has a slower roll-off linear trend compared with the "0.9/0.1". At BER = 1e-9, the required OSNRs for ratios of "0.8/0.2" and "0.9/0.1" are comparable, being approximately 24 dB. This is much lower than the required OSNR of approximately 32 dB for the case of "0.7/0.3" ratio, thus offering a gain of 6–8 dB. The NRZ, CSRZ and RZ modulation formats of DQPSK modulation are also investigated, as shown in Figure 8.26. The slope of the BER curves for DQPSK formats is steeper than that of the dual-level DPSK. However, the dual MSK format still performs better than that of the CSRZ DQPSK.

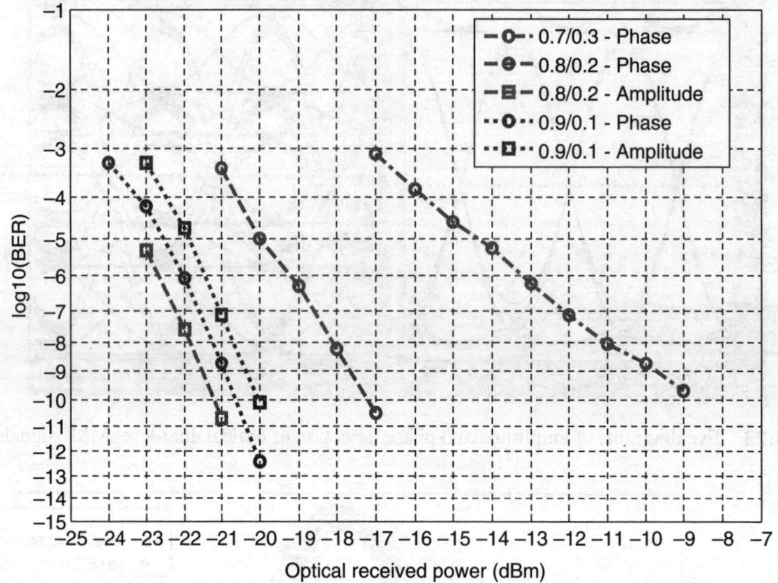

FIGURE 8.25 Receiver sensitivities of amplitude and phase detections for dual-level MSK format with splitting power ratios of "0.8/0.2", "0.9/0.1" and "0.7/0.3".

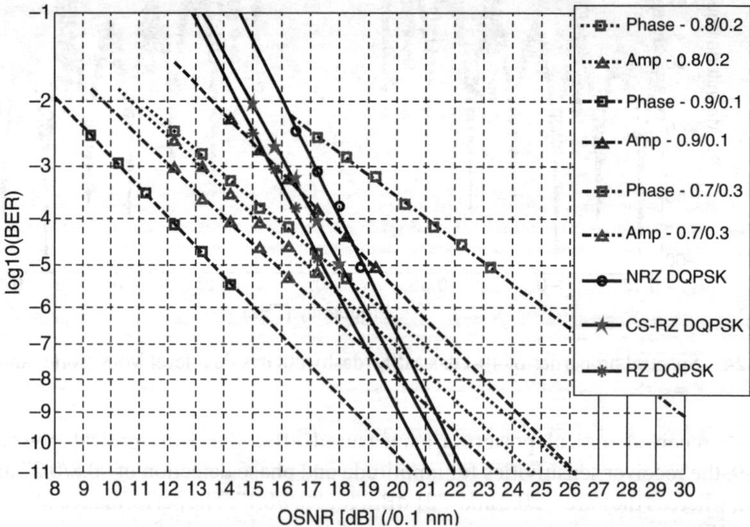

FIGURE 8.26 Performance for amplitude and phase detections of 80-Gb/s optical dual-level MSK format with splitting ratios of "0.8/0.2", "0.9/0.1" and "0.7/0.3", and DQPSK modulation of NRZ, CSRZ and RZ formats.

8.7.5 REMARKS

Spectral properties of the 80-Gb/s optical dual-MSK are similar to those of the 40-Gb/s optical MSK and better than the 40-Gb/s optical DPSK counterpart. This clearly indicates the high spectral efficiency for transmission of the proposed format. The obtained BER performances show the feasible receiver sensitivities and prove the potential for long-haul transmission of the optical 80-Gb/s dual-level MSK format. Finally, the generation scheme can be extended to generate 4-ary MSK signals which employs 3-bit-per-symbol scheme.

8.8   REMARKS

This chapter presents a simple receiver configuration for non-coherent detection of 40 Gb/s optical MSK signals with the utilization of the optical narrowband frequency discrimination receiver. The operational principles of the receiver are explained in detail. Optimum values for key parameters in receiver design, including the bandwidth and the center frequency of the optical discrimination filters and the optical delay line, are obtained. The proposed 40-Gb/s optical MSK receiver offers significant residual dispersion tolerance of up to ± 340 ps/nm for 1-dB penalty, and it is a promising candidate for future high-capacity optical transmission systems.

The non-coherent detection of 10-Gb/s optical MSK signals over an uncompensated 560-km SSMF optical link is possible. This transmission distance covers typical ring or mesh configuration of metropolitan optical networks.

The optical MSK narrowband frequency discrimination receiver significantly improves the residual chromatic dispersion tolerance up to $\pm 10{,}115$ ps/nm. The receiver is also shown to be robust to PMD. Numerical study on resilience of 40-Gb/s optical MSK transmission to SPM non-linearity has been conducted over 880 km SSMF dispersion-managed link and compared with DPSK and ASK counterparts. However, there is a trade-off between significant improvement in dispersion tolerance and non-linearity resilience. With careful power distribution consideration, optical MSK modulation format has been proved to be promising for long-haul and metropolitan transmission.

A dual-level optical MSK transmission system is also introduced as a multilevel MSK scheme. Spectral properties of the 80-Gb/s optical dual-MSK are similar to those of the 40-Gb/s optical MSK, and narrower than 40-Gb/s optical DPSK counterparts. This clearly indicates the high spectral efficiency for transmission of the multi-level format. The obtained BER performances show the feasible receiver sensitivities and prove the potential for long-haul transmission of the optical 80-Gb/s dual-level MSK format. The generation scheme can be extended to generate 4-ary MSK signals in which 3-bit-per-symbol can be used.

REFERENCES

1. Sakamoto, T., T. Kawanashi, and M. Izutsu. 2005. Optical minimum-shift keying with external modulation scheme. *Optics Express* 13 (20): 7741–47.
2. Ohm, M., and J. Speidel. 2004. Optical minimum-shift keying with direct detection (MSK/DD). In *Proceedings of SPIE*, vol. 5281, 150–61.
3. Huynh, T. L., L. N. Binh, K. K. Pang, and L. Chan. 2006. Photonic MSK transmitter models using linear and non-linear phase shaping for non-coherent long-haul optical transmission. In *SPIE Proceedings of of APOC'06*, 6353–85.
4. Mo, J., Y. J. Wen, Y. Dong, Y. Wang, and C. Lu. 2006. Optical minimum-shift keying format and its dispersion tolerance. Optical Fiber Conference, *OFC'06*, 5–10 March 2006, Annaheim CA, USA, paper JThB12.
5. Elrefaie, A. F., R. E. Wagner, D. A. Atlas, and A. D. Daut. 1988. Chromatic dispersion limitation in coherent lightwave systems. *IEEE Journal of Lightwaave Technology* 6 (5): 704–10.
6. Kaminow, I. P., and T. Li. 2002. *Optical fiber communications.* Vol. IVB. USA: Elsevier Science.
7. Binh, L. N., and T. L. Huynh. 2007. Linear and nonlinear distortion effects in direct detection 40Gb/s MSK modulation formats multi-span optically amplified transmission. *Optics Communications* 237 (2): 352–361.
8. Huynh, T. L., L. N. Binh, and K. K. Pang. 2006. Optical MSK long-haul transmission systems. In *SPIE Proceedings of Asia Pacific Optical Communications Conference*, September 2006, Kwanhju, Korea. *APOC'06* 6353–86, Thu9a

9. Sakamoto, T., T. Kawanishi, and M. Izutsu. 2005. Optical minimum-shift-keying with external modulation scheme. *Optic Express* 13 (20): 7741–47.

10. Binh, L. N., and T. L. Huynh.2007. Linear and nonlinear distortion effects in direct detection 40Gb/s MSK modulation formats multi-span optically amplified transmission. *Optics Communications* 273: 352–61.

11. Little, B. E. 2003. Advances in microRing resonator. In *Integrated photonics research conference 2003. Invited Paper*.

12. Proakis, J. G. 2001. *Digital communications*. 4th ed. New York: McGraw-Hill, 185–213.

13. Elrefaie, A. F., and R. E. Wagner. 1991. Chromatic dispersion limitations for FSK and DPSK systems with direct detection receivers. *IEEE Photonics Technology Letters* 3 (1): 71–73.

14. Alic, N., G. C. Papen, R. E. Saperstein, L. B. Milstein, and Y. Fainman. 2005. Signal statistics and maximum likelihood sequence estimation in intensity modulated fiber optic links containing a single optical preamplifier. *Optics Express* 13 (12): 4568–79.

15. Haunstein, H. 2004. PMD and chromatic dispersion control for 10 and 40 Gb/s systems. In *Proceedings of Optical Fiber Communications Conference, OFCO4*, Invited paper Thu3, March 2004, Annaheim CA, 1–3.

16. Agazzi, O. E., M. R. Hueda, H. S. Carrer, and D. E. Crivelli. 2005. Maximum-likelihood sequence estimation in dispersive optical channels. *Journal of Lightwave Technology* 23 (2): 749–62.

17. Proakis, J. G., and M. Salehi. 2002. *Communication systems engineering*. 2nd ed. NJ: Prentice Hall.

18. Proakis, J. G. 2001. *Digital communications*. 4th ed. New York: McGraw-Hill, 185–213.

19. Huynh, T. L., T. Sivahumaran, K. K. Pang, and L. N. Binh 2007. A narrowband filter receiver achieving 225 ps/nm residual dispersion tolerance for 40 Gb/s optical MSK transmission systems. Proc of OFC/NFOE'07, paper OThK3, 25–29 March 2007, Anaheim, CA, 1–3.

9 Partial Responses and Single-sideband Modulation Formats

9.1 PARTIAL RESPONSES: DUO-BINARY MODULATION FORMATS

9.1.1 INTRODUCTION

The demand for high-capacity long-haul telecommunication system has increased over recent years. To achieve high throughput of signals with minimum errors, different advanced modulation formats, such as ASK, PSK (coherent and differential in-coherent) and FSK have been proposed, and comparisons are made to determine which modulation format offers the best transmission performance. In countering performance degradation, modulation formats aim to narrow down the optical spectrum to enable close-channel spacing in the network. They increase symbol duration so that more uncompensated dispersion accumulates before ISI becomes significant. Furthermore, this format is more resilient to fiber non-linearities and optical signal distortion.

Modulation format is important in determining the performance of 40-Gb/s optical fiber communication systems. Duo-binary (DB) and continuous phase modulation (CPM) DB have been shown to offer high spectral efficiency [1,2]. DB modulation formats minimize intersymbol interference impairments in a controlled way instead of trying to eliminate it. It is possible to achieve a signaling rate equal to the Nyquist rate of 2 W symbols per second in a channel of bandwidth BW Hz. Optical DB technique has received much attention due to its high dispersion tolerance and high frequency-utilization efficiency by means of spectral narrowing. DBM format is similar to the non-return-to-zero (NRZ) format, with inclusion of phase coding. The phase characteristics of DBM signals compensate for the group velocity dispersion by reducing the spectral component in conventional NRZ modulation. ISI is reduced because bit patterns such as 101 are transmitted with the ones carrying opposite phase. Therefore, if pulses spread out into the zero time slot, due to dispersion, they tend to cancel each other out. The recovering of signals at the receiver is relatively simple. Furthermore, conventional direct detection receiver is applicable, hence simple receiver structure. There are two types of DBM schemes, which are constant phase and alternating phase in blocks of logics "1s".

This chapter presents the models for photonic transmission with optical channels operating under DB modulation format. This includes the development and implementation of the photonic transmitter, optical fiber propagation and the opto-electronic receiver. DBM encodes two-level electrical signal to three-level electrical signal before modulating the lightwave carrier. The transmitter of the Simulink® model consists of a DB encoder and a dual-drive MZIM. A baseband modulation is first implemented in the DB encoder, which encodes the binary signal into three levels signals: "1", "0" and "–1". MZIM is an electro-optic modulator that converts the electrical signal to an optical signal.

The DB or phase-shape binary modulation formats can be generated by modulating a dual-drive MZIM. Recent studies have shown that the driving voltages for the modulator can be reduced to generate variable pulse width DB optical signals. However, the pulse width of the

DB DPSK has not been thoroughly investigated under the alternating phase of the "1" coded bits. That means the "0" "π" "0" "π" phases of consecutive "1" are contrary to conventional duo-binary formats. We also present modeling performance of alternating phase DB modulation with a full-width half mark (FWHM) ratio with respect to the bit period of 100, 50 and 33% and compared with experimental transmission of carrier-suppressed DPSK over 50 km of SSMF and dispersion compensation. For the DB case, the transmission without dispersion compensation over the same SSMF length offers better performance for 50% FWHM DB modulation, and slightly worse for 33%. The transmission performance, the bit-error-rate versus receiver sensitivity, of these DB modulation formats are compared with the carrier-suppressed DPSK experimental transmission

Section 9.1.2 of this chapter describes the fundamental aspects of DBM format. It is then followed by the description of each component in 40-Gb/s DBM photonic transmission systems in Section 11.3. Section 11.4 is the implementation of the Simulink model of the communication system. Lastly, Section 14.5 gives simulated results. Finally, comparisons with theoretical analyses and other modulation formats are given.

9.1.2 THE DBM FORMATTER

Modulation formats aim to modulate one or more field properties to suit system needs. There are four types of field properties: intensity, phase, polarization and frequency. Symbols are constellated in one or more dimensions to carry more information and to travel further. Data modulation format is the information-carrying property of the optical field.

DBM format has become an attractive modulation format over recent years compared with other formats, such as non-return-to-zero (NRZ), and return-zero (RZ). This is because it can overcome the fiber chromatic dispersion in high-capacity transmissions. It is the characteristics of the DBM format that make it a preferred format.

DBM schemes can be described as correlative-level coding or partial-response signaling schemes. Correlative-level coding schemes means adding ISI to the transmitted signal in a controlled manner, and a signaling rate equal to the Nyquist rate of 2 W symbols per second in a channel of bandwidth W Hz can be achieved. "Duo" in the word duo-binary indicates the doubling of transmission capacity of a conventional binary system. DBM format is, in fact, NRZ modulation with an inclusion of phase coding. The one bits in the data input are phase-modulated. For instance, for a bit pattern of 101, this data will be transmitted with the ones carrying opposite phase, 0 and π. If the pulse of the one bits spread out to the zero time-slot in between, they cancel each other. This effect increases the dispersion tolerance, and allows the signal to be transmitted over a longer distance.

DB coding converts a two-level binary signal of 0's and 1's into a three level signal of "−1", "0" and "+1". This is done by first applying the binary sequence to a pulse-amplitude modulator to produce two level short pulses of amplitude of −1 and +1, with −1 corresponding to 0, and +1 corresponding to 1. This sequence is then applied to DB encoder to produce a three-level output of "−2", "0" and "2".

As shown in Figure 9.1 input sequence, $\{a_k\}$ of uncorrelated two-level pulses is transformed into $\{c_k\}$, which is a sequence of correlated three-level pulses. The correlation between adjacent pulses is equivalent to introducing ISI into transmitted signal in an artificial manner. The DB encoder is simply a filter involving a single delay element and summer (Figure 9.2). However, once errors are made, they tend to propagate through the output. This is because a decision made on the current input a_k depends on the decision made on the previous input a_{k-1}. Therefore, precoding is needed to avoid this error propagation phenomenon. Binary sequence, $\{b_k\}$ is converted into another binary sequence, $\{d_k\}$ by modulo-two addition, exclusive OR of b_k and d_{k-1}, as show in Figure 9.3

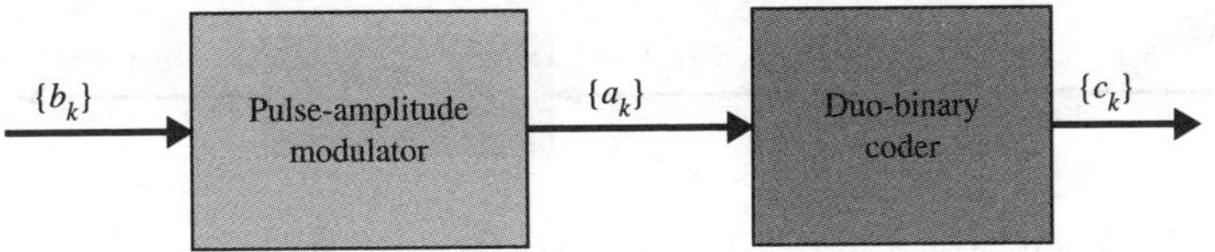

FIGURE 9.1 Brief overview of DB signaling.

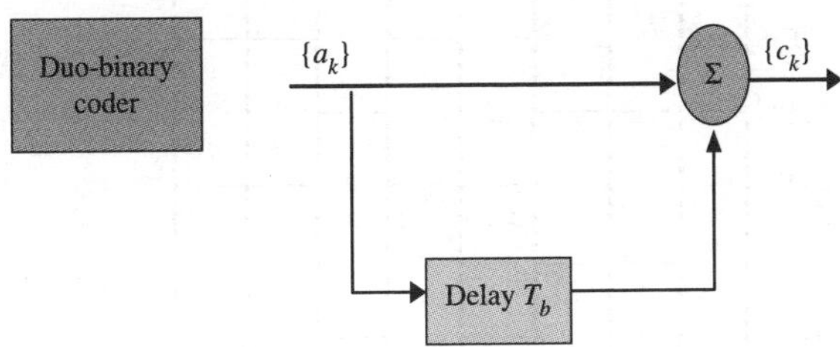

FIGURE 9.2 DB encoder: the block at the left is represented by the signal flow diagram shown on the right.

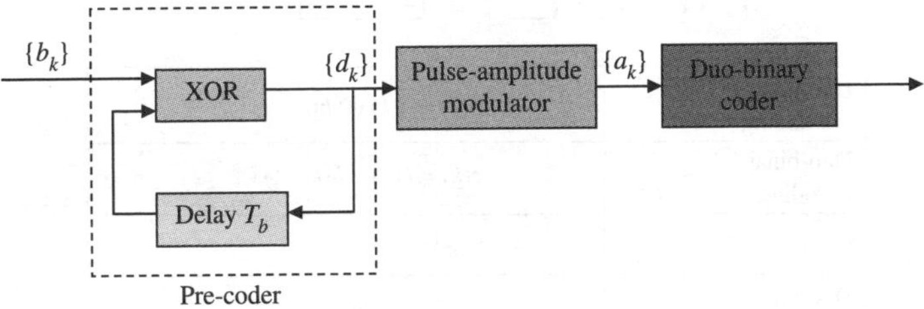

FIGURE 9.3 DBM scheme with pre-coder.

$$c_k = a_k + a_{k-1} \tag{9.1}$$

$$d_k = b_k \oplus d_{k-1}. \tag{9.2}$$

The three-level DB output, $\{c_k\}$ is then modulated into a two-level optical signal by an external modulator. The most commonly used external modulator is the MZIM. The optical DB signal has two intensity levels: "on" and "off". The "on" state signal can have one of two optical phases: 0 and π. The two "on" states correspond to the logic states "1" and "−1" of the DB-encoded signal, $\{a_k\}$ and the "off" state corresponds to the logic state "0". Figure 9.4a shows an example of the original binary signal, the DB-encoded signal, optical DB signal, and a summary of coding rule (Figure 9.4b).

9.1.3 40-Gb/s DB Optical Fiber Transmission Systems

Ultra-long terrestrial networks, transmitting signal at a bit rate of 40 Gb/s, has matured over recent years. Various advanced modulation schemes, such as RZ, NRZ, NRZ-DPSK, and RZ-DPSK have been proposed to achieve an extended reach and an improved capacity of the communication system.

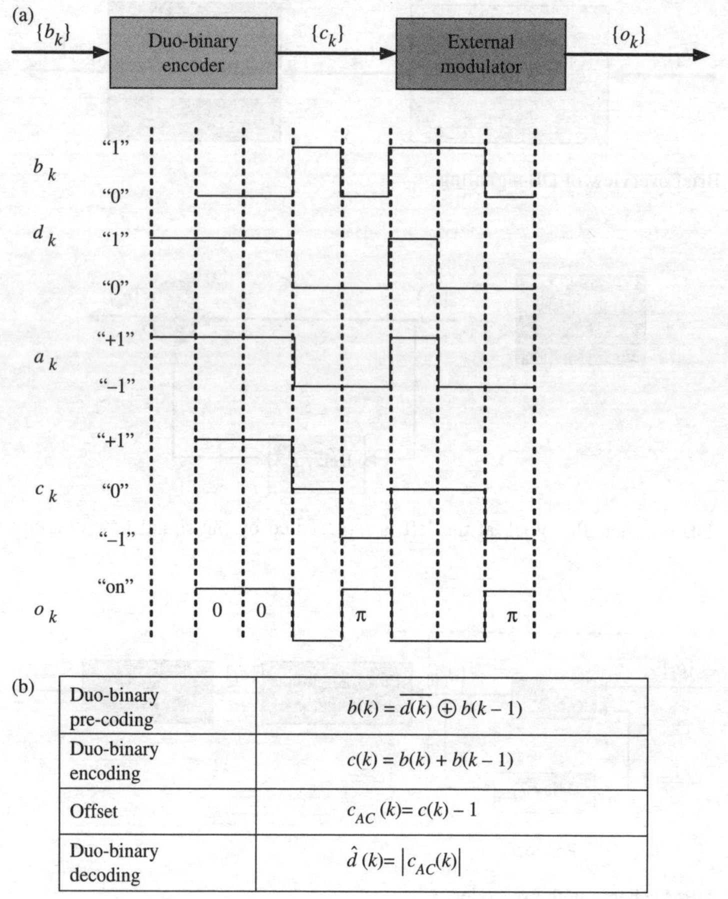

(b)	
Duo-binary pre-coding	$b(k) = \overline{d(k)} \oplus b(k-1)$
Duo-binary encoding	$c(k) = b(k) + b(k-1)$
Offset	$c_{AC}(k) = c(k) - 1$
Duo-binary decoding	$\hat{d}(k) = \left| c_{AC}(k) \right|$

FIGURE 9.4 (a) Example of original binary signal (b_k), pre-coded signal (d_k), DB encoded signal (c_k) and optical DB signal (o_k). (b) Summary of coding rule.

Figure 9.5 shows the typical DWDM optical fiber communication system. Signals are modulated at the transmitters, and are multiplexed together at the wavelength multiplexer before transmitting them into the fiber. The fiber link is divided into a number of spans. Each span consists of a SSMF and DCF. The EDFA is used to compensate for the optical power loss of the transmission span. At the end of the fiber, the signals are demultiplexed and detected at the receivers.

DBM format has become popular compared with other modulation formats because it extends the transmission distance as limited by fiber loss, without regenerative repeaters. It extends the dispersion limit without additional optical components, such as dispersion compensating fiber and DCF. Chromatic dispersion has become a main effect that limits the transmission distance. The optical three-level transmission can overcome this limitation because the narrowband signal has higher tolerance to chromatic dispersion compared with the broadband signal. Furthermore, DB optical fiber communication system can suppress stimulated Brillouin scattering (SBS).

The main modules of the communication system are the transmitter, optical fiber and the opto-electronic receiver, as shown in Figure 9.6. The transmitter consists of the DB encoder and the MZIM. A series of 0's and 1's is modulated under the DBM scheme. These three-level electrical data are then used to modulate the laser source, producing a two-level optical signal. This modulated

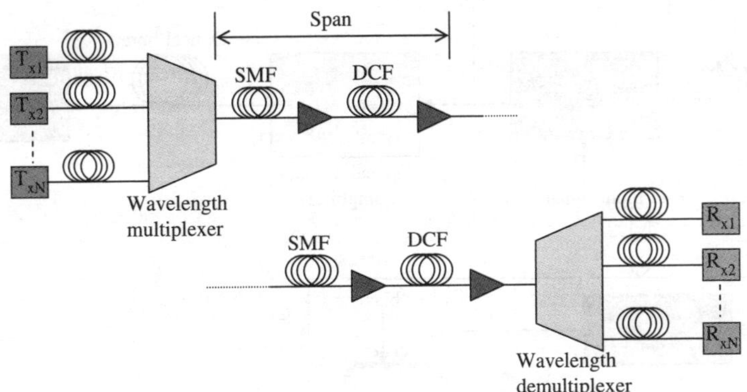

FIGURE 9.5   Ultra-long-haul fiber transmission.

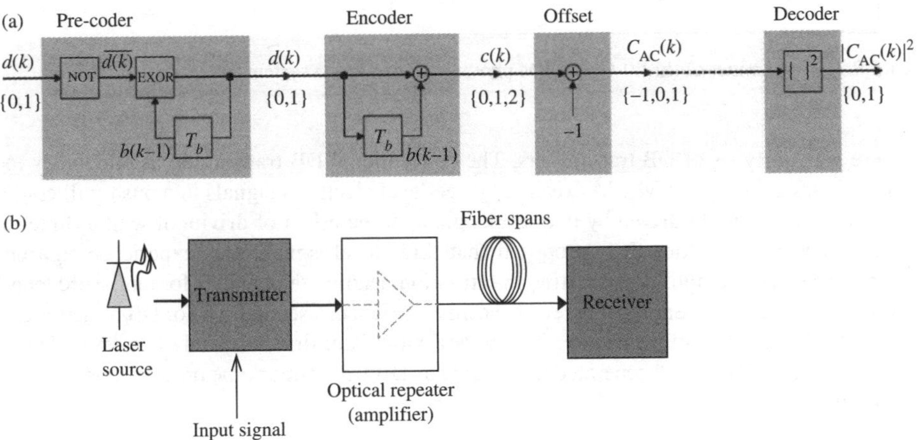

FIGURE 9.6   Main modules of a DBM optical communication system: (a) coder and decoder (b) generic transmission system.

signal is transmitted along an optical fiber transmission link toward the electro-optic receiver. The signal will be detected using a photodetector, which converts the two-level optical signal back to an electrical signal. Optical amplification can be done at some points along this transmission link to minimize the effect of fiber loss.

9.1.4   ELECTRO-OPTIC DUO-BINARY TRANSMITTER

The transmitter modulates and converts the incoming electrical signal into the optical domain. Depending on the nature of the signal, the resulting modulated light may be turned on and off, or may vary linearly in intensity between two levels. The output of the DB transmitter is the modulated lightwaves switched on and off at transitional instances of the input electrical signal. A general DBM transmitter is shown in Figure 9.7. It consists of a monochromatic laser source, a coder, and a photonic modulator. Binary data is encoded by the DB encoder. This resulting three-level electrical signal is converted into two-level an optical signal using the folding characteristic of an optical MZIM. It is then transmitted into the optical fiber.

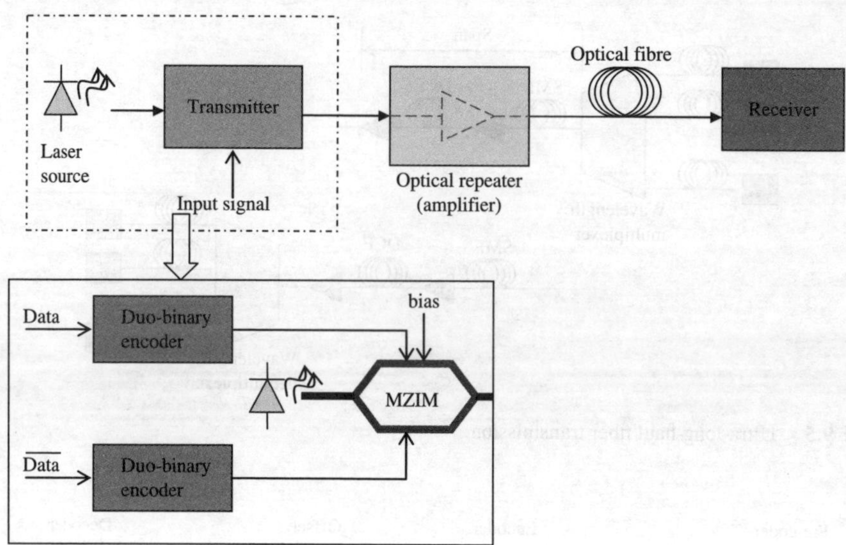

FIGURE 9.7 Transmitter of the 40-Gb/s DBM photonic transmission system.

There are two types of DB transmitters. The conventional DB transmitter, as previously mentioned, includes a dual-drive MZIM driven by three-level electrical signals in a push-pull configuration. MZIM is normally driven by two-level signal, so the effect of driving it with a three-level signal produces uncertainties. It is proposed that three-level signals may experience significant distortion in electrical amplifiers operating in saturation, leading to penalties for long word lengths. It may also cause the degradation of receiver sensitivity. Hence, a second type of DB transmitter has been proposed. This type of transmitter has the MZ modulator driven by only two-level electrical signals. The optical DB signal generated is the same as DB transmitter type one, i.e., constant phase in blocks of 1's.

9.1.5 THE DuoB ENCODER

The DB encoder encodes binary signals, which are a sequence of 0's, to a three-level electrical signal. The DB signal is a fundamental correlative coding in partial response signaling. A DB encoder consists of a pre-coder and a DB coder as shown in Figure 9.8. A pre-coder is used before DB coding to allow for easier recovery of binary data at the receiver, and to avoid error propagation. The pre-coder is a simple binary digital circuit that consists of an exclusive OR (XOR) and a 1-bit delay feedback. The DB coder is a filter consisting of a single delay element and an adder.

Binary data input is pre-coded, with initialization of the 1-bit delay to 0. The output of the DB pre-coder is, then, modulated by a pulse-amplitude modulator to produce a two-level electrical signal with an amplitude of −1 and 1. The DB signal is produced by adding data delayed by 1-bit period to the present data. This DB signal is a three-level electrical signal of −2, 0 and 2. Finally, it is converted to a level of −1, 0 and 1. The three-level is mapped into the optical domain by modulating amplitude and phase. The "+1" and "−1" levels have the same optical intensity, but opposite optical phase.

9.1.6 THE EXTERNAL MODULATOR

Although the electro-optic modulator has been described in Chapter 2, it is essential to revisit the operation of the MZIM for DB operation. In a MZIM, the input optical carrier is split into

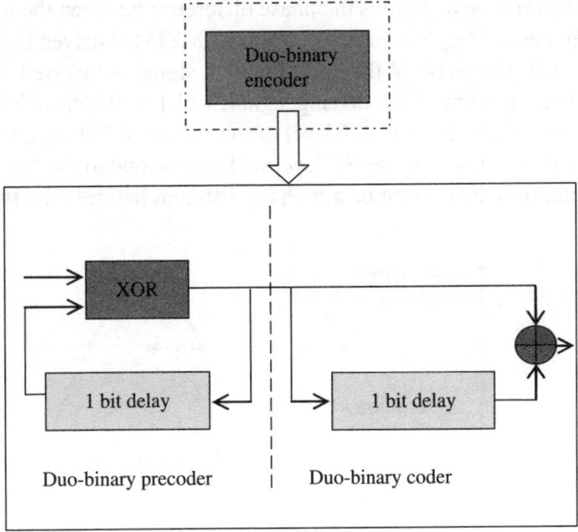

FIGURE 9.8 DBM encoder of the 40-Gbps DBM photonic transmission system.

two paths via a Y-junction. This Y-junction splits the input signal into $E_i/\sqrt{2}$ each. The resultant signal is

$$E_o = \frac{E_i}{2}\left[1 + \exp\left(j\pi\frac{V(t)}{V_\pi}\right)\right]$$

(9.3)

where V_π is the voltage to provide a π phase shift of each phase modulator and $V(t)$ is the driving voltage. The input and output relationship of this MZIM is shown in Figure 9.9. It is accompanied by a phase modulation of $\exp(j\varphi(t))$ with $\varphi(t) = \pi V(t)$. For $V(t)$ from 0 to V_π, E_o and E_i have the same phase, and as for $V(t)$ from V_π to $2V_\pi$, E_o and E_i have different phase.

MZIM can be single-drive or dual-drive. Single-drive x-cut $LiNbO_3$ MZM has no phase modulation along with the amplitude modulation. It follows the transfer characteristics shown in Figure 9.9. Dual-drive X-cut $LiNbO_3$ MZIM, on the other hand, has two paths phase modulated with opposite phase shifts in a push–pull operation. The V_π in Figure 9.9 is reduced by half in this case. For dual-drive y-cut $LiNbO_3$ MZIM, two paths are driven by complementary signal with V_1 equal to $-V_2$. The output electric of a dual-drive MZIM is

$$E_o = \frac{E_i}{2}\left[\exp\left(j\pi\frac{V_1}{V_\pi}\right) + \exp\left(j\pi\frac{V_2}{V_\pi}\right)\right].$$

(9.4)

DB optical signal is generated by driving a dual-drive MZIM with push–pull operation, as shown in Figure 9.10. One arm is driven by the DB signal and the second arm is driven by the inverted DB signal. Figure 9.11 shows the operation of the MZIM. The output electric field $E_o(t)$ can be expressed as

$$E_o(t) = E_i \cos\frac{\Delta\phi(t)}{2}\exp\left(-j\frac{\phi_0}{2}\right)$$

(9.5)

where E_i is the input electric field, $\Delta\varphi(t)$ is the phase difference between the lightwaves propagating in two optical waveguides, and φ_0 is a constant when the MZIM is driven in a push–pull operation. At point B of Figure 9.11, the phase of the output optical signal is inverted. The optical DB signal is dependent on the biasing point of the driving signal, which is the electrical DB signal. By biasing at point B in Figure 9.11, "−1" and "+1" level of the electrical DB signal will correspond to the "on" state of the optical signal, whilst the "0" level will correspond to the "off" state. To achieve the effect of carrier suppression, there must be a π phase different between the two arms.

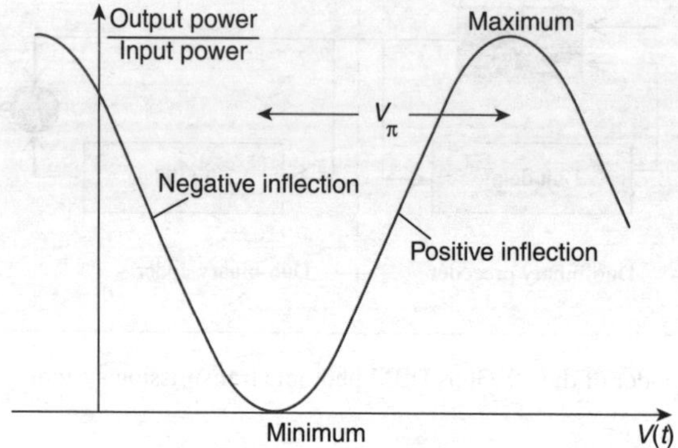

FIGURE 9.9 Input–output transfer characteristics of the MZIM.

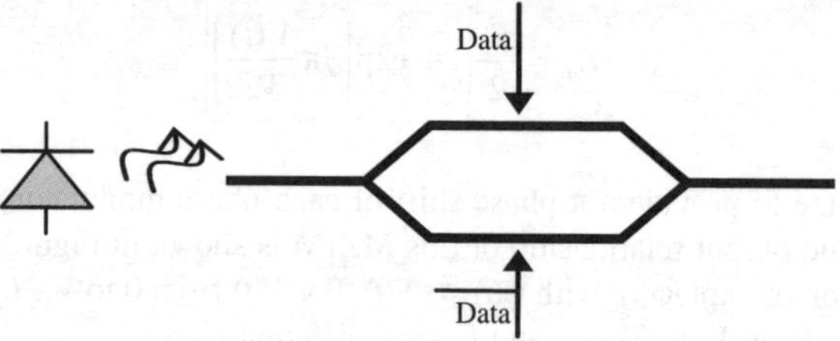

FIGURE 9.10 Dual-drive MZIM.

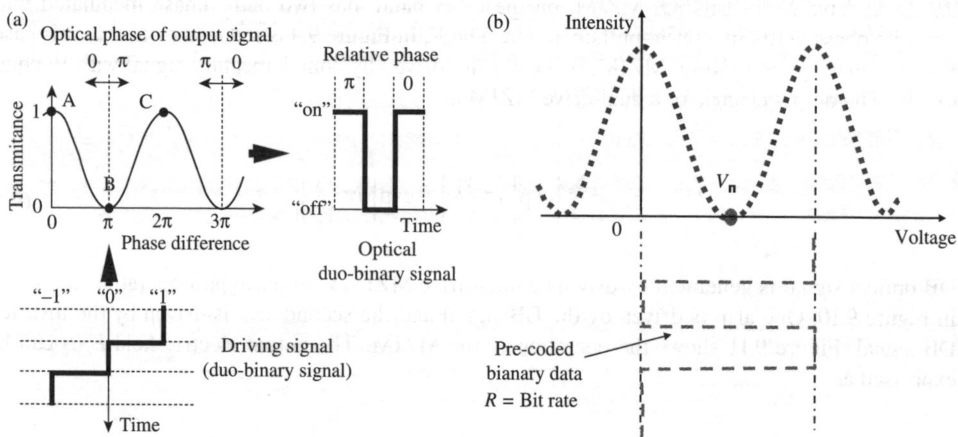

FIGURE 9.11 Driving operation of dual-drive MZM (a) duo-binary signaling (b) DPSK.

9.1.7 DuoB Transmitters and Pre-Coder

In general, the transmitter model consists of the DB coder and the MZIM. The DB coder encodes the incoming binary sequence of 0's and 1's to the DB electrical signal. This signal is then used to drive the arms of the dual-drive MZIM. One arm is driven by the DB signal, and the other arm is driven by the inverted DB signal. The Bernoulli binary generator generates a random sequence of binary electrical signals. It is set to generate the data at a rate of 40 Gb/s. This signal is encoded by the DB encoder, which consists of a DB pre-coder and a DB coder. The first output of the encoder is shifted up by 1 to produced levels of "0", "1" and "2". This electrical DB signal is sent to the phase shift block, as shown in Figure 9.12 and Figure 9.13, to represent these levels with a certain phase. This, in fact, represents the biasing point on the transmittance curve. For dual-drive MZIM, the driving signal is biased at $V_{\pi/2}$. The second output of the DB encoder is the inversion of output 1. This output is used to modulate the second arm of MZIM. The output 2 signal is shifted down by -1, to bias at the point $-V_{\pi/2}$ of the transmittance curve. MZIM is an amplitude modulator, accompanied by a shift of phase. This modulation is also called AM-PSK. This lightwave, which is the sine wave produced by the sine wave function, is modulated by the DB signal through the complex phase shift block. The input sine wave is shifted by the amount specified at the Ph input.

The DB coder consists of a pre-coder and a coder, as shown in Figure 9.14. The pre-coder is a differential coder, with a exclusive OR (XOR) gate and a 1-bit delay feedback path. The addition of -0.5 and division by 2 function as the amplitude modulator, which shifts levels of the signal from "0" and "1" to "-1" and "$+1$". The signal is then added to its one-bit delay to produce a three-level DB signal of level "-2", "0" and "$+2$", followed by a conversion to a level of "-1", "0" and "$+1$". The summation of the signal with its one-bit delay is the DB coder. For the second output, the output of the differential coder is inverted, before going through the same operation as Out1. Zero-order hold is placed before the output of the DB encoder functions to discretize the signals to have a fast-to-slow transition of signals. It holds and samples the signal before transmitting it out. If the signals are transmitted out without the zero-order hold, the transition to "0" level will be overseen. The signal will only have two levels, which are "-1" and "$+1$".

9.1.8 Alternative Phase DB Transmitter

Two types of DB transmitter model are proposed. The conventional DB transmitter, as mentioned previously, uses a dual-drive Mach–Zehnder modulator driven by three-level DB electrical signals. MZIM shown in Figure 9.11 is usually driven by a two-level electrical signal. In some cases, this three-level driving signal may experience significant distortion in electrical amplifiers operating at saturation. This may lead to penalties for long word lengths. There may be a degradation of receiver sensitivity. Due to these uncertainties, an alternative DB transmitter, as shown in Figure 9.15, is proposed. A MATLAB® Simulink® models for such modulator and a differential encoder are shown in Figure 9.16 and Figure 9.17 respectively. This second type of DB transmitter has the MZIM driven by two-level electrical signals. It consists of a differential encoder, a one-bit-period electrical time delay and a MZ modulator. One arm of the dual-drive MZ modulator is driven with the signal from the, whereas the other arm is driven by the one-bit-delayed inverted signal. Both DB transmitters produce the same result, which is constant phase in blocks of 1's.

9.1.9 Fiber Propagation

As described in Chapter 3, the fiber propagation model models the linear and non-linear dispersion effects that exist over the entire length of the optical fiber can be represented by the NLSE including all non-linear effects, given by

$$\frac{\partial A}{\partial z} = +\beta_1 \frac{\partial A}{\partial t} + \frac{j}{2}\beta_2 \frac{\partial^2 A}{\partial t^2} - \frac{1}{6}\beta_3 \frac{\partial^3 A}{\partial t^3} = j\gamma \, |A|^2 A \tag{9.6}$$

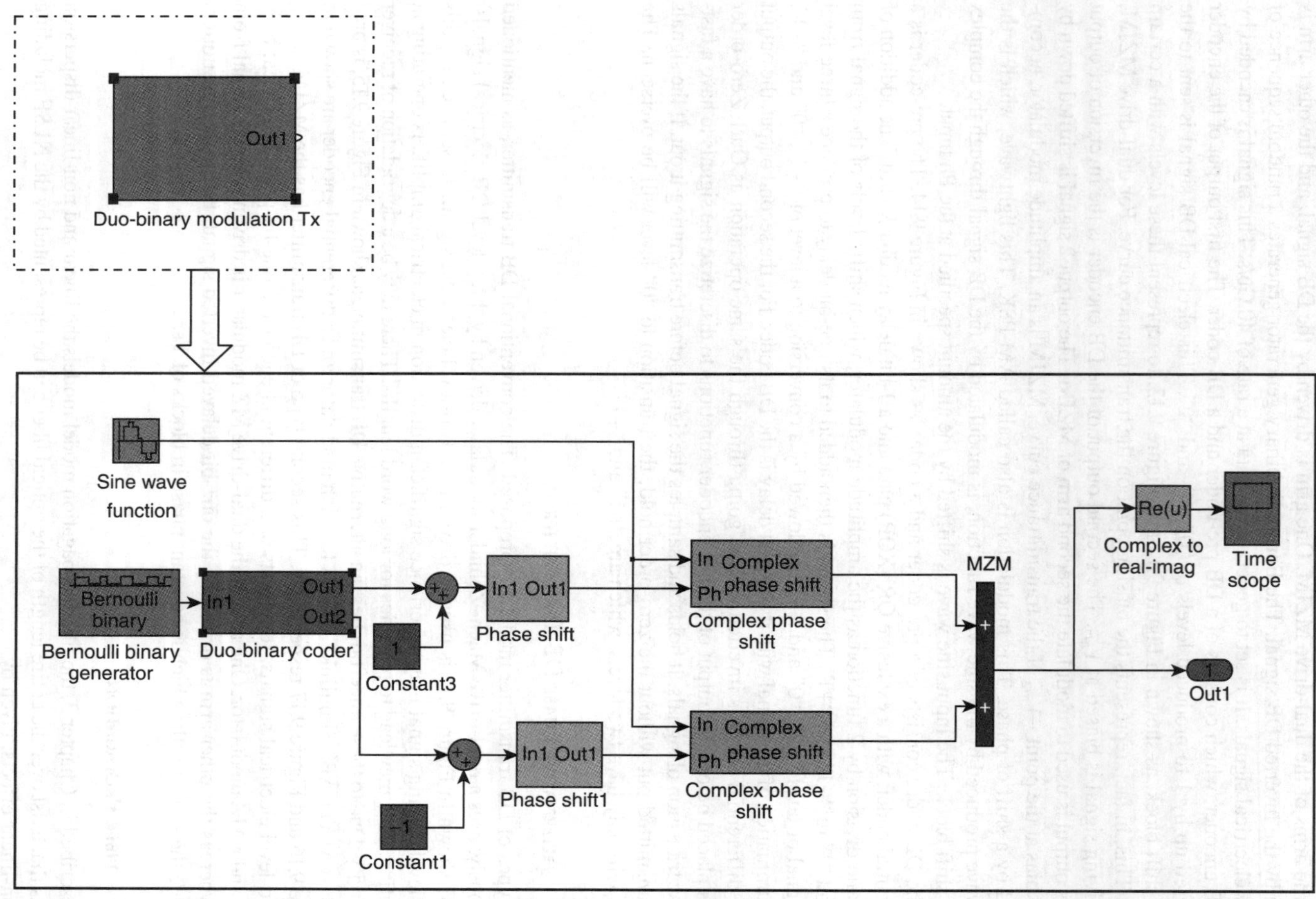

FIGURE 9.12 Conventional transmitter model of the 40-Gb/s DB optical fiber communication system.

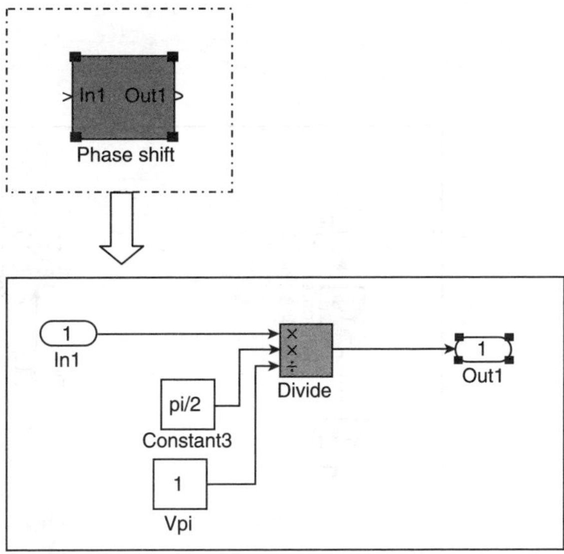

FIGURE 9.13 Phase-shift block of the transmitter model.

where β_1 is the lightwave group delay, β_2 and β_3 are the first- and second-order group velocity dispersion, γ is the non-linear coefficient and A is the pulse envelope. NLSE can be solved by the SSFM that integrates two main steps of split-step model, the non-linear effects acts alone, and the linear effects.

$$\frac{\partial A}{\partial t} = (L + N)A \tag{9.7}$$

where L and N represent linear and non-linear operators, respectively, which can be extracted from the NLSE.

The fiber propagation block, as shown in Figure 9.18, consists of Gaussian filter, a gain factor and the single-mode fiber (SMF) model (Figure 9.19). This model is based on the SSF method. It splits the fiber into a number of small sections, dz. All parameters needed for the SSF operations are concentrated into a matrix, before passing it to the MATLAB® function. The MATLAB function block solves the NLSE using the split-step Fourier method. Linear operation is implemented in all steps. When the peak power is greater than the non-linear (SPM) threshold, the non-linear operator is activated. In this fiber propagation model, buffer is attached at various points. Buffer is used to redistribute the input samples to a new frame size, in this case, a larger frame than the input frame size. Buffering to a larger frame size yields an output with a slower frame rate than the input. The "unbuffer" block unbuffers the frame-based input into a sample-based output. The buffer used in this model determines the number of bits sent into the fiber, which is the SMF block.

Probes are connected at two different points of the fiber propagation model. These probes determine the sampling time at these points. Sampling time of $2.5e^{-11}$ means that the signal is downsample to baseband. At baseband, the complex envelope of the signal will be extracted and this extracted signal will be transmitting through the fiber, which is represented by SMF in Figure 9.18. However, the phase contents of the signal are maintained and are transmitted through the fiber. This phase component of the DB signal is important because it increases the dispersion tolerance and allow the signals to travel a longer distance.

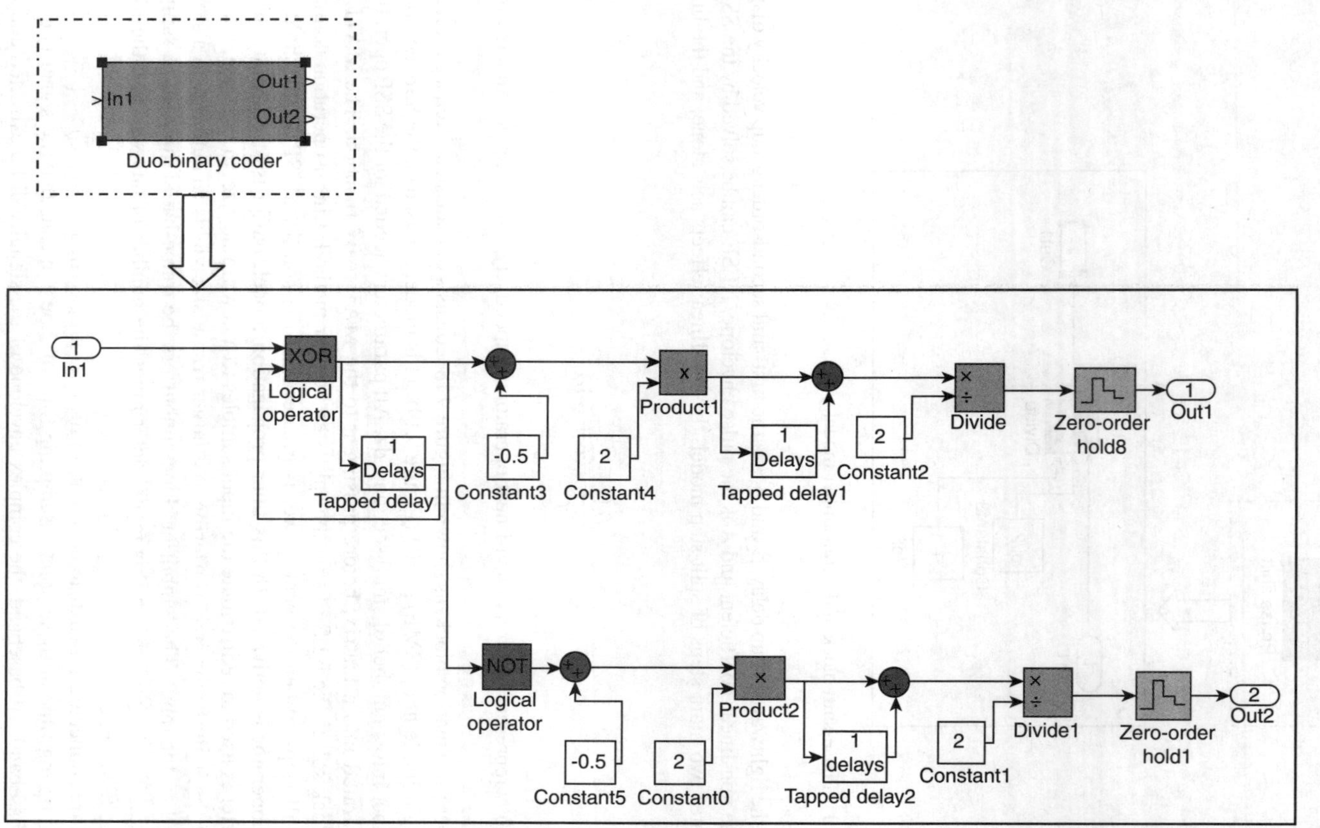

FIGURE 9.14 DB coder of the transmitter model.

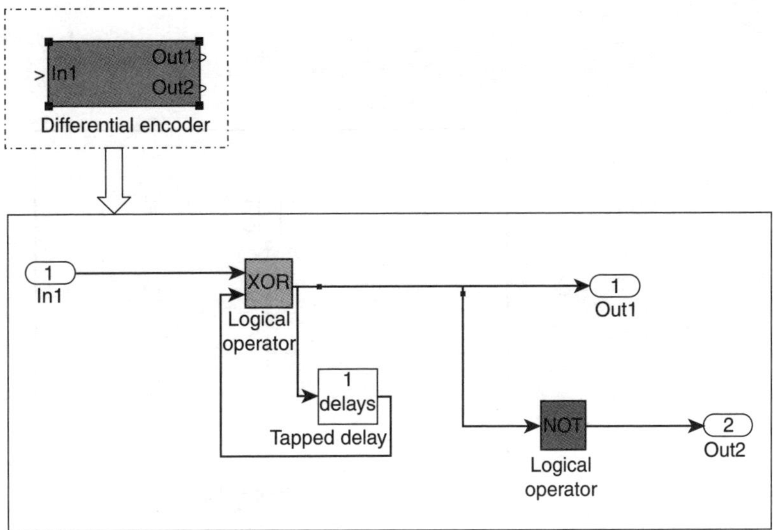

FIGURE 9.15 DB transmitter (type 2).

9.2 DUO-BINARY DIRECT DETECTION RECEIVER

The receiver model of the 40-Gb/s DB optical fiber communication system consists of a Gaussian filter and scopes to observe the performance of the system. The receiver is a conventional direct-detection type. Therefore, a demodulator or decoder is not needed. Power spectrum and eye diagram can be observed directly from this point. The Gaussian filter functions as a baseband filter. The absolute value of the incoming signal is taken because the DB receiver detects the intensity of the signal. The probe is used to determine the sampling time at that point. "Demodsignals" block collects all the data at this point and saves it in the workspace. These data are used to plot the histogram, which is used to determine the Q factor and BER of the system.

Demodulation is needed at the receiver, depending on the modulation format used. For instance, DPSK receivers consist of a Mach–Zehnder delay-interferometer (MZDI), which demodulates the incoming signal before detecting it using a photodetector. MZDI lets two adjacent bits interfere with each other at its output port. This interference creates the presence or absence of power at the output port, depending on whether the interference is constructive, or destructive. The preceding bit in the DPSK signal acts as a phase reference for demodulating the current bit. The DI output ports are detected by the balanced detectors. The optical DB signals can be demodulated into a binary signal with a conventional direct detection type optical receiver, as shown in Figure 9.21. A decoder is not required in this case. The received signal is directly detected by a photodiode operating as a square law detector. Optical DB signal consists of two states, "on" and "off". The photodiode works by detecting the incoming intensity of the signal. The recovery of the original electrical signal can be done by simply inverting the signal detected at the photodiode. This inversion is done within the circuit shown as the decision circuit in Figure 9.12. The signal is observed at this point of the system. Most commonly used parameters to test and observe the performance of a system are the power spectrum, eye diagram, Q factor, BER and received optical power. The Q-factor is the quality factor of the system, under the Gaussian noise distribution. It is determined by the mean voltage level and the standard deviation of the noise.

$$\delta = \frac{\mu_1 - \mu_0}{\sigma_1 + \sigma_0} \tag{9.8}$$

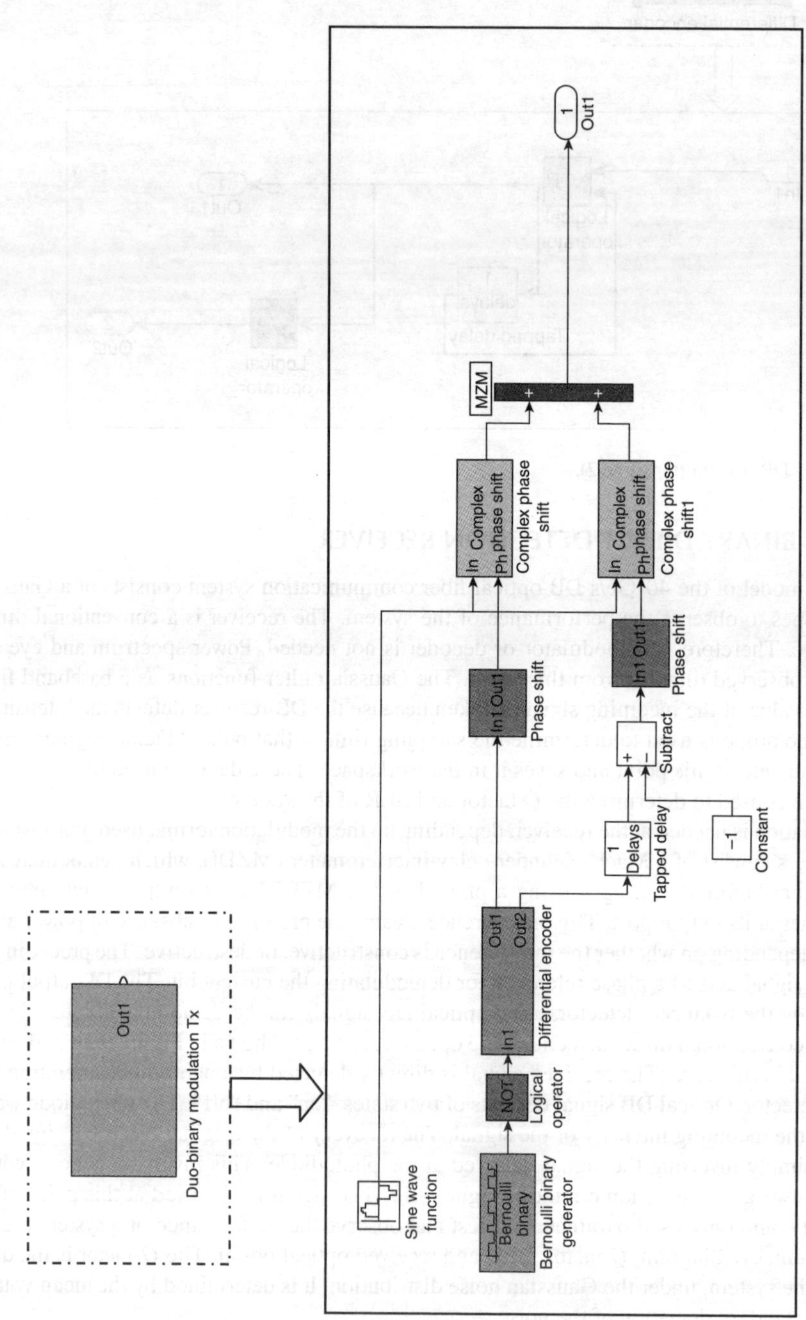

FIGURE 9.16 Simulink® model of MZIM.

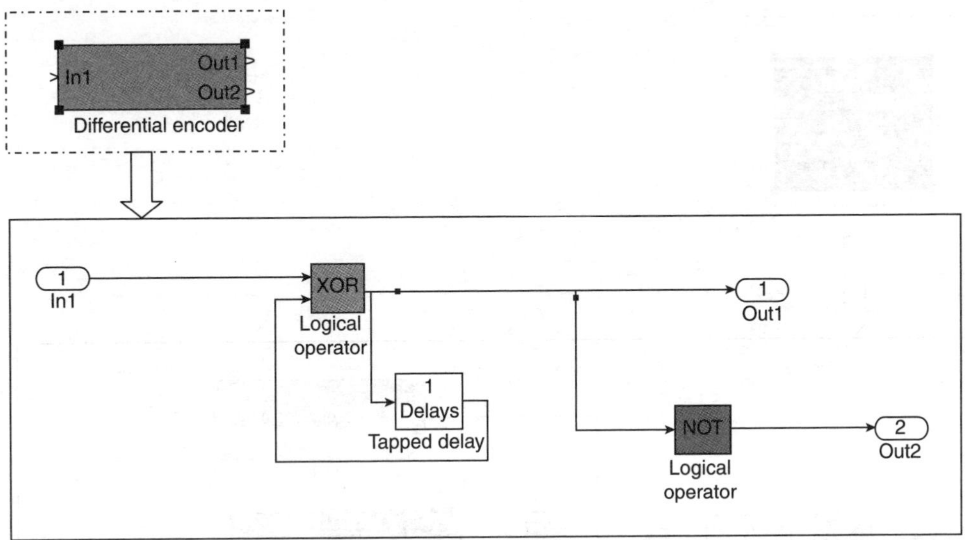

FIGURE 9.17 Differential encoder of type 2 duo-binary transmitter.

where μ_1 is the mean voltage level of the "1" received, μ_0 is the mean voltage level of "0" received, σ_1 is the standard deviation of noise of the "1" received and σ_0 is the standard deviation of the noise of the "0" received.

$$\text{BER} = \frac{1}{2}\operatorname{erfc}\left(\frac{\delta}{\sqrt{2}}\right). \tag{9.9}$$

The 40-Gb/s DB optical fiber communication system can be implemented on the Simulink® platform. Simulink has been chosen as the computer software for this development of the model because it consists of a variety of communication blocks which can assist in simplifying the process of implementing and improving of this model. Figure 9.22 shows the overall system of the 40-Gb/s DB optical fiber communication system. The main modules of this communication system are the DB transmitter, named as the DBM Tx, the fiber propagation and the receiver. The DB transmitter, in general, consists of the DB encoder and the MZIM. The fiber propagation block introduces the linear and non-linear dispersion effect that exists over the entire length of optical fiber to the signal. The receiver is of direct-detection type. Therefore, the output at the receiver is directly observed using scope, in the form of eye diagrams and power spectrum.

9.3 SYSTEM TRANSMISSION AND PERFORMANCE

The overall performance of this 40-Gb/s DB optical fiber communication system can be observed at the receiver of the system. The eye diagram and power spectrum are the parameters that are used to observe the performance of the system. Multiple debugging and testing processes have to be carried out to prove that the system is functioning as expected. Simulation using Simulink® has vastly reduced the time and difficulties involved in these processes.

The first step involved the testing of the transmitter. It is important to ensure that DB optical signal is generated at the output of the transmitter. The DB encoder is critical in this case. Its output is checked using the time scope to ensure the three-level electrical signal is produced. Power spectrum and eye diagram are connected at the output of the transmitter. These obtained results are compared

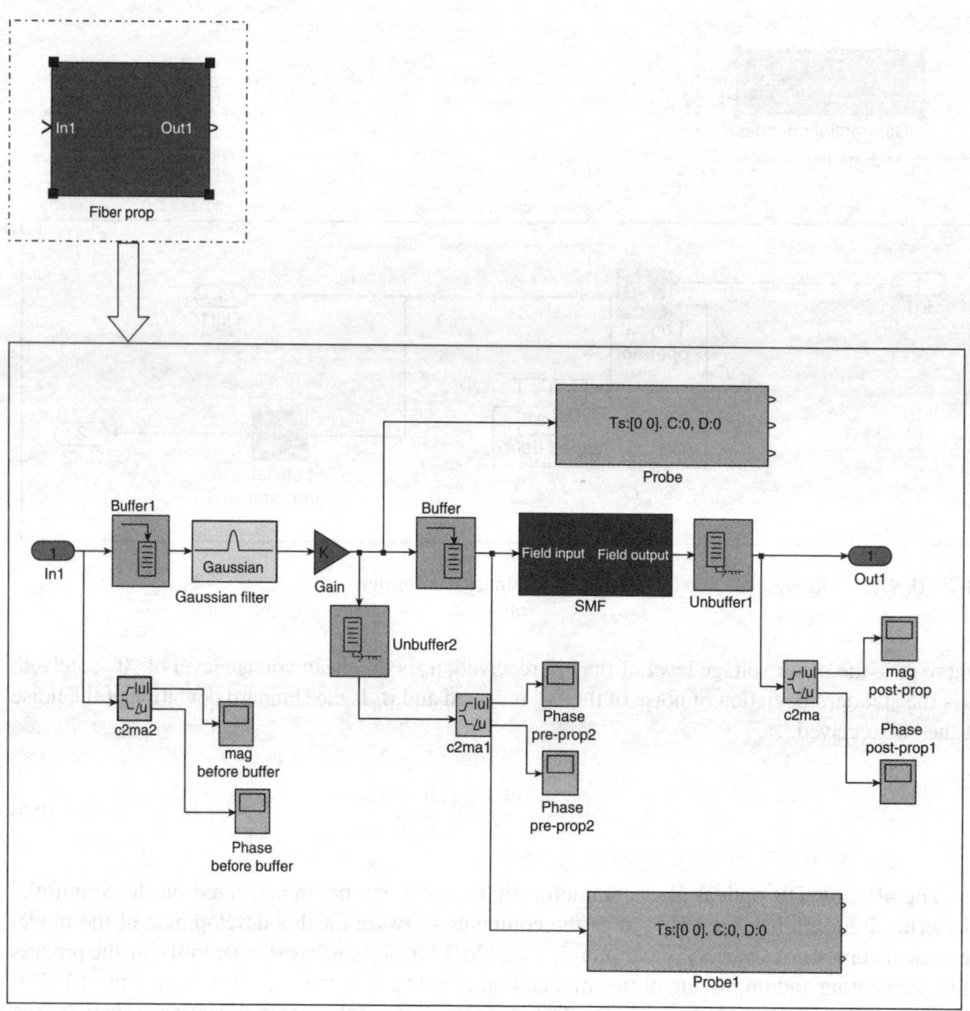

FIGURE 9.18 Fiber propagation model of the 40-Gbps DB optical fiber communication system.

with those obtained experimentally and theoretically to verify that the DB transmitter is generating the correct signals. The fiber propagation model is then connected to the output of the transmitter. The fiber propagation model will introduce the linear and non-linear effects of fiber depending on the distance the signal travels. The DB receiver, which is of conventional direct-detection type, is connected after fiber propagation. The signal is observed directly at the receiver. By observing the eye diagram at this point, the distance the signal can propagate without severe distortion can be estimated. The testing is started with a fiber length of 1 km. The distance is increased until the point when significant distortion to the eye diagram can be observed, and the "eye" of the eye diagram has closed.

9.3.1 THE DB ENCODER

The DB encoder is the first implemented model in this 40-Gb/s DB optical fiber communication system for generation of three-level DB electrical signals. Scopes probe at various points of the DB encoder, as shown in Figure 9.14. All time scopes are set to the same range to allow for comparisons of the bits within the same range. The temporal sequences at the outputs of the encoder for driving the

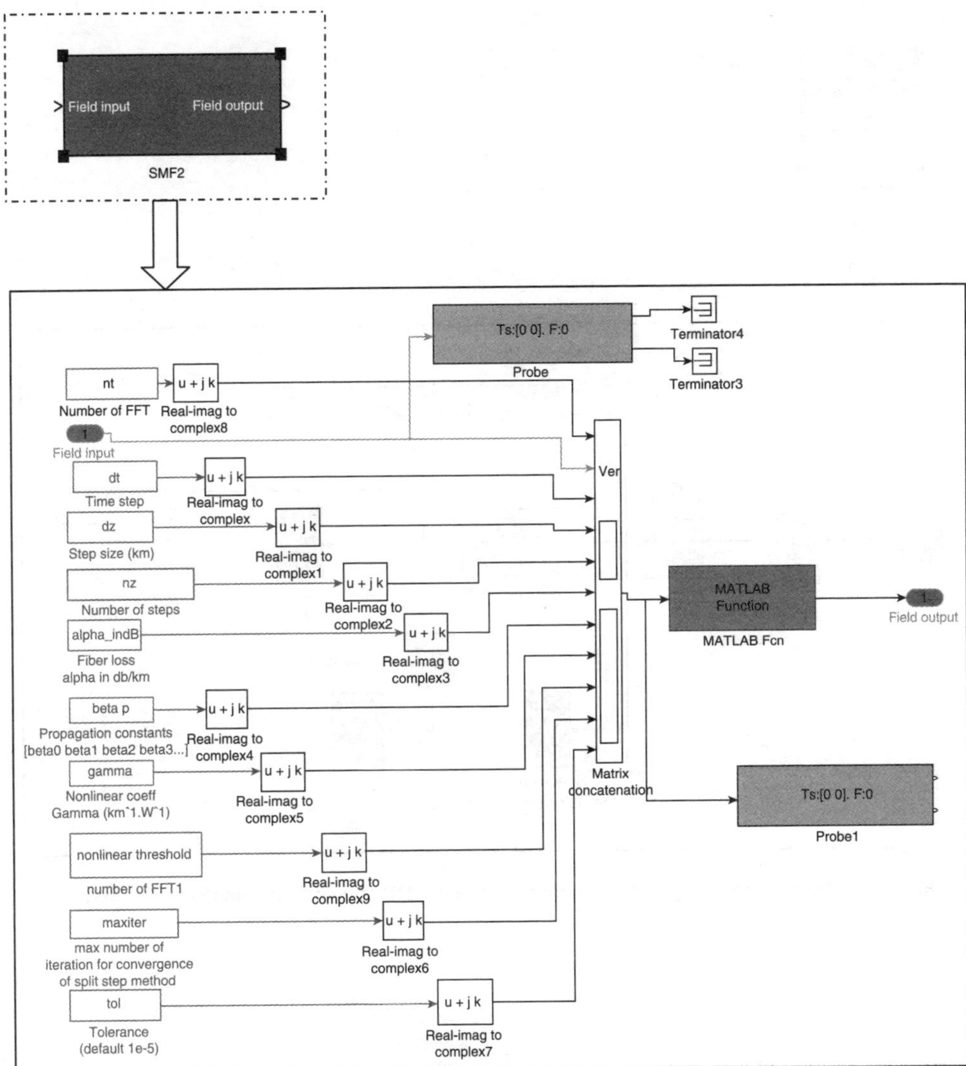

FIGURE 9.19 Single-mode fiber models using the SSFM.

MZIM are shown in Figure 9.23. Scope 1 shows the data generated by the Bernoulli Binary Generator. Scopes 2–4 show the output at each arm. A bit "0" is encoded as "+2" or "–2", while the bit "1" is encoded as 0. This agrees with the DB coding scheme. They also show three-level electrical signal. Scope 4 is the inverted version of scope, as expected, due to the NOT gate applied to the second arm.

9.3.2 The Transmitter

The transmitter includes the DB encoder and MZIM. After the output of the DB encoder is verified to produce the correct signal, the levels of the signal are represented with a phase. This phase is used to shift the laser source produced by sine wave function. The testing of the implemented DB transmitter includes observing the eye diagram and power spectrum at the output. Time scopes are attached at various points of the transmitter to observe the signal at these points. Spectrum scopes are connected at the output of the transmitter, which is after the summation of both arms of MZIM and to each arm of the MZIM to observe the power spectrum at these arms. Some adjustments must be made to the transmitter model in order to observe the power spectrum. The sine wave function

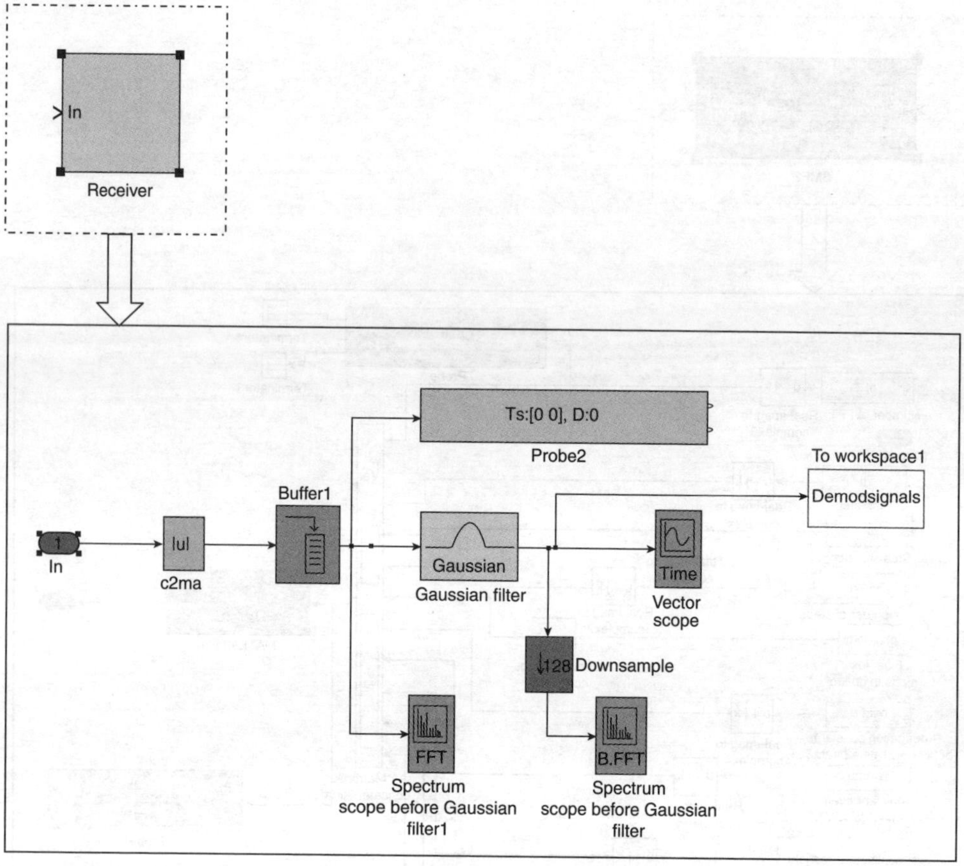

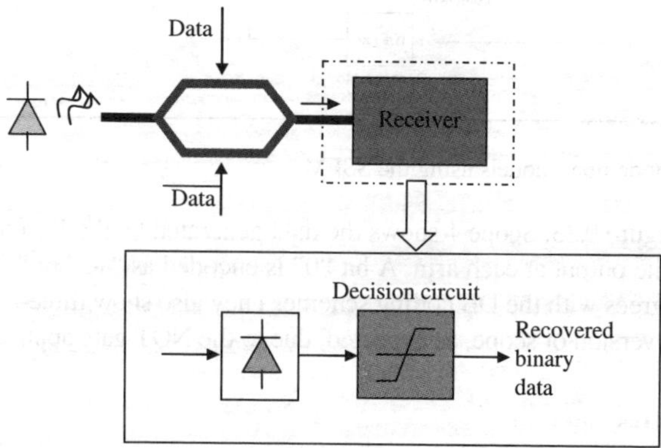

FIGURE 9.20 Direct detection optical receiver of the 40-Gb/s DB optical fiber communication system.

FIGURE 9.21 Receiver of the 40-Gbps DB photonic transmission system; 40-Gb/s DB transmission Simulink model.

is set to produce a wave of 200 GHz rather than 193 THz. This is to enable the observation of the spectrum centered at a lower frequency, and better spectral resolution. The zero-order hold which functions is to hold and sample the incoming signal is set to $1e^{-12}$, so that the x-axis of the spectrum scope is in the THz range. The obtained results, as shown in Figure 9.25, reflect on the results

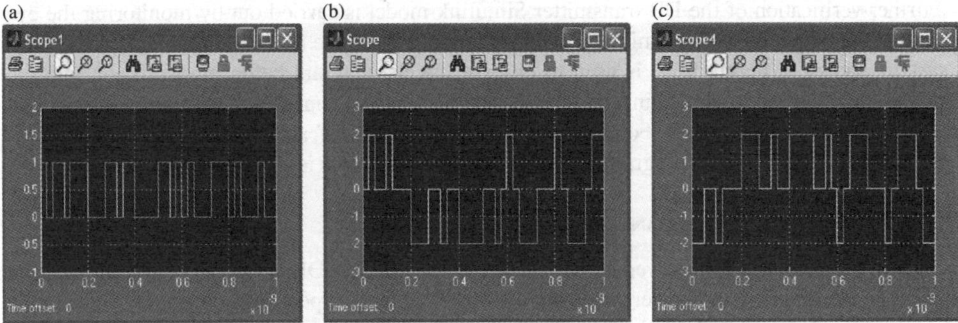

FIGURE 9.22　40-Gb/s DB optical fiber communication system Simulink model.

(a)　(b)　(c)

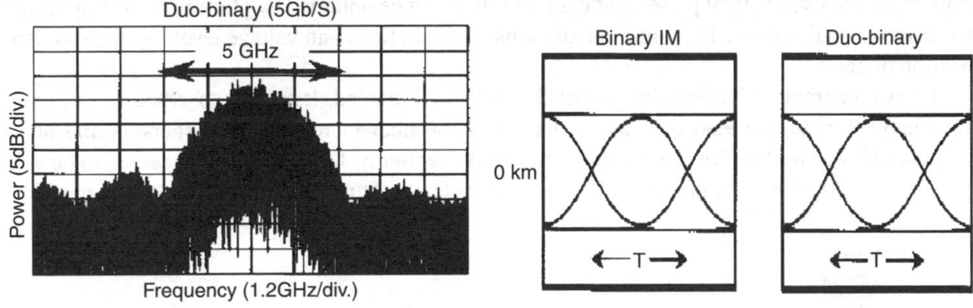

FIGURE 9.23　DB encoder outputs (a) Bernoulli binary sequence (b) and (c) DB electrical sequences for driving the electrodes of the MZIM.

FIGURE 9.24　DB power spectrum and eye diagram for 5-Gb/s DB optical fiber communication system. (From Frank, T., P. B. Hansen, T. N. Nielsem, and L. Eskildson, *IEEE Photonics Technology Letters*, 10 (4), 597–99, 1998. With permission.)

(a)　(b)　(c)

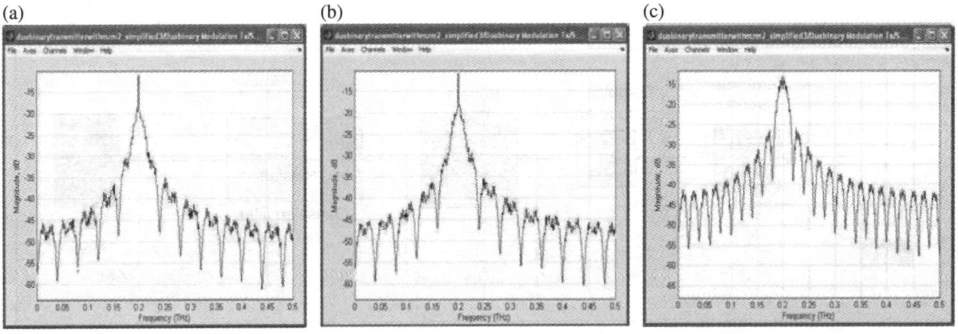

FIGURE 9.25　Power spectrum obtained from each arm of the MZIM, (a) and (b) and the output of DB transmitter (c).

expected from experiments and theories. The power spectrum for a 5-Gb/s DB optical fiber communication system is shown in Figure 9.24a. This is used to verify the power spectrum obtained from the Simulink® model. Spectrum (c) of Figure 9.25 corresponds to spectrum scope in Figure 9.18. The obtained spectrum reflects on that in Figure 9.25a the shape is approximately the same. The bandwidth of the obtained spectrum corresponds to the data rate, as in Figure 9.24a. The spectra obtained are centered at 200 GHz, which is the carrier frequency of the model. Spectrum Figure 9.25c shows the carrier-suppressed case, but not the other two spectra. This is as expected from the DBM format. Due to the π-phase difference in the two arms of MZIM, the output is expected to have its carrier suppressed. Figure 9.26 shows the Simulink block diagram of the monitoring of the signal and eye diagram before the transmission block of the 40-Gbps DB optical fiber communication system. Figure 9.27 shows the eye diagram obtained before propagation block.

Further verification of the DB transmitter Simulink model is carried out by monitoring the eye diagram at the output of the transmitter. The observation prior to the transmission block in the overall system, shown in Figure 9.22, is used to observe the eye diagram before transmission into the fiber. The eye diagram, as shown in Figure 9.28, obtained has an "open eye" with amplitude of 0.6. It reflects the one in Figure 9.24. For a range of 100 ps, four "eyes" are obtained, giving one "eye" 25 ps. This proves that the eye diagram is correct, since the data rate is 40 Gbps, one bit every 25 ps.

9.3.3 Transmission Performance

The overall performance is observed at the receiver of the 40-Gb/s DB optical fiber communication system. The eye diagram and power spectrum are observed by the spectrum scope and vector scope attached inside the receiver. Eye diagrams show the distortion and attenuation of signals at various length of the fiber. When the "eye" of the eye diagram closes, the signal is severely distorted and dispersed, and thus recovering the signal will be impossible. The eye diagram can also be used to calculate the received optical power. Q factor and BER can be obtained by plotting the histogram of the received signal, followed by some calculations to obtain the mean voltage level and the standard deviation of noise.

The eye diagrams of various lengths of the fiber are obtained. It can be observed from Figure 9.28, Figure 9.30 Figure 9.30 that, as the length of the fiber increases, the dispersion and noise increases with it. The "eye" of the eye diagram closes eventually. DB signals are expected to be able to travel up to 200 km without the "eye" closing completely. This has been proved by the model. It

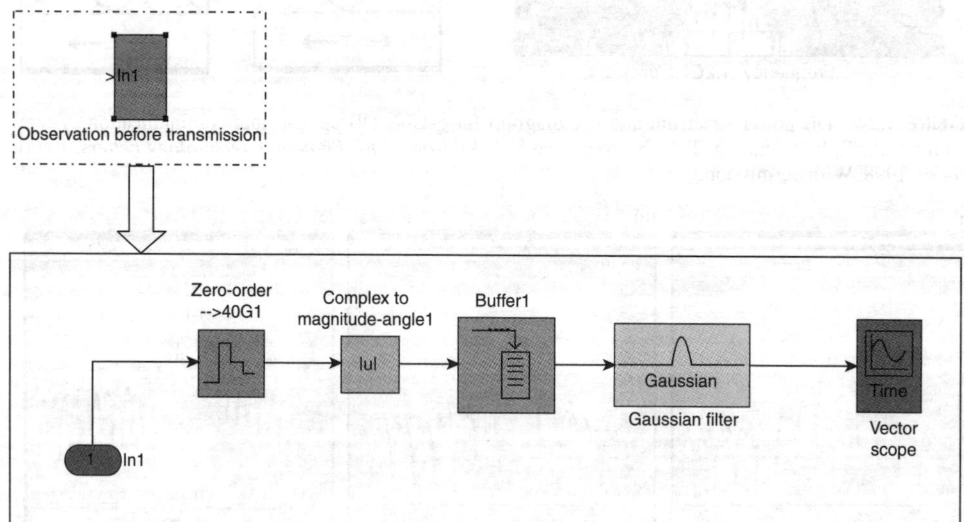

FIGURE 9.26 Signal and eye diagram monitored before transmission block of the 40-Gbps DB optical fiber communication system.

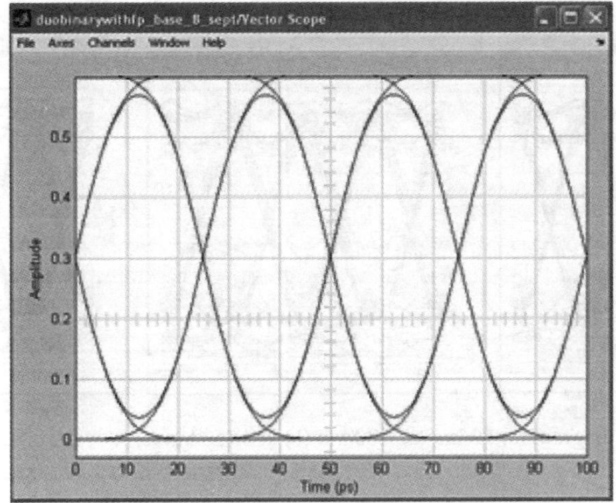

FIGURE 9.27   Eye diagram obtained before propagation block.

(a)   (b)   (c)

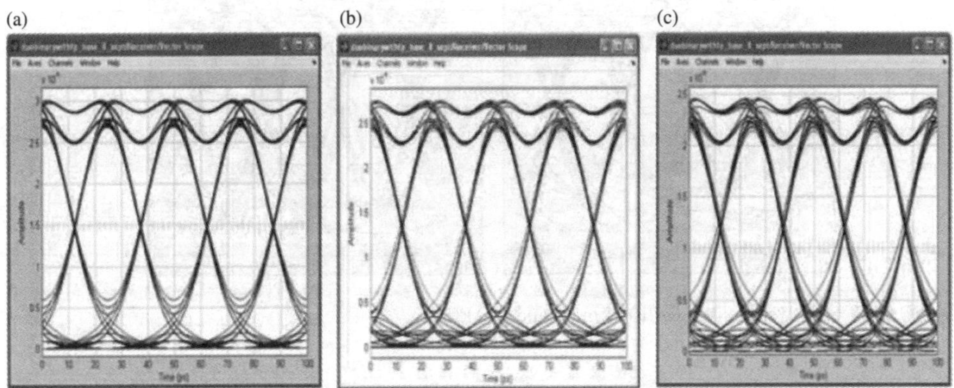

FIGURE 9.28   DB eye diagram for 1 km (a), 5 km (b) and 10 km (c).

can be observed that at the length of 250 km, as shown in Figure 9.30b, the "eye" of the eye diagram has yet to be fully closed. Dispersion effects can be observed and they progressively increase with the distance of the fiber.

Data obtained from the demodsignals block, shown in Figure 9.20, is used to plot the histogram which determines the Q factor and BER. Q factor is the quality factor of the system. From the Q factor, the BER can be calculated. It is expected that for a BER of 10^{-9}, the Q factor is approximately six. The histograms for 1 km, 5 km and 10 km are shown in Figure 9.32. The two points on the histogram are compared with the histogram at the receiver to obtain the mean value, μ_0 and μ_1 and the standard deviation of noise, σ_0 and BER; or bit error rate can be calculated from the Q factor found. Figure 9.32 shows the plot of BER versus distance. As the distance of propagation of signal increase, the bit error rate increases. This is as expected because the linear and non-linear effects of the fiber introduce noise and errors to the transmitting signal. Figure 9.33 is the plot of BER versus receiver sensitivity. The receiver sensitivity obtained of the region −23 dBm to −20 dBm. Experimental results of the transmission of CSRZ ASK and RZ ASK format are also included for system over 328 km SSMF and DCM modules, completely dispersion compensated with five EDFA modules integrated.

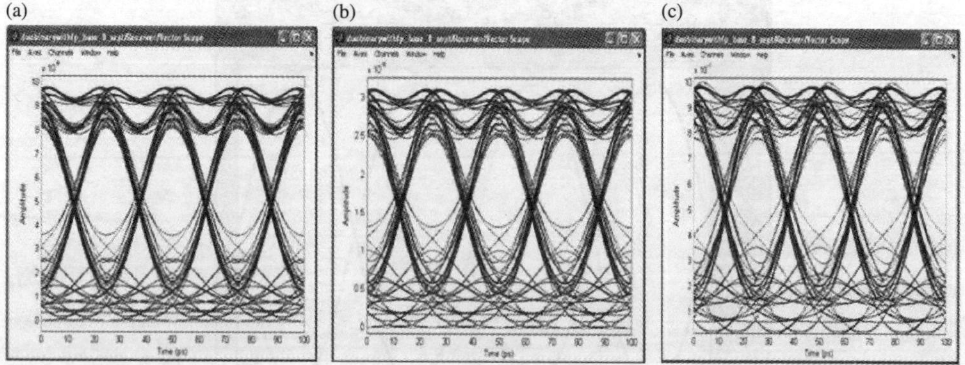

FIGURE 9.29 DB eye diagram for 50 km (a), 100 km (b) and 150 km (c).

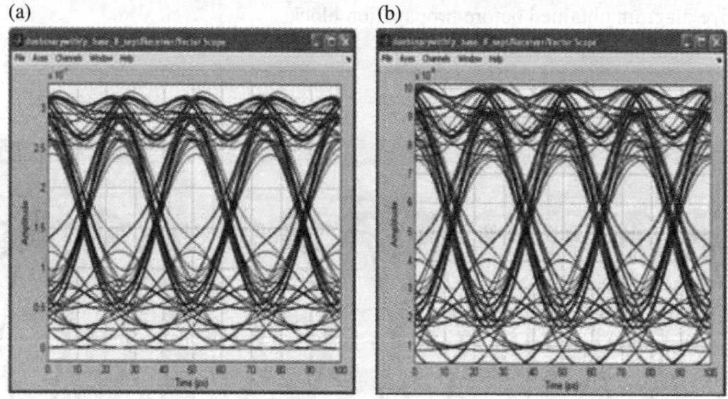

FIGURE 9.30 DB eye diagram for 100 km (a) and 150 km (b).

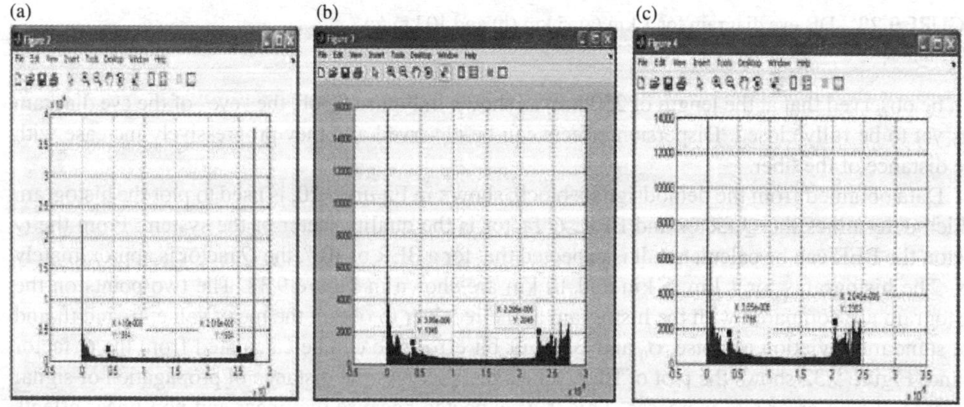

FIGURE 9.31 Histogram obtained from data demodsignals for 1 km (a), 5 km (b) and 10 km (c).

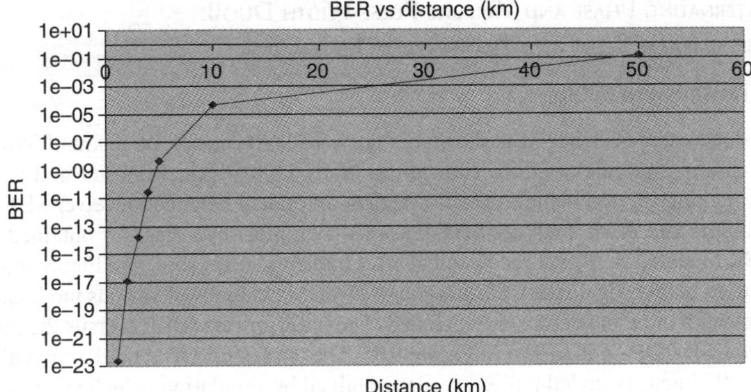

FIGURE 9.32   Plot of BER vs. distance (km) of SSMF.

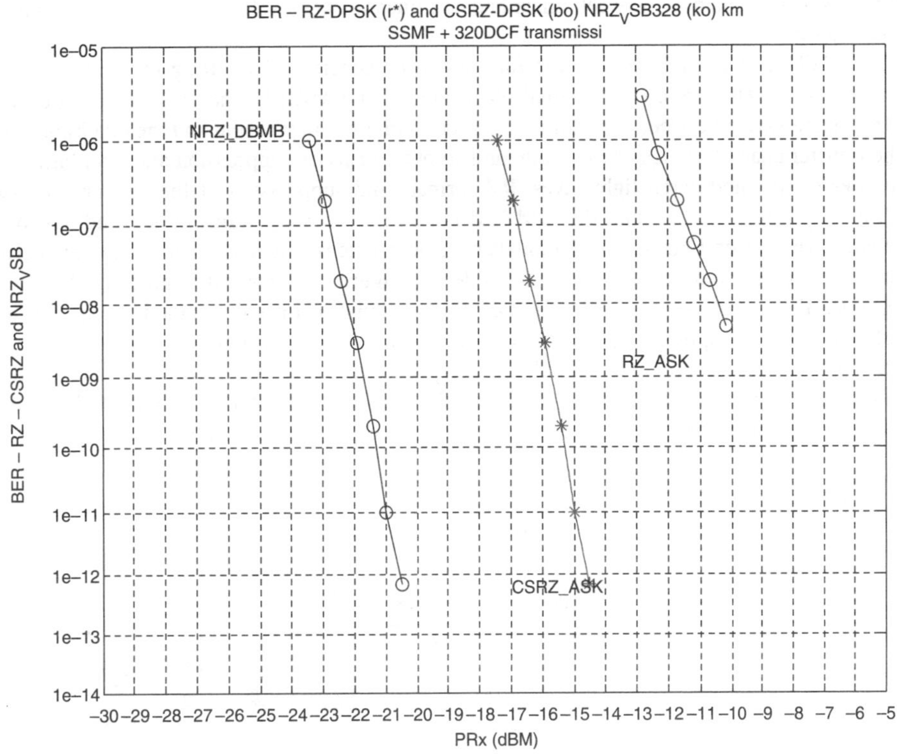

FIGURE 9.33   Plot of BER vs. receiver sensitivity (dBm) of simulated NRZ_DBM (black o) and experimental CSRZ_ASK, RZ_ASK. Experimental and simulation transmission system of 328 km (as shown in Chapter 5).

9.3.4 Alternating Phase and Variable Pulse Width DuoB: Experimental Setup and Transmission Performance

9.3.4.1 Transmission Setup

A transmission system is arranged as shown in Figure 9.34a. It consists of 50 km of SSMF and associate compensating module, a DPSK transmitter (SHF 5300), optical booster and pre-amplifiers, wavelength multiplexer and demultiplexer, a MZDI, an optical balanced receiver SHF-5008, clock recovery module and error analyzer SHF-EA 44A. A typical eye diagram obtained for 40-Gb/s CSRZ-DPSK modulation format recovered at after transmission is shown in Figure 9.34b. This test bed is also used to investigate the transmission of DWDM channels of various modulation formats. So the wavelength mux and demux are included. The bit-error-rate (BER) versus the receiver sensitivity for the CSRZ DPSK is shown in Figure 9.35. The curves for DB modulation with alternating phase of the "1" intensity coded pulses are also obtained by simulation, which are given in the next section.

DB optical transmission can be experimentally determined via simulation whose platform can be implemented on MATLAB® Simulink® for several modulation formats, especially the phase-modulated optical transmission. A typical system arrangement of optical transmitter using a dual-drive MZIM for generation of DPSK and alternating phase DB modulation for simulation is shown in Figure 9.36. The fiber propagation model follows the well-known split-step Fourier (SSF) method with provision for switching between linear and non-linear power level propagation so as to minimize computing time.

The FWHM of the DB modulation format can be generated by setting the amplitude of the swing voltage levels applied to the dual electrodes of the MZIM. The biasing condition of the MZIM can be varied between the minimum and maximum transmission and the quadrature points of the transfer characteristics of the modulator to obtain carrier suppression and alternating phase properties of the modulated lightwaves. RZ formats and suppression of the carrier can also be generated by using another dual-drive MZIM biased at minimum transmission and half bit rate frequency synthesizer. Our Simulink models have been extensively tested and the system performance agrees well between experiments and modeling. Figure 9.35 shows the agreement between the BER versus the receiver sensitivity for CSRZ DPSK modulation format obtained by experiment and simulation. Noises of optical amplifiers have also been taken into account in both cases. We selected CSRZ–DPSK format to compare with alternating phase DB modulation because it has been proven to offer superior performance as compared with RZ-DPSK and NRZ-DPSK. The MZIM is modulated and biased such that the width of the DB modulated pulses can be altered with ease.

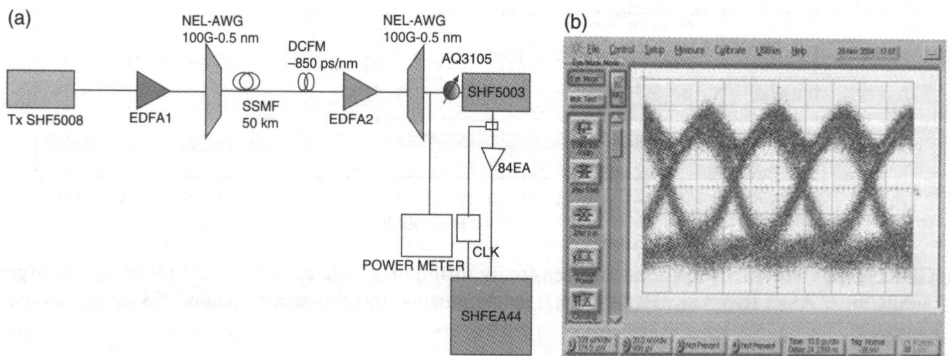

FIGURE 9.34 System test bed for CSRZ-DPSK transmission over 50-km SSMF, balanced receiver (a) schematic diagram of the test bed (b) dispersion compensated received eye diagram of CSRZ-DPSK.

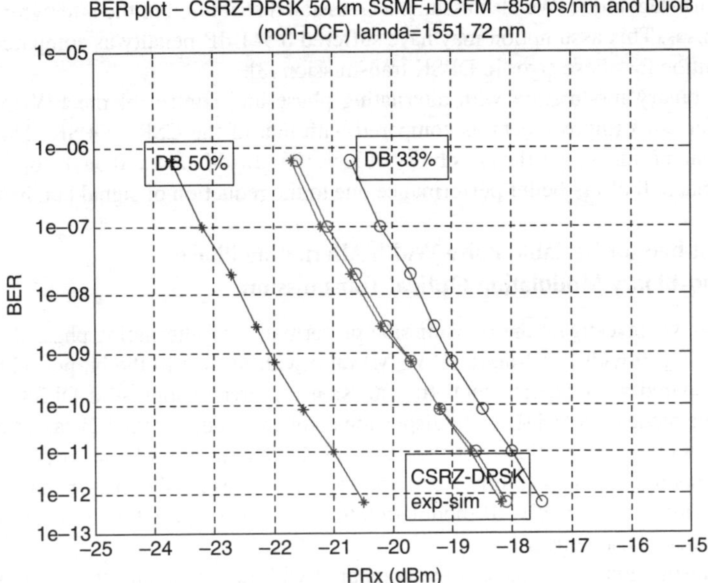

FIGURE 9.35   BER versus receiver sensitivity for 50-km transmission (a) red curve – CSRZ DPSK transmission with complete compensation (b) blue curves 50 and 33% FWHM alternating phase DB modulation.

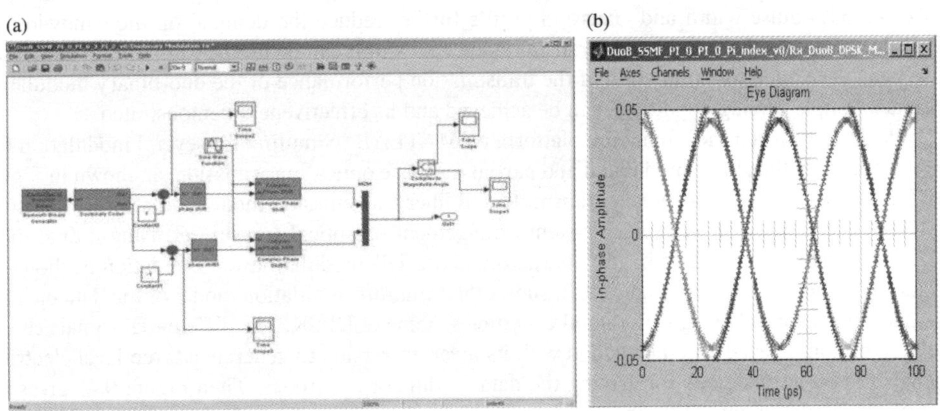

FIGURE 9.36   (a) Optical transmitter Simulink model for generation of alternating phase DB modulation format (b) non-dispersion compensating balanced received eye diagram for DB 33% FWHM.

Simulation results are obtained for alternating phase DB with a FWHM of 50 and 33% of the bit period for 40 Gb/s, as shown by the blue curves shown in Figure 9.35. An optical Gaussian filter type is also used at the output of the transmitter prior to transmission. We observe almost 2-dB better receiver sensitivity of the 50% FWHM alternating phase DB formats as compared with the CSRZ-DPSK. However, the 33% FWHM DB case offers 1-dB less sensitivity. This indicates the effectiveness of the duo-binary modulation. The 50-km SSMF without compensation is used for 40-Gb/s duo-binary transmission. A typical eye diagram obtained at the output of the balanced receiver for 33% FWHM DB is shown in Figure 9.36b. The simulation is conducted with 256 random bit pattern and several frames sufficient for measurement of the Q-factor without resorting to the Monte Carlo technique. We also assume a Gaussian distribution of the intersymbol interference and phase noises. This assumption may have suffered 0.5–1 dB penalty as compared with the chi-spare distribution for phase error in DPSK transmission [3].

The duo-binary modulation with alternating phase and control of the FWHM of the pulse sequence offer 2-dB improvement as compared with that of the CSRZ DPSK. The 50% FWHM with a Gaussian profile allows the possibility of lower bandwidth demand on the optical modulators [4]. This format still offers better performance due to the reduction of signal bandwidth.

9.3.4.2 Test-bed for Variable Pulse Width Alternating Phase Duo-Binary Modulation Optical Transmission

In this section, we investigate the transmission performance of alternating phase duo-binary (DB) modulation of a full-width half-mark (FWHM) ratio with respect to the bit period of 100, 50 and 33% and compared with experimental transmission of carrier-suppressed DPSK over 50 km of standard single mode fiber (SSMF) and dispersion compensation. For the DB case, the transmission without dispersion compensation over the same SSMF length offers better performance for 50% FWHM DB modulation, and slightly worse for 33%. The transmission performance, the BER versus receiver sensitivity, of these DB modulation formats are compared with the carrier-suppressed DPSK experimental transmission.

These transmission performances are compared with non-compensating 50-km SSMF transmission of duo-binary modulation with pulse width of 33% and 50%. The later format offers at least 2-dB improvement in terms of the BER and receiver sensitivity over that of CSRZ-DPSK. The 50% and 33% FWHM DB modulation schemes offer simpler driving circuitry for the optical modulators due to their lower swing voltage levels applied to the electrodes of the dual-drive interferometric modulator. This is very important when the bit rate is in the multi-GHz region. Furthermore, the 50% and 33% pulse width and Gaussian profile further reduce the demand on the bandwidth of optical modulators, i.e., $a \leq 30$ GHz transmittance bandwidth can be used. A Simulink® model has also been developed for simulation of the transmission performance of the duo-binary modulation formats. Simple driving conditions can be achieved and its effectiveness demonstrated.

We describe a generic simulation platform on MATLAB® Simulink for several modulation formats, especially the phase-modulated and partial response optical transmission, as shown in Figure 9.37. It consists of two duo-binary transmitters, a fiber transmission model and a direct detection opto-electronic receiver. A typical system arrangement of optical transmitter using a dual drive MZIM for generation of DPSK and alternating phase DB modulation for simulation is shown in Figure 9.38 and Figure 9.39. Figure 9.40 shows the Simulink simulation model of the data encoder for duo-binary format using differential encoding scheme of DPSK. The differential signals can be delayed or non-delayed and then added with its inverted version to generate a three-level electrical signals to two-level signals for driving the data modulator electrodes. Then Figure 9.41 gives the Simulink simulation model for the propagation of duo-binary-modulated carrier over a dispersion-compensated transmission link.

The fiber propagation model follows the well-known split-step Fourier method (SSFM) with provision for switching between linear and non-linear power level propagation to minimize computing time.

The pulse FWHM of the DB modulation format can be generated by setting the amplitude of the swing voltage levels applied to the dual electrodes of the MZIM. The biasing condition of the MZIM can be varied between the minimum and maximum transmission and the quadrature points of the transfer characteristics of the modulator to obtain carrier suppression and alternating phase properties of the modulated lightwaves. RZ formats and suppression of the carrier can also be generated by using another dual-drive MZIM biased at minimum transmission points of both sides of the voltage-optical intensity transfer curve of MZIM. This is the pulse carver that would be required to generate variable pulse width optical "clock pulses" before feeding into the data modulator.

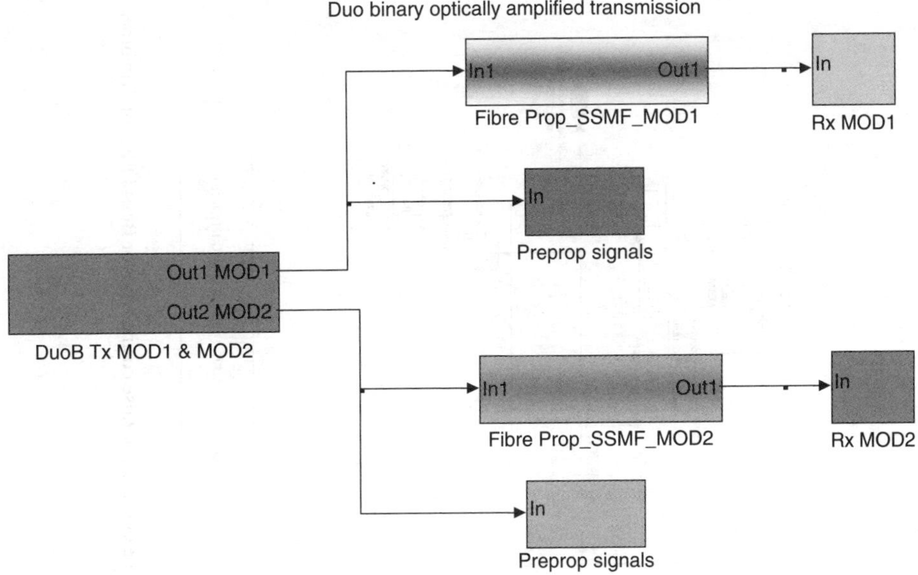

FIGURE 9.37 Generic simulation model for duo-binary variable pulse width transmission. The transmitter is shown on the far left and consistsing of two sub-system transmitters – one is the NRZ duobinary and the other a different variable pulse width duo-binary format for comparison. Each span consists of a transmission fiber (standard single-mode fibre SSMF) in association with a dispersion-compensating fibre and two in-line amplifiers (see Figure 9.41). Cascaded spans are identical with 100-km SSMF and 100-km DCF (negative dispersion factor and matched slope) and 20 dB gain plus 5 dB NF EDFAs.

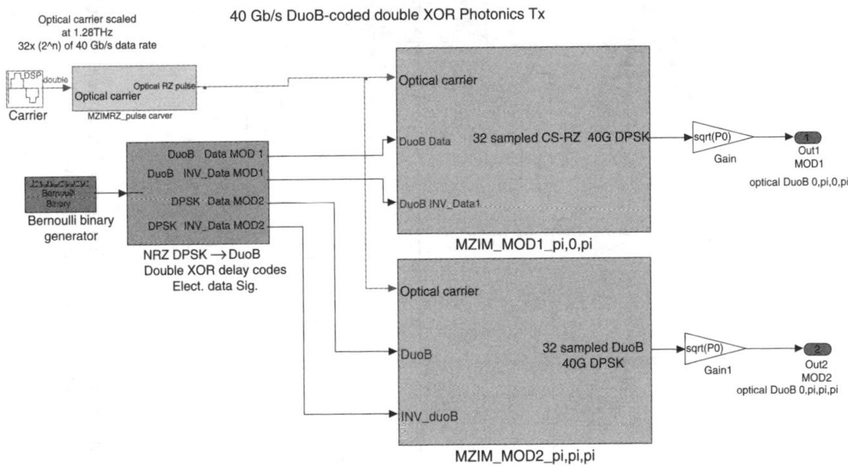

FIGURE 9.38 MATLAB® Simulink model showing the pulse carver or optical "clock sources" and the data dual-drive MZIM for generating of variable pulse width RZ-duo-binary transmitter.

Half bit rate frequency synthesizer is used as the driving source applied to the two electrodes for generation of RZ periodic pulse sequence. The optical spectrum of a lightwave 50% RZ duo-binary modulated signals (Figure 9.44) confirms the estimation of the modulation technique.

Simulink models have been extensively tested and the system performance agrees well between experiments and modeling. Figure 9.48 shows the agreement between the BER versus the receiver sensitivity for CSRZ DPSK modulation format obtained experimentally and by simulation. Noises

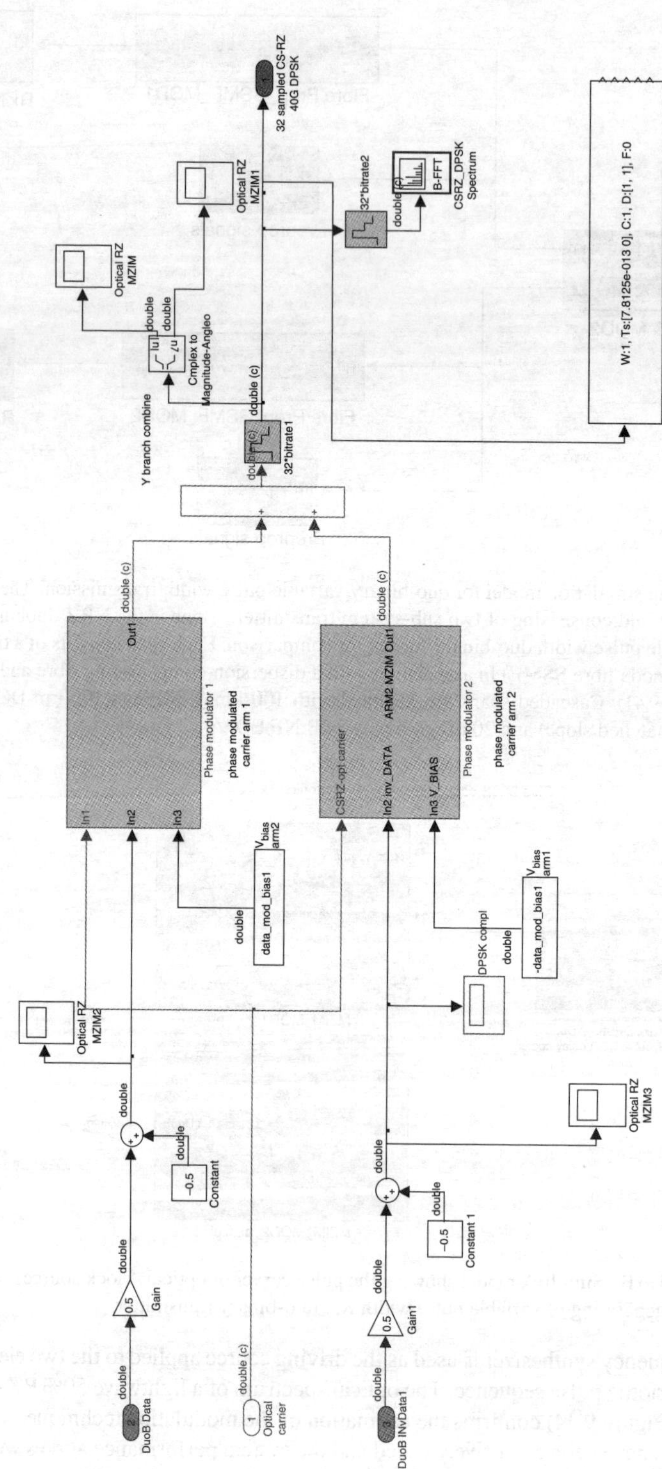

FIGURE 9.39 Electrical filter inserted before modulating the data optical modulator. Electrical filter can be Gaussian or raise cosine types or Bessel filter of fifth order can be inserted to reduce the required transmitting signal bandwidth.

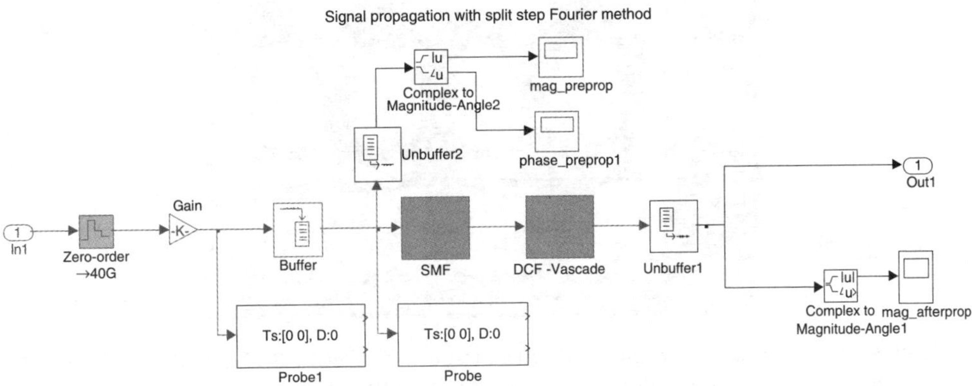

FIGURE 9.40 Simulink simulation model of the data encoder for duo-binary format using differential encoding scheme of DPSK. The differential signals can be delayed or non-delayed and then added with its inverted version to generate a three-level electrical signals to two-level signals for driving the data modulator electrodes.

FIGURE 9.41 Simulink simulation model for the propagation of duo-binary-modulated carrier over a dispersion- compensated transmission link.

of optical amplifiers have also been taken into account in both cases. We selected the CSRZ–DPSK format to compare with alternating phase DB modulation because it is proven in practice, offering superior performance as compared with RZ-DPSK and NRZ-DPSK. The MZIM is modulated and biased such that the width of the DB modulated pulses can be altered.

Simulation results are obtained for alternating phase DB with a FWHM of 50% and 33% of the bit period for 40-Gb/s, as shown by the blue curves indicated in Figure 9.48. An optical Gaussian filter type is also used at the output of the transmitter prior to transmission. We observe almost 2-dB better receiver sensitivity of the 50% FWHM alternating phase DB formats as compared with the CSRZ-DPSK [5]. However, the 33% FWHM DB case is 1-dB less sensitive. A single photodetector can be used to recover duo-binary-modulated lightwave signals. The simulation is conducted with 256 random bit pattern and several frames are sufficient for measurement of the Q-factor without resorting to the Monte Carlo technique. We also assume a Gaussian distribution of the inter-symbol interference (ISI) and phase noises. This assumption may have suffered 0.5–1 dB penalty as compared with the chi-spare distribution for phase error in DPSK transmission [6].

9.3.4.2.1 CSRZ DPSK Experimental Transmission Platform and Transmission Performance
A transmission system is arranged as shown in Figure 9.42a which shows the experimental and simulated system test bed for CSRZ-, NRZ- and RZ-DPSK and duobinary transmission over 328-km

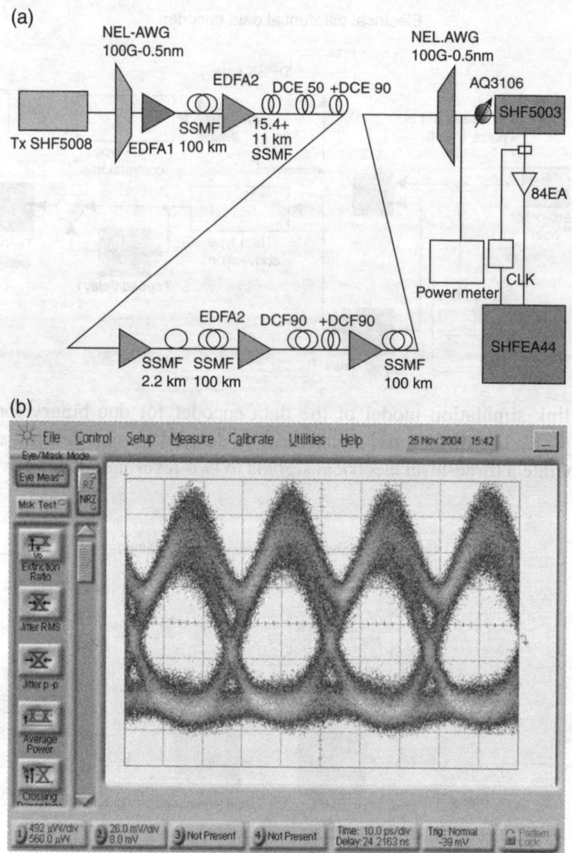

FIGURE 9.42 Experimental and simulated system test bed for CSRZ-, NRZ- and RZ- DPSK and duo-binary transmission over 328-km SSMF + 320-km DCF and associated EDFAs using balanced receiver (a) schematic diagram of the test bed (b) 34 ps/nm residual dispersion with dispersion compensated received eye diagram of CSRZ- DPSK.

SSMF +320-km DCF and associated EDFAs using balanced receiver. Figure 9.42b shows the received eye diagram of CSRZ-DPSK at the receiver with 34 ps/nm residual dispersion with dispersion compensated link. It consists of 50 km of SSMF and associate compensating module, a DPSK transmitter (SHF 5300), optical booster and pre-amplifiers, wavelength multiplexer and demultiplexer, MZIM, an optical balanced receiver SHF-5008, clock recovery module and error analyzer SHF-EA 44A. A typical eye diagram obtained for the 40-Gb/s CSRZ-DPSK modulation format recovered after transmission is shown in Figure 9.42b. This test bed is also used to investigate the transmission of DWDM channels of various modulation formats. So the wavelength mux and demux are included here. The bit-error-rate (BER) versus the receiver sensitivity for the CSRZ DPSK is shown in Figure 9.48. The curves for DB modulation with alternating phase of the "1" intensity coded pulses are also obtained by simulation.

The duo-binary simulation setup is structured close to the experimental setup for CSRZ DPSK transmission over 320-km SSMF optical fiber transmission multi-span links. Although the modulators employed in the SHF 5008 optical transmitter are single drive, the simulated modulators are dual-drive, and signaling and coding blocks are described in the previous section. The receiver model is simple as direct detection receiver circuitry is used, including a wideband photodetector and microwave amplifier. A Besssel fifth order or Gaussian (80% or duo-binary signal bandwidth) electrical filter can also be incorporated to minimize the noise contribution. Figure 9.43a shows

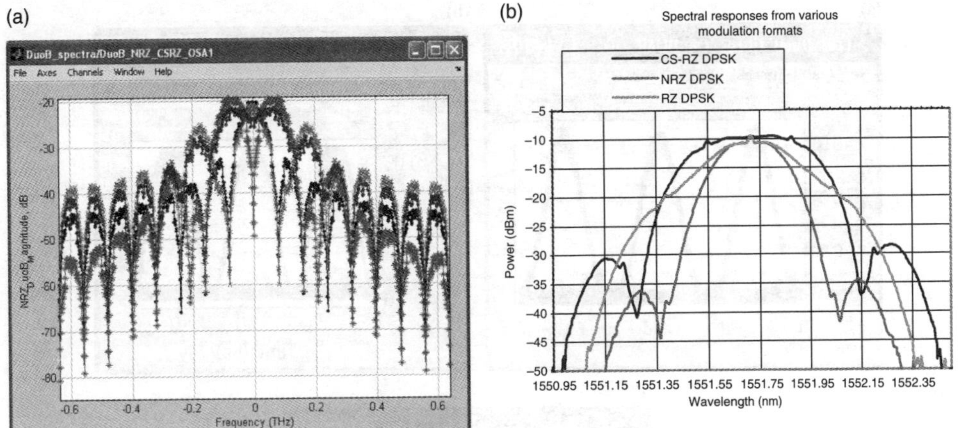

FIGURE 9.43   Optical spectrum as monitored at the output of (a) the 40-Gb/s duo-binary transmitter with formats NRZ (dotted-black), CSRZ 67% pulse width (red +), 50% RZ (blue dot) and RZ33% (magneta *). All normalized with respect to the optical carrier, which is set at 192.52 THz (b) Spectra of CSRZ-, NRZ- and RZ- DPSK modulation formats obtained by experiment.

optical spectrum as monitored at the output of the 40-Gb/s duo-binary transmitter with modulation formats NRZ, CSRZ 67% pulse width, 50% RZ and RZ33%. All normalized with respect to the optical carrier, which is set at 192.52 THz. Figure 9.43b shows the spectra of CSRZ-, NRZ- and RZ-DPSK modulation formats obtained by experiment.

The optical spectra of DPSK signals under different pulse width and pulse carver operations are shown in Figure 9.44b, while Figure 9.44a shows the simulation spectra of their counterparts under the duo-binary coded scheme. The spectra are centered at the optical carrier frequency. The red curve (+) clearly shows the suppression of the carrier, while the other duo-binary schemes show minimum carrier suppression, but substantial signal power level that indicates the specific characteristics of the di-phase modulation, especially when the data modulator is biased such that the phase difference between the two arms of the MZIM is π.

Figure 9.36 shows the simulated eye diagrams at the receiver (after the photodetector, including electronic amplifier noises) of the duo-binary coded sequence of NRZ (c) 50% RZ (b) and 33% RZ (a). Gaussian electrical filtering has been applied to the NRZ duo-binary and none applied to the other RZ formats of the scheme. No electrical filtering is used at the transmitter. The bit rate is 40 Gb/s.

The effects of electrical filtering at the transmitter on the received eye diagrams are shown in Figure 9.45a–d. A 30% bit rate (BT = 0.3) electrical Gaussian filter bandwidth would still sustain the eye opening after 320-km transmission, but substantial closure of the temporal eyes (Figure 9.45a and Figure 9.45b), i.e., the decision sampling time, must be very accurate to be error-free, even under fully compensated transmission. A 50% BT would, as observed under simulation, be much more tolerable to the sampling error. Figure 9.45c and Figure 9.45d shows such filtering effects would not exist for a raised cosine electrical filter with a 0.5 roll-off factor for 40 Gb/s CSRZ 67% duo-binary format transmission over 320 km 3-span optically amplified fully compensated transmission. Figure 9.47a shows the experimental BER versus receiver power for 320-km SSMF and 320-km (effective negative 320 km SSMF) DCF transmission using CSRZ DPSK modulation format with complete compensation and experimental RZ-DPSK and simulated duo-binary 67%, simulated duo-binary 50% FWHM alternating phase modulation. It is observed that 67% improves over 50% duobinary by 4.2 dB. Figure 9.47b shows the BER versus the residual equivalent SSMF length. The launched power is set at 0 dBm at the output of transmitter. The simulated BER for 40 Gb/s duo-binary 67% and 50% and 33% FWHM is also shown in Figure 9.47a.

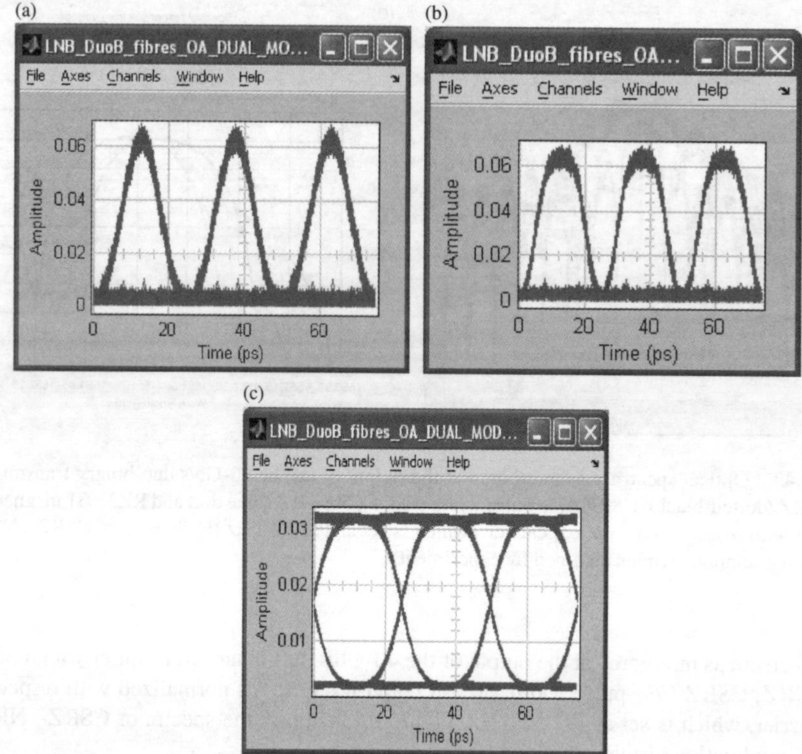

FIGURE 9.44 Received eyes after (3 spans) 320-km SSMF fiber fully compensated, including two in-line optical amplifiers (20-dB gain and 5-dB noise figure) transmission. (a) DB 33% FWHM (b) and (c) 50% duo-binary.

Fully compensated eye diagrams after transmission over the 320-km optically amplified multi-span SSMF transmission link are shown in Figure 9.48a and the dispersion tolerance versus the BER is shown in Figure 9.48b. The dispersion tolerance is similar for various pulse-width duo-binary signals. The 33% and 50% RZ-modified duobinary formats shown indicate error-free transmission with a residual distance of about 6 km duo-binary format, while experimentally we could obtain error-free transmission only after 3-km SSMF residual dispersion for the CSRZ 67% DPSK format.

Experimentally and under simulation, the launched power is adjusted and the BER versus the receiver sensitivity is obtained as shown in Figure 9.48. Included in this figure is the performance curve of the CSRZ DPSK transmission. The BER curves indicate that 50% RZ and 67% NRZ duo-binary formats outperform the CSRZ DPSK by at least 2 dB. Although these duo-binary performance curves are from simulation, we expect the comparison with those of experimental CSRZ DPSK would not be much different as we have taken into account the electronic noise and the signal-dependent shot noises in the models of the photodetector. In experimental and modeling cases, it is reasonable to assume that the PMD effects are negligible. We expect that the 33% RZ duo-binary performance would fall between these two formats. We have not sufficient time to conduct detailed transmission results for this format and these results will be reported in the near future.

The model represents exactly the photonic behavior of lightwave modulated modulation formats, i.e., the square law direct detection incorporating the equivalent electronic noises of 40-GHz receiver. Signal-dependent shot noise and equivalent electronic noise current are modeled. The bandwidth of the receiver is adjusted according to the effective bandwidth of the modulation scheme. This allows us to justify the comparison of practical implementation of the direct detection and balanced receivers. The pulse sequence at the receiver is integrated and an average power

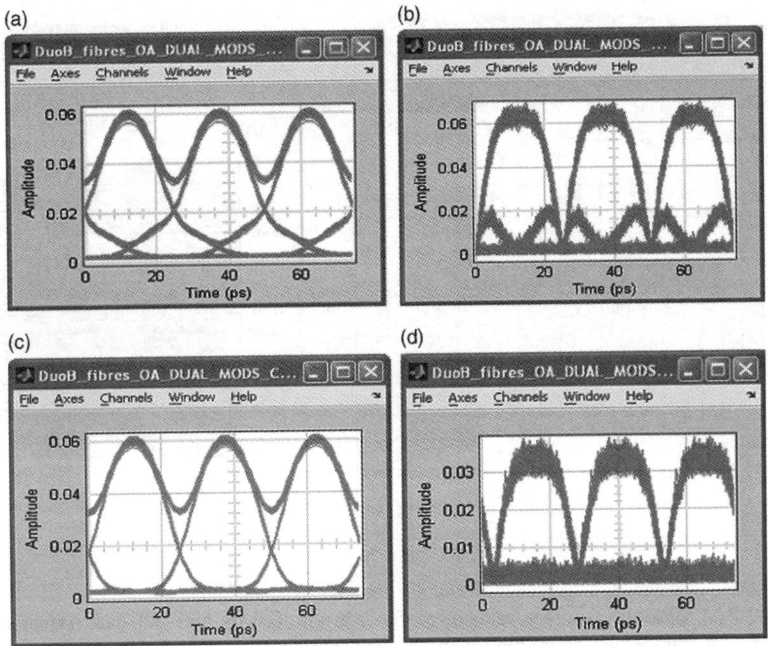

FIGURE 9.45   Received eyes after (~3 spans) 320-km SSMF fiber fully compensated, including two in-line optical amplifiers (20-dB gain and 5-dB noise figure) 40-Gb/s transmission. (a) CSRZ 67% duo-binary with the integration of Gaussian electrical filter with 30% bit rate (b) same as (a) but without filtering at the receiver (c) CSRZ 67% duo-binary with raised cosine filtering at transmitter and no filter at the receiver.

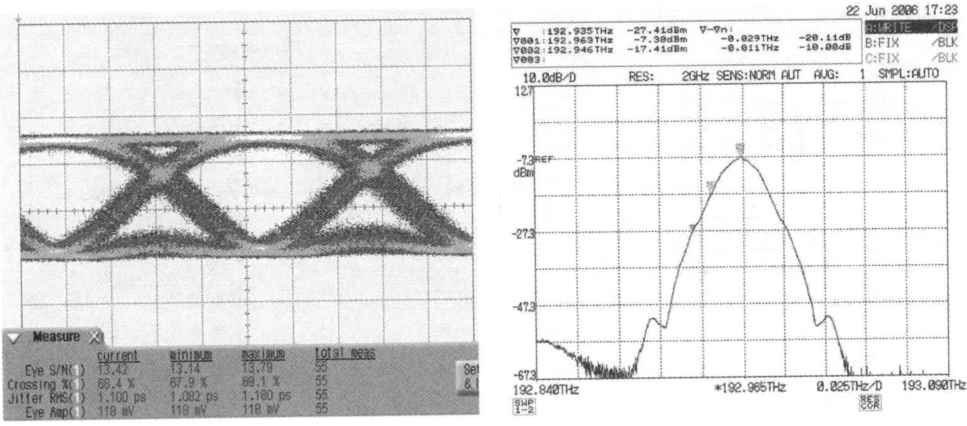

FIGURE 9.46   Duo-binary signals. (a) Eye diagram. (b) Measured optical spectrum. (Courtesy of SHF Ltd, Berlin, Germany).

obtained. Then its equivalent quantum shot noise is calculated over the electronic amplifier bandwidth. The equivalent electronic noise current at the input of the electronic amplifier is then added to these signal-dependent noises, and a signal-to-noise ratio obtained. Thus, the received optical power can be derived and plotted with respect to the BER obtained from processing the eye diagram obtained after each transmission. The simulation duration is set long enough so that the number of pulses transmitted and received is sufficiently high. Although we understand that the noise statistics of the phase modulated optical transmission is asymmetrical and may follow a Maxwellian

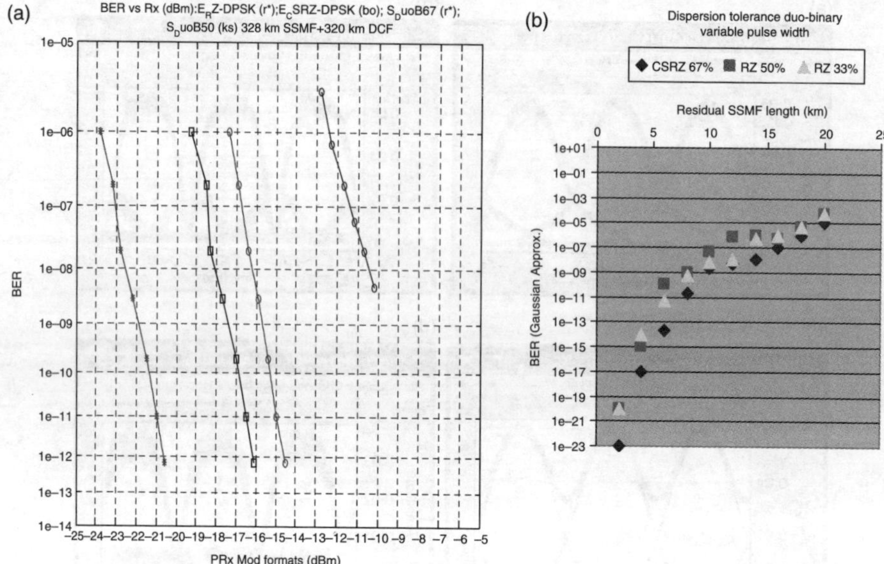

FIGURE 9.47 (a) BER versus receiver power for 320-km SSMF and 320-km (effective negative 320 km SSMF) DCF transmission (a) red (o) curve – **experimental** CSRZ DPSK transmission with complete compensation (b) **blue o** experimental RZ-DPSK. **Red (*)** curves simulated duo-binary 67%, **square black** simulated duo-binary 50% FWHM alternating phase modulation. 67 % improves over 50% duobinary by 4.2 dB. (b) Simulated dispersion tolerance for 40 Gb/s duo-binary 67% and 50% and 33% FWHM – BER versus residual equivalent SSMF length. Launched power 0 dBm at the output of transmitter.

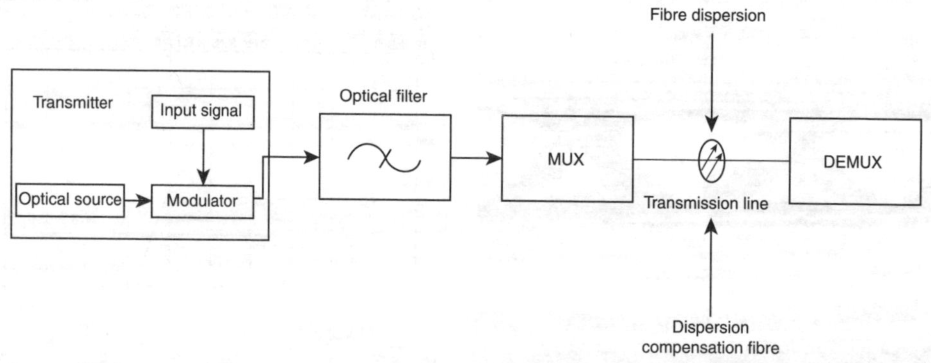

FIGURE 9.48 Model of the VSB transmission system (schematic).

distribution, in this article it is assumed that the noise statistics follow a Gaussian distribution. This assumption would suffer only about 0.5–1.0-dB difference from the true distribution.

The duo-binary modulation with alternating phase and control of the FWHM of the pulse sequence offer 5-dB improvement as compared with that of the CSRZ DPSK obtained experimentally. For the simulated duo-binary 50% pulse format, a 1.2-dB improvement on the receiver power is obtained. Equivalent noise current at the receiver of 2.5 pA/(Hz)$^{1/2}$ is used, which is compatible to that of the commercial 50-GHz DPSK optical receiver. The bandwidth of the receiver is adjusted to narrower for 67% duo-binary case, and hence less noises for the quantum signal-dependent shot noise and total equivalent electronic noise power. The 40-Gb/s duo-binary signals in the time domain and its spectrum are shown in Figure 9.46 using a filter of 17 GHz in the transmitter. The

3-dB bandwidth shows clearly the single sideband and embedded carrier. The 50% FWHM with a Gaussian profile allows the possibility of lower bandwidth demand on the optical modulators [4]. Still this format offers better performance due to the reduction of the signal bandwidth. All performance curves shown in Figure 9.48 were obtained when the modulation and the carrier power were operating under linear region with the non-linear threshold set at a power that would create a total phase change of 0.1π. A 4.2-dB improvement of 67% as compared with 50% duo-binary is due to the reduction of the electronic and quantum shot noise at the receiver.

9.3.5 Remarks

DBMs have been demonstrated via simulation using a Simulink® transmission model test bed. The system performance is error-free, with receiver sensitivity −60 dBm without receiver electronic noise. The non-dispersion compensation transmission can reach 9–10 km for 40 Gb/s, equivalent to 150 km for 10 Gb/s. The Simulink model is successfully developed. We have also demonstrated experimentally and in modeling the CSRZ DPSK transmission. These transmission performances are compared with non-compensating 50-km SSMF transmission of duo-binary modulation with pulse width of 33 and 50%. The latter format offers at least 2-dB improvement in terms of the BER and receiver sensitivity over that of CSRZ-DPSK. The 50 and 33% FWHM DB modulation schemes offer simpler driving circuitry for the optical modulators due to their lower swing voltage levels applied to the electrodes of the dual-drive MZIM. This is very important when the bit rate is in the multi-GHz region. Furthermore, the 50 and 33% pulse width and Gaussian profile will further reduce the demand on the bandwidth of optical modulators, that is, a 30 GHz or lower MZIM can be used. A Simulink model has also been developed for simulation of the transmission performance of the duo-binary modulation formats. Simple driving conditions can be achieved and its effectiveness demonstrated.

We have demonstrated experimentally and in modeling the CSRZ DPSK and duo-binary transmission. These transmission performances are compared with non-compensating 50-km SSMF transmission of duo-binary modulation with pulse width of 50 and 67% (NRZ). The latter formats (50 and 67%) offer at least 5-dB and 1.2 , respectively, in terms of the BER and receiver sensitivity over that of CSRZ-DPSK. The 67%, 50% and 33% FWHM DB modulation schemes offer simpler driving circuitry for the optical modulators due to their lower swing voltage levels applied to the electrodes of the dual-drive MZIM. Simpler and narrower bandwidth of the electronic amplifier for 67% duo-binary case would also improve the receiver sensitivity. We thus expect 33% duo-binary would be worse than CSRZ-DPSK and hence no simulation is conducted for this scenario. This is very important when the bit rate is in the multi-GHz region. Furthermore, the 67%, 50% pulse widths and Gaussian profile will further reduce the demand on the bandwidth of optical modulators, that is, a 30 GHz or lower MZIM can be used. Simulink models have also been developed for simulation of the transmission performance of the duo-binary modulation formats. The balanced receivers are modeled exactly as a real practical sub-system. This allows a fair comparison with the transmission performance of the CSRZ-DPSK formats. Simple driving conditions can be achieved and its effectiveness demonstrated.

We have not conducted the transmission of duo-binary format pulse sequences of variable pulse width over the multi-span optically amplified distance in the power region where the self-phase modulation effects may occur. However, it is expected that the CSRZ 67%, RZ 33% and RZ 50% duo-binary format would outperform their DPSK counterparts. It is also expected that the RZ 33% duo-binary would suffer much less non-linear distortion effects than its 50 and 67% RZ counterparts. These performances will be reported in the future. Similarly, the effects of electrical filtering at the transmitter will also be investigated.

9.4 DWDM VSB MODULATION-FORMAT OPTICAL TRANSMISSION

This section presents the transmission of optical multiplexed channels of 40 Gb/s using vestigial single side band modulation format over a long-reach optical fiber transmission system. The effects on the Q-factor of fibers dispersion, the pass band and roll-off frequency of the optical filters and the channel spacing are described. The performance of the optical transmission using low and non-zero dispersion fibers or/and dispersion compensation is discussed. It has been demonstrated that BER of 10^{-12} or better can be achieved across all channels, and minimum degradation of the channels can be obtained under this modulation format. Optical filters are designed with asymmetric roll-off bands. Simulations of the transmission system are also given and compared for channel spacing of 20, 30 and 40 GHz. It is shown that the passband of 28 GHz and 20-dB cut-off band performs best for 40-GHz channel spacing.

Optical communications have been extensively developed for ultra-high-capacity transport networks. However, the demand for high-speed communication system over ultra-long reach and ultra-long-haul offering greater capacity is expected to offer challenges for further technical development for band width-efficient networking. Global Internet traffic has been growing rapidly, typically doubling the backbone traffic each year. This drives a requirement for higher channel speeds per Internet port. As Internet backbone routers are currently moving to 10 Gbit/s and 40 Gb/s as a standard, it is foreseen that efficient transmission techniques for increasing the bit rate may be required in the near future.

The most common modulation formats for 40-Gb/s optical systems are the RZ, NRZ, CS-RZ, which can also be integrated with differential PSK coding [7–8]. The vestigial sideband (VSB) RZ modulation format can also be considered as the appropriate choice for Tb/s long-haul optical transmission systems due to its half band property, and hence is highly tolerant to dispersion and non-linearity. On the other hand, VSB with NRZ could provide higher spectral efficiency than the VSB-RZ format since NRZ occupies only half of the RZ bandwidth and requires a lower peak transmit power in order to maintain the same energy per bit. This would offer the same bit-error-rate (BER) as the RZ format.

This section gives a numerical simulation of the VSB-NRZ modulation format for long-haul optical fibers transmission system, and the effects of optical filters (OF) on its transmission performance. The OF used for eliminating unwanted optical side bands are critical in the generation of the desired format. The design of such filters is given so as to alter the pass- and roll-off bands of the filters to investigate their effects on transmission system performance. Furthermore, the channel spacing of the multiplexed channels is also important and is a major factor in the specification of OFs. Simulation is presented for eight channels DWDM of 40-GHz channel spacing optical fibers transmission system. The channels are transmitted at 40 Gb/s over a dispersion-managed 100-km span. The effects of these filters on back-to-back transmission and dispersion tolerance are the principal objectives of this work.

This section is thus structured as follows. Section 9.2.2 gives an overview of the simulation design and the transmission system. Sections 9.2.3 and 9.2.4 describe the optical transmitter and modulation format coding. The properties of the OF passband and the signal spectra of the VSB channels are also given. Section 9.2.5 presents the details of the design of the VSB filters and the DWDM, multiplexing and demultiplexing. Section 9.2.6 discusses the roles of the group velocity dispersion (GVD) factor of the transmission and dispersion compensating fibers. Section 9.2.7 gives the evaluation of the transmission performance using the received eye diagram and the Q-factor and the effects of several factors on the bit-error rate (BER). Finally, section 8 offers concluding remarks.

9.4.1 TRANSMISSION SYSTEM

The schematic diagram of the optical transmission employing VSB modulation format is shown in Figure 9.48. An optical source operating in the continuous mode is launched into an external optical modulator which is modulating the lightwave carrier via a random bit pattern generator and associated microwave power amplifiers. The external modulator, an X-cut $LiNbO_3$ MZ intensity modulator, is used to offer chirp-free modulated output. For 40 Gb/s, the stabilization and linearity of the external modulator is critical. Normally two modulators would be used, one for generating the required NRZ format and the other used for carrier-suppression and phase modulation if required, depending on the transmission format.

NRZ data format is used in this simulation. VSB modulation technique is used and generated by the use of an OF, which filters the unwanted sideband of the information channel. After filtering, a number of optical VSB channels can be multiplexed and then transmitted through a number of optically amplified dispersion-managed fibers spans. Each optically amplified dispersion-compensating in-line unit would consist of an in-line optical amplifier followed by a dispersion compensating module and then another optical amplification booster that would then enhance the total average optical power for transmission to the next span. Eye diagrams are obtained and the bit-error-rate (BER) can be deduced. For example, the back-to-back eye diagram is observed when dispersion compensation fibers are used with appropriate dispersion slope for equalization. Naturally, at the end of the transmission line, the multiplexed channels are separated via a demultiplexer. The principal objective of this article is to study the effects of the OF on VSB system performance, thus we do not include optical amplification noises in our modeling in the pre- and booster amplifiers located at each transmission span. RZ and the NRZ formats can be used. The main advantage of NRZ format is that it occupies half of the RZ bandwidth. VSB modulation allows a small amount of the unwanted sideband existing at the output of modulator depending on the roll-off passband of the filter. Instead of eliminating the entire second sideband (such as in SSB), the VSB modulation suppresses most of, but not all, the second sideband. Using this technique, the difficulty in generating a very sharp cut-off can be overcome. The VSB modulation can be implemented with OFs to eliminate most of the second sideband. The spectrum of a VSB signal can be obtained as in [9]

$$x_c(t) = \frac{1}{2}AE\cos(\omega_c - \omega_1)t + \frac{1}{2}A(1-E)\cos(\omega_c + \omega_1)t + \frac{1}{2}B\cos(\omega_c + \omega_2)t \qquad (9.10)$$

where E is magnitude of the optical field envelope, ω_c is the optical carrier radial frequency, ω_1 and ω_2 are the arbitrary radial frequencies of the signals. The signal can be demodulated by multiplying by $4\cos(\omega_c t)$ and applying the low-pass OF, leading to

$$E(t) = A\,E\cos\omega_1 t + A(1 - E)\cos\omega_1 t + B\cos\omega_2 t \qquad (9.11)$$
$$e(t) = A\cos\omega_1 t + B\cos\omega_2 t.$$

The basic characteristics of a VSB OF can be illustrated in Figure 9.49. The roll-off and passbands are asymmetric.

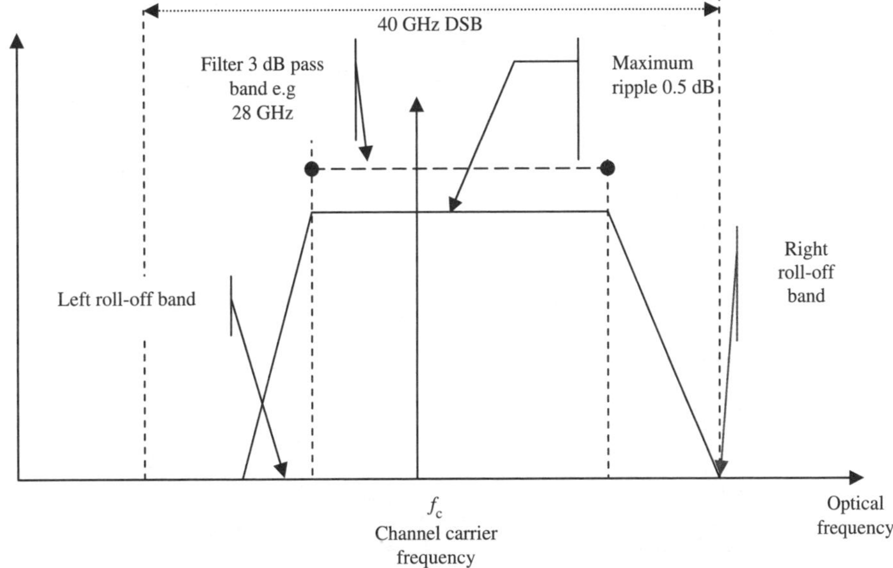

FIGURE 9.49 Spectral property of the VSB modulation format as compared with that of DSB (double sideband) (a) general frequency shifting and (b) details of optical VSB filter pass-band and roll-off bands.

9.4.2 VSB Filtering and DWDM Channels

To achieve the VSB modulation-formatted signals, OF is implemented. In this article, a number of low-pass elliptic filters (LPEF) are chosen because this filter type offers the steepest transition region between the passband and stop-band without suffering stability problems. The elliptic filter is a combination of the Chebyshev type I and Chebyshev type II, and exhibit some amplitude response ripples in the passband and stop-band. The main advantage of elliptic filters is that the width of the transition band is minimized for a finite ripple limit in the passband and a minimum attenuation in the stop-band. Furthermore, these filters can be implemented using planar circuit technology [10]. The spectral response of the optical low pass elliptic filter (LPEF) of order N is given by [10]

$$\left|H_{LP}(j\omega)\right|^2 = \frac{1}{1+\varepsilon^2 E_N^2\,(\omega)} \tag{9.12}$$

where $E_N^2(\omega)$ is the Chebyshev rational function and can be determined from the specified ripple characteristics. Similarly, the s-domain transfer function of the LPEF of order N can be obtained as

$$H_{LP}(s) = \frac{H_0}{D(s)} \prod_{i=1}^{r} \frac{s^2 + A_{0i}}{s^2 + B_{1i}s + B_{0i}} \tag{9.13}$$

where $s = j\omega$, $r = N - 1/2$ odd N and $r = N/2$ for even N. The definitions of all the bands of the filters are referred to in Figure 9.49. Figure 9.50 illustrates a typical response of the LPEF. The carrier wavelength of 1550 nm is used as the center wavelength corresponding to the optical frequency of 193.41 THz. The characteristic of the elliptic filter is designed with the following properties: passband ripple = 0.5 dB or less; minimum stop-band attenuation = 10 dB; passband region = 28 GHz; stop-band = 2 GHz and 20-dB cut-off band = 10 GHz and 20 GHz for the left and right sides,

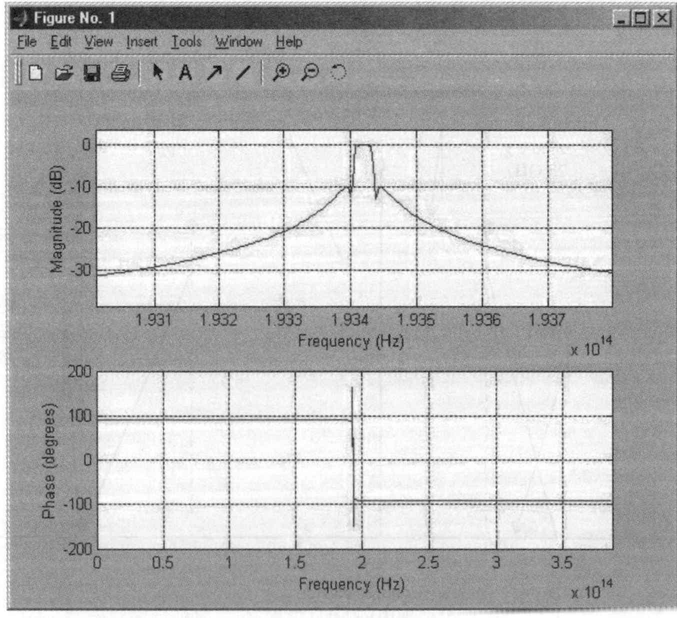

FIGURE 9.50 Frequency responses of the elliptic filter including the amplitude (upper curve) and phase (lower graph).

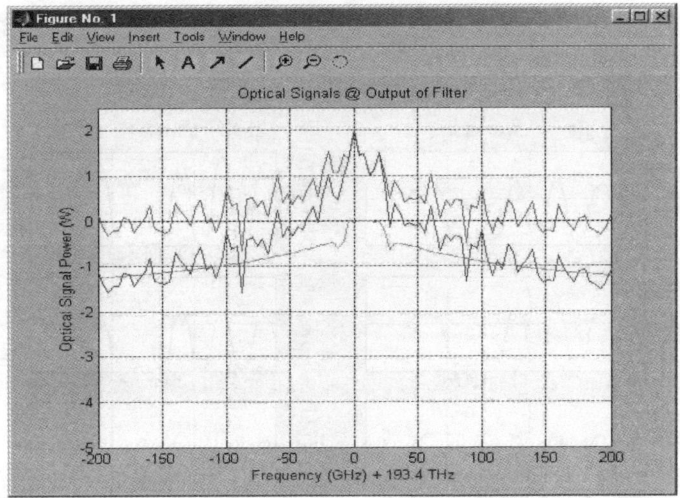

FIGURE 9.51 Filter characteristics (bottom) and signal spectra before filtering (most upper) and after filtering (middle curve). The filter frequency response is also included and indicated as the bottom graph.

respectively. The filter spectral characteristics are illustrated in Figure 9.50 and Figure 9.51. The eight multiplexed signals at the output of a LPEF filter are plotted in Figure 9.52. This type of OFs can be implemented with fibers Bragg gratings or multi-stage silica-on-silicon planar integrated MZDI. The MZDI can also be considered an all-pass (all-zero) filter that would offer no resonant peaks, and hence its group delay is almost constant, so dispersion would not be introduced by these OFs. In this work, we would take into account the dispersion characteristics deduced from the group delay of the design filter, and this would be compensated for by the dispersion-compensating module.

The optical transmission system is simulated for eight wavelength channels in which the 1550 nm wavelength is taken as the center wavelength and the frequency spacing of 20, 30 and 40 GHz spacing wavelength grid. For simplicity, we use 1550 nm as the center wavelength rather than the exact ITU spectral grid. After filtering out the unwanted sideband, the multi-channels are multiplexed and then propagated through the single-mode low non-zero dispersion-shifted fibers (NZDSF). The multiplexing is based on the wavelength division multiplexing (WDM) technique in which multiple optical carriers at different wavelength are modulated using independent electrical bit streams, and are then transmitted over the same fibers. Our design employs variable channel spacing of 20 GHz to 40 GHz. The signal spectra in the frequency and the time domain at the output of multiplexer are depicted in Figure 9.53. There are some ripples in the temporal signals. It is believed that these ripples appearing at the output of multiplexer are due to the cross talks generated in multi-channel transmission. At the output of the LPEF, the ripple is negligible as the channels are filtered independently of each other.

9.4.3 TRANSMISSION DISPERSION AND COMPENSATION FIBERS

Single-mode fibers (SMF), standard or NZ-DSF types, are naturally chosen as the transmission media in this simulation. This section briefly describes the design of the fibers so that their dispersion properties can be specified accurately instead of using the data provided by fiber manufacturers. Due to chromatic dispersion, the GVD is frequency-dependent. Consequently, different spectral components of the pulse travel at a different velocity and cause pulse dispersion that limits the performance of SMFs. Fiber dispersion consists of two components: material dispersion and waveguide dispersion. The transmission fibers and the matched dispersion compensating fibers are described in Ref [11] with matched dispersion factors and dispersion slopes for complete compensation of the channels across the waveband. A brief summary of the design is explained below.

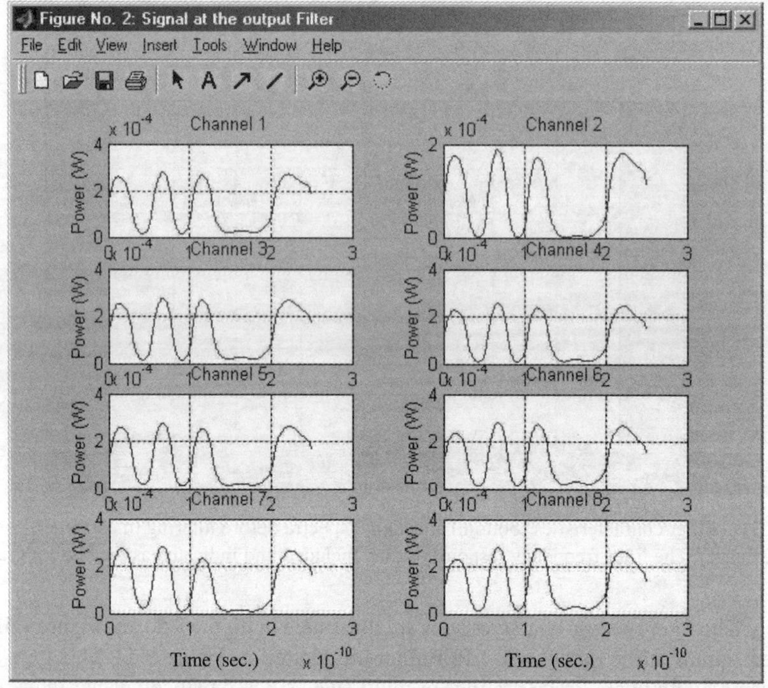

FIGURE 9.52   Signals of the DWDM channels at the output of the VSB filters.

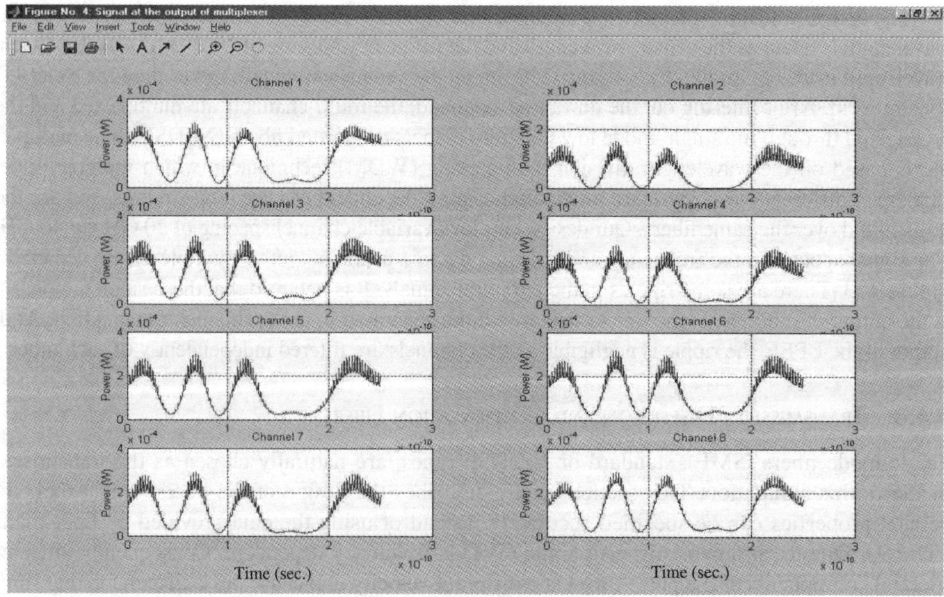

FIGURE 9.53   Time-domain signals of the channels at the outputs of the optical multiplexer.

For the transmission fiber, the material $M(\lambda)$ and $D_W(\lambda)$ waveguide dispersion factors can be estimated calculated by using

$$M(\lambda) = -\frac{\lambda}{c}\frac{d^2 n}{d\lambda^2} \quad D_w(\lambda) = \frac{n_2(\lambda)\Delta}{c\lambda} V \frac{d(Vb)}{dV^2}. \tag{9.14}$$

The waveguide-dependent factor can be approximated for when the V-parameter is constricted with the single mode range of 1.3 to 2.6 as

$$V\frac{d^2(Vb)}{dV^2} = 0.080 + 0.549(2.834 - V)^2. \tag{9.15}$$

Hence, the fibers total dispersion is given by

$$D_T = M(\lambda) + D_w(\lambda). \tag{9.16}$$

In order to obtain non-zero low total dispersion over a wide wavelength range, techniques for the design of dispersion-flattened fibers are employed. An optimized large effective area fiber with pure silica material for the cladding is designed for the transmission medium since it gives the best dispersion slope for matching with those of the dispersion compensation module. Sellmeier's constants for pure silica fibers are: $G_1 = 0.696750$ and $\lambda_1 = 0.069066e-6$; $G_2 = 0.408218$ and $\lambda_2 = 0.115662e-6$; $G_3 = 0.890815$ and $\lambda_3 = 9.9900559e-6$. The refractive index of pure silica can be expressed as

$$n(\lambda) = c_1 + c_2\lambda^2 + c_3\lambda^{-2} \tag{9.17}$$

if $c_1 = 1.45084$, $c_2 = -0.00343$ μm^{-2} and $c_3 = 0.00292$ μm^2, then the refractive index $[n(\lambda)]$ for pure silica fibers is 1.45. We have designed the transmission and dispersion-compensating optical fibers with specific dispersion factor 17 ps/nm/km and dispersion slope 0.3 ps/(nm^2.km) for transmitting several channels. The dispersion and dispersion slope values of the dispersion-compensating fiber are designed with a factor of five times of those of the transmission fiber with a residual dispersion of about 1.5 ps/nm/km for the highest and lowest wavelength channels. Fiber parameters are used in this simulation: fiber radius, $a = 1.6$ μm; relative refractive index difference, $\Delta = 0.0339$. The dispersion factors of the centered channel (only channel 5 is illustrated) are indicated in Figure 9.54. The pulse broadening after transmitting through the fiber length L can be expressed as

$$\Delta\tau = D(\lambda)\sigma_\lambda L \tag{9.18}$$

where $D(\lambda)$ is fiber spectral dispersion (ps/nm.km) and σ_λ is the signal bandwidth when the laser linewidth is much smaller than that of the VSB signals, and L is the fiber transmission distance. Figure 9.54 shows the signals at the output of the demultiplexer for each channel (channel 1 to channel 8) without dispersion compensation.

From Figure 9.54, the simulation results show that over a 100-km span transmission line, the eye of the transmitted data closes completely, as expected, due to fiber dispersion. The eye is still open at a BER of about 10^{-9} after 16-km designed fibers transmission. This is a significant improvement over the NRZ format.

We now turn to the dispersion-compensating fibers. Since the GVD limits the system performance, it is essential to implement a dispersion-compensating fiber (DCF). The condition of dispersion compensation can be expressed as

$$D_1 L_1 + D_2 L_2 + \Gamma = 0 \tag{9.19}$$

where D_1 is total dispersion of the fiber's transmission fibers section (ps/nm.km); L_1 is the fiber transmission distance (km); D_2 is the fiber dispersion-compensation distance (ps/nm.km); L_2 is the dispersion-compensation distance (km) and Γ is the dispersion factor contributed by the VSB

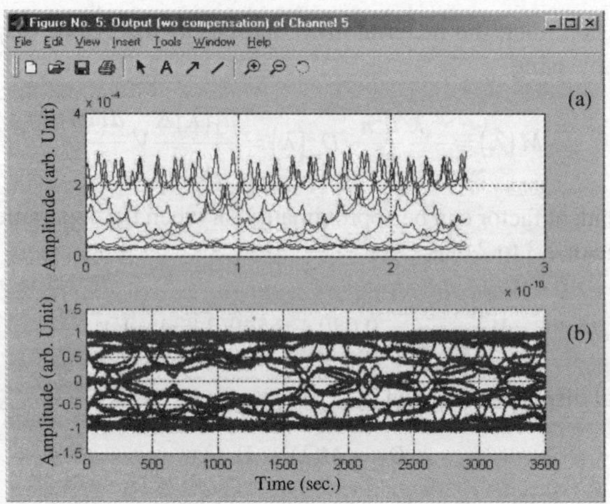

FIGURE 9.54   Signal outputs at the end of transmission fibers without dispersion compensation of the centered channel 5 at 1550 nm with $D_T = 2.7455e\text{-}6$ ps/(nm.km) (a) time sequence (b) eye diagram.

OF. Equation 9.10 shows that the DCF must have its GVD negative at 1550 nm to compensate for the positive dispersion of the transmission fibers and the OF. The parameters of the DCF can be optimized for minimum non-linear self-phase-modulation effect with: fiber radius $a = 1.7$ μm; relative refractive index difference $\Delta = 0.0243$. Figure 9.55 shows the transmitted data at the end of the dispersion-managed system with a dispersion compensation for channel 5 of the 8 channels. The effectiveness of dispersion-compensation fibers in the improvement of the BER as it recovers the quality of the data channels close to their original quality. The residual ripple of the compensated pulses clearly shows the effects of the filter passband on the time-domain pulses. These ripples contribute to the penalty on the eye closure and hence the Q-factor. An optical multiplexer then combines the modulated lightwaves to the transmitting fibers. At the receiver, the channels are separated into different channels by an optical demultiplexer. Figure 9.54 gives the simulation results of transmitted signals at the output of multiplexer without dispersion compensation, while Figure 9.55 shows the transmitted signals at the output of multiplexers with DCF in cascade with the transmission fibers. Optical amplifiers at the input and booster amplifier at the output are not included in this simulation.

9.4.4   Transmission Performance

The eye diagram is used to deduce the Q-factor and hence the system BER. The random digital optical pulse sequence suffer distortion by noise, pulse broadening and timing jitter errors introduced in the optically amplified fibers link. Assuming that the probability density function of the "1" and "0" are equal and Gaussian, the eye diagram can be used to estimate the Q-factor by

$$Q = \frac{\mu_1 - \mu_0}{\sigma_1 - \sigma_0} \tag{9.20}$$

where μ_1 and μ_0 are the means of the current at the output of the photodetector of the decision circuitry of the receiver at the sampling instant respectively for symbol "1" and "0". While σ_1 and σ_0 are the standard deviations of the current at the decision circuit input at the sampling instant respectively for symbol "1" and "0". Figure 9.56a and Figure 9.56b show the eye diagram of the transmitted pulse random sequence at 1550 nm (channel 5) with 40-GHz channel spacing which depict the eye diagram at the end of the transmission line without and with dispersion compensation fibers, respectively. The

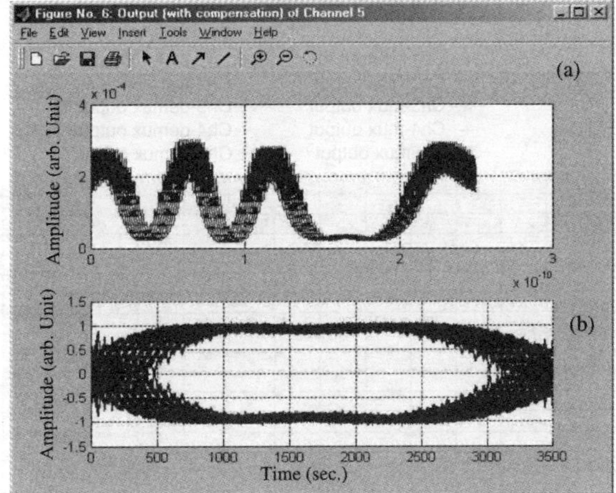

FIGURE 9.55   Signal outputs at the end of transmission line with dispersion compensation of the centre channel-channel 5 at 1550 nm with DCF dispersion factor of −2.7918e-6 ps/(nm.km) (a) temporal sequence (b) eye diagram.

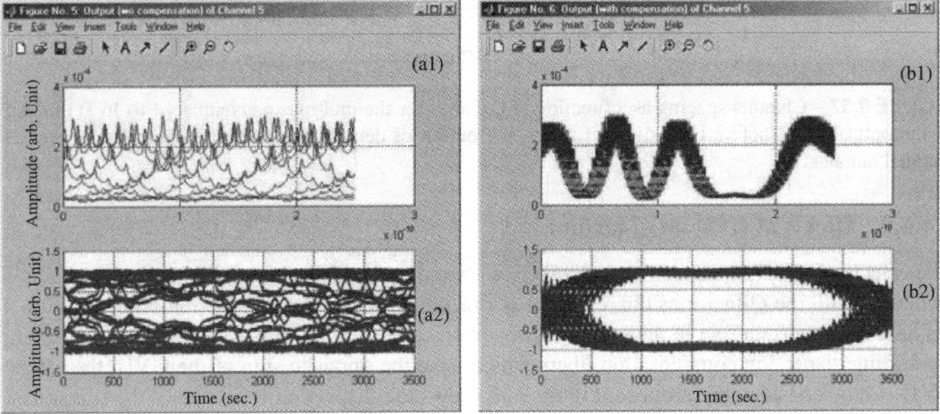

FIGURE 9.56   Signals received after transmission (a) without dispersion compensation and (b) with dispersion compensation. Upper curve (a1) and (b1) are temporal sequence and (a2) and (b2) correspond with eye diagrams.

eye diagram of data with dispersion compensation is wide open compared with that without dispersion compensation. The wide-open eye diagram shows that the system performance is good since the error has been compensated.

9.4.4.1   Effects of Channel Spacing on Q-Factor

The capacity of the WDM system depends on how close channels can be packed into the wavelength domain. The minimum channel spacing is limited by inter-channel crosstalk. The typical value of channel spacing should exceed four times that of the bit rate. The Q-factor is obtained for each channel of at the output of the multiplexer and at the output of the demultiplexer, at the end of the 16 channels DWDM 10km designed fibers transmission without dispersion compensation in Figure 9.57. The Q-factor at the output of demultiplexer decreases significantly for channel spacing 20 GHz and 30 GHz. While the Q-factor for the 40-GHz channel spacing decreases only slightly. This indicates the resilience of the VSB modulation format to chromatic dispersion.

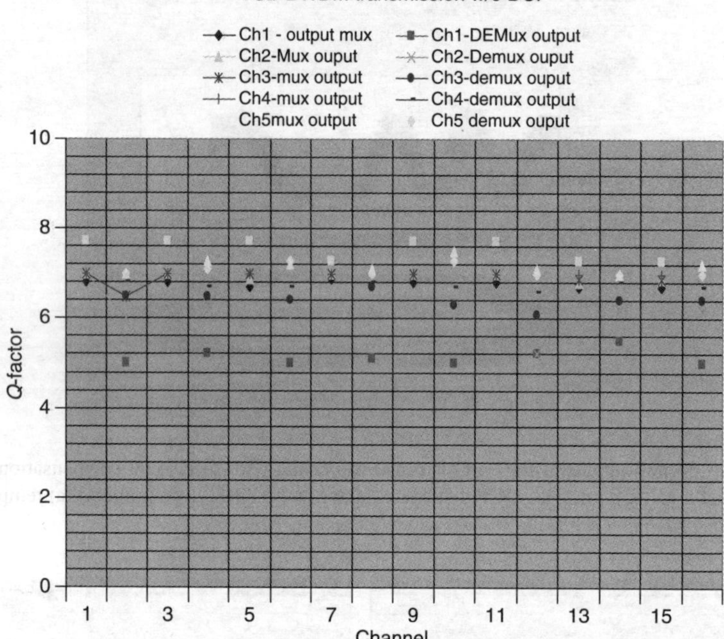

FIGURE 9.57   Channel spacing as a function of Q-factor for the multiplexed channels 1 to 16 at the output of the optical mux (before transmission) and at the output of demux (after transmission). Horizontal axis: channel number.

9.4.4.2   Effects of GVD on Q-factor

The effects of GVD on transmitted channels with and without dispersion-compensation fibers are evaluated with the Q-factor as the reference performance parameter. Dispersion-compensation fibers have an important role in optical transmission since the GVD limits transmission performance. By adding dispersion compensation fibers which have the opposite sign of the GVD, the effect of GVD can be considerably reduced. Figure 9.58 shows the dispersion tolerance of the transmission system, the DCF has significantly improved the Q-factor at the receiver to an equivalent BER close to an error-free 10^{-9} at a bit rate of 40 Gb/s. Shown also in this figure is the dispersion tolerance of RZ-DQPSK of 20 Gbauds/s. The performance of two modulation formats is very similar. However, the VSB format may offer an advantage of a simplified transmitter with additional optical filter, rather than another parallel transmitter arm for generation of the quadrature constellation in addition to the in-phase components [12].

9.4.4.3   Effects of Filter Pass-band on the Q-Factor

The effect of VSB filtering on Q-factor and its dispersion tolerance can be examined by varying the passband characteristic of the VSB filter and measuring the Q-factor for each lightwave channel at the output of demultiplexer, as shown in Figure 9.59. For 40-Gb/s NRZ data format, the system penalty is one unit of the Q-factor, which is equivalent to one decade of BER or about 10 dBQ when the passband of the OF is extended from 20 GHz to 24 GHz. This penalty could be due to the cut-off of the signal band by the roll-off band of the OF with more than half of the bandwidth eliminated. For the 28-GHz passband, the Q-factor is considerably increasing to 7.8 or BER $= 10^{-15}$, that is, error-free transmission when the sampling is at the center of the eye. This shows that the performance of VSB modulation format is better than double side band (DSB) modulation since the VSB format eliminates most (but not all) of the redundant sideband.

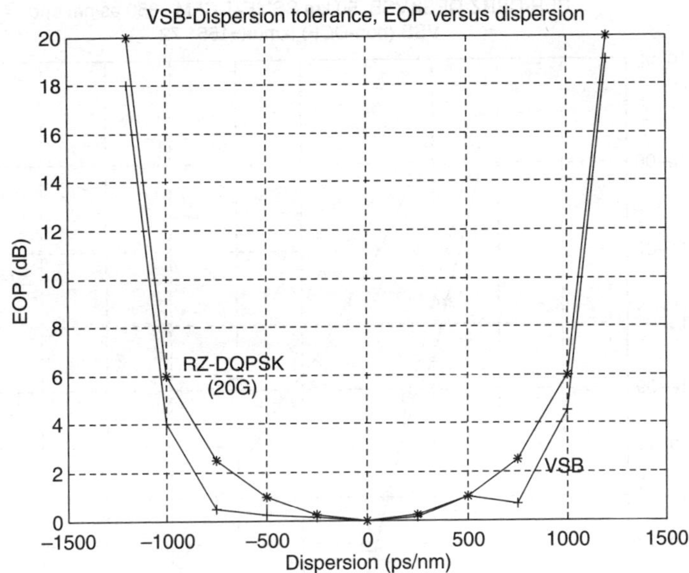

FIGURE 9.58 Dispersion tolerance of VSB format at 40 Gb/s (+) transmission as compared with 20 GBaud/s RZ-DQPSK (*). Eye opening penalty version dispersion factor at BER of 1e-9 of channel 5.

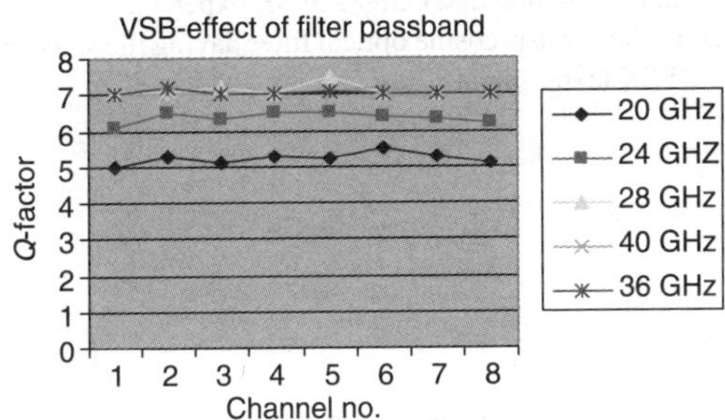

FIGURE 9.59 Variation of the Q-factor as a function of the VSB filter pass-band.

When dispersion-compensating fibers and pure gain optical amplifiers are incorporated in front and after the DCF, the transmission distance can be extended to 15 spans with a BER of 10^{-12} for all 16 wavelength channels.

Figure 9.60 shows the BER versus the receiver sensitivity of the VSB experiment simulation and measured experimental values of transmission of modulation format, the carrier-suppressed return-to-zero differential phase shift keying (CSRZ-DPSK) [11]. Figure 9.61 shows the simulink model of the SSB Hilbert transform phase-shift modulator. Figure 9.62 shows (a) the input RF signal (b) the Hilbert transform the input RF signal (c) Upper arm of dual electrode MZ modulator with delay 10 ns in Hilbert transform phase shift system (d) the lower arm of dual electrode MZ modulator with Hilbert transform phase shift, and (e) dual electrode MZ modulator single sideband system spectrum resulting with lower SB cancels out. Figure 9.63 shows the demodulator platform for SSB signals. Figure 9.64 shows (a) the demodulated signal in the time scope (b) the demodulated cosine signal time scope before the low pass filter (c) the demodulated signal time scope after the low pass filter (d) the spectrum of the demodulated signal, and (e) the demodulated eye diagram of NRZ-SSB modulation format with Q=8 & BER=10^{-15}. Optical filters with a raised cosine shape and that of the Chebychev type would offer nearly the same transmission performance of BER versus receiver sensitivity and about 2 dB at a BER of 10^{-9} worse than CS-RZ modulation format. This is

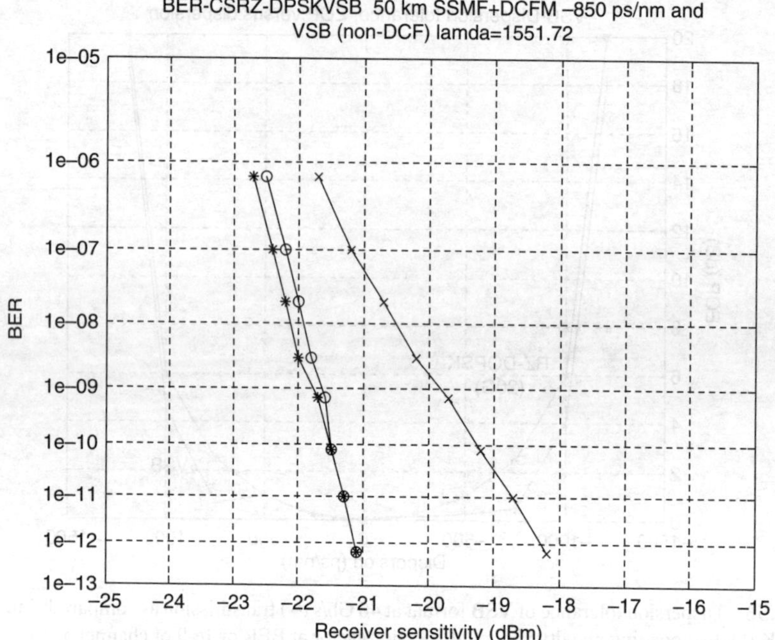

FIGURE 9.60 VSB experiment simulation and CSRZ-DPSK experimental values. Legend: (o) − VSB with Chebychev type optical filter; (*) VSB raise-cosine optical filter having the same cutoff passband as that of the Chebychev filter, (+) CSRZ-DPSK (experiment).

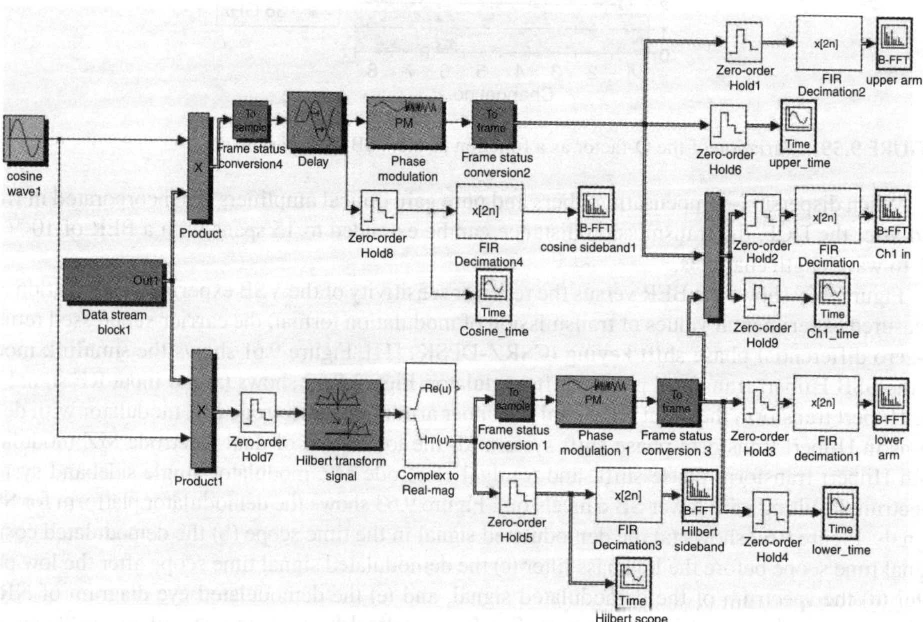

FIGURE 9.61 SSB Hilbert transform phase-shift modulator.

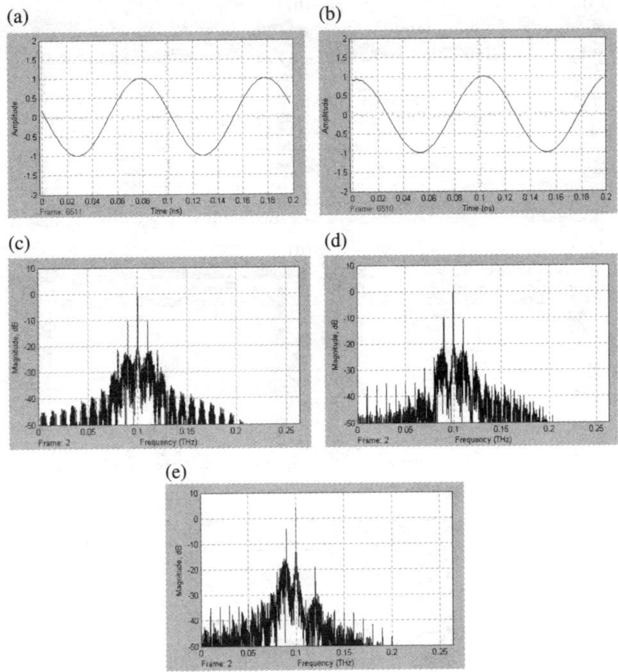

FIGURE 9.62 (a) Input RF signal time scope (cosine signal) (b) Hilbert transform the input RF signal $H\{x(t)\} = A\cos(2\pi f_o t - \pi/2) = A\sin((2\pi f_o t)$ (c) Upper arm of dual electrode MZ modulator with delay 10 ns in Hilbert transform phase shift system (d) Lower arm of dDual eElectrode MZ modulator with Hilbert transform phase shift $\theta = \pi/2$ (e) dual electrode MZ modulator single sideband system spectrum result with lower SB cancels out.

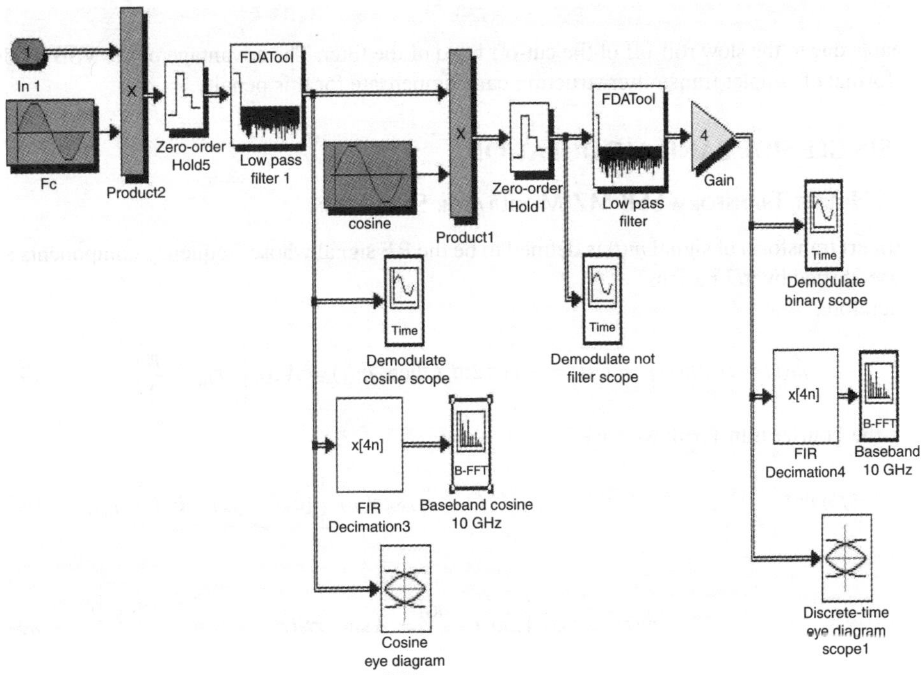

FIGURE 9.63 Single sideband demodulator platform.

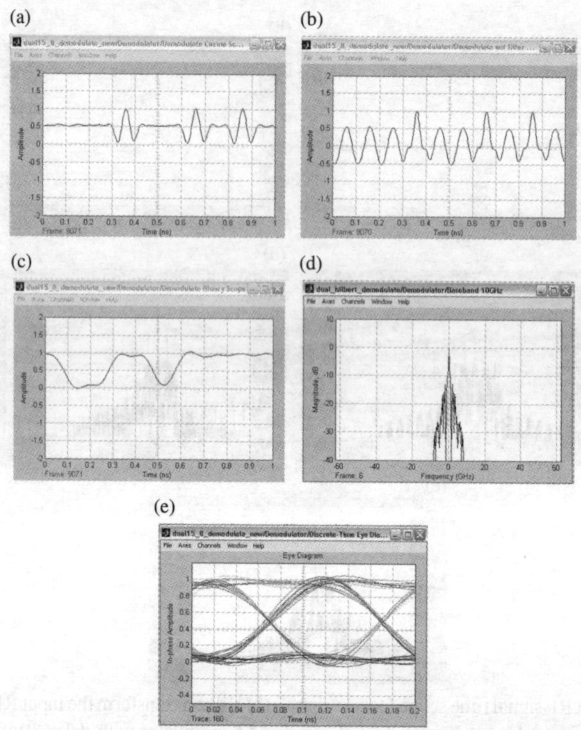

FIGURE 9.64 Demodulate F_c signal time scope (b) Demodulate cosine signal time scope before low pass filter (c) Demodulate signal time scope after Low Pass filter (d) Demodulate signal spectrum (e) Demodulated signal Eye Diagram NRZ-SSB with $Q = 8$ & BER $= 10^{-15}$.

reasonable due to the slow roll-off of the cut-off band of the filter. The advantage of the VSB modulation format of simpler transmitter structure can compensate for this penalty.

9.5 SINGLE SIDE BAND MODULATION

9.5.1 Hilbert Transform SSB MZ Modulator Simulation

The Hilbert transform of signal $m(t)$ is defined to be the RF signal whose frequency components are all phase-shifted by $\pi/2$ radians[*].

Therefore,

$$\hat{m}(t) = H\left\{m(t)\right\} \rightarrow m(t) = A\cos 2\pi f_o t \text{ then } \hat{m}(t) = A\cos\left(2\pi f_o t - \frac{\pi}{2}\right). \tag{9.21}$$

Taking the Fourier transform, we have

$$M(f) = -j\operatorname{sgn}(f)\frac{A}{2}[\delta(f + f_o) + \delta(f - f_o)]M(f) = -j\operatorname{sgn}(f)\frac{A}{2}[\delta(f + f_o) + \delta(f - f_o)] \tag{9.22}$$

Thus,

$$\hat{m}(t) = A\cos\left(2\pi f_o t - \frac{\pi}{2}\right) = A\sin 2\pi f_o t. \tag{9.23}$$

---

[*] Hilbert transform Single sideband, Digital and Analog Communication Systems, 6th edition, p. 312.

9.5.2 SSB DEMODULATOR SIMULATION

The modulated signal is:

$$u(t) = A_c/2 \left\{ \cos(\omega_c t + \gamma\pi + \alpha\pi\cos\omega_{rf}t) + \cos(\omega_c t + \alpha\pi\cos(\omega_{rf}t + \theta)) \right\}. \tag{9.24}$$

The demodulated signal is:

$$s(t) = A_c \cos 2\pi f_c t * u(t). \tag{9.25}$$

The low-pass filter is used to filter high frequency components in the signal. The only component left is the modulating signal (10 Gbps [binary*cosine signal]) after demodulation. It is required to multiply the demodulate signal by in-phase cosine signal and also there is high frequency signal, so that LPF is required to filter the high frequency components in the signal.

9.6 CONCLUDING REMARKS

We have presented the simulation results of 40-Gb/s DWDM optical transmission systems with NRZ-VSB modulation format. The effects of filtering on the Q-factor with respect to fiber dispersion and compensation, channel spectral spacing and the effects of symmetry and asymmetry of the filter pass and cut-off bands have been examined. We can draw the following conclusions: (i) The Q-factor at the output of demultiplexer decreases significantly for channel spacing of 20 GHz and 30 GHz due to noise and crosstalk interference. However, the Q-factor decreases slightly for 40-GHz channel spacing. A BER of 10–12 can be achieved over all channels with this channel spacing. For dense and super-dense WDM optical transmission, the demands on the roll-off, cut-off and the pass-band of VSB OFs are high; (ii) Fiber dispersion can reduce the system performance dramatically. Using dispersion-compensation fibers along the transmission line improves the system performance significantly. Specific designed fibers are used with deterministic dispersion are used; (iii) The 20 GHz and 24 GHz pass-band OFs give low performance. The low performance could be caused by the unstable system since about half of the bandwidth is eliminated. While, for the 28-GHz pass-band, the Q-factor is considerably improved to 7.5. Thus, the performance of VSB modulation format is comparable and marginally enhanced as compared with that of the double side band NRZ amplitude shift keying modulation format since the VSB format eliminates most, but not all, of the redundant sideband; (iv) Several types of OFs are examined, and the sharp roll-off band is expected to contribute to the improvement of the VSB transmission. However, the roll-off band must not be at least 20 dB below the power level of the centre of the spectrum of the signal and carrier frequency to avoid pulse ripples; and (v) The all-zero all pass optical filters can be designed and implemented in planar lightwave circuit technology to substitute for the two-poles elliptic filters to eliminate the filter contribution to the total dispersion of the transmission system.

Non-linear effects such as the self phase modulation, cross-phase modulation, Raman scattering and patterning effects are not considered in this chapter.

REFERENCES

1. Gene, J. M., R. Nieves, A. Buxens, C. Peucheret, J. Prat, and P. Jeppesen. Reduced driving voltage optical duobinary transmitter and its impact on transmission performance over standard single-mode fiber. *IEEE Photonic Technology Letters* 14 (6): 843–45.
2. Kim, Y., J. Lee, Y. Kim, and J. Jeon. 2004. Evaluation of transmission performance in cost-effective optical duobinary transmission utilizing modulator's bandwidth or low-pass filter implemented by a single capacitor. *Optical Fiber Technology* 10 (4): 312–24.
3. Ho, K. P. 2005. *Phase-modulated optical communications systems.* Berlin: Springer Verlag, 212.

4. Gnauk, H., and P.J. Winzer. 2004. Optical Phase-Shift-Keyed Transmission. *IEE Journal of Lightwave Technology* 23(1): 115–18.

5. Elrefaie, F., R. E. Wagner, D. A. Atlas, and D. G. Daut. 1988. Chromatic dispersion limitations in coherent lightwave transmission system. *IEE Journal of Lightwave Technology* 6 (5): 704–9.

6. Yonenage, K., and S. Kuwano. 1997. Dispersion-tolerant optical transmission system using DB transmitter and binary receiver. *IEE Journal of Lightwave Technology* 15 (8): 1530–37.

7. Bergano, N. S., et al. 2004. Chirped return to zero formats for ultra long-haul fibers communications. Presented at 2004 IEEE Workshop on Advanced Modulation Formats, San Francisco, CA, USA.

8. Tokle, T., et al. 2004. Transmission of RZ-DQPSK over 6500 km with 0.66 bits/Hz spectral efficiency. Presented at IEEE Workshop on Advanced Modulation Formats, San Francisco, CA, USA.

9. Ziemer, R. E., W. H. Tranter, and D. R. Fannin. 1989. *Signal and systems: Continuous and discrete.* New York: Macmillan.

10. Binh, L. N., et al. 2005 Photonic signal processing—Part II.2: Tuneable photonic filters using cascaded all-pole micro-rings and all-zero interferometers. Monash University, Clayton, Technical Report code MECSE-12-2005.

11. Binh, L. N., T. L. Huynh, K.-Y. Chin, and D. Sharma. 2004. Design of dispersion flattened and compensating fibers for dispersion-managed optical communications systems. *International Journal on Wireless and Optical Communications* 1: 1–21.

12. Serbay, M., et al. 2004. Comparison of six different RZ-DQPSK transmitter sert-ups regarding their tolerance towards fibers impairment in 8×40Gb/s WDM-systems. Presented at IEEE Workshop on Advanced Modulation Formats, San Francisco, CA, USA.

13. Kim, S. K., J. Lee, and J. Jeong. 2004. Transmission performance of 10-Gb/s optical DB transmission systems considering adjustable chirp of nonideal $LiNbO_3$ MZIMs due to applied voltage ratio and filter bandwidth. *IEEE Journal of Lightwave Technology* 19 (4): 465–70.

14. Zhang, S. Advanced Optical Modulation formats in High Speed Lightwave System. University of Kansas, 17–32.

15 Yonenaga, K., S. Kuwano, S. Norimatsu, and N. Shibata. 1995. Optical DB transmission system with no receiver sensitivity degradation. *Electronics Letters* 31 (4): 736–37.

16. Frank, T., P. B. Hansen, T. N. Nielsem, and L. Eskildson. 1998. DB transmitter with lower intersymbol interference. *IEEE Photonics Technology Letters* 10 (4): 597–99.

17. Ono, T., Y. Yuno, K. Fukuchi, T. Ito, H. Yamazaki, M. Yamaguchi, and K. Emura. 1998. Characteristics of optical DB signals in terabit/s capacity, high-spectral efficiency WDM systems. *IEEE Journal of Lightwave Technology* 16 (5): 788–97.

18. Miyamoto, Y., M. Yoneyama, T. Otsuji, K. Yonenaga, and N. Shimizu. 1999. 40G-bits/s TDM transmission technologies based on ultra-high-speed IC's. *IEEE Journal of Solid-State Circuits* 34 (9): 1246–53.

19. Kim, S., and J. Jeong. 2000. Transmission performance on frequency response of receivers and chirping shape of transmitters for 10 Gb/s $LiNbO_3$ modulator based lightwave systems. *Optics Communication* 175: 109–23.

20. Sinkin, O. V., J. Zweck, and C. R. Menyuk. 2001. Comparative study of pulse interactions in optical fiber transmission systems with different modulation formats. *Optics Express* 9 (4): 339–50.

21. Yoneyama, M., K. Yonenaga, Y. Kisaka, and Y. Miyamoto. 1999. Differential precoder IC modules for 20- and 40-Gbits/s optical DB transmission system. *IEEE Transactions on Microwave Theory and Techniques* 47 (12): 2263–70.

22. Ho, K. P., and J. M. Kahn. 2004. Spectrum of externally modulated optical signals. *IEEE Journal of Lightwave Technology* 22 (2): 658–63.

23. Haykin, S. 2001. *Communication systems.* 4th ed. Wiley.

24. Bolvin, D., et al. 2005. *IEEE Photonics Technology Letters* 17 (6): 1331.

25. Ohm, T. 2004. Comparison of different DQPSK transmitters with NRZ and RZ impulse shaping. Presented at IEEE Workshop on Advanced Modulation Formats, San Francisco, CA, USA.

26. Zhu, Y., et al. 2004. Highly spectral efficient transmission with CSRZ-DQPSK. Presented at IEEE Workshop on Advanced Modulation Formats, San Francisco, CA, USA.

10 Temporal Lens and Adaptive Electronic/Photonic Equalization

10.1 INTRODUCTION

For an optical pulse propagating through a SMF, its optical spectrum would travel down the fiber at a different speed due to the material and waveguide dispersion, as well as the polarization modal delay and naturally the non-linear phase effects if the intensity of the pulse is above the SPM threshold level. If the bit rate reaches 160 Gb/s, the compensation and equalization is preferred to be in an active mode, thus an optical phase modulator operating at very high speed should be used. An integrated optical phase modulator could change the phase (controlled shift on the phase of a light beam) of the optical signal by applying a driving traveling wave voltage V. When the electric field is generated and applied across an optical channel waveguide formed on an electro-optic substrate, the change in the refractive index would induce a change in the propagation constant of the propagating mode and create a phase shift for all traveling lights through that region [1]. The optical signals can be synchronized and chirped by an optical phase modulator to reverse the chirped effects of the fiber, hence equalization of the distortion of the signal envelop. This type of equalization can be implemented adaptively.

Furthermore, it is much more difficult to fully compensate higher-order dispersion such as third- or fourth-order dispersion in ultra-fast signal processing by using DCF fiber only. Mismatched compensation between SMF and DCF would be accumulated over long distance and cause serious distortion of received pulses. Therefore, another dispersion compensation scheme that can be implemented at the front end of the receiver in order to fully recover transmitted signals is reported in this article.

Currently, the remarkable progress of integrated photonic technology make possible the fabrication of active devices such as modulators and integrated transmitters which are commercially available for operating speed of up to several GHz. These include ultra-wideband optical phase modulators that can offer significant phase modulation over a very short period, enabling the alteration of the phase of the lightwave carrier imbedded within a very narrow-width pulse sequence. These modulators may enable of chirping the carrier frequency and hence dispersion in the negative or positive sense for equalization of the phase disturbance within an optical pulse due to the quadratic phase effects of single-mode optical fibers. The phase modulation for equalization can be implemented in an adaptive mode.

The propagation of an optical pulse through an optical fiber can be considered as the evolution of a pulse through a quadratic phase medium, that is, the amplitude remains invariant but its phase, the carrier phase, is altered following a quadratic function of its spectral components. This phenomenon is similar to the diffraction of a lightwave beam through a spatial slit. Alternatively, one could consider the fiber as a temporal lens that defocuses or focuses the guided lightwave beam depending on the sign of the dispersion factor of the fiber. Therefore, there is duality between the spatial and temporal domain of the propagation of lightwave beams or pulse sequence through a spatial lens or a guided wave optical device. The correction of the blurred optical pulses can be made via the modulation of the phase of the carrier within the pulse period whether in passive, active and adaptive modes. We demonstrate that it can be achieved by using an optical phase modulator, and that the performance of this on-line adaptive equalization is very effective for an ultra-high-speed optical transmission system, especially at 160 Gb/s.

The organization of this report is as follows. In Section 10.2, space–time duality and its, as well as the principles of temporal imaging, are described. That temporal imaging is necessary and plays a key part in long-haul transmission optic fiber is also discussed. Section 10.3 presents simulation results of the equalization of the impairment due to second-order and third-order dispersion of a SMF. Advantages and disadvantages between using sinusoidal and ideal parabolic driving voltage on an optical phase modulator for equalization is also given in detail. Section 10.4 discusses the simulation model by MATLAB® Simulink® for the transmission, modulation and equalization techniques for a 160-Gb/s transmission system. Section 10.5 outlines significant results of a single pulse transmission and equalization of the 160-Gb/s transmission system. BER characteristics and eye diagrams are illustrated together with the transmission performance. Finally, some concluding remarks are stated in Section 10.6.

10.2 SPACE–TIME DUALITY AND EQUALIZATION

The duality between the spatial domain and the time domain has been investigated, leading to the formation of temporal imaging and thus the analogy of a thin lens in the spatial domain and a quadratic phase modulation in the time domain [2]. Temporal imaging enables the expansion or compression of signals in time and their envelope profiles could be maintained in long-haul transmission [3,4]. Temporal imaging is based on broadening pulse width due to the dispersion factor of the fiber and spreading of a beam due to Fresnel diffraction under far-field consideration. When optical signals propagating through a single-mode optical fiber or are transmitted in the air, the pulse width of these signals are broadened due to the optical dispersion or diffraction effects. Consequently, the carrier frequencies of these signals would be chirped up or chirped down depending on transmission medium. Thus, the spectra of these waveforms would be broadened or compressed due to the changes of the phase of each frequency component. The Fresnel or Fraunhofer (far-field) diffraction were discussed by Papoulis in the 1960s [5], and signals dispersion is described by Agrawal [6,7]. We also further clarify these effects in this section.

Furthermore, a key element in a temporal imaging system is a time lens, which is considered as a quadratic phase modulation in the time domain. In addition, a dispersive element such as standard single-mode fiber (SSMF), dispersion-compensating fiber (DCF) or fiber Bragg grating (FBG) also performs an equivalent role of diffraction. A quadratic phase modulation could be produced using several methods, and a popular method for ultra-short-pulse compression or generation is an ultra-high speed electro-optic phase modulator [8,9]. Assuming that chirped signals are launched into a real lens (divergence or convergence lens) in space or a quadratic phase modulator in the time domain time. The phase modulator would chirp the frequencies of the imbedded lightwave carrier under the signal envelop in the opposite sign. Each frequency component of the linearly chirped signals which are generated by a parabolic phase modulator travel at different speeds due to the group velocity dispersion (GVD) in the fiber, hence the fiber acts as a time lens. Consequently, the frequency components of these signals are reorganized before they are again launched to free air or a dispersive element such as an optical fiber. By using a suitable length or dispersion before and after a space lens or a time lens, the output pulses can be recovered or even shortened compared with the original pulses.

The main purpose of this chapter is to study space–time duality and how to apply this property to design an adaptive equalization system for ultra-fast optical signal processing and equalization of distorted pulse sequence over an ultra-long-haul and ultra-high-speed optical fiber transmission system. The duality between spatial domain and time domain is proved and summarized. Furthermore, the broadening factor of a Gaussian pulse in transmission fiber depends on chirping factor.

The duality between light diffraction in the spatial domain and narrow-band dispersion in the time domain has been studied recently. This duality could be applied to expand or compress transmitted pulses in the time domain while the shapes of the pulses are maintained. This process can be termed "temporal imaging". The theory of temporal imaging is based on the duality between paraxial diffraction and narrow-band dispersion. This interesting duality is analyzed and discussed in more detail below. This section is devoted to reviewing an interesting analogy between paraxial diffraction of a light beam in spatial domain and the temporal dispersion of a narrow-band pulse in a dielectric [2,10], [11]. Furthermore, the space–time duality has also led to the employment of quadratic phase modulation devices such as optical adaptive equalizer [12] or electro-optic phase modulator [8] in the time domain, in the analogy of a thin lens in spatial domain [13] which is equivalent to a time lenses. In addition, a real-time Fourier transformation that uses a time lens would be considered as a temporal equivalence with the spatial Fourier transformation [14]. This equivalence can be used for implementation of an equalization system, which is introduced in the next section.

10.2.1   Space–Time Duality

Assuming the frequency spectrum of the wave is monochromatic. This assumption can be made due to the fact that the harmonic time variation and time derivatives become multiplicative constants in the wave equation.

10.2.1.1   Paraxial Diffraction

Maxwell's equations would be used to obtain a three-dimensional vector equation. The paraxial form of the Helmholtz equation and its solution is given as follows

$$\nabla^2 E(x,y,z) - j2k \frac{\partial E(x,y,z)}{\partial z} = 0 \tag{10.1}$$

and

$$\frac{\partial^2 E}{\partial x^2} + \frac{\partial^2 E}{\partial y^2} + \frac{\partial^2 E}{\partial z^2} - j2k \frac{\partial E}{\partial z} = 0. \tag{10.2}$$

The propagation direction is in the z-direction and polarization in the transverse plane of x- and y-directions. Owing to the paraxial approximation, the curvature of the field envelope in the direction or propagation (z-direction) is much less than the curvature of the transverse profile.

Therefore the term $|\partial^2 E/\partial z^2|$ could be concluded that it is much smaller compared to terms $|\partial^2 E/\partial x^2|$, $|\partial^2 E/\partial y^2|$, $|2k/(\partial E/\partial z)|$, therefore it might be eliminated. Finally, the electric field propagating down the z-axis can be found as

$$\frac{\partial^2 E}{\partial x^2} + \frac{\partial^2 E}{\partial y^2} - j2k \frac{\partial E}{\partial z} = 0$$

$$\therefore E_z = -\frac{j}{2k}\left(E_{xx} + E_{yy}\right). \tag{10.3}$$

These equations are parabolic and similar to the heat diffusion equation. Therefore, the behavior of diffraction and dispersion, which is spreading like a temperature distribution, can be derived by using the solutions of the diffusion problem [2].

10.2.1.2   The Governing Non-linear Schrödinger (NLS) Equation

An optical signal whose temporal envelope is in the $E(z,t) = u(z,t)e^{j\omega t}$ propagating through an optical fiber can be represented by the non-linear Schrödinger (NLS) equation

$$\frac{\partial u}{\partial z} + \frac{j\beta_2}{2}\frac{\partial^2 u}{\partial t^2} - \frac{\beta_3}{6}\frac{\delta^3 u}{\delta t^3} - j\gamma|u|^2 u = -\frac{\alpha}{2}u$$

$$\therefore \frac{\partial u}{\partial z} = -\frac{\alpha}{2}u - \frac{j\beta_2}{2}\frac{\partial^2 u}{\partial t^2} + \frac{\beta_3}{6}\frac{\delta^3 u}{\delta t^3} + j\gamma|u|^2 u \tag{10.4}$$

where the amplitude $u = u(z,t)$ is the complex envelope carried by the lightwaves of wavelength λ, along the propagation in z-axis, and t is the time variable. Pulse broadening is a result of the frequency dependence of λ. By using Taylor series to $\beta(\omega)$ around the carrier ω_c so that the second-, third- and higher-order dispersion can be obtained $\beta_n = d^n\beta/d\omega^n|_{\omega=\omega_c}$, $\beta_2 = -D\lambda^2/2\pi c$ is the group velocity dispersion co-efficient and $\beta_3 = (2\pi c/\lambda^2)^{-2}[S - (4\pi c/\lambda^3)\beta_3]$ the third-order dispersion factor and related to the dispersion slope S. $\gamma = n_2\omega/cA_{\text{eff}}$ is the non-linear co-efficient of an optical fiber with an effective area A_{eff} corresponding to a mode spot size r_0, with n_2 is the non-linear refractive index taking a typical value for silica based glass $2.6e - 20 m^2/W$. If the losses during the transmissions on the optical fiber can be ignored, and a value of β_3 (that is about $0.117\,\text{ps}^3/\text{km}$) is very small compared with a value of β_2 (about $21.6\,\text{ps}^2/\text{km}$). Therefore Equation 10.4 can be normalized to

$$\frac{\partial u}{\partial z} = -j\frac{\beta_2}{2}\frac{\partial^2 u}{\partial t^2} + j\gamma|u|^2 u. \tag{10.5}$$

10.2.1.3   Diffractive and Dispersive Phases

Different forms of diffusion could be modeled quantitatively using the diffusion equation, which went by different names depending on the physical situation. Steady-state thermal diffusion is governed by Fourier's law. In all cases of diffusion, the net flux of the transported quantity (atoms, energy, electrons) are equal to a physical property (diffusivity, thermal conductivity, electrical conductivity) multiplied by a gradient (concentration, thermal, electric field gradient) [2].

The governing equations of the paraxial equation and the non-linear Schrödinger equation follow the forms

$$E_z = -\frac{j}{2k}\left(E_{xx} + E_{yy}\right) \tag{10.6}$$

$$\frac{\partial u}{\partial z} = -\frac{j\beta_2}{2}\frac{\partial^2 u}{\partial t^2} + j\gamma|u|^2 u \tag{10.7}$$

and equation of heat diffusion for two-dimensional diffusion is

$$u_t = c(u_{xx} + u_{yy}). \tag{10.8}$$

Paraxial, NLS and heat diffusion equations are thus governed by the same parabolic differential equation, so they should have similar forms. Let k_x be the Fourier domain variable or the propagation constant in the x direction, and $U(k_x,k_y,0)$ is the initial Fourier spectrum. By applying the two-dimensional equation of heat diffusion to the two wave equations, solutions for two-dimensional diffraction and wave-propagation can be solved [2].

$$u(x,y,z) = \frac{1}{(2\pi)^2} \int\limits_{-\infty}^{\infty} \int U(k_x,k_y,0) e^{j(k_x^2+k_y^2)z/2k} e^{j(k_x x+k_y y)} dk_x dk_y \tag{10.9}$$

$$u(z,t) = \frac{1}{2\pi} \int\limits_{-\infty}^{\infty} U(0,\omega) e^{-j\frac{z}{2}\frac{d^2\beta}{d\omega^2}\omega^2} e^{j\omega t} d\omega. \tag{10.10}$$

Comparing Equation 10.9 and Equation 10.10, the envelope of the input pulse is similar to the paraxial diffraction. The terms (k_x,k_y) and x in spatial domain correspond to the terms ω and t in the time domain, respectively. In the other words, the beam displacement in spatial domain corresponds to a pulse delay in the time domain [15]. According to Equation 10.9 and Equation 10.10 the diffractive and dispersive phases are given as

$$\phi(t) = \frac{k_x^2 + k_y^2}{2k} z \tag{10.11}$$

$$\phi(\omega) = -\frac{d^2\beta}{2d\omega^2} \omega^2 z. \tag{10.12}$$

10.2.1.4 Spatial Lens

A lens has an important property, that is, phase retardation or delay due to the propagation velocity of an optical disturbance which is less than the velocity in the air [16]. Furthermore, a lens is said to be thin when a ray enters at a point on one side of a thin lens and emerges from the same point on the other side of a thin lens [17]. Therefore, a thin lens delays an incident wavefront by an amount proportional to the variation in thickness of the lens.

Recalling the incident monochromatic wavefront $u_i(x,y)$ at plane U_i emerging through the lens would give an output wavefront $u_o(x,y)$ at plane U_o. Thus, the output wavefront could be considered as the product of the input wavefront multiplied by the phase transform function of the lens ($u_o(x,y) = T(x,y) \times u_i(x,y)$) as shown in Figure 10.1a and Figure 10.1b. The effect of a thin lens would be described by a phase function as

$$T(x,y) = \exp\left[jkn\Delta t\right] \exp\left[-jk\frac{x^2+y^2}{2f}\right] \tag{10.13}$$

where f is the focal length of the lens. Relationship between the focal length, refractive index and two radius of lens is calculated $1/f = (n-1)((1/R_1)+(1/R_2))$. Equation 10.13 shows that a thin lens could produce a quadratic phase modulation in real spatial space. Additionally, the relevant spatial phase transformation of a thin lens can be found as a phase function given by

$$\phi(x,y) = k\frac{x^2+y^2}{2f}. \tag{10.14}$$

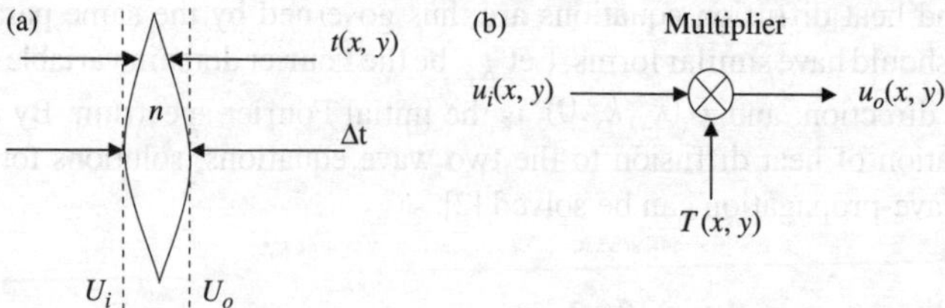

FIGURE 10.1　(a) The thickness function (b) Analogy system of a lens transformation.

10.2.1.5　Time Lens

A time lens is basically just a quadratic phase modulator in time which can be implemented using a traveling-wave electro-optic modulator driven at a microwave ω_m. However, it is difficult to generate a true quadratic phase modulation for an ideal time lens in practice. Therefore, the required modulation function could be approximated by a portion of a sinusoidal phase modulation [15]. Applying a sinusoidal wave with the local time variable t with a travelling wave electro-optic modulator driven at a modulation frequency ω_m as

$$V(t) = V_o \cos(\omega_m t). \tag{10.15}$$

By using the Taylor series expansion to expand Equation 10.15, the corresponding quadratic approximation around $t = 0$ is

$$V(t) \approx V_o\left(1 - \frac{1}{2}\omega_m^2 t^2\right). \tag{10.16}$$

By denoting $f_T = \omega_c/((\pi V_c/V_\pi)/\omega_m^2)$ as the equivalent focal time of the time lens, with ω_c is an optical carrier frequency. The approximate quadratic phase modulation of time lens in the time domain would follow a phase function of

$$\phi(t) = \frac{1}{2}\frac{\omega_c}{f_T}t^2. \tag{10.17}$$

10.2.1.6　Temporal Imaging

The properties of space–time duality are summarized in Table 10.1. The diffraction from an object from a lens to spatial image is dependent on the phase shift provided by lens [18]. Figure 10.2 shows a spatial imaging and temporal imaging systems. The dispersion effects from an object pulse to the lens and from the lens to the images are provided by two grating pairs which are installed before and after a quadratic phase modulator [10]. The former acts as a time lens that provides a time-varying phase shift.

　　　The equivalence between the diffraction–dispersion phase functions and space–time lens phase function between the spatial domain and the time domain leads to the relationship between conventional spatial imaging and temporal imaging configurations, as summarized in Figure 10.2a for spatial imaging and Figure 10.2b for temporal imaging. Let an unchirped pulse train enter temporal and spatial imaging. Firstly, these pulses would be distorted due to diffraction effect in space or dispersion effect in time. At this moment, the phase of each frequency component would be changed in the frequency domain and therefore a time delay of these frequency components would occur in the time domain. Consequently, the pulse shape of these pulses is blurred or broadened.

TABLE 10.1
Space–Time Duality

	Space–Time Duality	
	Space	**Time (Optical Fiber)**
Governing equations	$E_z = -\dfrac{j}{2k}\left(E_{xx} + E_{yy}\right)$	$\dfrac{\partial u}{\partial z} = -\dfrac{j\beta_2}{2}\dfrac{\partial^2 u}{\partial t^2} + j\gamma\left|u\right|^2 u$
Propagation variables	z	μ
Transverse variables	x, y	t
Fourier variables	k_x, k_y	ω
Diffractive/dispersive phases	$\phi(t) = \dfrac{k_x^2 + k_y^2}{2k}z$	$\phi(\omega) = -\dfrac{d^2\beta}{2d\omega^2}\omega^2 z$
Lenses/temporal Lens phase	$\phi(x,y) = k\dfrac{x^2 + y^2}{2f}$	$\phi(t) = \dfrac{1}{2}\dfrac{\omega_c}{f_T}t^2$
Spatial/temporal focal point	$f = \dfrac{1}{(n-1)\left(\dfrac{1}{R_1} - \dfrac{1}{R_2}\right)}$	$f_T = \dfrac{\omega_c}{\dfrac{\pi V_c}{V_\pi}\omega_m^2}$
Imaging condition	$\dfrac{1}{d_i} + \dfrac{1}{d_o} = \dfrac{1}{f}$	$\dfrac{1}{\mu_i\dfrac{d^2\beta_i}{d\omega^2}} + \dfrac{1}{\mu_o\dfrac{d^2\beta_o}{d\omega^2}} = -\dfrac{\omega_c}{f_\tau}$
Magnification factor M	$-\dfrac{d_o}{d_i}$	$-\dfrac{\mu_o\dfrac{d^2\beta_o}{d\omega^2}}{\mu_i\dfrac{d^2\beta_i}{d\omega^2}}$

These pulses are fed into a space lens or time lens. At this time, carrier frequency is chirped up or down depending on its applications. Lastly, this pulse train leaves a space lens or a time lens and is diffracted in the air or dispersed by the grating pair again. This diffraction or dispersion is necessary and very important. All the frequency components of these pulses must be reorganized again in order to compress the input pulses due to their different time delay. Moreover, in order to expand or compress a pulse, magnification parameter M should be modified. This magnification M could be changed by adjusting object–lens distance, lens–image distance or lens characteristic in space or changing grating pair separation distance or quadratic phase modulator characteristic in time.

In particular, when these input pulses are compressed, the bit period is shortened. In order to avoid this problem, two thin lenses are used and separated by a mask which has the role of the Fourier transform plane. In a practical setup, a pulse train is transmitted to the first lens and these pulses are diffracted. Masks, including amplitude mask and phase mark (called "spatial light modulators") to control the amplitude and phase of the pulse could be used to yield desired pulses. Lastly, these pulses are refocused by a second lens. As a result, a received pulse train still keeps its original bit rate, while the pulse width of each individual pulse is reduced.

10.2.1.7 Electro-optic Phase Modulator as a Time Lens

Electro-optic is an optical device which is used to modulate a beam and light. In this section, only phase modulation of a modulated beam is discussed. Electro-optic phase modulator lens has been

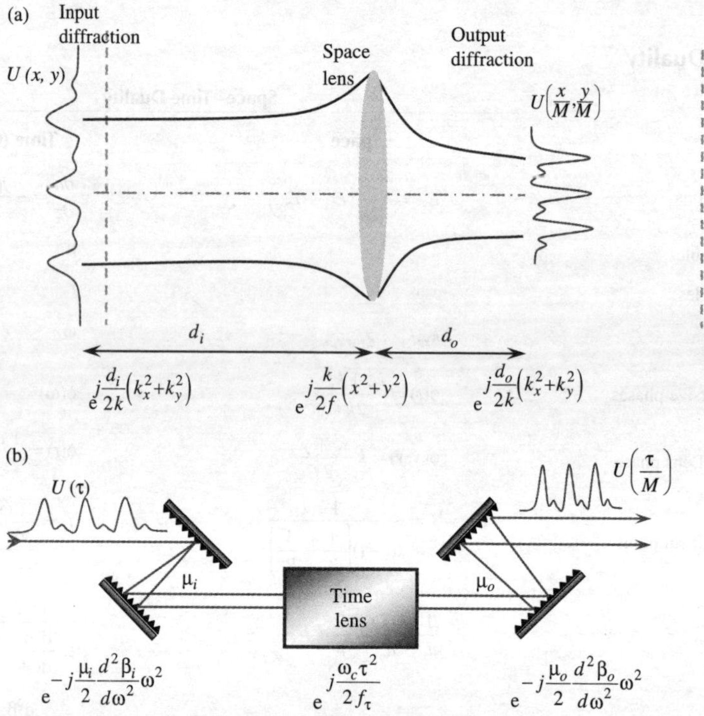

FIGURE 10.2 Configuration for (a) Spatial imaging and (b) Temporal imaging. (From Khayim, T., M. Yamauchi, D. S. Kim, and T. Kobayashi, *IEEE Journal of Quantum Electronics*, 35(10), 1412–18, 1999; Bennett, C. V., and B. H. Kolner. *IEEE Journal of Quantum Electronics*, 36(6), 649–55, 2000; and Yan, L. S., Q. Yu, T. Luo, and X. S. Yao, *IEEE Photonics Technology Letters*, 14(6), 858–1012, 2002. With permission.)

often used in pulse compression for ultra-short optical pulses, such as few pico-seconds or few hundred femto-seconds. Khayim et al. [8] introduced the relationship between the applied electric field corresponding to electro-optic lens as well as chirped optical carrier frequency, as shown in Figure 10.3. Furthermore, the focal length of the integrated guided wave electro-optic phase modulator can also be given as

$$f_\tau = -\frac{hd^2}{r_{33}n_e^3 V(t)l_o} \tag{10.18}$$

with $2d$ – The maximum width of the parabolic electrode; h – the spacing between the traveling wave electrode; l_o – the maximum length of the electrode; r_{33} – The electro-optic coefficient of the $LiNbO_3$ Z–cut X-prop substrate; n_e – the extra-ordinary index of refraction; and $V(t)$ – the time-variable applied voltage.

Equation 10.18 and Figure 10.3 show the convex or concave parabolic-shape of an applied electric field corresponding to the diverging or converging lens and the down-chirping or up-chirping of the optical carrier frequency, respectively. Depending on specific applications, the carrier frequency would be chirped up or chirped down by adjusting the applied driving voltage applied to the electro-optic modulator in the time domain that is equivalent to converging or diverging lens in the spatial domain.

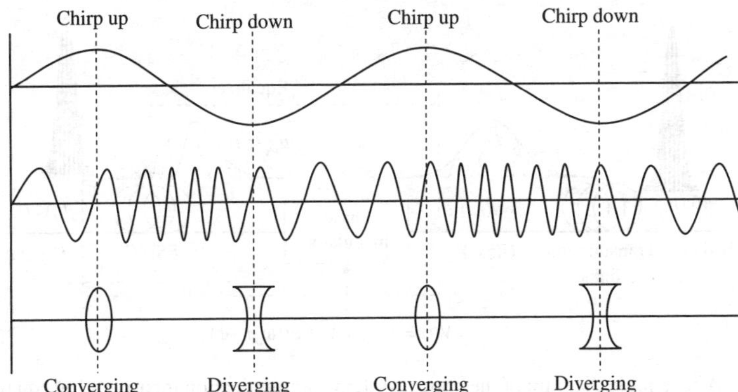

FIGURE 10.3 Relationship between phase modulator electric field, chirped carrier frequency and corresponding lens. (From Khayim, T., M. Yamauchi, D. S. Kim, and T. Kobayashi, *IEEE Journal of Quantum Electronics*, 35(10), 1412–18, 1999; Bennett, C. V., and B. H. Kolner. *IEEE Journal of Quantum Electronics*, 36(6), 649–55, 2000; and Yan, L. S., Q. Yu, T. Luo, and X. S. Yao, *IEEE Photonics Technology Letters*, 14(6) 858–1012, 2002. With permission.)

10.2.2 EQUALIZATION IN TRANSMISSION SYSTEM

The ultra-high-speed optical transmission system is considered to be the essential backbone of the next generation of ultra-high-speed networks. When the pulse width reaches less than a few picoseconds, not only the second-order but even higher-order dispersion must be taken into account. On the other hand, third-order and higher-order dispersion effects are not fully compensated by using traditional dispersion compensating devices such as DCF fibers. Moreover, the ultra-fast transmission signals are very sensitive to environmental effects (e.g., temperature, vibration). Thus, the received signals at the receiver end would be distorted dramatically, even if only a small outside effect affects the transmission link. Therefore, the tuneable adaptive equalizer system is used at the receiver end to recover higher-order dispersion which could not be compensated by DCF fiber, and to recover distorted signals to improve the bit-error-rate (BER) or reduce the eye-opening penalty (EOP).

The structure of an equalizer system based on spatial imaging and temporal imaging can offer important applications of the time lens optical system, such as compensating group velocity dispersion, third- and fourth-order dispersion, and reducing timing jitter for a pico-seconds pulse train as well as time reversal with magnification [19,20]. Numerous applications of time lens optical systems, temporal magnification and reversal optical data have used the analogy between spatial issue of diffraction and temporal issue of dispersion [21]. It showed that space–time duality had valuable affects on ultra-fast photonic signals processing, and its applications could be further explored in the mitigation of non-linearity impairments. Adaptive dispersion equalization would become a key technique for ultra-high-speed transmission systems and temporal imaging theorem could be used to create this adaptive equalizer. Jannson et al realized that the spatial Fourier transformations of temporal signals are equivalent to real-time Fourier transformation in dispersive optical fibers [14].

Initially, a temporal self-imaging effect in single-mode fiber and a time-domain Collett–Wolf equivalence theorem were studied and applied to transference and propagation of the information contained in periodic signals in fibers [22,23]. Consequently, this equalizer system model was built and developed based on the implementation of this well-known temporal imaging. In this case, a single-mode optical fiber acted as free space in the spatial domain and the adaptive equalizer

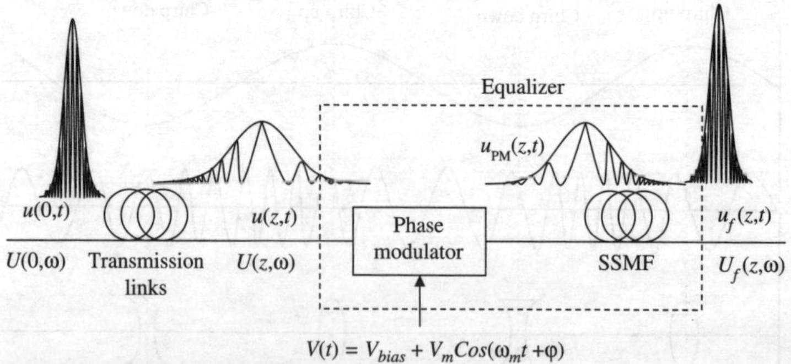

FIGURE 10.4   A schematic diagram of the optical system communication incorporating equalizer with sinusoidal driving voltage, 40-GHz clock recovered signals for phase modulation.

acted as a time lens in the time domain. Following this quadratic phase modulator is a dispersive element that is used to change the phase of each frequency component in frequency domain. Thus, different time delays appeared between these frequency components and the final pulse shape would be recovered. Standard SMF fiber link is used as a dispersive element (dispersion compensation fiber (DCF) or fiber Bragg grating (FBG) could be used instead of using standard SMF fiber). This equalizer system including quadratic phase modulator and SMF is used to provide the required quadratic phase modulation [13]. Further investigations, theoretical calculations and simulation results of transmission system using adaptive equalizer are discussed in more detail in later sections.

10.2.2.1   Equalization with Sinusoidal Driven-Voltage Phase Modulator

For the optical fiber communication system, the GVD has a significant role because it distorts signals dramatically, especially at ultra-high-speed. Also, the third-order dispersion effects or even fourth-order dispersion effects also become more significant [24]. A combination of different fibers, such as single-mode fiber, reverse-dispersion fiber and dispersion-compensating fiber, would fully compensate the second-order dispersion, but residual dispersion effects are present. With an ultrafast transmitted signal, e.g., 160 Gb/s, the system is very sensitive to environmental effects such as the PMD and fluctuation of the dispersion factor. For example, if there is a small change in temperature, there would be a mismatch in compensating between SMF and DCF fiber. This compensating mismatch would be accumulated over a long-haul transmission, and the received signals would be distorted randomly. Furthermore, for an installed transmission system, the third-order dispersion could not be eliminated by using DCF fiber for 160-Gb/s transmitted signals. Therefore, an adaptive equalizer which can be tuned to equalize any distortion is used. It could compensate second-, third- and even fourth-order dispersion. Moreover, it also eliminated the timing jitter effect, thus the BER at the receiver.

In order to investigate the effects of using the equalizer, some significant and necessary formulas are briefly given. In applying a sinusoidal driving voltage to the phase modulator, a linear chirp would be generated, hence amplitude distortion. The applied electric field would change the phase of each frequency component in the time domain, thus the signal spectrum is broadened in frequency domain. The transmitted signal at the output of the phase modulator is

$$u_{PM}(z,t) = u(z,t)e^{j\frac{1}{2}Kt^2}.$$ (10.19)

The signal at the input of the receiver end would be the convolution between the signal at the output of the phase modulator and the dispersive device, the standard single mode fiber and given as

$$u_f(z,t) = u_{\text{PM}}(z,t) * e^{-j\frac{t^2}{2D}} \tag{10.20}$$

with $D = \beta_2 L_{\text{PM}}$ and $|D| = T_0^2$. Assuming the chirping rate of the phase modulator $K = 1/D$, the received signal at the output of the equalizer can be obtained as

$$u_f(z,t) = \sqrt{\frac{j}{2\pi D}}\, A\sqrt{2\pi T_0^2}\, e^{-\frac{T_0^2}{2}\frac{t^2}{D^2}}\, e^{-j\frac{1}{2}Kt^2}\, e^{j\phi\left(\frac{t}{D}\right)}. \tag{10.21}$$

In order to recover the initial Gaussian pulse shape, condition $|D| = T_0^2$ should be applied into Equation 10.21 to obtain

$$u_f(z,t) = A\sqrt{\frac{j}{\text{sign}(\beta_2)}}\, e^{-\frac{t^2}{2T_0^2}}\, e^{j\left[-\frac{1}{2}Kt^2 + \phi\left(\frac{t}{D}\right)\right]}. \tag{10.22}$$

If a sinusoidal driving voltage $V(t) = V_{\text{bias}} + V_{\text{m}}\cos(\omega_{\text{m}}t + \varphi)$ is applied into the phase modulator to chirp carrier frequency of signals, then by using Taylor series expansion, and by the approximation quadratic driving voltage, phase of driving voltage would be

$$\phi(t) = e^{j\left[\frac{\pi}{V_\pi}(V_{\text{bias}} + V_{\text{m}}) - \frac{1}{2}\frac{\pi V_{\text{m}}}{V_\pi}\omega_{\text{m}}^2 t^2\right]}. \tag{10.23}$$

Comparing Equation 10.22 and Equation 10.23, in order to recover the Gaussian pulse at the end of the system, a chirping rate should be $|K| = |(\pi V_{\text{m}}/V_\pi)|/\omega_{\text{m}}^2$. This depends on the amplitude of driving voltage and modulation frequency $\omega_{\text{m}} = 1/T_o\sqrt{|(V_\pi/\pi V_{\text{m}})|}$. In a practical system, ω_{m} could be tuned in order to achieve an almost reconstructed signal. Apply this sinusoidal to a phase modulator in a 160-Gb/s system with a full width at half maximum (FWHM) of initial un-chirped $T_{\text{FWHM}} = 2\,\text{ps}$, then the corresponding input pulse half width at 1/e intensity point can be estimated as $T_o = T_{\text{FWHM}}/1.665 \approx 1.2012\,\text{ps}$. Thus, the necessary chirping rate needed to recover the distorted signal is $K = 1/T_o^2 = 6.944 \times 10^{23}\,\text{s}^{-2}$.

Assuming that $V_\pi = 3.5\,V$ and $V_{\text{bias}} = V_{\text{m}} = 7V$ in order to shift the sinusoidal wave up $V_{\text{m}}/2$. Therefore, if a 40-GHz clock signal is used, the chirping rate becomes $K = (\pi V_{\text{m}}/V_\pi)/\omega_{\text{m}}^2 = 3.969 \times 10^{23}$. Thus, two 40-GHz phase modulators must be employed to generate the required chirping rate. If the dispersion factor of standard SMF fiber $D = 17\,\text{ps/(nm-km)}$, then the GVD co-efficient would be $\beta_2 = -2.168e - 26s^2/\text{m}$. Thus, the length of single mode fiber to be inserted in the equalizer would be calculated as $L_{\text{PM}} = T_o^2/\beta_2 \approx 66.55\,\text{m}$.

10.2.2.2 Equalization with Parabolic Driven-Voltage Phase Modulator

The transmission system including a phase equalization sub-system is shown in Figure 10.5. Because a driving voltage has an ideal parabolic shape, it could create a parabolic phase. A single-mode fiber transfer function is $H(f) = e^{-j\alpha f^2}$ [25] which also has a parabolic phase. Therefore, this ideal parabolic driving voltage applied a parabolic electric field to a phase modulator, and it certainly could fully compensate any existed distortion after transmission link. If a parabolic driving voltage $V(t) = V_{\text{bias}} - V_{\text{m}}(\omega_{\text{m}}t)^2$ is applied into the phase modulator instead of using sinusoidal driving voltage, phase of the phase modulator would become

$$\phi(t) = e^{j\left[\frac{\pi}{V_\pi}V_{\text{bias}} - \frac{\pi V_{\text{m}}}{V_\pi}(\omega_{\text{m}}t)^2\right]}. \tag{10.24}$$

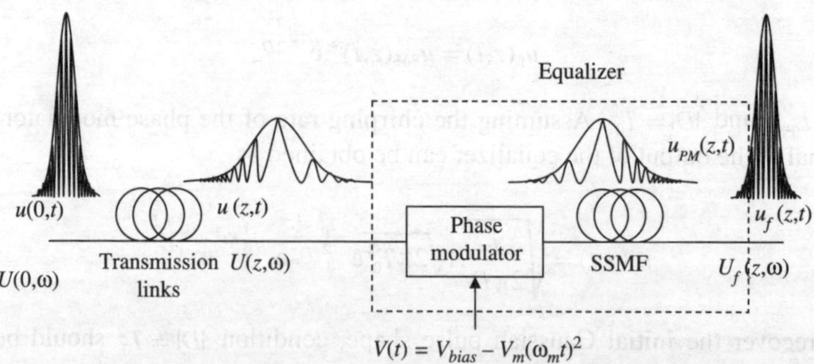

FIGURE 10.5   The overview of the optical system communication by using equalizer with ideal parabolic driving voltage and 40-GHz clock recovery to drive phase modulator.

A needed chirping rate for compensation at this time $|K| = 2|\pi V_m/V_\pi| f_m^2$ with a modulation frequency $f_m = 1/(2\pi T_o)/\sqrt{|V_\pi/2\pi V_m|}$ for 160 Gb/s bit rate system. The necessary parameters for compensation at this time are listed as: chirping rate $K = 6.944 \times 10^{23}\,\mathrm{s}^{-2}$, $V_\pi = 3.5\,\mathrm{V}$, $V_m = 6\,\mathrm{V}$, $V_{bias} = 2V_m = 12\,\mathrm{V}$, and SMF length in equalizer system is $L_{PM} \approx 66.55\,\mathrm{m}$. The modulation frequency can be estimated as $f_m = 1/(2\pi T_o)/\sqrt{|V_\pi/2\pi V_m|} \approx 40\,\mathrm{GHz}$.

In summary, if an ideal parabolic driving voltage is assumed, it is possible to achieve full compensation of all the distortions and only one phase modulator can be used in this case. However, it is very difficult to generate an ideal parabolic driving voltage in practice. The received signals of sinusoidal variation can be acceptable, which could be generated easily. Thus, sinusoidal signal driving-signals should be investigated in more detail in simulation and compared with the ideal parabolic waveform.

10.3   SIMULATION OF TRANSMISSION AND EQUALIZATION

10.3.1   SINGLE PULSE TRANSMISSION

Simulation results are carried out for dispersion due to the mismatched length between SMF and DCM of the transmission links, leading to distortion. Consequently, an adaptive equalizer should be installed at the end of the long-haul transmission link, especially at 160 Gb/s, in order to fully compensate all dispersion effects and timing jitter. Initially, a 2-ps width single pulse is launched into the optical fiber. The distorted output pulse is monitored after propagating through a transmission link. Then this distorted pulse is led to equalizer system to recover its original pulse width and shape.

10.3.1.1   Equalization of Second Order Dispersion

Figure 10.6 and Figure 10.7 show the equalization of the second-order dispersion effect for a 2-ps Gaussian pulse transmission system. Let T_o be a pulse half width at e^{-1} intensity point. By calculating the second-order dispersion length $L_{D2} = T_o^2/\beta_2 \approx 66.553\,\mathrm{m}$ and third-order dispersion length $L_{D3} = T_o^3/\beta_3 \approx 17.756\,\mathrm{km}$ [26] with a group velocity co-efficient $\beta_2 = -2.168\mathrm{e}-26\,\mathrm{s}^2/\mathrm{m}$ and third-order dispersion co-efficient $\beta_3 = 9.761\mathrm{e}-41\,\mathrm{s}^3/\mathrm{m}$, the output pulse is obviously broadened dramatically if a length of transmission link is greater than L_{D3} in order to investigate third-order dispersion effect. Table 10.2 tabulates the parameters of the equalizer required for the equalization of 160-Gb/s transmission due to second- and third-order dispersion to demonstrate the effectiveness of the equalizer in the refocusing or reshaping of the transmitted signals.

The simulated results tabulated in Table 10.2 show strong consistency between estimation and simulation results. Figure 10.6a and Figure 10.7a show the initial pulses before transmission through a 300-m optical fiber length under the two different waveforms of the applied voltage. These pulses would be spreading at the output of the receiver. The equalizer is inserted at the front-end of the receiver in order to recover these distorted pulses to their original shape. On the other hand, the pulse shape observed after an equalizer with a sinusoidal driving voltage is not as close to its original pulse shape with modulator driven by a parabolic driving voltage. The tails of recovered pulse is spreading over and can overlap with other tails from the adjacent pulses. This overlapping contributes to the penalty of the eye diagram. This effect can be observed due to the Taylor approximation as assumed

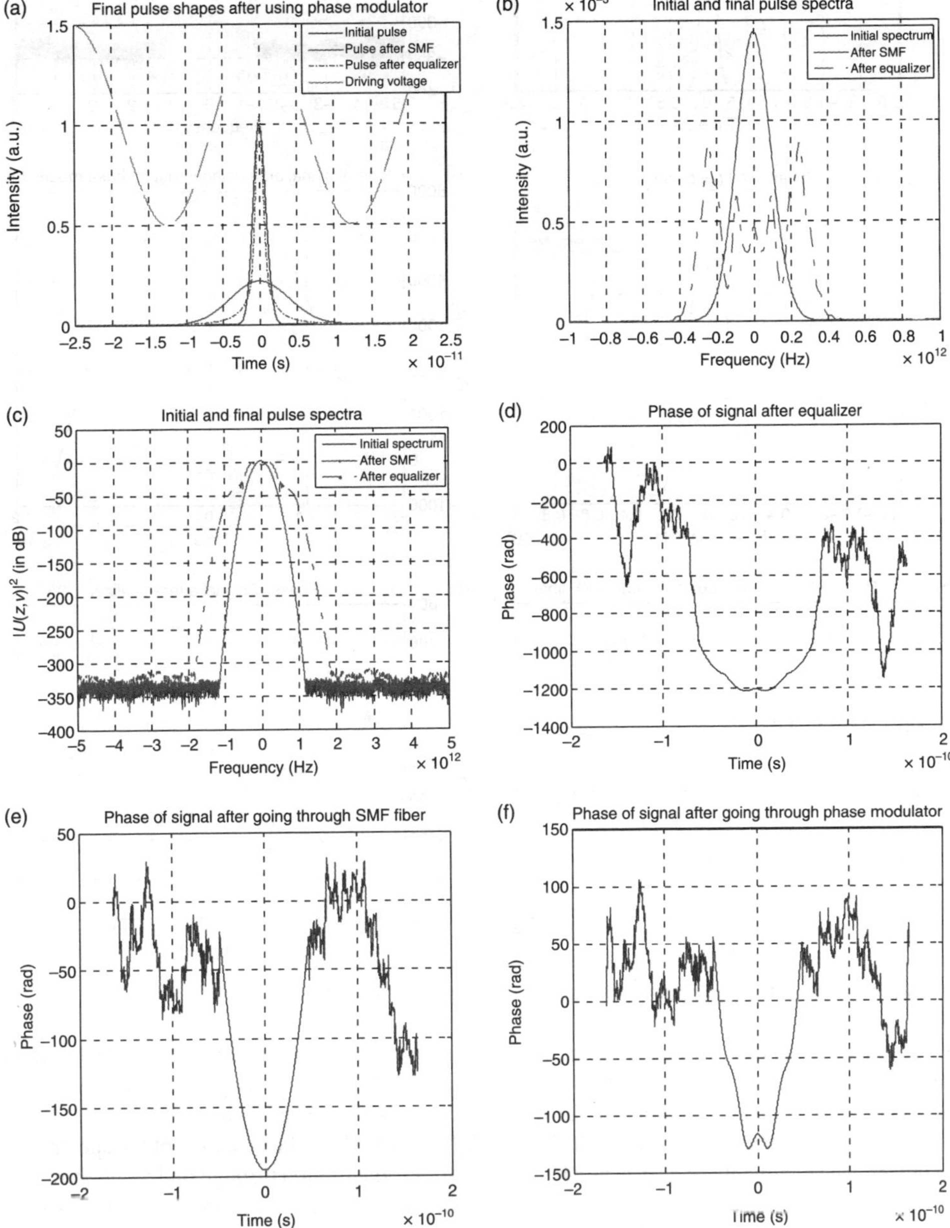

FIGURE 10.6   2-ps Gaussian pulse with sinusoidal applied driving voltage.

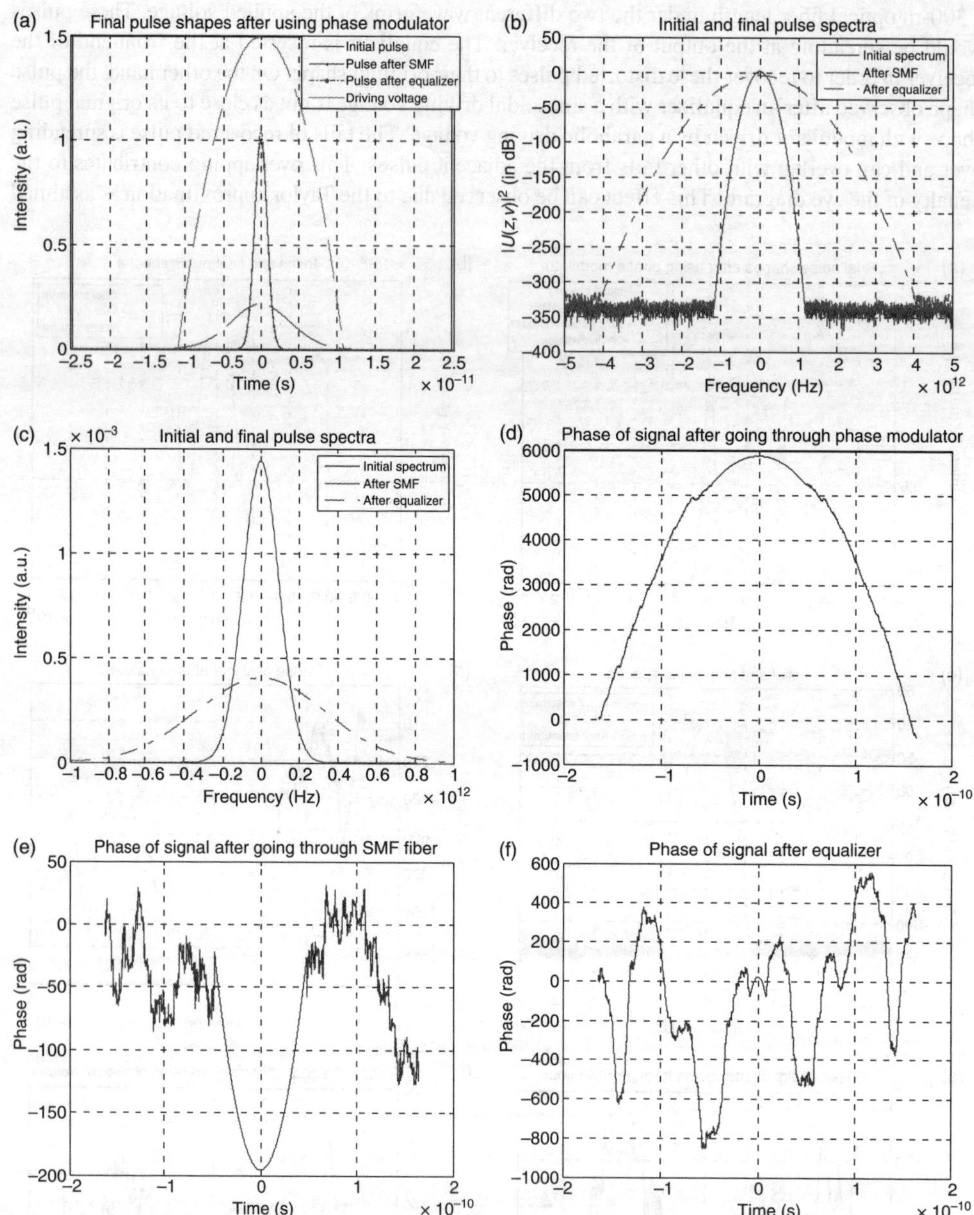

FIGURE 10.7　2-ps Gaussian pulse with parabolic applied driving voltage.

in Section 10.2.2.1. Due to the approximation of parabolic shape in the region $t = 0$, the Fourier-transformed pulses have prolonged tails at the wings of pulse. This overlapping penalty would be considered, discussed and investigated further in Section 10.3 and Section 10.4.

Figure 10.6d, Figure 10.6e, Figure 10.7d, and Figure 10.7e show the evolution of the spectrum of the pulses. The bandwidth of a spectrum before and after transmission through optical fiber remains the same because the phase of each frequency component is changed in the frequency domain. These changes created a time delay of each carrier frequency component in the time domain, thus the pulse is broadening. And carrier frequency is chirped down while transmitting through SMF

TABLE 10.2

Operational Parameters of the Equalization of the Second-Order Dispersion Effects on the Transmission of a Single Pulse

Parameters	Sinusoidal Driving Voltage	Parabolic Driving Voltage
β_2	$-2.168e-26 s^2/m$	$-2.168e-26 s^2/m$
β_3	$0 s^3/m$	$0 s^3/m$
$L_{Transmission}$	300 m	300 m
$L_{Equalizer}$	62 m	62 m
V_π	3.5 V	3.5 V
V_m	7 V	6 V
V_{bias}	7 V	12 V
φ	0	0
f_m	40 GHz	40 GHz
Number of PMs	2 phase modulators	1 phase modulator

fiber. By using a phase modulator, the carrier frequency under pulse envelope is chirped up. Spectrum is also broadening at this step because the phase of each frequency component is changed in the time domain, and it leads to a frequency delay in the frequency domain. Therefore, the envelope of these pulses is not changed, while the spectrum is expanded substantially.

Figure 10.6d, Figure10.6e, Figure 10.6f and Figure 10.7d, Figure 10.7e, and Figure 10.7f plot the phase of signal at the output of the SMF fiber, at the output of the phase modulator, and at the output of the equalizer, respectively. These phase plots indicated that the carrier frequency is chirped down or chirped up. Firstly, a pulse propagated through the SMF fiber, carrier frequency is chirped down, thus phase shape is a convex parabolic. But the applied electric field is approximately concave parabolic and carrier frequency would be chirped up. Only phase of these frequency components are changed in the time domain, so the pulse shape remains the same. Then carrier frequency is chirped down again when these pulses are transmitted through the SMF in equalizer. At this process, there is a re-organization of each frequency component due to the different time delay of different frequency component and the output pulse shape would be recovered. Furthermore, attention should be paid when there is a little concave parabolic shape around $t = 0$ area. It indicates that the carrier frequency under the pulse envelope is still chirped up, thus if SMF length in the equalizer is increased, the output pulse from equalizer would be even more compressed than the initial pulse.

In other words, equalizer is a tunable system which could modify the output pulse width and its shape accordingly to change some parameters in that equalizer such as voltage amplitude or SMF length.

10.3.1.2 Equalization of Third-Order Dispersion

Table 10.3 shows all the necessary parameters for third-order dispersion-compensation if sinusoidal or parabolic driving voltages are used to drive the phase modulator. In this case, an electric field applied to a phase modulator is shifted right by an angle of $\pi/24$ due to the following reasons. Firstly, the oscillated tail of this pulse should be covered by this driving voltage. Secondly, this distorted pulse is asymmetric and third-order dispersion "pulls" a pulse to right direction. If the driving voltage is still biased at the quadrature point as in the previous section, phase of some frequency components might be either unchanged or changed in the opposite direction. This is due to the spreading of the tail of the pulse outside the concave parabolic curve of the driving voltage.

As shown in Figure 10.8a and Figure 10.9a, the output pulse is almost fully recovered accordingly to different voltage waveforms. However, assuming a pulse train is transmitted, if a tail of this pulse is broadened to neighbor duration, it would be very difficult to be recovered. Figure 10.8b and Figure 10.9b are the spectral profiles corresponding to Figure 10.8a and Figure 10.6a, respectively. Distortion in the time domain is converted to that in the frequency domain by Fourier transform, which is discussed above.

10.3.2 Pulse Train Transmission

10.3.2.1 Second-Order Dispersion

Figure 10.10 and Figure 10.11 show the propagation of a pulse train with a bit sequence [1 0 0 1 1 1 0 1] with 2-ps pulse width through a 300-m and 600-m optical fiber, respectively. All other parameters remain the same for these two situations and they are listed in Table 10.3.

Figure 10.10a and Figure 10.11a show an initial and output pulse train after propagation through 300-m and 600-m standard SMF fiber, respectively. The distorted pulse train is recovered and

TABLE 10.3

Summary of Parameters for the Equalization of Third-Order Dispersion Under Single-Pulse Transmission

Parameters	Sinusoidal Driving Voltage	Parabolic Driving Voltage
β_2	$0\,\text{s}^2/\text{m}$	$0\,\text{s}^2/\text{m}$
β_3	$9.761\text{e}-41\,\text{s}^3/\text{m}$	$9.761\text{e}-41\,\text{s}^3/\text{m}$
$L_{\text{Transmission}}$	120 km	120 km
$L_{\text{Equalizer}}$	67 m	67 m
V_π	3.5 V	3.5 V
V_m	7 V	6 V
V_{bias}	7 V	12 V
φ	$-\dfrac{\pi}{24}$	$-\dfrac{\pi}{24}$
f_m	40 GHz	40 GHz
Number of PM	2 phase modulators	1 phase modulator

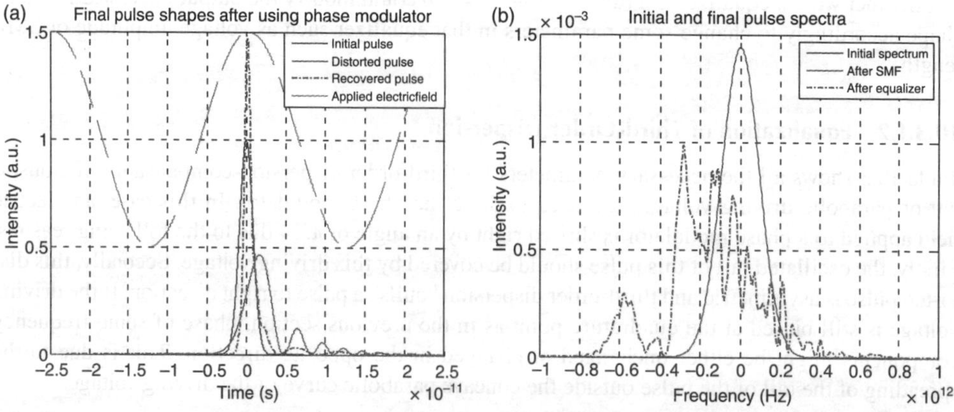

FIGURE 10.8 2-ps Gaussian pulse with sinusoidal applied driving voltage.

TABLE 10.4

Parameters for Equalization of Second-Order Elimination Under Pulse-Train Transmission

	β_2	β_3	$L_{\text{Equalizer}}$	V_π	$V_m = V_{\text{bias}}$	φ	f_m	Number of PM
Values	$-2.168e-26\,s^2/m$	$0\,s^3/m$	$67\,m$	$3.5\,V$	$7\,V$	π	$40\,GHz$	2

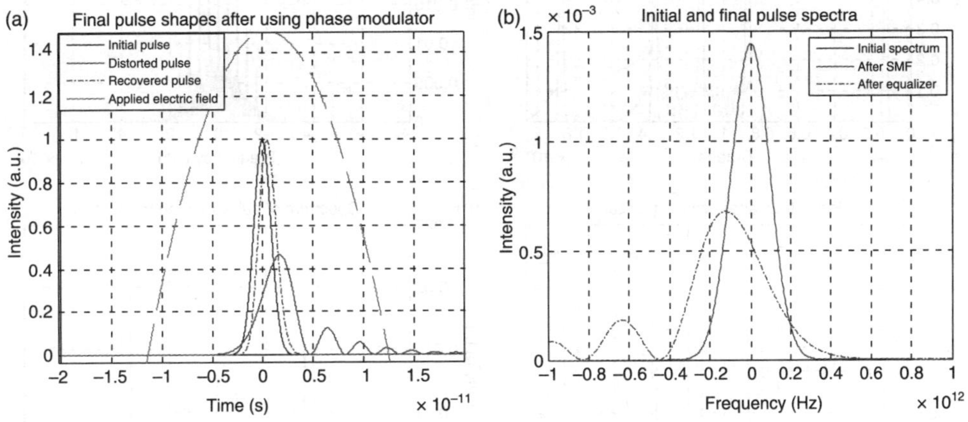

FIGURE 10.9   2-ps Gaussian pulse with parabolic applied driving voltage.

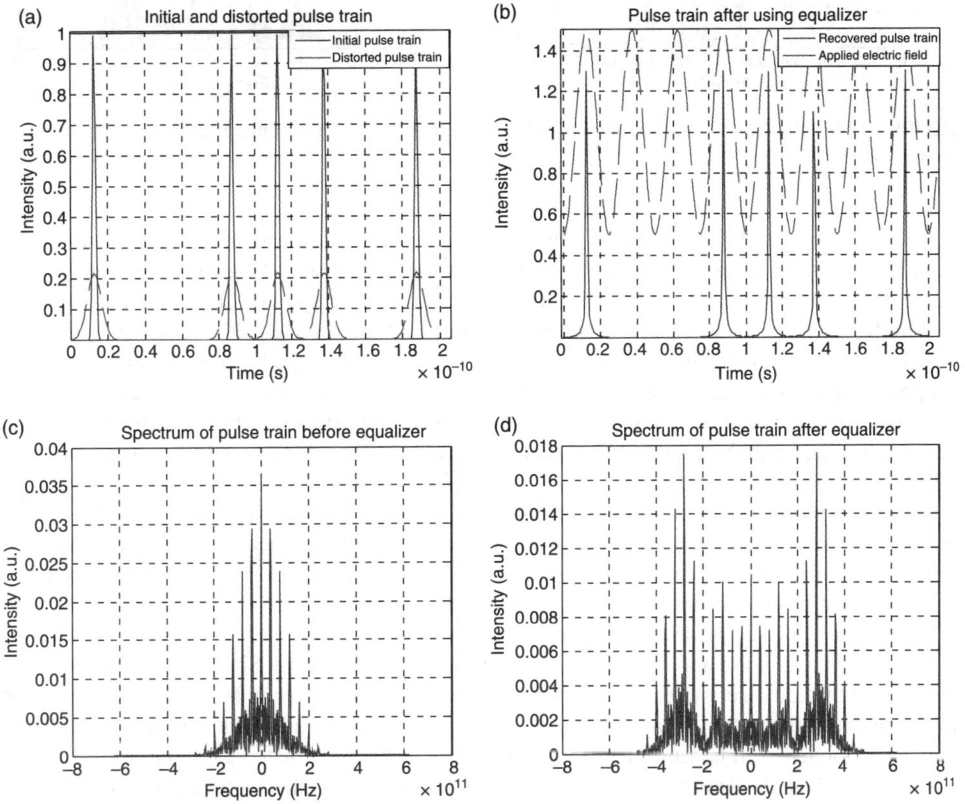

FIGURE 10.10   40-Gb/s with 2-ps Gaussian pulse propagating through 300-m optical fiber with sinusoidal driving voltage.

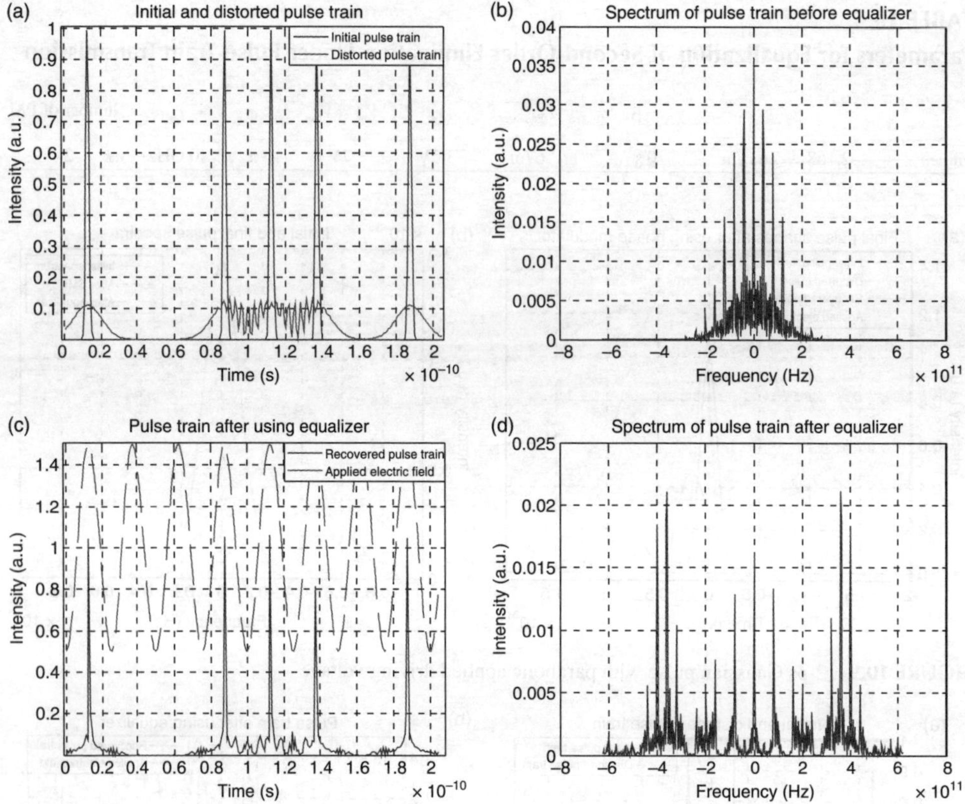

FIGURE 10.11 40-Gb/s with 2-ps Gaussian pulse propagating through 600-m optical fiber with sinusoidal driving voltage.

showed in Figure 10.10b and Figure 10.11b. The corresponding spectrum to that pulse train is plotted in Figure 10.10c–d and Figure 10.11c–d. Based on the transmission length of SMF fiber, there is a huge difference between Figure 10.10 and Figure 10.11 which should be discussed thoroughly.

Firstly, when this pulse train is transmitted through 300-m fiber link, although each pulse in this pulse train is dispersed and broadened, their tails are still not overlapped with each other. Therefore, compensating these dispersive pulses is not difficult. The spectrum of this signal is also monitored and it is obviously broadened, which satisfies the expectation explained above. Furthermore, there are several spikes, and the difference between each spike is about 40 GHz, which satisfies the theoretical calculation and expectation.

Since this pulse train is transmitted through 600 m of optical fiber, the output pulse train is distorted significantly and their tails overlapped and modulated to each other. The overlapping between neighboring pulses would lead to constructive or destructive effects corresponding to in-phase or out of phase of those carrier frequencies, respectively. Therefore, there is an oscillation when two tails of two neighboring pulses modulating to each other, as in Figure 10.11a. These oscillating tails would contribute to noise and increase the noise level of optical system. Since the output after passing through the equalizer system is plotted in Figure 10.11b, the output pulse train had a significant improvement in dispersion compensation, as well as noise reduction. Although total cancellation of noise is not achieved because the driving voltage is just approximated parabolic waveform, the existing noise level is also reduced significantly and this noise level after using equalizer is much less than before using it.

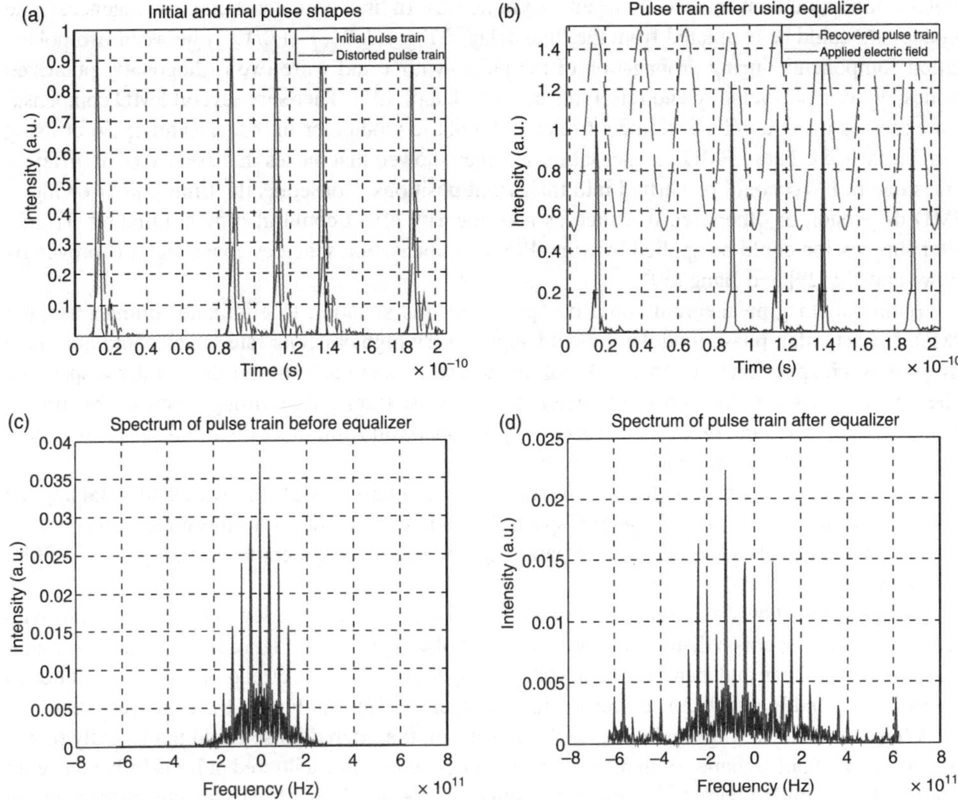

FIGURE 10.12 40-Gb/s with 2-ps Gaussian pulse propagating through 120-km optical fiber with sinusoidal driving voltage.

10.3.2.2 Equalization of Third-Order Dispersion

Figure 10.12a and Figure 10.12b showed the initial, distorted and recovered pulse train after transmission through 120-km transmission link, while Figure 10.12c and Figure 10.12d showed the spectrum corresponding to that pulse train. This distorted pulse train recovered by using equalizer system and this recovering phenomenon is discussed clearly in Section 10.1.2 above.

On the other hand, a pulse train is investigated because there are some effects which could not obtained or monitored by using a single pulse, such as a timing jitter effect. When there are neighboring pulses, two closed pulses would be pulled or pushed from their original position due to third-order dispersion, or non-linearity effects. Thus, this effect would contribute to bit-error-rate in transmission system. Because the receiver recovery clock is always sampled at the original instant of the pulse sequence in order to detect bit 1 or bit 0 when the pulses are jittering. Therefore, recovered signals may not be fully reconstructed.

In order to eliminate this problem, equalizer would be used to push or pull these pulses back to their initial positions as well as remove their distortion. Section 10.3.3 would discuss further with some simulation results about timing jitter elimination.

10.3.3 Equalization of Timing Jitter and Polarization-Mode Dispersion

Polarization mode dispersion (PMD) is a form of modal dispersion where two different polarizations of light in a waveguide, which normally travel at the same speed, travel at different speeds. Thus, it would become broader as the two components disperse along the fiber due to their different group velocity, random imperfections and asymmetries. In fiber with constant birefringence, pulse broadening could be estimated from the time delay $\Delta T = |(L/V_{gx}) - (L/V_{gy})|$ between two polarization components during propagation of the pulse, with x and y are two orthogonally polarized modes, V_g is group velocity and L is a transmission length [27]. There are several PMD compensation schemes using a polarization controller and a phase modulator in the transmitter [9] or using "time lens" [28]. These PMD compensation schemes showed that pulses that are placed at different times due to PMD could be shifted into their right positions. Consequently, timing jitter owing to PMD dispersion or others reasons such as non-linearity effect could also be eliminated. Timing jitter suppression could be applied by using PMD compensation schemes above and it is proven by Howe and Xu [29] and Jiang [30].

Temporal imaging theorem could be applied in this situation to eliminate timing jitter and PMD. In particular, phase modulator would apply a very high chirping rate which could dominate all existing chirping rate due to PMD, non-linear effect, and second- and third-order dispersion. Then these pulses with high chirped carrier frequency are transmitted through optical fiber link to compress their pulse and recover their original pulse shape and pull those distorted pulses by timing jitter or PMD back to their original positions.

Figure 10.13 shows timing jitter elimination by using phase modulation with sinusoidal applied voltage and ideal parabolic voltage and spectral profile corresponding to those recovered pulses. Table 3.5 indicates all necessary setting parameters in equalizer system for timing jitter and PMD elimination.

Figure 10.13a shows the timing shift in a range of about ± 4 ps is added to the input signals with PMD and TOD dispersion distortion. The recovered pulse is plotted in Figure 10.13 b and c with sinusoidal applied voltage and ideal parabolic voltage, respectively. The spectral profile are also plotted in Figure 10.13d and Figure 10.13e corresponding to Figure 10.13b and Figure 10.13c respectively.

When a driving voltage had a sinusoidal waveform, the jittered pulse could not be fully recovered its original pulse shape. With respect to parabolic modulation, a jittered pulse is fully recovered from TOD and pulled back to its original position. Therefore, it is difficult to fully reduce timing jitter and PMD under sinusoidal modulation, especially when jittered pulse is broadened within the whole curvature of the modulation. However, under a reasonable jitter or PMD, it is very possible to compensate and pull it back to its original form.

The ideal situation is to use ideal parabolic modulation to fully recover pulse. However, there is a trade-off between using parabolic and sinusoidal applied voltage. Firstly, it is very hard to generate an ideal parabolic in practice. Secondly, if it is possible to build a parabolic driving voltage, it would be expensive compared with very popular sinusoidal waveform. Thirdly, if jittered or PMD pulses are still in a reasonable range, they could be recovered and pulled back to their initial position with reasonable results. In conclusion, depended on using purpose in order to decide to use sinusoidal driving voltage or spending more money to create an ideal or closed to ideal parabolic modulation.

10.4 EQUALIZATION IN 160-Gb/s TRANSMISSION SYSTEMS

10.4.1 System Overview

10.4.1.1 System Configurations

Recently, standard single-mode fiber with a group velocity dispersion value of about 17 ps/(nmkm) was used widely in long-haul transmission. As discussed above, dispersion-compensating fiber is one of the dispersion management methods to recover signals after a period of distance. However, dispersion slope compensation and wavelength division multiplexed transmission is very hard to fully compensate, especially for ultra-fast transmitted signals. For example, it would return a significant variation in transmission performance for 160 Gb/s even though there is only a small change

TABLE 10.5
Parameters for Equalization of Third-Order Elimination Under Pulse-Train Transmission

	β_2	β_3	$L_{\text{Equalizer}}$	V_π	$V_m = V_{\text{bias}}$	φ	f_m	Number of PM
Values	$0\ \text{s}^2/\text{m}$	$9.761\text{e}-41\text{s}^3/\text{m}$	80 m	3.5 V	7 V	$5\pi/6$	40 GHz	2

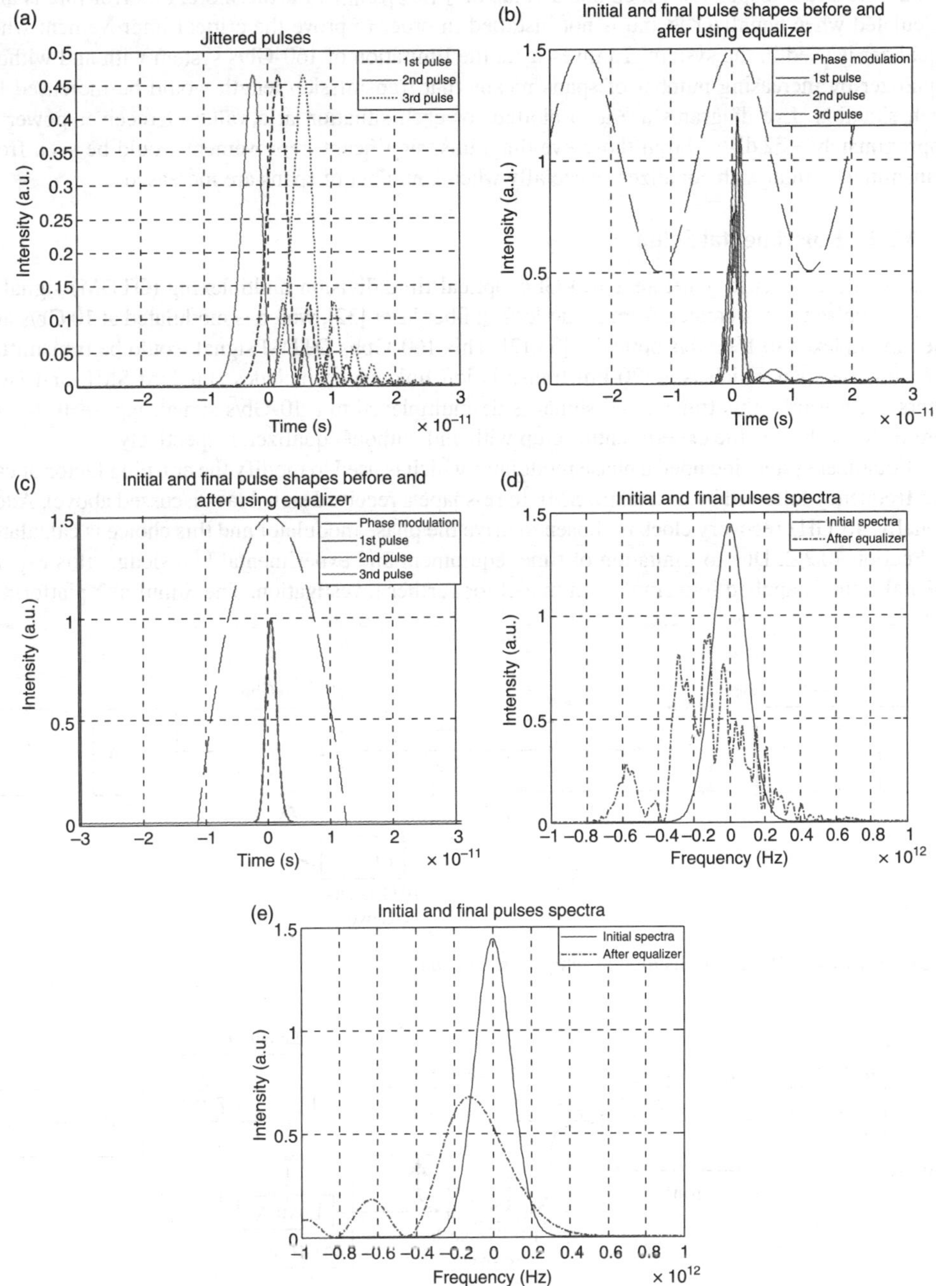

FIGURE 10.13   Investigation of three jittered 2-ps Gaussian pulses.

in dispersion due to environmental or setting-up effects. Therefore, the equalizer system which included two main devices, which are phase modulator followed by a dispersive element, is used before receiver. This equalizer is used to compensate not only group velocity dispersion, but also for third- or fourth-order dispersion [31,24], and timing jitter [30].

In this study, equalizer system is used for one-span 120 km transmission fiber (including standard SMF and DCF) to compensate a group velocity dispersion which had a value of 1.28 ps/nm and a third-order dispersion which had a value of 1.692 ps/nm^2. Furthermore, bit-error-rate is also calculated when equalizer is and is not installed in order to prove the critical improvement when equalizer is used in the system. To investigate the limitation of 160-Gb/s system with and without equalizer by increasing number of spans means that transmission length would be increased for each situation. Eye diagrams are also plotted for each situation at a value of receiver power of approximately -32 dBm. From those eye diagrams, significant improvement would be seen from transmitted system with equalizer, especially when a number of spans are increased.

10.4.1.2 Experimental Setup

First of all, in order to generate a 160-Gb/s optical time division multiplexing (OTDM) signal, a 10-GHz pulse train is created from mode-locked fiber laser [32], which is modulated at 10 Gb/s and then multiplexed to 160 Gb/s optically [33,12]. This 160-Gpbs OTDM signal would be transmitted through one span which is a 120-km transmission link which includes standard SMF and DCF fibers. Afterwards, this transmitted signal is de-multiplexed to a 10-Gb/s signal. Figure 10.14 and Figure 10.15 showed the experimental setup with and without equalizer, respectively.

Equalizer system included a phase modulator which is used to modify the chirping factor of carrier frequency followed by a standard SMF to re-shape a received signal (as discussed above). Additionally, 40-GHz recovery clock is chosen to drive the phase modulator and this choice is calculated in Section 10.2.2. Due to limitation of time, equipment and experimental knowledge, this experimental setup is applied into simulation model for further investigation. The Simulink® platform is

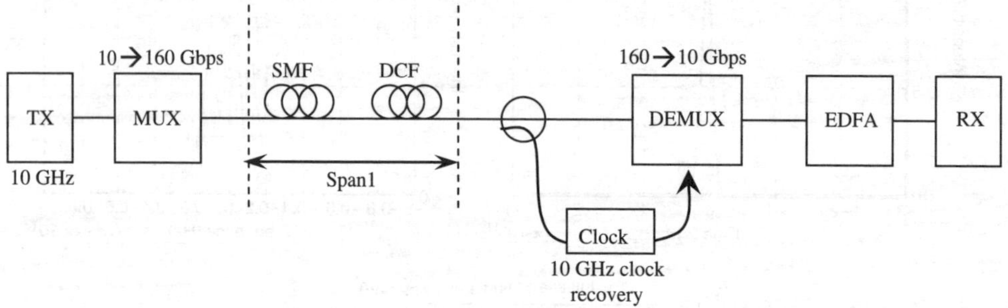

FIGURE 10.14   160-Gb/s transmission setup without equalizer.

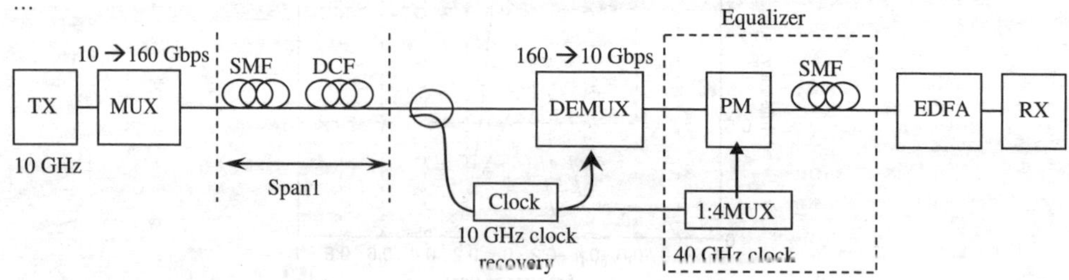

FIGURE 10.15   160-Gb/s transmission setup with equalizer. (From Bennett, C. V., and B. H. Kolner, *IEEE Journal of Quantum Electronics*, 36(4), 430–37, 2000; and Hirooka, T., and M. Nakazawa, *Journal of Lightwave Technology*, 24(7), 2530–40, 2006. With permission.)

chosen because of its many advantages. First of all, MATLAB® is the standard mathematical tool boxes in teaching and research laboratories. It is also very useful and powerful tool for simulation. Furthermore, several packages have already been developed for teaching and research purposes, and they are adapted and modified for this simulation purpose. Lastly, individual Simulink® block could be re-used and applied for other Simulink models without conflicts between these models.

Because of huge advantages of using Simulink model in MATLAB for running simulation, this platform has been chosen to build a 160-Gb/s transmitted signal system with and without equalizer based on a real experimental setup. Section 10.4.2 is devoted to introducing, building Simulink model, and setting parameters for this study.

10.4.2 SIMULATION MODEL OVERVIEW

10.4.2.1 SYSTEM OVERVIEW

Modeling of an optical communication system that satisfies some requirements, such as simplicity, accuracy in terms of phenomena and corroborating with experimental systems, is very important. Thus, this Simulink model is adapted and modified from existing Simulink blocks and they are used, modified and some new Simulink blocks can be developed. This would create a necessary 160-Gb/s system that is built-up based on a real experimental setup which was introduced briefly above.

The purpose of this Simulink model is to investigate GVD and TOD effects in ultra-fast signal processing. Only an equalization process in linear regime is considered at this stage, any non-linearity effects are negligible. Actually, non-linear effect might change the phase of each frequency component and it would create a chirp rate on optical carrier frequency. However, chirp rate which is created from non-linear effect is sufficiently small compared with the generated chirp rate from the phase modulator. Thus, it is reasonable to eliminate non-linearity effect at this stage. This section concentrates on explaining and developing a Simulink model which could be capable of simulating a dispersive compensation before and after using an adaptive equalizer. Some results are also demonstrated in this section.

First of all, a transmitter which could generate 160-Gb/s signals transmitted data through fibers at 10-dBm peak power. Although this launching power is high enough to have non-linearity effect during transmitting signals, it is still temporary ignored owing to the explanations stated above. Let us assume that only one span of transmission fiber which includes standard single-mode fiber and dispersion compensating fiber is used in these transmission links. When the signal is launched

---

TABLE 10.6
Summary of Timing Jitter Elimination Parameters

Parameters	Sinusoidal Driving Voltage	Parabolic Driving Voltage
β_2	$0\,s^2/m$	$0\,s^2/m$
β_3	$9.761e-41s^3/m$	$9.761e-41s^3/m$
$L_{\text{Transmission}}$	120 km	120 km
$L_{\text{Equalizer}}$	80 m	67 m
V_π	3.5V	3.5 V
V_m	7V	6 V
V_{bias}	7V	12 V
φ	$-\dfrac{\pi}{24}$	$-\dfrac{\pi}{24}$
f_m	40 GHz	40 GHz
Number of PM	2 phase modulators	1 phase modulator

through SMF fiber, GVD and TOD would affect the propagation signal and the output signal would be distorted dramatically. Although DCF is used for compensation, final output signal would not be fully recovered due to other factors, such as third-order dispersion, polarization-mode dispersion, and timing jitter. These problems are also analyzed and discussed in more detail in Section 10.2 above and Section 10.5 are devoted to show some simulation results about those dispersive effects and their elimination.

Because the 160 Gb/s source is so high that it has not been produced yet, high-speed optical time division multiplexing (OTDM) technique would be used to generate the high transmission rate. The 160 Gb/s or an even higher bit rate have already been introduced and generated by using a 10-GHz regenerative mode-locked fiber laser [34–36]. Consequently, this Simulink 160 Gb/s laser source is built based on this OTDM technique.

Initially, sixteen 10-Gb/s transmitters are used and these sixteen 10-Gb/s signals are multiplexed by using OTDM technique. This high-speed signal is transmitted to 120 km (one span) and several spans of transmission links at later stage. Because the transmitted signals are multiplexed before launching through optical fiber, they should be de-multiplexed back into 10-Gb/s signal. This de-multiplexed signal is fed into an equalizer system to recover distorted signals. Erbium-doped fiber is also used to amplify the output signal from equalizer before receiving this signal at a receiver end. Moreover, an error calculation block is also inserted right before the receiver block in order to count errors that would be used for calculating BER by using the Monte-Carlo method. Eye diagram has also been plotted on a pop-up window when the simulation has started.

10.4.2.2 Transmitter Block

The transmitter is a vital device in any communication system. An optical transmitter source is used to launch a laser beam into an optical fiber in transmission system. This laser beam carries transmitted data which is modulated under different modulation formats such as non-return-to-zero (NRZ), minimum-shift-keying (MSK), and phase-shift-keying (PSK). Return-to-zero (RZ) format is used in this investigation. Optical transmitter block is showed in Figure 10.4.

A pulse generator block is used to generate 10-GHz square wave with 2-ps pulse width. Parameters of pulse generator are set with (i) Pulse amplitude = 1; (ii) Pulse period = 100 ps; and (iii) Pulse width = 2 ps, i.e., about 2% of the pulse period

This square wave is fed into MUX block in order to generate 160-GHz square wave-transmitted signals. Figure 10.17 shows that 10-GHz square wave from pulse generator block is split into sixteen input square waves with different time delays by using transport delay block. This time delay could be created by using micro-ring with different lengths in the experimental setup. Furthermore, this MUX block also included sixteen Bernoulli random sources which are considered as sixteen different transmitting data. These sixteen different data signals are multiplied with those square waves above, and these multiplication results is added up to generate 160 Gb/s square wave signal afterwards. This output square wave signal has been filtered by using Gaussian filter to generate 160-Gb/s Gaussian-pulse-shape signal that is launched into transmission link under RZ fnormat at 10-dBm peak-launched-power. The 160 Gb/s Gaussian output multiplexed signal is show in Figure 10.18.

In order to compare transmitted and received signals for bit-error-rate calculation, a Goto tag is added. In the experimental setup, this multiplexed signal would be de-multiplexed into sixteen different data at the receiver end, and these data have been detected at the receiver. Assuming that the tenth data is picked up and investigated in this Simulink® model, the tenth signal is saved in *Goto* tag and sent it to *Error Calculation* block for comparison at the receiver end.

10.4.2.3 Transmission Link

The schematic diagrams, both in sub-system blocks and in Simulink®, of a 120-km transmission link including SMF-DCF fibers is described in Figure 10.15 and Figure 10.16, respectively. The 160-Gb/s RZ transmitted signals with Gaussian shape are launched into 98-km SSMF at 10-dBm peak

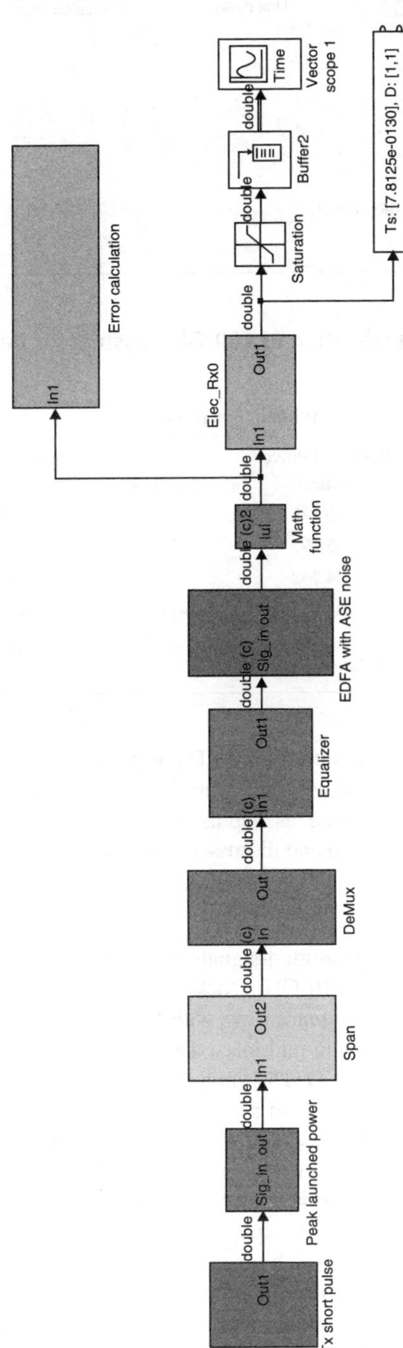

FIGURE 10.16 Overview of Simulink® model for an optical communication system.

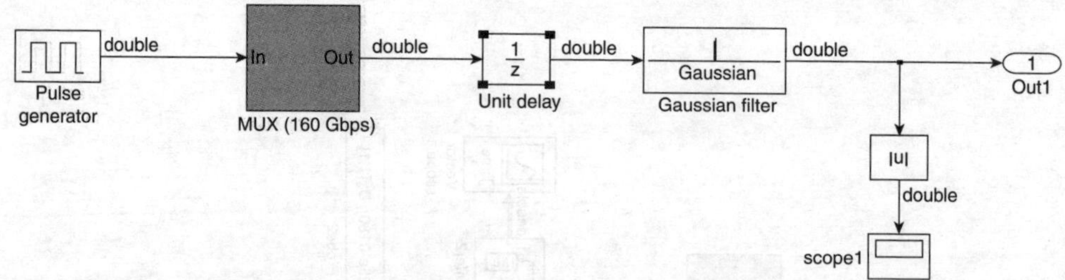

FIGURE 10.17   Transmitter Simulink model.

TABLE 10.7

BER Characteristic for GVD Equalization of 160-Gb/s System for Back-to-Back, with and without Equalizer

Back-to-Back		Without Equalizer		With Equalizer	
Received Power (dBm)	log(BER)	Received Power (dBm)	log(BER)	Received Power (dBm)	log(BER)
−37	−4.521	−36.375	−4.453	−36.745	−4.495
−36	−5.647	−35.193	−5.521	−35.498	−5.821
−35	−6.727	−34.152	−6.470	−34.437	−7.006
−34	−7.853	−33.012	−7.455	−33.414	−8.054
−33	−8.972	−31.983	−8.196	−32.356	−9.017
−32	−9.987				

power. Dispersion-compensating fiber is used for GVD compensation. Oscilloscopes and spectrum scopes are inserted in front of SMF, between SMF and DCF fiber and after DCF fiber in order to observe and measure the pulse broadening, as well as compensating factors during a transmission. Pulses and spectral profiles are analyzed and discussed in detail in the next section

10.4.2.4   De-multiplexer

At the receiver end, the 160-Gb/s transmitted signals are de-multiplexed to 10-Gb/s signal after passing through a de-multiplexer block; 10-GHz clock recovery has been applied to de-multiplexer block. This clock generated a 10-GHz square wave with 6.25-ps pulse width. The purpose of this clock recovery is illustrated in a simple de-multiplexed example in Figure 10.21.

We assumed that there is a 30-Gb/s OTDM multiplexed signal which needed to be de-multiplexed to 10-Gb/s signals at the receiver end of a transmission system. In the de-multiplexer block, clock one, two, and three are used to recover for first, second, and third data, respectively, by multiplying this multiplexed signal with each clock recovery. The products of this multiplication are sending data. Likewise, synchronization is also an important factor which must be taken into account. Therefore, a unit delay is added in this *DeMux* block for synchronization purposes, as shown in Figure 10.22. Pulse generator which acted as a clock recovery is used to generate a 10-GHz square wave with 6.25-ps pulse width. This square wave is delayed by feeding into unit delay block, and then multiplied with the input signal for the de-multiplexing process.

Figure 10.23 demonstrates an example of a successful de-multiplexed signal after passing through *DeMux* block. The higher-position scope (scope 1) showed the original multiplexed signal and the de-multiplexed output signal is plotted on the lower scope (scope 3). According to input signal, output signal is successfully de-multiplexed. In addition, synchronization issue is even much harder and more significant when third-order dispersion and time-jittering effects play a critical role

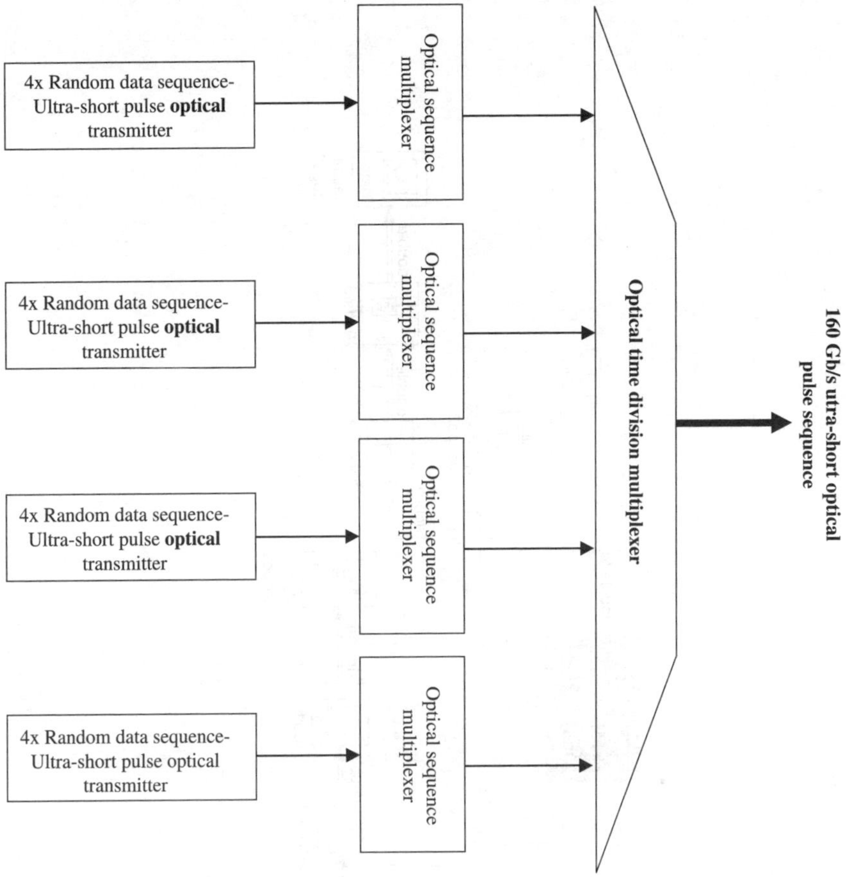

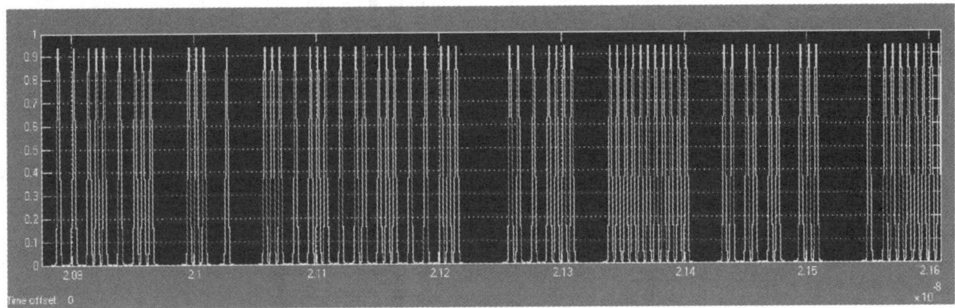

FIGURE 10.18 Multiplexer model with sixteen different sources.

FIGURE 10.19 160-Gb/s OTDM output signal.

during propagation through transmission link. Thus, a clock should be synchronized with suitable pulses accurately in order to sample correct bits.

10.4.2.5 Equalizer System

The equalizer system which has been invented according to temporal imaging theorem is introduced and discussed on working algorithms as well as some demonstrated simulation results in previous sections. This equalizer included a phase modulator followed by a dispersive element which

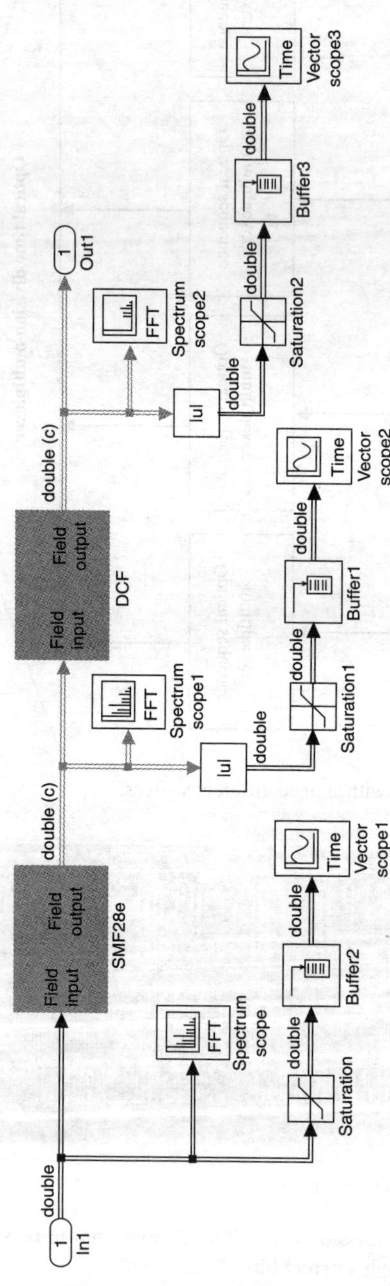

FIGURE 10.20 SMF-DCF fiber in one transmission span.

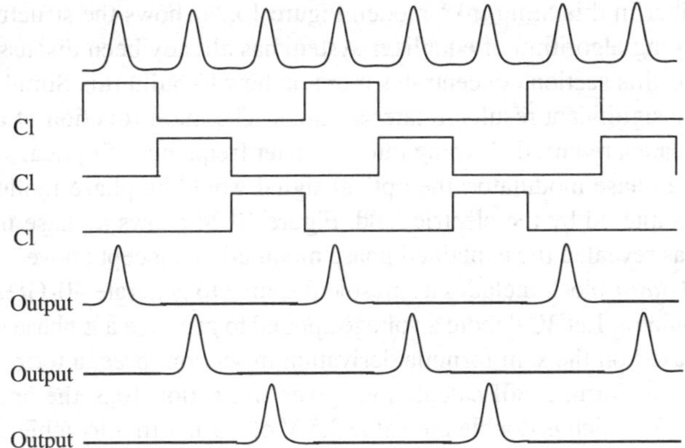

FIGURE 10.21 Time-domain output signals of multiplexers in Figure 10.18.

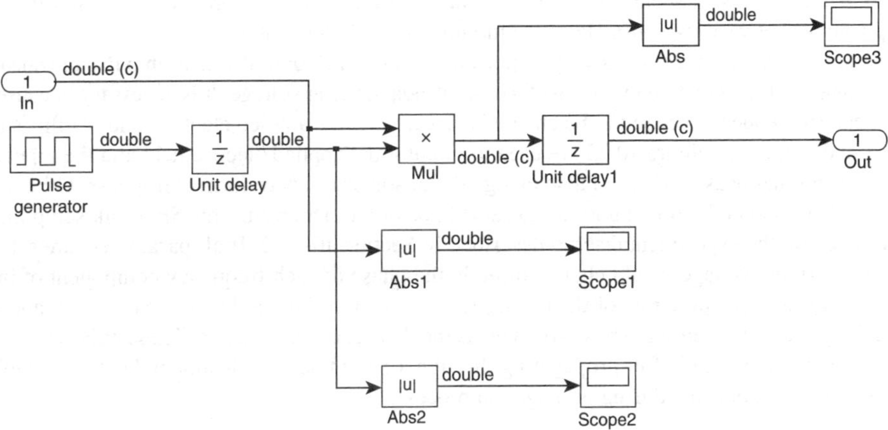

FIGURE 10.22 De-multiplexer block in detail.

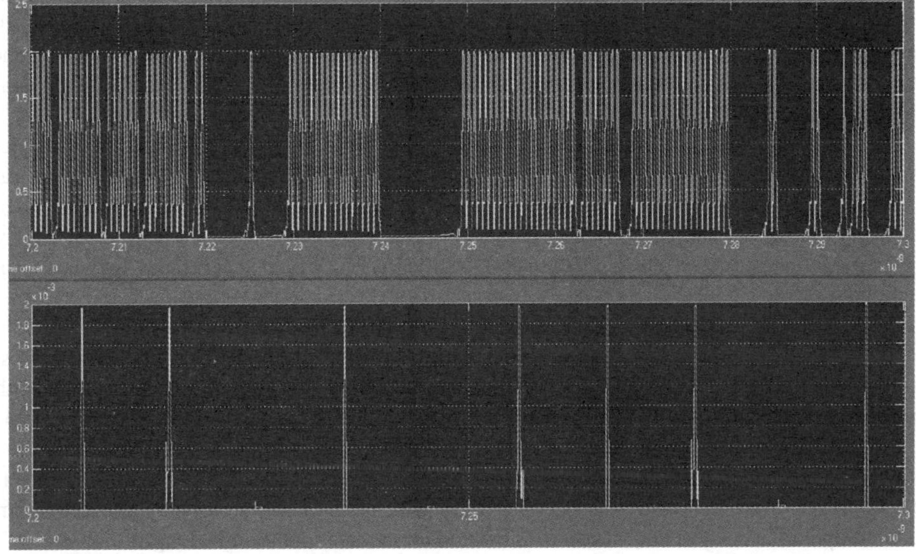

FIGURE 10.23 De-multiplexed signals.

is a standard SMF fiber in this Simulink® model. Figure 10.24 shows the structure of the sub-system model. The working algorithm of equalizer system has already been discussed in detail in the previous section, thus this section concentrates more on how to build this Simulink model, setting parameters and some significant results in later sections. The main function of a phase modulator in this model is to create a required chirping rate to carrier frequency of optical signal. By applying a driving voltage to a phase modulator, the optical signal would be phase-modulated because the optical path length is altered by the electric field. Figure 10.24 shows a phase modulator block in more detail, and it has revealed the explained phase modulation concept above.

This *Phase Modulator* block includes a sinusoidal source to generate 40-GHz sinusoidal waves to drive a phase modulator. Let V_π denote a voltage applied to generate a π phase shift on the optical carrier frequency. Based on the sam formula derivation in section three, a term π/V_π is multiplied into that sinusoidal waveform. In all calculations given in Section 10.3, the amplitude of driving voltage V_m is set to 7 V which is double the value 3.5 V of V_π in order to achieve a 2π phase shift on carrier frequency to guarantee the necessary chirp rate. Furthermore, two 40-GHz phase modulators are used in this situation to recover output signal as theoretical calculation in section three and simulation results in section four above. Thus, the multiplication product is a result of three multiplications between two 40-GHz phase modulator and input signal.

In addition, synchronization is very important because each output Gaussian pulse of transmitted data must be fully covered by this 40-GHz sinusoidal driving voltage. It is necessary because all optical carrier frequency needed to be chirped correctly by this applied electric field. At this stage, scope 1 and scope 2 in Figure 10.25 are used to monitor the applied electric field and the input signal to confirm that phases of each pulse of signal and sinusoidal waveform is in-phase. Otherwise, driving voltage should be tuned until an expected position is achieved. This Simulink setup model is also based on the experimental setup described in Section 10.4.1.2. If all parameters are set correctly, this driving voltage would change directly the phase of each frequency component of input signal. As a result, chirping rate of the carrier frequency of this signal had to be changed, and this carrier frequency has been set to be chirped up as the theoretical calculation. These pulses would be compressed and recovered after propagating through a SMF fiber in this model due to reorganization frequency components during propagation process.

10.4.2.6 Errors Calculation

Firstly, this *Errors Calculation* block is built up to calculate bit-error-rate (BER) based on the Monte-Carlo method, which is widely used class of computational algorithms for simulating the behavior of systems. Furthermore, Monte-Carlo method is used due to its useful for modeling phenomena with significant uncertainty in inputs. Figure 10.26 shows a BER calculation block using the Monte-Carlo method.

Threshold Time Decision Binary Detection (TTDBD) block is adapted and modified to detect between bit 1 and bit 0 of the received signal. The output from TTDBD is connected to error rate calculation (ERC) block to calculate BER and find delay (FD) block to set the time delay between signal from transmitter end (from binary data tag which is collected from transmitter block) and receiver end to integer delay block. The purpose of this FD block is to synchronize between transmitted and received signals.

Finally, computational delay in ERC block should be set with the same value as the value from FD block. For example, a number 1794 which is returned from FD block should be set for integer delay block. As a result, ERC block should also ignore the first 1794 samples at the beginning of the comparison in this block. It follows that this number should be set for computation delay parameter. Additionally, correlation window length (samples) parameter of FD block should be set sufficiently large so that the computed delay eventually stabilized at a constant value. However, there is a trade-off between reliability of the computed delay and processing time to compute the delay. Thus, a reasonable value for correlation window length should also be taken into account.

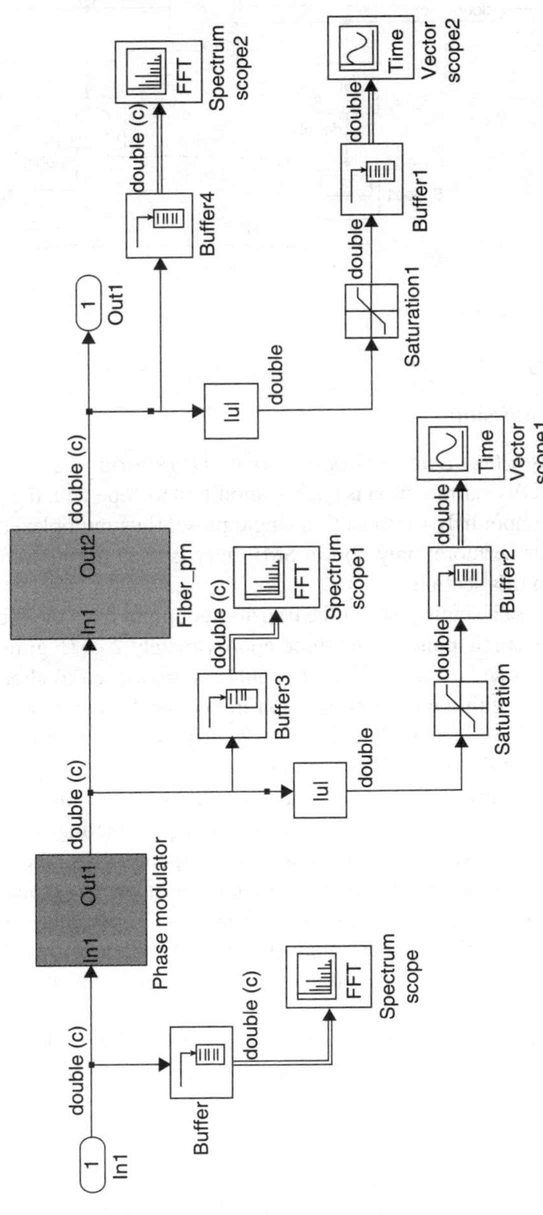

FIGURE 10.24 Equalizer system including phase modulator followed by a SMF.

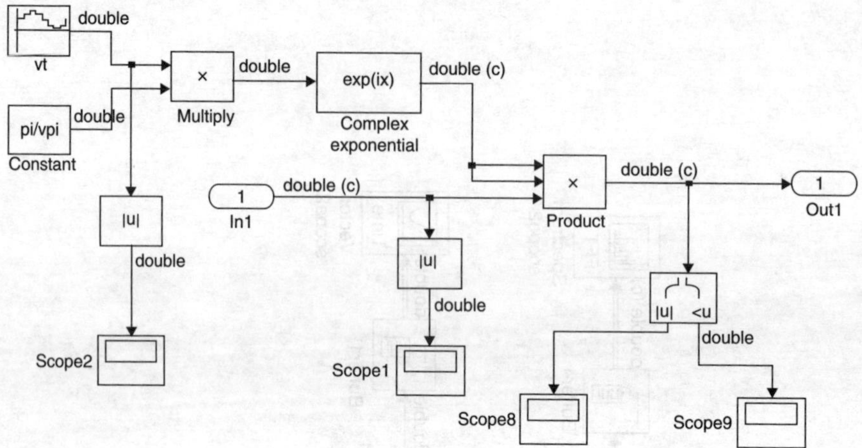

FIGURE 10.25 Phase modulator block in detail.

10.4.3 SIMULATION RESULTS

10.4.3.1 Single Pulse Transmission

10.4.3.1.1 Equalization of Distortion Due to Group Velocity Dispersion

The Simulink® model for this GVD elimination is slightly modified to cope with the purpose of this section. Transmitter is used temporarily to transmit a single pulse, thus multiplexer and de-multiplexer blocks are removed. Furthermore, only 300-m SMF fiber is used in this transmission link, DCF is not necessarily used in this section.

With respect to pulse shape of a single pulse of the transmitter output from the oscilloscope, full width at half maximum of this single pulse is measured approximately 2 ps (Figure 10.27a). After propagating through 300 m of standard SMF fiber, this pulse is broadened to about 9 ps (Figure 10.27b). This result is consistent with the simulation result in Section 10.3 and theoretical calculation. The final recovered pulse shape is plotted in Figure 10.27c with a pulse width value of 2 ps, but it had a longer tail compared with the original pulse shape.

Spectral profiles of a single pulse before and after using equalizer are shown in Figure 10.27d and Figure 10.27e. The spectra of a single pulse before and after propagating through SMF fiber remain the same because only the phase of each frequency components are changed in the frequency domain, and it would create different time delay for each frequency component in the time domain due to their different GVD during transmission in fiber link. These delays are a reason for pulse broadening. Compared with spectra after equalizer that is broadened dramatically, it indicated that the optical carrier frequency is chirped and it is up-chirped, which is consistent with theoretical calculation and expectation.

Lastly, all simulation results for a transmitted single pulse matched with the obtained results from Section 10.4 and expectation. Therefore, equalizer could eliminate a GVD effect during transmission system.

10.4.3.1.2 Third-Order Dispersion Elimination

Third-order dispersion played a significant part in ultra-high speed signals in the transmission system. From theoretical calculation and simulation results from Section 10.2 and Section 10.3 respectively, sinusoidal modulation would not fully recover distorted signals due to their asymmetric TOD. On the other hand, the oscillated tail of the recovered signal was reasonable reduction, thus the noise level created from this tail did not have a big affect on transmission system and its contribution to BER is acceptable. Figure 10.28a, Figure 10.28b, and Figure 10.28c showed

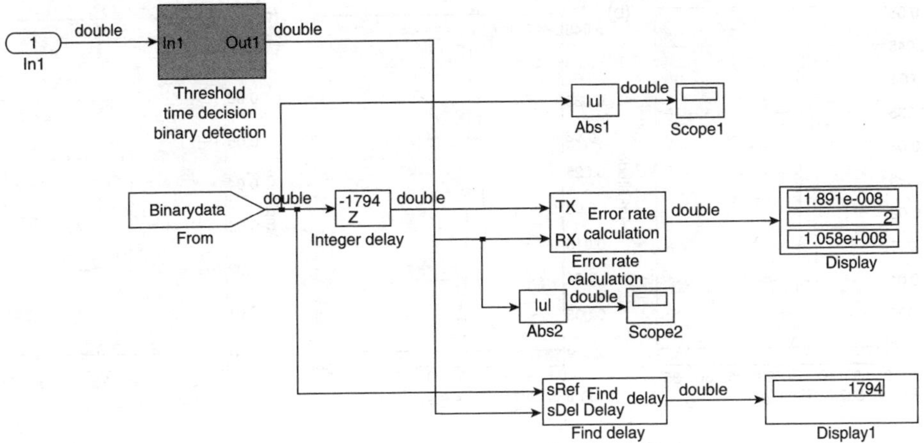

FIGURE 10.26 BER calculation block.

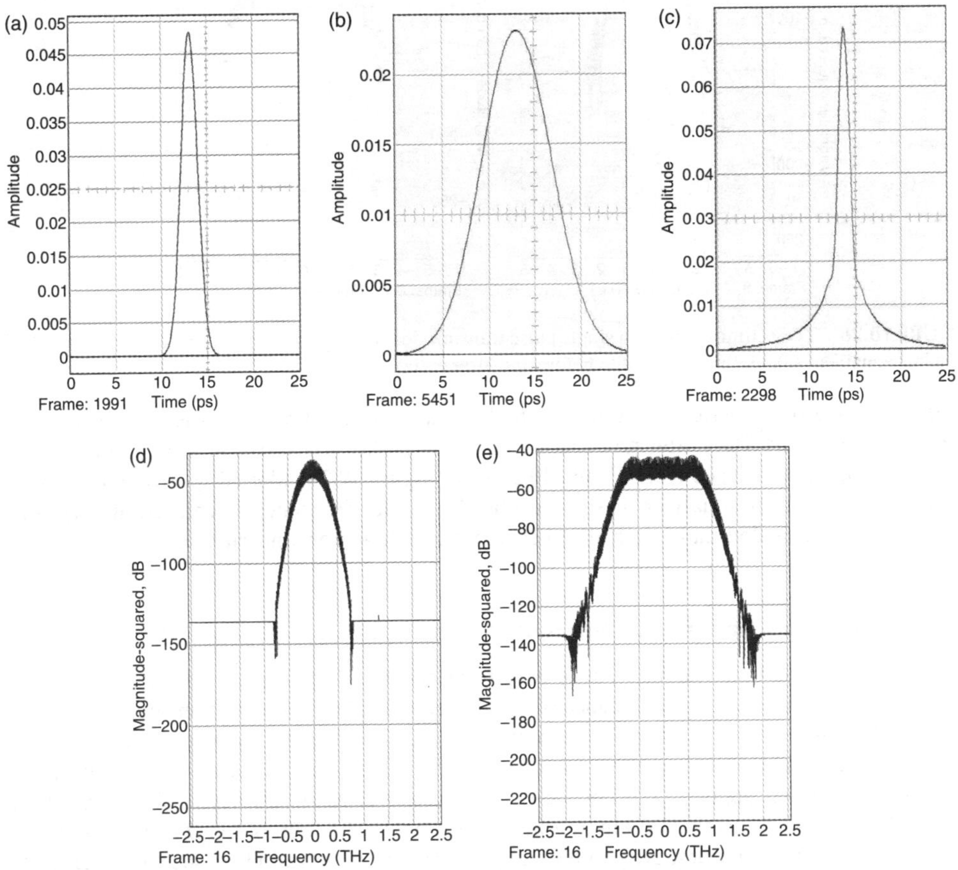

FIGURE 10.27 GVD elimination for a single pulse transmission (a) At a transmitter, (b) After 300-m SMF, (c) After equalizer, (d), (e) Spectral profile of a single pulse before and after equalizer.

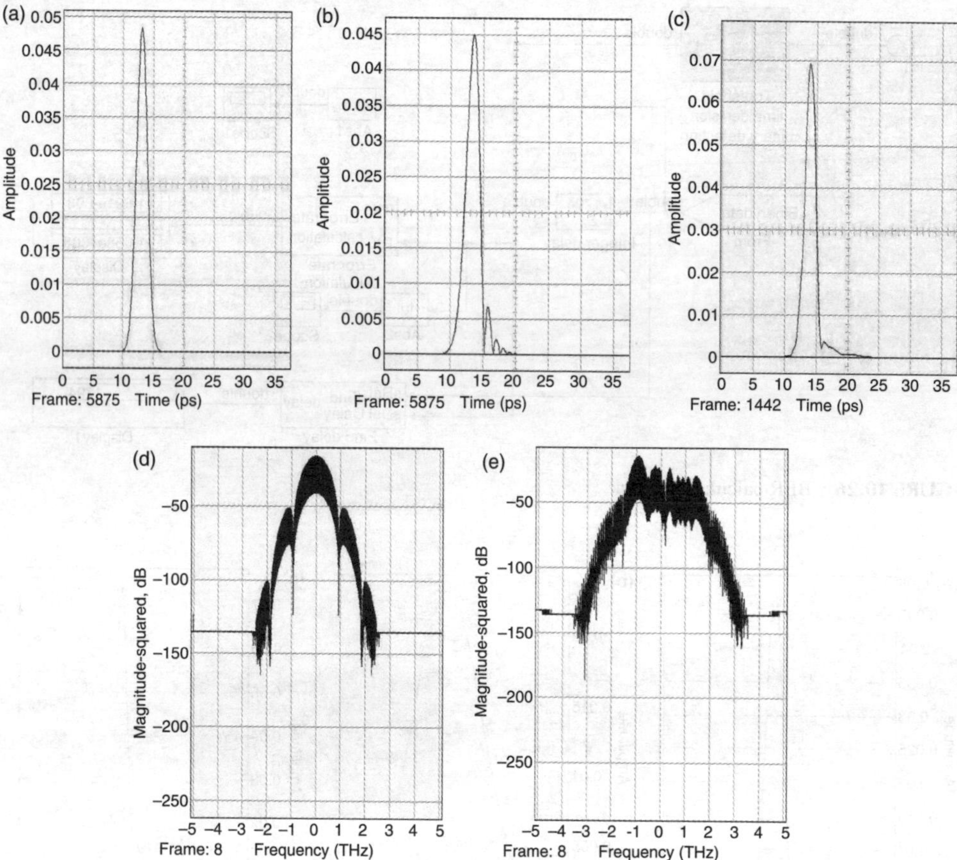

FIGURE 10.28   TOD elimination for a single pulse transmission (a) At a transmitter, (b) After 120-km SMF, (c) After equalizer, (d), (e) Spectral profile before and after equalizer.

an initial single pulse before being transmitted through optical fiber, distorted pulse which is affected critically by TOD after propagating through 120-km transmission link, and recovered pulse after equalizer, respectively. Figure 10.28d and Figure 10.28e plotted spectra of this single pulse before and after equalizer. It showed that the carrier frequency is chirped at very high chirping rate that is dominant over all other effects, such as non-linearity, second order, third order, jitter or PMD effects.

10.4.3.2   160-Gb/s Transmission and Equalization

Ultra-high-speed communication systems have been becoming increasingly popular because of the increase in the demand for high-capacity optical networks. Furthermore, transmission distance for this ultra-fast transmitted signal is improving and developing. In order to achieve a longer transmission link with high-speed transmission, different modulation formats such as differential phase-shift keying, is applied [37]. For the OTDM system using on-off keying format has been extended up to approximately 600-km transmission distance. The limit of the OTDM technique is due to its sensitivity properties, even with small perturbations in optical fibers such as GVD, TOD, PMD and timing jitter can distort signals dramatically [38,39]. In this section, some significant results are given for 120-km transmission link with BER characteristic. Transmission length is a parameter for evaluation of the equalized system and eye diagram is illustrated for the received power of about –32 dBm.

10.4.3.2.1 GVD Equalization of 120-km SSMF Transmission

Simulation results for GVD equalization are shown in Figure 10.30 for three situations: back-to-back (transmitter and receiver are connected directly together without transmission link), without equalizer and with equalizer. Then these data are plotted in Figure 10.29 with a GVD value of 1.28 ps/nm. TOD, non-linear effect, and timing jitter are set to zero in order to investigate GVD effect only. In this simulation setup, a 97.84-km standard SMF fiber with dispersion parameter $D = 17$ ps/(nm km) is used as transmission link. In order to have a second-order dispersion compensation, a 22.16-km DCF fiber length is used with dispersion parameter $D = -75$ ps/(nm km). Therefore, GVD value of 1.28 ps/nm is the mismatched dispersion.

With a low value of received power, there is an improvement of about 1/2 dBm between with and without an equalizer at a log(BER) value of -5. However, this difference is increased when the received power is increased. For example, at a log(BER) value of -8, this improvement is >1 dBm. This is further improved as a function of the receiver sensitivity. Thus using equalizer could improve long-haul transmission performance, especially at higher received power (or at higher signal-to-noise ratio).

Back-to-back BER characteristic is also plotted in Figure 10.29. Although the improvement of the EOP for using equalizer is still approximately half 1-dBm worse than back-to-back system, it still shows a improvement when the equalizer is used. This small residual penalty is due to the approximation of using sinusoidal waveform instead of an ideal parabolic waveform. If the received power is increased further, there would be a significant BER improvement and saturation at about -32 dBm. Thus, the error floor would appear following a diamond-trend at higher received power value.

Figure 10.30 shows eye diagrams corresponding to the individual curve of the BER transmission performance. For example, Figure 10.30a1 shows an eye diagram before de-multiplexing, Figure 10.30a2 and Figure 10.30a3 show eye diagrams after de-multiplexing at received power values of -36.375 dBm and -36.745 dBm without and with equalizer, respectively. Before de-multiplexing input signals, 160-Gb/s rate signals as still observed. The middle and right-hand-side eye diagrams are de-multiplexed 10-Gb/s signals at receiver end without and with equalizer, respectively.

Through the eye diagrams obtained for different received power levels, it is observed that the higher the received power, the greater the eye opening. The FWHM of distorted eye is about 3 ps, while the FWHM of recovered eye is only approximately 2 ps at about -32 dBm received power.

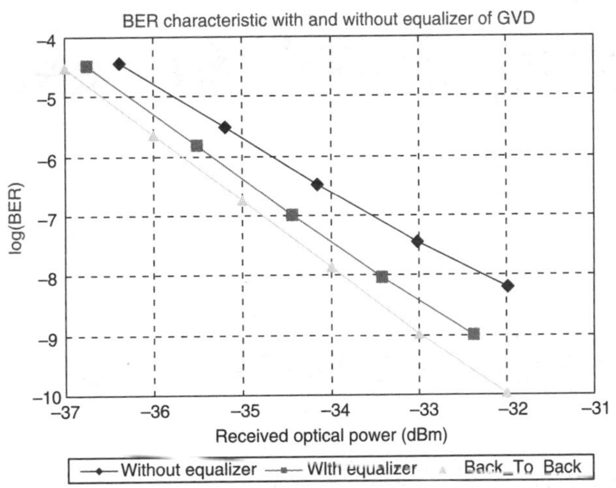

FIGURE 10.29 BER characteristic of 10-Gb/s de-multiplexing signal in GVD elimination.

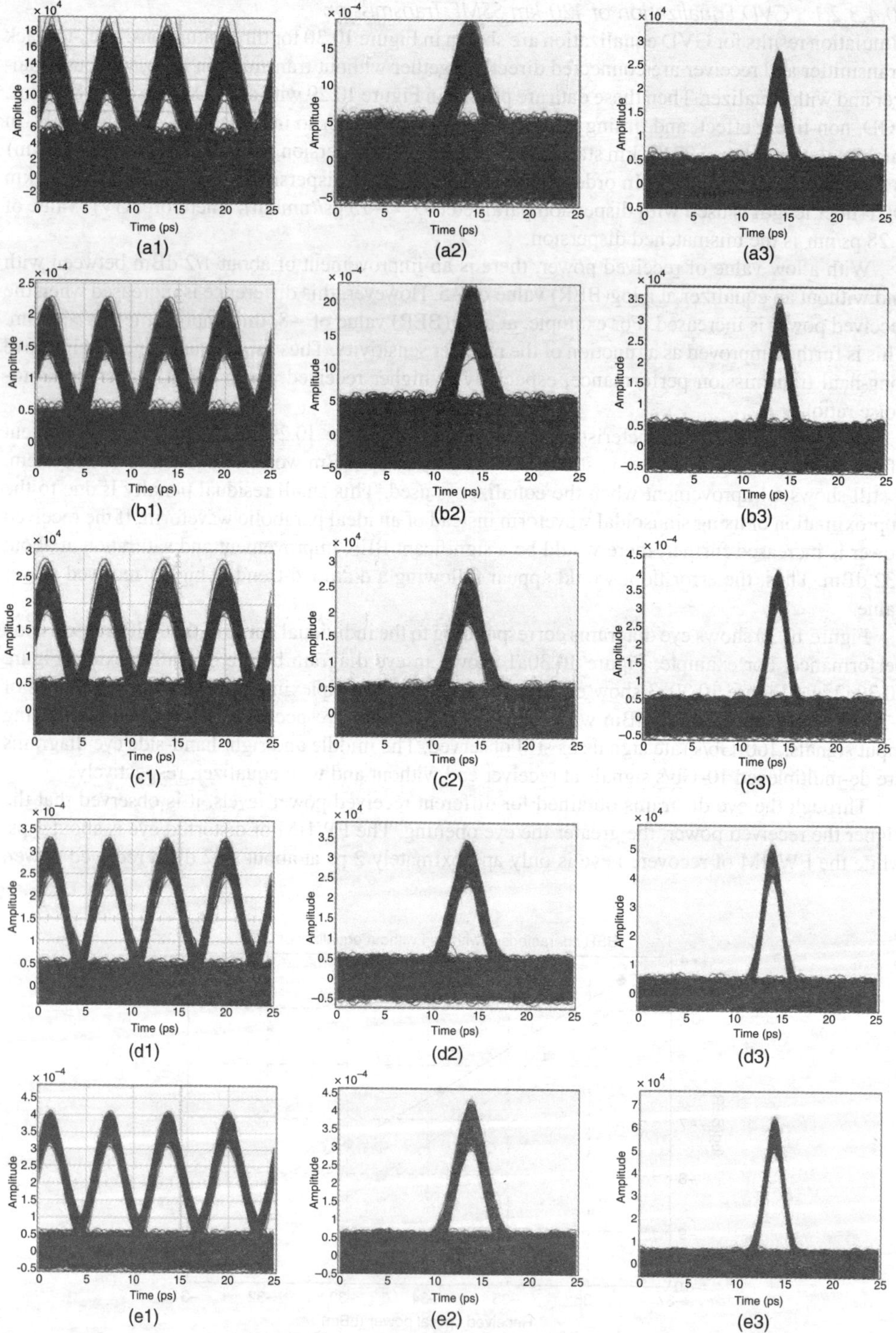

FIGURE 10.30 Eye diagrams for GVD with ranges of received power for three cases (from left to right): before de-multiplexing, after de-multiplexing without and with equalizer.

Thus, there is a significant improvement when equalizer is inserted, especially in long-haul transmission under accumulation of mismatched dispersion.

10.4.3.2.2 Third-Order Dispersion Equalization of 120 km SSMF Transmission

For TOD elimination, second-order dispersion is assumed to be fully compensated by DCF fiber. A 104.7-km standard SMF and a 15.3-km DCF fiber with dispersion slope values of $S = 0.06$ ps/(nm^{2}km) and 15.3 ps/(nm^{2}km), respectively, are used in each span of transmission system. Thus, the dispersion slope TOD is calculated with a value of about 1.692 ps/nm^2.

Table 10.8 and Figure 10.31 show the simulation results, and BER characteristic of TOD equalization scheme. Although the TOD effect for 120 km is not very significant compared with GVD effect as discussed in the previous section, equalizer still shows its usefulness in improving transmission performance. However, this BER characteristic when using equalizer still could not get closer to the back-to-back line (triangle-line). Error floor is possible due to the diamond-line for non-equalization at higher received power, while a square-line for using equalizer system still drops steeply. Consequently, the improvement between using and not using equalizer would be much greater than 1 dBm (as at about $\log(\text{BER}) = -8$ in Figure 10.31).

TABLE 10.8

BER Characteristic Summary of TOD Elimination for 160-Gb/s System for Back-to-Back, with and without Equalizer

Back-to-Back		Without Equalizer		With Equalizer	
Received Power (dBm)	log(BER)	Received Power (dBm)	log(BER)	Received Power (dBm)	log(BER)
−37	−4.521	−36.366	−4.673	−36.703	−4.526
−36	−5.647	−35.375	−5.521	−35.616	−5.708
−35	−6.727	−34.363	−6.570	−34.597	−6.820
−34	−7.853	−33.202	−7.489	−33.468	−7.984
−33	−8.972	−32.134	−8.296	−32.413	−9.056
−32	−9.987				

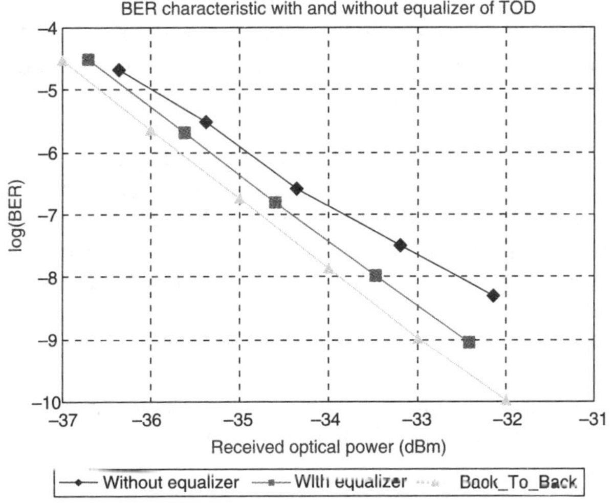

FIGURE 10.31 BER characteristic of 10-Gb/s de-multiplexing signal in TOD elimination.

In conclusion, equalizer offers significant improvement of the transmission performance in long-haul optical transmission. Furthermore, when signal to noise ratio is improved, there is a significant improvement between using and not using equalizer system. Some eye diagrams are also plotted in Figure 10.32.

Figure 10.32a1, Figure 10.32a2 and Figure 10.32a3 show eye diagrams for 160-Gb/s signals before de-multiplexed, 10-Gb/s signals after de-multiplexed not using equalizer, and 10-Gb/s signals after de-multiplexed using equalizer, respectively. From (a) to (e), eye diagrams for each received power values from simulation are shown. There is an improvement when a higher power is received at the receiver end. Furthermore, with the same received power values, eye diagrams in the TOD case are a bit bigger than eye diagrams in the GVD case. Therefore, a received BER from simulation also showed lower BER values in the TOD case for case using equalizer. Although BER of system remained pretty much the same as the GVD elimination above due to the approximation of sinusoidal applied voltage, it still confirmed its usefulness in compensating second-order and third-order effects (may also be timing jitter and PMD during transmission) through Figure 10.30 and Figure 10.32. Furthermore, the FWHM of distorted eye is about 2.8 ps, while FWHM of recovered eye is approximately 2.2 ps at received power value of about −32 dBm.

10.4.3.2.3 *Equalization of Third-Order Dispersion with Variable Fiber Lengths*

TOD critically contributes to the pulse distortion in a ultra-high-speed system in long-haul transmission. Unlike the effect of the GVD that could be fully compensated by using DCF fiber, it is very difficult to compensate TOD fully by using existing DCF. Therefor there is always distortion due to the mismatch between SMF and DCF. To compensate TOD, the equalizer offers some additional equalization potential that may be necessary at the end of transmission system to compensate the effects of the TOD.

Five pairs of eye diagrams of Figure 10.33 show the recovered signals after transmission through 240 km (two spans), 480 km (four spans), 720 km (six spans), 960 km (eight spans) and 1200 km (ten spans) optical links, respectively, at a received power value of approximately −32 dBm. All set parameters for equalizer system remain the same and inserted at the end of transmission link. Erbium-doped fiber with gain value of 30 dB an 5 dB noise value are used between each 120-km transmission span to compensate for the optical losses of transmitted signals due to attenuation. Figure 10.33a1–a5 indicated that TOD was accumulated over long transmission length and the received signals would be distorted rapidly. For two spans of transmission link, received signals have not been distorted too much, and the FWHM of an eye is measured to be about 2.6 ps. When a transmission link is increased up to 4 spans (480 km), third-order dispersive effect is seen clearly. But the eye is still widely opening and FWHM is about 3 ps at this time. On the other hand, when it is more than six spans, greater than 720-km fiber length, received signals are distorted dramatically, especially at their tails. These long oscillation tails would substantially affect neighboring pulses during transmission.

Figure 10.33b1–b5 show the TOD equalization. The eye diagrams of five situations from two spans to ten spans are also distorted since more than six spans are used in transmission link, but these distortions are not too much and the eye windows of these eye diagrams are still opening wide enough to have a reasonable BER. Furthermore, FWHM of an eye when two spans are used is about 2.2 ps, while this FWHM is approximately 3 ps after transmitting through ten spans (1200-km fiber length). The received results after using equalizer are reasonable and there is a huge improvement compared with signals before using equalizer.

Further investigations might be done by running simulation to count number of errors and calculated BER of this transmitted system for variable transmission length afterwards. This would give correct and exact estimations and calculations how much received signals could be improved by using the equalizer.

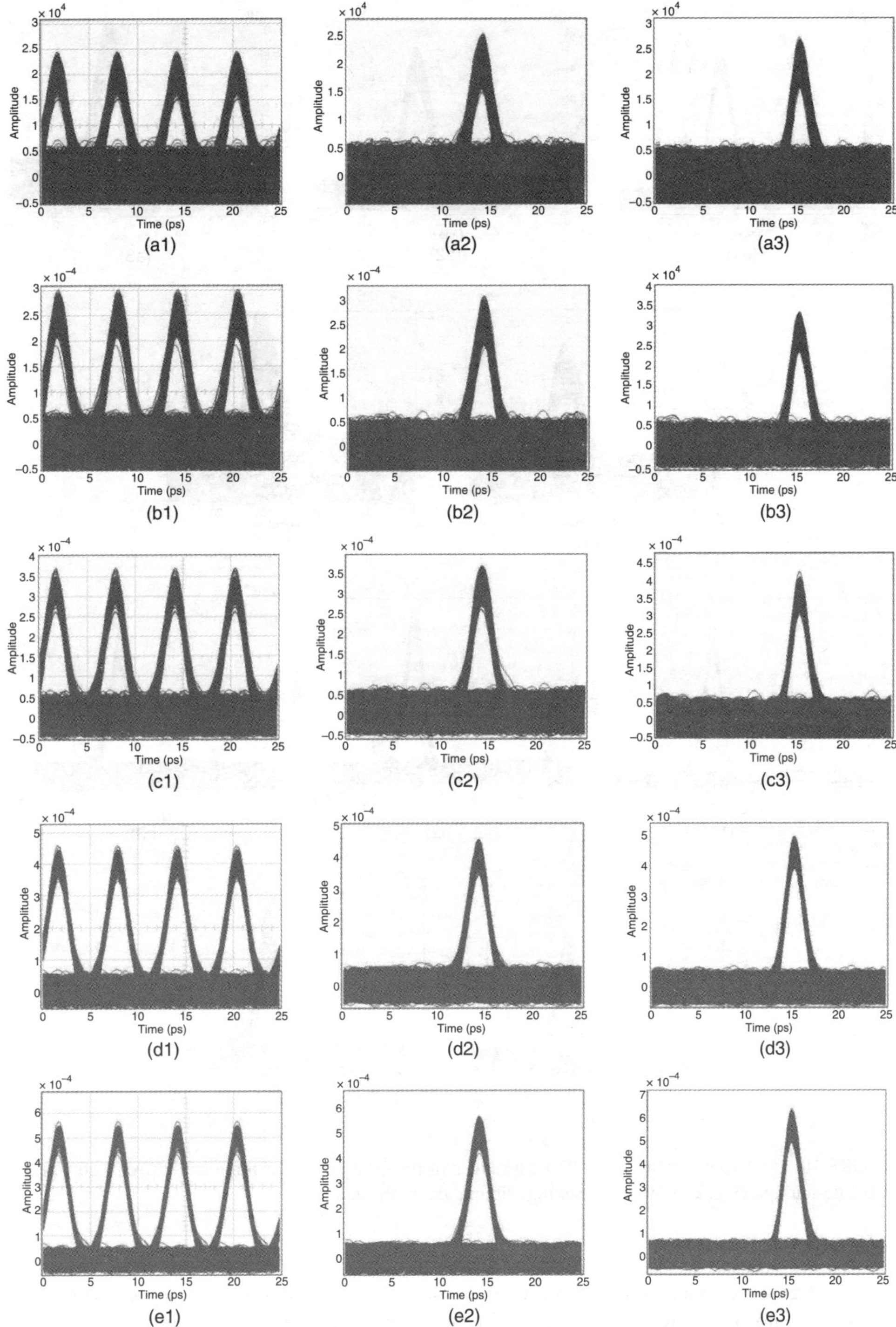

FIGURE 10.32 Eye diagrams for 160-Gb/s signals before de-multiplexing (left column) and after de-multiplexing with and without equalization (center and right columns).

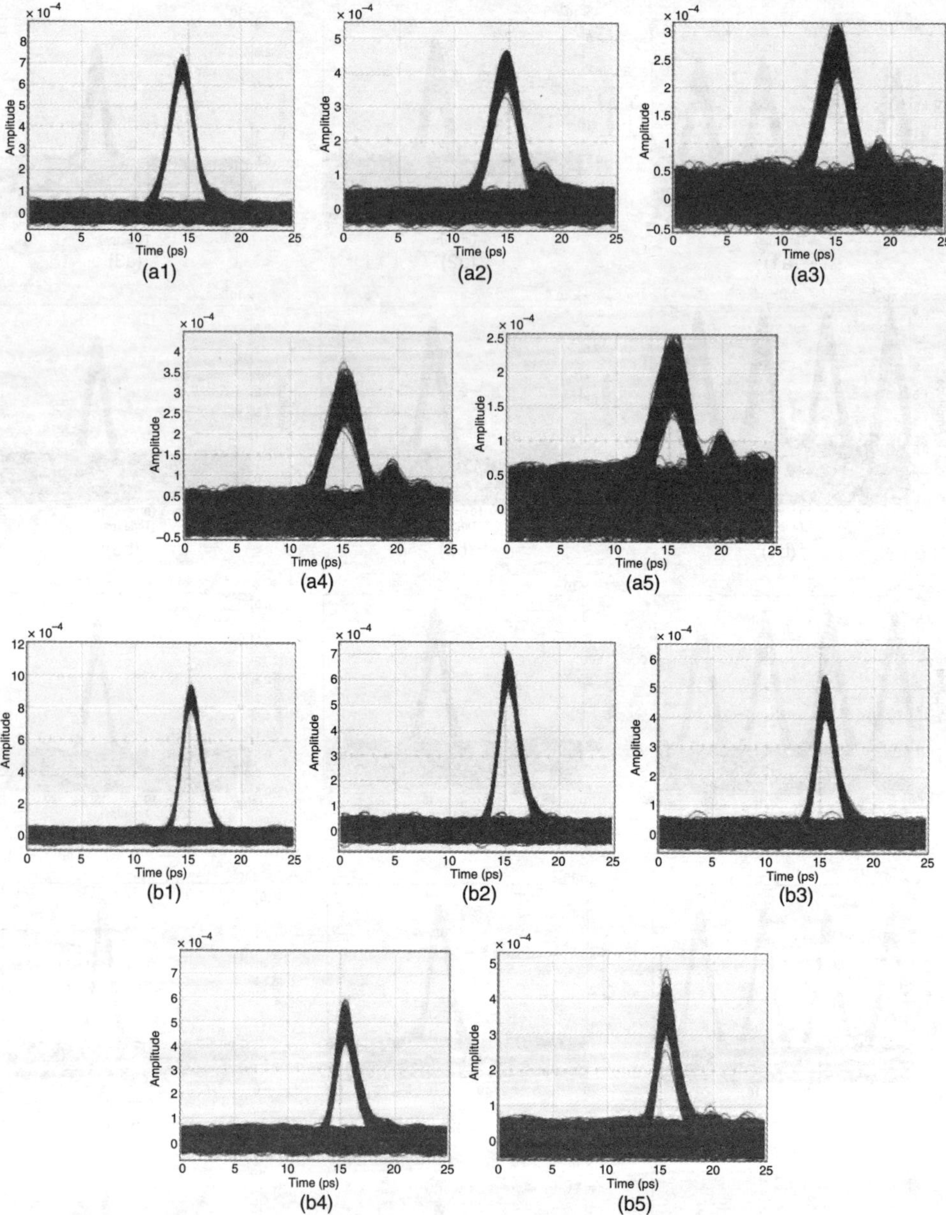

FIGURE 10.33 Eye diagrams for TOD with ranges of received power for three cases (from left to right): before de-multiplexing, after de-multiplexing without and with equalizer.

10.5 CONCLUSIONS

Photonic signal processing based on space–time duality is very important in ultra-high-speed transmission systems. In this report, foundation for the principles of temporal imaging and space–time duality have been studied and proven. The equivalence between paraxial diffraction in space and narrow-band dispersion in time can be used to construct an analogous approximation of the relative bandwidths of spatial and temporal Fourier spectra. By using the diffraction–dispersion analogy, a quadratic phase modulation acted as the time lens, and the focal length of the spatial and time lens had already been derived. The duality properties between the spatial and time domain has led to temporal imaging that would be very useful in future work, especially for designing adaptive equalizer by using phase modulation to expand or compress pulses in an ultra-high-speed communication system. By analyzing the impulse response of the temporal imaging system, it revealed the resolution which is related to the Fourier transform of the aperture function, and thus is the time-domain equivalent of the Fraunhofer diffraction pattern [2]. It is a very important result and solidified the nature of the space–time duality.

With respect to photonic signal processing, a single pulse transmission has been studied by propagating through SSMF with GVD and TOD effects without an initial chirp factor. The knowledge from simulation results for single pulse is then used for the design of the dispersion-compensating scheme that could eliminate not only second-order dispersion, but also third-order dispersion, polarization-mode dispersion and timing jitter.

The space–time duality and temporal imaging are studied and then applied in the design of an adaptive equalizer system that includes a phase modulator followed by a dispersive element that could recover the initial pulse shape from a distorted pulse. An efficient adaptive equalizer acted as a time lens in the time domain for optical pulse width compression by using a phase modulator and a dispersive element proposed, analyzed and proven by simulation results. By using the equalizer system, the GVD effect is fully compensated. Furthermore, TOD, PMD and timing jitter are also reduced significantly in ultra-high-speed transmission system using an parabolic-like sinusoidal voltage waveform in the driving of the phase modulator. It has also shown that an ideal parabolic phase modulation could fully recover original pulse.

Furthermore, this adaptive equalizer has also been applied in ultra-high speed transmission and equalization at 160 Gb/s. Higher bit rate such as 320 Gb/s or even higher might be applicable and would be investigated at a later stage. Evaluation of the improvement with and without using an equalizer is also made. By monitoring eye diagrams at the receiver end of this system, the signal eye diagram under non-equalization is significantly distorted, thus much higher BER is obtained in the transmission. On the other hand, when an adaptive equalizer is inserted, there is a critical improvement in BER as well as a wide opening eye in an eye diagram. The transmission length is increased up to 1200 km in order to investigate this limitation. We have shown agreement between theoretical and simulation results of the performance of the equalization technique for 160-Gb/s optical transmission system under the influence of GVD, TOD, PMD impairment and the elimination of the timing jitter not only with a single pulse, but also with 160 Gb/s random sequence data transmission. The non-linear phase shift and noises in the system might challenge the system stability and this is the topical development of the next phase.

REFERENCES

1. Lifante, G. 2003. *Integrated photonics: Fundamentals*. West Sussex: John Wiley & Sons.
2. Kolner, B. H. 1994. Space-time duality and the theory of temporal imaging. *IEEE Journal of Quantum Electronics* 30 (8): 1951–63.
3. Bennett, C. V., and B. H. Kolner. 2000. Priciples of parametric temporal imaging – part I: System configurations. *IEEE Journal of Quantum Electronics* 36 (4): 430–37.
4. Bennett, C. V., and B. H. Kolner. 2000. Principles of parametric temporal imaging – part II: System performance. *IEEE Journal of Quantum Electronics* 36 (6): 649–55.
5. Papoulis, A. 1968. *Systems and transforms with applications in optics*. New York: McGraw-Hill.
6. Adams, A. R., and Y. Suematsu. 1994. *Handbook of semiconductor lasers and photonic nitegrated circuits*. New York: Chapman and Hall.
7. Agrawal, G. P. 2001. *Nonlinear fiber optics*. 3rd ed. CA, USA: Academic Press.
8. Khayim, T., M. Yamauchi, D. S. Kim, and T. Kobayashi. 1999. Femtosecond optical pulse generation from a CW laser using an electroopitc phase modulator featuring lens modulation. *IEEE Journal of Quantum Electronics* 35 (10): 1412–18.

9. Yan, L. S., Q. Yu, T. Luo, and X. S. Yao. 2002. Compensation of higher order polarization-mode dispersion using phase modulation and polarization control in the transmitter. *IEEE Photonics Technology Letters* 14 (6): 858–1012.

10. Treacy, E. B. 1969. Optical pulse compression with diffraction gratings. *IEEE Journal of Quantum Electronics* QE-5: 454–58.

11. Papoulis, A. 1994. Pulse compression, fiber communication, and diffraction: A unified approach. *Journal of Optical Society of America A* 11 (1): 3–13.

12. Hirooka, T., M. Nakazawa, F. Futami, and S. Watanabe. 2004. A new adaptive equalization scheme for a 160-Gb/s transmitted signal using time-domain optical fourier transformation. *IEEE Photonics Technology Letters* 16 (10): 2371–73.

13. Azana, J., and M. A. Muriel. 2000. Real-time optical spectrum analysis based on the time-space duality in chirped fiber gratings. *IEEE Journal of Quantum Electron.* 36 (5): 517–26.

14. Jannson, T. 1983. Real-time fourier transformation in dispersive optical fibers. *Optic Letters* 8 (4): 232–34.

15. Godil, A. A., B. A. Auld, and D. M. Bloom. 1994. Picosecond time-lenses. *IEEE Journal of Quantum Electron* 30 (3): 827–37.

16. Yu, F. T. S., and I. C. Khoo. 1990. *Principles of optical engineering.* New York: John Wiley & Sons.

17. Goodman, J. W. 1993. *Introduction to Fourier optic.* New York: McGraw Hill.

18. Kauffman, M. T., A. A. Godil, B. A. Auld, W. C. Banyai, and D. M. Bloom. 1993. Applications of time lens optical systems. *Electronics Letters* 29 (3): 268–69.

19. Kolner, B. H., and M. Nazarathy. 1989. Temporal imaging with a time lens. *Optic Letters* 14 (12): 630–32.

20. Kolner, B. H., and M. Nazarathy. 1990. Temporal imaging with a time lens: Erratum. *Optic Letters* 15 (11): 655.

21. Bennet, C. V., R. P. Scott, and B. H. Kolner. 1994. Temporal magnification and reversal of 100 Gb/s optical data with an un-conversion time micro-scope. *Applied Physics Letters* 65 (20): 2513–15.

22. Jannson, T., and J. Jannson. 1981. Temporal self-imaging effect in single-mode fibers. *Journal of Optical Society of America* 71 (11): 1373–76.

23. Saleh, B. E. A., and M. I. Irshid. 1982. Collett-Wolf equivalence theorem and propagation of a pulse in a signel mode optical fiber. *Optic Letters* 7 (7): 342–43.

24. Yamamoto, T., and M. Nakazawa. 2001. Third- and fourth-order active dispersion compensation with a phase modulator in a terabit-per-second optical time-division multiplexed transmission. *Optic Letters* 11 (9): 647–49.

25. Elrefaie, A. F., R. E. Wagner, D. A. Atlas, and D. G. Daut. 1988. Chromatic dispersion limitations in coherent lightwave transmission systems. *Journal of Lightwave Technology* 6 (5): 704–9.

26. Li, X., X. Chen, and M. Qasmi. 2005. A broad-band digital filtering approach for time-domain simulation of pulse propagation in optical fiber. *Journal of Lightwave Technology* 23 (2): 864–75.

27. Agrawal, G. P. 2002. *Fiber-optic communication systems.* 3rd ed. New York: John Wiley & Sons.

28. Romagnoli, M., P. Franco, R. Corsini, A. Schiffini, and M. Midrio. 1999. Time-domain Fourier optics for polarization-mode dispersion. *Optic Letters* 24 (17): 1197–99.

29. Howe, J. V., and C. Xu. 2006. Ultrafast optical signal processing based upon space-time dualities. *Journal of Lightwave Technology* 24 (7): 2649–62.

30. Jiang, L. A., M. E. Grein, H. A. Haus, E. P. Ippen, and H. Yokoyama. 2003. Timing Jitter eater for optical pulse trains. *Optic Letters* 28: 78–80.

31. Hellstrom, E., H. Sunnerud, M. Westlund, and M. Karlsson. 2003. Third-order dispersion compensation using a phase modulator. *Journal of Lightwave Technology* 21 (5): 1188–97.

32. Nakazawa, M., E. Yoshida, and Y. Kimura. 1994. Ultrastable harmonically and regeneratively mod-elocked polarization-maintaining erbium fiber ring laser. *Electronics Letters* 30: 1603–4.

33. Hirooka, T., and M. Nakazawa. 2006. Optical adaptive equalization of high-speed signals using time-domain optical Fourier transformation. *Journal of Lightwave Technology* 24 (7): 2530–40.

34. Nakazawa, M., T. Yamamoto, and K. R. Tamura. 2000. 1.28Tbits/s-70km OTDM transmission using thrid- and fourth-order simultaneous dispersion compensation with a phase modulator. *Electronics Letters* 36 (24): 2027–29.

35. Kawanishi, S., H. Takara, T. Morioka, O. Kamatani, K. Takiguchi, T. Kitoh, and M. Saruwatari. 1996. Single channel 400Gbit/s time-division-multiplexed transmission of 0.98 ps pulses over 40 km employing dispersion slope compensation. *Electronics Letters* 32: 916–18.

36. Diez, S., R. Ludwig, and H. G. Wener. 1998. All-optical switch for TDM and WDM/TDM systems demonstrated in a 640 Gbit/s demultiplexing experiment. *Electronics Letters* 34: 803–5.
37. Daikoku, M., T. Miyazaki, I. Morita, H. Tanaka, F. Kubota, and M. Suzuki. 2005. 160 Gbit/s-based field transmission experiments with singlepolarization RZ-DPSK signals and simple PMD compensator. In *European conference optical communication (ECOC 2005), Paper We2.2.1,* Sep. 2005, Glasgow, Scotland.
38. Lehmann, G., W. Schairer, H. Rohde, E. Sikora, Y. R. Zhou, A. Lord, D. Payne, et al. 2004. 160 Gbit/s OTDM transmission field trial over 550 km of legacy SSMF. In *European conference optical communication (ECOC 2004), Paper We1.5.2,* Sep. 2004, Stockholm, Sweden.
39. Hirooka, T., K. I. Hagiuda, T. Kumakura, K. Osawa, and M. Nakazawa. 2006. 160-Gb/s–600-km OTDM transmission using time-domain optical Fourier transformation. *IEEE Photonics Technology Letters* 18 (24): 2647–49.

36. Otani, T., K. Ludwig, and H. G. Weber. 1998. All-optical demultiplexer for TDM and WDM/TDM systems demonstrated in a 640 Gbit/s demultiplexing experiment. *Electron. Lett.* 34: 60–[illegible].

37. Bilenca, M., T. Nuccio, F. Morichetti, H. Banks, H. Kuboto, and A. [illegible]. 2005. 160 Gbit/s-based field demonstration experiments with shelf polarization [illegible] DPSK signals and sample PMD compensator. In *European Conference on Optical Communication (ECOC 2005)*. Paper We2.2.1, Sep. 2005, Glasgow, Scotland.

38. Fukuchi, O., W. Schairer, H. Rohde, [illegible] V. R. Zhou, A. Lord, D. Payne, et al. 2004. 160 Gbit/s OTDM transmission field trial over 550 km of legacy SSMF. In *European conference optical communication (ECOC 2004)*. Paper We1.5, Sep. 2004, Stockholm, Sweden.

39. Hirooka, K., T. Nagpua, T. Kamitani, K. Ogawa, and M. Nakazawa. 2006. 160 Gbit/s OTDM transmission using time-domain optical Fourier transformation. *IEEE Photonics Technology Letters* 18 [illegible] 2006-44.

11 Electronic Signal Processing in Optical Transmission Systems

11.1 INTRODUCTION

With the processing of digital signal processors and the decrease of the cost of the electronic processing systems, electronic signal processing is becoming very attractive to mitigate various impairments that severely affect optical transmission. Equalization and compensation of the impairment effects of the transmission of optical signals through optically amplified systems have been of principal concern to transmission engineering. Compensation of linear and non-linear dispersion represents a typical and important application of the processing. Over the years the compensation has been implemented in long-haul transmission with the insertion of dispersion compensating fibers (DCF) at the end of a transmission span (e.g., standard SMF or LEA etc.) or an FBG with phase fluctuation. These compensating devices would normally suffer appreciable attenuation and hence an additional optical amplifier must be used to compensate for these lost factors. Thus additional optical amplifiers are necessary for long-haul transmission systems. Unavoidable accumulated noises of the ASE noises sources from these amplifiers will shorten the transmission distance. Furthermore, the narrow core of a DCF renders it more susceptible to fiber non-linearity effects. The DCF is also polarization sensitive.

On the other hand, electrical equalization can offer the advantages of more flexibility, a lower cost, and a smaller size through integration within the transceiver electronics, especially those that are programmable using digital signal processing. Thus if one could compensate or equalize the pulse sequence in the electrical domain either by pre-distortion or post-compensation or a combination of both processes, then the elimination of these amplifying processes can be reduced and thus longer transmission reached.

Furthermore optical networks with different transmission paths carrying lightpaths of different distance due to different number of hops would suffer dispersion and non-linear effects due to the nature of ultra-high speed at 10 Gb/s or higher. The 10 Gb/s Ethernet will appear in networks in the near future and there is an urgent demand that inexpensive techniques should be available for network implementation. These techniques must be very adaptive and tunable with control signals. Electronic equalization can offer these solutions for high speed optical networks. Table 11.1 shows the penalty of the eye opening with respect to the bit rate and the length of SSMF when no dispersion compensation is used. This dispersion tolerance becomes very critical when the bit rate is increased. Thus the implementation of equalization is very important in the electronic domain, the digital signal processing, so that it can be tuned or controlled and adapted to the number of hops of the lightwave channels travelling over the network nodes. These digital signal processing systems for equalization would enhance the received opening eye diagram. The eye opening penalty of some advanced modulation formats due to the distortion by the chromatic dispersion is shown in Table 11.1.

Naturally the dispersion-limited distance can be improved using optical pre-distortion techniques such as the control of the chirp of the laser to 200 km SSMF [1] for 10 Gb/s or the technique of dispersion-supported transmission to 250 km SSMF [2]. But these techniques require high complexity of the optical transmitters.

For conventional direct-detection receivers, the linear distortion that is induced by chromatic dispersion in the optical domain is transformed into a non-linear distortion in the electrical signal,

TABLE 11.1

Eye Opening Penalty, Dispersion Tolerance and Equivalent Length of SSM for Different Bit Rate

Modulation Technique	Direct modulation (DM) External modulation (EM)	2 dB Penalty ps/nm/km (SSMF)	
DML (2.5 Gb/s)	DM	1000–2000	60–120
DML (10 Gb/s) (eye opening and distortion)	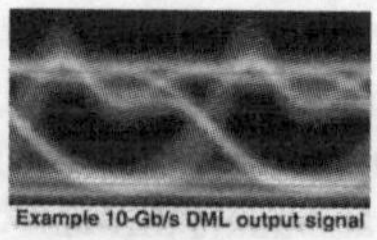	50–200	3–12
EML (10 Gb/s)	EM	400–2000	25–120

Example 10-Gb/s DML output signal

which explains why only limited performance improvements can be achieved by using a linear baseband equalizer with only one baseband received signal. This also explains why non-linear techniques, such as decision feedback equalization (DFE) and maximum-likelihood sequence detection (MLSE), are more effective in combating chromatic dispersion in direct-detection receivers. Strictly speaking the MLSE system attempts neither to compensate the chromatic dispersion effects nor to equalize them, but duly account for these effects in the data detection process.

On the other hand, in a coherent detection system, chromatic dispersion is linear in the electrical signal at the receiver, which is why fractionally spaced equalizers with "complex coefficients" can achieve a performance that is limited only by the number of taps used within the equalizer accounting only for the fiber chromatic dispersion. Recently with the availability of narrow linewidth and tunable lasers and ultra-high speed receivers, coherent lightwave systems have attracted significant attention. However they have not reached the practical and installation stage so far. This is attributed mainly to the success of wavelength-division multiplexing (WDM) technology with the advent of erbium-doped fiber amplifiers (EDFAs) in the late 1990s. Another reason is the complexity of coherent transmitters and receivers.

The detrimental effect of chromatic dispersion in SSMF can also be reduced by using reduced-bandwidth modulation formats such as optical duobinary signals, single sideband or vestigial sideband, which can extend the reach of 10 Gb/s systems to distances in excess of 200 km, compared with approximately 80 km for conventional NRZ OOK. In this chapter, optical duobinary signalling is combined with linear electrical pre-equalization scheme [3] to demonstrate the extension of the reach of a 10-Gb/s system, to several hundred kilometers without any optical chromatic dispersion compensation. In fact, the system reach is influenced by other fiber impairments other than chromatic dispersion. The equalization is based on exploiting the linearity of a coherent lightwave system to move the equalization process to the transmitter, where the data, especially the carrier phase is still undisturbed. More specifically, the duobinary signal is pre-equalized using two tunable T/2-spaced finite-impulse response (FIR) filters. The outputs of the FIR filters then modulate two optical carriers that are in phase quadrature. Thus, the advantages of coherent receiver equalization are still maintained while utilizing a conventional non-coherent direct-detection receiver. Although this chapter focuses on duobinary modulation, this choice is principally driven by the reduced-bandwidth advantage, compared with conventional NRZ OOK modulation, and the ability to use a conventional direct-detection receiver. In principle, electrical pre-equalization can also be applied to other modulation formats for potential improvements.

Duobinary offers the most effective mitigation of the linear dispersion due to its single sideband property and direct detection at the receiver. Duobinary signalling can also be combined with a proposed electrical pre-equalization scheme to extend the reach of 10-Gb/s signals that are transmitted over standard single-mode fiber. The proposed scheme is based on pre-distorting the duobinary signal using two T/2-spaced finite-impulse response (FIR) filters. The outputs of the FIR filters then modulate two optical carriers that are then applied to the two parallel MZIM to generate duobinary optical signals. This equalization and duobinary modulation formats for extending the transmission reach are demonstrated in the first section of this chapter with an adaptation of the algorithm given by El Said et al. [3].

Another equalization system that would be implemented in the electrical domain are the MLSE equalizers. They have recently received much interest as an effective solution to overcome the severe distortion caused by inter-symbol interference (ISI) of the optical channels and to improve the signal to noise ratio performance, thus extending the reach of the optical transmission systems and improving signal to noise ratio performance. Most of the studies on the use of non-linear MLSE equalizers have concentrated on amplitude or differential phase modulation formats [4–6]. However there are not many reports on the performance of MLSE equalizers for optical transmission systems employing minimum shift keying modulation (MSK) format.

A narrowband filter receiver with a breakthrough dispersion tolerance limits for the detection of non-coherent 40 Gb/s optical MSK transmission system has been proposed and explained in detail in Chapter 8. The performance of this receiver was found to be limited by ISI caused by the differential group delay of the optical channel and the narrow bandwidth of the optical detection filter. In order to combat the ISI, in this chapter, two non-linear equalizers are proposed and integrated with the narrowband filter receiver for detection of non-coherent 40 Gb/s optical MSK signals. The performance of the proposed schemes are numerically evaluated. The proposed schemes extend the reach of dispersion tolerance of the proposed narrowband optical MSK receiver to ± 952 ps/nm and ± 884 ps/nm at BER $= 1e-9$ or equivalently 52 km and 56 km of SSMF with required OSNR of 19 dB and 23.5 dB, respectively. To the best of the authors knowledge, these dispersion tolerance limits have not been reached before.

Recently, MLSE equalizer for optical transmission systems employing minimum shift keying modulation (MSK) format has been reported by Alic et al. [4]. The uncompensated distance can equivalently reach up to approximately 960 km standard single mode fiber (SSMF) for 10 Gb/s transmission. The results are comparable to the distance of 1040 km SSMF as reported by Haunstein [5] for IMDD systems. In Haunstein's study, up to 8192 trellis states are used [5]. Longer reach for IMDD systems require a very high number of states which is not even feasible in simulation. Due to large bandwidth optical filters being used in IMDD systems, noise is not greatly suppressed, leading to high values of required optical signal to noise (OSNR) ratios. In addition, noise distribution in IMDD systems no longer follows a Gaussian profile, which increases the complexity of the equalizer in order to estimate the noise distribution for optimal performance of the Viterbi algorithm. These issues can be effectively mitigated with the introduction of narrowband optical filtering in an optical MSK transmission system.

In this section, we numerically described the possibility of transmission over 1472 km SSMF optical link without inline dispersion compensation for 10 Gb/s optical MSK signals. The achievement is enabled by the integration of 1024-state Viterbi-MLSE equalizer with the non-coherent narrowband frequency discrimination receiver. The proposed receiver mitigates the fiber inter-symbol interference (ISI) effectively and greatly reduces the noise floor, thus enabling low OSNR for receiver sensitivity. The simulation results also show significant OSNR improvements with two and four samples per bit over the conventional one sample per bit. Finally, MLSE equalizer in our proposed scheme can operate optimally due to Gaussian noise distribution [6]. Thus, the complexity of the MLSE equalizer is reduced compared to the IM/DD systems.

This chapter is organized as follows: The next two sections give a general overview and detailed schemes of equalization whereby the equalizers are placed at the receiver, the transmitter and a share between transmitter and receiver. Essential expressions of the sampled impulse responses of the equalizer sub-systems, the FFE, DFE and combined non-linear and linear equalizers are summarized from Franceschini et al. [7]. The minimization of the square error in the FFE and DFE is also integrated to optimize the equalization process. The doubling sampling or partial response equalization is also included. Two special cases of equalization are described in detail. The duobinary and MSK modulation formats are described in Sections 5 and 6, 7 and 8 using double sampling and MLSE techniques, respectively. Section 9 then gives some degree of uncertainty in the equalization process and the gain in the eye opening penalty, the number of taps or length of the templates in MLSE schemes are obtained. Then finally concluding remarks are given.

11.2　ELECTRONIC DIGITAL PROCESSING EQUALIZATION

Electronic equalization can be implemented with the possibilities of (i) integration of digital signal processing (DSP) at the receive side, or post-equalization; (ii) placing the equalization DSP at the transmit side, or pre-distortion, of the optical signals by modification of the electrical signals driving the external modulator; or (iii) sharing the equalization function at both the receive-side and transmit-side, i.e using both pre-distortion and post-equalization.

Figure 11.1 shows the generic arrangement of the equalizers at both the transmit-side and receive-side. The equalization processes can occur by using the pre-distortion at the transmitter (Tx) only or post compensation at the receive-side (Rx) only or sharing between the pre-distortion and post compensation at both the Tx and Rx side. Figure 11.2 shows the equalization processing both in the optical and electronic domain at the receiver and transmitter sides. The optical equalization works with the field of lightwaves while the electronic processor works with the electrical signals obtained from the conversion of the optical intensity of the signals, except for the case of coherent reception, the recovered signals are directly related to the phase in the electrical signals. A typical layered optical network using IP over WDM is shown in Figure 11.5 in which the equalizer is used to equalize any distortion residual from the optical compensator, for example phase ripple of the phase group delay of the fiber Bragg grating (FBG) dispersion compensator, or different travelling hops of the light paths. It is noted that electronic equalizers do not suffer the problems of over dispersion as in the case of optical compensator, because the electrical type deals with the intensity while with optical, the field and over or under compensation occurs due to the variation of the residual dispersion in different lengths of hops of the lightpaths. Figure 11.6 also shows a typical light of a wavelength channel travelling through the network. The insert of this figure extends the dispersion tolerance with and without equalization. Optical equalizer can be shared by the channels of different wavelengths while electronic equalizer is normally integrated into a receiver for a specific channel.

When the equalizer is placed at the receiver then this post-equalization would take advantage of the knowledge of all distortion effects and deal with the signals dependent on the intensity and equalized signals can be achieved. If any unexpected distortion occurs this post equalization can be adaptive to equalize the pulse sequence. On the other hand, an equalizer placed at the transmit-side is processing the lightwave by creating a pre-distortion on the phase of the carrier. Note that the channel is purely phase distortion and thus the sequence can be electrically pre-distorted before driving the external modulator to chirp the phase in such a way that opposes the chirping by the group velocity of the optical fiber.

The post equalization processing usually requires a number of samples to ensure sufficient data for its equalization technique. The higher the number of samples the better the equalized sequence. However for long-haul optical transmission systems the pulse may be very dispersive and the samples may suffer significant distortion. In order to assist the post equalizer a pre-distortion compensator may be placed at the transmitter to share the equalization functionality with the post equalizer to extend the reach of the optical transmission system.

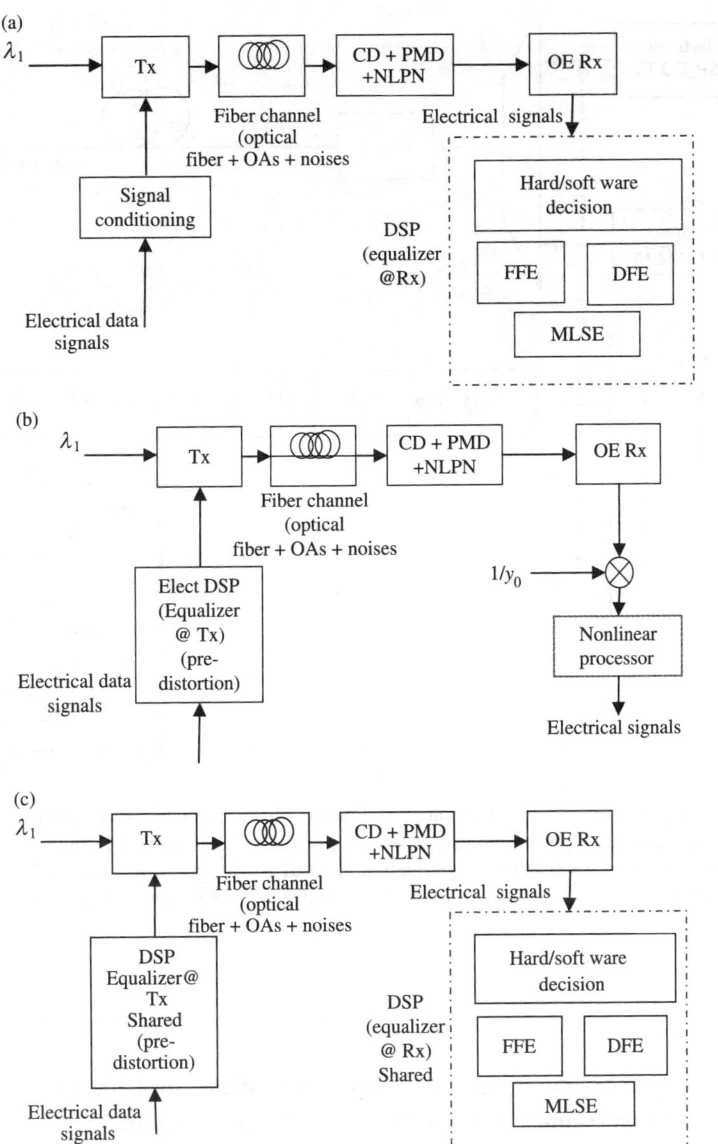

FIGURE 11.1 Arrangement of electronic digital signal processing for equalization in optical transmission systems. Electronic DSP for pre-distortion, electronic DSP at the Rx is for post-equalization and shared pre-distortion and post-equalization at the Tx and Rx. (a) Equalization at the transmit-side. (b) Equalization at the receive-side and (c) equalization shared between Tx and Rx sides.

The benefit of equalization does also depend on the modulation format as the algorithm must be adapted to the mode of modulation, for example amplitude, phase or frequency shift keying. Table 11.2 shows the improvement factors of the equalization techniques on different modulation formats using amplitude shift keying. The duobinary is a tri-level modulation format using both amplitude and phase to represent the levels. However its detection is intensity-based and thus included here.

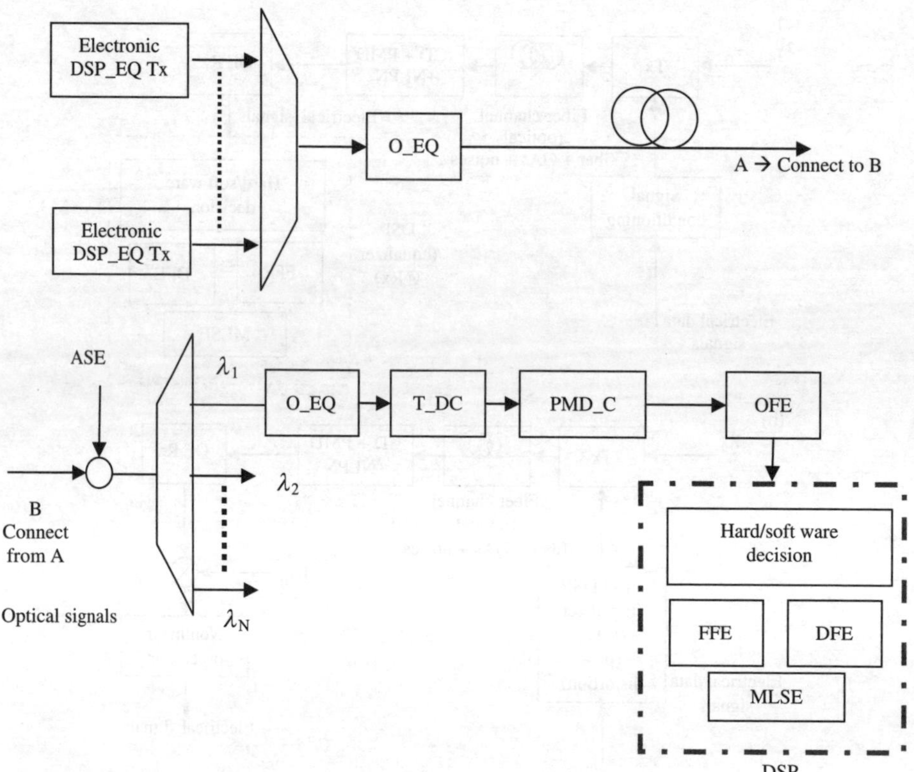

FIGURE 11.2 Typical structure and position of the DSP equalizer of a receive-side equalized optical system. Both optical and electronic equalization processors are integrated. Legend: O_EQ, optical equalizer; T_DC, tunable dispersion compensator; PMDC, polarization mode dispersion compensation; OFE, optical to electrical converter; DFE, decision feedback equalizer; FFE, feed forward equalizer; MLSE, maximum likelihood sequence estimator.

TABLE 11.2

Dispersion Tolerance and Equivalent Length of SSM for Different Modulation Formats and Equalization Techniques and Improved Factors

| | Measured/Simulated 10.7 Gb/s at 2 dB Penalty BER $=10^{-9}$ | | | |
	No Electronic Dispersion Compensation (EDC)	Equalization FFE/DFE	MLSE	Improved Factor
Modulation Technique				
Chirp free NRZ-ASK	50 km SSMF	80	130	X2.6
Chirp NRZ-ASK	80	130	150	
Duobinary	200	200	260	→x1.3
Chirp managed DM ASK	200	270	300	
Improved factor	x2.4		x2.3	

11.3   SYSTEM REPRESENTATION OF EQUALIZED TRANSFER FUNCTION

11.3.1   GENERIC EQUALIZATION FORMULATION

11.3.1.1   Signal Representation and Channel Pure Phase Distortion

Given that the single mode fiber is purely phase distortion, the equalization is dealing with the processing of sampled amplitude or phase of the recovered signals pending the modulation formats, whether its amplitude, phase or frequency is altered. This section gives a brief strategy of development of the requirement on the transfer function of the equalized system and associate filters. It is essential that the impulse response of the channel (the single mode optical fiber) must be obtained in order to specify the equalization sub-systems. Thus, this section gives an analytical derivation of the pure distortion channel. Both the impulse and step responses of the fiber channel are analytically derived.

Let the impulse response of the channel (the single mode optical fiber) is $y(t)$ then the z-transform of the impulse response with a sampling time equals to the bit period is given by

$$y(nT) = [y_0 \ y_1 \ ... \ y_g 0 \ 0 \ ... \ 0]$$

$$Y(z) = y_0 + y_1 z^{-1} \ ... + y_g z^{-g}$$

$$(11.1)$$

with $n \geq 2g + 1$ and y_i are complex-valued.

Under a purely phase distortion then the channel lost no energy and hence the conservation of energy requires that

$$\sum_{i=0}^{n} |y_i|^2 = 1 \tag{11.2}$$

with the total energy contained in the impulse assumed to be unity.

Now consider the input signals having complex-valued sampled components, such as the QAM optical signals that comprise two DPSK in phase quadrature or two AM signals in quadrature, one of which carries the data symbols $\{s_{1,i}\}$. Each AM signal is associated with the corresponding modulator and demodulator, to recover back to the base band channel. For advanced modulation formats, the detection can be incoherent or coherent. Under the coherent detection the impulse response takes the amplitude and preserves its phase. While for incoherent or direct detection the impulse response is the result of the beating of different components of the response in the photodetection process and its response follows intensity-based square law detection. In this case the response can be estimated by considering the Gaussian distribution with the pulse spreading factor.

The input signals can now be written as

$$s_i = s_{0,i} + j s_{1,i} \tag{11.3}$$

with $j = (-1)^{1/2}$ and both $s_{0,i}$ and $s_{1,i}$ can take on independently any one of the m different values

$$-(m-1)k, -(m-3)k, \ldots, (m-1)k. \tag{11.4}$$

An element of the QAM signal carrying the data symbol s_i is itself the sum of two double sideband suppressed carrier AM signals elements in phase quadrature, carrying the data symbols $s_{0,i}$ and $s_{1,i}$.

The output signals can be real or complex-valued. The sequence of the signals at the output of the channel is, of course, given by the sequence of samples at the output of the samplers. The real and imaginary parts of this sequence are samples of the respective baseband signals at the outputs of the two constituent parallel channels that make up the linear baseband channel. It can now be seen that if the sampled impulse response of the linear baseband channel is real, then there is no coupling between the two parallel channels. The distortion of these two channels is the same and can be determined as for ASK signals. On the other hand if they are complex-valued, the coupling is introduced between the two constituent channels by the imaginary parts of the components of the sampled impulse response. The impulse response of a single mode optical fiber is purely complex-valued and will be described in the next section.

When the channel introduces pure phase distortion, it introduces a unitary transformation into the transmitted signals [8], such that the received signal-elements corresponding to the resultant (complex-valued) transmitted signal-elements are orthogonal. An orthogonal transformation is a special case of unitary transformation where all signals are real-valued.

Under coherent detection then the detected signals are given by

$$r_i = s_i y_0 + w_i. \tag{11.5}$$

Where we have assumed, without lost of generality, that the impulse response takes only a non-zero value y_0 and w_i is the noise contributed to the signal. The noise is statistically independent Gaussian random variables with zero mean and variance σ^2. Thus we have

$$r_i = s_{0,i} y_0 + j s_{1,i} y_0 + w_i. \tag{11.6}$$

Obviously the received signals must be detected with both the real and imaginary parts. Thus in the absence of noise we have:

$$\left| r_i \right|^2 = s_{0,i}^2 + s_{1,i}^2. \tag{11.7}$$

Since the impulse response is of purely phase distortion then the absolute value of components of the sampled time response is unity and thus when the input signals are convolved with the impulse response, the phase of the components of the received signals are rotated by a fixed amount of phase.

This is equivalent to reversing the components of the impulse response pivoting around the central value at $t = 0$ by the complex conjugates, thus the impulse sequence can be written as:

$$y_i = y_i^*$$

$$y(nT) = \left[y_0 \; y_1 \ldots 0 \ldots y_m^* \; y_{m-1}^* \ldots y_1^* \right]. \tag{11.8}$$

Alternatively we can state that whenever there is a reversal of the complex-conjugate of a sequence pivoted around its components at $t = 0$ then the sequence represents a pure phase distortion. Thus when the sequence convolves with its original sequence gives a sequence all of whose components are zero except for the component at time $t = 0$ which is unity.

11.3.1.2 Equalizers at Receiver

11.3.1.2.1 Zero Forcing Equalization

The equalizer shown in Figure 11.3 is a decision feedback equalizer (DFE) which is a non-linear equalizer in contrast to the linear filter. A linear filter usually consists of a transversal feed forward filter (FFE) that delays the signal by time interval T and then tapped and multiplied with a certain set of coefficients and then sum up all those signals to form an equalized output. The non-linear DFE employed in this chapter is structured with both linear FFE and non-linear DFE. A zero-forcing equalizer operates as follows.

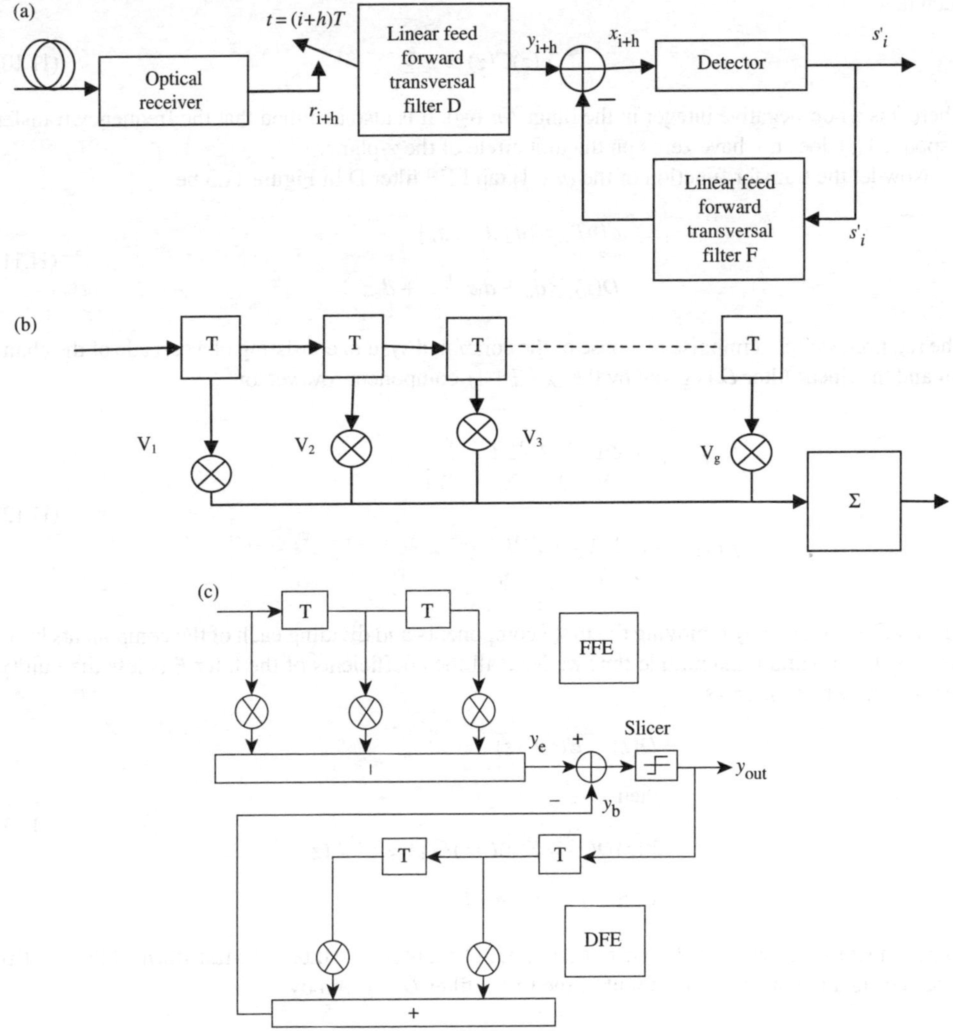

FIGURE 11.3 Schematic of a non-linear feedback equalizer following an optical receiver with (a) a feed forward transversal filter at the input and at the feed back path (FFE-DFE). (b) Linear feed forward transversal filter (FFE). (c) Combined FFE and DFE equalization-slicer is the data decision.

The linear FFE partially equalizes the signal by setting to zero all components of the sampled channels impulse response preceding that of the largest magnitude, without changing the relative values of the remaining components. The non-linear equalizer then completes the equalization process to give accurate equalization of the channel. The equalizer is non-linear because the detector (or slicer) is a non-linear device.

Now let y_l be the largest magnitude component of the impulse response (Equation 11.1) then the z-transform of the linear FFE of $(m+1)$-taps is given by

$$C(z) = c_0 + c_1 z^{-1} + \ldots + c_m z^{-m}.$$

(11.9)

Such that

$$Y(z)C(z) \approx z^{-h} \tag{11.10}$$

where h is a non-negative integer in the range $(m+g)$. It is also assumed that the frequency transfer response $Y(z)$ does not have zeros on the unit circle of the z-plane.

Now let the transfer function of the $(n+1)$ tap FFE filter D in Figure 11.3 be

$$d(nT) = [d_0 \ d_1 \ ... \ d_n]$$

$$D(z) = d_0 + d_1 z^{-1} ... + d_n z^{-n} \tag{11.11}$$

The required sampled impulse response of the combined system consisting of a cascade of the channel and the linear filter D is given by the $(g-l+1)$-component row vector

$$E = \left[1 + \frac{y_{l+1}}{y_l} \ \frac{y_{l+2}}{y_l} \ \frac{y_{l+3}}{y_l} \ ... \ \frac{y_g}{y_l} \right]$$

$$E(z) = \left[1 + \frac{y_{l+1}}{y_l} z^{-1} + \frac{y_{l+2}}{y_l} z^{-2} + \frac{y_{l+3}}{y_l} z^{-3} \ ... \ \frac{y_g}{y_l} z^{-g+1} \right]. \tag{11.12}$$

Clearly E is obtained by removing the first l components and dividing each of the components by y_l. As y_l is the maximum magnitude thus we have all the coefficients of the filter E is less than unity. $D(z)$ can now be written as

$$D(z) = E(z)C(z)$$

then

$$Y(z)D(z) = Y(z)C(z)E(z) \approx z^{-h}E(z) \tag{11.13}$$

with $\quad n = m+g-l.$

The z-transform of the $(i+1)$th transmitted signal-element is $s_i z^{-i}$ thus the z-transform of the $(i+1)$th received signal-element at the output of the linear filter D is given by:

$$s_i z^{-i-h} E(z) = s_i z^{-i-h} \left(1 + \frac{y_{l+1}}{y_l} z^{-1} + \frac{y_{l+2}}{y_l} z^{-2} + \frac{y_{l+3}}{y_l} z^{-3} \ ... + \frac{y_g}{y_l} z^{-g+1} \right). \tag{11.14}$$

It is thus observed that whenever there is a non-zero term there would be intersymbol interference (ISI) between the signal elements at the output of the linear filter. This ISI is then removed in the non-linear equalizer that equalizes the z-transform $z^{-h}E(z)$ as given above. Clearly the non-linear DFE follows that of the Figure 11.2 with the linear FFE with the number of taps of $(g-l)$ with the tap gain coefficients of (instead of the tap coefficients indicated in the Figure 11.2b).

$$\frac{y_{l+1}}{y_l}, \frac{y_{l+2}}{y_l}, \frac{y_{l+3}}{y_l} \ ... \ \frac{y_g}{y_l}. \tag{11.15}$$

The input to the substractor is thus, at the instant $(i+h)T$ is

$$v_{i+h} = s'_i + \frac{y_{l+1}}{y_l} s'_{i-1} + \frac{y_{l+2}}{y_l} s'_{i-2} + \frac{y_{l+3}}{y_l} s'_{i-3} + ... + \frac{y_g}{y_l} s'_{i-g+l} + u_{i+h} \tag{11.16}$$

where s'_{i-j} is the detected value of s_{i-j} at the output of the equalizer. With the correct detection of $s_{i-1}, s_{i-2}\cdots \ldots, s_{i-g+1,}$ the signals at the detector input becomes

$$x_{i+h} \simeq s_i + u_{i+h}. \tag{11.17}$$

When $x_{i+h} > 0$, $s_i' = k$ and when $x_{i+h} < 0$ $s_i' = -k$, especially for the balanced receiver detection of DPSK or DQPSK or QAM constellation. The noise component u_{i+h} is given by

$$u_{i+h} = \sum_{j=0}^{n} w_{i+h-j} d_j. \tag{11.18}$$

The noise components are statistically independent Gaussian variables with zero mean and a variance, thus we have to total variance the noise component of the detected signal is

$$\eta = \sigma^2 \sum_{j=0}^{n} d_j^2 = \sigma^2 |D|^2 \tag{11.19}$$

with D is the Euclidean length of the vector D. Thus the probability of error of the detection of s_i given the correct detection of $s_{i-1,} s_{i-2,}\ldots \ldots s_{i-g+l}$ can be approximately given as

$$P_e = \int_{k}^{\infty} \frac{1}{\sqrt{2\pi\eta^2}} e^{-\frac{u^2}{2\eta^2}} du = Q\left(\frac{k}{\eta}\right) = Q\left(\frac{k}{\sigma|D|}\right) \tag{11.20}$$

with k is the average total amplitude of the signal at the output of the receiver. The equalizer presented in this section can be optimized to reduce the probability of error further, for example the use of two cascaded transfer functions to replace the equivalent transfer function of the fiber, the linear equalizer takes on one part and then another by the non-linear equalizer. One could thus reduce the number of taps of the linear FFE. The least number of taps the better the demands puts on the electronic processor operating very high speed.

11.3.1.2.2 Equalization that Minimizes Mean Square Error (MSE Equalizer)

Both FFE and DFE can equalize and minimize the error contributed by the noise and distortion that would generally offer a more effective degree of equalization than an equalizer that minimizes the peak distortion. Such an equalizer minimizes the mean difference between the actual and ideal sampled values at its output for a given number of taps for the case of FFE. For DFE the minimization of the error occurs at the point where the feedback linear FFE and that at the output of the linear filter is shown as in Figure 11.3.

Under linear equalization, we recall that the channel sampled transfer function given in Equation 11.1 of the sampled impulse response of the channel and the equalizer is given the $(m+g+1)$ row vector

$$E = [e_0 e_1 \ldots\ldots\ldots e_{m+g}]. \tag{11.21}$$

The ideal value of the sampled impulse response of the channel and the equalizer is given by $(m+g+1)$-component row vector

$$E_h = \left[\underbrace{0..0...0...0}_{h} 1\, 0\, ...0\right]. \tag{11.22}$$

With h is an integer and is within the range $(m+g)$. The minimization is thus conducted so that the expected mean value

$$\mathrm{E}\left[\left(x_{i+h}-s_i\right)^2\right]=k^2\left|E-E_h\right|^2+\sigma^2\left|C\right|^2. \tag{11.23}$$

Where $k^2|E-E_h|^2$ and $\sigma^2|C|^2$ are the mean square error terms in the received signals x_{i+h} due to ISI and due to the Gaussian noise respectively, C is the sampled impulse response of $(m+1)$-component row of the FFE denoted by

$$C=[c_0 c_1 \ldots \ldots \ldots c_m]. \tag{11.24}$$

Under non-linear equalizer DFE, the minimization occurs at the error substractor as shown in Figure 11.3. The mean square error of the equalized signals needs to be minimized as the expected value of the mean square expressed as

$$\varepsilon = \mathrm{E}[(x_{i+h}-s_i)^2]. \tag{11.25}$$

Given that the data symbols $s_{i-1}, s_{i-2}, \ldots \ldots s_{i-\mu}$ have been correctly detected. This is implemented by adjusting the FFE filters D, the FFE before the DFE and F as the FFE in the feedback path of the DFE. The sampled impulse responses of these filters are expressed as

$$D=[d_0 d_1 \ldots \ldots \ldots d_n] \tag{11.26}$$

$$E=[e_0 e_1 \ldots \ldots \ldots e_{n+g}] \tag{11.27}$$

$$F=[f_o f_1 \ldots \ldots \ldots f_\mu] \tag{11.28}$$

$$S=[s_i s_{i-1} s_{i-2} \ldots \ldots \ldots s_{i-n-g}]. \tag{11.29}$$

With E as the sampled impulse of the fiber channel and the linear equalizer FFE D, S is the sampled signals correctly detected at the output of the equalizer. The signals at the input of the detector in Figure 11.3 at time $t=(i+h)T$ are given by

$$x_{i+h}=\sum_{j=0}^{n+g}s_{i+h}e_j-\sum_{j=1}^{\mu}s_{i-j}f_j+u_{i+h}=S_{i+h}E^T-S_{i+h}F_0^T+u_{i+h}$$

with

$$F_0=\left[\underbrace{0\ldots0}_{h} f_1 f_2 \ldots f_\mu\right]. \tag{11.30}$$

The mean square error can thus be written as [9]

$$\varepsilon=k^2\left|B\right|^2+\sigma^2\left|D\right|^2$$

with

$$B=E-F_0-E_h=[b_0 \ b_1 \ldots \ldots b_{n+g}] \tag{11.31}$$

and

$$E_h=\left[\underbrace{0\ldots0}_{h} \ 1\,0\,0\right].$$

The data symbols are assumed to be statistically orthogonal with zero mean and variance k^2. The DFE feedback non-linear equalizer is assumed to have μ taps in the filter F. This is the lowest number of taps whereby exact decision-directed cancellation can be achieved of all ISI components in x_{i+h}, involving data symbols $s_{i-1}, s_{i-2}, \ldots s_{i-\mu}$ that have been detected.

After a number of manipulation steps the mean square error can be obtained as:

$$\varepsilon = k^2 \left[1 - E_h Z^T \left(ZZ^T + \frac{\sigma^2}{k^2} I \right)^{-1} ZE_h^T \right]$$

with   (11.32)

$Z = [(n+1)x(n+g+1)]$ derived from

$\left[Y_c \right] = $ convolution matrix of the sampled impulse response.

The matrix Z can be written as

$$Z = \begin{bmatrix} y_0 & y_1 & y_2\ldots & \ldots 0\ldots 0\ldots 0 \\ 0 & y_0 & y_1\ldots & \ldots 0\ldots 0\ldots 0 \\ 0 & 0 & y_2\ldots & \ldots 0\ldots 0\ldots 0 \\ \ldots & \ldots & \ldots & \ldots 0\ldots 0\ldots 0 \\ 0 & 0 & 0 & \ldots y_0\, y_1 & y_2 & 0\ldots & 0 \\ 0 & 0 & 0 & 0\, y_0 & y_1 & y_2\ldots & 0 \\ 0 & 0 & 0 & 0\, 0 & y_0 & y_1\ldots & 0 \\ 0 & 0 & 0 & 0\, 0 & 0 & y_0\ldots & 0 \end{bmatrix}.$$ (11.33)

11.3.1.2.3   Non-linear Equalization that Minimizes Signal to Noise Ratio

An optimum DFE can be considered as a DFE that further minimizes the BER in the detected signals for a given DFE structure. As previously assumed the z-transform of the fiber channel in intensity has no zeros or roots on the unit circle of the z-plane. Thus we have the sampled impulse response of the channel and a FFE of the DFE is given by

$$Y(z)C(z) \simeq z^{-h} \quad \text{or} \quad C(z) \simeq z^{-h} Y^{-1}(z).$$ (11.34)

Then the FFE D of the DFE can be composed of the FFE C and a filter E under the constraint that

$$Y(z)D(z) \simeq z^{-h} E(z)$$

                                                                                        (11.35)

$$\text{where} \quad E(z) = [e_0\, e_1 \ldots e_{n-m}].$$

The non-linear equalizer of Figure 11.3 equalizes $E(z)$ so that the linear FFE of the feedback path F has $(n-m)$ taps with the gains equal, respectively, to the components $e_i, e_2, \ldots e_{n-m}$ of the vector E. The signals at the output of the linear filter D is thus given by, at the instant $t = (i+h)T$

$$v_{i+h} \simeq s_i + s_{i-1}e_1 + s_{i-2}e_2 + \ldots + s_{i-n+m}e_{n-m}.$$ (11.36)

With the corrected detection of the input signals $s_{i,t}$ then the equalized signals at the output of the DFE can be written as

$$x_{i+h} \simeq s_i + u_{i+h} \tag{11.37}$$

So when

$$x_{i+h} = \begin{cases} > 0....s'_i = k \\ < 0.....s'_i = -k. \end{cases}$$

The noise components u_{i+h} is a Gaussian random variable as assumed previously with zero mean and variance of η given as

$$\eta = \sigma^2 |D|^2. \tag{11.38}$$

Thus the BER can be obtained as

$$P_e = \int\limits_k^\infty \frac{1}{\sqrt{2\pi\eta^2}} e^{-\frac{u^2}{2\eta^2}} \, du = Q\left(\frac{k}{\eta}\right) = Q\left(\frac{k}{\sigma|D|}\right). \tag{11.39}$$

So, it is necessary to minimize the Euclidean length of the linear FFE filter or find the minimum number of taps of the filter that still allow the maximizing of the signal to noise ratio. Thus the DFE should be adjusted to as to minimize the vector length D. This minimization can be implemented as follows. This minimization can be done at the linear filter D of the DFE such that

$$|D| = |G_0 - LM| \tag{11.40}$$

With

$$G_0 = \begin{bmatrix} 0 & c_0 & c_1 & c_2...c_m & 0...0 \end{bmatrix}$$

$$M = \begin{bmatrix} 0 & c_0 & c_1 & c_2...0 & 0 \\ & & 0 & c_0...0 & 0 \\ ... & ... & ... & ... & ... \\ & & & c_m...0 & 0 \\ & & & c_{m-1}...0 & 0 \\ & & & c_{m-2} & c_{m-1} & c_m \end{bmatrix} \tag{11.41}$$

$$L = \begin{bmatrix} -e_1 & -e_2... & -e_{n-2-m} & -e_{n-1-m} & -e_{n-m} \end{bmatrix}.$$

The task is now adjusting L so that minimization of the vector length D happens.

11.3.1.2.4 Case Study of Equalization Schemes

This section gives a case study of direct detection with the impulse response of the single mode fiber or the residual dispersion of a dispersion managed transmission system. The sampled impulse response is a real impulse response after the photodetector and electronic preamplifier. This impulse response follows a Gaussian distribution with the broadening of the impulse at the e^{-1} value of the maximum peak by an amount given by

$$\Delta\tau = D.L.B_{3dB} \tag{11.42}$$

where the $B_{3\,dB}$ is taken as the 3 dB bandwidth of the power spectra of the signals under different modulation formats.

A typical dispersive impulse at the instant $t=iT$ can be given by

$$r_i = 0.3s_i + s_{i-1} + w_i \tag{11.43}$$

where the data symbol $\{s_i\}$ are statistically independent and equally likely to have either value $\pm k$ and $\{w_i\}$ and are statistically independent Gaussian random variable with zero mean and variance of σ^2. Effectively the impulse response of the fiber transmission disperses to 30% of the peak amplitude at the output of the electronic preamplifier. We can compare the tolerance to noises of the following equalization processes (a) A simple threshold level detector; (b) a linear FFE; (c) a purely NL DFE; (d) linear and non-linear equalization of two factors of the channel; and (e) optimum DFE.

11.3.1.2.4.1 Non-equalization Direct Detection

At a high SNR practically all the errors occur when $s_{i+1}=-s_i$ that is when $r_{i+1}=0.7s_i+w_i$. Thus the BER can be derived as

$$P_e = \int_{0.7k}^{\infty} \frac{1}{\sqrt{2\pi\sigma^2}} e^{-\frac{u^2}{2\sigma^2}} du = Q\left(\frac{0.7k}{\sigma}\right). \tag{11.44}$$

(a) Linear equalization

The sampled impulse response of the fiber channel is given by the two components sequence $Y=[0.3\ 1]$ so that the z transform of the sequence is given by $Y(z)=0.3+z^{-1}$. Now assume that a 6-tap linear FFE is designed for the equalization so as to minimize the peak distortion is $M(z)=1+0.3\,z^{-1}$. The equalizer with a 6-tap FFE can be found by taking a long division of $(1+0.3\,z^{-1})^{-1}$. The long division gives

$$N(z)=1-0.3z^{-1}+0.09z^{-2}-0.027z^{-3}+0.0081z^{-4}-0.00243z^{-5}. \tag{11.45}$$

Thus, the $C(z)$ of the FFE can be obtained by reversing the order of the $N(z)$ as

$$C(z)=1.z^{-5}-0.3z^{-4}+0.09z^{-3}-0.027z^{-2}+0.0081z^{-1}-0.00243. \tag{11.46}$$

Hence, the z-transform of the cascaded channel and the FFE is thus simply given as

$$C(z)Y(z)=0.000729+z^{-6}\simeq z^{-6}. \tag{11.47}$$

Thus, the equalized pulse at the instant $t=(i+6)T$ is

$$x_{i+6}=1w_{i+1}-0.3w_{i+2}+0.09w_{i+3}-0.027w_{i+4}+0.0081w_{i+5}-0.00243w_{i+6}. \tag{11.48}$$

The Gaussian noise can be estimated as

$$\eta=1w_{i+1}-0.3w_{i+2}+0.09w_{i+3}-0.027w_{i+4}+0.0081w_{i+5}-0.00243w_{i+6}. \tag{11.49}$$

Thus, the noise variance is obtained as

$$\eta^2=\sigma^2(1-0.3+0.09-0.027+0.0081-0.00243)=1.0989\sigma^2. \tag{11.50}$$

Thus, the probability of error is given by

$$P_e = \int_k^\infty \frac{1}{\sqrt{2\pi\eta^2}} e^{-\frac{u^2}{2\eta^2}} \, du = Q\left(\frac{k}{\eta}\right) = Q\left(\frac{k}{1.0989\sigma}\right) = Q\left(\frac{0.9539k}{\sigma}\right).$$

(11.51)

(b) NL DFE

On the receive of the signals r_i the NL DFE determines s'_{i-1} to generate the equalized signals $x_i = r_i - s'_{i-1}$. With the correct detection of s_{i-1} we have

$$x_i = 0.3s_i + w_i.$$

(11.52)

And s_i is now detected from x_i as follows.

$$x_{i+h} = \begin{cases} > 0 \ldots s'_i = k \\ < 0 \ldots s'_i = -k. \end{cases}$$

(11.53)

An error occurs here in the detection of s_i when w_i has a magnitude greater than 0.3 k and the opposite sign to s_i. Thus the probability of error of the equalized signals is given by

$$P_e = \int_k^\infty \frac{1}{\sqrt{2\pi\eta^2}} e^{-\frac{u^2}{2\eta^2}} \, du = Q\left(\frac{k}{\eta}\right) = Q\left(\frac{0.3k}{\sigma}\right).$$

(11.54)

(c) Linear and non-linear equalization

The linear and non-linear equalization of the two factors here would degenerate into a linear equalizer since the impulse response z-transform of $Y(z)$ does not have zero in the outside region of the unit circle in the z-plane. Thus the probability of error is the same as in the case (b) that is

$$P_e = Q\left(\frac{0.9539k}{\sigma}\right).$$

(11.55)

(d) Optimum DFE

The first FFE of the non-linear equalizer peforms an orthogonal transformation on the receive signals such that the z-transform of the channel and the filter becomes:

$$Y(z)D(z) \simeq z^{-h}(1 + 0.3z^{-1}).$$

(11.56)

The filter does not change the SNR and the noise property or any distortion of the amplitude of the received signal. Its output signal at $t = (i+h)T$ is

$$v_{i+h} = s_i + 0.3s_{i-1} + u_{i+h}$$

(11.57)

where u_i is a Gaussian random variable of zero mean and a variance of σ^2. Thus the probability of error is

$$P_e = \int_k^\infty \frac{1}{\sqrt{2\pi\sigma^2}} e^{-\frac{u^2}{2\sigma^2}} \, du.$$

(11.58)

On the receive of v_{i+h} the non-linear filter has formed $0.3s'_{i-1}$ which is subtracted from v_{i+h} to give the equalized signal

$$x_{i+h} = v_{i+h} - 0.3s'_{i-1}.$$ (11.59)

Then fed to the detector or decision device, the correct detected signal then the equalized signal becomes

$$x_{i+h} = s_i + u_{i+h} \quad \text{when} \quad x_{i+h} = \begin{cases} > 0...s'_i = k \\ < 0...s'_i = -k. \end{cases}$$ (11.60)

An error occurs in the detection of s_i from x_{i+h} when u_{i+h} has a magnitude greater than k and opposite in sign to s_i so that regardless of the sign of s_i the probability of detection of s_i can be approximately given as

$$P_e = \int_k^\infty \frac{1}{\sqrt{2\pi\sigma^2}} e^{-\frac{u^2}{2\sigma^2}} du = Q\left(\frac{k}{\sigma}\right).$$ (11.61)

Thus, from the above analytical estimation of the probability of error as a function of the SNR k/σ we could see that the resilience to noise of the optimum DFE is best with about 0.4 dB while the pure non-optimized non-linear DFE it is worst with -10 dB and -2.7 dB for direct detection without equalization and 0 dB for linear FFE.

11.3.1.2.5 *Maximum Likelihood Sequence Estimation*

The maximum likelihood sequence estimation (MLSE) is considered as the detection of the transmitted sequence with minimum probability of error and it depends on the coding and modulation formats of the signals. This method is further explained in details elsewhere in this chapter.

11.3.1.3 Equalizers at the Transmitter

Consider the arrangement so that the equalizer acting as a pre-distortion and placed at the transmitter. The non-linear equalizer in this case converts the sequence of data symbols $\{s_i\}$ into a non-linear channel $\{f_i\}$ at the output of the transmitter, normally an optical modulator driven by the output of the non-linear electrical equalizer as shown in Figure 11.1b and re-modeled as shown Figure 11.4, Figure 11.5 and Figure 11.6. These symbols are sampled and transmitted as sampled impulses $\{f_i\delta(t-iT)\}$. The SNR in this case is given by $E[f_i^2]/\sigma^2$ with the energy is the expected average energy of the pre-distorted signal sequence and must be equal to k^2. A non-linear processor is necessary at the output before the final recovery of the system. This processor performs a modulo-m operation on the received sequence $x[\text{modulo}]-\text{m}=x-jm$ with j an appropriate integer. The z-transform $F(z)$ at the output of the non-linear distortion equalizer is given by

$$F(z) = M\left[S(z) - F(z)(y_0^{-1}Y(z) - 1\right]$$ (11.62)

where $S(z)$ is the z-transform of the sampled input signals and $Y(z)$ is the z-transform of the channel transfer function. Then the z-transform at the output of the non-linear processor at the receiver can be written as:

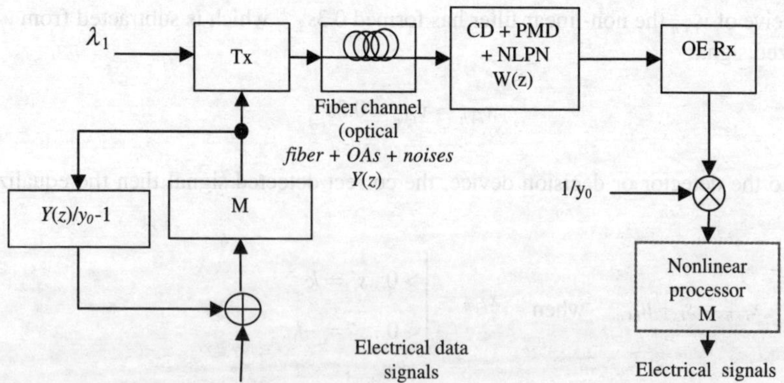

FIGURE 11.4 Schematic of a non-linear feedback equalizer at the transmitter.

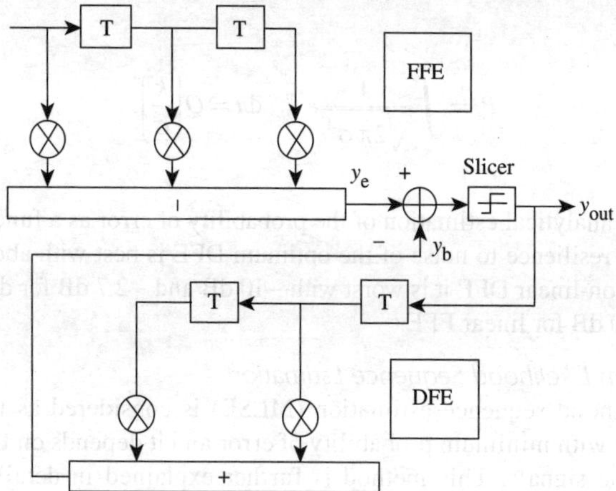

FIGURE 11.5 Typical layered graph model for IP over WDM network with optical receivers and equalizers.

$$X(z) = M\left[F(z)Y(z) + W(z)\right] = M\left[F(z)(y_0^{-1}Y(z) - y_0^{-1}W(z)\right]$$

$$\text{or } X(z) = M\left[S(z) - y_0^{-1}W(z)\right]$$

that is

$$x_i = M\left[y_0^{-1}r_i\right] = M\left[s_i + y_0^{-1}w_i\right].$$

(11.63)

With x_i denoting the signal at the output of the non-linear processor and clearly all ISI has been eliminated except the noise contribution.

The non-linear processor M operates as a modulo-m as

$$M[q] = [(q + 2k)\text{modulo} - 4k] - 2k = q - 4jk$$
$$-2k \le M[q] \le 2k.$$

(11.64)

Thus, an error would happen when

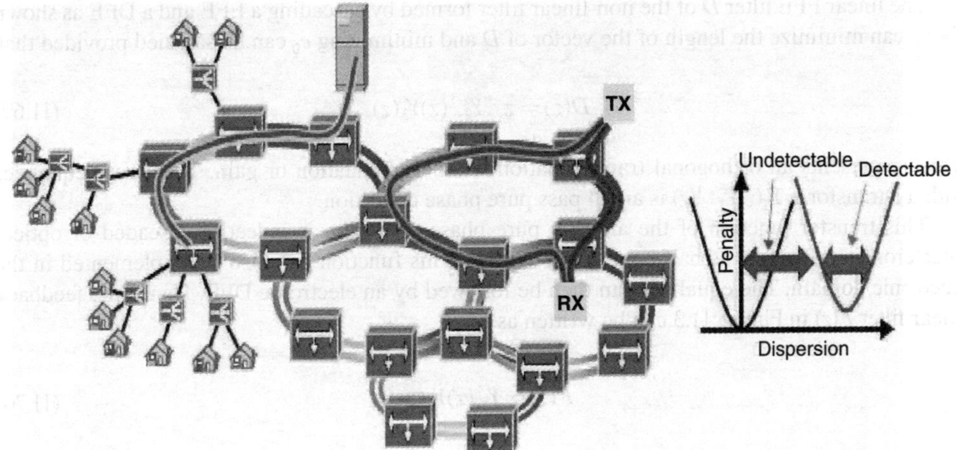

FIGURE 11.6 Typical lightpaths of optical networks and dispersion tolerance under the case of without and with electrical equalization. (From Alic, N., G. C. Papen, R. E. Saperstein, L. B. Milstein, and Y. Fainman, *Optics Express* 13 (12), 4568–79, 2005. With permission.)

$$(4j-3)k \le y_0^{-1}w_i \le (4j-1)k \quad \forall j. \tag{11.65}$$

The probability of error is then given as

$$P_e = 2 \int_k^\infty \frac{1}{\sqrt{2\pi y_0^{-2}\sigma^2}} e^{-\frac{w^2}{2y_0^2\sigma^2}} \, dw = 2Q\left(\frac{k|y_0|}{\sigma}\right). \tag{11.66}$$

11.3.1.4 Equalization Shared Between Receiver and Transmitter

Long-haul transmission requires the equalization so that the extension of the transmission as long as possible. Equalization at the receiver can be supplemented with that at the transmitter to further increase the reach.

The equalizer at the transmitter of the above section can be optimized by inserting another filter at the receiver. Consider the channel transfer function $Y(z)$ that can be written as a cascade of two linear transfer functions $Y_1(z)$ and $Y_2(z)$ as

$$Y(z) = Y_1(z)Y_2(z)$$

$$\text{with} \begin{cases} Y_1(z) = 1 + p_1 z^{-1} + \dots + p_{g-f} z^{-g+f} \\ Y_2(z) = q_0 + q_1 z^{-1} + \dots + q_f z^{-f}. \end{cases} \tag{11.67}$$

$Y_1(z)$ has all the zero inside the unit circle and $Y_2(z)$ all outside the unit circle. Then we could find a third system with a transfer function that has all the coefficients in reverse of those of $Y_2(z)$ given as

$$Y_3(z) = q_f + q_{f-1}z^{-1} + \dots + q_0 z^{-f}. \tag{11.68}$$

With the reverse coefficients, the zeroes of the $Y_3(z)$ now lie inside the unit circle.

The linear FFE filter D of the non-linear filter formed by cascading a FFE and a DFE as shown above can minimize the length of the vector of D and minimizing e_0 can be satisfied provided that

$$D(z) = z^{-h} Y_2^{-1}(z) Y_3(z) \tag{11.69}$$

which represents an orthogonal transformation without attenuation or gain. The linear equalizer with a z-transform $Y_2(z)Y_3^{-1}(z)$ is an all pass pure phase distortion.

This transfer function of the all pass pure phase distortion is indeed a cascaded of optical interferometers, known as half-band filters [10,11]. This function can also be implemented in the electronic domain. The equalizer can then be followed by an electronic DFE. Hence the feedback linear filter $F(z)$ in Figure 11.3 can be written as

$$F(z) = Y_1(z) Y_3(z). \tag{11.70}$$

The probability of error can be similarly evaluated and given by

$$P_e = \int_k^\infty \frac{1}{\sqrt{2\pi y_0^{-2}\sigma^2}} e^{-\frac{w^2}{2y_0^2\sigma^2}} \, dw = Q\left(\frac{k|q_f|}{\sigma}\right). \tag{11.71}$$

With q_f, the first coefficient of $Y_3(z)$.

11.3.1.5 Performance of FFE and DFE

The performance of linear equalizer using FIR filter with different order is shown in Figure 11.18 with the penalty of eye opening versus the dispersion factor [12]. The higher the order of the FIR the better the equalization of the distorted impulse, or the longer the length of the SSMF that the signals can propagate. A 7th order FIR filter can equalize the distorted signals over 150 km with 1 dB EOP. While the receiver without equalization would suffer an EOP of at least 12 dB.

Figure 11.20 shows the EOP versus the fibre length (SSMF) for the case of non-linear equalizer with a FFE and a DFE with the orders of 2nd and 3rd, respectively. With 1 dB EOP the FFE and DFE combined non-linear filter can extend the transmission of 10 Gb/s NRZ-ASK to 150 km and 300 km with non-linear correction.

Figure 11.20 shows the EOP of modulation formats of NRZ-ASK and NRZ duo-binary using FIR of number of taps of 3 and 16 with the distortion of signals. Up to 400 km SSMF transmission is possible with the equalized system for 1 dB EOP.

In the next section, the impulse response of the fiber is described. It will be proven that it is a purely imaginary response because the fiber channel is a purely phase distortion medium. The impulse response related to the intensity distortion will be found with Gaussian-shape approximation, and the broadening of the impulse can be estimated. A detailed description follows of the equalization, using MLSE for continuous phase FSK or optical MSK modulation schemes.

11.3.2 Impulse and Step Responses of the Single Mode Optical Fiber

The treatment of lightwaves through single mode fiber has been well documented as shown in Chapter 4. In the following section, we shall restrict our study to the linear region of the media. Furthermore, the delay term in the transmittance function for the fiber is ignored, as it has no

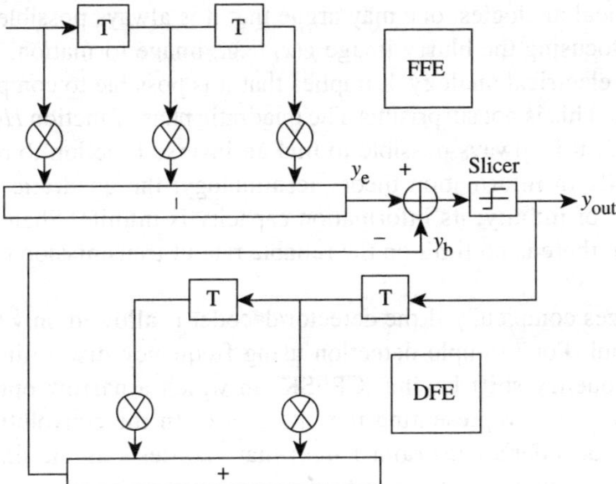

FIGURE 11.7 Representation of the quadratic-phase transmittance function in frequency and time domain.

bearing on the size and shape of the pulses. We can thus model the fiber simply as a quadratic phase function. The input–output relationship of the pulse can therefore depicted in Figure 11.7.

Equation 11.1 expresses the time-domain impulse response $h(t)$ and the frequency domain transfer function $H(\omega)$ as a Fourier transform pair:

$$h(t) = \sqrt{\frac{1}{j4\pi\beta_2}}\, \exp\left(\frac{jt^2}{4\beta_2}\right) \quad \leftrightarrow \quad H(\omega) = \exp(-j\beta_2\omega^2) \tag{11.72}$$

where β_2 is known as the group velocity dispersion (GVD) parameter. The input function $f(t)$ is typically a rectangular pulse sequence; the parameter β_2 is proportional to the length of the fiber. The output function $g(t)$ is the dispersed waveform of the pulse sequence. The electro system of Figure 11.7 is an exact analogy of diffraction in optical systems (see item 1, Table 2.1, p. 14, A. Papoulis [13]). Thus, the quadratic phase function also describes the diffraction mechanism in one-dimensional optical systems, where distance x is analogous to time t. The establishment of this analogy affords us to borrow many of the imageries and analytical results that have been developed in the diffraction theory. Thus, we may express the step response $s(t)$ of the system $H(\omega)$ in terms of Fresnel cosine integral $C(t)$ and Fresnel sine integral $S(t)$ as follows:

$$s(t) = \int_0^t \sqrt{\frac{1}{j4\pi\beta_2}}\, \exp\left(\frac{jt^2}{4\beta_2}\right) dt$$

$$= \sqrt{\frac{1}{j4\pi\beta_2}}\left[C\left(\sqrt{1/4\beta_2}\,t\right) + jS\left(\sqrt{1/4\beta_2}\,t\right)\right] \tag{11.73}$$

$$C(t) = \int_0^t \cos\left(\frac{\pi}{2}\tau^2\right) d\tau$$

with

$$S(t) = \int_0^t \sin\left(\frac{\pi}{2}\tau^2\right) d\tau. \tag{11.74}$$

Using the electro-optical analogies, one may argue that it is always possible to restore the original pattern $f(x)$ by refocusing the blurry image $g(x)$ (e.g., image formation, item 5, Table 2.1, A. Papoulis [13]). In the electrical analogy, it implies that it is possible to compensate the quadratic phase media perfectly. This is not surprising. The quadratic phase function $H(\omega)$ in Equation 11.72 is an all-pass function, it is always possible to find an inverse function to recover $f(t)$ from $g(t)$. Express this differently in information theory terminology: the quadratic phase channel has a theoretical bandwidth of infinity; its information capacity is infinite. Shannon's channel capacity theorem states that there is no limit on the reliable rate of transmission through the quadratic phase channel.

The picture changes completely if the detector/decoder is allowed only a finite time window to decode each symbol. For example detection using frequency discrimination techniques for continuous phase frequency shift keying (CPFSK) in which a narrow optical filter is used to extract the carrier frequency representing the bits 1 or 0. In the convolutional coding scheme for example, it is the decoder's constraint length that manifests as the finite time window. In adaptive equalization schemes, it is the number of equalizer's delay elements, which determines the decoder window length. Since the transmitted symbols have already been broadened by the quadratic phase channel, if they are next gated by a finite time window, the information received could be severely reduced. The longer the fiber, the more the pulses are broadened, the more uncertain it becomes in the decoding. It is the interaction of the pulse broadening on one hand, and the restrictive detection time window on the other, that gives rise to the finite channel capacity.

Figure 11.8a shows the impulse response of Gaussian pulse transmission of 100 Gb/s pulse sequence through a SMF of length $L = 200$ km, dispersion $= 0$e-6 s/m^2, (a) dispersion slope $S = +0.06$e$+5$ s/m^3 while Figure 11.8b shows response over $L = 2$ km of the same dispersion factor and dispersion slope. It is noted from Equation 11.72 that the impulse response is purely phase function and thus phase distortion. The oscillation of the tail of the impulse response for long fiber indicates the phase chirping of the lightwave carrier in the advanced region of time-dependent pulses. While over a short length of fiber the Gaussian pulse remains with minimum change of the phase underneath its envelope. Figure 11.8c shows the impulse tail oscillation on the other side when it propagates through a dispersion-compensated 200 km fiber span with a total residual dispersion of -8.5 ps/nm that is typically optically amplified fiber transmission system.

The step response (Equation 11.73) consists of the in phase and quadrature phase components. Figure 11.9a, Figure11.9b and Figure11.9c shows the pulse and step responses, its frequency spectrum and step response after propagating through a 1 km SSMF. The over shooting of the edge of the pulse with the chirping of the lightwave carrier occurs mostly near the edge which indicates that for the detection of the frequency shift keying modulation (FSK) format would significantly influence the bit error rate ad the receiver sensitivity of the optical systems.

11.4 ELECTRICAL LINEAR DOUBLE SAMPLING EQUALIZERS FOR DUOBINARY MODULATION FORMATS FOR OPTICAL TRANSMISSION

In general, chromatic dispersion is a time-varying impairment mainly because of temperature variations. Implications can be quite critical for 40-Gb/s systems that have tight chromatic dispersion tolerances. For 10-Gb/s systems, the chromatic dispersion tolerance is far less stringent than 40-Gb/s systems (approximately 1000 ps/nm for a 1 dB power penalty). A variation of approximately 0.15 ps/nm/km over the temperature range from 40 to 60°C, compared with a typical value of 17 ps/nm/ km, the variation can be neglected. Therefore, it can be assumed, for our immediate purposes, that chromatic dispersion is a static impairment that can be accurately modeled, and hence static equalizers can be utilized. The ideal zero-forcing equalizer is simply the inverse of the channel transfer

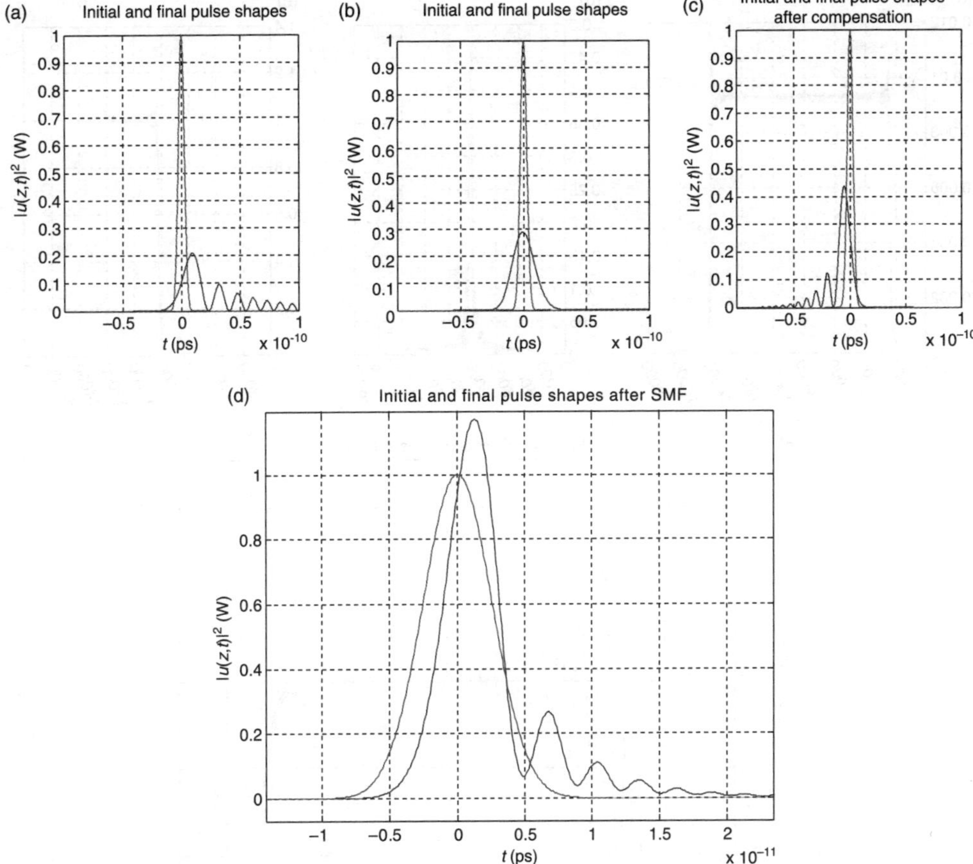

FIGURE 11.8   Gaussian pulse transmission of 100 Gb/s pulse sequence through a SMF of length $L=2$ km, dispersion $=0$e-6 s/m^2, (a) dispersion slope $S=+0.06$e$+5$ s/m^3. Fiber length $=200$ km. (b) $L=2$ km dispersion $=17$e-6 s/m^2 and $S=+0.06$e$+5$ s/m^3. (c) Over 200 km SSMF and completely compensated with a mismatch of -8.5 ps/nm dispersion. (d) Enlarged responses with a dispersion slope of DCF $=0.3$e$+5$ s/m^3. Note the oscillation of the tail of the impulse response.

function given in Chapter 3. Thus, the required equalizer transfer function for the equalizer $H_{eq}(f)$ is given by

$$H_{eq}(f) = H_f^{-1}(f) = e^{+j\alpha f^2}$$

with   (11.75)

$$\alpha = \pi DL \frac{\lambda}{c^2}.$$

However, this transfer function does not maintain conjugate symmetry, that is

$$H_{eq}(f) \neq H_{eq}^{-1}(-f).$$ (11.76)

Thus, the impulse response of the equalizing filter is complex as shown in the previous section (Equation 9.2). Consequently, this filter can not be realized by a baseband equalizer using only one

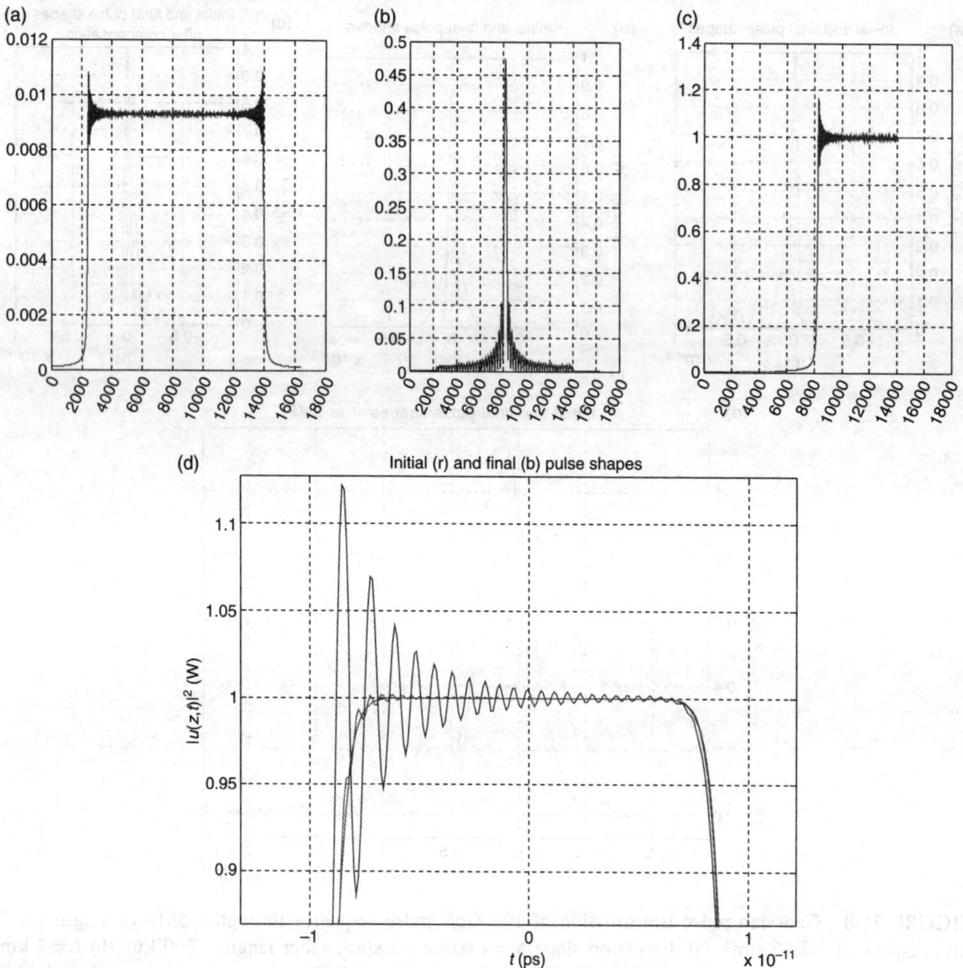

FIGURE 11.9 Rectangular pulse transmission of through a SMF, near field response (a) Pulse response. (b) Spectrum of pulse in (a) and (c) step response. Note horizontal axis unit normalized with time unit. (d) Enlargement of chirping of pulse at the edge due to dispersion slope.

baseband received signal, which explains the limited capability of linear equalizers that are used in direct-detection receivers to mitigate the chromatic dispersion.

On the other hand, it also explains why fractionally spaced linear equalizers, especially the double sampling type that are used within a coherent receiver can potentially extend the system reach to distances that are only limited by the number of equalizer taps. However, it is important to note that the non-ideal effects such as the laser phase noise and the fiber non-linearity, which eventually set a limit on the maximum achievable transmission distance that should also be taken into account. With the equalizer in a coherent receiver, most of the complexity can be avoided at the receiver by an equalization process at the transmitter. The required filter would thus have complex coefficients, the pre-distorted data have to modulate two optical carriers that are in phase quadrature.

Although the concept of using two optical carriers that are in phase quadrature is not a conventional one, its practical feasibility was shown in Chapter 4 where two orthogonal optical carriers were used to obtain an optical differential-quadrature phase-shift-keying (oDQPSK) transmission system. To determine the required equalizer coefficients, the simulation setup in Figure 11.10 can be used, where an adaptive algorithm is used to adaptively adjust the FIR filter coefficients to their optimum values. The adaptive algorithm adjusts the FIR filter taps to compensate not only for chromatic dispersion but other linear dispersion effects such as PMD and any other phase-delay type filter such as phase ripples of chirp FBGs. Once the filter taps converge to their optimal values, the linearity of the system allows transfer of the filter to the transmitter side, where it acts as a pre-distortion filter. The proposed pre-distortion scheme is depicted in Figure 11.10. Figure 11.11 shows the optical transmitter for duo-binary is formed using two parallel MZMs and a π/2 optical phase shift. The MZMs are biased for minimum optical transmission and driven by the pre-equalized data from the FIR filters. The duobinary filters can be implemented using analog filters. The analog filters also account for the limited bandwidth of the signal path between the pre-equalizer and the MZM electrodes. The tap spacing of the FIR filter can be as high as a whole bit period as proposed in Figure 11.12. However, reducing the tap spacing to half the bit period doubles the frequency band over which equalization can be applied, which translates into a substantial improvement in signal quality, but at the expense of a higher clock speed. Alternatively, the FIR coefficients can be computed mathematically using, for example, the minimum mean square error (MSE) criterion. This is accomplished by solving the Wiener–Hopf equation [3]. The signal input and tap-weight vectors can be defined as shown in Figure 11.12 and written as

$$s^H(n) = \left[s^*(n)...s^*(n-1)...s^*(n-2)...s^*(n-N+1) \right]^H$$

$$s^T(n) = \left[s(n)...s(n-1)...s(n-2)...s(n-N+1) \right]^T$$

(11.77)

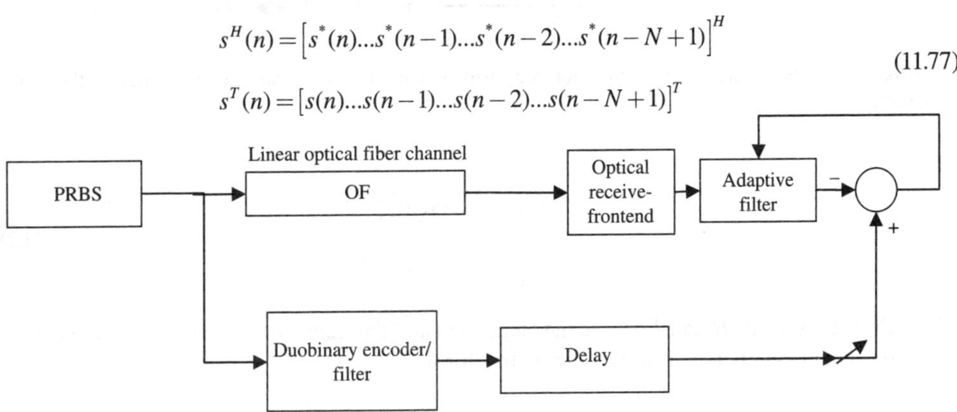

FIGURE 11.10 Schematic of the pre-distortion equalizer for linear distortion of optical fiber transmission.

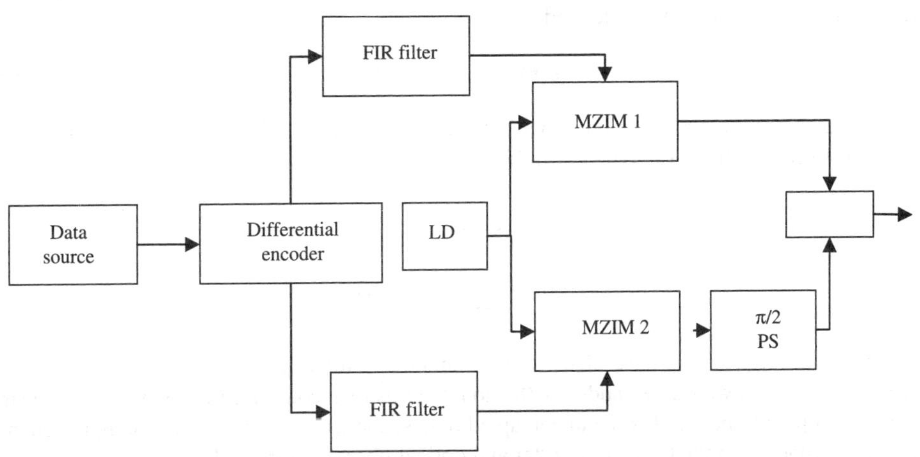

FIGURE 11.11 Schematic of optical modulation structure for generation of NRZ duobinary format. Biasing voltages and microwave amplifiers are not shown for MZIMs.

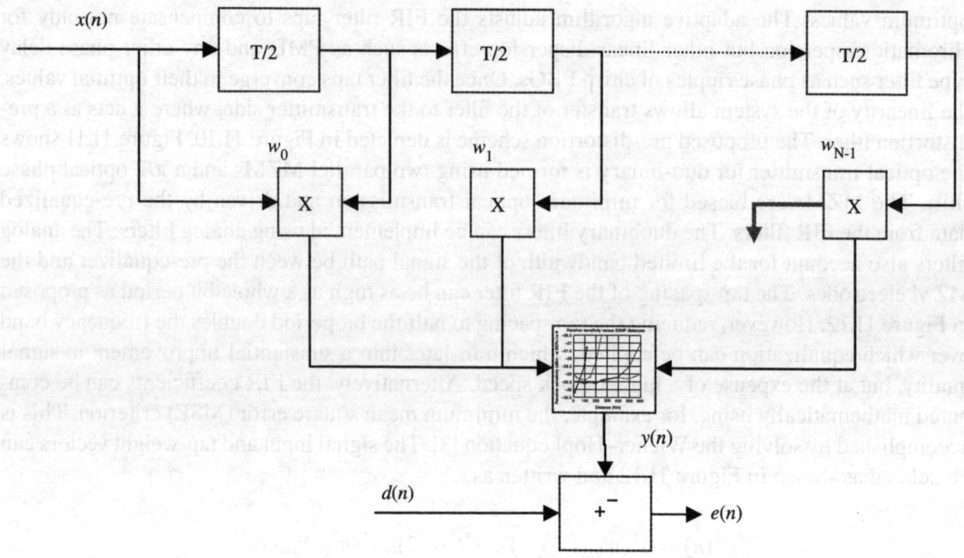

FIGURE 11.12 Structure of FIR filter for pre-distortion of the electrical driving signal at the optical transmitter.

and

$$c^T = \left[c_0 ... c_1 ... c_2 ... c_N \right]^T$$

$$c^H = \left[c_0^* ... c_1^* c_2^* ... c_N^* \right]^H$$

(11.78)

where the superscripts, H and T denote the Hermitian and the transpose operation respectively. The equalizer output can then be expressed in matrix form

$$y(n) = c^T s(n).$$

(11.79)

Note that the samples of the input, $s(n)$ and output, $y(n)$ of the equalizer, are at $T/2$ intervals. However, in the optimization of the equalizer tap weights, only the output samples at T intervals are of concern. Thus, the error signal is defined as

$$e(n) = d(n) - y(2n).$$

(11.80)

Figure 11.12 shows the proposed pre-equalization scheme and the fractionally spaced FIR filter, thus the performance function to be minimized is written as

$$J = E\left[e^2(n) \right]$$

$$d(n) = \sum_i \gamma_i w(n - i)$$

(11.81)

where γ_i is the impulse response of the digital duobinary filter. This impulse response if under numerical simulation would be padded with enough zeros to account for the delay of the transmitter filter and optical fiber, and is the uncorrupted data sequence. The fractionally sampled channel response (includes both the transmitter filter and optical fiber) is expressed as

$$y(nT) = \left[y_0 ... y_1 ... y_2 ... y_{M-1} \right]^T$$

(11.82)

where h is a sufficiently large integer such that the values of h_i for $i > (M-1)$ are negligible. The optimum tap weights are obtained by solving the Wiener–Hopf equation [3], which yields the optimum tap-weight vector, given by

$$c_0 = R^{-1}p$$

with

$$R = E[s(2n)s^H(2n)] \quad \text{and} \quad p = E[d^*(n)s(2n)] \tag{11.83}$$

and

$$s(2n) = Hw(n) + v(2n)$$

$$w(n) = \left[w(n)...w(n-1)...w\left(n - \frac{N}{2} + 1\right) \right]^T \tag{11.84}$$

$$\text{with} \quad H = \begin{bmatrix} y_0 & y_2 & y_4 & y_6 & \cdots & y_{M-2} & 0 & \cdots \\ 0 & y_1 & y_3 & y_5 & \cdots & y_{M-3} & y_{M-1} & \cdots \\ 0 & 0 & y_2 & y_4 & \cdot\cdot & y_{M-4} & y_{M-2} & \cdots \\ \cdot & \cdot & \cdot & \cdot & \cdot & \cdot & \cdot & \cdot \\ \cdot & \cdot & \cdot & \cdot & \cdot & \cdot & \cdot & \cdot \end{bmatrix}^T$$

$w(n)$ is a vector of noise samples. Note that the matrix H is a circulant matrix with alternating rows in which a row is formed by shifting one column of the row two levels above it. The noise, taken to be white noise, is added to avoid excessive filter gains at the frequencies where the magnitude of the channel frequency response is very small. Equation 11.14 may also be rewritten as

$$d(n) = \gamma^T w(n) \tag{11.85}$$

where $d(n)$ is a column vector of the target impulse response. From Equation 11.14, and using the fact that takes the values of $+1$ and -1 only we can obtain

$$R = E[Hss^H H^H] + \sigma^2 I = HIH^H + \sigma^2 I = HH^H + \sigma^2 I \tag{11.86}$$

where σ is the variance of the added noise, and I is an identity matrix. Similarly

$$P = E[Hw^H H^H \gamma] = HI\gamma = H\gamma. \tag{11.87}$$

Thus, the desired optimum $T/2$-spaced FIR equalizer can thus be written as

$$c_0 = [HH^H + \sigma^2 I]^{-1} H\gamma. \tag{11.88}$$

Figure 11.13 shows significant improvement of the sensitivity of the linear equalizer in the 10 Gb/s duobinary modulation format under the conditions as shown in Table 11.3. Recently the fundamental limits of the direct detection of duobinary modulation signals has been studied [7].

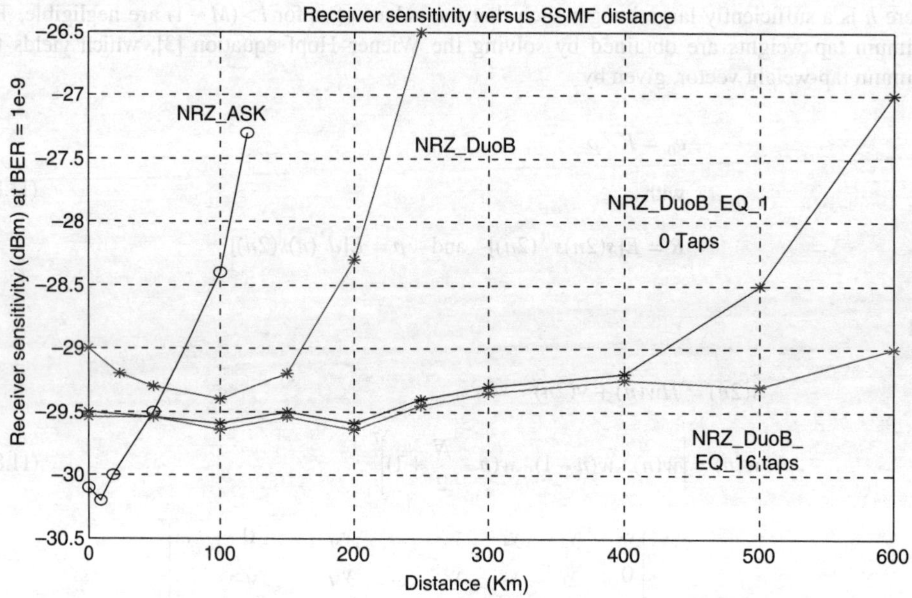

FIGURE 11.13 Receiver sensitivity versus fiber SSMF distance for conventional NRZ-ASK, NRZ duobinary, and pre-distortion equalized duobinary of different number of taps of linear equalization system. (From Alic, N., G. C. Papen, and Y. Fainman. Performance of maximum likelihood sequence estimation with different modulation formats. *Proceedings of LEOS'04,* 49–50, 2004. With permission.)

TABLE 11.3

Key Simulation Parameters used in the LE for Duobinary Modulation Format

Optical amplifier gain 20 dB, ASE noise $n_{sp} = 2$	Narrowband Gaussian filter: $B = 5.2$ GHz or $BT = 0.13$
Operating wavelength: $\lambda = 1550$ nm	Constant delay: $t_d = |2\pi f_D \beta_2 L|$
Bit rate: $R = 10$ Gb/s	Pre-amp EDFA of the OFDR: $G = 15$ dB and NF $= 5$ dB
SSMF fiber: $|\beta_2| = 2.68$e-26 or $|D| = 17$ ps/nm/km, dispersion slope 0.072 ps/(nm².km), 80 μm² effective diameter	$i_d = 10$ nA
Attenuation: $\alpha = 0.25$ dB/km	$N_{eq} = 20$ pA/(Hz)$^{1/2}$
Laser linewidth	4 MHz
Optical filter bandwidth and duobinary bandwidth	100 and 3.5 GHz (3rd order filter Bessel type)

11.5 NON-LINEAR MLSE EQUALIZERS FOR MSK OPTICAL TRANSMISSION SYSTEMS

11.5.1 NON-LINEAR MLSE

Maximum likelihood sequence estimation (MLSE) is a well-known technique in communications for equalization and detection of the transmitted digital signals. The concept of MLSE is discussed in brief. A MLSE receiver determines a sequence b as the most likely transmitted sequence when the conditional probability $Pr(y|b)$ is maximized where y is the received output sequence. If the received signal y is corrupted by a noise vector $\boldsymbol{n}$ which is modeled as a Gaussian source (i.e., $y = b + n$), it is shown that the above maximization operation can be equivalent to the process of minimization of the Euclidean distance d [14]:

$$d = \sum_k |y_k - b_k|^2. \tag{11.89}$$

MLSE can be carried out effectively with the implementation of Viterbi algorithm based on state trellis structure.

11.5.2 Trellis Structure and Viterbi Algorithm

11.5.2.1 Trellis Structure

Information signal b is mapped to a state trellis structure by a finite state machine (FSM) giving the mapping output signal c as shown in Figure 11.14. The FSM can be a convolutional encoder, a trellis-based detector or as presented in more detail later on, the fiber medium for optical communications.

A state trellis structure created by the FSM is illustrated in Figure 11.15. At the epoch n^{th}, the current state B has two possible output branches connecting to states E and F. In this case, these two branches correspond to the two possible transmitted symbols of 0 or 1, respectively. This predefined state trellis applies to the 1-bit-per-symbol modulation formats such as binary ASK or DPSK formats. In cases of multi-bit modulation per symbol, the phase trellis has to be modified. For example, in the case of DQPSK, there are four possible branches leaving the current state B and connect the next states, corresponding to the possible input symbols of 00, 01, 10 and 11.

Furthermore for a simplified explanation, several assumptions are made. These assumptions and according notations can be described as: (i) Current state B is the starting state of only two branches which connect to states E and F, denoted as: b_{BE} and b_{BF}. In general, $b_{\text{B}*}$ represents all the possible branches in the trellis which starts from state B. In this case, these two branches correspond to the

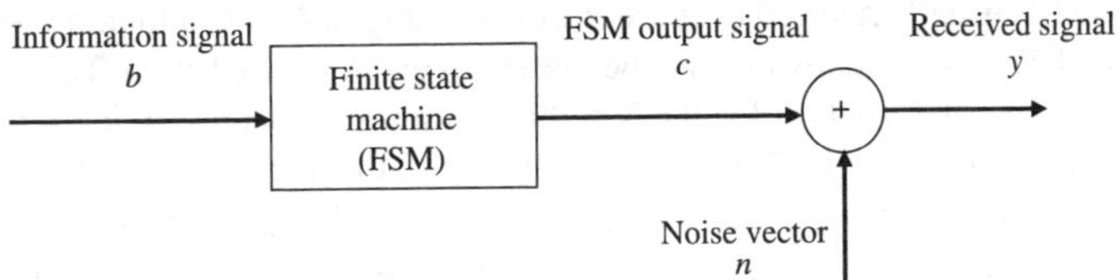

FIGURE 11.14 Schematic of the MLSE equalizer as a finite state machine.

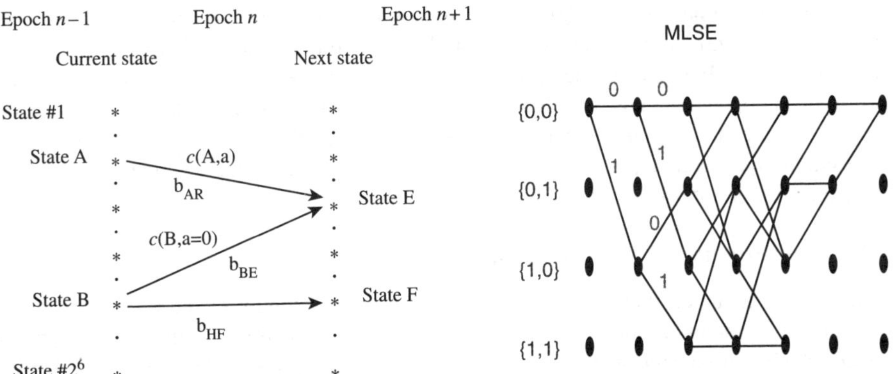

FIGURE 11.15 State trellis of a finite state machine. (a) Trellis path. (b) Tracing of possible maximum likelihood estimation with assigned bit pattern.

possible binary transmitted information symbols of 0 or 1, respectively. The notation $c(B, a = 0)$ represents the encoded symbol (at the FSM output) for the branch BE, b_{BE} which starts from state B and corresponds to a transmitted symbol $a = 0$. Similar conventions are applied for the case of branch BF, b_{BF}, i.e., $c(B, a = 1)$ and the all the branches in the trellis. (ii) There are only two possible output branches ending at state E (from state A and state B). They are denoted as: b_{AE} and b_{BF}. In general, b_{*E} represents all the possible branches in the trellis which end at state E. The number of states 2^δ in the trellis structure is determined by the number of memory bits, δ also known as the constraint length. Figure 11.15b also shows a possible trace of the trellis with the assigned states of the bit pattern.

11.5.2.2 Viterbi Algorithm

The principles of the Viterbi algorithm follow two principal phases.

11.5.2.2.1 Calculation of State and Branch Metrics

At a sampling index n, epoch n, a branch metric is calculated for each of the branches in the trellis. For example, considering the connecting branch between the current state B and the next state E, which corresponds to the case when 0 symbol is transmitted, the branch metric BM_{BE} is calculated as $BM_{BE}(n) = |_{(n)} - c_{BE}|^2$

The state metric $SM_E(n)$ the state E is calculated as the sum between the metric of the previous state, that is state B and the branch metric obtained above. The calculation of state metrics $SM_E(n)$ of state E is given by

$$SM_E(n) = \min_{\forall B \to E} \left\{ SM_B(n-1) + BM_{BE}(n) \right\}. \tag{11.90}$$

The branch giving the minimum state metric $SM_E(n)$ for state E is named as the *preferred path*.

11.5.2.2.2 Trace-back Process

The process of calculating state and branch metrics continues along the state trellis before terminating at epoch N_{trace} which is usually referred to as the trace-back length. A rule of thumb is shown by Proakis [14,15], the value of N_{trace} is normally taken to be five times the sequence length. This value comes from the results showing that the solution for the Viterbi algorithm converges giving a unique path from epoch 1 to epoch N_{trace} for the MLSE detection. (i) At N_{trace}, the terminating state with the minimum state metric and its connecting *preferred path* are identified. (ii) The previous state is then identified as well as its previous *preferred path*. (iii) The trace-back process continues until reaching the epoch 1.

11.5.3 OPTICAL FIBER AS A FINITE STATE MACHINE

The finite state machine show in Figure 11.15 is now replaced with the optical channel based on optical fiber. It is significant to understand how the state trellis structure is formed for an optical transmission system. In this case, the optical fiber involves all the ISI sources causing the waveform distortion of the optical signals, which includes CD, PMD and filtering effects. Thus, the optical fiber FSM excludes the signal corruption by noise as well as the random-process non-linearities. The intesymbol interference (ISI) caused by dispersion phenomena and non-linear phase and ASE and non-linear effects has been described in Chapters 3 and 8.

11.5.3.1 Construction of State Trellis Structure

At epoch n, it is assumed that the effect of ISI on an output symbol of the FSM c_n is caused by both executive δ pre-cursor and δ symbols on each side. In this case, c_n is the middle symbol of a sequence which can be represented as $c = (c_{n-\delta}, ..., c_n, ..., c_{n+\delta})$ $b \in \{0,1\}; n > 2\delta + 1$. Unlike the conventional FSM used in wireless communications, the output symbol c_n of the optical fiber FSM is selected as the middle symbol of the sequence c. This is due to the fact the middle symbol gives the most reliable picture about the effects of ISI in optical fiber channel.

A state trellis is defined with a total number of $2^{2\delta}$ possible data sequences. At epoch n, the current state B is determined by the data sequence $(b_{n-2\delta+1},...,b_{n-\delta},...b_n)$ $b \in \{0,1\}; n \geq 2\delta+1$ where b_n is the current input symbol into the optical fiber FSM. Therefore, the state trellis structure can be constructed from the optical fiber FSM and ready to be integrated with the Viterbi algorithm, especially for the calculation of a branch metric.

The MLSE non-linear equalizer can effectively combat all the above ISI effects and it is expected that the tolerance to both CD and PMD of optical MSK systems with OFDR receiver can be improved significantly.

11.6 MLSE EQUALIZER FOR OPTICAL MSK SYSTEMS

11.6.1 CONFIGURATION OF MLSE EQUALIZER IN OFDR

Figure 11.16 shows the block diagram of narrowband filter receiver integrated with non-linear equalizers for the detection of 40 Gb/s optical MSK signals. Two narrowband filters are used to discriminate the USB and the LSB frequencies that correspond to logic "1" and "0" transmitted, respectively. A constant optical delay line which is easily implemented in integrated optics is introduced on one branch to compensate the differential group delay $t_d = 2\pi f_d \beta_2 L$ between f_1 and f_2 where $f_d = f_1 - f_2 = R/2$, β_2 represents group velocity delay (GVD) parameter of the fiber and L is the fiber length. If the differential group delay is fully compensated, the optical lightwaves in two paths arrive at the photodiodes simultaneously. The outputs of the filters are then converted to electrical domain through the photodiodes. These two separately detected electrical signals are sampled before being fed as the inputs to the non-linear equalizer.

11.6.2 MLSE EQUALIZER WITH VITERTIC ALGORITHM

At epoch k, it is assumed that the effect of ISI on an output symbol of the FSM c_k is caused by both executive δ pre-cursor and δ post-cursor symbols on each side. First, a state trellis is constructed with 2δ states for both detection branches of the OFDR. A lookup table per branch corresponding to symbols 0 and 1 transmitted and containing all the possible $2^{2\delta}$ states of all 11 symbol-length possible sequences is constructed by sending all the training sequences incremented from 1 to $2^{2\delta}$.

The output sequence $c(k) = f(b_k, b_{k-1},...,b_{k-2\delta-1}) = (c_1,c_2,...,c_{2\delta})$ is the non-linear function representing the ISI caused by the δ adjacent pre-cursor and post-cursor symbols of the optical fiber FSM. This sequence is obtained by selecting the middle symbols c_k of 2δ possible sequences with length of $2\delta + 1$ symbol intervals.

The samples of the two filter outputs at epoch k y_k^i, $i=1,2$ can be represented as $y^i(k) = c^i(k) + n_{ASE}^i(k) + n_{Elec}^i(k)$, $i=1,2$ Here, n_{ASE}^i n_{Elec}^i represent the amplified spontaneous emission (ASE) noise and the electrical noise, respectively.

In linear transmission of an optical system, the received sequence y_n is corrupted by ASE noise of the optical amplifiers, n_{ASE} and the electronic noise of the receiver, n_{Elec}. It has been proven that the calculation of branch metric and hence state metric is optimum when the distribution of noise follows the normal/Gaussian distribution, i.e., the ASE noise and the electronic noise are collectively modelled as samples from Gaussian distributions. If noise distribution departs from the Gaussian distribution, the minimization process is suboptimum.

The Viterbi algorithm sub-system is implemented on each detection branch of the OFDR. However the MLSE with Viterbi algorithm may be too computationally complex to be implemented at 40 Gb/s with the current integration technology. However, there have been commercial products available for 10 Gb/s optical systems. Thus a second MLSE equalizer using technique of reduced-state template matching is presented in the next section.

In an optical MSK transmission system, narrowband optical filtering plays the main role in shaping the noise distribution back to the Gaussian profile. The Gaussian-profile noise distribution is verified in Figure 11.17. Thus, branch metric calculation in the Viterbi algorithm which

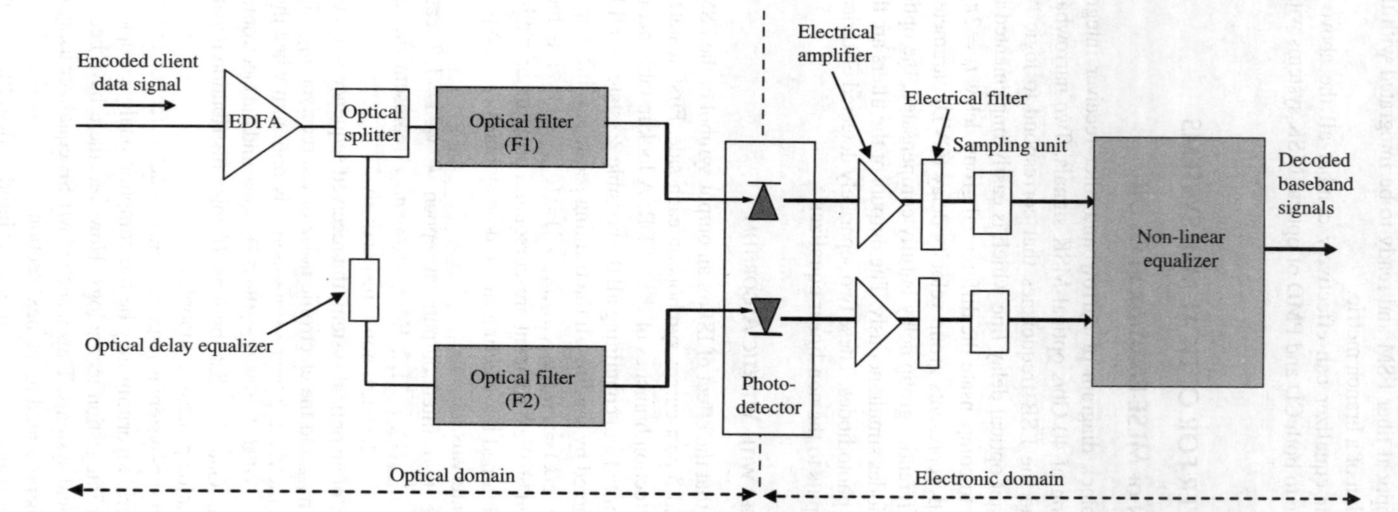

FIGURE 11.16   Block diagram of narrowband filter receiver integrated with non-linear equalizers for the detection of 40 Gb/s optical MSK signals. (From Alic, N., G. C. Papen, R. E. Saperstein, L. B. Milstein, and Y. Fainman, *Optic Express*, 13(12), 4568–79, 2005. With permission.)

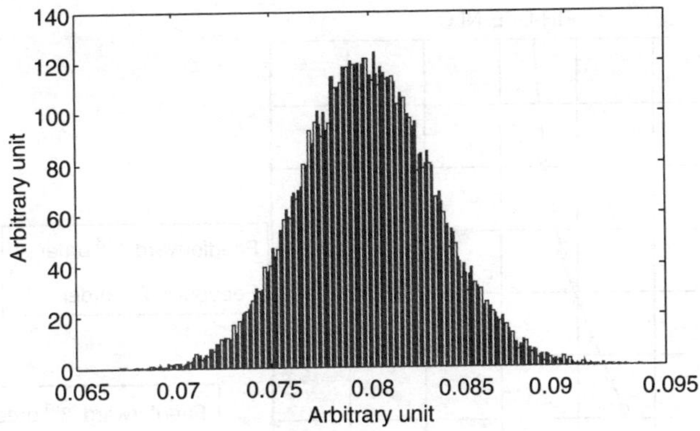

FIGURE 11.17   Noise distribution following Gaussian shape due to narrowband optical filtering.

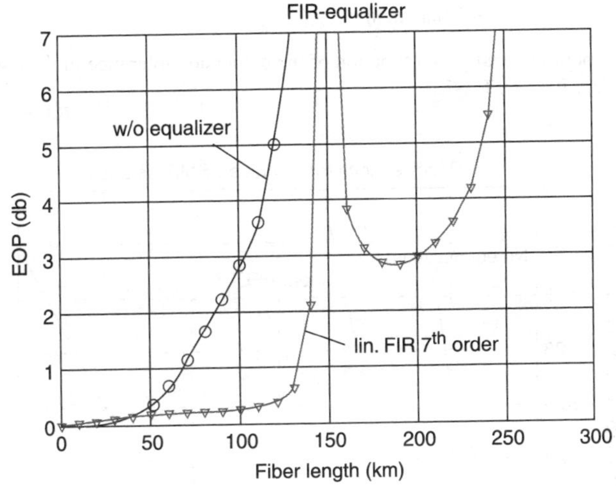

FIGURE 11.18   Eye opening penalty as a function of the dispersion tolerance in kms of SSMF without and with linear equalizer FIR.

is based on minimum Euclidean distance over the trellis can achieve the optimum performance. Also, the computational effort is less complex than ASK or DPSK systems due to the issue of non-Gaussian noise distribution. Figure 11.18 shows the performance improvement with the eye opening penalty as a function of the dispersion tolerance in kms of SSMF without and with linear equalizer FIR for ASK modulation format. Similarly, Figure 11.19 and Figure 11.20 shows the eye opening penalty as a function of the dispersion tolerance in kms of SSMF without and with using a non-linear equalizer FFE-DFE-NLC and non-linear equalizer FFE-DFE respectively.

11.6.3   MLSE EQUALIZER WITH REDUCED-STATE TEMPLATE MATCHING

The modified MLSE is a single shot template matching algorithm. First a table of $2^{2\delta+1}$ templates, $\mathbf{g}^k, k = 1,2,\cdots,2^{2\delta+1}$ corresponding to the $2^{2\delta+1}$ possible information sequences, I^k of length $2^{\delta+1}$ is constructed. Each template is also a vector of size $2^{\delta+1}$ which is obtained by transmitting the corresponding information sequence through the optical channel and obtaining the $2^{\delta+1}$ consecutive received samples. At each symbol period, n the sequence, $\hat{I}_n$ with the minimum metric is selected as $c(k) = \arg\min_{b(k)}\{m(b(k), y(k))\}$. The middle element of the selected information sequence

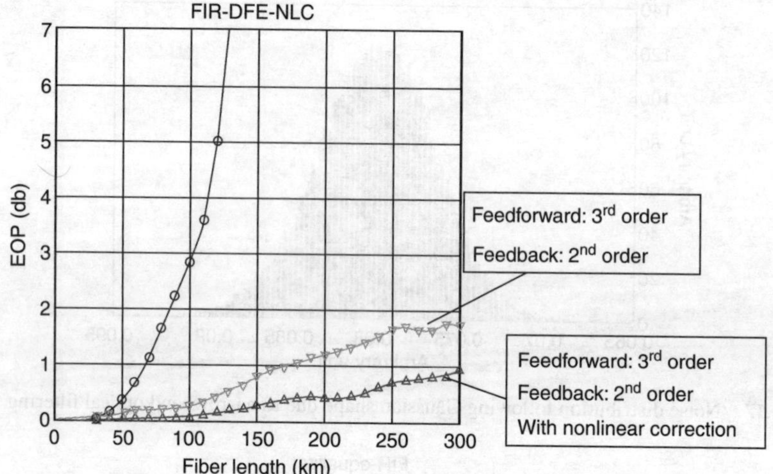

FIGURE 11.19　Eye opening penalty as a function of the dispersion tolerance in kms of SSMF without and with non-linear equalizer FFE-DFE-NLC.

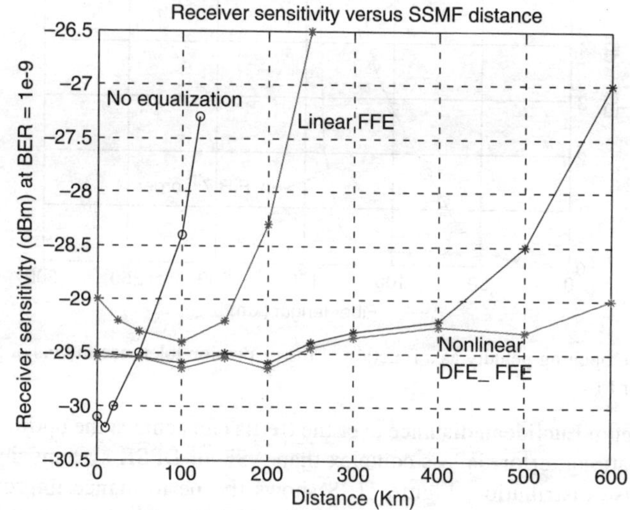

FIGURE 11.20　Eye opening penalty as a function of the dispersion tolerance (kms) of SSMF without and with linear and non-linear equalizer FFE-DFE.

is then output as the nth decision, $\hat{I}_n$ i.e., $\hat{I}_n = \hat{\mathbf{I}}_n(\delta+1)$. The minimization is performed over the information sequences, I^k which satisfy the condition that the $\delta{-}2$ elements are equal to the previously decoded symbols $\hat{I}_{n-\delta}, \hat{I}_{n-(\delta-1)}, \cdots \hat{I}_{n-2}$ and the metric, $m(I^k, r_n)$ is given by $m(I^k, r_n) = \{w \bullet (g^k - r_n)\}^T \{w \bullet (g^k - r_n)\}$ where w is a weighting vector which is chosen carefully to improve the reliability of the metric. The weighting vector is selected so that when the template is compared with the received samples, less weighting is given to the samples further away from the middle sample. For example, we found through numerical results that a weighting vector with elements $\mathbf{w}(i) = 2^{-|i-(\delta+1)|}, i = 1, 2, \cdots, 2\delta+1$ gives good results. Here, $\bullet$ represents Hadamard multiplication of two vectors, $(.)^T$ transpose of a vector and $|.|$ is the modulus operation.

11.7   MLSE SCHEME PERFORMANCE

11.7.1   Performance of MLSE Schemes in 40 Gb/s Transmission

Figure 11.21 shows the simulation system configuration used for the investigation of the performance of both the above schemes when used with the narrowband optical Gaussian filter receiver for the detection of non-coherent 40Gb/s optical MSK systems. The input power into fiber (P_0) is -3 dBm which is much lower than the non-linear threshold power. The EDFA2 provides 23 dB gain to maintain the receiver sensitivity of -23.2 dBm at BER $= $ 1e-9. As shown in Figure 11.21, the optical received power (P_{Rx}) is measured at the input of the narrowband MSK receiver and the OSNR is monitored to obtain the BER curves for different fiber lengths. Length of SSMF is varied from 48 km to 60 km in a step of 4 km to investigate the performance of the equalizers to the degradation caused by fiber cumulative dispersion. The narrowband Gaussian filter with the time bandwidth product of 0.13 is used for the detection filters.

Electronic noise of the receiver can be modelled with equivalent noise current density of electrical amplifier of 20pA/√Hz and dark current of 10 nA. The key parameters of the transmission system are given in Table 11.4.

The Viterbi algorithm used with the MLSE equalizer has a constraint length of 6 (that is 2^5 number of states) and a trace back length of 30. Figure 11.22 shows the BER performance of both non-linear equalizers plotted against the required optical signal to noise ratio (OSNR). The BER performance of the optical MSK receiver without any equalizers for 25 km SSMF transmission is also shown in Figure 11.22 for quantitative comparison. The numerical results are obtained via Monte Carlo simulation (triangular markers as shown in Figure 11.22) with the low BER tail of the curve linearly extrapolated. The OSNR penalty (at BER $= $ 1e-9) versus residual dispersion corresponding to 48 km, 52 km, 56 km and 60 km SSMF are presented in Figure 11.23. The MLSE scheme outperforms the modified MLSE schemes especially at low OSNR. In the case of 60 km SSMF, the improvement at BER $= $ 1e-9 is approximately 1 dB and in the case of 48 km SSMF, the

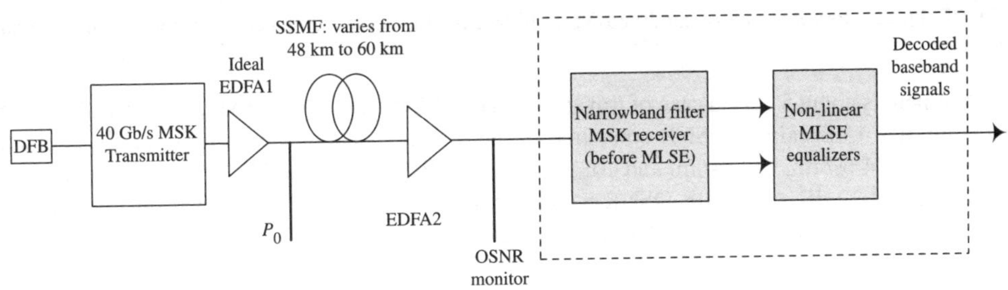

FIGURE 11.21   Simulation setup for performance evaluation of MLSE and modified MLSE schemes for detection of 40 Gb/s optical MSK systems.

TABLE 11.4
Key Simulation Parameters used in the MLSE MSK Modulation Format

Input power: $P_0 = -3$ dBm	Narrowband Gaussian filter: $B = 5.2$ GHz or $BT = 0.13$
Operating wavelength: $\lambda = 1550$ nm	Constant delay: $t_d = \lvert 2\pi f_D \beta_2 L \rvert$
Bit rate: $R = 40$ Gb/s	Pre-amp EDFA of the OFDR: $G = 15$ dB and NF $= 5$ dB
SSMF fiber: $\lvert \beta_2 \rvert = 2.68$e-26 or $\lvert D \rvert = 17$ ps/nm/km.	$i_d = 10$ nA
Attenuation: $\alpha = 0.2$ dB/km	$N_{eq} = 20$ pA/(Hz)$^{1/2}$

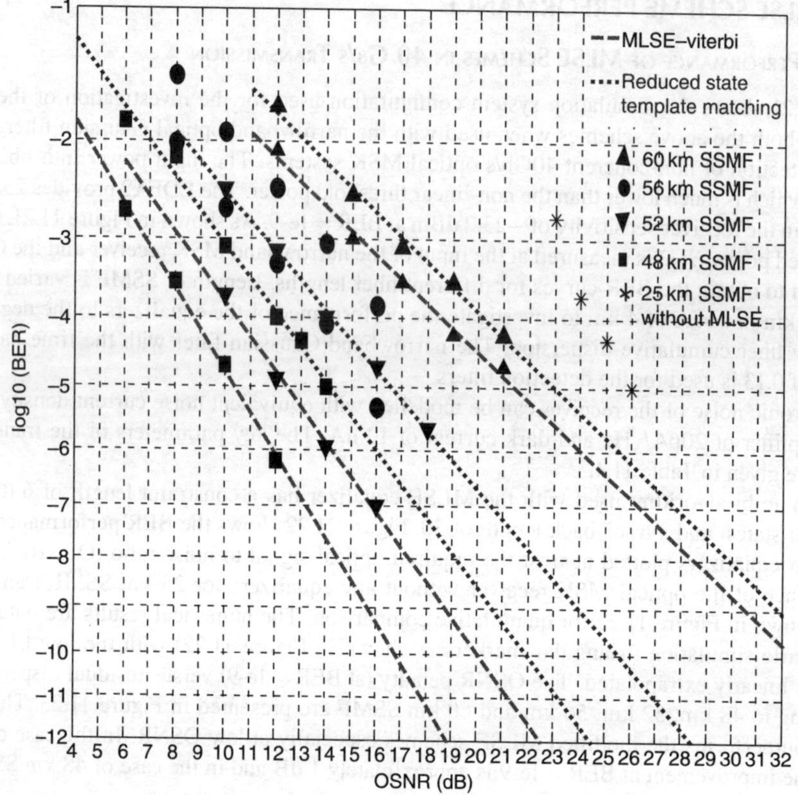

FIGURE 11.22 Performance of modified Verterbi-MLSE and template matching schemes, BER versus OSNR.

improvement is about 5 dB. In case of transmission of 52 km and 56 km SSMF, MLSE with Viterbi algorithm has 4 dB gain in OSNR compared to the modified scheme. With residual dispersion at 816 ps/nm, 884 ps/nm, 952 ps/nm and 1020 ps/nm, at BER of 1e-9, the MLSE scheme requires 12 dB, 16 dB and 19 dB and 26 dB OSNR penalty, respectively while the modified scheme requires 17 dB, 20 dB, 23 dB and 27 dB OSNR, respectively.

11.7.2 TRANSMISSION OF 10 GB/S OPTICAL MSK SIGNALS OVER 1472 KM SSMF UNCOMPENSATED OPTICAL LINK

Figure 11.24 shows the simulation setup for 10 Gb/s transmission of optical MSK signals over 1472 km SSMF. The receiver employs an optical narrowband frequency discrimination receiver integrated with a 1024-state Viterbi-MLSE post-equalizer. The input power into fiber (P_0) is −3 dBm lower than the fiber non-linear threshold power. The optical amplifier EDFA1 provides an optical gain to compensate the attenuation of each span completely. The EDFA2 is used as a noise loading source to vary the required OSNR. The receiver electronic noise is modeled with equivalent noise current density of the electrical amplifier of 20 20pA/√Hz and dark current of 10 nA for each photodiode. A narrowband optical Gaussian filter with two-sided bandwidth of 2.6 GHz (one-sided BT = 0.13) is optimized for detection. A back-to-back OSNR = 8 dB is required for BER at 1e-3 for each branch. The correspondent received power is −25 dBm. This low OSNR is possible due to large suppression of noise after being filtered by narrowband optical filters. A trace back length of 70 is used in the Viterbi algorithm. Figure 11.24 shows the simulation results of BER versus the required OSNR for 10Gb/s optical MSK transmission over 1472 and 1520 km SSMF uncompensated optical links with

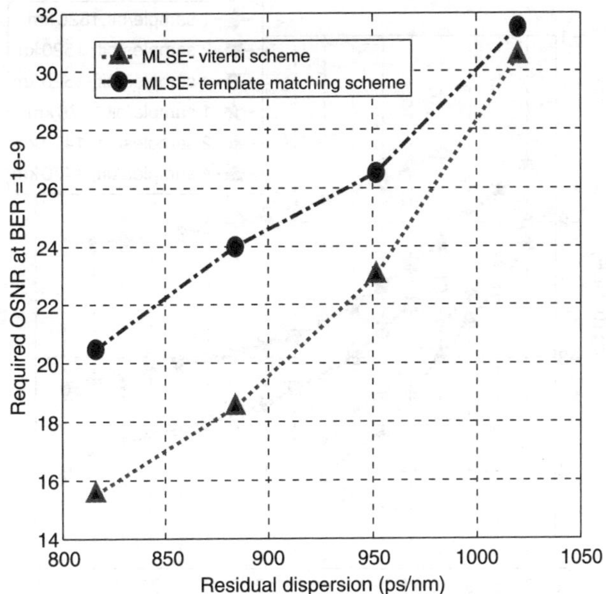

FIGURE 11.23 Required OSNR (at BER = 1e-9) versus residual dispersion in Viterbi-MLSE and template-matching MLSE schemes.

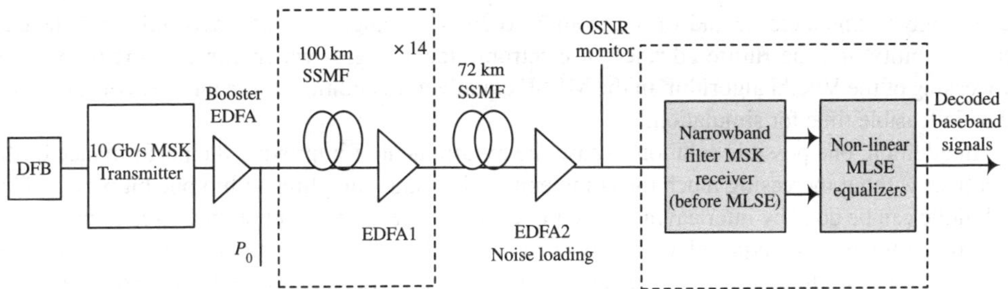

FIGURE 11.24 Simulation setup for transmission of 10 Gb/s optical MSK signals over 1472 km SSMF with MLSE-Viterbi equalizer integrated with the narrowband optical filter receiver.

one, two and four samples per bit, respectively. 1520 km SSMF with one samples per bit is seen as the limit for 1024-state Viterbi algorithm due to the slow roll-off and the error floor. However, two and four samples per bit can obtain error values lower than the FEC limit of 1e-3.

Thus, 1520 km SSMF transmission of 10G/bs optical MSK signals can reach the error-free detection with use of a high performance FEC. In case of 1472 km SSFM transmission, the error events follow a linear trend without sign of error floor and therefore, error-free detection can be comfortably accomplished.

Figure 11.25 also shows the significant improvement in OSNR of two and four samples per bit over 1 sample per bit counterpart with values of approximately 5 dB and 6 dB respectively. In terms of OSNR penalty at BER of 1e-3 from back-to-back setup, four samples per bit for 1472 km and 1520 km transmission distance suffers 2 dB and 5 dB penalty, respectively.

11.7.3 PERFORMANCE LIMITS OF VITERBI-MLSE EQUALIZERS

The performance limits of Viterbi-MLSE equalizer to combat ISI effects are investigated against various SSMF lengths of the optical link. The number of states used in the equalizer is incremented

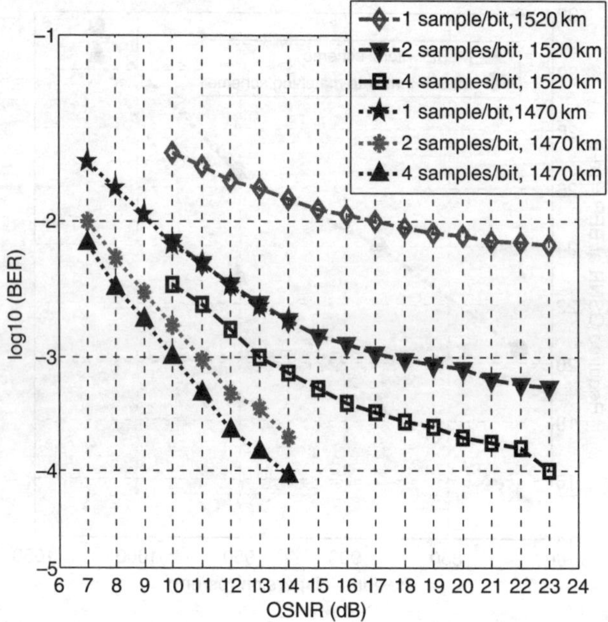

FIGURE 11.25   BER versus required OSNR for transmission of 10 Gb/s optical MSK signals over 1472 and 1520 km SSMF uncompensated optical link.

accordingly to this increase and varied from 2^6 to 2^{10}. This range was chosen as reflecting the current feasibility and the future advance of electronic technologies which can support high-speed processing of the Viterbi algorithm in the MLSE equalizer. In addition, these numbers of states also provide feasible time for simulation.

In addition, one possible solution to ease the requirement of improving the performance of the equalizer without increasing much the complexity is by multi-sampling within one bit period. This technique can be done by interleaving the samplers at different time. Although, a greater number of electronic samplers are required, they only need to operate at the same bit rate as the received MSK electrical signals. Moreover, it will be shown later on that there is no noticeable improvement with more than two samples per bit period. Hence, the complexity of the MLSE equalizer can be affordable while improving the performance significantly.

Figure 11.26 shows the simulation setup for 10 Gb/s optical MSK transmission systems with lengths of uncompensated optical links varying up to 1472 km SSMF. In this setup, the input power into fiber (P_0) is set to be -3 dBm, thus avoid the effects of fiber non-linearities. The optical amplifier EDFA1 provides an optical gain to compensate the attenuation of each span completely. The EDFA2 is used as a noise loading source to vary the required OSNR values. Moreover, a Gaussian filter with two-sided 3 dB bandwidth of 9 GHz (one-sided BT = 0.45) is utilized as the optical discrimination filter because this BT product gives the maximized detection's eye openings (refer to Section 6.4). The receiver electronic noise is modelled with equivalent noise current density of the electrical amplifier of $20pA/\sqrt{Hz}$ and a dark current of 10 nA for each photodiode. A back-to-back OSNR = 15 dB is required for a BER of 1e-4 on each branch and the correspondent receiver sensitivity is -25 dBm. A trace back length of 70 is used for the Viterbi algorithm in the MLSE equalizer. Figure 11.26, Figure 11.27 and Figure 11.28 show the BER performance curves of 10 Gb/s OFDR-based MSK optical transmission systems over 928 km, 960 km, 1280 km and 1470 km SSMF uncompensated optical links for different number of states used in the Viterbi-MLSE equalizer. In these figures, the performance of Viterbi-MLSE equalizers is given by the plot of the BER versus the required OSNR for several detection configurations: balanced receiver (without the equalizer), the conventional single-sample per bit

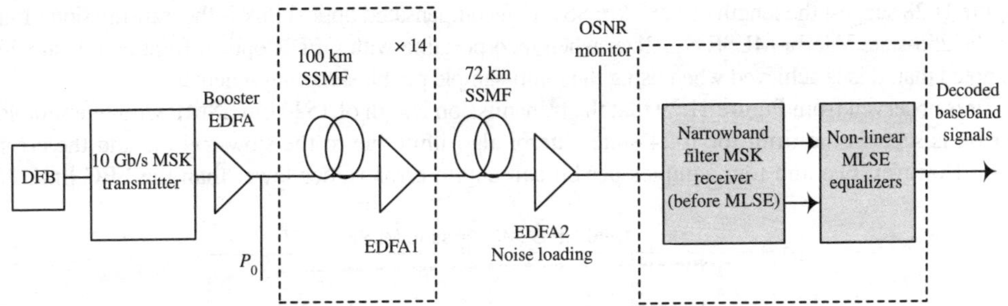

FIGURE 11.26 Simulation setup of OFDR-based 10 Gb/s MSK optical transmission for study of performance limits of MLSE-Viterbi equalizer.

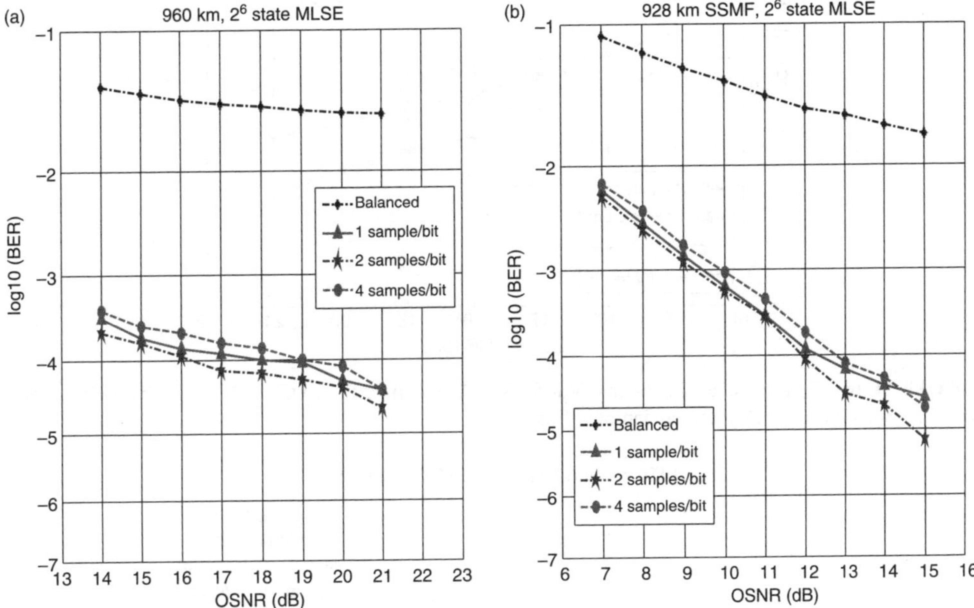

FIGURE 11.27 Performance of 64-state Viterbi-MLSE equalizers for 10 Gb/s OFDR-based MSK optical systems over (a) 960 km and (b) 928 km SSMF uncompensated link.

sampling technique and the multi-sample per bit sampling techniques (two and four samples per a bit slot). The simulation results are obtained by the Monte Carlo method.

The significance of multi-samples per bit-slot in improving the performance for MLSE equalizer in cases of uncompensated long distances is shown. It is found that the tolerance limits to the ISI effects induced from the residual CD of a 10-Gb/s MLSE equalizer using 2^6, 2^8 and 2^{10} states are approximately equivalent to lengths of 928 km, 1280 and 1440 SSMF, respectively. The equivalent numerical figures in the case of 40 Gb/s transmission are corresponding to lengths of 62 km, 80 km and 90 km SSMF, respectively.

In the case of 64 states over 960 km SSMF uncompensated optical link, the BER curve encounters an error floor which can not be overcome even by using high performance FEC schemes. However, at 928 km, the linear BER curve indicates the possibility of recovering the transmitted data with the use of high performance FEC. Thus, for 10 Gb/s OFDR-based MSK optical systems, a length of 928 km SSMF can be considered as the transmission limit for the 64-state Viterbi-MLSE equalizer. Results shown in

Figure 11.28 suggest the length of 1280 km SSMF uncompensated optical link is the transmission limit for the 256-state Viterbi-MLSE equalizer when incorporating with OFDR optical front-end. It should be noted that, this is achieved when using the multi-sample per bit sampling schemes.

It is observed from Figure 11.29 that the transmission length of 1520 km SSMF with one sample per bit is seen as the limit for 1024-state Viterbi algorithm due to the slow roll-off and the error floor. However, two and four samples per bit can obtain error values lower than the FEC limit of

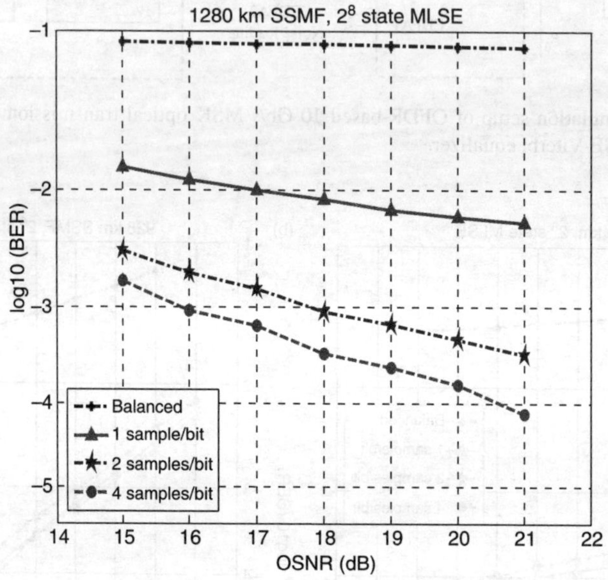

FIGURE 11.28 Performance of 256-state Viterbi-MLSE equalizer for 10 Gb/s OFDR-based MSK optical systems over 1280 km SSMF uncompensated link.

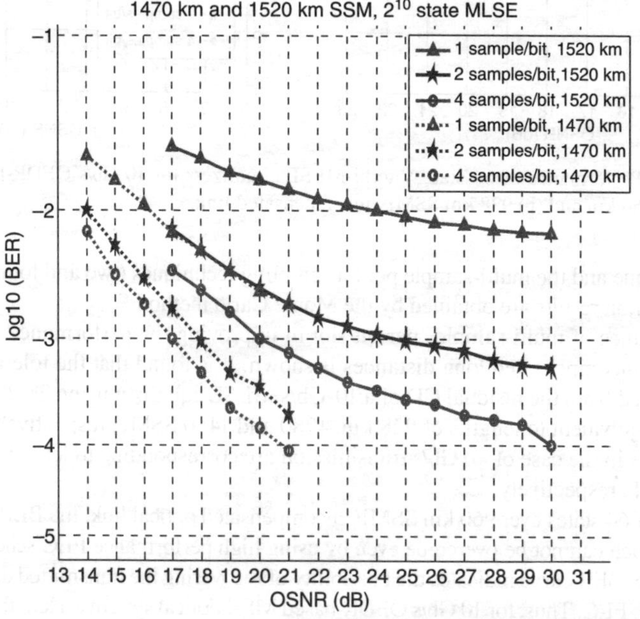

FIGURE 11.29 Performance of 1024-state MLSE equalizer for 10 Gb/s optical MSK signals versus required OSNR over 1472 and 1520 km SSMF uncompensated optical link.

1e-3. Thus, 1520 km SSMF transmission of 10 G/bs optical MSK signals can reach the error-free detection with use of a high performance FEC. In case of 1472 km SSFM transmission, the error events follow a linear trend without sign of error floor and therefore, error-free detection can be achieved. Figure 11.29 also shows the significant OSNR improvement of the sampling techniques with two and four samples per bit compared to the single-sample per bit counterpart. In terms of OSNR penalty at BER of 1e-3 from back-to-back setup, four samples per bit for 1472 km and 1520 km transmission distance suffers 2 dB and 5 dB penalty, respectively.

It is found that for incoherent detection of optical MSK signal based on the OFDR, MLSE equalizer using 2^4 states does not offer better performance than the balanced configuration of the OFDR itself, i.e., the uncompensated distance is not over 35 km SSMF for 40 Gb/s or 560 km SSMF for 10 Gb/s transmission systems, respectively. It is most likely that in this case, the severe ISI effect caused by the optical fiber channel has spread beyond the time window of five bit slots (two precursor and two post-cursor bits), the window which 16-state MLSE equalizer can handle.

The trace-back length used in the investigation is chosen to be 70 which guarantees the convergence of the Viterbi algorithm. The longer the trace-back length, the larger memory is required. With state-of-the-art technology for high storage capacity nowadays, memory is no longer a big issue. Very fast processing speed at 40 Gb/s operations hinders the implementation of 40 Gb/s Viterbi-MLSE equalizers at the mean time. Multi-sample sampling schemes offer an exciting solution for implementing fast signal processing process. With the advance of the semiconductor industry, this challenge may also be overcome in the near future. At the present, the realization of Viterbi-MLSE equalizers operating at 10 Gb/s has been commercially demonstrated [16].

11.7.4 Viterbi-MLSE Equalizers for PMD Mitigation

Figure 11.30 shows the simulation test-bed for the investigation of MLSE equalization of the PMD effect. The transmission link consists of a number of spans which are comprised of 100 km SSMF (with $D = +17$ ps/nm.km, $\alpha = 0.2$ dB/km) and 10 km DCF (with $D = -170$ps/nm.km, $\alpha = 0.9$ dB/km). Input power into each span (P_0) is -3 dBm. The EDFA1 has a gain of 19 dB, hence providing input power into the DCF to be -4 dBm which is lower than the non-linear threshold of the DCF. The 10 dB gain of EDFA2 guarantees the input power into next span unchanged of -3 dBm value. An OSNR of 10 dB is required for receiver sensitivity at BER of 1e-4 in case of back-to-back configuration. Considering the practical aspect and complexity a Viterbi-MLSE equalizer for PMD equalization, a small number of 4-state bits, or effectively 16 states were chosen for the Viterbi algorithm in the simulation study.

The performance of MLSE against the PMD dynamic of optical fiber is investigated. BER versus required OSNR for different values of normalized average differential group delay (DGD) $\Delta\tau$ are shown in Figure 11.31. The mean DGD factor is normalized over one bit period of 100 ps and 25 ps for 10 and 40 Gb/s bit rate, respectively. The numerical studies are conducted for a range of values from 0 to 1 of normalized DGD $\Delta\tau$ which equivalently corresponds to the instant delays of up to 25 ps or 100 ps in terms of 40 Gb/s and 10 Gb/s transmission bit rate, respectively.

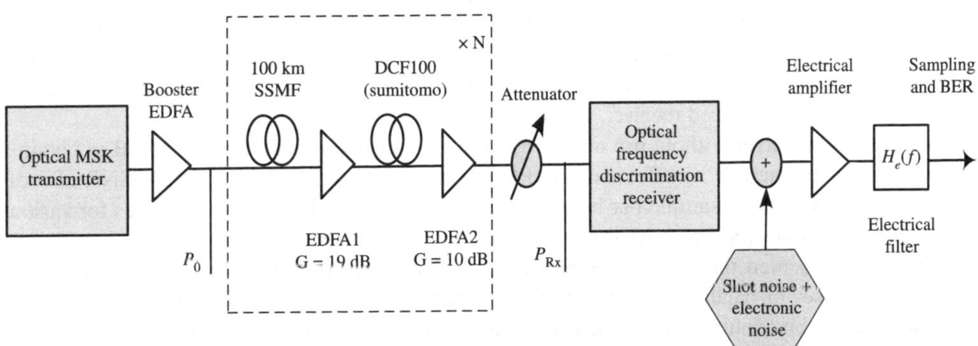

FIGURE 11.30   Simulation test-bed for investigation of effectiveness of MLSE equalizer to PMD.

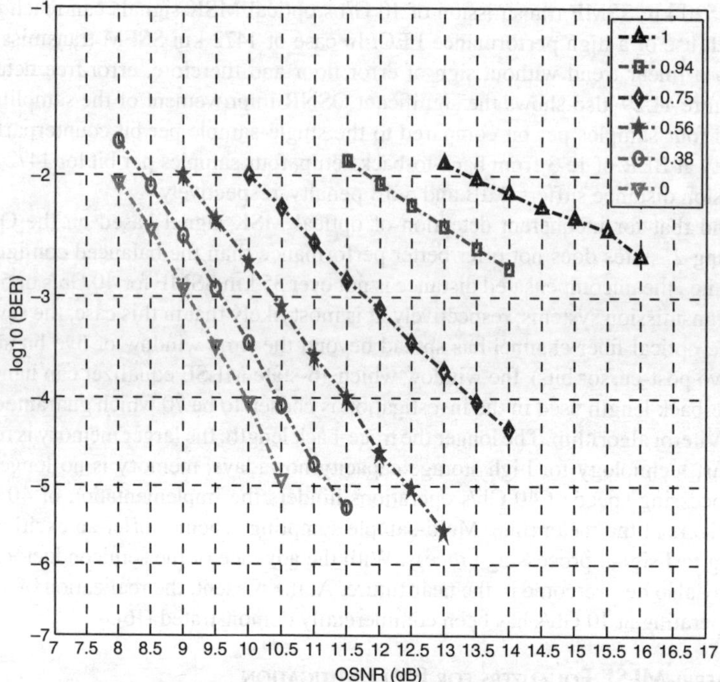

FIGURE 11.31 Performance of MLSE equalizer versus normalized Δτ values for one sample per bit. (From Sivahumaran, T., T. L. Huynh, K. K. Pang, and L. N. Binh, Nonlinear equalizers in narrowband filter receiver acheiving 950 ps/nm residual dispersion tolerance for 40 Gb/s optical MSK transmission systems. In *Proceedings of OFC'07, Paper OThK3,* Annaheim, CA, 2007. With permission.)

The advantages of using multiple samples per bit over the conventional single sample per bit in MLSE equalizer are also numerically studied and the Monte Carlo results for normalized DGD values of 0.38 and 0.56 are shown in Figure 11.32 and Figure 11.33. However, the increase from two to four samples per bit does not offer any gain in the performance of the Viterbi-MLSE equalizer. Thus, two samples per bit are preferred to reduce the complexity of the equalizer. Here and for the rest of this chapter, multi samples per bit implies the implementation of two samples per bit. Performance of MLSE equalizer for different normalized DGD values with two samples per bit is shown in Figure 11.34. From Figure 11.34, the important remark is that 16-state MLSE equalizer implementing two samples per bit enables the optical MSK transmission systems achieving a PMD tolerance of up to one bit period at BER = 1e-4 with a required OSNR of 8 dB. This delay value starts introducing a BER floor indicating the limit of the 16-state MLSE equalizer. This problem can be overcome with a high performance FEC. However, the DGD mean value of 0.94 can be considered as the limit for an acceptable performance of the proposed MLSE equalizer without aid of high performance FEC. Figure 11.35 shows the required OSNR for MLSE performance at 1e-4 versus normalized Δτ values in configurations of balanced receiver, one sample per bit and two samples per bit.

A balanced receiver without use of the MLSE equalizer requires an OSNR = 5 dB to obtain a BER = 1e-4 when there is no effects of the PMD at all compared to the required OSNR of 3 dB and 1 dB in cases of one sample per bit and two samples per bit. The OSNR penalties for various normalized Δτ values for the above three configurations are shown in Figure 11.36.

It can be observed that the OSNR penalties (back-to-back) of approximately 3 dB and 1 dB apply to the cases of balanced receiver and one sample per bit, respectively with reference to the two samples per bit conturation. Another important remark is that 16-state Viterbi-MLSE equalizer

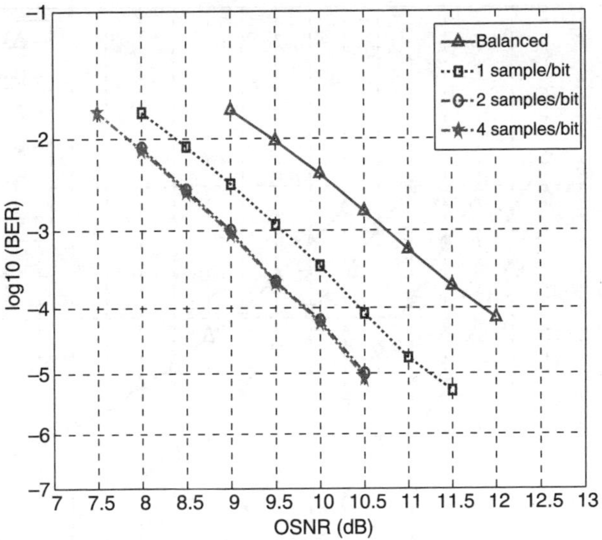

FIGURE 11.32 Comparison of MLSE performance for configurations of one, two and four samples per bit with normalized $\Delta\tau$ value of 0.38.

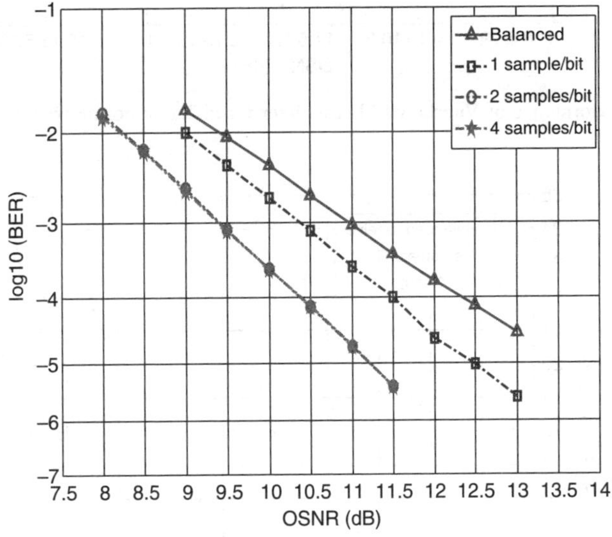

FIGURE 11.33 Performance of Viterbi-MLSE equalizer for configurations of one, two and four samples per bit with a normalized $\Delta\tau$ value of 0.56.

which implements two samples per bit enables the optical MSK transmission systems achieving a PMD tolerance of up to one bit period at a BER of 1e-4 with a power penalty of about 6 dB. Moreover, the best 2-dB penalty occurs at 0.75 for the value of normalized $\Delta\tau$. This result shows that the combination of OFDR-based MSK optical systems and Viterbi-MLSE equalizers, particularly with the use of multi-sample sampling schemes was found to be highly effective in combating the fiber PMD dynamic impairment and better than recently reported PMD performance for OOK and DPSK modulation formats [17].

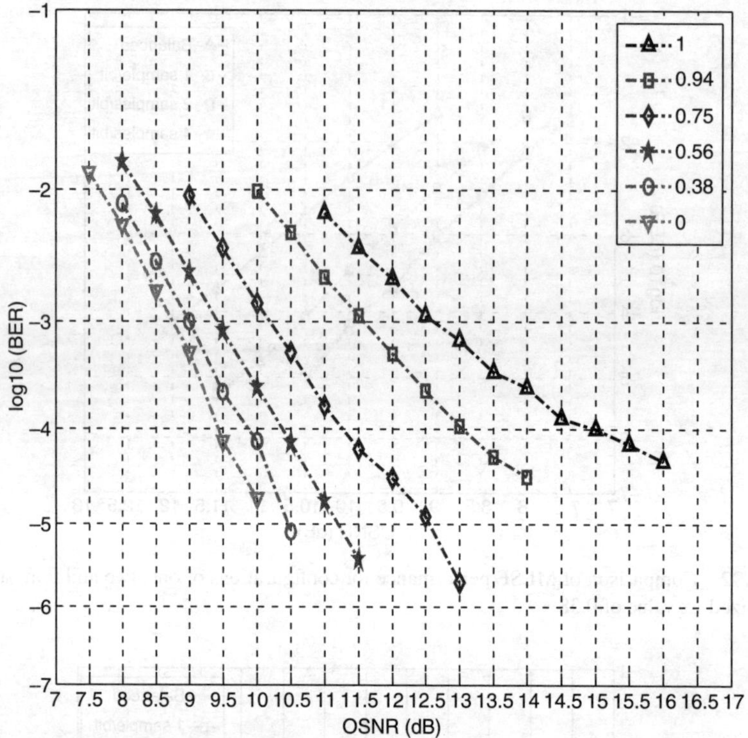

FIGURE 11.34 Performance of Viterbi-MLSE equalizer for different normalized $\Delta\tau$ values with two and four samples per bit.

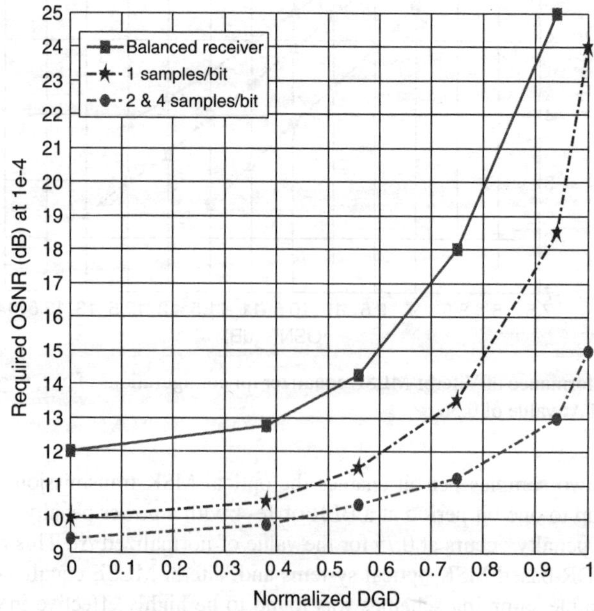

FIGURE 11.35 Required OSNR for MLSE performance at 1e-4 versus normalized $\Delta\tau$ values in configurations of balanced receiver and one and two samples per bit.

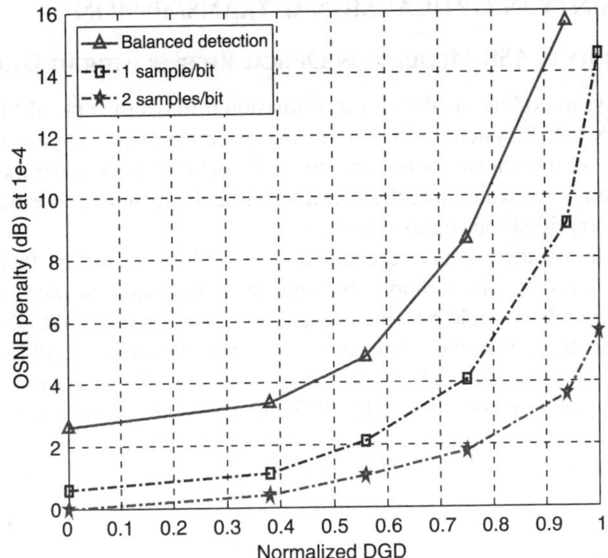

FIGURE 11.36   OSNR penalties of MLSE performance at 1e-4 for various normalized $\Delta\tau$ values in configurations of balanced receiver, single-sample, and two samples per bit sampling schemes.

11.7.5 ON THE UNCERTAINTY AND TRANSMISSION LIMITATION OF EQUALIZATION PROCESS

The fundamental limitations of a quadratic phase media on the transmission speed are formulated in signal space for significant applications in signal equalization. These limitations are quantitatively derived for both coherent and incoherent optical systems. This section investigates the dispersive effect of optical fiber on the transmission speed of the media. The main effect of dispersion is pulse broadening (and frequency chirping) as the signal pulses propagate through the quadratic phase media. This causes intersymbol interference (ISI) and thereby limiting the speed of transmission. There are various ways of mitigating the harmful effect of ISI. One way is to use optical fiber with anti-dispersive property to equalize the dispersion. In practice, this optical scheme reduces signal level and thus requires amplification. The amplifier introduces noise, which in turn limits the channel speed (Shannon's information capacity theorem). Another way of combating ISI is by digital electronic equalization means such as maximum likelihood sequence estimation (MLSE) which is applicable for both coherent and incoherent detection schemes [18–22]. These reported results show clearly that there is a fundamental limit on the achievable transmission speed for a given fiber length.

This section studies the limitation of quadratic phase channel, the single mode optical fiber, in terms of signal space. The signal space is chosen so that it consists of binary signals in 8-bit block code. As the 8-bit symbols propagate down the quadratic phase channel, naturally the waveform patterns of the symbols would become less distinctive, hence more difficult to discriminate between symbols. How do we quantify the detrimental effect of the quadratic phase channel? Although it is obvious that the more information we have to hand the more accurately one would be able to obtain the BER. We approach this problem entirely from the digital communication perspective, and the results so derived are the fundamental limits imposed by the quadratic phase channel. It is virtually independent of the detection scheme used. Two mechanisms that limit the transmission speed are explained. One is brought about by the finite time window available for detection. In all practical schemes, the decoder must decode each symbol within a finite time. The other is brought about by not using the phase information for detection. Depending on the complexity of the detection scheme chosen, it is quite often that the phase information may be lost inadvertently when the optical signal is converted into the electrical signal.

11.8 UNCERTAINTY IN OPTICAL SIGNAL TRANSMISSION

11.8.1 UNCERTAINTY IN ASK MODULATION OPTICAL RECEIVER WITHOUT EQUALIZATION

There are many ways of coding the data for transmission through the optical fiber. For simplicity, we first choose a NRZ-ASK system for this study and the transmission rate is operating at 40 Gb/s. Later in this section we investigate the uncertainty of the data recovery under the MSK modulation and the use of equalization technique, the maximum likelihood sequence estimation (MLSE) in long-haul optically amplified transmission system.

The NRZ-ASK transmitted signal is organized into 8-bit blocks and thus, there are 2^8 (256) possible pattern symbols. As the 8-bit symbols propagate down the quadratic phase channel, the waveform patterns of the symbols would become less and less distinctive. It makes the task of decoding more and more difficult and uncertain. How do we measure this uncertainty? In the field of digital communications, the uncertainty of MLSE detection is measured in symbol error rate (SER) which is a function of the signal to noise ratio (SNR). SNR, in turn, depends on the distances between signal constellation, and it is the shortest distance that dominates the error probability. The SER and the Q-factor are expressed as a function of the normalized shortest Euclidean distance metric d_{norm} between all possible received symbol pair combinations as:

$$\text{SER} = Q\left(\sqrt{\text{SNR}}\right) = Q\left(\sqrt{d_{norm}^2 \times \text{SNR}_{nofiber}}\right) \qquad 0 < d_{norm} \le 1 \qquad (11.91)$$

where $\text{SNR}_{nofiber}$ is the optical SNR (OSNR) in the case of back to back setup i.e., without fiber. From Equation 11.91, the logarithmic scale of the d_{norm}^2 represents the OSNR penalty as expressed in the following equation:

$$d_{norm[dB]}^2 = \text{OSNRPenalty} = \text{SNR}_{[dB]} - \text{SNR}_{norfiber[dB]}. \qquad (11.92)$$

Figure 11.37 depicts the penalty in the discrimination property with the OSNR penalty being plotted against the length of the standard single mode fiber (SSMF) (the dispersion factor $D = 17$ ps/(nm. km)) for the case of coherent detection schemes i.e., both the magnitude and the phase information of the received waveform are used.

The family of curves shown in Figure 11.37 corresponds to a range of different detection window widths; from a window of 8 bit periods (the minimum detection width) to 9, 10, and 32 bit periods. As expected, the discrimination property is significantly improved as the detection window length

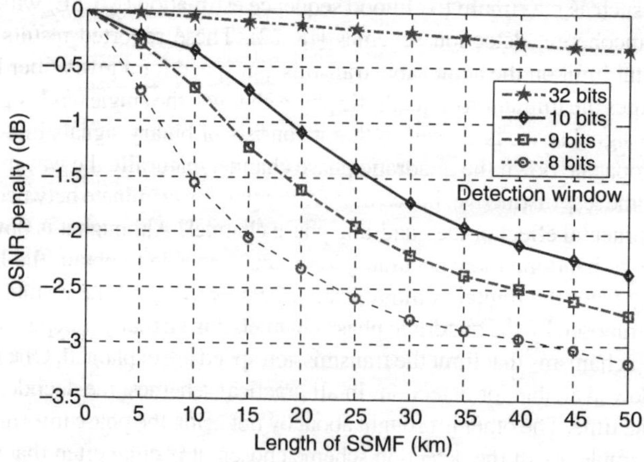

FIGURE 11.37 Discrimination penalty versus transmission distance at 40 Gb/s when using both magnitude and phase information.

increases. In the case of 32 bit window length, there is insignificant penalty over the fiber lengths of up to 50 km SSMF. This result re-affirms the previous discussion that the channel capacity is infinite, i.e., there is no uncertainty if the detection time window is long enough. At the fiber distance of 5 km, a penalty of about 0.7 dB occurs if the 8-symbol detection width is used. Depending on the system noise level, this signal level may or may no be acceptable for a reliable data transmission.

We are now considering incoherent optical transmission systems that utilize a photodiode to convert the optical signals to electrical signals. The use of photodiodes as such could result in the loss of the phase information of the signal. Furthermore, electrical equalization techniques based on the MLSE principles employ only the magnitude information [18,23,24]. To model this incoherent property, the distance metric utilizes the absolute magnitude of the received waveform only. The discrimination penalty versus the length of SSMF is depicted Figure 11.38.

The family of curves in Figure 11.38 also corresponds to a range of different detection window widths (8, 9 10 and 32 bit periods). When the length of the fiber is less than about 15 km, there is no appreciable improvement by choosing a larger detection window. However, when the length of the fiber increases further, the deterioration is quite severe and having a larger detection window helps enormously in reducing this deterioration. At the SSMF length of 5 km, the incoherent detection suffers a penalty of approximately 3.5 dB compared to the 0.7 dB penalty for the coherent detection and thus, leading to an improvement of about 3 dB when using the coherent schemes with the availability of both amplitude and phase information. This agrees with the popularly known advantage on the receiver sensitivity of coherent detections over the incoherent counterparts. In addition, comparing the two sets of results in Figure 11.38 and Figure 11.37 shows clearly the extent of improvement possible if a coherent scheme is used instead of a incoherent scheme. This is achieved at the expense of greater system complexity on the receiver structure of coherent optical systems.

11.8.2 Uncertainty in MSK Optical Receiver with Equalization

The equalization processing system is placed at the output of the optical receiver shown in Figure 11.30. The equalization is completed in the electronic domain, i.e., the optical impulse and pulses have been converted into the electrical domain and the phase chirping has already converted to amplitude distortion. However the broadening of the pulse influence the sampled values obtained from the broadened pulses for the processing of the equalization system.

The narrowband filter receiver in Figure 11.14 is now integrated with the MLSE non-linear equalizers for the detection of 40 Gb/s optical MSK signals. Two narrowband filters are used to discriminate the USB and the LSB frequencies that correspond to logic 1 and 0 transmitted, respectively. A constant optical delay line which is easily implemented in integrated optics is introduced on

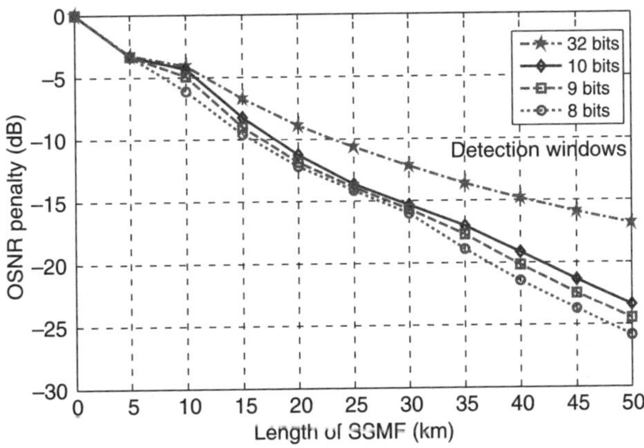

FIGURE 11.38 Discrimination penalty versus transmission distance at 40 Gb/s when considering only the magnitude information.

one branch to compensate the differential group delay $t_d = 2\pi f_d \beta_2 L$ between the 1 and 0 logic pulses of frequencies f_1 and f_2 where $fd = f_1 - f_2 = R/2$ β_2 represents group velocity delay (GVD) parameter of the fiber and L is the fiber length. If the differential group delay is fully compensated, the optical lightwaves in two paths arrive at the photodiodes simultaneously. The outputs of the filters are then converted to electrical domain through the photodiodes. These two separately-detected electrical signals are sampled before being fed as the inputs to the non-linear equalizer. The MLSE equalizer requires values of the samples in both advanced and later events. This is the impact of the impulse response of the fiber on this equalization and the number of events required for the detection and equalization is very critical. Figure 11.29 shows the performance of the optical frequency discrimination integrated with an MLSE equalizer in a long-haul optically amplified fiber transmission systems of about 1500 km when the number of one, two and four samples per bit is used in the equalization processing system. The time distance between the samples is controlled so that they are within the reasonable time slot under the effects of chirping of the phase of the lightwave.

11.9 ELECTRONIC DISPERSION COMPENSATION OF MODULATION FORMATS

Accepting the MLSE as the pattern comparison and decision made based on the matching, the equalization techniques using FFE, FFE-DFE, non-linear equalization and FFE result in different improvements in the eye opening of digital modulation formats of either amplitude shift keying of phase shift keying.

Simulations are conducted for NRZ formats with the modulation of ASK, duobinary, DPSK and SSB. The transmission setup parameters are optically amplification used for compensate all losses due to the fiber, no dispersion compensation fibers are used. The bit rate is 10 Gb/s and optical filter of 50 GHz 3 dB passband is used before the opto-electronic detection. Transmission fibers are SSMF with 17 ps/nm/km dispersion factor is assumed. The OSNR is plotted against the linear dispersion factor and are shown in Figure 11.39, Figure 11.41, and Figure 11.42 for NRZ-ASK, NRZ-duobinary (50%), NRZ-DPSK and NRZ SSB formats. It notes that (i) moderate improvement of the OSNR for NRZ-ASK (Figure 11.39); (ii) Significant improvement of OSNR for SSB format (see Figure 11.42); (iii) Mild improvement for DPSK format; and (iv) moderate improvement of the duobinary format.

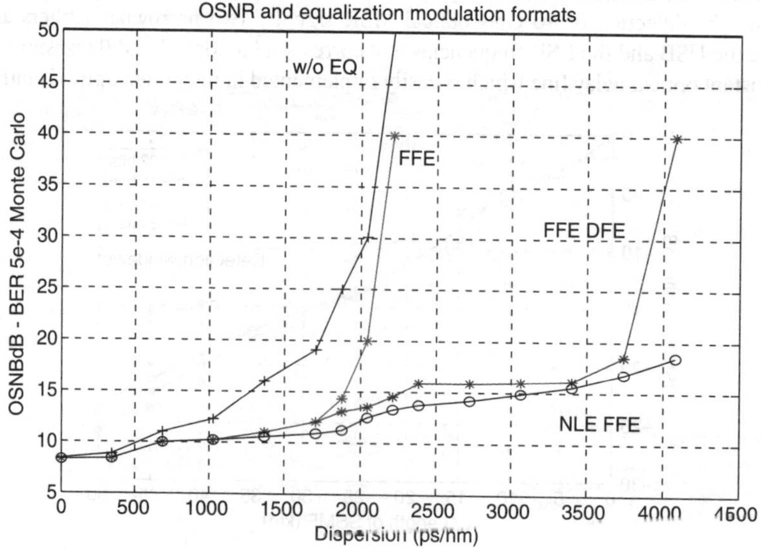

FIGURE 11.39 OSNR required for BER 5e-4 (Monte Carlo simulation) versus linear CD under different equalization schemes for NRZ-ASK modulation format.

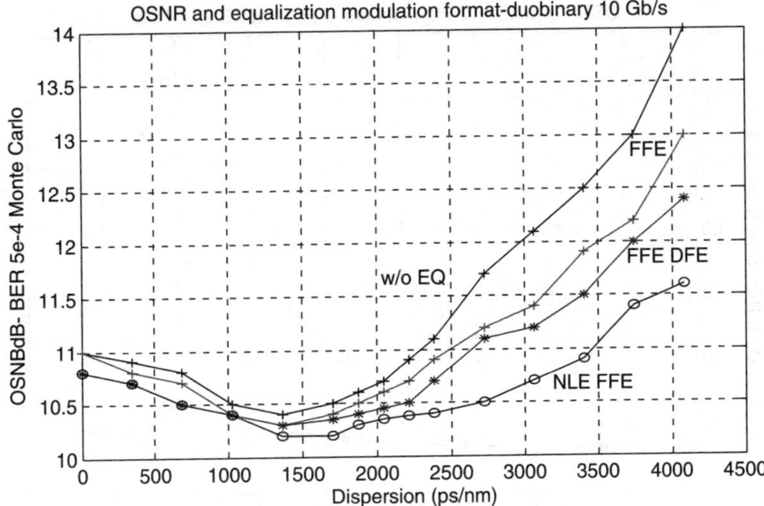

FIGURE 11.40 OSNR required for BER 5e-4 (Monte Carlo simulation) versus linear CD under different equalization schemes for NRZ-duobinary modulation format.

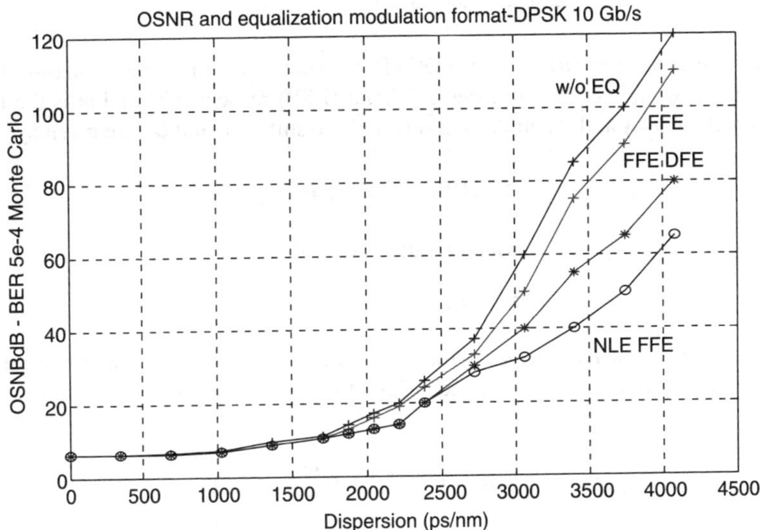

FIGURE 11.41 OSNR required for BER 5e-4 (Monte Carlo simulation) versus linear CD under different equalization schemes for NRZ-DPSK modulation format.

Thus effectively we note that significant improvement of the eye opening is achieved for single side band signals. So why this is so? In order to explain these results, refer to recent clarifications by Rosenkranz and Xia [25], and earlier attempts by Sieben et al. [26]. The behavior of the signals received after the photodetector, a square law detection device, can be explained using the phasor representation given in Figure 11.43.

Considering a monochromatic wave of frequency ω_1 of the double side band of the NRZ-ASK modulated carrier, then under ASK modulation the complex field envelope of the received signals entering the photodetector is given by

$$s_{ASK}(t) = 1 + A\frac{m}{2}e^{j(\omega_1 t - \phi_A)} + B\frac{m}{2}e^{j(-\omega_1 t - \phi_B)}. \tag{11.93}$$

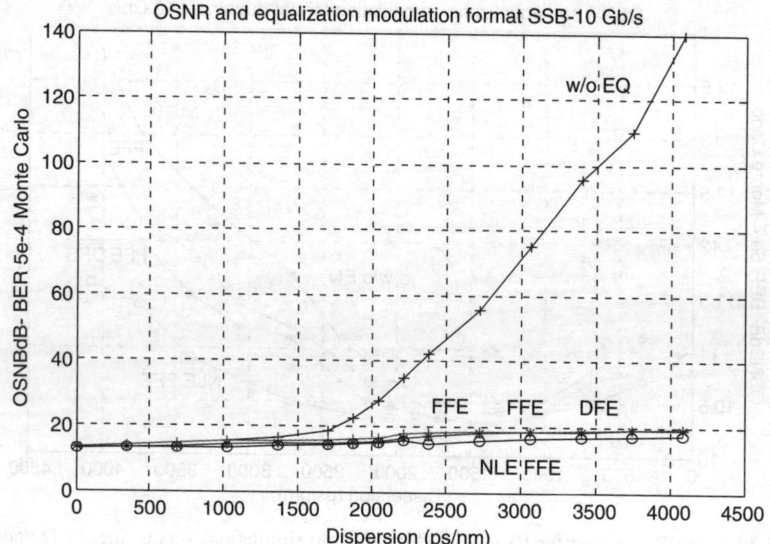

FIGURE 11.42　OSNR required for BER 5e-4 (Monte Carlo simulation) versus linear CD under different equalization schemes for NRZ-SSB modulation format.

The first term represents the carrier term at DC after the detection, the second and third terms come from the upper sideband (USB) and lower sideband (LSB) respectively, and m is the modulation index. The coefficients A and B and the additional phase shifts, ϕ_A and ϕ_B, are given by

$$A = \left| H(\omega_1) \right| = B = \left| H(-\omega_1) \right|$$

$$\text{with} \quad H(\omega) = \exp(-j\beta_2\omega^2) \tag{11.94}$$

$$\phi_A = \phi_B = \beta_2\omega_1^2 L/2.$$

With $H(\omega)$ is the fiber transfer function in the frequency domain. Thus the electronic current generated by the photodetector is given by the squaring of the magnitude of the optical field and can be written as

$$\left| s_{ASK}(t) \right|^2 = \left| 1 + A\frac{m}{2}e^{j(\omega_1 t - \phi_A)} + B\frac{m}{2}e^{j(-\omega_1 t - \phi_B)} \right|^2$$

$$\tag{11.95}$$

$$= 1 + \frac{m^2}{2} + 2m\cos\phi_A\cos\omega_1 t + \frac{m^2}{2}\cos 2\omega_1 t.$$

We note from this equation that there are three terms: a DC term, a first harmonic term (linear term) and a second harmonic term which represents the distortion due to dispersion.

For the DPSK, duobinary and SSB modulation formats we can follow a similar analysis to obtain the electronic currents generated after the photodetector

$$\left| s_{DuoB}(t) \right|^2 = \left| s_{DuoB}(t) \right|^2 = \frac{m^2}{2}(1 + 2m\cos\omega_1 t)$$

$$\tag{11.96}$$

$$\left| s_{SSB}(t) \right|^2 = (1 + \frac{m^2}{4} + m\cos(\omega_1 t - \phi_A).$$

Obviously the currents are linear and no non-linear term existed.

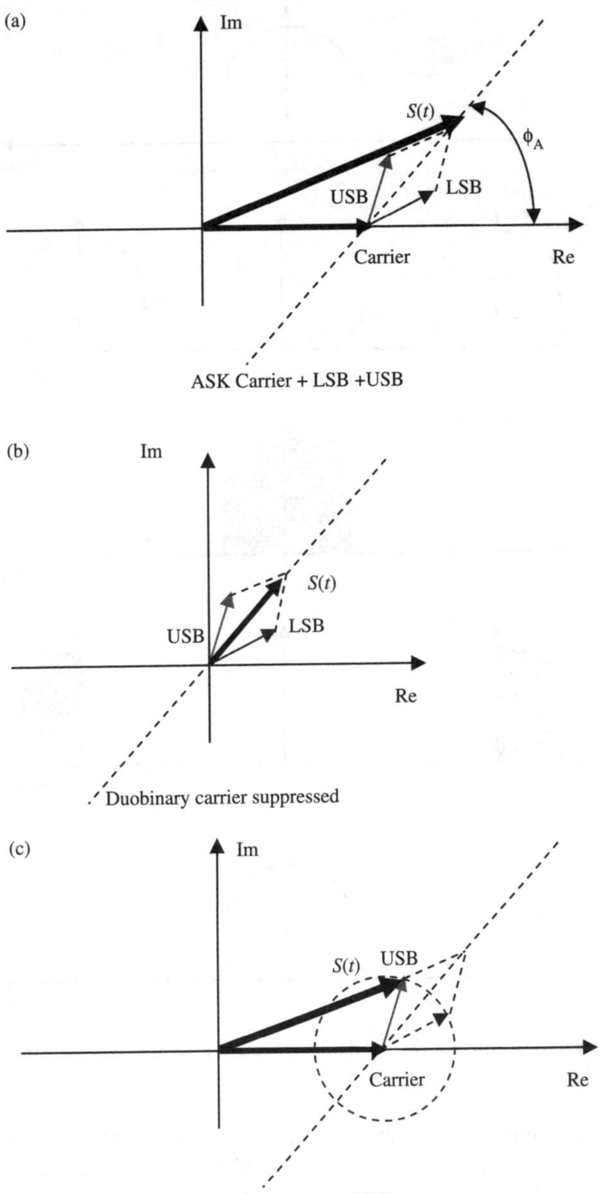

FIGURE 11.43 Phasor representation of optical signals under different modulation formats (a) ASK, with carrier suppression and lower side band (LSB) and upper side band (USB) (c) SSB with carrier suppressed plus USB.

A linear equalizer equalizes the signal waveform by minimizing the ISI due to the linear part of Equation 11.95 and Equation 11.96. When the phase term $\cos(\phi_A)$ becomes smaller after a propagation length of about 100 km (or 1700 ps/nm) the linear term diminishes significantly and the linear equalization would have no effect. The non-linear distortion term becomes significant and takes over the linear term as we can observe from Figure 11.39.

It is noted that the linear part comes from the carrier component. While for DPSK and duobinary the carrier is embedded into the signal band and plays no role. This is mentioned before that it is

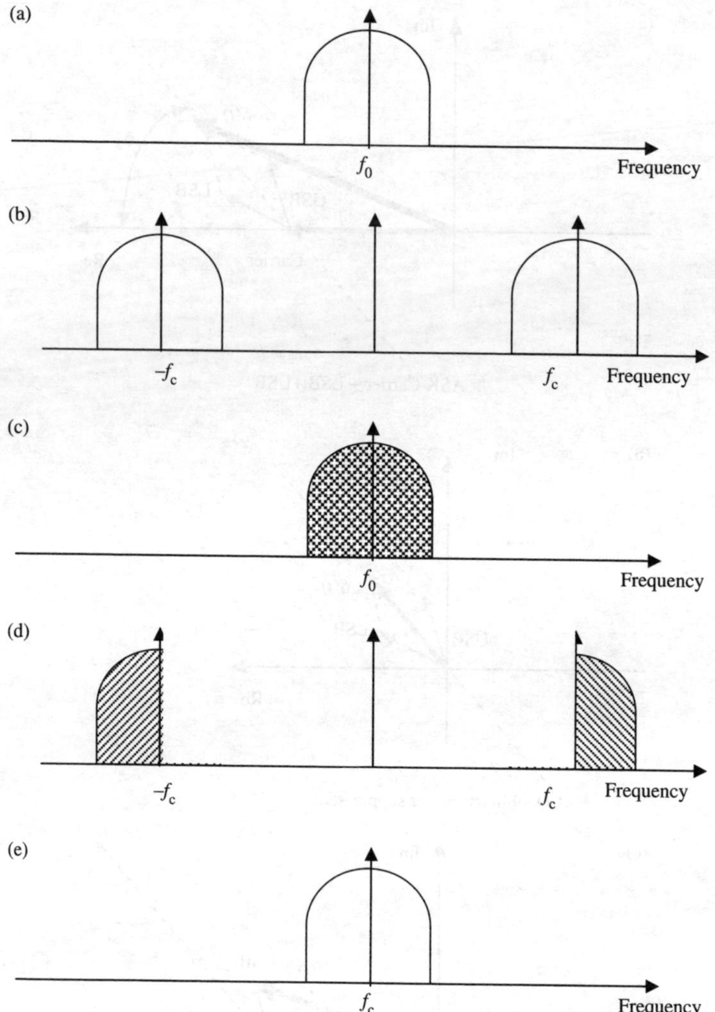

FIGURE 11.44 Spectral illustration of DSB and SSB operation after the lightwave modulated signals propagating through a fiber length.

due to the effective sampling of the waveform by the switching on and off of the amplitude in ASK. Thus from Equation 11.96 the linear term disappear and hence the linear equalizer is not effective as observed from Figure 11.40 and Figure 11.41.

Now then for SSB modulation we could se that there is a DC term and a linear part. Thus the linear FFE would perform very well and equals that of the non-linear FFE and DFE system as observed in Figure 11.42.

We can visualize the effects of pure phase distortion channel on modulation schemes with DSB and SSB as shown in Figure 11.44. For DSB modulation the spectrum shown in Figure 11.44a in the baseband is shifted to the passband (Figure 11.44b). After propagating through the fiber the spectrum amplitude remains the same but its phase altered by the fiber phase distortion by an amount $-\beta_2 L f_2$.

So when the signals are recovered in the photodetector the double side bands interfered with each other with the overlapping (hatched area) in Figure 11.44c. Thus distortion is generated by both the in band distortion and the interference between the two side bands. On the other hand for SSB, this recovered process would not result in any interference. So distortion could only come from the pure distortion within the bands and no distortion due to spectral overlapping. Thus linear equalization works well for SSB linear distortion.

11.10 CONCLUSIONS

We have described the equalization of real and imaginary signals using FFE, DFE and FFE, linear and non-linear equalizers that are placed at either the transmitters, the receivers or shared between receivers and transmitters. The effects of equalizers on modulation formats are identified.

For phase shift keying and duobinary the carrier is embedded within the signal band and thus only non-linear frequency doubling terms exists in the electronic current generated by the photo-detector. Thus non-linear equalization works well and linear FFE contributes no effect when the linear term diminished after certain lengths of fiber. The electronic current generated under the SSB modulation format, on the other hand exhibits only the linear term and no frequency doubling term (no non-linear term), thus the linear equalizer FFE works as well as the non-linear equalizer. Thus only FFE is required as the implementation of the FFE is much simpler at high speed. We thus expect that linear FFE also works well for MSK modulation formats.

Supplementing these investigations to prove the effects of the carrier in modulation formats, two case studies have been conducted in detail. Firstly, a study of a linear electrical equalizer for duobinary modulation formats has been conducted and two non-linear equalizers are proposed. MLSE using Viterbi algorithm and the modified MLSE scheme, which are integrated with the narrowband optical filter receiver proposed elsewhere [27–30] for non-coherent detection of 40 Gb/s optical MSK signals. This chapter has presented the non-linear MLSE equalizers, especially when integrated with Viterbi algorithm, for digital photonic communications in general and for OFDR-based MSK optical systems. Backgrounds for MSLE and Viterbi algorithm are briefly given and the construction of the state trellis structure for use in optical communications has been explained. It has also been shown that noise after tight optical filtering follows the Gaussian distribution and can be minimized. Thus, the Viterbi algorithm can obtain its optimum performance without requiring more complex algorithms for estimating the exact noise distribution. Two MLSE electronic equalizers based on Viterbi procedures and template-matching algorithm, are investigated and formulated for OFDR-based MSK optical transmission systems. In these two schemes, OFDR serves as the optical front-end. The proposed equalizers significantly extend the reach of uncompensated link. With a reasonably small number of states (64 states), the uncompensated optical link can equivalently reach up to approximately 960 km SSMF for 10 Gb/s transmission or 60 km SSMF for 40 Gb/s. The results are comparable to the distance of 1040 km SSMF as reported previously [31] for OOK/IM systems, in which up to 8192 trellis states are required.

The performance of the Viterbi-MLSE equalizers for OFDR-based optical MSK transmission systems in terms of transmission limits and PMD mitigation was also numerically investigated. Finally, this chapter has proven the improvement in performance of Viterbi-MLSE equalizers by using multi samples per bit compared to the conventional single sample bit. The novel scheme in which Viterbi-MLSE equalizers are incorporated with OFDR-based MSK optical transmission systems provides high robustness to fiber PMD dynamic and better performance compared to other modulation formats.

Finally, we investigate quantitatively the transmission limits due to the combination of the dispersive effect of the quadratic phase media and restrictiveness of a finite detection and equalization window. The parameter used to quantify these limits is the least Euclidean distance between all symbol combinations and this distance corresponds to the penalty in the system's OSNR. For the coherent detection, both phase and magnitude information of the received signals is used in the determination of the Euclidean distance. While only the magnitude information is used in the incoherent detection. It is found that coherent systems provide a much higher degree of discrimination that is significant for the signal equalization at the cost of the receiver's complexity.

It is no doubt that the equalizers must be adaptive for automatic control of the adjustment of the equalization due to the nature of the dispersion fluctuation of CD and PMD as well as the non-linear effects of signals transmitted over the long-haul distance. In practice the equalizers described in this chapter must be adaptive. One such example is the equalization of channels that have been propagation through different of hops of DWDM optical networks.

Electronic equalization emerges as an important technique for mitigation of impairments of optical signals propagating through optical channels and will remain as a most probable candidate for reshaping the distorted signals. The progresses in the development of high speed digital electronic processors will assist practical implementation of these equalizers in optical transmission systems and networks.

Sieben et al. [27] has recently reported the electrical equalization of OFDM signals over long-haul transmission. The integration of multi-carrier modulation and electrical equalization techniques would enable the extension of transmission beyond the dispersion limit in single mode optical fiber communications systems [31–46]. However the double sideband nodulation of OFDM [47] to the optical passband may face non-linearity distortion and the equalizer may have to be non-linear equalizer and thus difficulty in the implementation unless it is implemented after the recovery of bit sequence after the PFDM decoder as described in Chapter 10.

REFERENCES

1. Matsui, Y., D. Mahgerefteh, X. Zheng, C. Liao, Z.F. Fan, K. McCallion, and P. Tayebati. 2006. Chirp-managed directly modulated laser (CML). *IEEE Photonics Technology Letters* 18 (2): 385–87.
2. Wedding, B., B. Franz, and B. Junginger. 1994. 10-Gb/s optical transmission up to 253km via standard single mode fiber using the method of dispersion supported transmission. *Journal of Lightwave Technology* 12: 1720–27.
3. El Said, M. M., J. Stitch, and M. I. Elmasry. 2005. An electrically pre-equalized 10 Gb/s duobinary transmission system. *IEEE Journal of Lightwave Technology* 23 (1): 388–400.
4. Alic, N., G. C. Papen, R. E. Saperstein, L. B. Milstein, and Y. Fainman. 2005. Signal statistics and maximum likelihood sequence estimation in intensity modulated fiber optic links containing a single optical preamplifier. *Optics Express* 13 (12): 4568–79.
5. Haunstein, H. 2004. PMD and chromatic dispersion control for 10 and 40Gb/s systems. *Proceedings of Optical Fiber Conference OFC'04, Invited Paper, ThU3*.
6. Agazzi, O. E., M. R. Hueda, H. S. Carrer, and D. E. Crivelli. 2005. Maximum-likelihood sequence estimation in dispersive optical channels. *Journal Lightwave Technology* 23 (2): 749–62.
7. Franceschini, M., G. Bongioni, G. Ferrari, R. Rahedi, F. Mehli and A. Castoldi. 2007. Fundamental limits of electronic signal processing in direct detection optical communications. *IEEE Journal of Lightwave Technology* 25 (7): 1742–52.
8. Clarke, A. P. 1985. *Equalizers for high speed modems*. London: Pentech Press.
9. Clarke, A. P. 1985. *Equalizers for digital modem*. London: Pentec Press.
10. Jinguji, K., and M. Oguma. 2000. Optical half-band filters. *IEEE Journal of Lightwave Technology* 18 (2): 252–59.
11. Binh, L. N., and V. A. T. Tran. 2004. Design of photonic half-band filters using multirate DSP technique, Technical report MECSE-27-2004, http://www.ds.eng.monash.edu.au/techrep/reports/2004/MECSE-27-2004.pdf, accessed December 2007.
12. Rosenkranz, W. *Lecture notes on optical transmission II*. Kiel Germany: Technical Faculty, University of Kiel.
13. Papoulis, A. 1968. *Systems and transforms with applications in optics*. Kiel, Germany: McGraw Hill.
14. Proakis, J. G., and M. Salehi. 2002. *Communication systems engineering*. 2nd ed. Englewood Cliffs, NJ: Prentice Hall.
15. Proakis, J. G. 2001. *Digital communications*. 4th ed. New York: McGraw-Hill.
16. Core Optics AG. 2007. 10 Gbit/sec Digital Equalization Chip (DECS), http://www.coreoptics.com/product/prod_ic.phpUSA, accessed December 2007, Nurnberg, Germany.
17. Sivahumaran, T., T. L. Huynh, K. K. Pang, and L. N. Binh. 2007. Non-linear equalizers in narrowband filter receiver achieving 950 ps/nm residual dispersion tolerance for 40Gb/s optical MSK transmission systems. *Proceedings of OFC'07, Paper OThK3*, Annaheim, CA, USA.
18. Poggiolini, P., G. Bosco, M. Visintin, S. J. Savory, Y. Benlachtar, P. Bayvel, and R. I. Killey. 2007. MLSE-EDC versus optical dispersion compensation in a single-channel SPM-limited 800 km link at 10 Gbit/s. *ECOC2007, Paper 1.3*, Berlin, Germany.

19. Alic, A., G. C. Papen, and Y. Fainman. 2004. Performance of maximum likelihood sequence estimation with different modulation formats. *Proceedings of LEOS'04*, Orlando, FL, USA, 49–50.

20. Sivahumaran, T., T. L. Huynh, K. K. Pang, and L.N. Binh. 2007. Non-linear equalizers in narrowband filter receiver achieving 950 ps/nm residual dispersion tolerance for 40Gb/s optical MSK transmission systems. *Proceedings of the Optical Fiber Communications Conference, OFC'07, Paper OThK3*, Anaheim, CA, 1–3.

21. Curri, V., R. Gaudino, A. Napoli, and P. Poggiolini. 2004. Electronic equalization for advanced modulation formats in dispersion-limited systems. *Photonics Technology Letters* 16 (11): 2556–58.

22. Katz, G., D. Sadot, and J. Tabrikian. 2005. Electrical dispersion compensation equalizers in optical long-haul coherent-detection system. *Proceedings of ICTON'05, Paper We.C1.5*, 2005, 7th International Conference on Transparent Optical Networks, 3–7 July 2005, Barcelona, Spain.

23. Crivelli, D. E., H. S. Carrer, and M. R. Hueda. 2004. On the performance of reduced-state viterbi receivers in IM/DD optical transmission systems. *Proceedings ECOC'04, Paper WE4.P.083*, Stockholm, Sweden.

24. Haunstein, H. 2004. PMD and chromatic dispersion control for 10 and 40Gb/s systems. *Proceedings of OFC'04, Invited Paper, ThU3*, Los Angeles, CA, USA.

25. Rosenkranz, W., and C. Xia. 2007. Electrical equalization for advanced optical communications systems. *International Journal of Electronics and Communications* 61: 153–57.

26. Sieben, M., J. Conradi, and D. E. Dodds. 1999. Optical single sideband transmission at 10Gb/s using only electrical dispersion compensation. *IEEE Journal of Lightwave Technology* 17 (10): 1742–49.

27. Huynh, T. L., T. Sivahumaran, K. K. Pang, and L. N. Binh. 2007. A narrowband filter receiver achieving 225 ps/nm residual dispersion tolerance for 40 Gb/s optical MSK transmission systems. *Proceedings of OFC'07*, Anaheim, CA.

28. Sivahumaran, T., T. L. Huynh, K. K. Pang, and L. N. Binh. 2007. Non-linear equalizers in narrowband filter receiver achieving 950 ps/nm residual dispersion tolerance for 40 Gb/s optical MSK transmission systems. *Proceedings of OFC'07, Paper OThK3*, Anaheim, CA.

29. Huynh, T. L., T. Sivahumaran, L. N. Binh, and K. K. Pang. 2007. Narrowband frequency discrimination receiver for high dispersion tolerance optical MSK systems. *Proceedings of Coin-AOFT International Conference, Paper TuA1-3*, 25–27 June 2007, Melbourne, Australia.

30. Huynh, T. L., T. Sivahumaran, L. N. Binh, and K. K. Pang. 2007. Sensitivity improvement with offset filtering in optical MSK narrowband frequency discrimination receiver. *Proceedings of Coin-ACOFT International Conference, Paper TuA1-5*, 25–27 June 2007, Melbourne, Australia.

31. Savory, S. J., Y. Benlachtar, R. I. Killey, P. Pavel, G. Bosco, and P. Poggiolini. 2007. IMDD transmission over 1040 km of standard single-mode fiber at 10Gb/s using a one-sample-per-bit reduced complexity MLSE receiver. *Proceedings of the Optical Fiber Communications Conference, OFC'07, Paper OThK2*, Anaheim, CA, 1–3.

32. Secondini, M., E. Forestieri, and G. Prati. 2007. A theoretical comparison of robustness in combating CD and PMD in fiber-optic systems. *IEEE Photonics Technology Letters* 19 (7).

33. Heffner, B., T. Schmidt, R. Saunders, R. Hui, D. Richards, and G. Nicholl. 2007. 43 Gb/s Adaptive polarization mode dispersion compensator field trial. *Proceedings of OFC'07, Paper NTuA1*, Anaheim, CA.

34. Technical Specifications of OTS 4540 PMD Compensator, StrataLight Communications, http://www.stratalight.com/ots.php, accessed December 2007.

35. Buchali, F. 2006. Electronic dispersion compensation for enhanced optical transmission. *Proceedings of OFC'6, Paper OWR5*, Anaheim, CA.

36. Xia, C., and W. Rosenkranz. 2006. Electrical dispersion compensation for different modulation formats with optical filtering. *Proceedings of OFC6, Paper OWR2*, Anaheim, CA.

37. Klekamp, A., F. Buchali, M. Audoin and H. Bülow. 2006. Non-linear limitations of electronic dispersion pre-compensation by intra-channel effects. *Proceedings of OFC'06, Paper OWR1*, Anaheim, CA.

38. Kupfer, T., S. Langenbach, N. Stojanovic, S. Gehrke, and J. Whiteaway. 2007. Performance of MLSE in optical communication systems. *ECOC2007, Paper 1.1* Berlin, Germany.

39. Agrawal, G. P. 2001. *Non-linear fibre optics*. 3rd ed. CA: Academic Press.

40. Papoulis, A. 1968. *Systems and transforms with applications in optics*. McGraw Hill.

41. Haunstein, H. 2004. PMD and chromatic dispersion control for 10 and 40Gb/s systems. *Proceedings OFC'04, Invited Paper, ThU3*.

42. Crivelli, D. E., H. S. Carrer, and M. R. Hueda. 2004. On the performance of reduced-state viterbi receivers in IM/DD optical transmission systems. *Proceedings of ECOC'04, Paper WE4.P.083*.

43. Alic, N., G. C. Papen, and Y. Fainman. 2004. Performance of maximum likelihood sequence estimation with different modulation formats. *Proceedings of LEOS'04* 49–50.
44. Sivahumaran, T., T. L. Huynh, K. K. Pang, and L. N. Binh. 2007. Non-linear equalizers in narrowband filter receiver achieving 950 ps/nm residual dispersion tolerance for 40Gb/s optical MSK transmission systems. *Proceedings of OFC'07, Paper OThK3*, Anaheim, CA.
45. Agrawal, G. P. 2001. *Non-linear fibre optics*. 3rd ed. CA: Academic Press.
46. Papoulis, A. 1968. *Systems and transforms with applications in optics*. McGraw Hill.
47. Schmidt, B., A. J. Lowery, and J. Armstrong. 2008. Experimental demonstrations of electronic dispersion compensation for long-Haul transmission using direct-detection optical OFDM. *IEEE Journal of Lightwave Technology* 26 (1): 196–202.

12 Optical Soliton Transmission

12.1 INTRODUCTION

A fascinating manifestation of the fiber non-linearity occurs in the anomalous dispersion regime where the fiber can support optical solitons through interplay between the dispersive and non-linear effects. The term *soliton* refer to special kinds of waves that can propagate undistorted over long distances and remain unaffected after collision with each other. Solitons have been studied extensively in several fields of physics. In the context of optical fibers, solitons are not only of fundamental interest but also has the potential application in the field of optical fiber communications. The word *soliton* was coined [1] in 1965 to describe the particle-like properties of pulse envelopes in dispersive non-linear media.

The basic concepts behind fiber solitons and its basic propagation equation, known as the non-linear Schrodinger equation are introduced in Section 10.3 and Section 10.4. The numerical approach using *Beam Propagation Method* (BPM) to solve the Schrodinger equation is discussed in detail in Section 10.3. The analytical approach using *Inverse Scattering Method* (ISM) to solve the Schrodinger equation is discussed in detail in Section 10.4. The results obtained through these two approaches are compared and verified.

The results of the numerical approach are discussed in Section 10.5. In this Section, the behavior of fundamental and higher order solitons are observed. In Section 10.6, soliton interactions are simulated and discussed. Techniques for controlling the soliton interaction effect are then suggested.

Section 10.7 revisits the ISM for simulation of optical solitons, in particular the interaction between sequential pulses for optical communications systems. Steps for interaction and reconstruction of solitons are described and proven to be the most accurate technique for design and control of optical solitons pulses.

All the results of our synthesis for optical solitons propagation and detection techniques are described for both bright and dark solitons and summarized with suggestions in Section 10.9.

12.2 FUNDAMENTALS OF NON-LINEAR PROPAGATION THEORY

Similar to all electromagnetic phenomena, the propagation of optical fields either in linear or non-linear regimes in fibers is governed by Maxwell's equations. From Maxwell's equations, one can obtain [2],

$$\nabla \times \nabla \times E = -\frac{1}{c^2}\frac{\partial^2 E}{\partial t^2} - \mu_0 \frac{\partial^2 P}{\partial t^2} \tag{12.1}$$

where E is the electric field and P is the induced polarization. The non-linear effects in the optical fibers involve the use of short pulses with widths ranging from 10 ns to 10 fs. When such optical pulses propagate inside the fiber, both dispersive and non-linear effects influence their shape and spectrum. Thus, Equation 12.1 can be further developed [2] to cater for these effects,

$$\nabla^2 E - \frac{1}{c^2}\frac{\partial^2 E}{\partial t^2} = -\mu_0 \frac{\partial^2 P_L}{\partial t^2} - \mu_0 \frac{\partial^2 P_{\mathrm{NL}}}{\partial t^2} \tag{12.2}$$

where P_L and P_{NL} are the linear part and non-linear part of the induced polarization. Using the method of separation of variables to solve Equation 12.2 and taking perturbation theory into consideration, we could obtain the following results,

$$\frac{\partial A}{\partial z} + \beta_1 \frac{\partial A}{\partial t} + \frac{j}{2}\beta_2 \frac{\partial^2 A}{\partial t^2} + \frac{\alpha}{2}A = j\gamma |A|^2 A \qquad (12.3)$$

where the envelope A is a slowly varying function of z. β_1 and β_2 are the first and second order derivatives of β with respect to optical frequency ω as defined in Chapter 3. γ is the fiber non-linear coefficient and α is the intrinsic fiber attenuation express in dB/km. Taking higher order dispersion effect, stimulated inelastic scattering effect and Raman gain effect into consideration together with perturbation approach leads to three additional terms in Equation 12.3. The generalized propagation equation takes the form,

$$\frac{\partial A}{\partial z} + \frac{\alpha}{2}A + \frac{j}{2}\beta_2 \frac{\partial^2 A}{\partial T^2} - \frac{1}{6}\beta_3 \frac{\partial^3 A}{\partial T^3} = j\gamma \left\{ |A|^2 A + \frac{2j}{\omega_0}\frac{\partial}{\partial T}\left(|A|^2 A\right) - T_R A \frac{\partial |A|^2}{\partial T} \right\} \qquad (12.4)$$

where ω_0 is the carrier frequency and T_R is related to the slope of the Raman gain. Notice that β_1 has disappeared because of the transformation T, that is the observation frame is situated on the soliton itself, which is given by,

$$T = t - \beta_1 z. \qquad (12.5)$$

Before beginning to simulate solitons evolution using either the numerical or analytical approach, a few important terms of solitons systems need to be clarified. It is useful to write Equation 12.4 in a normalized form by introducing:

$$\tau = \frac{t - \beta_1 z}{T_0} \qquad (12.6)$$

$$\xi = \frac{z}{L_D} \qquad (12.7)$$

$$U = \frac{A}{\sqrt{P_0}} \qquad (12.8)$$

where T_0 is the soliton pulse width, P_0 is the soliton peak power, and the dispersion length, L_D is defined as [3]:

$$L_D = \frac{T_0^2}{|\beta_2|} \qquad (12.9)$$

where β_2 is obtained from the total dispersion factor of the optical fiber which is given by

$$\beta_2 = -\left(\frac{D_T \lambda^2}{2\pi c}\right). \qquad (12.10)$$

The order of the soliton, N is defined as

$$N^2 = \frac{\gamma P_0 T_0^2}{|\beta_2|}.$$ (12.11)

The fundamental soliton corresponds to $N=1$. Thus, in order to maintain $N=1$, the soliton pulse needs to have its peak power so that the square root of Equation 12.5 will be an integer value. The fundamental soliton has a secant shape and its initial amplitude given by

$$U(0, \tau) = \text{sech}(\tau).$$ (12.12)

This pulse shape is to be launched into the fiber. Its shape remains unchanged during the propagation. However for $N>1$, such input shape is recovered at each *soliton period* defined as

$$z_0 = \frac{\pi}{2} L_D = \frac{\pi}{2} \frac{T_0^2}{|\beta_2|}.$$ (12.13)

The soliton period z_0 and the soliton order N play an important role in the theory of optical solitons.

12.3 NUMERICAL APPROACH

12.3.1 Beam Propagation Method

Equation 12.4 is a non-linear partial differential equation that does not generally lend itself to analytic solutions except for some specific cases in which the inverse scattering method (Section 10.4) can be employed. A numerical approach is therefore often necessary for an understanding of the non-linear effects in optical fibers. Split Step Fourier Model (SSFM) described in Chapter 3 is used to solve the pulse-propagation in the non-linear guided wave medium. Thus, the evolution of solitons can be observed at any time span along the propagation distance. The beam propagation method using split step method has been presented in detail in Chapter 3. The SSFM is extensively implemented in this chapter for soliton transmission.

12.3.2 Analytical Approach—Inverse Scattering Method

The inverse scattering method is similar in spirit to the beam propagation method that is commonly used to solve linear partial differential equations. The incident field at $z=0$ is used to obtain the initial scattering data whose evolution along z is easily obtained by solving the linear scattering problem. The propagated field is reconstructed from the evolved scattering data. The details of the inverse scattering method are available in many texts [1,2], we describe only briefly how this method is used to solve the Schrodinger equation in Equation 12.4.

Using the normalized parameters in Equation 12.6 to Equation 12.13, the parameter N can be eliminated from Equation 12.4 by defining

$$u = NU = \sqrt{\left(\frac{\gamma T_0^2}{|\beta_2|}\right)} A$$ (12.14)

and the non-linear Schrodinger equation becomes

$$j\frac{\partial u}{\partial \xi} + \frac{1}{2}\frac{\partial^2 u}{\partial \tau^2} + |u|^2 u = 0.$$ (12.15)

In the inverse scattering method, the scattering problem associated with Equation 12.21 is

$$\frac{\partial v_1}{\partial \tau} + j\zeta v_1 = uv_2$$ (12.16)

$$\frac{\partial v_2}{\partial \tau} - j\zeta v_2 = -u^* v_1$$ (12.17)

where v_1 and v_2 are the amplitudes of the waves scattering in a potential $u(\xi,\tau)$ and ζ is the eigenvalue. The soliton order is characterized by the number N of poles or eigenvalues ζ_j ($j=1$ to N). The general solution is [4]

$$u(\xi,\tau) = -2\sum_{j=1}^{N}\lambda_j^* \psi_{2j}^*$$ (12.18)

where

$$\lambda_j = \sqrt{c_j}\,\exp\!\left(j\zeta_j\tau + j\zeta_j^2\xi\right)$$ (12.19)

and ψ_{2j} is obtained by solving the linear set of the following equations:

$$\psi_{1j} + \sum_{k=1}^{N}\frac{\lambda_j\lambda_k^*}{\zeta_j - \zeta_k^*}\psi_{2k}^* = 0$$ (12.20)

$$\psi_{2j}^* - \sum_{k=1}^{N}\frac{\lambda_k\lambda_j^*}{\zeta_j^* - \zeta_k}\psi_{1k} = \lambda_j^*.$$ (12.21)

The eigenvalues ζ_j is given by

$$\zeta_j = i\eta_j$$ (12.22)

and Equation 12.25 becomes

$$\lambda_j = \sqrt{c_j}\,\exp\!\left(-\eta_j\tau - jn_j^2\xi\right).$$ (12.23)

Assuming the soliton is symmetrical about $\tau = 0$, the residues are related to the eigenvalues by the relation

$$c_j = \frac{\displaystyle\prod_{k=1}^{N}(\eta_j + \eta_k)}{\displaystyle\prod_{k\neq j}^{N}|(\eta_j - \eta_k)|}.$$ (12.24)

The fundamental soliton corresponds to the case of a single eigenvalue η_1 which has a value of 0.5 for $N=1$. In general, for $N>1$, the eigenvalues are given by

$$\eta_i = \frac{2i-1}{2} \tag{12.25}$$

where $i=1, 2, 3,..., N$.

12.3.2.1 Soliton $N=1$ by Inverse Scattering

Fundamental soliton has only one eigenvalue which is shown in Equation 12.27. Thus, we obtain $\eta_1 = 0.5$. The residue which is related to the eigenvalue can be found by using Equation 12.26. Then, the eigenvalue and its residue are substituted in Equation 12.25 in order to obtain the eigenfunction λ. The complex eigenvalue can be obtained by using Equation 12.28. Eigenpotential ψ_{21} shown in Equation 12.26 and Equation 12.27 need to be solved simultaneously and the process of solving this problem is described in the following. From Equation 12.26 and Equation 12.27 we get

$$\psi_{11} + \frac{\lambda_1 \lambda_1^*}{\zeta_1 - \zeta_1^*} \psi_{21}^* = 0 \tag{12.26}$$

$$\psi_{21}^* = \lambda_1^* + \frac{\lambda_1 \lambda_1^*}{\zeta_1^* - \zeta_1} \psi_{11}. \tag{12.27}$$

Substituting Equation 12.28 into Equation 12.29 we obtain

$$\psi_{21}^* = \frac{\zeta_1 - \zeta_1^*}{\lambda_1 \lambda_1^*} \left[\frac{\lambda_1 \lambda_1^{*2} \left(\zeta_1^* - \zeta_1 \right)}{\left(\zeta_1^* - \zeta_1 \right)\left(\zeta_1 - \zeta_1^* \right) + \left(\lambda_1 \lambda_1^* \right)^2} \right]. \tag{12.28}$$

Thus, by substituting the eigenpotential and eigenfunction in Equation 12.24, we can obtain the fundamental soliton function $U(\xi,\tau)$.

12.3.2.2 Soliton $N=2$ by Inverse Scattering

All the steps of solving the non-linear Schroedinger equation are the same as listed in Section 10.4.1 except that the eigenvalues, eigenfunction and the eigenpotential have to be redefined. Solving the eigenvalues and the corresponding eigenfunction is straight forward. However, solving the eigenpotentials requires more effort. Again, from Equation 12.26 and Equation 12.27 we obtain the following expressions

$$\psi_{11} + \left(\frac{\lambda_1 \lambda_1^*}{\zeta_1 - \zeta_1^*} \psi_{21}^* + \frac{\lambda_1 \lambda_2^*}{\zeta_1 - \zeta_2^*} \psi_{22}^* \right) = 0 \tag{12.29}$$

$$\psi_{21}^* - \left(\frac{\lambda_1 \lambda_1^*}{\zeta_1^* - \zeta_1} \psi_{11} + \frac{\lambda_1 \lambda_2^*}{\zeta_1^* - \zeta_2} \psi_{12} \right) = \lambda_1^* \tag{12.30}$$

$$\psi_{12} + \left(\frac{\lambda_2 \lambda_1^*}{\zeta_2 - \zeta_1^*} \psi_{21}^* + \frac{\lambda_2 \lambda_2^*}{\zeta_2 - \zeta_2^*} \psi_{22}^* \right) = 0 \tag{12.31}$$

$$\Psi_{22}^* - \left(\frac{\lambda_2 \lambda_1^*}{\zeta_2^* - \zeta_1} \Psi_{11} + \frac{\lambda_2 \lambda_2^*}{\zeta_2^* - \zeta_2} \Psi_{12} \right) = \lambda_2^*. \qquad (12.32)$$

There are four unknowns and four equations, thus it is possible to solve them simultaneously. We can represent Equations 12.31 to Equation 12.34 in a matrix form

$$
\begin{pmatrix} \Psi_{11} \\ \Psi_{12} \\ \Psi_{21}^* \\ \Psi_{22}^* \end{pmatrix} =
\begin{pmatrix}
0 & 0 & \dfrac{\lambda_1 \lambda_1^*}{\zeta_1 - \zeta_1^*} \Psi_{21}^* & \dfrac{\lambda_1 \lambda_2^*}{\zeta_1 - \zeta_2^*} \Psi_{22}^* \\[2mm]
0 & 0 & \dfrac{\lambda_2 \lambda_1^*}{\zeta_2 - \zeta_1^*} \Psi_{21}^* & \dfrac{\lambda_2 \lambda_2^*}{\zeta_2 - \zeta_2^*} \Psi_{22}^* \\[2mm]
-\dfrac{\lambda_1 \lambda_1^*}{\zeta_1^* - \zeta_1} \Psi_{11} & -\dfrac{\lambda_1 \lambda_2^*}{\zeta_1^* - \zeta_2} \Psi_{12} & 0 & 0 \\[2mm]
-\dfrac{\lambda_2 \lambda_1^*}{\zeta_2^* - \zeta_1} \Psi_{11} & -\dfrac{\lambda_2 \lambda_2^*}{\zeta_2^* - \zeta_2} \Psi_{12} & 0 & 0
\end{pmatrix}
\begin{pmatrix} \Psi_{11} \\ \Psi_{12} \\ \Psi_{21}^* \\ \Psi_{22}^* \end{pmatrix} +
\begin{pmatrix} 0 \\ 0 \\ \lambda_1^* \\ \lambda_2^* \end{pmatrix} \cdot (12.33)
$$

Thus, by substituting the eigenpotential and eigenfunction in Equation 12.24, we can obtain the second order of soliton function $U(\xi,\tau)$.

The results of both beam propagation (numerical approach) and inverse scattering method (analytical approach) are shown in Figure 12.1 and Figure 12.2 for fundamental soliton and soliton of order 2, respectively. From the results, we notice that the soliton pulse is recovered at one soliton period. The different pulse widths observed is due to different setting of the initial soliton pulse width. The results of the numerical approach agree with the analytical results and hence we conclude that our numerical algorithm in synthesizing solitons evolution has been a success.

12.4 FUNDAMENTAL AND HIGHER ORDER SOLITONS

12.4.1 Soliton Evolution for $N = 1, 2, 3, 4$ and 5

The evolutionary process of optical soliton is shown in Figure 12.1, Figure 12.2 and Figure 12.3 for fundamental, second, third, fourth and fifth order of soliton, respectively. Periodic evolution of soliton is very much depending on the soliton period z_o and the soliton order N. As we can observe in those figures above, the soliton pulse begins to evolve and its original shape will be recovered at every soliton period.

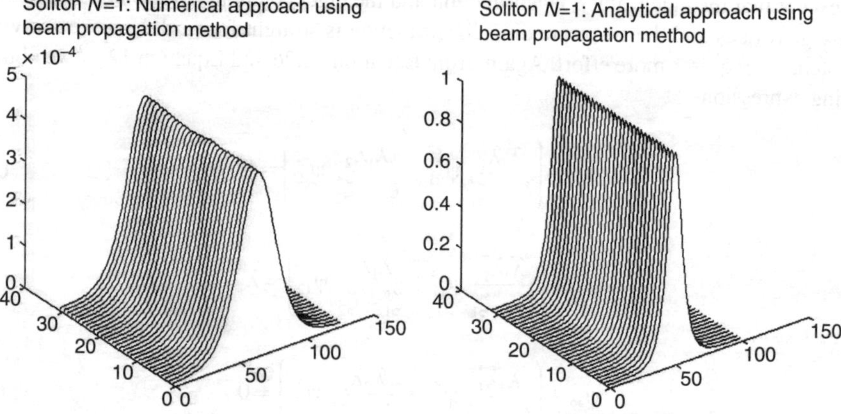

FIGURE 12.1 Comparison between numerical and analytic approach for soliton $N=1$.

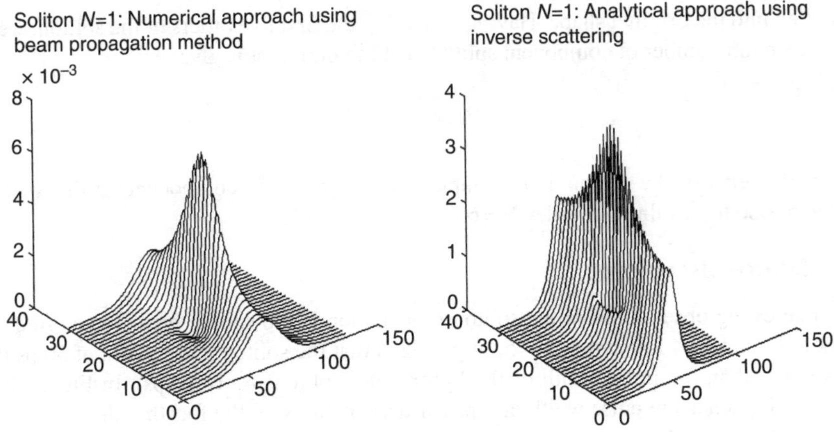

FIGURE 12.2   Comparison between numerical and analytic approach for soliton $N=2$.

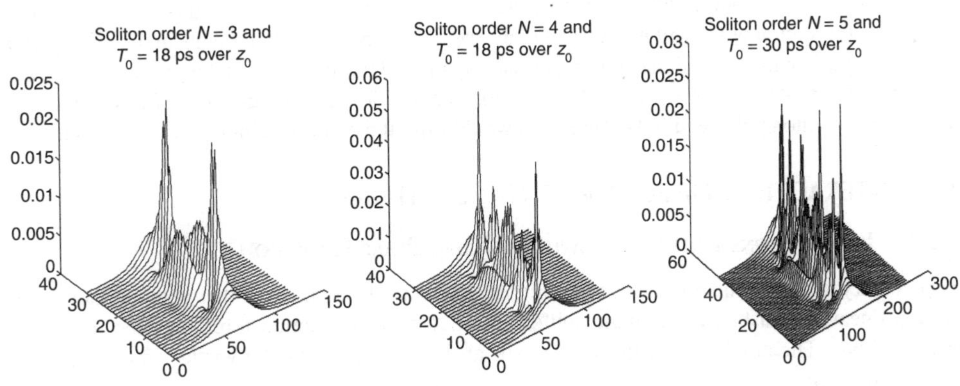

FIGURE 12.3   Soliton evolution for $N=3$, 4 and 5.

For fundamental soliton, the pulse shape remains the same throughout the propagation distance. However for soliton $N>1$, the soliton pulse will be evolved into various shapes periodically depending on the order of the soliton. Eventually, these distorted pulses will come back to the shape of the fundamental soliton at every soliton period. As for soliton $N=2$, the pulse begins to contract and the narrowing effect starts to take place and causes the peak to increase gradually. In the midpoint of the soliton period, the soliton peak has increased by three times (Figure 12.2, numerical results) and there are two minor components on each side of the main component of the soliton pulse.

For soliton $N=3$, as the pulse propagates along the fiber, it first contracts to a fraction of its initial width, splits into two components, and then mergers again to recover its original shape at the end of the soliton period. As for the soliton $N=4$, the fundamental pulse shape splits into two components, and then further splits into three components. After reaching the midpoint of the soliton period, the three components pulse begins to merge into two components and finally recovers to its original shape at the end of the soliton period.

Each plot is taken at certain distance of propagation. Starting at 0 km, we have the fundamental soliton shape. Then it splits into two components at 400 km, further splitting into three components occurs at 600 km and when it reaches the midpoint of the soliton period which is at 1210 km, it splits into four components. After the midpoint, the four components begin to merge into three, and into two, and then into the original shape. Having analyzing the first five orders of the soliton, we notice that the maximum number of component split (N_c) at the midpoint is given by

$$N_c = N - 1 \tag{12.34}$$

where N is the order of the soliton. For example, they are five split components at the midpoint of the soliton period for a soliton of order $N = 6$.

12.4.2 SOLITON BREAKDOWN

Another interesting observation of the soliton propagation is the soliton breakdown effect. Referring back to Figure 12.4 for soliton $N = 5$, we realized that the soliton pulse width of 30 ps is used, which is larger than the pulse width of the lower order solitons (only 18 ps). In Figure 12.5, we purposely assign a narrow pulse width of 9 ps (instead of 18 ps) to the fourth order soliton that the soliton breakdown may be seen.

This proves that we need a substantial soliton energy to support higher order soliton in order for it to recover back to its own shape. Since the pulse width is related to the total power contained in the soliton, we need to increase the pulse width for higher order transmission. Otherwise, soliton breakdown will occur where the soliton pulse will not recover to its original shape. At certain points in the propagation distance, the pulse will split and may not merge back again. All these effects can be seen in Figure 12.5. Hence, there is a trade off between the order of soliton and the bit rate. For high bit rate transmission, we need a narrow soliton pulse width. However this would limit the order of soliton that we can use.

12.5 INTERACTION OF FUNDAMENTAL SOLITONS

12.5.1 TWO SOLITONS INTERACTION WITH DIFFERENT PULSE SEPARATION

It is necessary to determine how close two solitons can come closer to each other without interacting. The non-linearity that bounds a single soliton introduces an interaction force between the neighboring solitons. The amplitude of the soliton pair at the fiber input can be written in the following normalized form:

$$u(0, \tau) = \operatorname{sech}(\tau - q_0) + r \operatorname{sech}\{r(\tau + q_0)\} \exp(j\theta) \tag{12.35}$$

where q_0 is the initial pulse separation and r is the relative amplitude. Thus, soliton interaction can be studied by solving Equation 12.37 by using the numerical approach—beam propagation method (BPM) discussed in Section 10.3. Figure 12.6 shows the interaction of a soliton pair at different pulse separation. Figure 12.7 displays the evolution pattern showing periodic collapse of a soliton pair for various pulse separations.

The periodic collapse of neighboring solitons is undesirable from the system standpoint. One way to avoid the interaction problem is to increase the separation such that the collapse distance, Z_p is much larger than the transmission distance L_T. From the results in Figure 12.7, we can measure the collapse distance Z_p at each different pulse separation q_0 and thus draw a rough estimation by interpolating these data. All the data are tabulated in Table 12.1 and the interpolated curve is shown in Figure 12.8.

Thus, the curve shown in Figure 12.1 is useful as a guide line for selection of optimum pulse separation given a certain transmission distance. Pulse separation has to be minimized in order to achieve high bit rate transmission.

12.5.2 TWO SOLITONS INTERACTION WITH DIFFERENT RELATIVE AMPLITUDE

Shown in Figure 12.9 is the two solitons interaction with different relative amplitude. Referring to Equation 12.37, we can set different relative amplitudes to the two pulses and the effect of setting asymmetrical pair of soliton is investigated in the following.

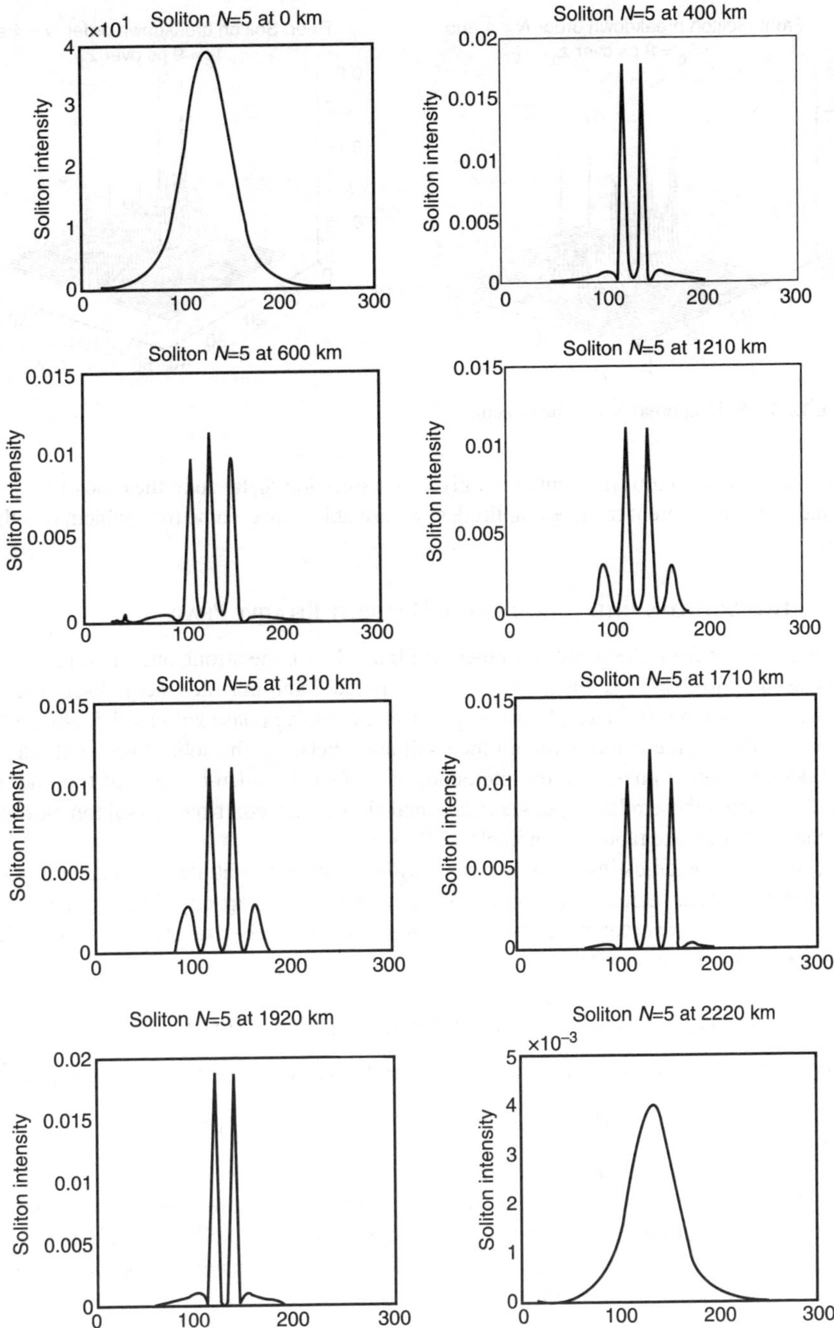

FIGURE 12.4 Evolution of solution of order 5–*N*=5.

From Figure 12.9, we observe that by setting a slight difference between the amplitudes of the soliton pulses, the soliton collapse problem is effectively solved. Another observation we obtain is that as the relative amplitude increases, the periodical attraction between the two pulses happens more frequently. For example in the case of $r = 1.1$, the two pulses come closer to each other in three

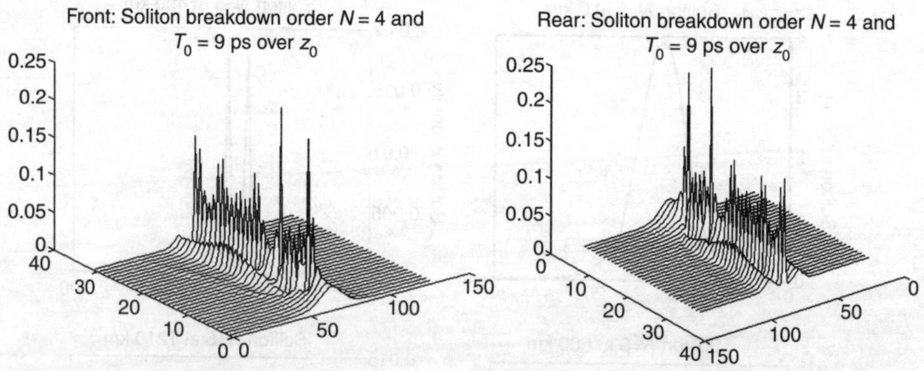

FIGURE 12.5   Soliton breakdown phenomena.

incidents, compared to four incidents for a given transmission distance in the case of $r = 1.3$. Thus, by setting slightly different relative amplitudes, we are able to overcome the problem of soliton pair collapse.

12.5.3   TWO SOLITONS INTERACTION UNDER DIFFERENT RELATIVE PHASE

As we can observe from the results obtained in Figure 12.10, the strong attractive force turns into a repulsive force for $\theta \neq 0$ that the soliton pair is separated and may not merge back. Figure 12.11 shows the case for three solitons interaction at different relative phase with $r = 1.5$. Notice that relative phase of 180 degree could maintain the oscillation between the soliton pair without merging. The 90 degree relative phase soliton interaction is in fact the mirror image of the one with 270 degree. Apart from those relative phases mentioned above, they contribute to soliton pair repulsion. These characteristics are tabulated in Table 12.2.

In conclusion, by adjusting the pulse separation, relative amplitude and relative phase of the soliton pair, we could increase the bit rate or the transmission capacity. The problems of soliton pair collapsing can be overcome effectively by varying those three parameters so that stable soliton transmission system can be achieved.

12.5.4   THREE SOLITON INTERACTION WITH DIFFERENT RELATIVE PHASE

The amplitude of three solitons at the fiber input can be written in the following normalized form:

$$u\,(0,\tau) = \sec h\,(\tau) + r \sec h\{r\,(\tau - q0)\} \exp{(j\theta)}. \qquad (12.36)$$

As for the case of three solitons interaction, the characteristics observed are pretty much the same as those that have been discussed for the soliton pair in terms of the relative phase. In Figure 12.12, we notice that only at the relative phase of 180 degree, the three solitons would not merge. The severe oscillation of the three solitons would create severe problems for the receiving/detection terminal. Thus, we must stabilize the oscillations by introducing a relative amplitude of $r = 1.5$ for the first and the third pulse shown in the next section and in Figure 12.13.

12.5.5   THREE SOLITONS INTERACTION WITH DIFFERENT RELATIVE PHASE AND $r = 1.5$

From Figure 12.13 and Figure 12.14, we can see that for relative phase less than 180 degree, the sides solitons are repelling each other away from the centre soliton. The repulsive force reaches its maximum at relative phase of 90 degree. Notice that the repulsive force exerted in the case of 45 degree and 135 degree are the same. When the relative phase is adjusted to 180 degree, there is neither attractive nor repulsive force at all (same case when the relative phase $= 0$).

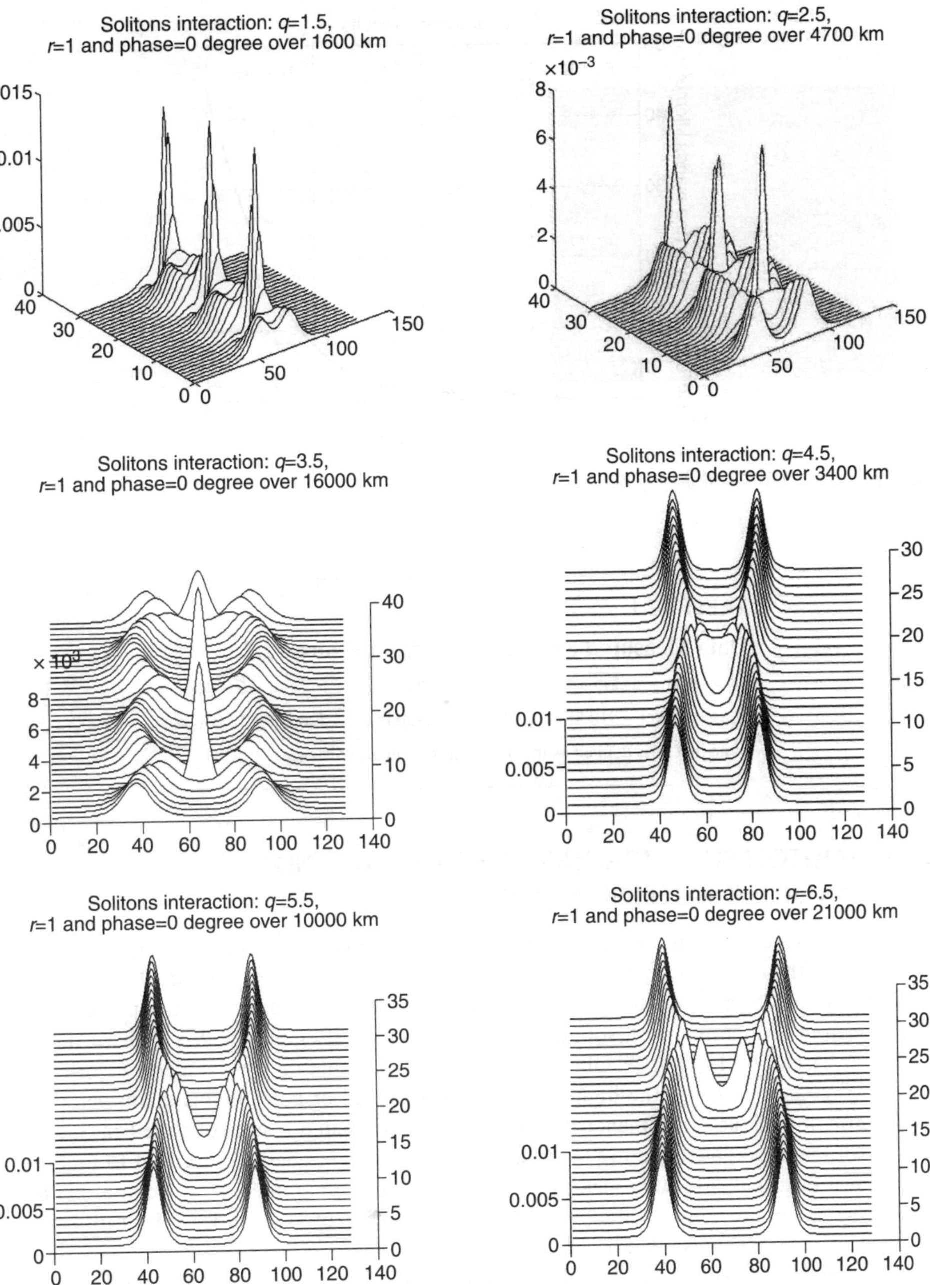

FIGURE 12.6 Two-soliton interaction with pulse separation as a parameter at different transmission distance, as indicated in the graphs.

When the relative phase is greater than 180 degree, various interesting behavior can be observed. In Figure 12.14, we see that the sides solitons begin to merge together with the centre soliton at relative phase = 200 degree. In fact, these solitons merge at the propagation distance of approximate to 14,500 km (Figure 12.14). When the relative phase = 225 degree, the three solitons will be scattered apart at propagation distance of about 10,000 km (Figure 12.14). Higher relative phase would still merge the solitons and there is no incidence of the repulsive force.

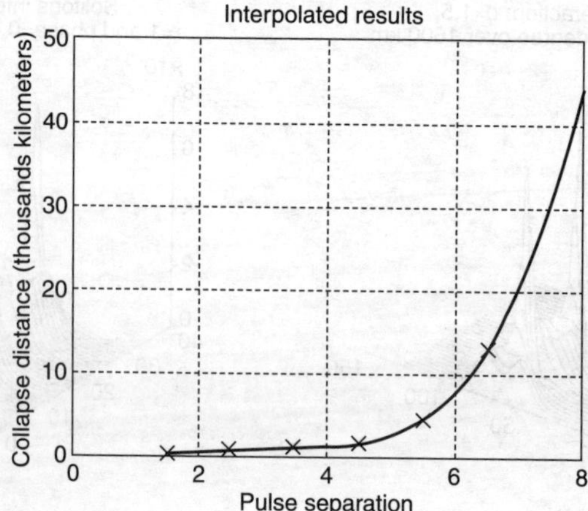

FIGURE 12.7 Interpolated data from Table 12.1.

TABLE 12.1

Data Measured and Collected from Figure 12.7

Q_0	1.5	2.5	3.5	4.5	5.5	6.5
Z_p	310 km	910 km	1,100 km	1,700 km	4,500 km	13,000 km

Note that Z_p is defined as the first solitons collapse distance.

12.6 SOLITON PULSE TRANSMISSION SYSTEMS AND INVERSE SCATTERING METHOD

This section deals with the design of a 4-bit soliton pulse train communication system. The behaviour of the optical solitons is studied via the inverse scattering method. Of particular interest is the interaction between the neighboring solitons and its effect on the pulse separation. The design of an optimal 4-bit soliton pulse train is then attempted and is then verified with a simulator based on MATLAB®. Obtained results indicate that the system design works well but the question as to whether the highest bit-rate is achieved or not remains unanswered. Also, a re-examination of the phenomenon observed by the simulated results obtained in Section 10.6 on the soliton breakdown due to "insufficient energy content" has been proven to be false. The breakdown occurs due to improper adjustment of sampling parameters to accommodate higher frequency components. A counterexample is shown whereby a 4th order soliton did not break down even at narrow pulse widths.

We recall here again the wave equation governing optical soliton propagation, Equation 12.3. If this equation is normalized to a simpler form and the third order term neglected (if the operating wavelength is substantially far way from the zero dispersion wavelength of the optical fiber), then

$$j\frac{\partial U}{\partial \xi} - \mathrm{sgn}(\beta_2)\frac{1}{2}\frac{\partial^2 U}{\partial \tau^2} + N^2 |U|^2 U = 0. \qquad (12.37)$$

N denotes the order of the soliton solution and it also indicates how many solitons are present in the solution.

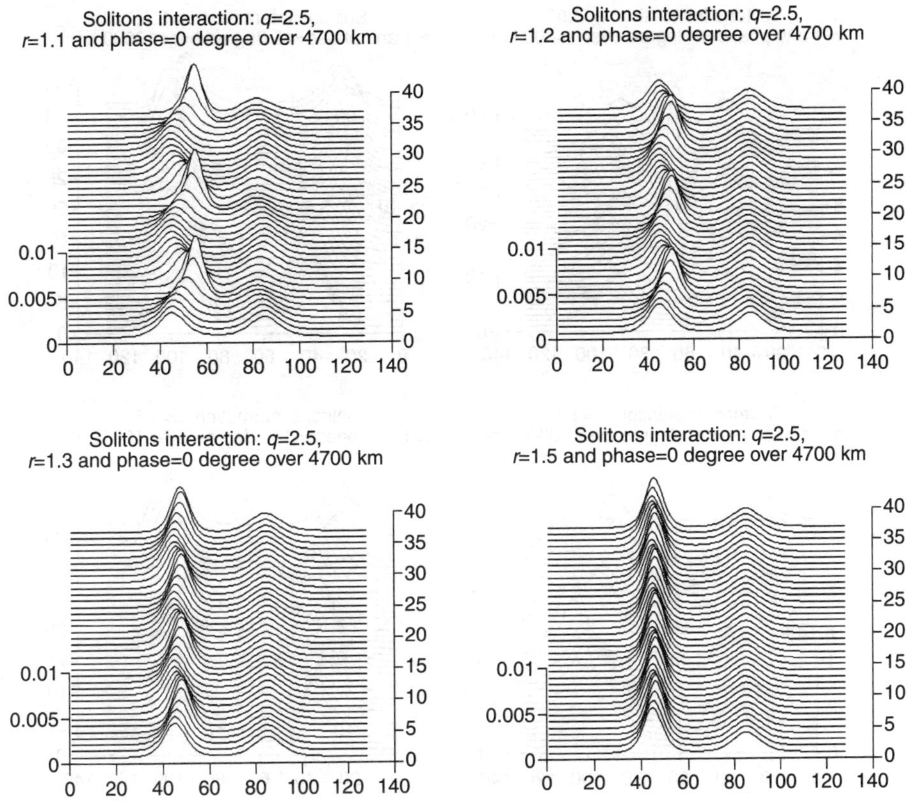

FIGURE 12.8 Two-soliton interaction with the amplitude of the pulses as a parameter at different transmission distance, as indicated in the graphs.

12.6.1 INVERSE SCATTERING METHOD REVISITED

The inverse scattering method was developed to deal with many non-linear wave equations encountered in the field of physics. One such equation studied was the non-linear Schroedinger equation (this equation differs slightly from Equation 12.39) by the roles of t and x

$$iq_t + q_{xx} + \chi |q|^2 q = 0. \tag{12.38}$$

A particular solution of this equation is the soliton wave

$$u(x,t) = \sqrt{2\chi}\eta \, \frac{\exp\left[-4j\left(\xi^2 - \eta^2\right)t - 2j\xi x + j\varphi\right]}{\cosh\left[2\eta(x - x_0) + 8\eta\xi t\right]} \tag{12.39}$$

where the constant η characterizes the amplitude as well as the width of the soliton, ξ the velocity of the soliton (with respect to the group velocity), x_0 the initial position of the soliton centre and φ, its initial phase. For a more general solution, we turn to the inverse scattering method.

Given initial boundary constraints, the aim of the inverse scattering method is to reconstruct the potential u in the non-linear Schroedinger equation from the asymptotic data which results from scattering processes. An outline of the method is given below:

The presented route shows that the inverse scattering method is broken up into three subproblems. This seems long and convoluted since from Figure 12.15, there is a more direct approach.

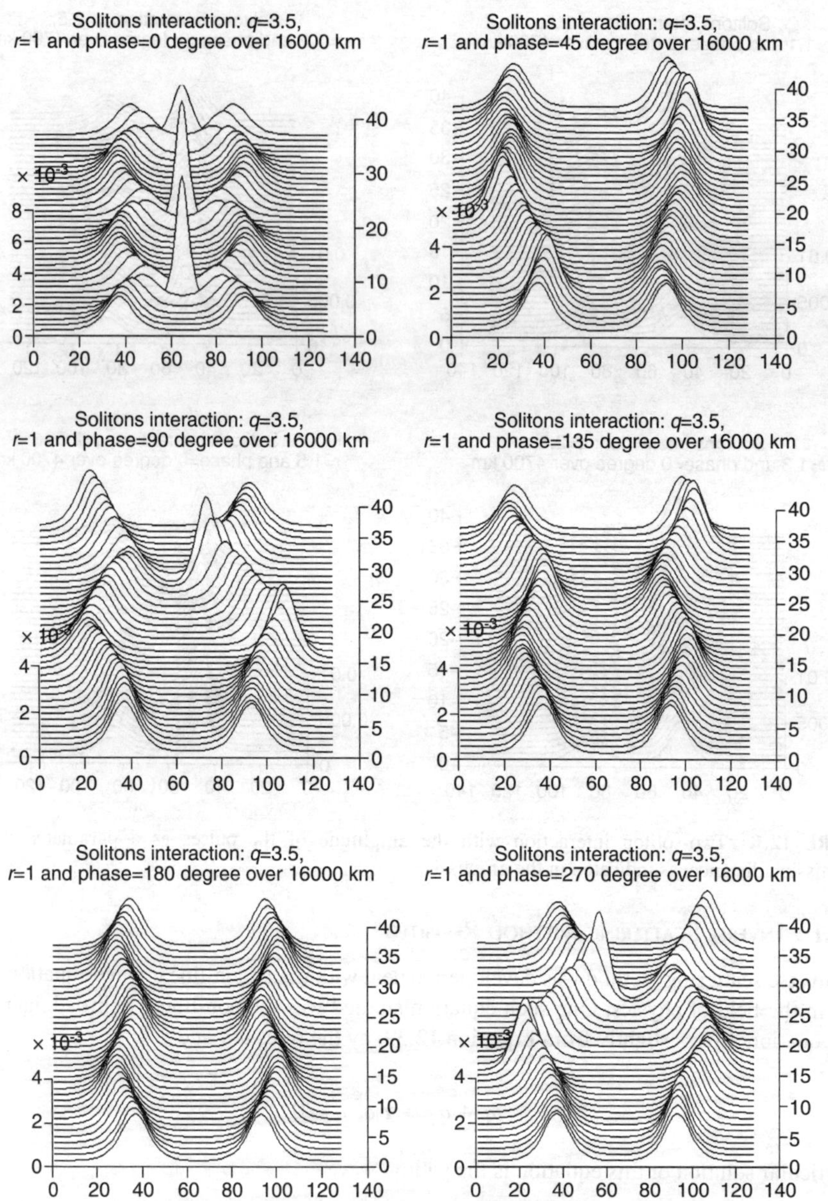

FIGURE 12.9 Two-soliton interaction with the relative carrier phase (greater than $\pi/2$) of the pulses as a parameter at different transmission distance, as indicated in the graphs.

However, one must note that the three steps involved in the inverse method deals with linear processes only and this is the advantage of the inverse scattering method.

Step 1: Direct scattering

The scattering data $S_\pm(0)$ corresponding to the initial function $F(x)$ is first obtained by solving the Schroedinger equation for the Jost solutions, which are solutions with the asymptotic form $\exp(-ikx)$ and $\exp(ikx)$ as $x \to \pm\infty$.

Step 2: Evolution of the scattering data

Once the scattering data is obtained for $t = 0$, the evolution of the scattering data is determined uniquely for arbitrary time t by

$$C_1\left(\hat{L}^A\right)u_t + C_2\left(\hat{L}^A\right)u_x = 0, \quad x \in \Re, t \geq 0 \tag{12.40}$$

where C_1, C_2 are real analytic functions and L is basically a reformulation of the Schroedinger equation as a linear operator.

Step 3: Inverse spectral transform
The potential $u(x,t)$ is then reconstructed from the scattering data $S_\pm(t)$, usually by solving some system of linear integrals equations. Returning to the non-linear Schroedinger equation, Equation 10.40, we can associate with this equation the Lax pair

$$\hat{L} = i\begin{bmatrix} 1+p & 0 \\ 0 & 1-p \end{bmatrix}\frac{\partial}{\partial x} + \begin{bmatrix} 0 & u^* \\ u & 0 \end{bmatrix}, \quad \chi = \frac{2}{1-p^2}$$

$$\hat{A} = -p\begin{bmatrix} 1 & 0 \\ 0 & 1 \end{bmatrix}\frac{\partial^2}{\partial x^2} + i\begin{bmatrix} \dfrac{|u|^2}{1+p} & ju_x^* \\ -ju_x & \dfrac{-|u|^2}{1-p} \end{bmatrix} \tag{12.41}$$

and the L operator evolves in time according to the commutation relation

$$\frac{\partial \hat{L}}{\partial t} = \left[\hat{A}, \hat{L}\right] \tag{12.42}$$

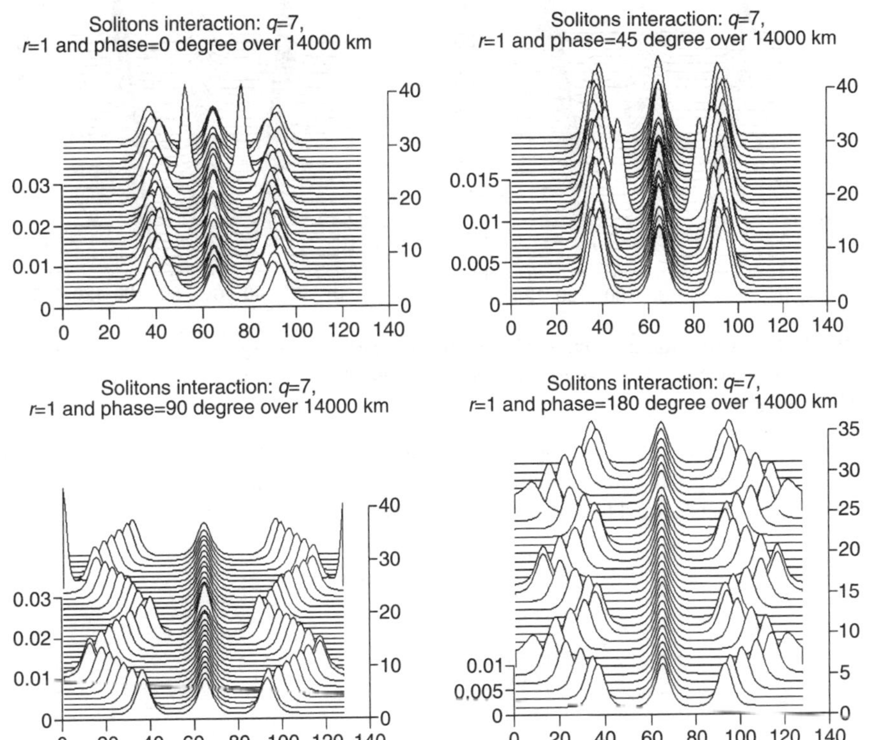

FIGURE 12.10 Two-soliton interaction with the relative carrier phase (less than $\pi/2$) of the pulses as a parameter at different transmission distance, as indicated in the graphs.

TABLE 12.2

The Characteristic of Soliton Pair Due to the Relative Phase Angle

Relative Phase	Observation and Effect
0 degree	Soliton pair is merging periodically
180 degree	Soliton pair is oscillating periodically without merging
Others	Soliton pair repulsion

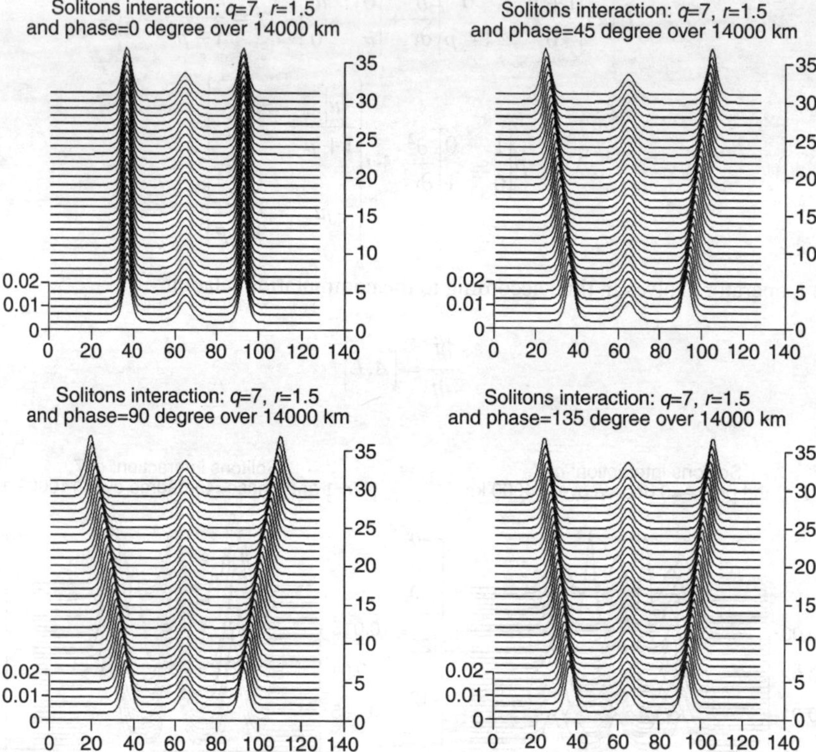

FIGURE 12.11 Two-soliton interaction with the relative carrier phase (greater than $\pi/2$) of the pulses and a pulse as the parameter with a separation temporal spacing of $r = 1.5$ at different transmission distance, as indicated in the graphs.

Here L and A are linear differential operators containing the sought function $u(x,t)$ in the form of a coefficient. By appropriate transformations of the eigenfunctions and rescaling of the eigenvalues, the pair could be transformed to resemble the form (see Anderson and Lisak [5]).

$$\hat{L} = j \begin{bmatrix} \dfrac{\partial}{\partial x} & -U(x,t) \\ R(x,t) & -\dfrac{\partial}{\partial x} \end{bmatrix}$$

$$\hat{A} = \begin{bmatrix} A(x,t,k) & B(x,t,k) \\ C(x,t,k) & -A(x,t,k) \end{bmatrix} + f(k)\mathbf{I} . \tag{12.43}$$

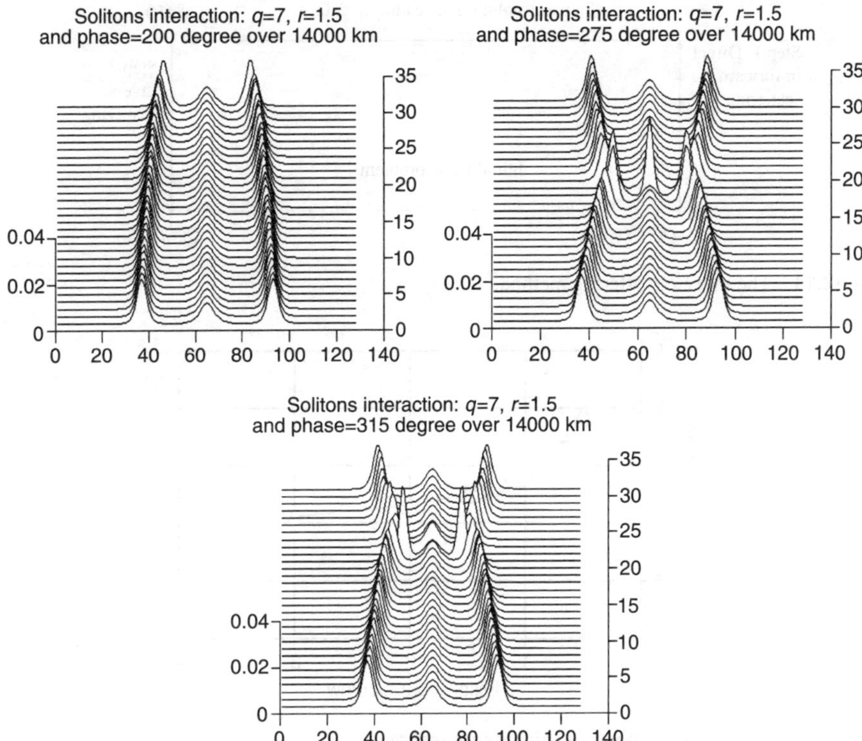

FIGURE 12.12 Three solitons interaction at different relative phase with r=1.5.

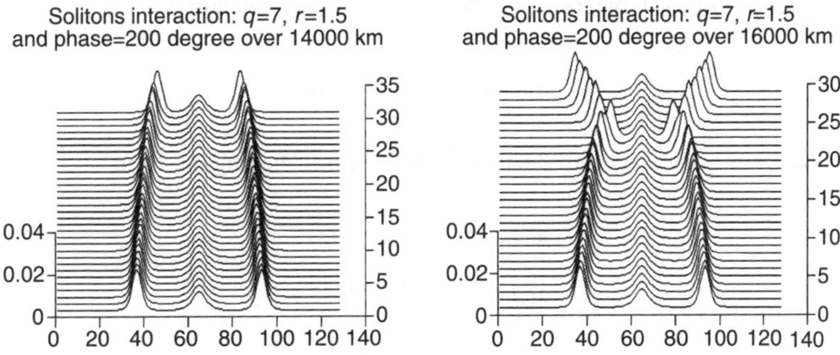

FIGURE 12.13 Three-soliton interaction at a distance of 14,000 km and 16,000 km with a relative phase difference of adjacent pulses of 200 degrees.

The key observation of Ablowitz, Kaup, Newell and Segur is that a set of non-linear equations can be obtained from the transformed Lax pair

$$A_x - UC + RB = 0$$

$$U_t - B_x - 2AU + 2jkB = 0 \qquad (12.44)$$

$$R_t - C_x + 2AR + 2jkC = 0.$$

This set of equations, in turn, can generate several classes of non-linear wave equations. A technique for obtaining some of the associated equations is to substitute polynomials in k for A, B, C and solve

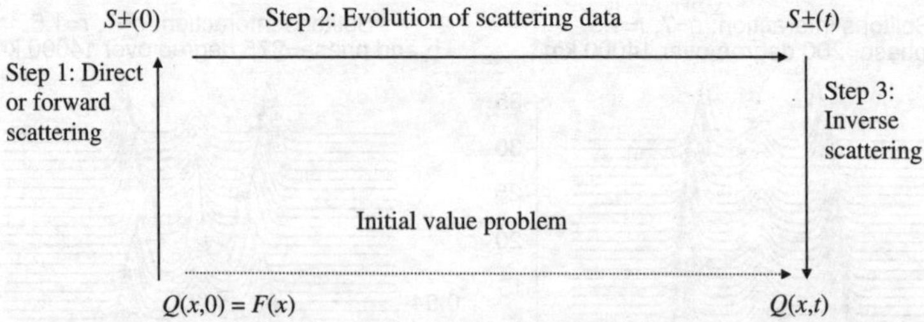

FIGURE 12.14 The inverse scattering method.

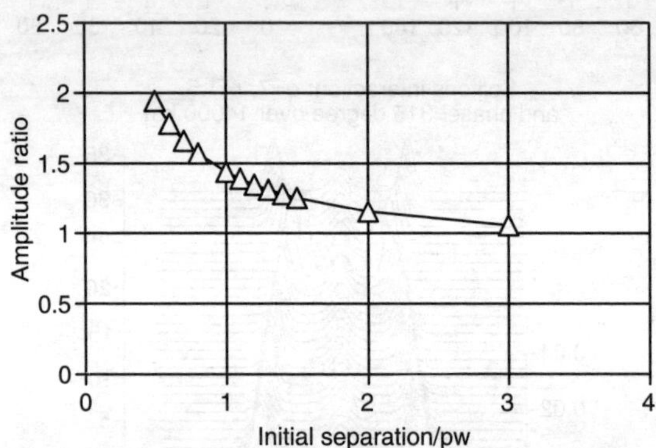

FIGURE 12.15 Amplitude ratio versus initial pulse separation.

recursively. This technique, reproduces besides the non-linear Schroedinger equation, several other important equations. For example, one such equation is the *KdV* equation

$$U_t + 6UU_x + U_{xxx} = 0 \tag{12.45}$$

and this equation was generated with the polynomials A, B, C having the form

$$A = -4ik^3 + 2jU - U_x,$$

$$B = 4Uk^2 + 2jU_x k - 2U^2 - U_{xx}, \tag{12.46}$$

$$C = -4k^2 + 2U, \quad R = -1.$$

Following the scheme as presented in Figure 12.16, we proceed with Step 1, the direct scattering problem.

Step 1: Direct scattering problem
Let us return to the L operator associated with the non-linear Schroedinger equation. We consider the system of equations

$$\hat{L}\psi = \lambda\psi, \quad \psi = \begin{Bmatrix} \psi_1 \\ \psi_2 \end{Bmatrix}. \tag{12.47}$$

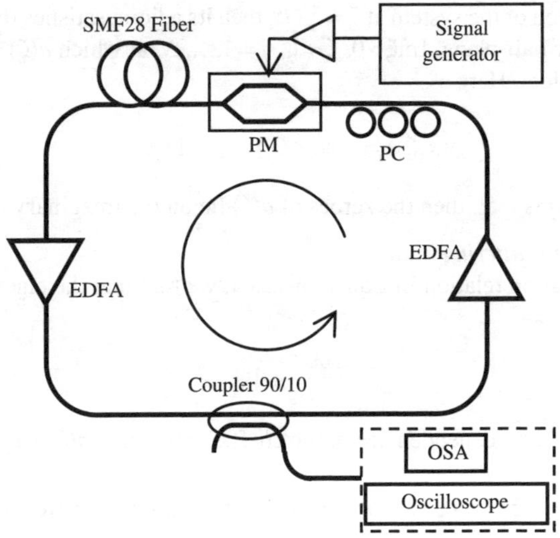

FIGURE 12.16 Experimental setup of the FM mode-locked fiber laser. PM, phase modulator; PC, polarization controller; OSA, optical spectrum analyzer.

Making the change of variables,

$$\psi_1 = \sqrt{1-p}\,\exp\!\left(-j\frac{\lambda x}{1-p^2}\right)v_2, \quad \psi_2 = \sqrt{1+p}\,\exp\!\left(-j\frac{\lambda x}{1-p^2}\right)v_1 \tag{12.48}$$

the equation can be rewritten in the following form

$$\frac{\partial v_1}{\partial t} + j\zeta v_1 = qv_2, \quad \frac{\partial v_2}{\partial t} - j\zeta v_2 = -q^* v_1,$$

$$q = \frac{ju}{\sqrt{1-p^2}}, \quad \zeta = \frac{\lambda p}{1-p^2}. \tag{12.49}$$

We define for real $\zeta = \xi$, the Jost functions φ and ψ, as solutions of Equation 12.47, with asymptotic values

$$\varphi \rightarrow \begin{bmatrix} 1 \\ 0 \end{bmatrix} \exp(-j\xi x), \quad x \rightarrow -\infty$$

$$\psi \rightarrow \begin{bmatrix} 0 \\ 1 \end{bmatrix} \exp(j\xi x), \quad x \rightarrow +\infty. \tag{12.50}$$

The solution ψ and its adjoint,

$$\bar{\psi} = \left\{ \begin{array}{c} \psi_2^* \\ -\psi_1^* \end{array} \right\} \tag{12.51}$$

form a complete system of solutions, and therefore

$$\varphi = a(\xi)\bar{\psi} + b(\xi)\psi. \tag{12.52}$$

Note that if ψ is a solution of the system at $\zeta=\xi+i\eta$ then its adjoint satisfies the system at $\zeta^*=\xi-i\eta$. The points of the upper half-plane, $\text{Im}\zeta>0$, $\zeta=\zeta_j$, $j=1,\ldots,N$, at which $a(\zeta)=0$, correspond to the eigenvalues of the problem. Here

$$\varphi(x,\zeta_j)=c_j\psi(x,\zeta_j), \quad j=1,\ldots,N. \tag{12.53}$$

It can be shown that if q is real, then the zeroes of $a(\zeta)$ lie on the imaginary axis.

Step 2: Evolution of the scattering data
Revisiting the commutation relation in Equation 12.44, we find that the eigenfunctions ψ obey the equation

$$i\frac{\partial\psi}{\partial t}=\hat{A}\psi. \tag{12.54}$$

From this equation, the time dependencies of the coefficients $a(\xi,t)$, $b(\xi,t)$, $c_j(t)$ are found to be

$$a(\xi,t)=a(\xi), \quad b(\xi,t)=b(\xi,0)\exp(4i\xi^2 t), \quad c_j(t)=c_j(0)\exp(4i\zeta^2 t). \tag{12.55}$$

Step 3: Inverse scattering problem
    The task is now to reconstruct $u(x,t)$ from the scattering data $a(\xi)$, $b(\xi,0)$ for $-\infty<\xi<+\infty$ and $c_j(0)$, which we have used to denote $b(\zeta_j,0)$. The eigenfunctions are related to the residues at the poles ζ_j and the coefficients $a(\xi)$, $b(\xi)$ via the following system of equations. See Agrawal [2] for complete derivation.

$$\phi_1-c(x,\xi)\frac{1+J}{2}\phi_2^*=-c(x,\xi)\sum_{k=1}^{N}\frac{\exp(-i\zeta_k^* x)}{\xi-\zeta_k^*}c_k^*\psi_2^*(x,\zeta_k),$$

$$c^*(x,\xi)\frac{1-J}{2}\phi_1+\phi_2^*=c^*(x,\xi)+c^*(x,\xi)\sum_{k=1}^{N}\frac{\exp(i\zeta_k x)}{\xi-\zeta_k}c_k\psi_1(x,\zeta_k),$$

$$\psi_1(x,\zeta_j)\exp(-i\zeta_j x)+\sum_{k=1}^{N}\frac{\exp(-i\zeta_k^* x)}{\zeta_j-\zeta_k^*}c_k^*\psi_2^*(x,\zeta_k)=\frac{1}{2}J\phi_2^*, \tag{12.56}$$

$$-\sum_{k=1}^{N}\frac{\exp(i\zeta_k x)}{\zeta_j^*-\zeta_k}c_k\psi_1(x,\zeta_k)+\psi_2^*(x,\zeta_j)\exp(i\zeta_j^* x)=1+\frac{1}{2}J\phi_1$$

where $(J\phi)(\xi)=\dfrac{1}{\pi i}\displaystyle\int_{-\infty}^{+\infty}\frac{\phi(\xi')}{\xi'-\xi}d\xi'$, $\phi(\xi)=\dfrac{b(\xi)}{a(\xi)}\exp(i\xi x)\psi(x,\xi)$.

The next step is to find an expression for q as u is related to q by a simple constant as in Equation 12.50. This can be done by comparing the asymptotic forms of the above system of equations

$$\begin{bmatrix}\psi_2(x,\zeta)\\-\psi_1(x,\zeta)\end{bmatrix}\exp(-i\zeta x)=\begin{bmatrix}1\\0\end{bmatrix}+\frac{1}{\zeta}\left[\sum_{k=1}^{N}c_k^*\exp(-i\zeta_k^* x)\psi^*(x,\zeta_k)+\frac{1}{2\pi i}\int\phi^*(\xi)d\xi\right]$$

$$+O\left(\frac{1}{\zeta^2}\right), \quad \zeta\to\infty \tag{12.57}$$

and the differential equation in Equation 12.51.

$$\psi(x,\zeta)\exp(-i\zeta x) = \begin{Bmatrix} 0 \\ 1 \end{Bmatrix} + \frac{1}{2i\zeta}\left[\begin{matrix} q(x) \\ \int_x^\infty |q(s)|^2\, ds \end{matrix}\right] + O\left(\frac{1}{\zeta^2}\right). \tag{12.58}$$

Thus, we get

$$q(x) = -2i\sum c_k^* \exp(-i\zeta_k^* x)\psi_2^*(x,\zeta_k) - \frac{1}{\pi}\int \phi_2^*(\xi)d\xi,$$

$$\tag{12.59}$$

$$\int_x^\infty |q(s)|^2\, ds = -2i\sum c_k \exp(i\zeta_k x)\psi_1(x,\zeta_k) + \frac{1}{\pi}\int \phi_1(\xi)d\xi.$$

Once a and b are obtained for an initial value, $u(x, t=0)$, then we can get $u(x,t)$ at an arbitrary instant through the above Equation 12.57 to Equation 12.61. We now try to find an explicit formula for general solution of the non-linear Schrodinger equation.

12.6.2 N-Solitons Solution (Explicit Formula)

We consider the inverse scattering problem with the assumption $b(\xi,t)=0$. Then $\phi(\xi)\equiv 0$ and the system of Equation 12.58 reduces to

$$\psi_{1j} + \sum_{k=1}^{N} \frac{\lambda_j \lambda_k^*}{\zeta_j - \zeta_k^*}\, \psi_{2k}^* = 0,$$

$$-\sum_{k=1}^{N} \frac{\lambda_k \lambda_j^*}{\zeta_j^* - \zeta_k}\, \psi_{1k} + \psi_{2j}^* = \lambda_j^*, \tag{12.60}$$

$$\psi_j = \begin{Bmatrix} \psi_{1j} \\ \psi_{2j} \end{Bmatrix} = \sqrt{c_j}\,\psi(x,\zeta_j), \quad \lambda_j = \sqrt{c_j}\,\exp(i\zeta_j x).$$

Subsequently, the potential is also simplified to

$$q(x) = -2i\sum_{k=1}^{N} \lambda_k^* \psi_{2k}^*, \quad \int_x^\infty |q(s)|^2\, ds = -2i\sum_{k=1}^{N} \lambda_k \psi_{1k}. \tag{12.61}$$

To see how a soliton can arise from this system of equations, we examine the case $N=1$ and $a(\zeta)$ has only one zero in the upper-half plane. The following system of equations is obtained.

$$\psi_1 + \frac{|\lambda|^2}{2i\eta}\psi_2^* = 0, \quad \frac{|\lambda|^2}{2i\eta}\psi_1 + \psi_2^* = \lambda^*. \tag{12.62}$$

This system describes a soliton with an amplitude η, the initial center and phase given by

$$x_0 = \frac{1}{2\eta}\ln\frac{|\lambda(0)|^2}{2\eta}, \quad \varphi = -2\arg\lambda(0). \tag{12.63}$$

Returning to the N-soliton case; in order to obtain the explicit solution for q and ultimately u, we need to solve the above system of equations, Equation 12.62 for ψ_j. If we represent the above equations in matrix form, then the potential $u(x,t)$ can be shown to be [2]:

$$\left|u(x,t)\right|^2 = \sqrt{2\chi}\,\frac{d^2}{dx^2}\ln\det\|\mathbf{A}\| = \sqrt{2\chi}\,\frac{d^2}{dx^2}\ln\det\|\mathbf{B}\mathbf{B}^* + 1\|,$$

$$(12.64)$$

$$\mathbf{B}_{jk} = \frac{\sqrt{c_j c_k^*}}{\zeta_j - \zeta_k^*}\exp\left[i\left(\zeta_j - \zeta_k^*\right)x\right].$$

The above result is a formula for reconstructing the potential $u(x,t)$. At the present moment, it does not describe a solution to any real physical problem because we have not specified the initial values or boundary conditions. Therefore let us examine the example from Satsuma and Yajima [6] where the initial potential has the form of a soliton, $A.\mathrm{sech}x$.

Example: $U(x,\tau) = A\mathrm{sech}x$

Let us turn our attention to the non-linear Schroedinger equation. Note that in this example, the equation is slightly different from Equation 12.39 and Equation 12.40.

$$ju_t = \frac{1}{2}u_{xx} + \left|u\right|^2 u.$$

$$(12.65)$$

We now consider the associated eigenvalue equation

$$jv_x + Uv = \zeta\sigma_3 v,$$

$$v = \begin{pmatrix} v_1 \\ v_2 \end{pmatrix}, \quad \sigma_3 = \begin{pmatrix} 1 & 0 \\ 0 & -1 \end{pmatrix}, \quad U = \begin{pmatrix} 0 & u \\ u^* & 0 \end{pmatrix}.$$

$$(12.66)$$

Note that if u is a solution of Equation 12.67, then the eigenvalue ζ is independent of time and v evolves in time according to

$$iv_t = Av,$$

$$A = \begin{pmatrix} 1 & 0 \\ 0 & 1 \end{pmatrix}\left(\frac{1}{2}\frac{\rho^2-1}{\rho^2+1}\frac{\partial^2}{\partial x^2} - i\zeta\frac{\partial}{\partial x} + C\right) + \frac{1}{\rho^2+1}\begin{pmatrix} \rho^2\left|u\right|^2 & iu_x \\ -j\rho^2 u_x^* & -\left|u\right|^2 \end{pmatrix}$$

$$(12.67)$$

where C is a constant independent of x. Proceeding with Step 1, we eliminate v_2 in Equation 12.68, and using the initial potential $u(x,t) = A.\mathrm{sech}x$, we have

$$s(1-s)\frac{d^2}{ds^2}v_1 + \left(\frac{1}{2} - s\right)\frac{d}{ds}v_1 + \left[A^2 + \frac{\zeta^2 + j\zeta(1-2s)}{4s(1-s)}\right]v_1 = 0, \quad s = \frac{1-\tanh x}{2}.$$

$$(12.68)$$

Further transformation of the dependent variable v_1 into $s^\alpha(1-s)^\beta\omega_1$ reduces the above equation to the hypergeometric functions,

$$v_{11}(s) = s^{j\zeta/2}\left(1-s\right)^{-j\zeta/2}F\left(-A, A, j\zeta + \frac{1}{2}; s\right),$$

$$(12.69)$$

$$v_{12}(s) = s^{\frac{1}{2} - j\zeta/2}\left(1-s\right)^{-j\zeta/2}F\left(\frac{1}{2} - j\zeta + A, \frac{1}{2} - j\zeta - A, \frac{3}{2} - j\zeta; s\right),$$

The solutions for v_2 can be obtained by replacing ζ with $-\zeta$ in the above equation,

$$v_{21}(s) = s^{-j\zeta/2}(1-s)^{j\zeta/2} F\left(-A, A, -j\zeta + \frac{1}{2}; s\right),$$

$$v_{12}(s) = s^{\frac{1}{2}+j\zeta/2}(1-s)^{j\zeta/2} F\left(\frac{1}{2}+j\zeta+A, \frac{1}{2}+j\zeta-A, \frac{3}{2}+j\zeta; s\right),$$

$$\tag{12.70}$$

From the asymptotic requirements of Equation 12.52, we can find expressions for the coefficients

$$\psi = \begin{pmatrix} \dfrac{A}{\xi+j/2} v_{12} \\ v_{21} \end{pmatrix}, \quad \bar{\psi} = \begin{pmatrix} v_{21}^{*} \\ -\dfrac{A}{\xi-j/2} v_{12}^{*} \end{pmatrix},$$

$$a(\xi) = \frac{\left[\Gamma\left(-j\xi+\dfrac{1}{2}\right)\right]^2}{\Gamma\left(-j\xi+A+\dfrac{1}{2}\right)\Gamma\left(-j\xi-A+\dfrac{1}{2}\right)},$$

$$\tag{12.71}$$

$$b(\xi) = \frac{j\left[\Gamma\left(j\xi+\dfrac{1}{2}\right)\right]^2}{\Gamma(A)\Gamma(1-A)} = j\frac{\sin(\pi A)}{\cosh(\pi\xi)}.$$

The eigenvalues ζ_r are obtained from the zeroes of $a(\zeta)$.

$$\zeta_r = j\left(A-r+\frac{1}{2}\right)$$

$$\tag{12.72}$$

r must be positive integers satisfying $A - r + \frac{1}{2} > 0$. Step 2 gives

$$\lambda_k = \sqrt{\frac{b(\zeta_k)}{\partial a(\zeta_k)/\partial t}}\,\exp\left[j\zeta_k x - j\zeta_k^2 t\right], \quad c(x,\xi) = \frac{b(\xi)}{a(\xi)}\exp\left[2j\xi x - 2j\xi^2 t\right]. \tag{12.73}$$

Note the slight difference in the above expressions when compared with Equation 12.24. Step 3 is an application of Equations 12.58 to Equation 12.63.

12.6.3 Special Case $A = N$

When $A = N$, a positive integer then by Equation 12.73, the coefficients $a(\zeta)$ reduces to

$$a(\zeta) = \prod_{r=1}^{N} \frac{(\zeta-\zeta_r)}{\zeta-\zeta_r^{*}}$$

$$\tag{12.74}$$

while $b(\xi) = 0$. At the eigenvalues,

$$b(\zeta_k) = j(-1)^{k-1}.$$

$$\tag{12.75}$$

The coefficients $c(x,\xi)$ defined by Equation 12.42 and subsequently $\phi_1 = \phi_2^* = 0$. Substituting the results in Equation 12.58, we can solve the equations for ψ_{k1} and ψ^*_{k2}. Explicit solutions for the cases $N=1$ and $N=2$ are given below:

$$u(x,t) = \exp(-jt/2)\operatorname{sech} x, \quad N=1$$

$$u(x,t) = 4\exp(-jt/2)\frac{\cosh 3x + 3\exp(-4jt)\cosh x}{\cosh 4x + 4\cosh 2x + 3\cos 4t}, \quad N=2. \tag{12.76}$$

Of interest is the long-term behavior of the soliton and this will be explored in the next section.

12.6.4 N-Soliton Solutions (Asymptotic Form as $\tau \to \pm\infty$)

The long-term behavior of the N-soliton solution is studied here for the special case where there are no two solitons with the same velocity i.e., no ξ_j are the same. It is found that the N-soliton solution breaks up into diverging solitons as $t \to \pm\infty$. To verify this for the case $t \to +\infty$, let us arrange the ξ_j in decreasing order, $\xi_1 > \xi_2 > \ldots > \xi_N$. From Equation 12.57 we have

$$\lambda_j(x,t) = \lambda_j(0)\exp\left[-\eta_j(x+4\xi_j t) + j\left(\xi_j x + 2(\xi_j^2 - \eta_j^2)t\right)\right],$$

$$|\lambda_j(x,t)| = |\lambda_j(0)|\exp(-\eta_j y_j), \quad y_j = x + 4\xi_j t. \tag{12.77}$$

Let us consider the reference frame where $y_m = $ constant as $t \to +\infty$ i.e., we are in interested in a moving reference frame of one of the solitons. Then

$$y_j \to 0, \quad |\lambda_j| \to 0 \text{ for } j < m,$$

$$y_j \to \infty, \quad |\lambda_j| \to \infty \text{ for } j > m, \tag{12.78}$$

It follows from the system of Equation 12.62 that $\psi_{1j}, \psi_{2j} \to 0$ when $j < m$ i.e., the effect of these solitons are made negligible. The number of equations is therefore reduced to $2(N-m-1)$

$$\psi_{1m} + \frac{|\lambda_m|}{2i\eta_m}\psi_{2m}^* = -\lambda_m \sum_{k=m+1}^{N}\frac{1}{\zeta - \zeta_k^*}\phi_{2k}^*,$$

$$\frac{|\lambda_m|^2}{2i\eta_m}\psi_{1m} + \psi_{2m}^* = \lambda_m^* + \lambda_m^*\sum_{k=m+1}^{N}\frac{1}{\zeta^* - \zeta_k}\phi_{1k},$$

$$\sum_{k=m+1}^{N}\frac{1}{\zeta - \zeta_k^*}\phi_{2k}^* = -\frac{\lambda_m^*}{\zeta_j - \zeta_m^*}\psi_{2m}^*,$$

$$\sum_{k=m+1}^{N}\frac{1}{\zeta^* - \zeta_k}\phi_{1k} = -1 - \frac{\lambda_m}{\zeta^* - \zeta_m}\psi_{1m}. \tag{12.79}$$

Solving first for ϕ_{1k} and ϕ_{2k}^*, we obtain

$$\phi_{1k} = a_k + \frac{2i\eta_m}{a_m} \frac{a_k}{\zeta_k - \zeta_m} \psi_{1m}\lambda_m,$$

$$\phi_{2k}^* = -\frac{2j\eta_m}{a_m^*} \frac{a_k^*}{\zeta_k^* - \zeta_m^*} \psi_{2m}^*\lambda_m^*, \tag{12.80}$$

$$a_k = \frac{\prod\limits_{p=m+1}^{N} \zeta_k - \zeta_p^*}{\prod\limits_{m<p\neq k}^{N} \zeta_k - \zeta_p}, \quad a_m = 2j\eta_m \prod\limits_{p=m+1}^{N} \frac{\zeta_m - \zeta_p^*}{\zeta_m - \zeta_p}.$$

After some tricky manipulation (see Dodd et al. [5] for details)

$$\psi_{1m} + \frac{\left|\lambda_m^+\right|^2}{2i\eta_m} \psi_{2m}^* = 0,$$

$$\frac{\left|\lambda_m^+\right|^2}{2i\eta_m} \psi_{1m} + \psi_{2m}^* = \left(\lambda_m^+\right)^*, \quad \lambda_m^+ = \lambda_m \prod\limits_{p=m+1}^{N} \frac{\zeta_m - \zeta_p}{\zeta_m - \zeta_p^*}. \tag{12.81}$$

The system coincides with the system from Equation 12.64 and describes a soliton with a displaced position of the centre x_0^+ and phase φ^+,

$$x_{0m}^+ - x_{0m} = \frac{1}{\eta_m} \sum\limits_{p=m+1}^{N} \ln\left|\frac{\zeta_m - \zeta_p}{\zeta_m - \zeta_p^*}\right| < 0,$$

$$\varphi_m^+ - \varphi_m = -2 \sum\limits_{p=m+1}^{N} \arg\left(\frac{\zeta_m - \zeta_p}{\zeta_m - \zeta_p^*}\right). \tag{12.82}$$

The calculations are similar for the case $t \to -\infty$. In the reference frames $y = x + \xi t$, where ξ does not coincide with any of the ξ_m as $t \to \pm\infty$, the reduced system tends to zero, hence proving the asymptotic breakdown of the N-soliton solution into solitons. When t changes from $-\infty$ to $+\infty$, the corresponding changes in the soliton centers and phases are

$$\Delta x_{0m} = x_{0m}^+ - x_{0m}^- = \frac{1}{\eta_m}\left(\sum\limits_{k=m+1}^{N} \ln\left|\frac{\zeta_m - \zeta_k}{\zeta_m - \zeta_k^*}\right| - \sum\limits_{k=1}^{m-1} \ln\left|\frac{\zeta_m - \zeta_k}{\zeta_m - \zeta_k^*}\right|\right),$$

$$\tag{12.83}$$

$$\Delta\varphi_m = \varphi_m^+ - \varphi_m^- = 2\sum\limits_{k=1}^{m-1} \arg\frac{\zeta_m - \zeta_k}{\zeta_m - \zeta_k^*} - 2\sum\limits_{k=m+1}^{N} \arg\frac{\zeta_m - \zeta_k}{\zeta_m - \zeta_k^*}.$$

Formula can be interpreted by assuming that the solitons interact with each other pairwise. In each interaction, the faster soliton moves forward by the first product in formula and the slower one shifts backwards by the second term in formula. The total soliton shift is equal to the algebraic sum of its shifts during the paired collisions or interactions. The potential thus takes the form

$$u(x,t) = \sqrt{2\chi} \sum\limits_{j=1}^{N} \eta_j \operatorname{sech}\left[2\eta_j(x - x_{0j}) + 8\eta_j\xi_j t\right] \exp\left[-4j\left(\xi_j^2 - \eta_j^2\right)t - 2j\xi_j x + j\varphi_j\right] \tag{12.84}$$

which is the sum of all the individual modal solitons. Note that this conclusion was arrived at employing the assumption that $b(\xi)$ vanishes. This may not always be true but Satsuma and Yajima [6] showed using the methods of complex analysis that the long-term solution still break down into individual solitons.

12.6.5 BOUND STATES AND MULTIPLE EIGENVALUES

From Equation 12.86, one can see that the rate of separation of a pair of solitons is proportional to the difference between the values of the parameters ξ. At equal values of ξ, the solitons do not separate but form a bound state. Consider the bound state of N solitons with initial potential $N\mathrm{sech}x$ as discussed earlier and suppose that their velocities ξ_j are all zero. Then the state of the soliton varies with time according to $\exp(-4j\eta_j^2 t)$ from the application of Equation 12.86. Thus, the state is a periodic variation with frequency $4\eta_j^2$. For a bound state of N solitons, there are N frequencies but we find that the behavior of the solitons is characterized by all the possible frequency differences or beat frequencies. For $N=2$, the bound state is characterized by the beat frequency $\omega=4(\eta_1^2-\eta_2^2)$. For $N=3$ or higher, the lowest beat frequency will give the main periodic variation. Refer to Figure 12.2 and Figure 12.16 to see the effect of the periodic variation. The solitons appear to diverge and then coalesce together. This process repeats itself with a frequency determined by the lowest beat frequency.

To study the effect of multiple eigenvalues, which is particularly important if we intend to study trains of equal sized solitons for communication purposes. Naturally, we would like to use Equation 10.83 or Equation 10.86 that have been developed earlier. Unfortunately, the formulas fail for solitons with equal amplitudes. Agrawal [2] has worked out an approximation for the limiting case as one eigenvalue approaches another and it was found that the distance between the two solitons with equal amplitudes increases with time like $\ln(4\eta^2 t)$. Another recourse would be to turn to perturbation methods as developed by Lisak, Anderson, Karpman and Solov'ev [7–9].

12.7 INTERACTION BETWEEN TWO SOLITONS IN AN OPTICAL FIBER

We now return to the non-linear Schrodinger equation describing the non-linear propagation of light waves in optical fibers. The motivation for this study is the eventual design of a soliton communication system. One needs to know how two solitons affect one another. The aim is to find suitable parameters for the soliton separation, widths, phases or speeds so that the solitons do not diverge too much or coalesce together. Also, the shape and size of the solitons should change negligibly otherwise they may not be recognized by the soliton detector.

Desem and Chu [10] gave an exact derivation of a two-soliton solution with arbitrary initial phase and separation

$$q(x,\tau) = \frac{|\alpha_1|\cosh\left(a_1+i\theta_1\right)e^{j\phi_2} + |\alpha_2|\cosh\left(a_2+i\theta_2\right)e^{j\phi_1}}{\alpha_3\cosh a_1\cosh a_2 - \alpha_4\left[\cosh\left(a_1+a_2\right)-\cos\left(\phi_2-\phi_1\right)\right]},$$

$$\phi_{1,2} = \left[\frac{\left(\eta_{1,2}^2-\xi_{1,2}^2\right)x}{2}-\tau\xi_{1,2}\right]+\left(\phi_0\right)_{1,2}, \quad a_{1,2} = \eta_{1,2}\left(\tau+x\xi_{1,2}\right)+\left(a_0\right)_{1,2},$$

$$(12.85)$$

$$|\alpha_{1,2}|e^{i\theta_{1,2}} = \pm\left\{\left[\frac{1}{\eta_{1,2}}-\frac{2\eta}{\Delta\xi^2+\eta^2}\right]\pm j\frac{2\Delta\xi}{\Delta\xi^2+\eta^2}\right\}, \quad \alpha_3 = \frac{1}{\eta_1\eta_2}, \quad \alpha_4 = \frac{2}{\eta^2+\Delta\xi^2}$$

$$\zeta_{1,2} = \frac{\xi_{1,2}+j\eta_{1,2}}{2}, \quad \Delta\xi = \xi_2-\xi_1, \quad \eta = \eta_1+\eta_2.$$

The authors were interested in the initial condition of the form

$$q(0,\tau) = \mathrm{sech}\left[\tau-\tau_0\right]+e^{j\theta}A\,\mathrm{sech}\left[A(\tau+\tau_0)\right]. \qquad (12.86)$$

12.7.1 SOLITON PAIR WITH INITIAL IDENTICAL PHASES

Launching the solitions with equal phases, that is $\theta = 0$ and hence $\xi_1 = \xi_2 = 0$, will result in a bound system. If $\theta \neq 0$, then $\xi_1 \neq \xi_2$ and from the previous section on the asymptotic behavior of a general soliton solution, the solitons separate eventually. The resulting equation upon substituting $\theta = 0$ is simplified to

$$q(\tau,x) = Q\left\{\eta_1 \operatorname{sech} \eta_1(\tau + \gamma_0)e^{j\eta_1^2 x/2} + \eta_2 \operatorname{sech} \eta_2(\tau - \gamma_0)e^{j\eta_2^2 x/2}\right\},$$

$$Q = \frac{\eta_2^2 - \eta_1^2}{\eta_1^2 + \eta_2^2 - 2\eta_1\eta_2[\tanh a_1 \tanh a_2 - \operatorname{sech} a_1 \operatorname{sech} a_2 \cos \psi]}, \tag{12.87}$$

$$a_{1,2} = \eta_{1,2}(\tau \pm \gamma_0), \quad \psi = \frac{\left(\eta_2^2 - \eta_1^2\right)x}{2}.$$

The two-soliton solution (Equation 12.89) describes the interaction of two solitons with unequal amplitudes ($\eta_1 \neq \eta_2$) bound together by the non-linear interaction. Here, owing to the slight variation in the coefficients of the non-linear Schroedinger equation in the optical fiber as compared to the one studied in the inverse scattering techniques, the beat frequencies are now

$$\omega = \left|\frac{\eta_2^2 - \eta_1^2}{2}\right| \tag{12.88}$$

and the two soliton pulses undergo an interaction which is periodic with $2\pi/\omega$.

12.7.2 SOLITON PAIR WITH INITIAL EQUAL AMPLITUDES

Here, the solitons are launched with equal amplitudes. The eigenvalues are

$$\eta_{1,2} \approx 1 + \frac{2\tau_0}{\sinh 2\tau_0} \pm \operatorname{sech}(\tau_0) \tag{12.89}$$

and are obtained by equating the first few terms of the Taylor series of Equation 12.88 to those of Equation 12.89 and applying the first conserved quantity.

$$\int_{-\infty}^{\infty} |q|^2 d\tau = 2(\eta_1 + \eta_2). \tag{12.90}$$

Referring to Figure 12.16. The period of oscillation in terms of the initial separation τ_0 is

$$\frac{\pi \sinh 2\tau_0 \cosh \tau_0}{2\tau_0 + \sinh 2\tau_0}. \tag{12.91}$$

As the initial separation τ_0 increases, the distance of coalescence increases exponentially. Therefore, to minimize interaction, the solitons should be widely separated.

12.7.3 SOLITON PAIR WITH INITIAL UNEQUAL AMPLITUDES

The eigenvalues for the case of unequal amplitudes are

$$\eta_{1,2} \approx \frac{A+1}{2} + \frac{2\tau_0\sqrt{A}}{\sinh 2\tau_0\sqrt{A}} + \left(\sqrt{A} - 1\right)\operatorname{sech} A\tau \pm \left[\frac{A-1}{2} + \operatorname{sech}\left(A_0\right)\right] \tag{12.92}$$

and the separation parameter is now

$$\gamma_0 \approx \tau_0 - \left(1 - \frac{2\,\mathrm{sech}\,A\tau_0}{A}\right)\left(\frac{1+A}{2A}\right)\ln\left(\frac{1+A}{A-1}\right).\qquad(12.93)$$

One can see from Equation 12.94 that as the initial separation τ_0 is increased, the eigenvalues η_1, η_2 approach A and 1, respectively. This implies that the oscillation period will no longer be governed by the separation but by the amplitude ratio A. Another key observation as $|\gamma_0|$ increases by increasing A, the value of the coefficient of $\cos\psi$ in Equation 12.92 is decreased. This means that as the amplitude ratio A is increased, the interaction is minimized. It can be seen that the interaction is still periodic but at no stage do they coalesce to form one pulse. An approximate expression for the minimum separation $\tau_{\min}$, in units of the pulse width of the unity-amplitude solution is given by

$$\tau_{\min} = \tau_0 - \frac{A+1}{4A\left[1+4A\,\mathrm{sech}\,A\tau_0\right]}\ln\left[1+4\left(\frac{A+1}{A-1}\right)^2\exp\left(\frac{-4A\tau_0}{A+1}\right)\right].\qquad(12.94)$$

This finding suggests that by employing two solitons with unequal amplitudes, we would achieve a much more stable wave formation for long distances as well as a higher bit rate since the pulses can be launched closer together (we can admit a higher bit rate because there is a minimum separation for the unequal amplitudes case whereas for the case of two solitons with equal amplitudes and zero phase difference, the soliton pulses do coalesce together, resulting in $\tau_{\min} = 0$).

A possibly troubling factor in employing unequal amplitudes is the fact that the amplitudes employed are not integers but can have fractions. Satsuma and Yajima [6] studied the behavior of non-integral amplitudes and concluded that if $A = N+\alpha$, $|\alpha| < \frac{1}{2}$, then for large times

$$\frac{\|u\|_\infty}{\|u\|_0} = 1 - \frac{\alpha^2}{\left(N+\alpha\right)^2}.\qquad(12.95)$$

Numerical simulations show that the solitons oscillates around the limiting value and slowly settles towards the steady or limiting value.

12.7.4 Design Strategy

Using the findings by Desem and Chu [10], we consider two designs in which equiphase solitons correspond to the transmission of four digital "1" signals as this is the worst case scenario which gives the most interaction. As mentioned earlier, a pair of non-equiphase solitons will result in an unbound system and the solitons will diverge. This is detrimental to the carrier system as the bit rate should be kept roughly constant. Also, the unequi-amplitude scheme is implemented as suggested by Desem and Chu [10] because it promises very little interaction among the soliton pulses.

Working in units of pulse width ($\pi\omega = 2$ for a unity-amplitude soliton), let us impose a design constraint that the minimum separation $2\tau_{\min}$ should not be less than $0.7 \cdot 2\tau_0$ ($2\tau_0$ is the initial separation). The reason is that we do not want the pulses to be too close together. Using Equation 10.96, we obtain for various values of τ_0, the corresponding A amplitude ratios and it is plotted in Figure 12.17.

Another constraint that needs to be imposed is that amplitude ratio should not be too large otherwise the smaller pulse may be mistaken by the detector for a "0" pulse as the detector may have a certain threshold. Let us assume that the threshold is $0.7 \cdot A_{\mathrm{peak}}^2$ where A_{peak} is the largest pulse amplitude (it is assumed that the detector detects the energy or intensity of the wave). This means the largest amplitude ratio possible is 1.2 corresponding to $\tau_0 = 1.75$ pw for the 1st design and $A = 1.1$, $\tau_0 = 2.5$ pw for the 2nd design. So, the 1st design seems superior than the 2nd design.

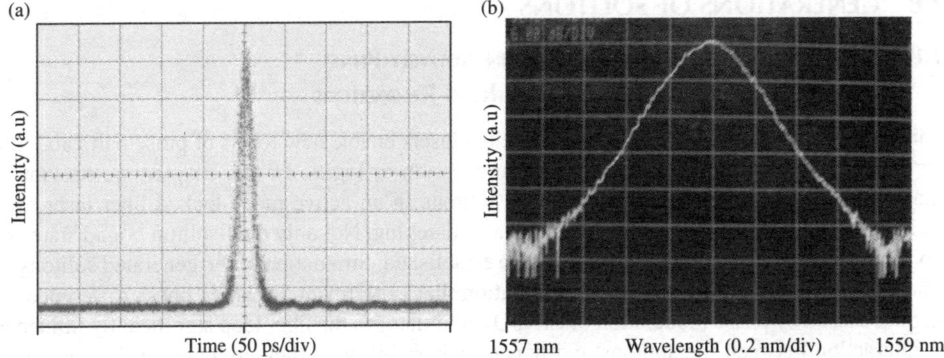

FIGURE 12.17   (a) The oscilloscope trace and (b) optical spectrum of a single soliton.

Next, one has to check for the variation in the amplitude of the soliton pulses. The constraint is that the amplitude of any soliton pulse should not be less than ½ the amplitude of another pulse. One way is applying perturbation theory [8,9]. The other method is to substitute the values of A and τ_0 into Equation 10.94 and Equation 10.95. Then, to find the variation of the soliton peaks, we let $\tau = -\gamma_0$ and $\tau = \gamma_0$ in Equation 12.89 to locate the peaks approximately. To estimate the extremum of the variation of the amplitude, we let $\cos\psi = \pm 1$. Note that this scheme gives a rough estimate only and no definite conclusion can be reached from this result. The MATLAB® program for the investigation of the 4-bit soliton pulse train is developed. It includes an automatic checker which alerts the user if (i) the pulses can no longer be distinguished from one another i.e., if we expect a signal of four "1"s, the pulses should be sufficiently far apart so that four "1"s can be distinguished. This happens when one pulse merges with another; (ii) pulse separation does not match the specification: 0.7/bit rate < pulse separation > 1.3/bit rate (the solitons drifts too much: drift >0.35/bit rate; and the eye window is less than half the maximum pulse amplitude.

Finally, we have yet to ascertain whether the maximum range of the 4-bit soliton pulse train. From study on two-solitons interaction, the range can be extended indefinitely for non-equi-amplitude soliton pulses. However, in the transmission scheme that I designed, there are scenarios where there may be interaction between equi-amplitude solitons.

So, the maximum range is determined by the interaction between the equi-amplitude solitons. The solitons coalesce at intervals defined by the soliton period. The formula for the soliton period is given by Desem and Chu [10],

$$x_0 = \frac{\pi \sinh(2\tau_0)\cosh(\tau_0)}{2\tau_0 + \sinh(2\tau_0)}. \tag{12.96}$$

Substituting the above value of $\tau_0 = 3.5$ pw $= 7$, we find that $x_0 = 1722.57$. Assuming that the coalescing process behaves like a cosine function, then if we demand that the separation should not be less than the 0.7 times the initial separation, then we have the maximum range is equal to $\cos^{-1}(0.7)/(\pi/2) \times 1722.57 \sim 872$. This is of course in normalized units. In real units and using the normalization constants as used in the MATLAB program, we obtain a value of $872 \times 157 = 58,404$ km. It turns out that for longer transmission distances, the second design is better as there is less chance of equi-amplitude soliton interaction. But of course, the bit-rate allowed is lower.

In summary, the design process follows the steps (i) Specify the minimum separation and amplitude threshold for a "1". (ii) Use initial relative phases of zero between neighboring solitons. This is to prevent solitons spreading apart. Use the equations given by Desem and Chu [10] to obtain a suitable pulse separation to and amplitude ration A; and (iii) check for variation of soliton amplitudes. Determine maximum range of transmission. Simulate with numerical method to confirm.

12.8 GENERATIONS OF SOLITONS

12.8.1 GENERATION OF BOUND-SOLITONS IN ACTIVELY PHASE MODULATION MODE-LOCKED FIBER RING RESONATORS

Bound solitons generated in actively mode-locked lasers enable new forms of pulse pairs and multiple pairs or groups of solitons in optical transmission or logics. In this chapter, we present the generation of stable bound states of multiple solitons in an active mode-locked fiber laser using continuous phase modulation for wideband phase matching. Not only dual-soliton bound states but also triple- and quadruple-soliton pulses can be established. Simulation of the generated solitons are demonstrated. We have also proven by simulation that experimental relative phase difference and chirping caused by phase modulation of $LiNbO_3$ modulator in the fiber loop significantly influences the interaction between the solitons and hence their stability as they circulate in the anomalous path-averaged dispersion fiber loop.

12.8.1.1 Introduction

Mode-locked fiber lasers are considered as important laser source for generating ultrashort soliton pulses. Recently, soliton fiber lasers have attracted significant research interests with experimentally demonstration of bound states of solitons as predicted in some theoretical works [11,12]. These bound-soliton states have been observed mostly in passive mode-locked fiber lasers [13–16]. There are, however, few reports on bound solitons in active mode-locked fiber lasers. The active mode locking offers significant advantage in the control of the repetition rate that would be critical for optical transmission systems. Observation of bound soliton pairs was first reported in a hybrid frequency modulation (FM) mode-locked fiber laser [17], in which a regime of bound-soliton pair harmonic mode locking at 10 GHz could be generated. There are, however no reports on multiple bound soliton states. Depending on the strength of soliton interation, the bound solitons can be classified into two categories: loosely bound solitons and tightly bound solitons which are determined by the relative phase difference between adjacent solitons. The phase difference may take the value of π or $\pi/2$ or any value depending on the fiber laser structures and mode locking conditions.

In this chapter, we report the bound states of multiple solitons in an active mode-locked fiber laser using continuous phase modulation or FM mechanism. By tuning the parameters for the phase matching of the lightwaves circulating in the fiber loop, not only that we could observe the dual-soliton bound state but also the triple- and quadruple-soliton bound states. Relative phase difference and chirping caused by phase modulation of $LiNbO_3$ modulator in the fiber loop significantly influences the interaction between the solitons and hence their stability as they circulate in the anomalous path-averaged dispersion fiber loop.

12.8.1.2 Formation of Bound States in a FM Mode-locked Fiber Laser

Although the formation of the stable bound soliton states are determined by the Kerr effect an anomalous averaged dispersion regime has been discussed in some configurations of passive mode-locked fiber lasers [14–16], it can be quite distinct in our active mode-locked fiber laser with the contribution of phase modulation of the $LiNbO_3$ modulator in stabilization of bound states. The formation of bound soliton states in a FM mode-locked laser can experience through two stages which include a process of pulse splitting and stabilization of muti-soliton bound states in the presence of a phase modulator in the cavity of mode-locked laser.

The first stage is splitting of a single pulse into multi pulses which occurs when the power in the fiber loop increases above a certain mode locking threshold [17,18]. At higher power, higher order solitons can be excited and in addition the accumulated non-linear phase shift in the loop is so high that a single pulse breaks up into many pulses [19]. The number of split pulses depends on the optical power in the loop, so there is a specific range of power for each splitting level. The fluctuation of pulses may occur at regions of power where there is a transition from the lower splitting level to the higher. Moreover, the chirping caused by phase modulator in the loop also makes the process of pulse conversion from a chirped single pulse into multi-pulses taking place more easily [20,21].

After splitting into multi pulses, the multi-pulse bound states are stabilized subsequently through the balance of the repulsive and attractive forces between neighboring pulses during circulating in a fiber loop of anomalous-averaged dispersion. The repulsive force comes from direct soliton interaction depending on the relative phase difference between neighboring pulses [22] and the effectively attractive force comes from the variation of group velocity of soliton pulse caused by the frequency chirping. Thus, in an anomalous average dispersion regime, the locked pulses should be located symmetrically around the extreme of positive phase modulation half-cycle, in other words the bound soliton pulses acquire an up-chirping when passing the phase modulator. In a specific mode-locked fiber laser setup, beside the optical power level and dispersion of the fiber cavity, the modulator-induced chirp or the phase modulation index determine not only the pulse width but also the time separation of bound-soliton pulses at which the interactive effects cancel each other [23,24].

The presence of a phase modulator in the cavity to balance the effective interactions among bound-soliton pulses is similar to the use of this device in a long-haul soliton transmission system to reduce the timing jitter, for this reason the simple perturbation theory can be applied to understand the role of phase modulation on the mechanism of bound solitons formation. A multi-soliton bound state can be described as following:

$$u_{bs} = \sum_{i=1}^{N} u_i(z,t) \tag{12.97}$$

and

$$u_i = A_i \operatorname{sech}\left\{A_i\left[(t - T_i)/T_0\right]\right\} \exp(j\theta_i - j\omega_i t) \tag{12.98}$$

where N is number of solitons in the bound state, T_0 is pulse width of soliton and A_i, T_i, θ_i, ω_i represent the amplitude, position, phase and frequency of soliton, respectively. In the simplest case of multi-soliton bound state, N is equal 2 or we consider the dual-soliton bound state with the identical amplitude of pulse and the phase difference of π value ($\Delta\theta = \theta_{i+1} - \theta_i = \pi$), the ordinary differential equations for the frequency difference and the pulse separation can be derived by using the perturbation method.

$$\frac{d\omega}{dz} = -\frac{4\beta_2}{T_0^3} \exp\left[-\frac{\Delta T}{T_0}\right] - 2\alpha_m \Delta T \tag{12.99}$$

$$\frac{d\Delta T}{dz} = \beta_2 \omega \tag{12.100}$$

where β_2 is the averaged group-velocity dispersion of the fiber loop, ΔT is pulse separation between two adjacent solitons ($T_{i+1} - T_i = \Delta T$) and $\alpha_m = m\omega_m^2/(2L_{cav})$, L_{cav} is the total length of the loop, m is the phase modulation index. Equation 12.100 shows the evolution of frequency difference and position of bound solitons in the fiber loop in which the first term on the right hand side represents the accumulated frequency difference of two adjacent pulses during a round trip of the fiber loop and the second one represents the relative frequency difference of these pulses when passing through the phase modulator. At steady state, the pulse separation is constant and the induced frequency differences cancel each other. On the other hand, if setting Equation 12.99 to zero we have

$$-\frac{4\beta_2}{T_0^3}\exp\left[-\frac{\Delta T}{T_0}\right]-2\alpha_m\Delta T=0 \qquad (12.101)$$

or

$$\Delta T\exp\left[\frac{\Delta T}{T_0}\right]=-\frac{4\beta_2}{T_0^3}\frac{L_{\text{cav}}}{m\omega_m^2}. \qquad (12.102)$$

We can see the effect of phase modulation to the pulse separation through Equation 12.102, in addition β_2 and α_m must have opposite signs which mean that in an anomalous dispersion fiber loop with negative value of β_2 the pulses should be up chirped. With a specific setup of FM fiber laser, when the magnitude of chirping increases, the bound pulse separation decreases subsequently. The pulse width also reduces according to the increase in the phase modulation index and modulation frequency, so that the ratio $\Delta T/T_0$ can not change much. Thus, the binding of solitons in the FM mode-locked fiber laser is assisted by the phase modulator. Bound solitons in the loop experience periodically the frequency shift and hence their velocity in response to changes in their temporal positions by the interactive forces in equilibrium state.

12.8.1.3 Experimental Setup and Results

Figure 12.18 shows the experimental setup of the FM mode locked fiber laser. Two erbium-doped fiber amplifiers (EDFA) pumped at 980 nm are used in the fiber loop to control the optical power in the loop for mode locking. Both are operating in saturated mode. A phase modulator driven in the region of 1 GHz modulation frequency assumes the role as a mode locker and controls the states of locking in the fiber ring. At input of the phase modulator, a polarization controller (PC) consisting of two quarter-wave plates and one half-wave plate is used to control the polarization of light which relates to the non-linear polarization evolution and influences to multi-pulse operation in formation of bound soliton states. A 50-m Corning SMF-28 fiber is inserted after the phase modulator to ensure that the average dispersion in the loop is anomalous. The fundamental frequency of the fiber loop is 1.7827 MHz that is equivalent to the 114 m total loop length. The outputs of the mode locked laser from the 90:10 coupler are monitored by an optical spectrum analyzer (HP 70952B) and an oscilloscope (Agilent DCA-J 86100C) of an optical bandwidth of 65 GHz.

Under normal conditions, the single pulse mode locking operation is performed at the average optical power of 5 dBm and modulation frequency of 998.315 MHz (the harmonic mode locking at the 560th order) as shown in Figure 12.19. The narrow pulses of 8–14 ps width depending on the RF driving power of the phase modulator are observed on the oscilloscope. The measured pulse

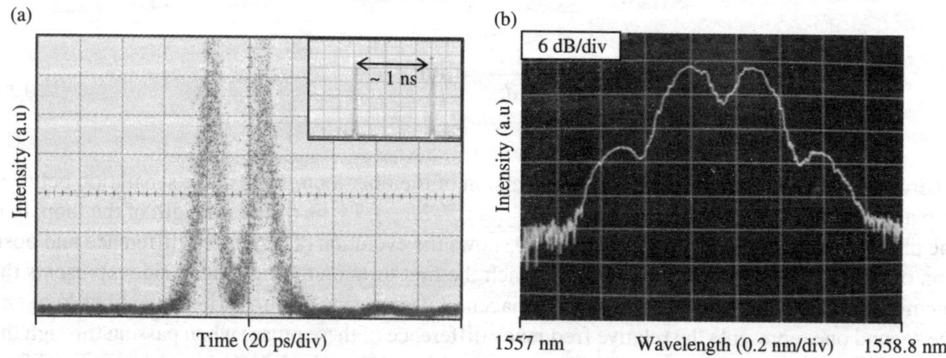

FIGURE 12.18 (a) The oscilloscope trace (inset: The periodic bound soliton pairs in time-domain) and (b) optical spectrum of dual-soliton bound state.

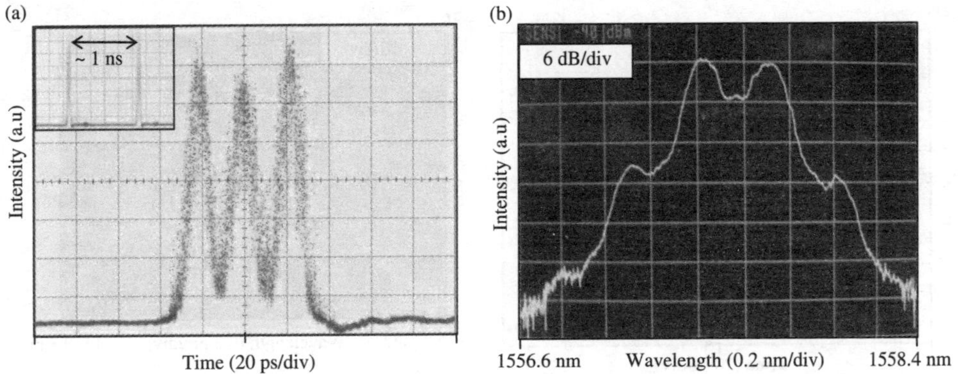

FIGURE 12.19 (a) The oscilloscope trace (inset: The periodic groups of triple bound solitons in time-domain) and (b) optical spectrum of triple-soliton bound state.

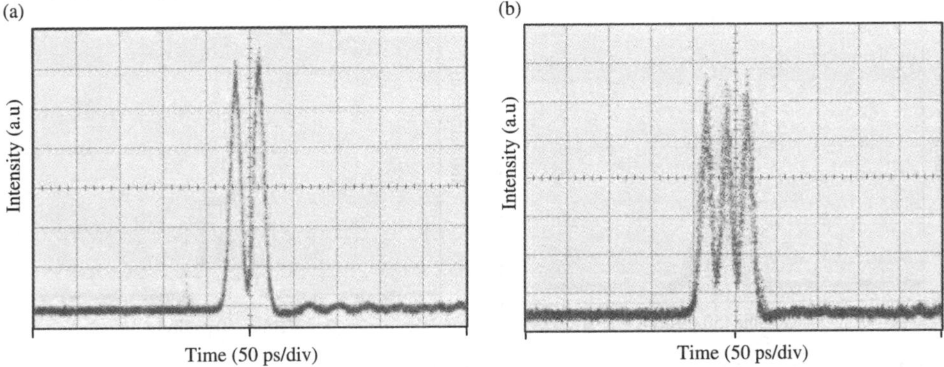

FIGURE 12.20 The oscilloscope traces of the dual-soliton bound state (a) and the triple-soliton bound state (b) after propagating through 1 km SMF.

spectrum has spectral shape of a soliton rather than a Gaussian pulse. By adjusting the polarization states of the PC wave plates at higher optical power, the dual-bound solitons or bound soliton pairs can be observed at average optical power circulating inside the fiber loop of about 10 dBm. Figure 12.20a shows the typical time-domain waveform and (b) the corresponding spectrum of the dual-bound soliton state. The estimated FWHM pulse width is about 9.5 ps and the temporal separation between two bound pulses is 24.5 ps which are correlated exactly to the distance between two spectral main lobes of 0.32 nm of the observed spectrum. When the average power inside the loop is increased to about 11.3 dBm and a slight adjustment of the polarization controller is performed, the triple-bound soliton state occurs as shown in Figure 12.21 with that the FWHM pulse width and the temporal separation of two close pulses are slightly less than 9.5 ps and 22.5 ps, respectively. Insets in Figure 12.20a and Figure 12.21a show the periodic sequence of bound solitons at the repetition rate equal exactly to the modulation frequency. This feature is quite different from that in a passive mode-locked fiber laser in which the positions of bound solitons is not stable and the direct soliton interaction causes a random movement and phase shift of bound soliton pairs [14]. On the other hand, it is advantageous to perform a stable periodic bound soliton sequence in a FM mode-locked fiber laser.

The symmetrical shapes of optical modulated spectrum in Figure 12.20b and Figure 12.21b indicate clearly that the relative phase difference between two adjacent bound solitons is of π value. At the center of spectrum there is a dip due to the suppression of the carrier with π phase difference

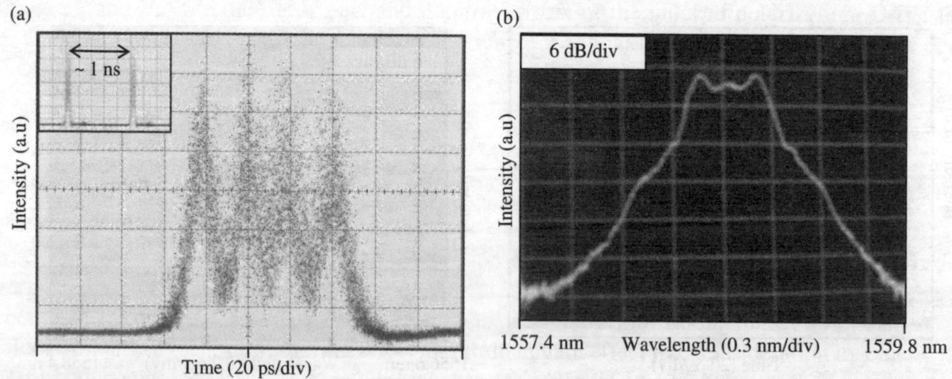

FIGURE 12.21   (a) The oscilloscope trace and (b) optical spectrum of quadruple-soliton bound state.

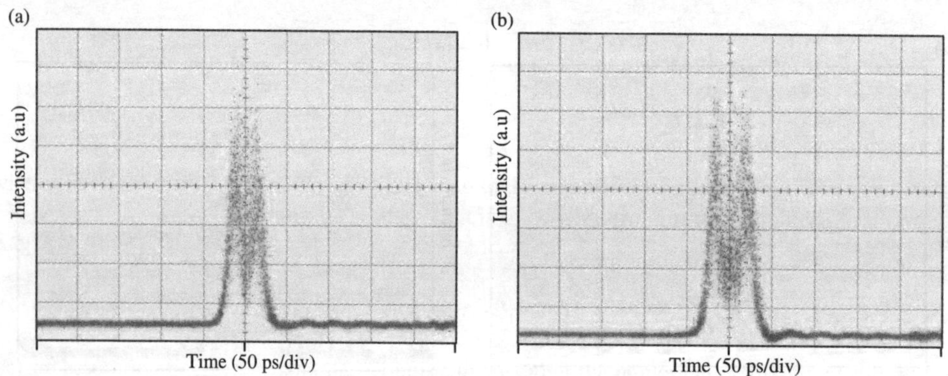

FIGURE 12.22   The oscilloscope traces of (a) the dual-soliton bound state at optical power threshold of 8 dBm and (b) the triple-soliton bound state at 10.7 dBm.

in case of dual-soliton bound state. While there is a small hump in case of triple-soliton bound state because three soliton pulses are bound together with the first and the last pulses in phase and out of phase with the middle pulse. The shape of spectrum will change which can be symmetrical or asymmetrical when this phase relationship varies. We believe that the bound state with the relative phase relationship of π between solitons is the most stable in our experimental setup and this observation also agrees with the theoretical prediction of stability of bound soliton pairs relating to photon-number fluctuation in different regimes of phase difference.

Similar to a single soliton, the phase coherence of bound solitons is still maintained as a unit when propagating through a dispersive medium. Figure 12.22 shows the dual-soliton bound state and triple-soliton state waveforms on the oscilloscope after they propagate through 1 km SMF fiber. There is also no change in their observed spectral shapes in both cases.

To prove that the multi-soliton bound operation can be formed in an FM mode-locked fiber by operating at a critical optical power level and phase modulator-induced chirp in a specific fiber loop of anomalous average dispersion, we increased the average optical power in the loop to maximum of 12.6 dBm and decrease the RF driving power of 15 dBm. It is really amazing when the quadruple-soliton state is generated as observed in Figure 12.23. The bound state occurring at lower phase modulation index is due to maintaining a small enough frequency shifting in a wider temporal duration of bound solitons to hold the balance between interactions of the group of four solitons in the fiber loop. However the optical power can still not be enough to stabilize the bound state as well as the larger noise from EDFAs at higher pump power, so that the time-domain waveform of

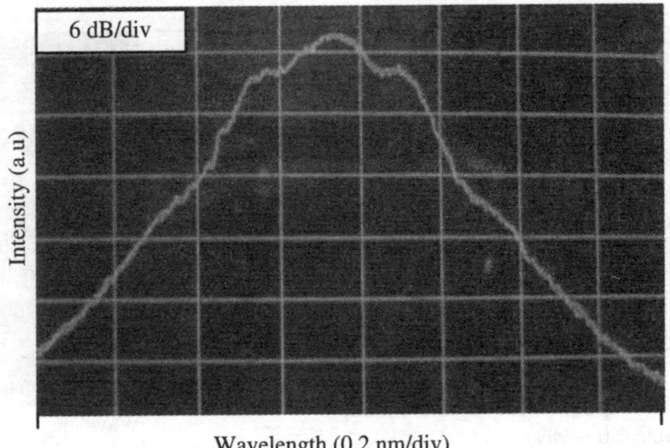

FIGURE 12.23   The spectrum of the bound soliton state with in phase pulses.

quadruple-soliton bound state is much noisier and its spectrum only can be clearly seen as two main lobes being inversely proportional to the temporal separation of pulses which is about 20.5 ps in our experimental setup. The results show the pulse separation reduces when the number of pulses in the bound states is larger.

Thus through the experimental results, both the phase modulation index and the cavity's optical power influence the existence of the multi-soliton bound states in the FM mode-locked fiber laser. The optical power determines the number of initially split pulses and maintains the pulse shape in the loop. Experimental observations show that there is a threshold of optical power for each bound level. At threshold value, the bound solitons show strong fluctuations in amplitude and oscillations in position considered as a transition state between different bound levels and the collisions of adjacent pulses even occur as shown in Figure 12.24. The phase difference of adjacent pulses can also change in these unstable states which means that the neighbouring pulses is not out of phase anymore but in phase as seen through the measured spectrum in Figure 12.25. Although the decrease in phase modulation index is required to maintain the stability at higher bound soliton level, it increases the pulse width and reduces the peak power. So the waveform seen as noisier and more sensitive to ambient environment and its spectrum is not strongly modulated as shown in Figure 12.26.

12.8.1.4   Simulation of Dynamics of Bound States in a FM Mode-locked Fiber Laser

12.8.1.4.1   Numerical Model of a Fm Mode-locked Fiber Laser
To understand the dynamic of FM bound soliton fiber laser, simulation technique is used to see what is going on in the FM mode-locked fiber loop. A recirculation model of the fiber loop is used to simulate propagation of the bound solitons in the fiber cavity. The simple model consists of basic components of a FM mode-locked fiber laser as shown in Figure 12.27, in other words, the cavity of FM mode-locked fiber laser is modeled as a sequence of different elements. The optical filter has a Gaussian transfer function with 2.4 nm bandwidth. The transfer function of phase modulator is $u_{out} = u_{in} \exp[jm\cos(\omega_m t)]$, m is the phase modulation index and $\omega_m = 2\pi f_m$ is angular modulation frequency, assumed to be a harmonic of the fundamental frequency of the fiber loop. The pulse propagation through the optical fibers is governed by the non-linear Schrodinger equation [22].

$$\frac{\partial u}{\partial z} + j\frac{\beta_2}{2}\frac{\partial^2 u}{\partial T^2} - \frac{\beta_3}{6}\frac{\partial^3 u}{\partial T^3} + \frac{\alpha}{2}u = j\gamma|u|^2 u \qquad (12.103)$$

where u is the complex envelop of optical pulse sequence, β_2 and β_3 account for the second- and third-order fiber dispersions, α and γ are the loss and non-linear parameters of the fiber respectively. The amplification of signal including the saturation of the EDFA can be represented as [22]

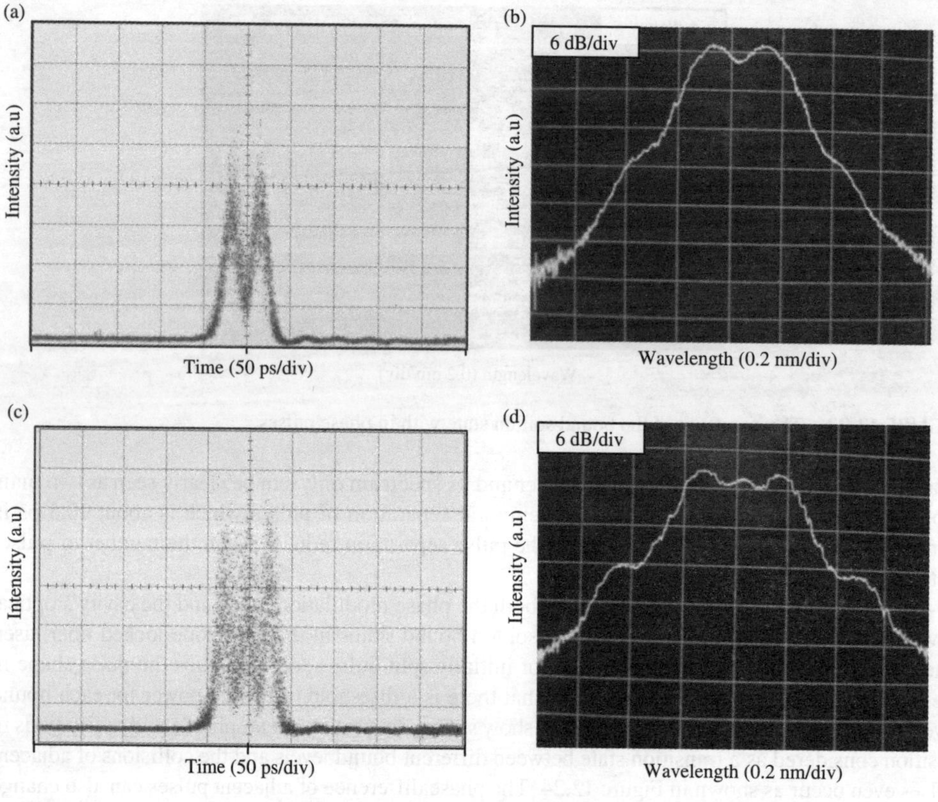

(a) 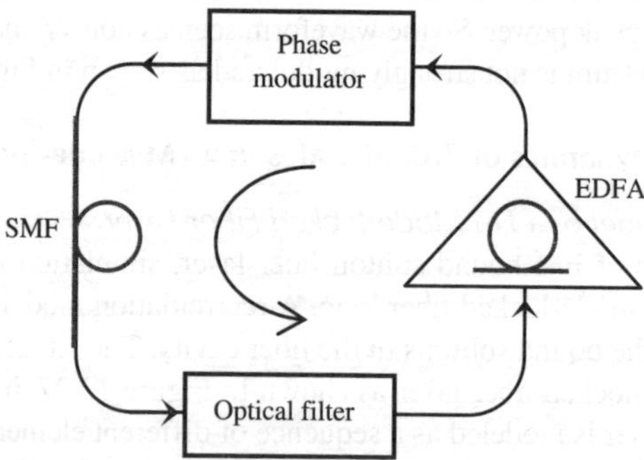

FIGURE 12.24   The oscilloscope traces of (a) the dual-soliton bound state and (c) the triple-soliton bound state and (b and d) their spectra, respectively at low RF driving power of 11 dBm.

FIGURE 12.25   A circulating model for simulating the FM mode-locked fiber ring laser. SMF, standard single mode fiber; EDFA, erbium-doped fiber amplifier.

$$u_{\text{out}} = \sqrt{G}\,u_{\text{in}} \tag{12.104}$$

with

$$G = G_0 \exp\left(-\frac{G-1}{G}\frac{P_{\text{out}}}{P_{\text{sat}}}\right) \tag{12.105}$$

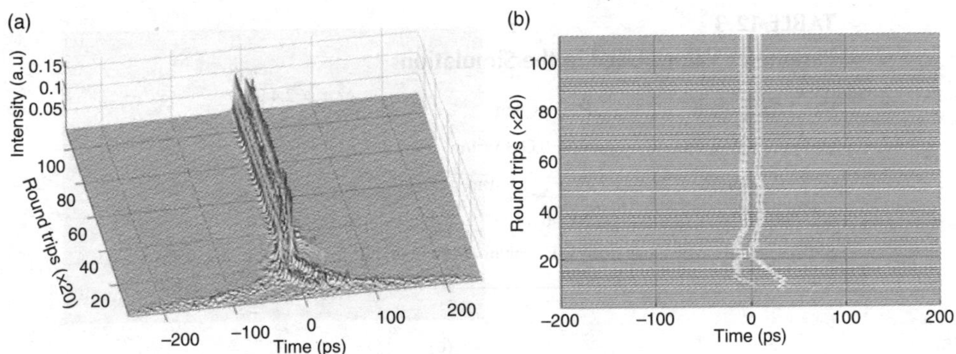

FIGURE 12.26   (a) Numerically simulated dual-soliton bound state formation from noise, (b) the evolution of the formation process in contour plot view.

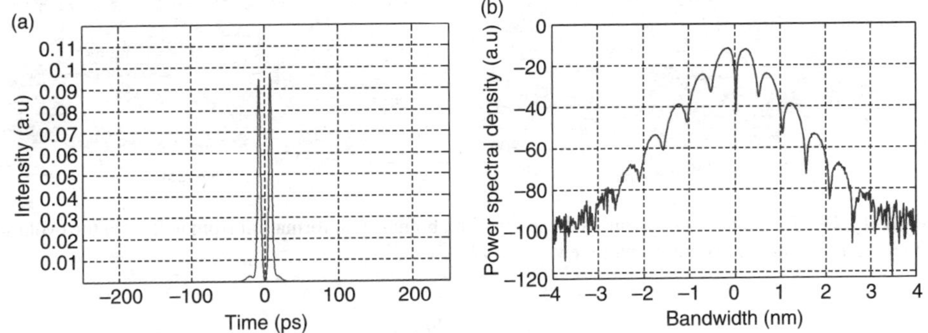

FIGURE 12.27   (a) The waveform, and (b) the corresponding spectrum of simulated dual-soliton bound state at the 2000th roundtrip.

where G is the optical amplification factor, G_0 is unsaturated amplifier gain, P_{out} and P_{sat} are output power and saturation power, respectively.

The difference of dispersion between the SMF fiber and the erbium-doped fiber in the cavity arranges a certain dispersion map and the fiber loop gets a positive net dispersion or an anomalous average dispersion which is important in forming the 'soliton-like' pulses in a FM fiber laser. Basic parameter values used in our simulations are listed in Table 12.3.

12.8.1.4.2   Simulation of the Formation Process of the Bound Soliton States
Firstly, we simulate the formation process of bound states in the FM mode-locked fiber laser whose parameters are those employed in the experiments described above. The lengths of the Er-doped fiber and SMF fiber are chosen to get the cavity's average dispersion $\overline{\beta}_2 = -10.7$ ps^2/km. The Schrodinger equation is applied with initial conditions that the random optical noises as the initial amplitudes of the waves which are then circulating around the ring via this propagation equation. The optical power is built up with the phase matching conditions and hence the formation of the solitons. Figure 12.28 shows a simulated dual-soliton bound state building up from initial Gaussian-distributed noise as an input seed over the first 2000 round-trips with P_{sat} value of 10 dBm and G_0 of 16 dB and m of 0.6π. The built-up pulse experiences transitions with large fluctuations of intensity, position and pulse width during the first 1000 roundtrips before reaching to the stable bound steady state. Figure 12.29 shows the time-domain waveform and spectrum of the output signal at the 2000th round trip.

The bound states with higher number of pulses can be formed at higher gain of the cavity, hence when the gain G_0 is increased to 18 dB, which is enhancing the average optical power in the

TABLE 12.3

Parameter Values Used in the Simulations

$\beta_2^{SMF} = -21a$ ps^2/km	β_2^{ErF} ps^2/km	$\Delta\lambda_{filter} = 2.4$ nm
$\gamma^{SMF} = 0.0019$ W^{-1}/m	$\gamma^{ErF} = 0.003$ W^{-1}/m	$f_m \approx 1$ GHz
$\alpha^{SMF} = 0.2$ dB/km	$P_{sat} = 7 \div 13$ dBm	$m = 0.1\pi \div 1\pi$
$L_{cav} = 115$ m	NF $= 6$ dB	$\lambda = 1558$ nm

SMF, standard single mode fiber; ErF, erbium-doped fiber; NF, noise figure of the EDFA.

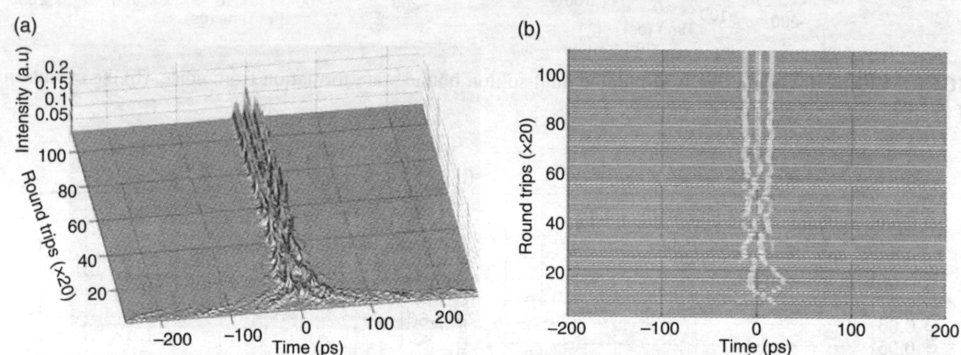

FIGURE 12.28   (a) Numerically simulated triple-soliton bound state formation from noise, (b) the evolution of the formation process in contour plot view.

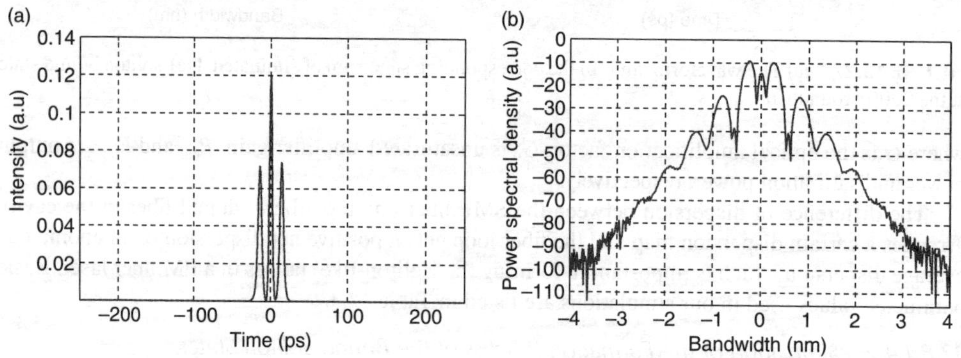

FIGURE 12.29   (a) The waveform, and (b) the corresponding spectrum of simulated triple-soliton bound state at the 2000th roundtrip.

loop, the triple-soliton bound steady-state is formed from the noise seed via simulation as shown in Figure 12.30. In the case of higher optical power, the fluctuation of signal at initial transitions is stronger and it needs more round trips to reach to a more stable three pulses bound state. The waveform and spectrum of the output signal from the FM mode-locked fiber laser at the 2000th round trip are shown in Figure 12.31a and Figure 12.31b, respectively.

Although the amplitude of pulses is not equal which indicates the bound state can require a larger number of round trips before the effects in the loop balance, the phase difference of pulses accumulated during circulating in the fiber loop is approximately a π value that is indicated by strongly modulated spectra. In particular from the simulation result, the phase difference between adjacent pulses is 0.98π in case of the dual-pulse bound state and 0.89π in case of the triple-pulse

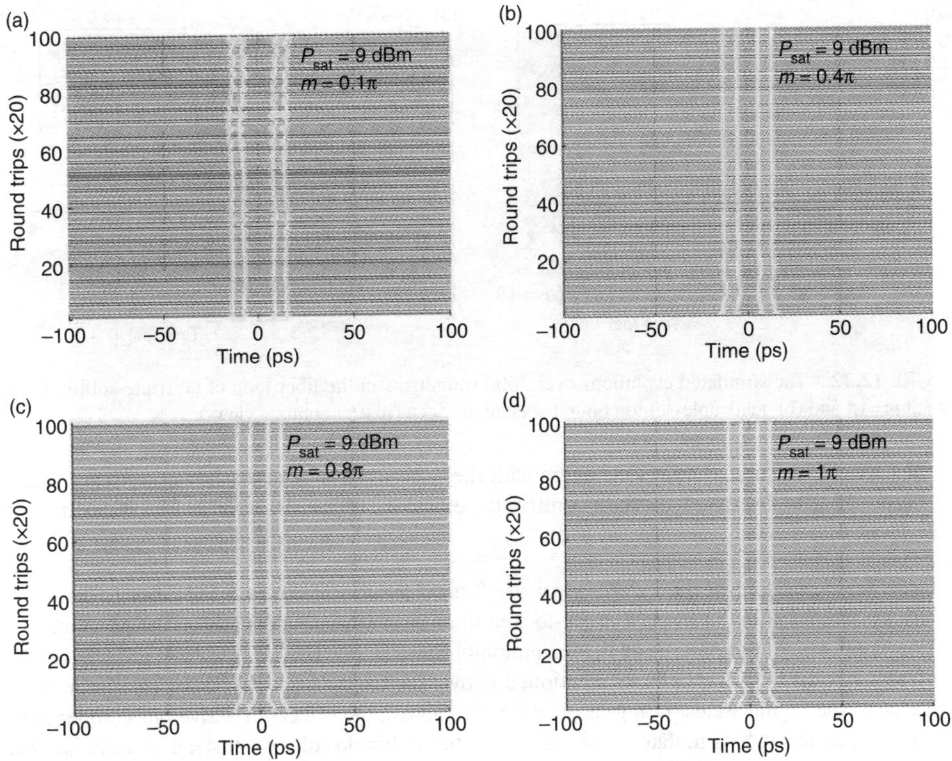

FIGURE 12.30 The simulated evolution of dual-soliton bound state over 2000 roundtrips in the fiber loop at different phase modulation indexes (a) $m=0.1\pi$, (b) $m=0.4\pi$, (c) $m=0.8\pi$, (d) $m=1\pi$.

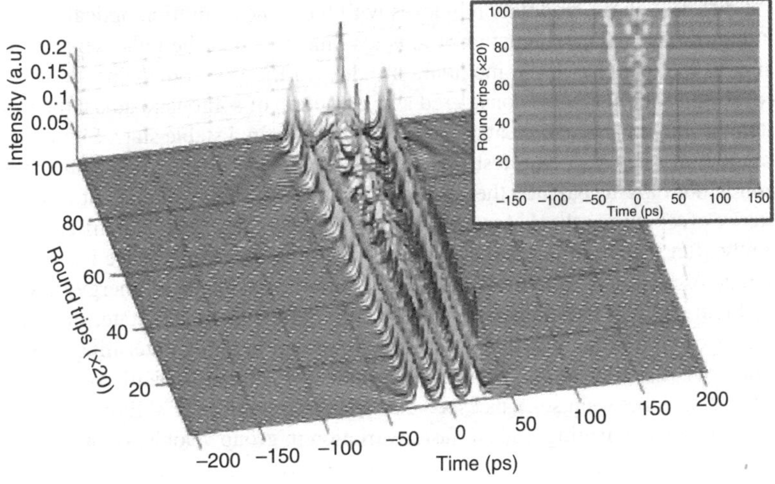

FIGURE 12.31 The simulated evolution of triple-soliton bound state over 2000 roundtrips at the low phase modulation index $m=0.1\pi$.

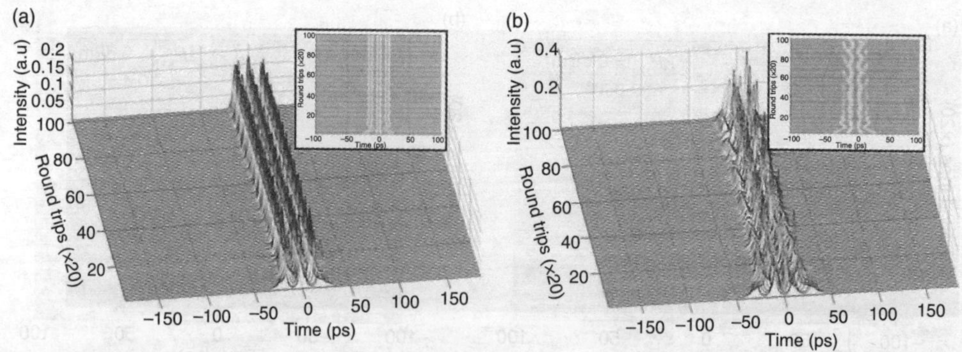

FIGURE 12.32 The simulated evolutions over 2000 roundtrips in the fiber loop of (a) triple-soliton bound state at $m=1\pi$ and (b) quadruple-soliton bound state at $m=0.7\pi$ (insets: contour views).

bound state. These simulation results agree with the experimental results (shown in Figure 12.20b and Figure 12.21b) discussed above to confirm the existence of multi-soliton bound states in a FM mode-locked fiber laser.

12.8.1.4.3 Simulation of the Evolution of the Bound Soliton States in a FM Fiber Loop

Stability of bound states in the FM mode-locked fiber laser strongly depends on the parameters of the fiber loop which also determine the formation of these states. Beside the phase modulation and group velocity dispersion (GVD) as mentioned above, the cavity's optical power also influence to existence of the bound states. Using the same model above, the effects of active phase modulation and optical power can be simulated to see the dynamics of bound solitons. Instead of the noise seed, the multi-soliton waveform following Equation 12.97 and Equation 12.98 are used as an initial seed for simplification of our simulation processes. Initial bound solitons are assumed to be identical with the phase difference between adjacent pulses of π value.

Firstly, the effect of the phase modulation index on the stability of the bound states is simulated through a typical example of the evolution of dual-soliton bound state over 2000 roundtrips in the loop at different phase modulation indexes with the same saturation optical power of 9 dBm as shown in Figure 12.32. The simulation results also indicate that the pulse separation decreases corresponding to the increase in the modulation index. In the first roundtrips, there is a periodic oscillation of bound solitons that is considered as a transition of solitons to adjust their own parameters to match the parameters of the cavity before reaching a final stable state. Simulations in other multi-soliton bound states also manifest this similar tendency. The periodic phase modulation in the fiber loop is not only to balance the interactive forces between solitons but also to retain the phase difference of π between them. At too small modulation index, the phase difference changes or reduces slightly after many round trips or in other words, the phase coherence is looser. This leads to the amplitude oscillation due to the alternatively periodic exchange of energy between solitons as shown in Figure 12.33. The higher the number of solitons in bound state, the more sensitive it is to the change in phase modulation index. The modulation index determines time separation between bound pulses, yet at too high modulation index, it is more difficult to balance the effectively attractive forces between solitons especially when the number of solitons in the bound state is larger. The increase in chirping leads to faster variation in group velocity of pulses when passing the phase modulator which can create the periodic oscillation of pulse position in the time domain evolution. Figure 12.34 shows the evolution of the triple-soliton bound state at the m value of 1π and the quadruple-soliton bound state at the m value of 0.7π. In case of quadruple-soliton bound state, solitons oscillate strongly and tend to collide together.

Another factor is the optical power of the fiber loop, it plays an important role not only in determination of multi-pulse bound states as in simulation of previous section, but also in stabilization

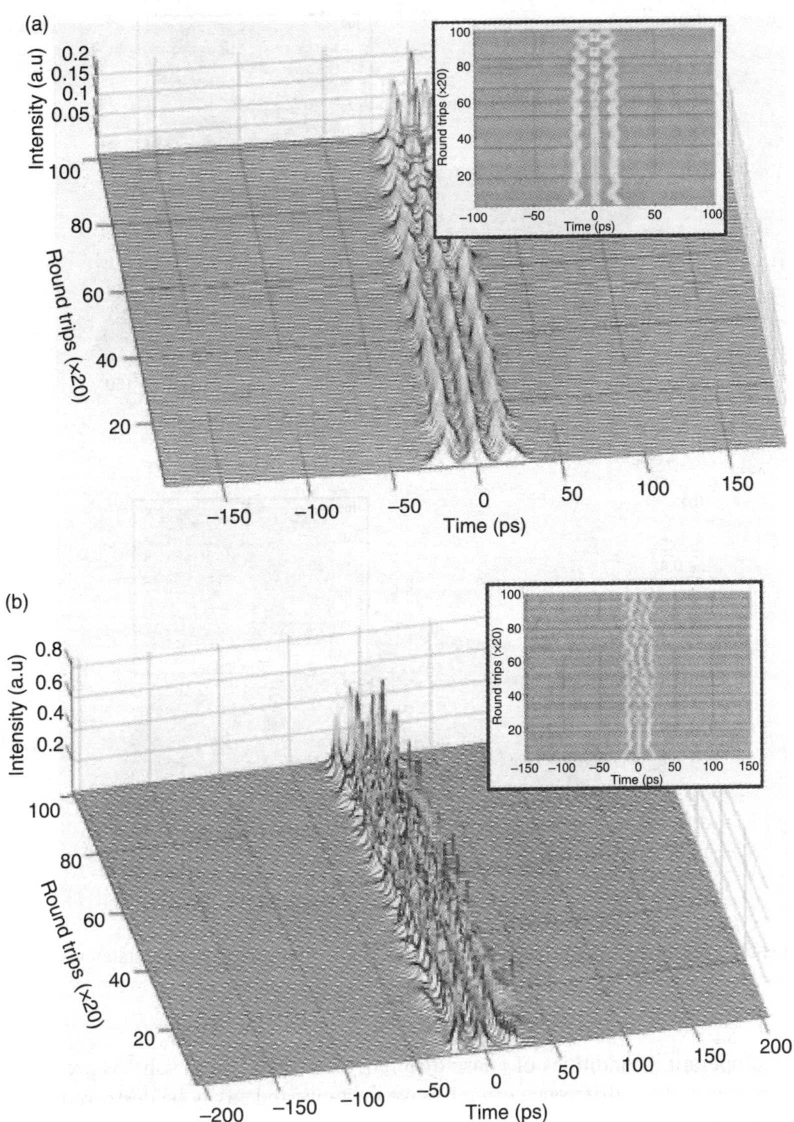

FIGURE 12.33 The simulated behaviors of triple-soliton bound state at unoptimized power (a) $P_{sat}=9$ dBm, $m=0.7\pi$ and (b) $P_{sat}=13$ dBm and $m=0.6\pi$ (insets: contour views).

of the bound states circulating in the loop. As mentioned in Section 12.8.2.3, each bound state has a specific range of operational optical power. In our simulations, dual- and triple- and quadruple-soliton bound states are in stable evolution in the loop when the saturated power P_{sat} is about 9, 11 and 12 dBm, respectively. When the optical power of the loop is not within these ranges, the bound states become unstable and they are more sensitive to the change of phase modulation index. At a power lower than the threshold, the bound states are out of bound and switched to lower level of bound state. In contrast, at high power level region, the generated pulses are broken into random pulse trains or decay into radiation. Figure 12.35 shows unstable evolution of the triple-soliton bound states under unoptimized operating conditions of the loop as an example.

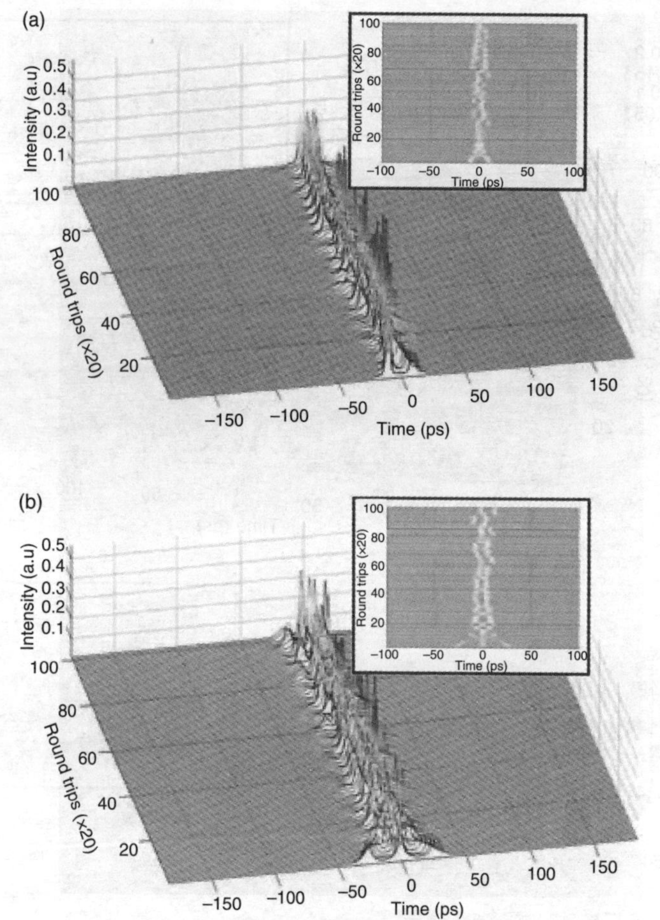

FIGURE 12.34 The time-domain behaviors of dual- and triple-soliton bound states respectively, within phase pulses.

Different operating conditions of phase difference between bound solitons are also simulated. The states of non-π phase difference often behave unstably and easily be destroyed as displayed in Figure 12.36. The simulation results agree well with the experimental observations as discussed above.

12.8.2 Active Harmonic Mode-Locked Fiber Laser for Soliton Generation

12.8.2.1 Experiment Setup

Figure 12.37 shows the schematic diagram of the active mode-locked fiber laser (MLFL). The fiber ring configuration incorporates an isolator to ensure unidirectional lasing. The gain media is obtained by using a 16 m erbium-doped fiber operating under bi-directional pump condition. The pump sources are two 980 nm laser diodes coupled to the ring through two 980/1550 WDM couplers. A FOL0906PRO-R17-980 laser diode from FITEL is used for the forward pump source and a COMSET PM09GL 980 nm laser diode for the backward pump source. A 18 m long dispersion shifted fiber (DSF) is used for shortening the locked pulse width. A 2 nm bandpass multi-layer thin-film optical filter is used to select the operation wavelength of the laser and moderately

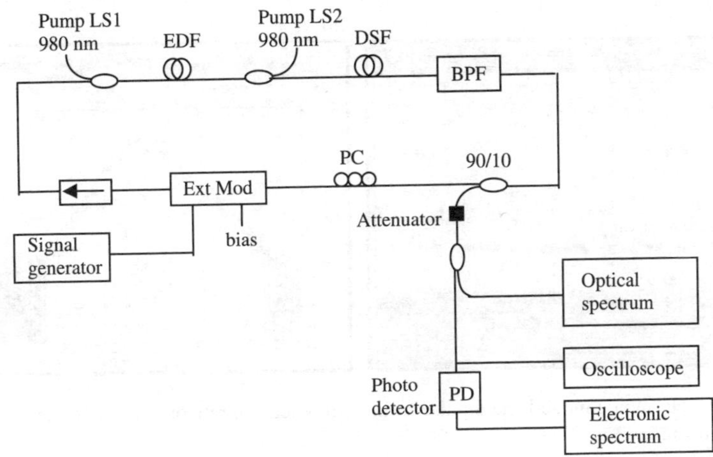

FIGURE 12.35 Active harmonic mode-locked fiber laser experiment setup.

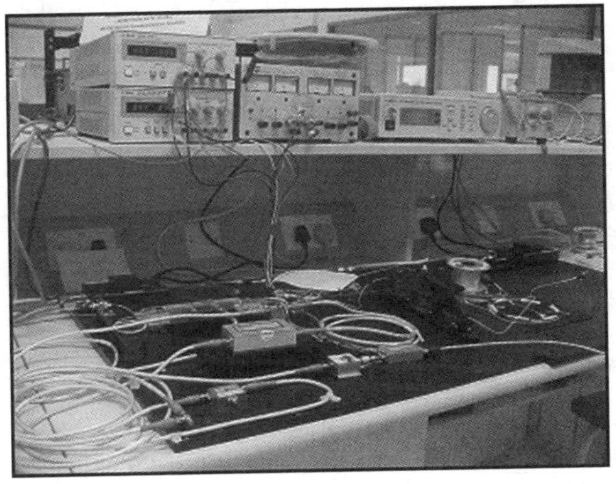

FIGURE 12.36 Active harmonic mode-locked fiber laser.

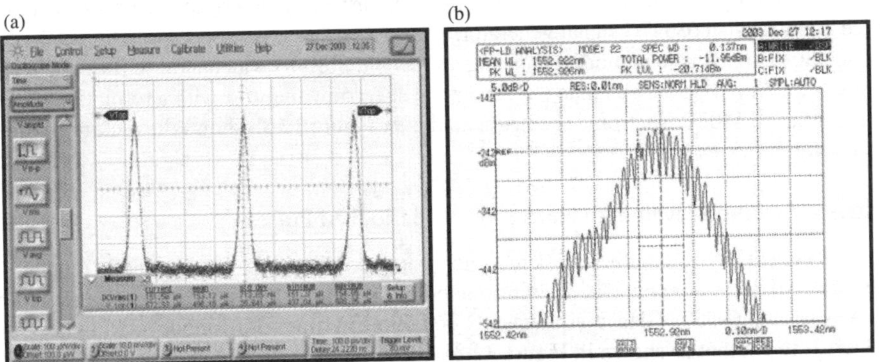

FIGURE 12.37 Harmonic mode-locked pulse train with repetition rate of 2.657 GHz (a) and its optical spectrum (b).

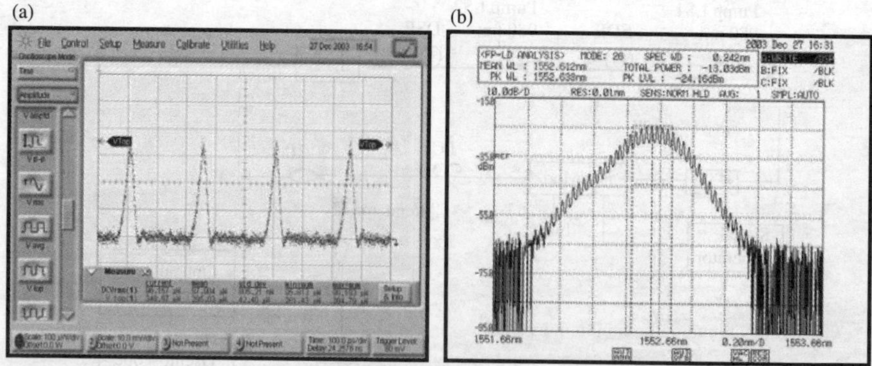

FIGURE 12.38 Harmonic mode-locked pulse train with repetition rate of 4.000 GHz; (a) temporal profile, (b) optical spectrum.

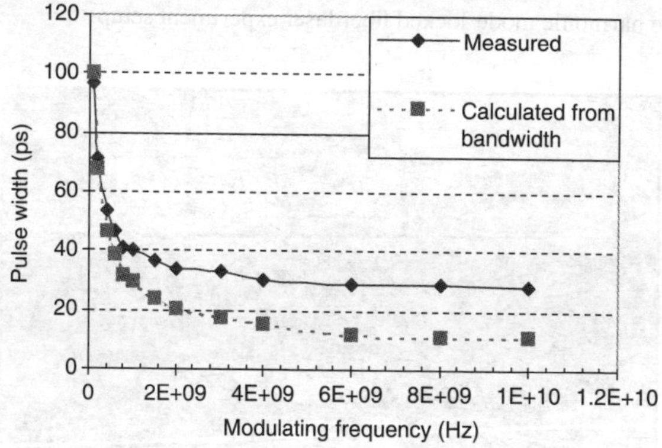

FIGURE 12.39 Pulse width of the HMLFL for various modulation frequencies.

accommodate the bandwidth of the output pulses. A polarization controller is also employed to maximize the coupling of lightwaves from fiber to the diffused waveguides of the Mach–Zehnder intensity modulator (MZIM). Output pulse trains are extracted via a 90/10 coupler. Mode locking is obtained by inserting into the ring a 20 GHz 3-dB Mach–Zehnder intensity modulator that periodically modulates the loss of the lightwaves traveling around the ring. The modulator is biased at the quadrature point with a voltage of −1.5 V and then driven by superimposing a sinusoidal signal derived from Anritsu 68347C signal generator.

The output signal is then monitored using an Agilent 86100B wideband oscilloscope with optical input sampling module, an Ando 6317B optical spectrum analyzer with a resolution of 0.01 nm and an Agilent E4407B electronic spectrum analyzer. Figure 12.38 shows the picture of the whole system setup for experiment.

12.8.2.2 Tunable Wavelength Harmonic Mode-locked Pulses

The total optical cavity length is 55.4 m which corresponding to a fundamental frequency f_R of 3.673 MHz (the measurement of f_R is presented in Section 12.8.6). When the modulation frequency is set at 2.695038 GHz or equivalently equal to 733 times of f_R, mode locking laser pulse is obtained at the output as shown in Figure 12.39a. The interval between adjacent pulses is 371 ps, corresponding to a repetition rate of 2.695 GHz. The locked pulses are formed with clear pedestal, with slightly fluctuated amplitude. The output pulse amplitude is measured and a peak power is recorded at 9.9 mW.

The pulse width is also measured and a value of 37 ps full width at half maximum (FWHM) is obtained. However, as this value is taken from the trace of pulse on the oscilloscope the rise time of the oscilloscope is included in the measured value. Since the rise time of the oscilloscope is 20 ps, which is comparable to the measured value, it can not be ignored. Besides that the rise time of the photodiode also contributes to the total measured value. Therefore the actual pulse width is much smaller than the measured value.

Figure 12.39b shows the optical spectrum of the output pulses. The FWHM bandwidth is 0.137 nm with central wavelength at 1552.926 nm. The time bandwidth product (TBP) was determined as:

$$\text{TBP} = T_{\text{FWHM}} \times BW \tag{12.106}$$

$$\text{TBP} = 37 \text{ ps} \times 17.125 \text{ THz} = 0.634.$$

The large value obtained here again confirms that the measured value of pulse width is mainly contributed by the rise times of oscilloscope and photodiode. The accurate pulse width should be obtained from using an optical auto-correlator.

Alternatively one can take advantage of the characteristic of the AM mode-locked fiber laser to approximate the pulse width. The pulses generated from AM mode-locked fiber lasers are nearly transform limited [13–16]. This means that the TBP takes the value of 0.44, assuming Gaussian shape pulse. Therefore the pulse width can be estimated from:

$$T_{\text{FWHM}} = \text{TBP}/BW \tag{12.107}$$

$$T_{\text{FWHM}} = 0.44/17.125 \text{ THz} = 26 \text{ ps.}$$

It can be seen from Figure 12.39b that the pulse train spectrum has sidelobes with a separation of 0.022 nm, which corresponds to a longitudinal mode separation of 2.695 GHz. This spectrum profile is typical for the mode-locked laser and usually referred as the mode-locked structure spectrum [17,18,24].

The above spectrum profile can be explained by taking the Fourier transform of the periodic Gaussian pulse train:

$$p(t) = e^{-t^2/2T_0^2} * \text{III}\left(\frac{t}{T}\right) \tag{12.108}$$

where T_0 is the width of the Gaussian pulse, T is the pulse repetition period and $\text{III}(t/T)$ is the comb function given by

$$\text{III}\left(\frac{t}{T}\right) = \sum_{n=-\infty}^{\infty} \delta(t - nT) \tag{12.109}$$

$$p(\omega) = F(p(t)) = T_0 \sqrt{2\pi} e^{-T_0^2 \omega^2/2} \sum_{n=\infty}^{\infty} \delta(\omega - n\Omega) \tag{12.110}$$

where $\Omega = 2\pi/T$; $P(\omega)$ has the structure of a train of Dirac function pulses separated by the repetition frequency with a Gaussian envelope. The laser wavelength is tunable over the whole C-band by tuning the central wavelength of the thin film bandpass filter inserted in the loop.

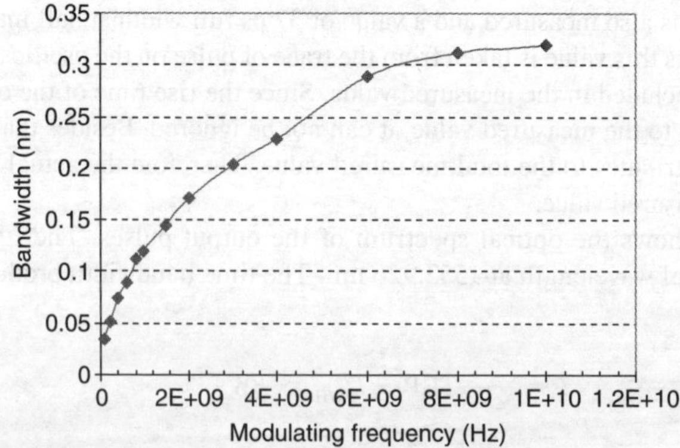

FIGURE 12.40 Bandwidth of the HMLFL for various modulation frequencies.

The repetition rate of the laser pulses can be increased by increasing the modulation frequency. Figure 12.40 shows the output pulses with the repetition rate of 4 GHz and its optical spectrum.

12.8.2.3 Measurement of the Fundamental Frequency

In active harmonic mode-locked fiber laser (HMLFL) the pulses are locked into one of the harmonic of the fundamental frequency f_R by tuning the modulation frequency f_m so that $f_m = k f_R$. In practice, the laser is mode-locked if f_m is within the locking range around the harmonic frequency, that is $f_m = k f_R \pm \Delta f$. If f_m falls outside the locking range the laser becomes unlocked, that is no pulse trains are observed. If f_m is adjusted until reaching the next harmonic frequency $(k+1)f_R$, the HMLFL is locked again into the next longitudinal resonant mode of the ring. By measuring those locked modulation frequencies, one can determine the fundamental frequency of the ring. The fundamental frequency is determined by:

$$(4018635400 - 3981902400)/10 \text{ Hz} \leq f_R \leq (4018636000 - 3981901400)/10 \text{ Hz}$$

$$3673300 \text{ Hz} \leq f_R \leq 3673460 \text{ Hz}$$

or

$$f_R = 3673380 \pm 80 \text{ Hz}.$$

12.8.2.4 Effect of the Modulation Frequency

Increasing modulation frequency not only generates higher repetition rate mode-locked pulses but also improves other parameters of the pulses such as bandwidth, pulse width.

The solid curve shown in Figure 12.41 indicates the relationship between the pulse width of the generated pulse train and modulation frequency. The pulse width does not change so much when the modulation frequency increase above 2 GHz. This is because the pulse width in this region is small comparable to the rise time of the oscilloscope. The actual values should be smaller than the measured values. This is confirmed when examining the equivalent bandwidth of the pulse trains.

Figure 12.42 shows the pulse train bandwidth versus the modulation frequency. The modulation frequency increases as the bandwidth increases. Unlike the behavior of pulse width, the pulse train bandwidth continues to increase even when the modulation frequency is above 2 GHz. This indicates that the actual pulse width should be smaller than its measured value. The pulse width is calculated and plotted as shown by the dash curve in Figure 12.41.

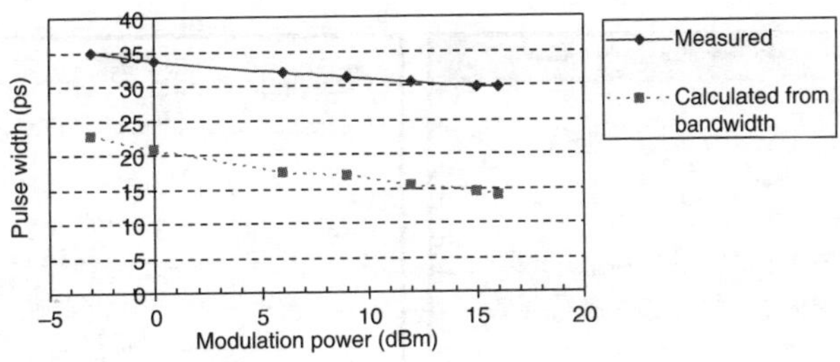

FIGURE 12.41 Pulse width of the HMLFL for various modulation powers.

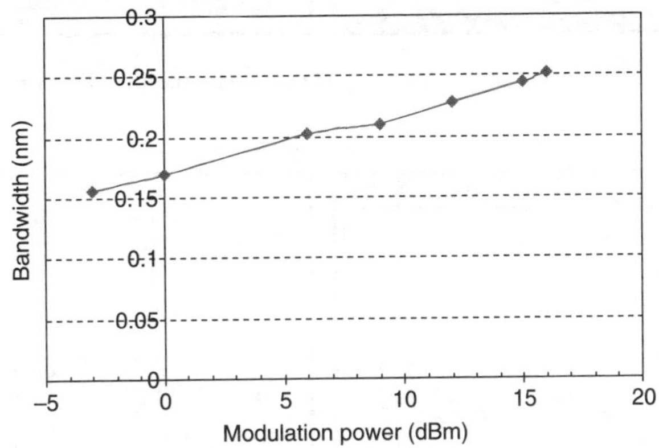

FIGURE 12.42 Bandwidth of the HMLFL for various modulation powers.

12.8.2.5 Effect of the Modulation Depth/Index

The same setup shown in Figure 12.37 is used to study the effect of modulation depth on the performance of the mode-locked pulse trains. The settings of the cavity are as following: (i) the total pump power is 274 mW; and (ii) the modulator is biased at the quadrature point. The modulation depth is varied by changing the signal amplitude (from the signal synthesizer) applied to the MZIM. The pulse width and bandwidth of the output pulse trains are measured for different values of modulation depth.

Figure 12.43 shows the relationship between pulse width and modulation amplitude/power. As the modulation power increases the pulse width slightly decreases. Therefore the pulse can be shortened by increasing the modulation power (or the modulation depth). However the pulse shortening effect of modulation depth is not as strong as that of modulation frequency. In addition, the modulation depth is limited to a maximum value of 1.

The relationship between pulse width and the modulation depth is verified by examining the effect of modulation power on the bandwidth. When the modulation power increases the bandwidth decreases as shown in Figure 12.44. This is consistent with the decrease of pulse width shown in Figure 12.43.

It is found that there is a threshold power of the modulation for locking. When the modulation power is reduced to −4 dBm corresponding to a modulation depth of 0.08, mode locking does not occur and pulse train cannot be obtained.

12.8.2.6 Effect of Fiber Ring Length

The same setup as in Figure 12.37 has been used to study the effect of the fiber ring length on the mode-locked pulse trains. The DSF fiber has been extended from 18 to 118 m. A total pump

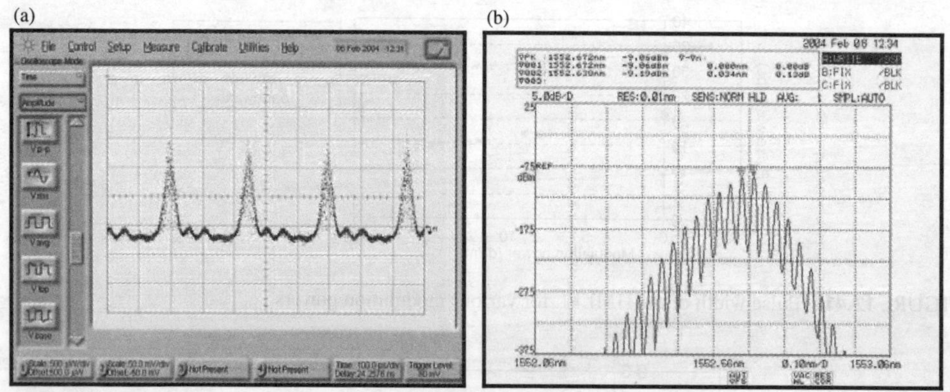

FIGURE 12.43   Locked pulses (a) and its optical spectrum (b) for the 18 m DSF ring laser.

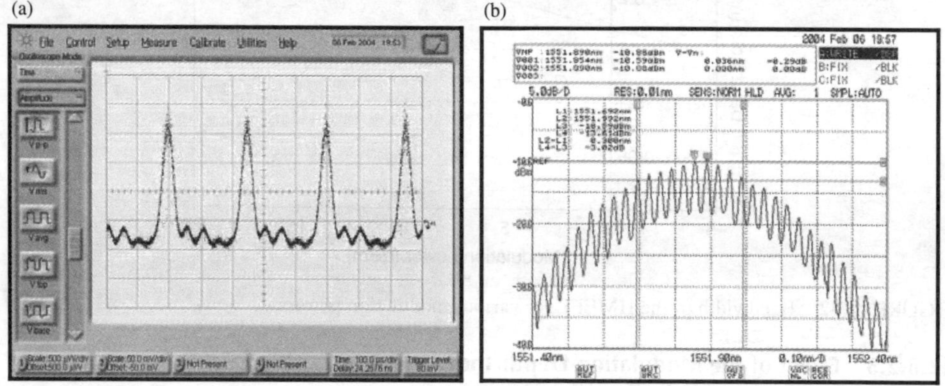

FIGURE 12.44   Locked pulses (a) and its optical spectrum (b) for the 118 m DSF ring laser.

power of 274 mW is used. The pulse width and bandwidth have been measured and compared with those described in Section 12.8.5. Figure 12.45 and Figure 12.46 show the mode-locked pulse temporal train and its optical spectrum for the 18 m DSF ring laser and the 118 m DSF ring laser, respectively.

The pulse trains are very much similar for both cases. Their pulse widths (FWHM) are nearly equal, 33 and 29 ps. However the pulse optical spectra are quite different. The longer ring laser has a bandwidth of 3 nm, larger than that of the shorter one, 0.184 nm. The broadening of the bandwidth of the pulse train of the 110 m ring length indicates that the pulse width is shorter than what it is observed by the sampling oscilloscope. The discrepancy here is again due to the affect of the photodiode and oscilloscope's rise time on the observed pulse.

The shortening of the pulse in the longer length ring can be due to the self phase modulation (SPM) effects or the non-linear induced phase in the fiber. The high intensity pulse traveling in the fiber suffers the non-linear effect which increases the peak of the pulse, shortens the pulse width and hence the bandwidth broadening.

To verify that the pulse shortening is due to the non-linear effect, the two different length rings are studied under lower pump power. It is found that when the pump power is reduced to 50 mW, the pulse width and bandwidth of the laser does not change regardless of the fiber length.

One can estimate the effective non-linear lengths of the fiber for the pump power of 274 and 50 mW. The peak pulse powers are 0.37 and 0.027 W, respectively. The non-linear lengths can be calculated as [22]:

(a)   (b)

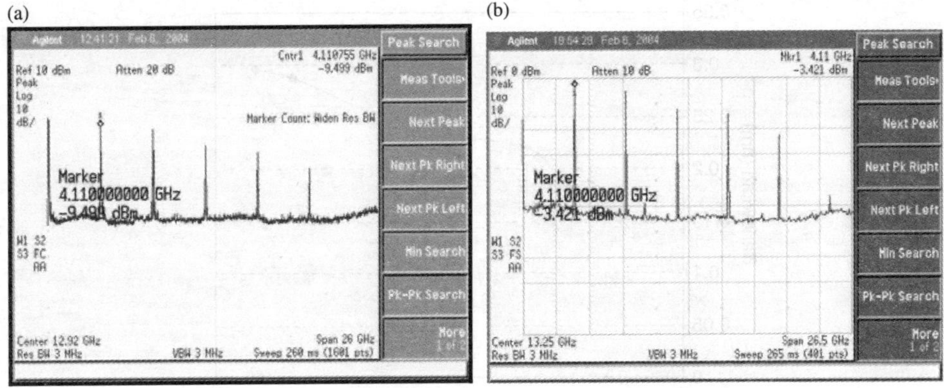

FIGURE 12.45   RF spectrum of HML laser with 18 m DSF (a), and 118 m DSF (b).

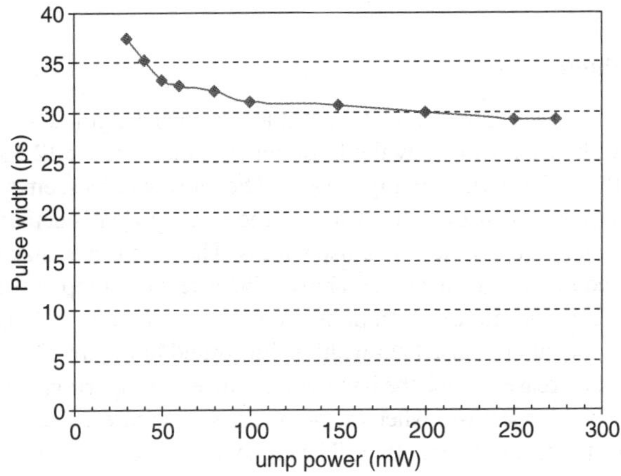

FIGURE 12.46   Pulse width of the HMLFL for various pump powers.

$$L_{\mathrm{NL}}(P_p = 274 \text{ mW}) = \frac{1}{\gamma P_0} = \frac{1}{2x10^{-3}x0.37} = 1351\,(m) \tag{12.111}$$

$$L_{\mathrm{NL}}(P_p = 27 \text{ mW}) = \frac{1}{\gamma P_0} = \frac{1}{2x10^{-3}x0.027} = 18518\,(m). \tag{12.112}$$

It can be seen that the effective non-linear length of the 118 m long fiber ring is comparable to that of the physical ring length when pumped at high power. Therefore the non-linear effect plays a significant role in this case. When the laser is pumped at lower power, the effective non-linear length is much longer than the laser ring length, hence the SPM effect is minute and can not be observed. However, one can predict the existence of the SPM effect if the fiber length is increased to be comparable with the non-linear length.

It is also noticed that the super-mode noise was reduced in the case of longer fiber ring. The pulses are more temporally stable and less fluctuating. To verify this, the RF spectrum of the lasers are recorded and compared. It can be seen from Figure 12.47 that the longer length laser has a lower noise floor than the shorter one.

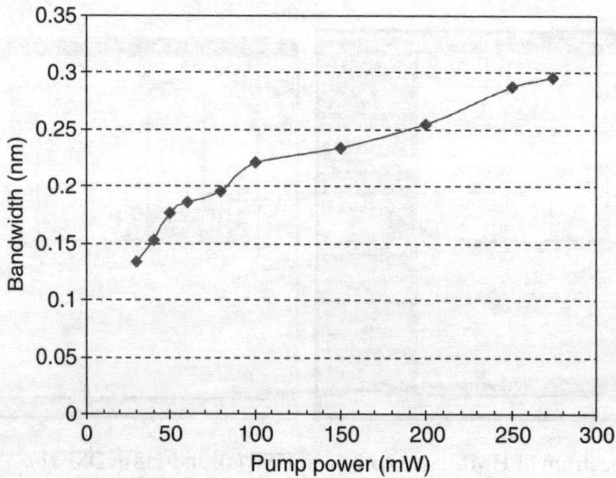

FIGURE 12.47   Bandwidth of the HMLFL for various pump powers.

12.8.2.7   Effect of Pump Power

The laser is configured as in Figure 12.37 except that the fiber length is 118 m. Different power levels are pumped into the EDFA to study the laser performance. Figure 12.48 shows the variation of the laser pulse widths under various pump powers. The pulse widths seem to be slightly affected by the change of pump power. However, the affect of the pump power is actually stronger since the pulse widths measured here are not the true pulse widths. They are larger than their true values due to the rise time of the equipment as discussed above. The affect of pump power on the pulse width can be drawn from the relationship between pump power and output pulse bandwidth.

The dependence of the output pulse bandwidth is illustrated in Figure 12.49. As the pump power increases the bandwidth increases. Since the bandwidth is inverse proportional to the pulse width, one can infer that the pulse width decreases when the pump power increases. The cause of the decrease in pulse width is attributed to the non-linear effect. The higher the peak power the stronger the non-linear effect, hence the shorter pulse width. It is also noticed that the pulse shortening effect of the pump power is only observed when the fiber length is long comparable to the effective non-linear length as discussed in Section 12.8.9.

12.9   REMARKS

We have presented the propagation of solitons through optical fibers through the numerical approach. The time dependent beam propagation method, the split step Fourier model, is employed to simulate the propagation of the soliton. Our algorithm has been verified by the analytical approach using the inverse scattering method. As both of these approaches produce consistent results, we further our investigation in interaction of two and three solitons by using the numerical approach.

We can conclude that there are three effective methods to overcome the undesirable effect of soliton collapse in our optical soliton transmission systems, which are, the method of setting the optimum pulse separation, relative amplitude and relative phase. Pulse separation method could be used if we are not too concerned about the maximum achievable bit rate. Relative phase method is only useful at 0 and 180 degrees. However, one can use the repulsive nature of the solitons by setting the relative phase to certain values in order to compensate the attractive nature of the solitons. However this method is difficult to implement in practice. Therefore, relative amplitude method would be the best in overcoming this problem. Only a slight difference in the amplitude of the solitons would prevent the solitons from collapsing.

Thus, we have to implement a device at the transmitting terminal to encode each binary digit with alternate value of amplitudes. This can be carried out by using an external modulator to modulate each pulse with different amplitude alternately: (i) The simulation assumes a zero-loss fibers. It would be very useful to take into account the losses due to propagation in the optical fiber, this could be implemented using distributed Raman amplification. (ii) Also, simulation model assumes that the parameters can be controlled exactly. It would be more accurate if the simulation also considers jittering effects in the phase, amplitudes as well as the initial separation of the solitons. (iii) Finally, the simulation should also consider the effects of the erbium-doped fiber amplifier scheme to boost signals as well as Raman amplification.

We have also experimentally demonstrated and simulated stable multi-soliton bound states which have been generated in an active mode-locked fiber laser under phase matching via optical phase modulation. It is believed that this stable existence of multi-soliton bound states is effectively supported by the phase modulation in an anomalous-dispersion fiber loop. Simulation results have confirmed the existence of multi-soliton bound states in the FM mode-locked fiber laser. Created bound states can be easily harmonic mode locked to generate periodically multi-soliton bound sequence at high repetition rate in this type of fiber laser that is much more prominent than those by passive types.

The pulse train of very short pulse width at high repetition rate, up to 10 GHz, generated from a HMLFL has also been demonstrated. The characteristics of the pulse train such as pulse width and bandwidth have been studied by varying the settings of the cavity. It is found that as the modulation frequency increases the pulse becomes shorter and its corresponding bandwidth increases. The increase of the modulation depth also makes the pulse shorter but its effect on the pulse is not as strong as modulation frequency. High pump power is also found to help shorten the pulse width in the long cavity laser.

REFERENCES

1. Zabusky, N., and M. D. Kruskal. 1965. *Physical Review Letters* 15: 240.
2. Agrawal, G. P. 1992. *Nonlinear fiber optics.* Boston, USA: Academic Press.
3. Agrawal, G. P. 2001. *Nonlinear fiber optics.* Boston, USA: Academic Press.
4. Zaharov, V. E., and A. B. Shabat. 1972. Exact theory of two-dimensional self-focusing and one-dimensional self-modulation of wave in nonlinear media. *Soviet Physics JETP* 34 (1): 62–69.
5. Dodd, R. K., J. C. Eilbeck, J. D. Gibbon, and H. C. Morris. 1982. *Solitons and nonlinear wave equations.* Boston, USA: Academic Press.
6. Satsuma, J., and N. Yajima. 1974. Initial value problems of one-dimensional self-modulation of nonlinear waves in dispersive media. *Progress of Theoretical Physics Supplement* 55: 284–306.
7. Anderson, D., and M. Lisak. 1986. Bandwidth limits due to mutual pulse interaction in optical soliton communication systems. *Optics Letters* 10 (3): 174–76.
8. Bondeson, A., M. Lisak, and D. Anderson. 1979. Soliton perturbations: A variational principle for soliton parameters. *Physica Scripta* 20: 479–85.
9. Karpman, V. I., and V. V. Solov'ev. 1981. A perturbational approach to the two-soliton systems. *Physica 3D* 3 (3): 487–502.
10. Desem, C., and P. L. Chu. 1987. Reducing soliton interaction in single-mode optical fibers. *IEE Proceedings* 134 (Pt J, 3): 145–51.
11. Malomed, B. A. 1991. Bound solitons in coupled nonlinear Schrodinger equation. *Journal of Physical Review A* 45: R8321–23.
12. Malomed, B. A. 1991. Bound solitons in the nonlinear Schrodinger–Ginzburg–Landau equation. *Journal of Physical Review A* 44: 6954–57.
13. Tang, D. Y., B. Zhao, D. Y. Shen, and C. Lu. 2002. Bound-soliton fiber laser. *Journal of Physical Review A* 66: Doc ID 033806, 1–6.
14. Gong, Y. D., D. Y. Tang, P. Shum, C. Lu, T. H. Cheng, W. S. Man, and H. Y. Tam. 2002. Mechanism of bound soliton pulse formation in a passively mode locked fiber ring laser. *Optical Engineering* 41 (11): 2778–82.
15. Grelu, P., F. Belhache, and F. Gutty. 2003. Relative phase locking of pulses in a passively mode-locked fiber laser. *Journal of Optical Society of America B* 20: 863–70.

16. Zhao, L. M., D. Y. Tang, T. H. Cheng, H. Y. Tam, C. Lu. 2007. Bound states of dispersion-managed solitons in a fiber laser at near zero dispersion. *Applied Optics* 46: 4768–73.

17. Hsiang, W. W., C. Y. Lin, and Y. Lai. 2006. Stable new bound soliton pairs in a 10 GHz hybrid frequency modulation mode locked Er-fiber laser. *Optics Letters* 31: 1627–29.

18. Doerr, C. R., H. A. Hauss, E. P. Ippen, M. Shirasaki, and K. Tamura. 1994. Additive-pulse limiting. *Optics Letters* 19: 31–33.

19. Davey, R., N. Langford, and A. Ferguson. 1991. Interacting solitons in erbium fiber laser. *Electronics Letters* 27: 1257–59.

20. Krylov, D., L. Leng, K. Bergman, J. C. Bronski, and J. N. Kutz. 1999. Observation of the breakup of a prechirped N-soliton in an optical fiber. *Optics Letters* 24: 1191–93.

21. Prilepsky, J. E., S. A. Derevyanko, and S. K. Turitsyn. 2007. Conversion of a chirped Gaussian pulse to a soliton or a bound multisoliton state in quasi-lossless and lossy optical fiber spans. *Journal of Optical Society of America B* 24: 1254–61.

22. Agrawal, A. P. 1992. *Fiber-optic communication systems*. New York, USA: Wiley.

23. Gordon, J. P. 1983. Interaction forces among solitons in optical fibers. *Optics Letters* 8 (10): 596–8.

24. Hasegawa, A. 1989. *Optical solitons in fibers*. Berlin, Germany: Spinger.

13 OFDM Optical Transmission Systems

13.1 INTRODUCTION

Orthogonal frequency division multiplexing (OFDM) is a transmission technology that is primarily known from wireless communications and wired transmission over copper cables 44. It is a special case of the widely known frequency division multiplexing (FDM) technique for which digital or analog data is modulated onto a certain number of carriers and transmitted in parallel over the same transmission medium. The main motivation for using FDM is the fact that due to parallel data transmission in frequency domain, each channel occupies only a small frequency band. Signal distortions originating from frequency-selective transmission channels, the fiber chromatic dispersion, can be minimized. The special property of OFDM is characterized by its very high spectral efficiency. While for conventional FDM, the spectral efficiency is limited by the selectivity of the bandpass filters required for demodulation, OFDM is designed such that the different carriers are pair wise orthogonal. In this way, the sampling point, the inter-carrier interference (ICI), is suppressed although the channels are allowed to overlap spectrally.

13.1.1 PRINCIPLES OF oOFDM

13.1.1.1 OFDM as a Multi-carrier Modulation Format

13.1.1.1.1 Spectra
OFDM is a multi-carrier transmission technique and uses multiple frequencies to simultaneously transmit multiple signals in parallel form. Each sub-channel is assigned a sub-carrier which is within the range allowable for an optical channel within the DWDM optical transmission system. Unlike the normal frequency division multiplexing technique in which the spectra of the sub-channels are separated by a guard band such that there is no overlapping between them, in OFDM the term "orthogonality" comes from the property that adjacent channels are orthogonal, that is they are perpendicular or the dot product of the channel is zero as shown in Figure 13.1a and Figure 13.1b, respectively. This allows the overlapping of the spectra of adjacent channels without creating any cross talk between them [1].

The entire bandwidth allocated for a wavelength channel may be occupied. The data source is distributed over all sub-carriers. Thus each sub-carrier transports a small amount of the information. Thus by this lowering of the bit rate per sub-carrier channel, the ISI due to the distortion of the channel can be significantly reduced.

13.1.1.1.2 Orthogonality
The complex baseband OFDM signal $s(t)$ can be written as

$$s(t) = \sum_{k} \sum_{n=0}^{N-1} a_n(k) g_n(t - kT) \tag{13.1}$$

where k is the time index, N is the total number of sub-carriers, $T = NT_s$ is the stretched OFDM symbol period as a result of the conversion from serial to parallel with T_s is the sampling period. $a_n(k)$ is the kth data symbol of the nth sub-carrier and $g_n(t)$ is the baseband data pulse given by

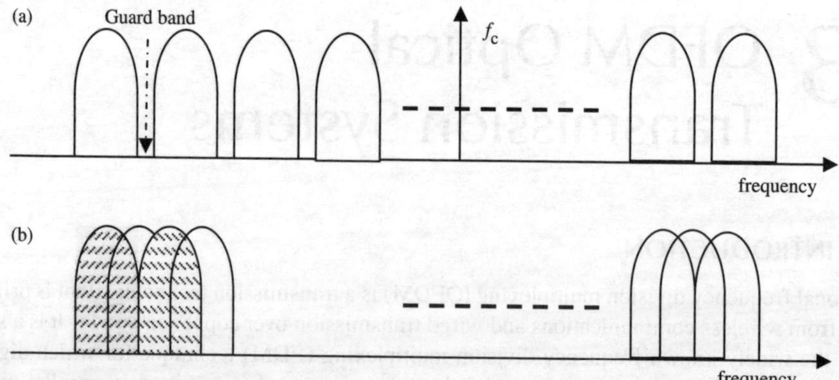

FIGURE 13.1 Multi-carrier modulation technique. (a) Spectrum of FDM sub-carriers. (b) Spectrum of OFDM sub-carriers.

$$g_n(t) = e^{j2\pi f_n} g(t)$$

$$f_n = f_c \pm n\Delta f. \tag{13.2}$$

This includes the sub-carrier component of frequency shifted from the central carrier frequency by n amounts of the frequency spacing between the sub-carriers. $g(t)$ is the pulse shaping function which normally is very close to Gaussian after propagating through the optical MZIM.

The orthogonality of the channels requires:

$$\int_{-\infty}^{\infty} f_n(t) f_m^*(t) dt = \frac{1}{T} \int_{-T/2}^{T/2} f_n(t) f_m^*(t) dt \begin{cases} 1 & n=m \\ 0 & n \neq m \end{cases} = \delta_0[n-m]. \tag{13.3}$$

The asterisk (*) indicates the complex conjugate. Thus it requires that the selection of the carrier frequency and the pulse shaping function in such a way that the orthogonality can be achieved. Block diagram of the sampled discrete time model of OFDM system using N-point FFT and IFT is shown in Figure 13.2. The resultant analog signals at the output of the D-to-A convertor used for modulating the optical modulator is shown in Figure 13.3. While Figure 13.4 shows the frequency domain representation of an OFDM symbol. Figure 13.5 shows a schematic diagram of an optical OFDM transmitter incorporated in a long-haul optical transmission system using multilevel modulation formats.

13.1.1.1.3 Sub-carriers and Pulse Shaping

For orthogonality, each sub-carrier should take an integer number of cycles over a symbol period T, and the number of cycles between adjacent sub-carriers differs by exactly unity. That means that the frequencies of the sub-carriers are multiples of each other.

The pulse shapes of the signals can take the form of rectangular or sinc function and can be written as

$$g(t) = \frac{\dfrac{4\alpha t}{T}\left\{\cos(1+\alpha)\dfrac{\pi t}{T} + \sin(1+\alpha)\dfrac{\pi t}{T}\right\}}{\dfrac{\pi}{T}\left(1 - \left[\dfrac{4\alpha t}{T}\right]^2\right)} \tag{13.4}$$

where α is the roll off factor, similar to the raise cosine function.

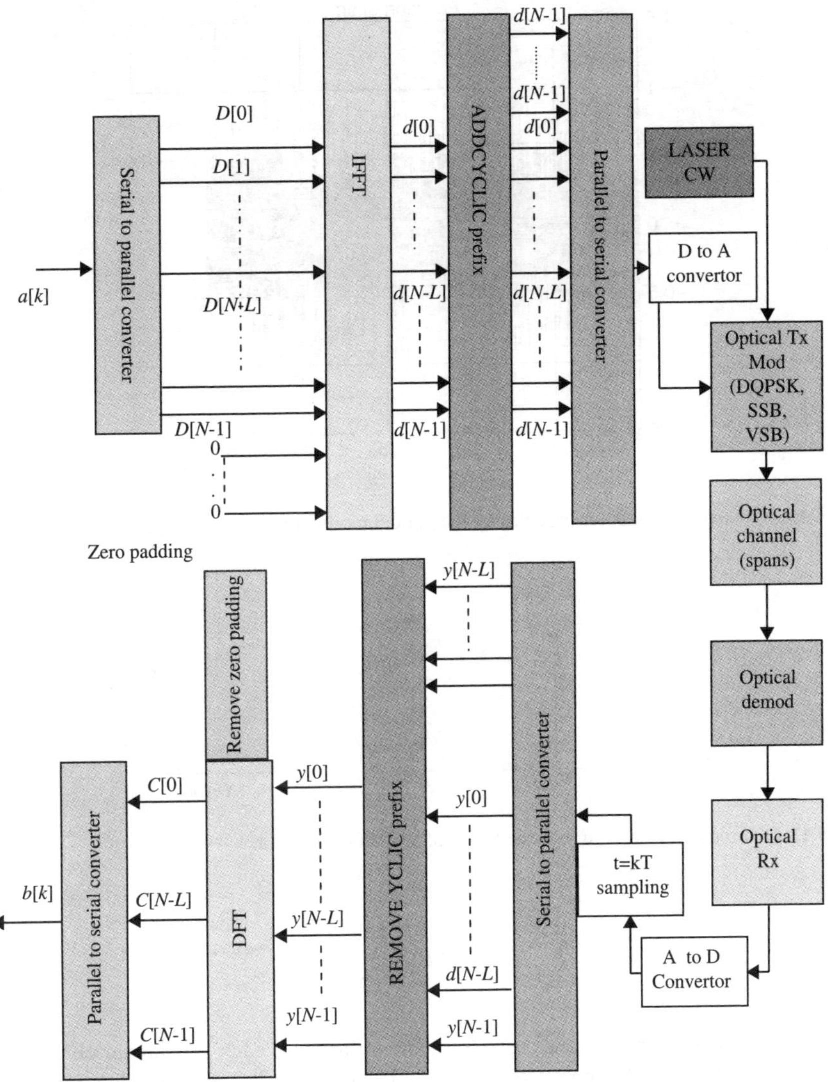

FIGURE 13.2 Block diagram of the sampled discrete time model of OFDM system using N-point FFT and IFFT.

For the rectangular pulse we have

$$g(t) = \begin{cases} 1 & |t| \le \dfrac{T}{2} \\ 0 & \text{otherwise.} \end{cases} \tag{13.5}$$

But this pulse shape would suffer some error.

The orthogonality can be achieved by setting $\Lambda f = 1/T$. For simplicity we can assume that the signal would take the form with the same shape function for all sub-carrier channels

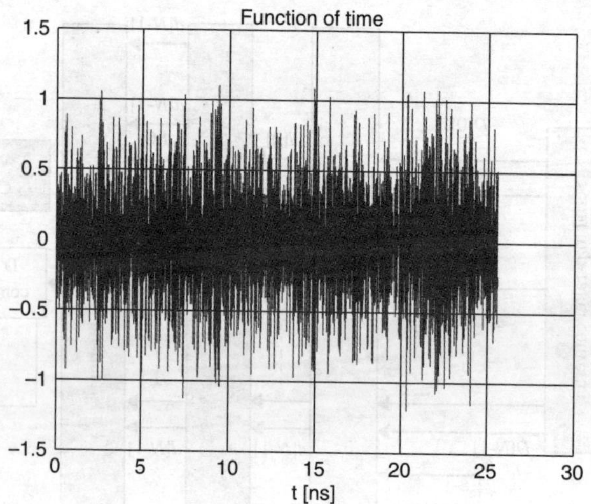

FIGURE 13.3 Analog signals for modulating the optical modulator shown in Figure 13.2.

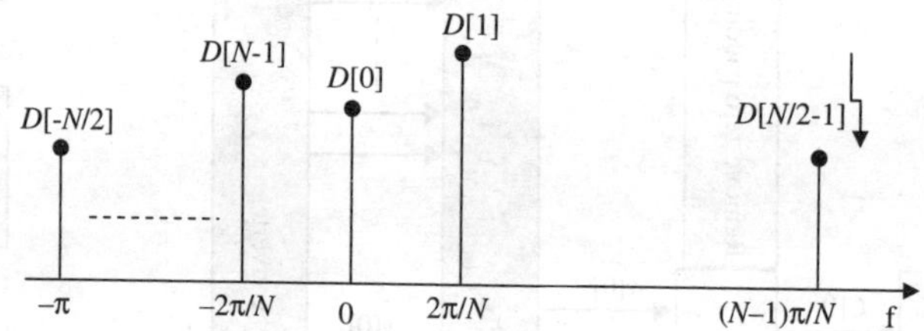

FIGURE 13.4 Frequency domain representation of an OFDM symbol.

$$s(t) = \sum_{n=0}^{N} a_n g_n(t - kT) = \sum_{n=0}^{N} a_n g(t) e^{j2\pi(f_c + n\Delta f)} \tag{13.6}$$

with N as the number of bits of the serial data stream fed into each sub carrier channel from the output of the serial to parallel converter as shown in Figure 13.2. Taking the Fourier transform of Equation 13.6 we have

$$S(f) = \sum_{n=0}^{N} a_n G(f - (f_c - n\Delta f)) \tag{13.7}$$

which is the summation of all sinc functions if the pulse shape is of rectangular shape. These spectral functions sinc would cross over at the null of each other and their peaks. On the other hand if the temporal function of the data of each sub-carrier channel then the frequency spectrum of the channel would be rectangular and thus they can be separated. Thus we can have

$$g(t) = \frac{1}{T}\operatorname{sinc}\left(\pi\frac{t}{T}\right) = \Delta f \operatorname{sinc}(\pi\Delta f t) \rightarrow G(f) = \operatorname{rect}\left(\frac{f}{\Delta f}\right) \tag{13.8}$$

with $\Delta f = 1/T$.

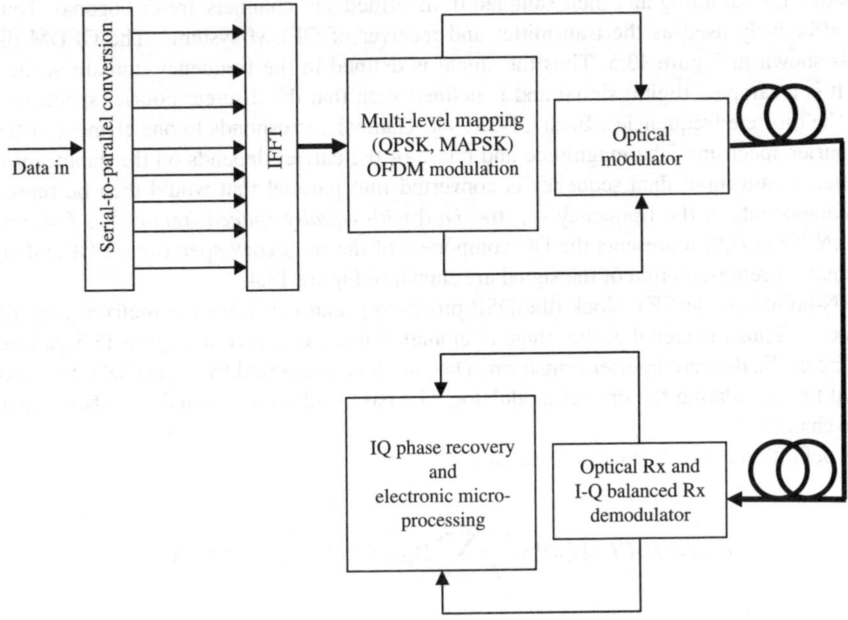

FIGURE 13.5 Schematic diagram of optical OFDM transmitter of a long-haul optical transmission system using multilevel modulation formats.

Therefore the spectrum of OFDM signal takes the form

$$S(f) = \sum_{n=0}^{N-1} a_n \text{rect}\left(\frac{f - f_c + n\Delta f}{\Delta f}\right) \tag{13.9}$$

which is made up by a number of rectangular spectra separated by Δf.

In practice the availability of rectangular time pulse shape would not be possible but with certain raise cosine feature, while for sinc shape pulses can be generated without difficulty by using a transmission filter in the electrical domain. The spectrum of OFDM signal is thus the summation of the sinc shape spectra of all the sub-channels. If the number of sub-carrier channel is sufficiently high then one would obtain a flat rectangular spectrum as shown in Figure 2.38e.

The orthogonal sub-channels can be generated using IFFT or FFT normally available from digital signal processors.

13.1.1.1.4   OFDM Receive

In a conventional FDM system the sub-channels are normally separated by filtering using banks of filters and demodulators. For orthogonal signals the sub-channels can be separated by correlation techniques [2,3].

13.1.2   FFT and IFFT Based OFDM Principles

To recover or demodulate the sub-carrier channels by using a bank of filters or local oscillator to tune and separate the sub-channels is very costly, in particular when there are a large number of sub-channels. Recently the speed of digital signal processing has increased significantly. Certainly the FFT and IFFT are the main features of these processors. The main feature of the FFT and

IFFT is that the sampling and their sampled transformed sub-channels are orthogonal. Thus they can be effectively used as the transmitter and receiver of OFDM systems. The OFDM discrete model is shown in Figure 13.5. Thus the signal is defined in the frequency domain at the transmitter. It is a sampled digital signal and is defined such that the discrete Fourier spectrum exists only at the discrete frequencies. Each OFDM sub-channel corresponds to one element of this discrete Fourier spectrum. The magnitude and phase of the carrier depends on the data transmitted. The serial-stream input data sequence is converted into parallel that would then be represented as the components of the frequency spectra, $D[n]$ *with equally spaced frequencies* $f_n=2\pi n/N$ for $n=0,1...N$. Thus $D[0]$ represents the DC component of the frequency spectrum of OFDM signals. The frequency representation of the signal are shown in Figure 13.4.

An N-point inverse FFT block (the DSP processor) then generates the prefixed time domain components. Thus the signal at this stage is in analog form as shown in Figure 13.5 padded with zeros for a cyclic discrete Fourier transform. They are then converted from parallel into serial form and used for modulating the optical modulator. The passband optical signals are then transmitted over the channel.

The definition of the N-point inverse DFT is

$$d[k] = IDFT\{D[n]\} = \frac{1}{N}\sum_{n=0}^{N-1} D[n]e^{j2\pi k/N} \quad n,k = 0,1...N-1 \tag{13.10}$$

with N is the symbol length. After the pulse shaping of a rectangular function of

$$\mathrm{rect}\left[\frac{k}{N}\right] = \begin{cases} 1 & \text{for } k = 0,1...N-1 \\ 0 & \text{otherwise.} \end{cases} \tag{13.11}$$

The sequence $d[n]$ in the time domain is given by

$$d[k] = \frac{1}{N}\sum_{n=0}^{N-1} D[n]e^{j2\pi k/N}\,\mathrm{rect}\left[\frac{k}{N}\right] \quad n,k = 0,1...N-1$$

$$= \frac{1}{N}\sum_{n=0}^{N-1} D[n]g_n[k] \tag{13.12}$$

$$\text{with } g_n[k] = e^{j2\pi k/N}\mathrm{rect}\left[\frac{k}{N}\right]$$

$g_n[k]$ can be considered as the baseband pulse which is considered to be orthogonal and satisfies the condition.

$$g_n[k] = e^{j2\pi k/N}\mathrm{rect}\left[\frac{k}{N}\right]. \tag{13.13}$$

Thus, a rectangular or sinc pulse shaper can be inserted between the serial to parallel and the IFFT blocks and then multiplied by the sub-carrier frequency and then superimposed and multiplied by a factor $1/N$.

At the receiver, after the photodetection and electronic pre-amplification, the noisy time domain OFDM samples are converted to its corresponding frequency domain symbols by an N-point FF. The coefficients of an N-point DFT are given by

$$C[n] = DFT\{y[k]\} = \sum_{n=0}^{N-1} y[k]e^{j2\pi k/N} \quad n,k = 0,1...N-1$$

$$= \sum_{-\infty}^{+\infty} y[k]e^{j2\pi k/N} \mathrm{rect}\left[\frac{k}{N}\right]. \tag{13.14}$$

An N-point DFT is equivalent to a bank of N-orthogonal "matched filters" matched to the corresponding "baseband pulses" in IDFT, followed by samplers that sample once per N symbols. The impulse of the matched filter can be written as

$$g_n^*[-k] = e^{j2\pi k/N} \mathrm{rect}\left[\frac{-k}{N}\right]. \tag{13.15}$$

This follows the logical sequence of

$$C[n] = y[k]*g_n^*[-k] = y[n]*e^{j2\pi k/N}\mathrm{rect}\left[\frac{k}{N}\right] \quad n,k = 0,1...N-1$$

$$= \sum_{-\infty}^{+\infty} y[k]\,e^{j2\pi k/N}\mathrm{rect}\left[\frac{-(k-u)}{N}\right]. \tag{13.16}$$

After sampling at the instant $k=0$ the frequency spectral component $C[n]$ becomes

$$C[n] = \sum_{-\infty}^{+\infty} y[u]\,e^{j2\pi u/N}\mathrm{rect}\left[\frac{u}{N}\right]$$

$$\text{with } u \to k \to \quad C[n] = \sum_{-\infty}^{+\infty} y[u]e^{j2\pi u/N}\mathrm{rect}\left[\frac{k}{N}\right] = \sum_{k=0}^{N-1} y[k]e^{j2\pi k/N} \tag{13.17}$$

which is the DFT of $y[k]$.

13.2 OPTICAL OFDM TRANSMISSION SYSTEMS

Orthogonality is achieved by placing the different RF-carriers onto a fixed frequency grid and assuming rectangular pulse shaping. For OFDM, the signal can be described as the output of a discrete inverse Fourier transform block using the parallel complex data symbols as input. This property has been one of the main driving aspects for OFDM in the past since modulation and demodulation of a high number of carriers can be realized by simple digital signal processing (DSP) instead of using many local oscillators in transmitter and receiver. Recently, OFDM has become attractive for digital optical communications [4,5]. Using OFDM appears to be very attractive since the low bandwidth occupied by a single OFDM channel increases the robustness towards fiber dispersion, drastically allowing the transmission of high data rates of 40 Gb/s and hundreds of kilometers without the need for dispersion compensation [2]. In the same way as for modulation formats like DPSK or DQPSK that were introduced in recent years, the challenge with OFDM for optical system engineers is to adapt a classical technology to the special properties of the optical channel and the requirements of optical transmitters and receivers.

However Hanzo et al. [3] shows that an optical SSB modulation can assist in combating fiber dispersion. SSB can be achieved by driving the MZIM with two microwave signals $\pi/2$ phase shift with each other or by optical filtering (VSB) as show in Chapters 2 and 9. However the true SSB transmitter using dual drive and $\pi/2$ phase shift is preferred so as to preserve the energy contained within the bands of the signals. In optical SSB the phase information can be preserved after the square-law detection of the photodetector and the CD is limited by reducing the optical spectral bandwidth by a factor of 2 [6].

Two approaches have been reported recently. An intuitive approach introduced by Lorente et al. [7] makes use of the fact that the wavelength-division multiplexing (WDM) technique already realizes data transmission over a certain number of different carriers. By means of special pulse shaping and carrier wavelength selection, the orthogonality between the different wavelength channels can be achieved resulting in the so-called orthogonal WDM technique (OWDM). However, in this way the option of simple modulation and demodulation by means of discrete Fourier transforms (DFT) cannot be utilized as this kind of digital signal processing is not available in the optical domain.

An alternative method [5] is the generation of an electrical OFDM signal by means of electrical signal processing followed by modulation onto a single optical carrier [8,9]. This approach is known as optical OFDM (oOFDM). Here, the modulation is a two-step process: first, the electrical OFDM signal already is a broadband bandpass signal which is then modulated onto the optical carrier. Second, to increase data throughput, oOFDM can be combined with WDM resulting in multi-Tb/s transmission systems as shown in Figure 13.5. Nevertheless, oOFDM itself offers different options for implementation. An important issue is optical demodulation that can be realized either by means of direct detection (DD) or coherent detection (CD) including a local laser oscillator. DD is preferable due to its simplicity. However, for DD the optical intensity has to be modulated. Due to the fact that the electrical OFDM signal is quasi-analog with zero mean and high peak-to-average ratio, the majority of the optical power has to be wasted for the optical carrier. That means there is an additional DC-value of the complex baseband signal, resulting in low receiver sensitivity. In addition, for chromatic dispersion (CD), the bandwidth efficiency is twice as high as for DD because pure intensity modulation inherently generates a double-sideband signal. For CD, a complex optical I-Q modulator composed of two real modulators in parallel followed by superposition with $\pi/2$ phase shift allows for transmission of twice as much data within the same bandwidth. For intensity modulation, the bandwidth efficiency may be increased by suppressing one of the redundant sidebands resulting in oOFDM with single-sideband (SSB) transmission. First the serial data in can be converted to parallel streams. These parallel data sequences are then mapped to QAM constellation in the frequency domain, then using IFFT converted it back to the time domain. The time domain signals are in I- and Q-components which are then fed into an I-Q-optical modulator. This optical modulation can be DQPSK or any other multilevel modulation sub-system. At the end of the optical fiber transmission, I and Q components are detected either by direct detection or coherent detection. For coherent detection a 2×4 90 degree hybrid coupler is used to mix the polarized optical fields of the local oscillator and that of the received signals. The outputs of the couplers are then fed into balanced optical receivers. The mixing of the local laser source and that of the signals preserves the phase of the signals which are then processed by a high speed electronic digital processor. For direct detection, I and Q components are detected differentially, the amplitude and phase detection are then compared and processed similarly as for the coherent case.

In order to show the robustness of oOFDM towards fiber dispersion and also fiber non-linearity, numerical simulations are carried out for a data stream of 42.7 Gb/s data rate. The number of OFDM channels can be varied between $N_{min}=256$ and $N_{max}=2048$. A guard interval of 12 ns can be inserted, a strategy belonging inherently to OFDM technology that ensures orthogonality of the different channels in case of a transmission channel with memory. For the optical modulation, intensity modulation using a single Mach-Zehnder modulator in conjunction with SSB filtering and direct detection was implemented. The non-linear optical transmission channel consisted of eight 80 km non-DCF spans of SSMF. MZIM with linearization should be used [10]. As a criterion for

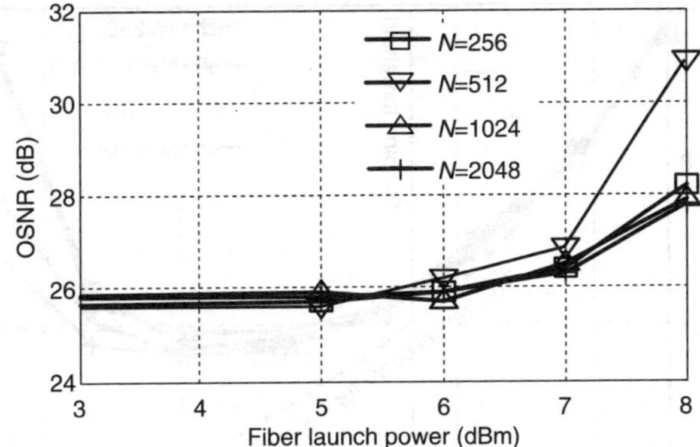

FIGURE 13.6 Simulation result for 42.7 Gb/s oOFDM transmission over 640 km of standard single-mode fiber; OSNR required for BER $= 10^{-3}$ as function of fiber launch power. (From Leibrich, J., Abdulamir Ali, and W. Rosenkranz, Optical OFDM as a promising technique for bandwidth-efficient high-speed data transmission over optical fiber. In *Proceedings of the 12th International OFDM-Workshop 2007, InOWo 2007*, 29–30 August 2007, Hamburg, Germany. With permission.)

performance, required OSNR for a BER of 10^{-3} (Monte Carlo) is measured. Using FEC after decoding the BER is equivalent to an error rate below 10^{-9} depending on the specific code.

Figure 13.6 shows the required OSNR as a function of fiber launch power for different values of N. Transmission is error free (1e-9) over 640 km of SSMF or a dispersion factor of about 1100 ps/nm without dispersion compensation. It can be explained by the fact that even for the lowest value of $N_{min}=256$, each sub-channel occupies a bandwidth of approximately 42.7 GHz/256=177 MHz resulting in high robustness towards fiber dispersion.

The principal difficulties of oOFDM are that the pure delay is due to the variation of the refractive index of the fiber with respect to the optical frequency lead to bunching of the sub-channels and hence the increase of the optical power, thus unexpected SPM may occur in a random manner.

13.2.1 Impacts on Non-linear Modulation Effects on Optical OFDM

The MZM is based on a quadrature point where the power transfer characteristic is linearizable. By means of low modulation depth, the non-linear distortions due to the MZM can be considered to be arbitrarily small, hence leading to low ratio of useful power to carrier power, and thus low sensitivity. Therefore the modulation depth is a compromise between these constraints.

Figure 13.7 shows results for the BER versus the OSNR under back-to-back transmission with the BER obtained by Monte-Carlo simulation [11,12]. The OSNR for BER=10^{-3} is plotted versus normalized driving voltage as shown in Figure 13.7. The normalization is performed such that minimum and maximum optical output power can be obtained for instantaneous input voltages of –0.5 and 0.5, respectively. Beyond these values, clipping occurs and inter-modulation distortion also exists, due to MZM characteristics. The OFDM signal is analog having nearly Gaussian amplitude distribution [13]. In good approximation, the peak voltage is within an interval from plus to minus the triple of the RMS voltage. This relation is used to create the lower from the upper of the two horizontal axes. The clipping threshold obtained for a peak voltage of ±0.5 is also given. Finally, the impact of MZM non-linearity, which within the range of acceptable sensitivity obviously does not depend significantly on N, is identified by means of the dashed line.

Figure 13.7 depicts the non-linear resilience of the fiber transmission link including 8×80 km SSMF without dispersion compensation. In order to investigate the impact of linear factor one

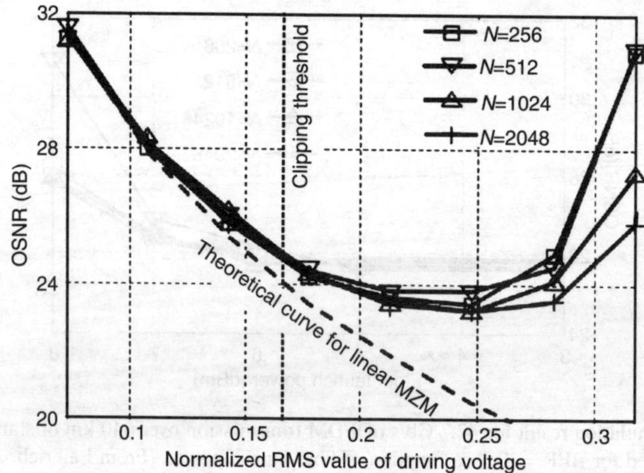

FIGURE 13.7 Required OSNR for BER $= 10^{-3}$ as function of driving voltage swing fed to MZIM electrodes. (From Leibrich, J., Abdulamir Ali, and W. Rosenkranz, Optical OFDM as a promising technique for bandwidth-efficient high-speed data transmission over optical fiber. In *Proceedings of the 12th International OFDM-Workshop 2007, InOWo 2007*, 29–30 August 2007, Hamburg, Germany. With permission.)

by one, the MZM is driven in the quasi-linear range with a normalized effective voltage swing of ≈ 0.15 resulting in a back-to-back OSNR of ≈ 25.5 dB. This value is significantly above those reported elsewhere [14], which is due to incoherent detection and higher bandwidth. For low values of launch power, Figure 13.7 shows the robustness of oOFDM towards fiber dispersion, as the OSNR penalty achieved with ≈ 11000 ps/nm accumulated dispersion is negligible. In the non-linear regime, however, penalty increases rapidly. Beyond 8 dBm launch power at the fiber input, the BER does not fall below 10^{-3} due to strong signal distortion. The optical power consists of a strong DC component and a weaker AC component carrying the OFDM-signal. Since only the AC component results in signal distortion, the acceptable launch power found in this contribution is higher than for coherent detection.

Figure 13.8 shows the eye opening penalty of modulation formats NRZ-ASK and OFDM with a guard interval of 25, 50 and 100 ps. It shows clearly the superiority of the OFDM over ASK at 42.7 Gb/s bit rate [15].

Except for the value for $N = 512$ at a launched power of 8 dBm, attributed to limited simulation accuracy, the result does not depend on N. Increasing N decreases the separation between the sub-channels, and fiber non-linearity is expected to cause strong XPM and FWM. However, with increasing N the power per sub-channel is decreased. Apparently, these two aspects have cancelled each other out so that equal performance for all N is achieved.

The impact of non-linear modulator characteristics and Kerr effect shows that for an uncompensated oOFDM 8×80 km fiber link with a varying number N of sub-carriers, both impairments are independent of N. Obviously, oOFDM is quite robust towards the specific non-linear impairments in fiber-optic transmission systems.

13.2.2 DISPERSION TOLERANCE

The variation of the eye opening penalty as a function of the ratio of the guard interval and the bit period with the fiber SSMF length as a parameter is obtained as shown in Figure 13.9. This indicates that the guard interval is very critical for different length of the fibers, and thus the bit rate and the number of sub-carriers.

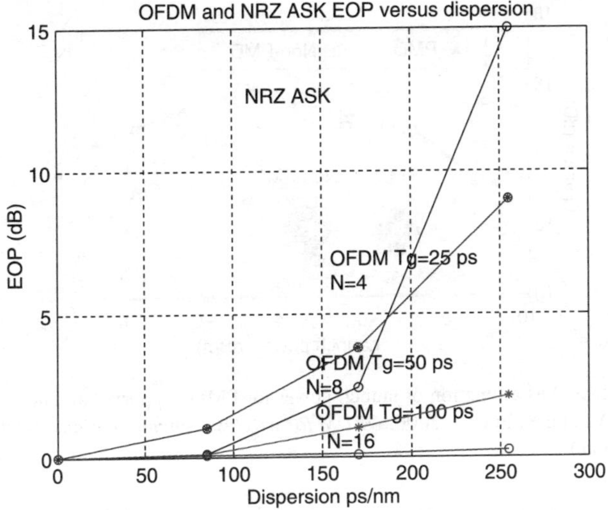

FIGURE 13.8 Eye opening penalty of OFDM as compared with NRZ ASK with different guard interval time of 25, 50 and 100 ps for 42.7 Gb/s bit rate.

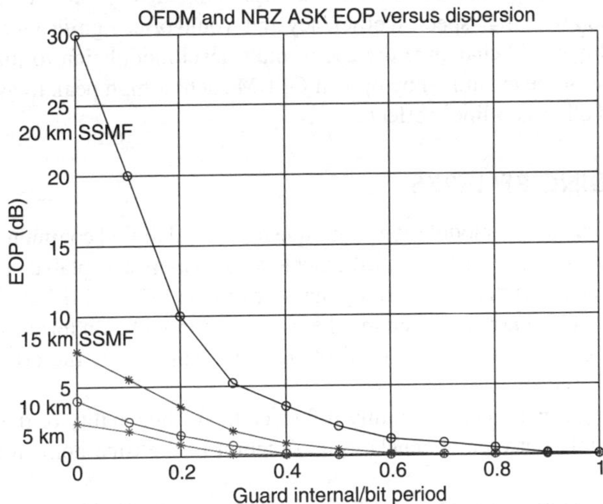

FIGURE 13.9 Eye opening penalty of OFDM against the ratio of the guard interval and bit period with fiber length as a parameter, $N = 8$.

13.2.3 RESILIENCE TO PMD EFFECTS

Jansen et al. [14] report the first experiment of PMD impact on fibre non-linearity in coherent optical OFDM (CO-OFDM) systems. The optimal Q value at 10.7 Gb/s has been improved by 1 dB after introduction of DGD of 900 ps with the variation of the Q-value against the launched power as shown in Figure 13.10.

13.3 OFDM AND DQPSK FORMATS FOR 100 GB/S ETHERNET

For 100 Gb/s Ethernet the decomposition of the bit rate to symbol rates by multi-level modulation formats are described in Chapter 7. The OFDM offers significant lower symbol rate due to the

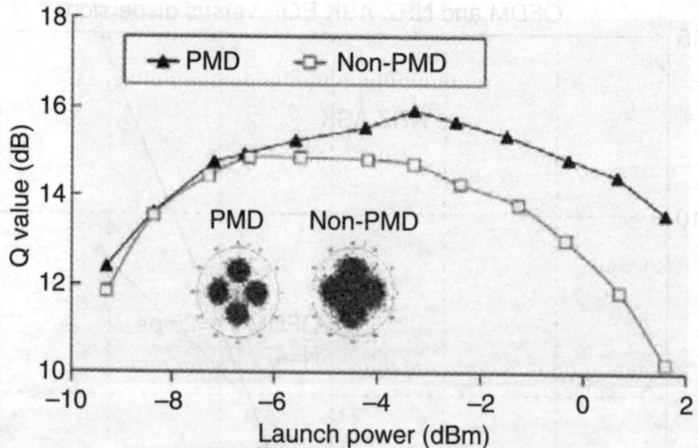

FIGURE 13.10 System Q as a function of launch power for PMD-supported and non-PMD-supported systems. (From Nee, R. V., and R. Prasad. 2000. *OFDM for wireless communications*. Norwood, USA: ARtech House. With permission.)

decomposition into several sub-channels. Lowery et al. [8] reported the transmission performance of RZ-DQPSK formats over 75 km of SSMF as well as RZ-ASK with OSNR of 22 dB and 36 dB, respectively.

On the other hand the OFDM can combine a large number of parallel data streams into one broadband data stream with high spectral efficiency and would offer significant reduction of bit rate into symbol rate. This would challenge the use of multi-level modulation formats. However there are still several issues to be examined by optical OFDM such as high peak to average carrier ratio or FWM effects and other non-linear effects.

13.4 CONCLUDING REMARKS

OFDM is strongly suitable as a modulation technique for optical digital communications, especially for lowering the ultra-high speed 100 Gb/s Ethernet to a much more acceptable rate per sub-channel. As a multi-carrier channel with natural orthogonality from the FFT and IFFT it is much more efficient than the use of single carrier as seen in other advanced modulation formats described in most chapters of this book. By closely spacing all of the sub-channels via the orthogonality one can increase the spectral efficiency.

The pulse shaping function can certainly influence the orthogonality of the sub-channels and minimize the cross talk between channels. Sinc function in the time domain would be the best choice.

The guard bands influences the transmission performance of OFDM transmission.

OFDM is shown to be more resilient to non-linearity and PDM as well as CD. Coherent OFDM offers higher receiver sensitivity or lower eye opening penalty than the direct detection OFDM without much more complexity as the processing of the received optical sequence can be implemented in the electronic domain.

REFERENCES

1. Nee, R. V., and R. Prasad. 2000. *OFDM for wireless communications*. Norwood, USA: ARtech House.
2. Hanzo, L., S. X. Ng, T. Keller, and W. Webb. 2004. *Quadrature amplitude modulation: From basics to adaptive trellis coded, turbo-equalized space-Tme coded OFDM, CDMA and MC-CDMA systems*. 2nd ed. Chichester, England: IEEE Press, Wiley.
3. Hanzo, L., M. Münster, B. J. Choi, and T. Keller. 2003. *OFDM and MC-CDMA for broadband multi-user communications, WLANs and broadcasting*. San Francisco, CA, USA: Wiley.

4. Llorente, R., J. H. Lee, R. Clavero, M. Ibsen, and J. Martí. 2005. Orthogonal wavelength-division-multiplexing technique feasibility evaluation. *Journal of Lightwave Technology* 23: 1145–51.
5. Leibrich, J., Abdulamir Ali, and W. Rosenkranz. 2007. Optical OFDM as a promising technique for bandwidth-efficient high-speed data transmission over optical fiber. In *Proceedings of the 12th International OFDM-Workshop 2007, InOWo 2007*, 29–30th August 2007, Hamburg, Germany.
6. Lowery, A. J., and J. Armstrong. 2006. Orthogonal frequency division multiplexing for dispersion compensation in long haul WDM systems. *Optics Express* 14 (6): 2079–84.
7. Sieben, M., J. Conradi, and D. E. Dodds. 1999. Optical single-sideband transmission at 10 Gb/s using electrical dispersion compensation. *IEEE Journal of Lightwave Technology* 17 (10): 2059–68.
8. Lowery, A. J., L. Du, and J. Armstrong. 2006. Orthogonal frequency division multiplexing for adaptive dispersion compensation in long haul WDM systems. In *Proceedings OFC 2006, Paper PDP39*, March, Anaheim, USA.
9. Shieh, W., and C. Athaudage. 2006. Coherent optical orthogonal frequency division multiplexing. *Electronics Letters* 42: 587–89.
10. Meng, X. J., A. Yacoubian, and J. H. Bechtel. 2001. Electro-optical pre-distortion technique for linearization of Mach-Zehnder modulators. *Electronics Letters* 37 (25): 1545–47.
11. Abdulamir Ali, J. Leibrich, and W. Rosenkranz. 2007. Impact of nonlinearities on optical OFDM with direct detection. In *Proceedings of ECOC 2007*, Berlin, Germany.
12. Abdulamir Ali. 2007. Orthogonal frequency division multiplexing in optical transmission systems with high spectral efficiency. Master Thesis Dissertation, Technische Facultaet, University of Kiel, Kiel, Germany.
13. Hanzo, L., M. Münster, B. J. Choi, and T. Keller. 2003. *OFDM and MC-CDMA for broadband multiuser communications, WLANs and broadcasting*. San Francisco, CA, USA: Wiley.
14. Jansen, S. L., I. Mortita, N. Takeda, and H. Tanaka. 2007. 20-Gb/s OFDM transmission over 4,160 km SSMF enabled by RF-pilot tone phase noise compensation, in Tech. Digest of the Conference on Optical Fiber Communication, (Optical Society of America, 2007), Postdeadline Paper PDP15.
15. Simulation developed under MOVE_IT (MATLAB) and simulation platform provided by M. Ali of Technische Facultaet, Universit of Kiel.
16. Shieh, W., and C. Athaudage. 2006. Coherent optical orthogonal frequency division multiplexing. *Electronic Letters* 42: 587–89.

14 Comparison of Modulations Formats for Long-Haul Optically Amplified Transmission Systems and Networks

14.1 IDENTIFICATION OF MODULATION FEATURES FOR COMBATING IMPAIRMENT EFFECTS

A sequence of information is used to modulate the lightwave for transmission over the optical fiber. The guided optical medium exerts variable delay difference of the lightwaves of the two sidebands of the passband signals, hence linear chromatic dispersion. Furthermore the random variation of the core geometry and stresses along the length of the fiber links the polarization mode dispersion that is created and thus reduces the low corner frequency of the fiber. The intensity of the optical signals or wavelength channels would also create non-linear phase effects on the lightwave channels and thus the non-linear threshold power level would be the limit of lightwave average power.

Modulation formats would be best to reduce these linear and non-linear impairments. In order to identify the roles of the digital signals and modulation techniques as well as the pulse formats to combat these undesirable effects, the spectral of the passband signals must be identified for amplitude, phase or frequency modulation.

One can identify the following features of modulation formats.

14.1.1 BINARY DIGITAL OPTICAL SIGNALS

The amplitude modulation at high speed exhibits the rise and fall of the intensity or optical field at the edge of the bit period, these rises would act like the sampling of a continuous wave and thus peaks of the spectrum of the passband signals. These peaks would contribute to the total average signal power and hence are limited by the non-linear SPM effects. These peaks would be about 6 dB in power higher than those under the phase modulation.

Phase shift keying is a discrete phase shift keying or non-continuous transition of the phase of the carrier at the boundary of the bit period. The amplitude of phase shift keying is constant. This constant envelope gives a smooth spectral property. Thus there are no spectral peaks due to the sampling of the switching of the amplitude in ASK signals, hence higher energy concentration in the spectra. Depending on the phase shift the roll off the spectra is sharp or slow and the sidelobes of the spectra are suppressed.

Continuous phase shift keying or continuous phase frequency shift keying offer a smooth spectrum similar to the discrete phase shift keying, sharp roll off and high suppression of the sidelobes of the spectrum. The detection of the logical bits can be either phase detection with balanced receiver or frequency discrimination using two narrow band optical filters with appropriate optical delay line.

Vestigial and single side band modulation formats offer narrower band when compared with DSB method of the signals for transmission to combat linear impairments. However the partial response modulation technique such as duobinary offers truly single side band signals with alternating phase shift to effectively combat the linear dispersion impairments and non-linear effects. Furthermore it can be detected directly and uses standard optical receivers.

All modulation techniques can also be applied in optical coherent detection systems to achieve high improvement of SNR, especially when integrated with the estimation of the phase of the demodulated signals using digital signal processors.

14.1.2 M-ARY DIGITAL OPTICAL SIGNALS

QPSK, DQPSK are the typical examples of the enhancement of transmission capacity with the use of distribution of signals over two $\pi/2$ phase shifts of the two DPSK constellations. M-ary signals can be formed by employing both amplitude levels and phase states such an ADPSK, M-ary Star-QAM. Provided that the detection of optical signals does not suffer losses the M-ary modulation signals exhibit an equivalent lower symbol rate for speed signals (e.g., 100 Gb/s). The major problem of the M-ary signals is the transition edges of the signals between the levels that create narrower eye opening in the time domain, the temporal eye.

14.1.3 MULTI SUB-CARRIER DIGITAL OPTICAL SIGNALS

Orthogonal frequency division multiplexing offers the distribution of signal energy into sub-carriers within the signal band. These sub-signals are combined and demodulated after propagation. Thus a high frequency signal can be lowered into several low frequency signals to combat the linear and non-linear impairments. They are then optically modulated for transmission using the digital modulation techniques, e.g., DPSK, DQPSK etc. described above. This modulation technique also takes advantage of the high speed signal processing systems and thus offers much flexibility in the tuning of the signals to the transmission media.

14.1.4 MODULATION FORMATS AND ELECTRONIC EQUALIZATION

In Chapter 11 it is identified that the optical carrier contributes to the linear and non-linear dispersion of the channel on the amplitude or phase modulation. Under discrete or continuous phase modulation the carrier is embedded in the signal band and contributes insignificantly to the linear and non-linear distortion of the data pulse sequence. The electronic equalization including linear and non-linear equalizers performs well when the passband signals exhibit single side band as their time domain signals are linear with respect to the frequency term and thus the LE with FFE as a linear filter performs best. It is thus expected that VSB is also behaving well under the electronic equalization. Further the MSK under the detection of optical filtering offers significant improvement in the resilience to CD, this modulation format would also be most suitable under linear equalization.

Thus, for long-haul optically amplified transmission, it is expected that future research will focus on demonstrating digital phase modulation and optical filtering, as well as the integration of electronic equalization. Because of the current interest in 100 Gb/s optical Ethernet, the electronic processing system that follows the optical receiver is probable and will be available in the near future.

Under the intense interest in the delivery of 10 Gb/s Ethernet and thus 100 Gb/s Ethernet, the electronic processing of the optical receivers and electronic equalization and optical modulators would not work very well at this speed. To solve this bottle neck multi-level modulation formats are expected to contribute due to its effectiveness in lowering the transmission symbol rate down to the matured electronic processing speed. However multi-level signaling would suffer a residual amount of optical carrier in the signal spectrum that may cause difficulty in the electronic equalization and high pattern dependency at the detection site.

14.2 AMPLITUDE, PHASE AND FREQUENCY MODULATION FORMATS

14.2.1 ASK AND DPSK & DPSK AND DQPSK

14.2.1.1 Dispersion Sensitivity of Different Modulation Formats of ASK and DPSK

ASK and DPSK modulation with formats NRZ, RZ, CSRZ at 40 Gb/s bit rate are measured under an experimental test bed shown in Figure 14.1 sent via SMF first them amp and then 1.2 nm filter to ensure no non-linearities. 'Cheated clock' was used, i.e., directly from PRBS generator to Error Analyzer for non-DPSK formats. The optoelectronic receiver is a 45 GHz Discovery Semiconductor type. PRBS bit pattern generator and error analyzer are an SHF Ltd. Product [1]. Optical amplifiers are included so as to boost the signal power to sufficient average power to meet the demand of the sensitivity of the receiver. Optical attenuator is used to vary the power into the receiver and a 10-dB coupler is used to tap the signal in front of the receiver and feed into an optical power meter for measuring the sensitivity in dBm. The optical transmitter is setup so that the biasing of the data modulator can be tuned to different biasing points so that RZ, CSRZ formats can be generated. We have measured the penalty of the eye monitored by an Agilent sampling oscilloscope for different ASK and DPSK modulation formats.

14.2.2 NRZ-ASK AND NRZ DPSK

Figure 14.2 shows the receiver sensitivity of the ASK and DPSK modulation with format NRZ. It is observed that the DPSK signals behave more linearly than that of the ASK. The carrier of the ASK modulation has contributed to the non-linear behavior due to the shift of the carrier as pointed out in Chapter 11 and it is expected that the DSPK formats would perform slightly better in the receiver sensitivity when the dispersion length is not of sufficient length. At 4 km the distortion for 40 Gb/s becomes serious and it is observed that DSPK has a 1-dB better sensitivity at 1e-11 BER as compared with that of NRZ ASK as shown in Figure 14.2a and Figure 14.2b (brown square).

Figure 14.3b and Figure 14.3c show the BER versus receiver sensitivity using one single filter and two filters. This is typical when the optical signals are transmitted with multiplexer and demultiplexer in an optical transmission systems. The DPSK formats behave more resilient to the dispersion with very much compatible receiver sensitivity. The setup of the experiment is shown in Figure 14.3a.

The effects of filter cascading can be observed from the measure of the BER versus the receiver sensitivity of 40 Gb/s CSRZ-DPSK modulation format under the multiplexing and demultiplexing by cascading two NEL 100GHz AWG's cf one AWG (3 dB bandwidth is 0.5 nm) and an optical filter with 1.2 nm passband which is a tunable filter, no fibre is used and thus only two AWGs offering narrow passband to pass 40G CS RZ DPSK sufficiently as observed in Figure 14.3. The transmission

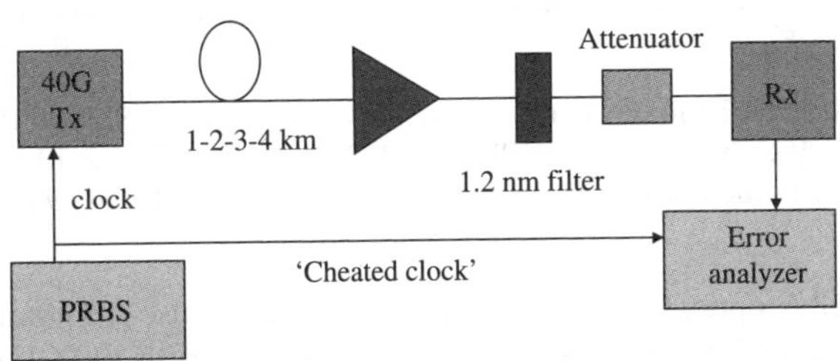

FIGURE 14.1 Experimental test bed for ASK and DSPK modulation formats. 'Cheated clock' is derived from the PRBS pattern generator.

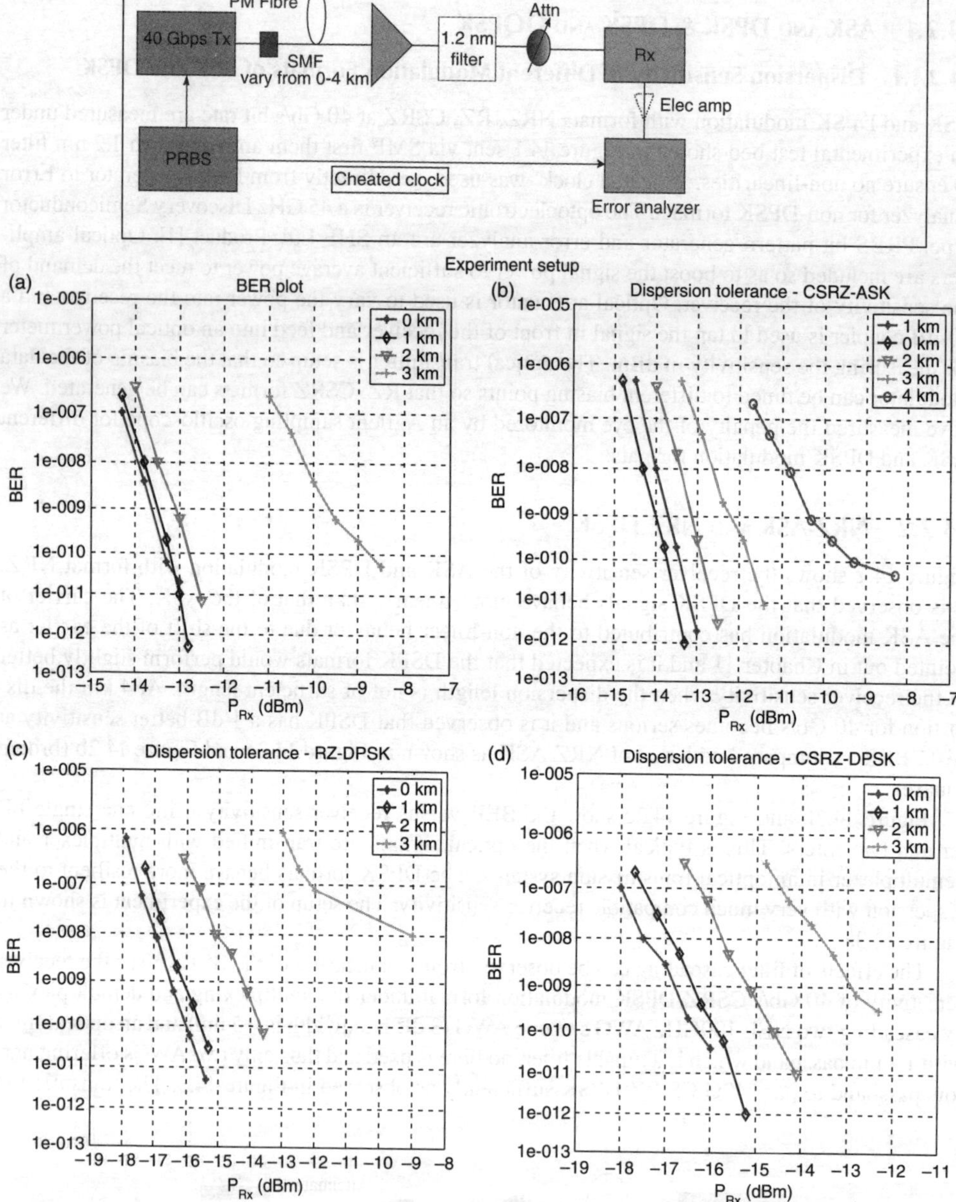

FIGURE 14.2 Experimental BER versus receiver sensitivity of (a) RZ-ASK and (b) CSRZ-ASK (c) RZ-DPSK (d) CSRZ-DPSK (d) under dispersion of SSMF of 1, 2, 3 and 4 km. Due to 10 dB coupler the Rx power level should be read with an addition −10 dB, that is the label is −5, −1…−10 and −9 dBm.

of 40 Gb/s over 10 Gb/s optical transmission infrastructure is also important for upgrading the transmission rate of one or few wavelength channels to 40 Gb/s. With 100 GHz spacing and 0.5 nm passband filters in the network, the BER has been measured versus the receiver sensitivity of 40 and 10 Gb/s of adjacent wavelength channels indicate that the CSRZ DPSK behaves very well and suffers no penalty as shown in Figure 14.14.

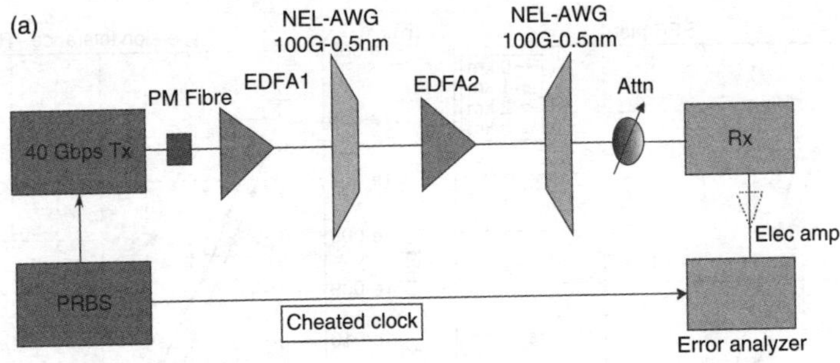

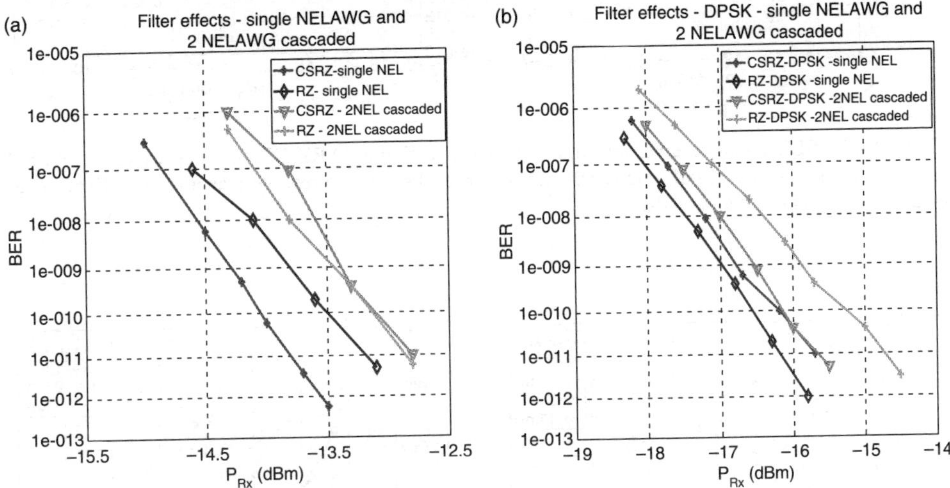

FIGURE 14.3 Experimental BER versus receiver sensitivity of (b) ASK and (c) DPSK with formats CSRZ and RZ with single filter , and duo-filters with the experimental setup shown in (a).

We thus expect that optical MSK channel would also perform better under conventional 10 Gb/s transmission systems.

14.2.3 RZ-ASK AND RZ-DPSK

The transmission performance of the RZ formats is shown in Figure 14.4 for ASK and DPSK modulation are also measured and the receiver sensitivity versus the BER of the same experimental platform. It is observed that an improvement of the receiver sensitivity of about 2, 2 and 4 dB for 1, 2 and 3 SSMF transmission respectively, at a BER of 1e-11 and 1 dB at 4 km of SSMF at BER of 1e-6. The RZ format would exhibit a switching or sampling of the waveform and thus spikes separated at 40 GHz is expected in the DPSK spectrum but at the carrier frequency. At short lengths below 3 km the contribution of the non-linear phase distortion is not high enough so that the improvement of the receiver sensitivity can be achieved while at 4 km SSMGF of 40 Gb/s the non-linear distortion takes over the linear distortion and so not much improvement in the eye diagram can be observed.

14.2.4 CSRZ ASK AND CSRZ-DPSK

As shown in Figure 14.5a there is a ~3–4 dB improvement of the CSRZ-DPSK over CSRZ-ASK formats. It is understood that a complete removal of the carrier would offer no linear distortion but only linear and much lower than that suffered by the CSRZ ASK. Thus there is a better improvement of CSRZ DPSK than CSRAZ-DPSK.

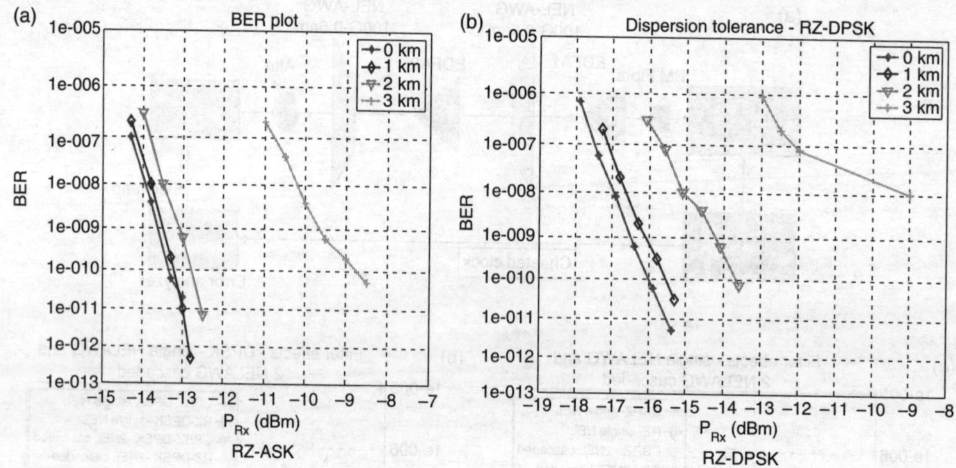

FIGURE 14.4   BER versus receiver sensitivity of (a) RZ-ASK and (b) RZ-DPSK under dispersion of SSMF of 1, 2, 3 and 4 km. Due to 10 dB coupler the Rx power level should be read with an addition –10 dB, that is the label is –15, –14…–10 and –9 dBm.

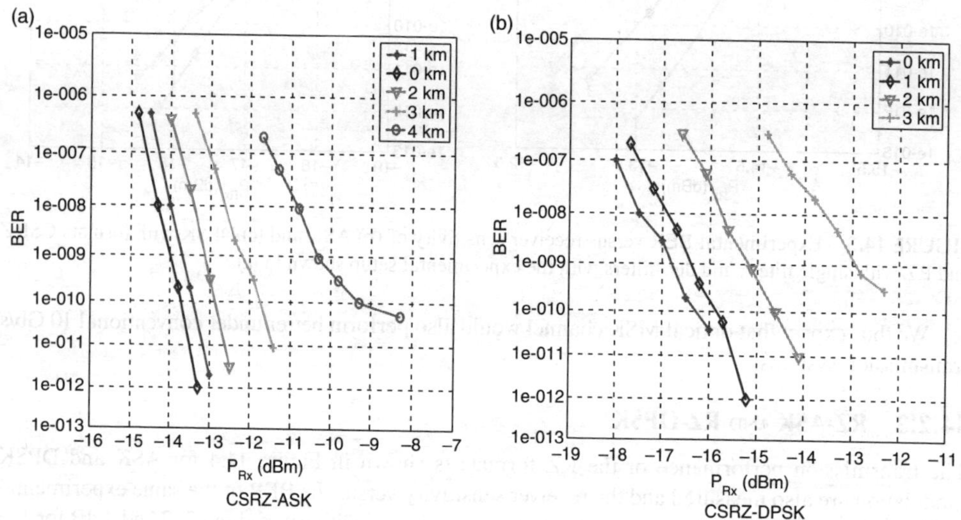

FIGURE 14.5   BER versus receiver sensitivity of (a) CSRZ-ASK and (b) CSRZ-DPSK under dispersion of SSMF of 1, 2, 3 and 4 km. Due to 10 dB coupler the Rx power level should be read with an addition –10 dB, that is the label is –15, –14…–10 and –9 dBm.

14.2.5   ASK and DPSK Spectra

The spectra of different formats on ASK and DPSK modulation as shown in Figure 14.6, the spikes due to the on–off of the ASK indicate the non-linear distortion in the waveform under high dispersion effects than for DPSK modulation whose time domain waveform has a constant envelop. The energy concentration of the signals over the spectrum of DPSK modulation allows better performance than ASK. Furthermore the linear equalization works for the modulation formats with discrete phase shift keying using non-coherent detection and naturally balanced receiver for an additional 3 dB improvement.

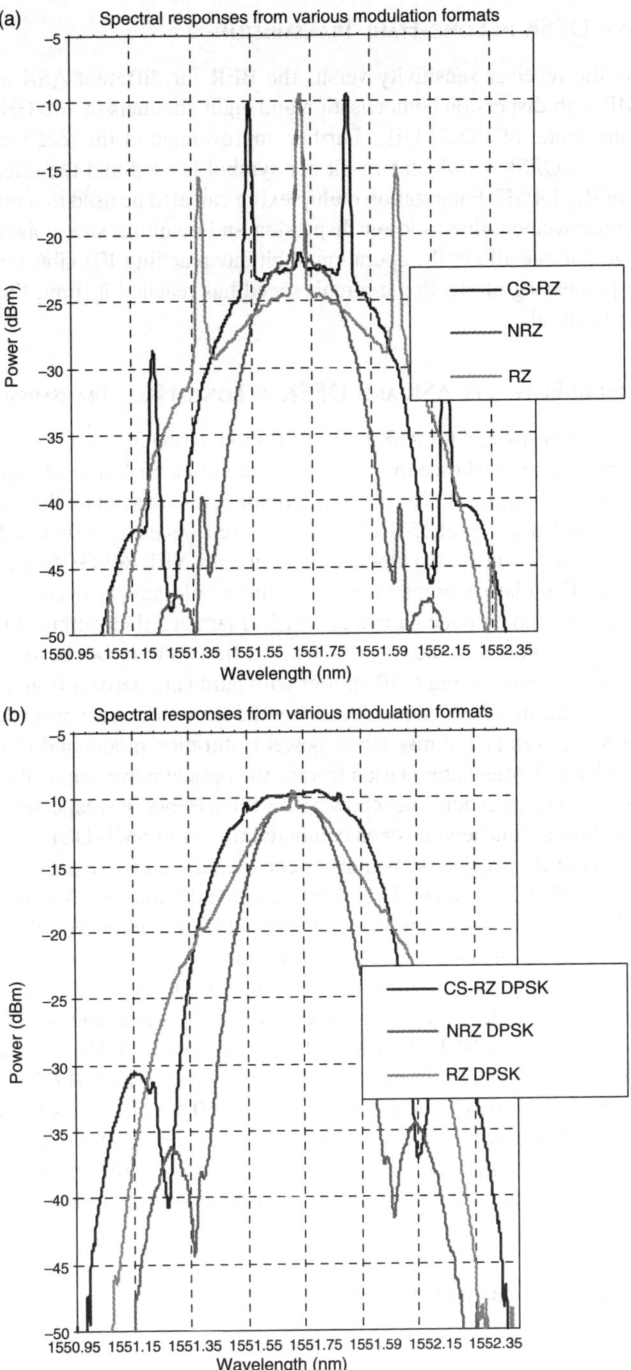

FIGURE 14.6 Optical spectra of (a) ASK and (b) DPSK modulation with different modulation formats.

14.2.6 ASK AND DPSK IN LONG-HAUL TRANSMISSION

Figure 14.7 shows the receiver sensitivity versus the BER for different ASK and DPSK formats over 320 km SSMF with dispersion compensation and eight channels of 100 GHz spacing with the matched filter at the center of 192.33 GHz. Further improvement of the receiver sensitivity can be expected for 40 Gb/s DQPSK in which two-bit per symbol is used and the effective bandwidth is about half of that of RZ-DPSK. Polarization multiplexing can also be used to reduce the symbol rate and hence better improvement due to lower dispersion and quantum shot noises. Furthermore the reduction of the symbol rate allows the operation of bit rate reaching 107 Gb/s for 10 Gb/s Ethernet. At this speed the processing of electronic circuit speed has reached it limit, thus the reduction of the symbol rate is essential.

14.2.7 NON-LINEAR EFFECTS IN ASK AND DPSK IN LONG-HAUL TRANSMISSION

Non-linear effects are investigated for ASK and DPSK modulation formats. The measurement on the transmission system is setup as shown in Figure 14.1 but with a 50-km SSMF spool and dispersion compensators and optical amplifiers. The photonic components are set with the following conditions. (i) Laser source : tunable laser set at 1551.72 nm for centred wavelength (lamda 5) and 1548.51 nm for lamda 1–1560 nm for lamda 16. (ii) Optical filters–muxes NEL-AWG=frequency spacing 100G 3 dB BW of 0.5 nm. Circulating property so spectrum would appear cyclic (note the spectrum), note of the channel input and output e.g input at port 5 of lamda 5 then output at port 8. Input lamda 1 at port 1 then output at port 8 (lamda 1). (iii) Optical transmitters can be set at CSRZ-DPSK or RZ- DPSK. (iv) Clock recovery using 6 dB splitter with hardware derived from the PRBS BPG (bit pattern generator). (v) Tuning of MZ phase decoder at the Rx is necessary when lamda channels are changed. (vi) EDFA 1 driven at 178 mW pump power (saturation mode) and EDFA2 driven at 197 mW (saturation mode). (vi) Attenuator is used to vary the optical power input at the receiver.

It is noted that DCF has lower non-linear power threshold. Hence, it is expected that non-linearities start their effects at lower launched power as compared to 50 km SMF-DCF.

The measurement of the receiver sensitivity is conducted with (i) Tuning the wavelength of the laser source: lamda 1=1548.51 nm, lamda 5=1551.72 nm and lamda 16=1560.61 nm (according to ITU grid standard). This is to show that different channels through the Mux/Demux with different dispersion characteristics, thence it is important to do mobbing for achievement of complete dispersion compensation. (ii) Mod formats implemented are: RZ-DPSK and CSRZ-DPSK. (iii) Power launched into the Sumitomo DCFM was measured to be 7.1 dBm. (iv) 40GHz clock Recovery unit was utilized. (v) Use two NELAWG Mux/Demux filters 100GHz spacing–3dB bandwidth 0.45 nm→alike the currently deployed real 100 GHz spacing system. (vi) DCF module for dispersion compensation is: Sumitomo DCFM (–850 ps/nm @ 1550 nm). (vii) Power used in the BER plot is the reading power at the power meter. The received power can easily be calculated by: Received Power = Reading power+10 dB (1:10 coupler)–0.7 (insertion loss of the coupler).

The following section give the transmission performance of the transmission system for ASK and DPSK formats.

14.2.7.1 Performance of DWDM RZ-DPSK and CSRZ-DPSK

The BER under low launched average power is measured versus the receiver sensitivity for different wavelength channels. Three channels are selected with wavelengths of the center C-band region and two other at the extreme, the shortest and longest. This is to ensure that the error free is received for all channels of the system. The BER versus the receiver sensitivity is shown in Figure 14.7.

14.2.7.2 Non-linear Effects on CSRZ-DPSK and RZ-DPSK

The receiver sensitivity is plotted against the BER for different launched power for RZ-DPSK and CSRZ-DPSK formats and is shown in Figure 14.8. It is observed that the CSRZ is more resilient to

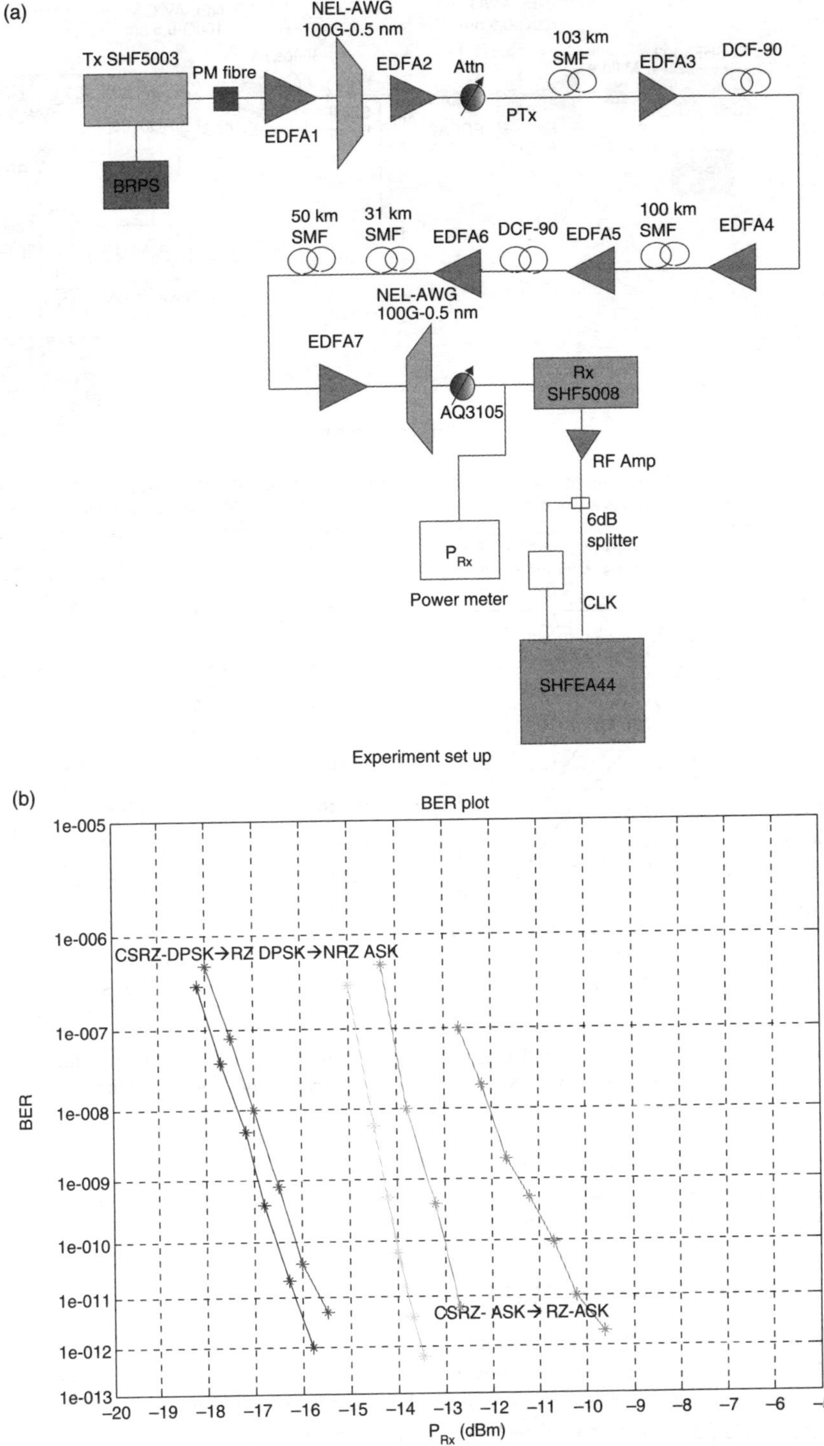

FIGURE 14.7 BER versus receiver sensitivity of 230 km transmission with optical amplifiers and dispersion compensating modules, 40 Gb/s bit rate and DWDM 8 channels of 100 GHz spacing and only one optical filter NEL array waveguide demultiplexer is used. (a) Experiment setup (b) BER of ASK and DPSK modulation formats.

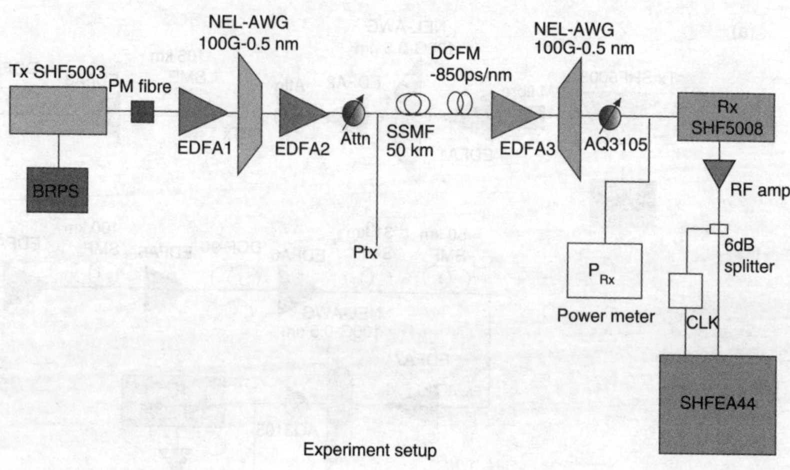

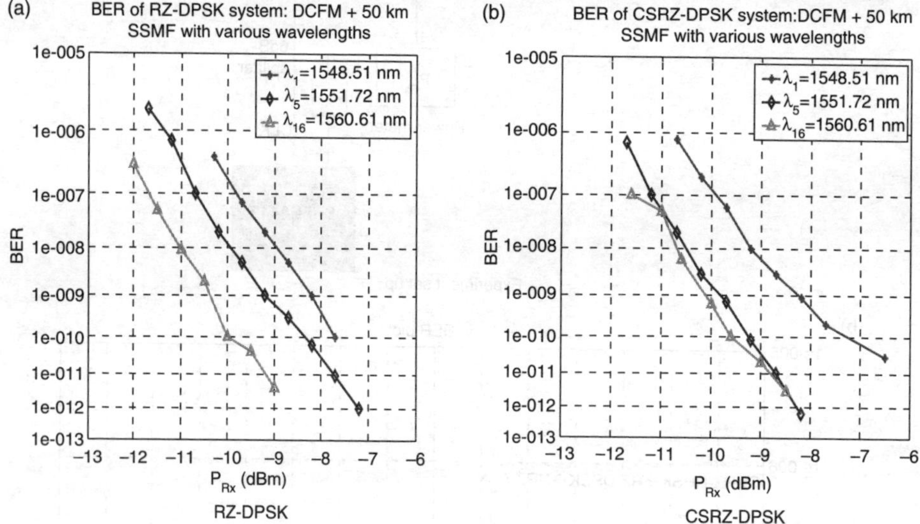

FIGURE 14.8 BER versus receiver sensitivity with different wavelength channels as a parameter of 50 km transmission with optical amplifiers and dispersion compensating modules, 40 Gb/s bit rate for (a) RZ-DPSK and (b) CSRZ-DPSK modulation formats. Note the channels are center and lowest and longest region.

the non-linear effects at a launched power of 11 dBm while the RZ-DPSK suffers at this power level. This can be due to the suppression of the carrier that allows the total average power lower than that of RZ-DPSK in which the carrier is still embedded in this format.

14.2.7.3 Non-linear Effects on CSRZ-ASK and RZ-ASK

The ASK modulation with formats RZ and CSRZ does not suffer non-linear phase distortion due to the SPM up to 11 dBm of launched power but the slope of the BER curve indicates it suffers from the non-linear phase distortion due to second harmonics as explained above. The CSRZ-ASK offers at least 1 dB better than that of the RZ-ASK, though the BER slope is nearly the same. This is explained in Chapter 11 as being due to the second harmonic distortion of the modulation. When the average power is increased to 13 dBm the distortion is severed and error free could not be achieved. The non-linear SPM effects can be observed in Figure 14.9 in which the BER is plotted versus the receiver sensitivity with the launched power as a parameter over 50 km transmission with optical

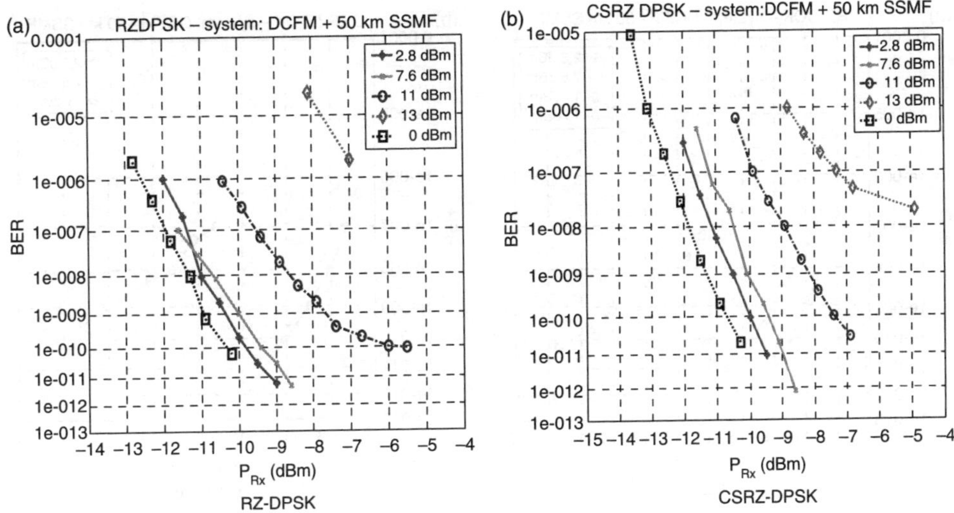

FIGURE 14.9 BER versus receiver sensitivity with launched power as a parameter of 50 km transmission with optical amplifiers and dispersion compensating modules, 40 Gb/s bit rate for (a) RZ-DPSK and (b) CSRZ-DPSK modulation formats. Experiment setup shown in Figure 14.8.

amplifiers and dispersion compensating modules. The bit rate is 40 Gb/s and the modulation formats are RZ-DPSK and CSRZ-DPSK. The experimental setup is shown in Figure 14.8. Furthermore, Figure 14.10 shows the SPM effects on phase shift keying modulation formats. The BER versus receiver sensitivity with launched power as a parameter of 50 km and 230 km transmission with optical amplifiers and dispersion compensating modules. The bit rate is 40 Gb/s and the modulation formats included in the figure are RZ-ASK, CSRZ-ASK, and CSRZ-DPSK. The experimental setup is also shown in Figure 14.10. The wavelength channel under experimental transmission reported in Figure 14.9, Figure 14.10 and Figure 14.11 is 1550.93 nm. Figure 4.11 shows the measured BER versus receiver sensitivity for two formats CSRZ DPSK and RZ DPSK of lightwave channel at different wavelength of 1551.72 nm. These channels are multiplexed and transmitted in the same fiber.

14.2.8 Continuous Phase Versus Discrete Phase Shift Keying

The continuous phase modulation offers much narrower bandwidth in the passband spectrum as compared to the discrete phase shift keying modulation formats due to the continuity of the carrier phase at the boundary of the bit period. That means there is chirp modulating of the carrier from one bit period to the other depending on the differential bits of the input data sequence. Furthermore the carrier is imbedded in the signal band with obviously no peaks of the spectrum of continuous phase shift keying formats.

When the separation between the frequencies of the continuous phase modulation equals to a quarter of the bit rate then orthogonality is achieved. This orthogonality allows improvement in the detection of the MSK signals either by balanced phase detection or by frequency discrimination. If there is photonic processing system that would distinguish between the orthogonal channels then the MSK signals would offer much ease of implementation in the photonic domain.

14.2.9 Multi Sub-carrier Versus Single/Dual Carrier Modulation

The lowering of the bit rate can be significantly reduced with multi-carriers especially when they are orthogonal and generated by the electronic processor of IFFT and FFT in which the nature of orthogonality can be provided via the discrete Fourier transform, the OFDM. Combining with DQPSK the scheme offers better resilience to non-linearity and lowering the dispersion effects.

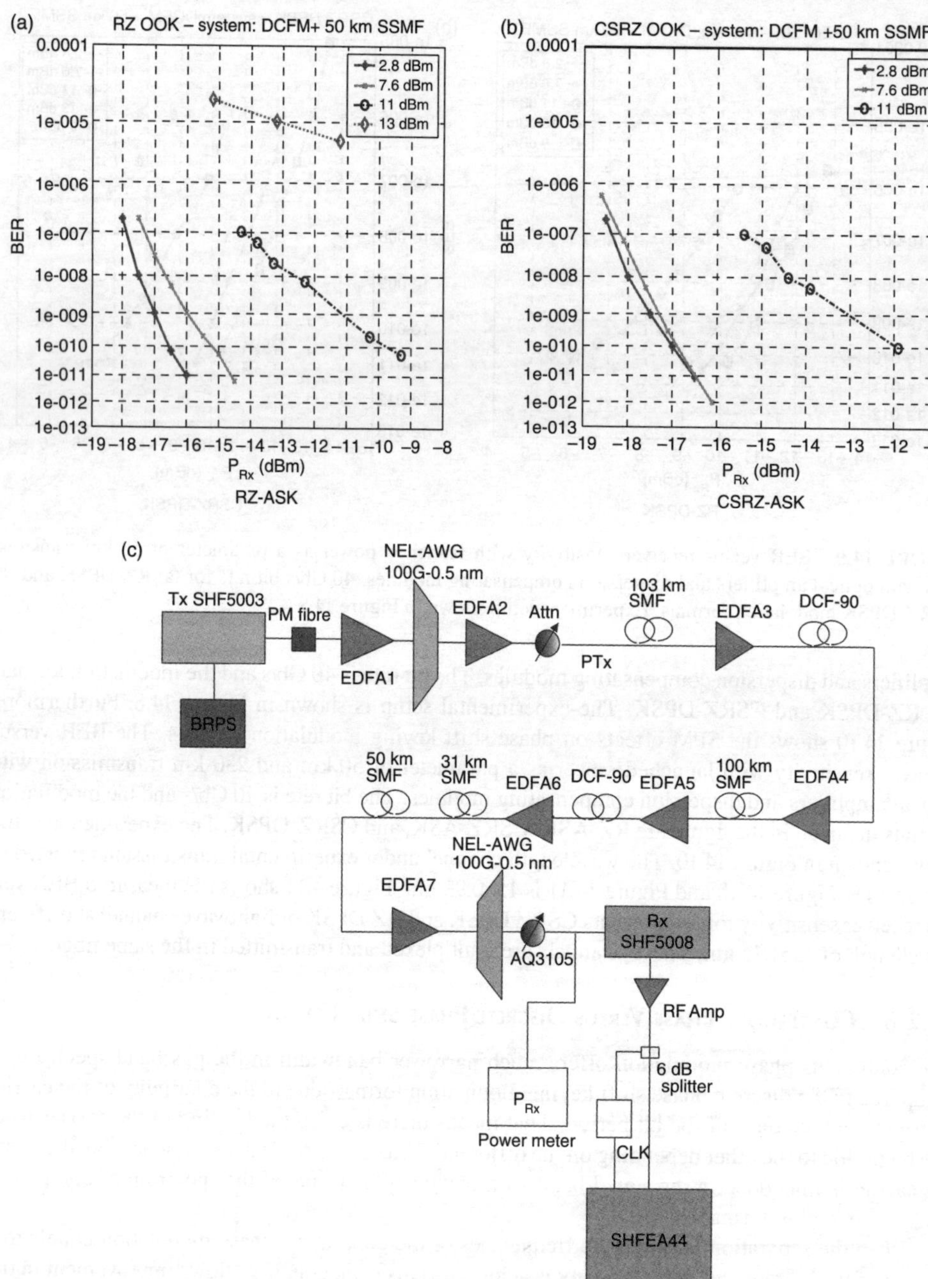

FIGURE 14.10 BER versus receiver sensitivity with launched power as a parameter of 50 km and 230 km transmission with optical amplifiers and dispersion compensating modules, 40 Gb/s bit rate for (a) RZ-ASK and (b) CSRZ-ASK (d) CSRZ-DPSK modulation formats. The transmission experiment setup is as shown in Figure 14.8 and (c) of this figure.

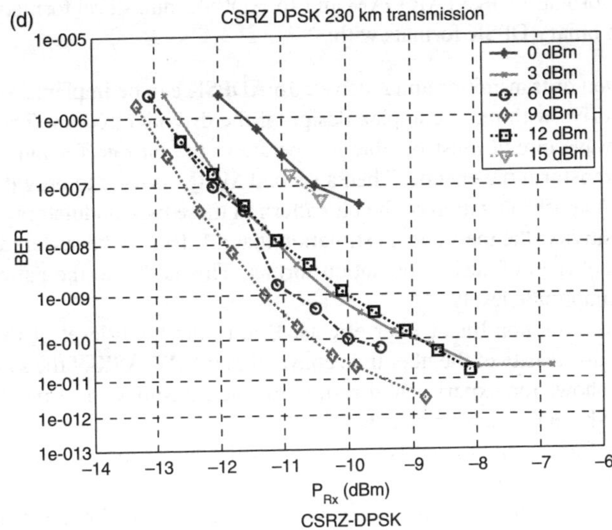

FIGURE 14.10 (*Continued*)

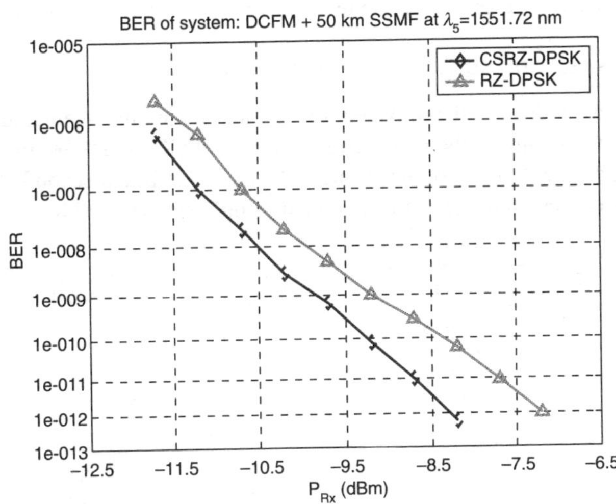

FIGURE 14.11 Measured BER versus receiver sensitivity for two formats CSRZ DPSK and RZ DPSK at wavelength 1551.72 nm.

However the high ratio of the peak power and average power makes this scheme unattractive. Further the FWM effect may be very serious amongst sub-carriers as they are very close to each other and phase matching can easily occur.

14.2.10 Multi-level Versus Binary or IQ Modulation

16 Star ASK and MADPSK multilevel formats offer the lowering of symbol rates and polarization multiplexing will allow the symbol rate reduced the bit rate of 100 Gb/s to around 12.5 Gb/s making use of digital electronic signal processing attractive for both coherent and incoherent transmission and detection. The pattern dependency is one of the major problems of the multi-level amplitude and phase modulation.

The multilevel formats of 16 STAR QAM and 16-ADPSK multi-level formats are characterized and compared with binary DPSK formats with:

1. Implementation of transmitter and receiver: 16 ADPSK can be implemented by using one MZIM dual drive with more complex design of the driving circuitry. The speed of electronic microwave circuit must be able to operate at the bit rate for multiplexing the bit pattern and transform the encode. Then a second MZIM is used to provide the amplitude modulation. A synchronization of the bit pattern in these two modulators is essential.
2. Receiver sensitivity: the theoretical estimation that a 1 dB may be suffered from the same bit rate (symbol rate) of ASK modulation format. This is due to the pattern dependency between the amplitude levels.
3. Robustness towards non-linear fiber effects: Due to the multi-level of the amplitudes it may suffer non-linearity of the fiber than conventional NRZ-ASK of the same symbol rate. Figure 14.12 shows the experimental results for the transmission of modulation formats ASK and DPSK under filtering effects.
4. Dispersion tolerance: multi-level is expected to offer better dispersion tolerance than its DPSK binary counterparts.
5. Spectral efficiency: The spectral efficiency of multilevel would be higher than the binary DPSK of the same symbol rate due to its narrower passband property.

Coherent detection of 16 STAR QAM may offer much better improvement, in the order of 10 to 20 dB if electronic signal processing can be used for the phase estimation of the detected signals.

14.2.11 SINGLE SIDEBAND AND PARTIAL RESPONSE MODULATION

In Chapter 11, we have addressed the impact of the carrier on the fundamental and higher order harmonic distortion in the nature of the modulation schemes, ASK and DPSK and the SSB as well as the duobinary formats. SSB and duobinary suffer the least non-linear second harmonic distortion. Furthermore the duobinary modulation offers a true single sideband with reduction of the carrier

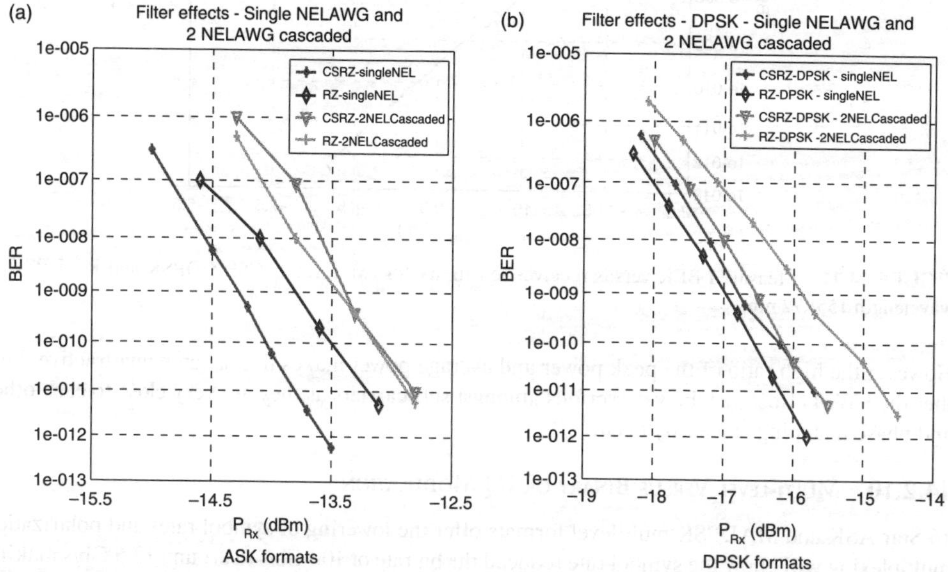

FIGURE 14.12 Measured BER versus receiver sensitivity for (a) ASK and (b) DPSK formats under multiple filtering effects—typical in all-optical networking.

plus its ability for direct detection without resorting to the balanced receiver structure. Thus these types of modulation formats would be considered to be the best binary modulation formats as they permit the reduction of linear CD effects and the electronic dispersion compensation at the receiver and shared pre-distortion and post-compensation (as described in Chapter 11).

14.3 MODULATION FORMATS AND ALL-OPTICAL NETWORKING

14.3.1 ADVANCED MODULATION FORMATS IN LONG-HAUL TRANSMISSION SYSTEMS

Over the years, especially in the last ten years, research works have concentrated on theory and simulation and experiments under various modulation formats. It is noted that along the transmission systems there must be optical add/drop multiplexers (OADM) and then fed into optical cross connect (OXCs) for routing to further transmission. Thus we expect the optical signals must be resilient to optical filtering cascaded along the transmission path. Optical filtering can thus assist the electronic filtering of the optical equalization at the receiver or as pre-distortion at the transmitter. The design of optical filtering can be integrated with these electronic processing sub-systems.

Coherent transmission and detection techniques have became reality with the availability of narrow band lasers for use as local oscillator. The increase of high bit rate also reduce the demand of the LO laser linewidth. Furthermore the processing of mixed signal detection and electronic processing for phase estimation of the received electronic signals will support the implementation and installation of coherent transmission in all-optical networks and systems.

14.3.2 ADVANCED MODULATION FORMATS IN ALL-OPTICAL NETWORKS

Future transparent networks would expect to be all optical and thus be spectrally efficient through the transmission systems, optical add/drop multiplexers, and optical cross connects (OXCs) for optical routing. In such networks the advanced modulation formats that offer narrow optical passband would minimize the effects of tight filtering at OADM and OXC sites. Formats, especially phase shift keying or continuous phase, such as SSB and VSB, MSK, RZ-DPSK and RZ-DQPSK and duobinary (as an exception) are shown to offer excellent transmission performance and adaptive to the electronic dispersion compensation placed at the receiver (see Chapter 11). Optical pre-filtering does generate the formats with truly single side band such as SSB, VSB and duobinary signals. In optical networks, cascading of several of these optical sub-systems is necessary and modulation format signals must be resilient to optical filtering cascading [2]. The complexity of network structures in next generation optical networks requires flexibility and adaptability of optical signals and spectral efficiency as well as ease of electronic equalization. Advanced modulation formats could offer to resolve a number of challenging issues which are to be integrated in the design and arrangement of OADM and OXCs in future networks. Monitoring of modulation formats, wavelength channels and eye distortion for adaptive equalization are also important for all-optical networking.

The effects of filter cascading can be observed from the measure of the BER versus the receiver sensitivity of 40 Gb/s CSRZ-DPSK modulation format under the multiplexing and demultiplexing by cascading two NEL 100 GHz AWG's cf one AWG (3 dB bandwidth is 0.5 nm) and an optical filter with 1.2 nm passband which is a tunable filter, no fibre is used and thus only two AWGs offering narrow passband to pass 40 G CS RZ DPSK sufficiently as observed in Figure 14.13. The transmission of 40 Gb/s over 10 Gb/s optical transmission infrastructure is also important for upgrading the transmission rate of one or few wavelength channels to 40 Gb/s. With 100 GHz spacing and 0.5 nm passband filters in the network, the BER has been measured versus the receiver sensitivity of 40 Gb/s and 10 Gb/s of adjacent wavelength channels indicate that the CSRZ DPSK behaves very well and suffers no penalty as shown in Figure 14.14. We thus expect that optical MSK channel would also perform better under conventional 10 Gb/s transmission systems.

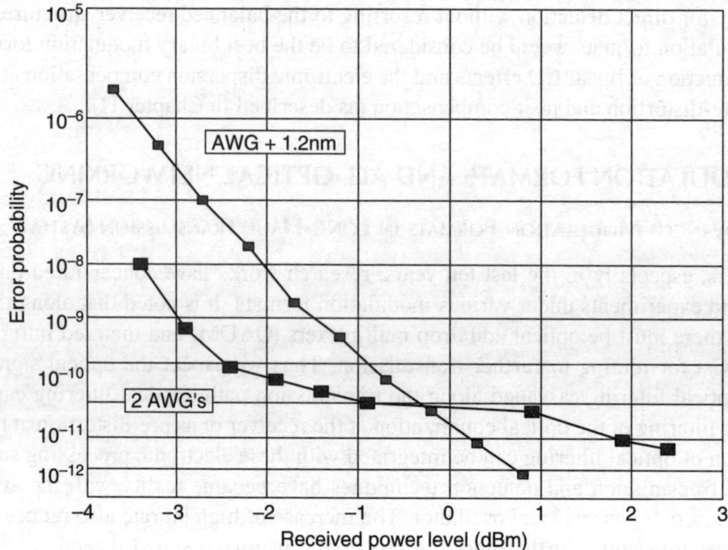

FIGURE 14.13 Measured BER versus receiver sensitivity for CSRZ_DPSK under cascading of two optical filters (AWGS) and one AWG. Horizontal axis must add −10 dB.

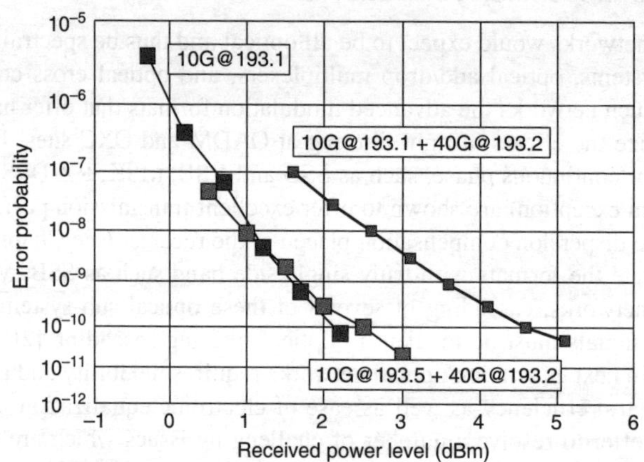

FIGURE 14.14 Measured BER versus receiver sensitivity for 40 Gb/s CSRZ_DPSK and 10 Gb/s under normal 10 Gb/s optical transmission setup.

14.4 HYBRID 40 GB/S OVER 10 GB/S OPTICAL NETWORKS: 328 KM SSMF+DCF FOR 320 KM TX: IMPACT OF ADJACENT 10 G/40 G CHANNELS

An experimental testbed is setup for a 320 km transmission performance with a 10 G NRZ-ASK and a CS-RZ DPSK 40 G channel to test adjacent and non-adjacent channel performance with a 100 GHz AWG mux and a 1.2 nm tunable filter at the input of the Rx. Significant penalty when adjacent 40 G channel switched on due to the narrow width of the tunable filter. Non-adjacent channel too far off to have any impact as observed from the results are shown in Figure 14.13, Figure 14.14, and Figure 14.15.

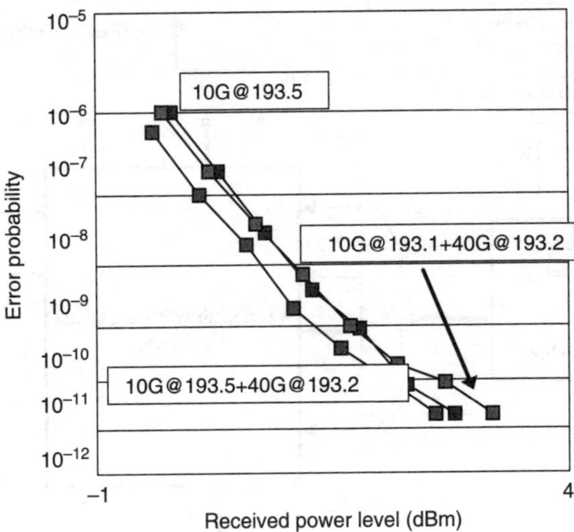

FIGURE 14.15   320 km transmission 40 G impact on 10 G channel: Probability of error versus receiver sensitivity (dBm). Effects of 40 G (CS_RZ DPSK) with 10 G (NRZ-ASK) channel simultaneously transmitted for NRZ ASK and CS-RZ DPSK formats. Blue dots for 1.2 nm thin film filter and dots for 0.5 nm AWG filter (demux with 100 GHz spacing).

Results show that there is no appreciable impact on 10 G signal of an adjacent 40 G signal. For the case of two 100 GHz AWG's used as the mux and Rx filter for CS RZ DPSK transmission over 320 km. We now measure the performance of the 40 G stream transmitted simultaneously with adjacent and non-adjacent 10 G channel. Note problem with using 2 AWG's at 40 Gb/s as there is significant error floor. However one can see that, as expected, 10 G adjacent or non-adjacent has no impact on 40 G signal. A small improvement with 40 G channel is probably due to less noise from EDFA.

14.5   ULTRA-FAST OPTICAL NETWORKS

14.5.1   OTDM NETWORKING

Future large scale deployment of broadband connection in the access area will require transport and add and drop of high speed capacity information with bit rate reaching 1600 Gb/s or even higher to 170 Gb/s including forward error coding.

Optical time division multiplexing (OTDM) should be employed to multiplex real time domain signal channels to higher rat, for example 4 × 40 Gb/s to 160 Gb/s. In previous chapters of this book most of our studies are based on the base rate of 40 Gb/s or 10 Gb/s with different modulation formats of amplitude, or phase or frequency of the lightwave carriers. This section addresses a number of technological issues that would enable the transmission of ultra-high speed 160 Gb/s signals over systems and networks.

160 Gb/s is normally composed of time-interleaved RZ coded lower bit rates, generally at 10 Gb/s or 40 Gb/s usually generated by mode-locked fiber lasers [3]. The transmission of this ultra-high bit rate signals requires precise compensation of the dispersion effects using optical phase modulation as addressed in Chapter 10.

To recover the lower rate channel further there needs to be time division demultiplexing and adding or dropping other channels. This enables the insertion and extraction of lower bit rate channels. In order to achieve the insertion and extraction, two principal photonic devices are required, a photonic clock recovery operating at the lower rate of the channel and an ultra-fast gating device usually an electro-absorption modulator (EAM) operating in a short time window within which other pulse sequence would be isolated so as no interference occurs. A contrast ration of at least 20 dB is necessary.

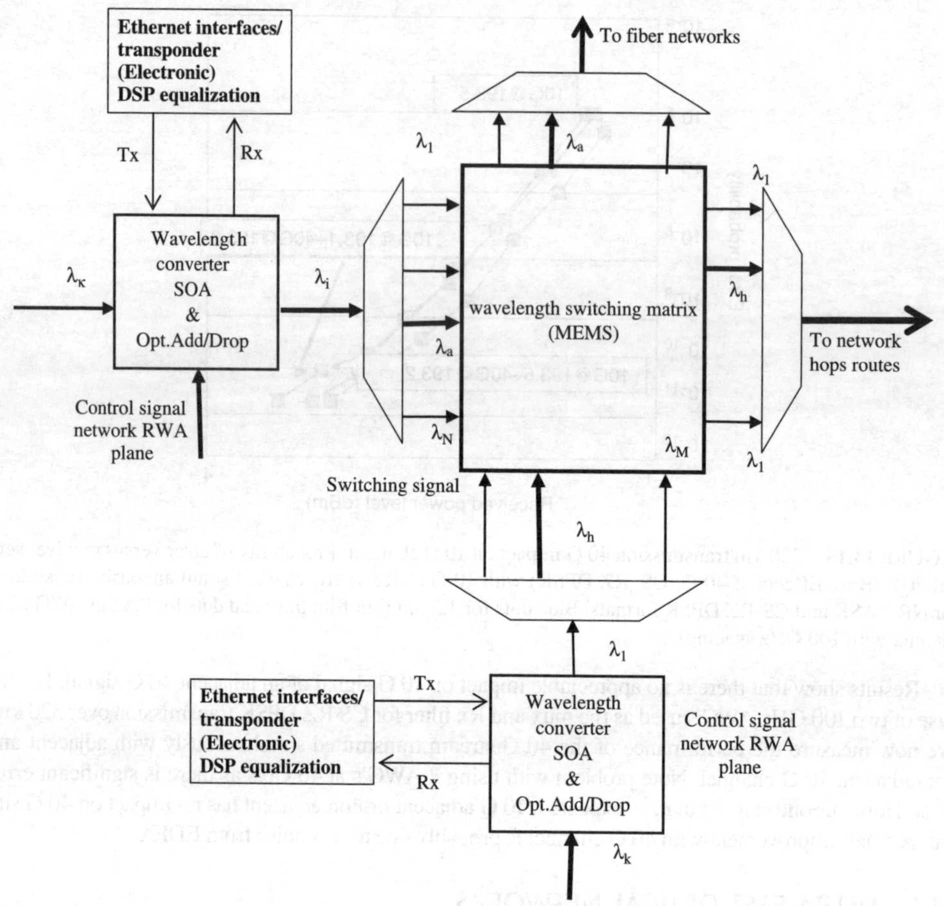

FIGURE 14.16 Detailed structure of an optical node for ultra-high speed optical Ethernet. Wavelength conversion and routing by an SOA and demultiplexer and switching matrix for routing a specific wavelength to a designated output port for further routing.

Photonic ultra-fast gate can be based on a number of operating principles such as phase shifting in interferometric type gate/switch, frequency shifting like parametric frequency conversion or four wave mixing or gain/absorption variation by electrical or optical excitation. The gating function can be provided by EAM [4], SOA—semiconductor amplifier-based ultra fast nonlinear interferometers [5,6], SOA based interferometers, Sagnac interferometers incorporating highly non-linear fifers [7–9]. Optical or electrical clock pulses are required depending on the operating domain of the gate.

Wavelength conversion for ultra-fast routing of these OTDM channels is also required in order to avoid the congestion at nodes in the networks as shown in Figure 14.16. The EAM seems to be the simplest photonic gate for such network node. The design and demonstration of such network nodes are critical for future deployment of ultra-fast optical networking.

REFERENCES

1. http://www.SHF.de, accessed December 2007.
2. Raybon, G. 2001. Performance of advanced modulation format in optically routed networks. *Proceeding of Optical Fiber Conference 2006, Paper OThR1, OFC 2006*, Anaheim CA USA, 1–3.

3. Winzer, P.J., G. Raybon, and M. Duelk. 2005. 107-Gb/s optical ETDM transmitter for 100G Ethernet transport. *Proceedings of the 31st European Conference on Optical Communication, 2005. ECOC 2005.* Vol. 6, 1–2.
4. Lam, H. Q., P. Shum, L. N. Binh, and Y. D. Gong. 2007. Mode-locked fiber lasers, polarization switching in an active harmonic mode locked fiber laser. *Conference on Optical Internet COIN-ACOFT*, June 24–26, Melbourne, Australia.
5. Ellis, et al. *IEEE Electronics Letters* 34 (18): 1788
6. Schubert, C. et al. *IEEE Electronics Letters* 39 (19): 1074.
7. Turkiewicz, J. et al. ECOC'03, PD Th4.4.5, 84.
8. Heid, M. et al. ECOC'02, We8.4.3.
9. Watanabe, S., R. Okabe, F. Futami, R. Hainberger, C. Schumidt-Langhorst, C. Schubert, and H. G. Weber. 2004. Novel fiber Kerr-switch with parametric gain: demonstration of optical demultiplexing and sampling up to 640 Gb/s. In *Proc. European Conference on Optical Communication (ECOC 2004)*, Postdeadline paper Th4.1.6, Stockholm Sweden.

Index